Mathematical Tools for Physicists

Edited by
Michael Grinfeld

Related Titles

Goodson, D.Z.

Mathematical Methods for Physical and Analytical Chemistry

2011

ISBN: 978-0-470-47354-2

Also available in digital formats

Trigg, G.L. (ed.)

Encyclopedia of Applied Physics

The Classic Softcover Edition

2004

ISBN: 978-3-527-40478-0

Gläser, M., Kochsiek, M. (eds.)

Handbook of Metrology

2010

ISBN: 978-3-527-40666-1

Stock, R. (ed.)

Encyclopedia of Applied High Energy and Particle Physics

2009

ISBN: 978-3-527-40691-3

Bohr, H.G. (ed.)

Handbook of Molecular Biophysics

Methods and Applications

2009

ISBN: 978-3-527-40702-6

Andrews, D.L. (ed.)

Encyclopedia of Applied Spectroscopy

2009

ISBN: 978-3-527-40773-6

Masujima, M.

Applied Mathematical Methods in Theoretical Physics

2nd Edition

2009

ISBN: 978-3-527-40936-5

Also available in digital formats

Willatzen, M., Lew Yan Voon, L.C.

Separable Boundary-Value Problems in Physics

2011

ISBN: 978-3-527-41020-0

Also available in digital formats

Mathematical Tools for Physicists

Edited by
Michael Grinfeld

Second Edition

WILEY-VCH
Verlag GmbH & Co. KGaA

The Editor

Dr. Michael Grinfeld
University of Strathclyde
Mathematics and Statistics
Glasgow
United Kingdom
m.grinfeld@strath.ac.uk

Cover
"Skeleton and Pore Partition", x-ray computed tomography image of a limestone core from Mt Gambier in South Australia. Courtesy of Olaf Delgado-Friedrichs

Library of Congress Card No.: applied for

British Library Cataloguing-in-Publication Data
A catalogue record for this book is available from the British Library.

Bibliographic information published by the Deutsche Nationalbibliothek
The Deutsche Nationalbibliothek lists this publication in the Deutsche Nationalbibliografie; detailed bibliographic data are available on the Internet at <http://dnb.d-nb.de>.

Print ISBN: 978-3-527-41188-7
ePDF ISBN: 978-3-527-68426-7
ePub ISBN: 978-3-527-68427-4
Mobi ISBN: 978-3-527-68425-0

Cover-Design Grafik-Design Schulz, Fußgönheim, Germany
Typesetting Laserwords Private Limited, Chennai, India
Printing and Binding Markono Print Media Pte Ltd, Singapore

Printed on acid-free paper

Contents

List of Contributors

Chris Athorne
University of Glasgow
School of Mathematics and Statistics
15 University Gardens
Glasgow G12 8QW
UK

David A.W. Barton
University of Bristol
Department of Engineering Mathematics
Merchant Venturers Building
Woodland Road
Bristol BS8 1UB
UK

Kurt Binder
Johannes Gutenberg University of Mainz
Department of Physics, Mathematics and Informatics
Institute of Physics
Staudingerweg 7
55128 Mainz
Germany

Lyonell Boulton
Heriot-Watt University
Department of Mathematics
School of Mathematical and Computer Sciences
Colin MacLaurin Building
Riccarton
Edinburgh EH14 4AS
UK

James R. Cruise
Heriot-Watt University
Department of Actuarial Mathematics and Statistics
and the Maxwell Institute for Mathematical Sciences
Edinburgh EH14 4AS
UK

James H. Davenport
University of Bath
Department of Computer Science
Claverton Down
Bath
Bath BA2 7AY
UK

Marcelo Epstein
University of Calgary
Department of Mechanical and Manufacturing Engineering Schulich School of Engineering
2500 University Drive NW
Calgary
Alberta, T2N 1N4
Canada

Ernesto Estrada
University of Strathclyde
Department of Mathematics and Statistics
26 Richmond Street
Glasgow G1 1XQ
UK

Pavel Exner
Czech Technical University
Doppler Institute
Břehová 7
11519 Prague
Czech Republic

and

Department of Theoretical Physics
Nuclear Physics Institute
Academy of Sciences of the
Czech Republic
Hlavní 130
25068 Řež
Czech Republic

Robert Gilmore
Drexel University
Department of Physics
3141 Chestnut Street
Philadelphia
PA 19104
USA

Ostap Hryniv
Durham University
Department of Mathematical Sciences
South Road
Durham DH1 3LE
UK

Gabriel J. Lord
Heriot-Watt University
Department of Mathematics
Maxwell Institute for Mathematical
Sciences
Edinburgh EH14 4AS
UK

James Murdock
Iowa State University
Department of Mathematics
478 Carver Hall
Ames
IA 50011
USA

Des McGhee
University of Strathclyde
Department of Mathematics and
Statistics
26 Richmond Street
Glasgow G1 1XH
Scotland
UK

Rainer Picard
Technische Universität Dresden
FR Mathematik
Institut für Analysis
Willersbau
Zellescher Weg 12-14
01069 Dresden
Germany

Vanessa Robins
The Australian National University
Department of Applied Mathematics
Research School of Physics and
Engineering
Canberra ACT 0200
Australia

Tomáš Roubíček
Charles University
Mathematical Institute
Sokolovská 83
Prague 186 75
Czech Republic

and

Institute of Thermomechanics
of the ASCR
Prague
Czech Republic

Sascha Trostorff
Technische Universität Dresden
FR Mathematik
Institut für Analysis
Willersbau
Zellescher Weg 12-14
01069 Dresden
Germany

Andrew R. Wade
Durham University
Department of Mathematical Sciences
South Road
Durham DH1 3LE
UK

Marcus Waurick
Technische Universität Dresden
FR Mathematik
Institut für Analysis
Willersbau
Zellescher Weg 12-14
01069 Dresden
Germany

Preface

The intensely fruitful symbiosis between physics and mathematics is nothing short of miraculous. It is not a symmetric interaction; however, physical laws, which involve relations between variables, are always in need of rules and techniques for manipulating these relations: hence, the role of mathematics in providing tools for physics, and hence, the need for books such as this one. Physics, in its turn, supplies motivation for the development of mathematically sound techniques and concepts; it is enough to mention the importance of celestial mechanics in the evolution of perturbation methods and the impetus Dirac's delta "function" gave to the work on generalized functions. So perhaps a sister volume to the present compendium would be *Physical Insights for Mathematicians*. (From the above, one might get the impression that physics and mathematics are two separate, even opposing, domains. This is of course not true; there is a well-established field of mathematical physics and many mathematicians working in fluid and continuum mechanics would be hard-pressed to separate the mathematics and the physics elements of their seamless activity.)

The present book is intended for the use of advanced undergraduates, graduate students, and researchers in physics, who are aware of the usefulness of a particular mathematical approach and need a quick point of entry into its vocabulary, main results, and the literature. However, one cannot cleanly single out parts of mathematics that are useful for physics and ones that are not; for example, while the theory of operators in Hilbert spaces is undoubtedly indispensable in quantum mechanics, an area as abstruse as category theory is becoming increasingly popular in cosmology and has found applications in developmental biology.[1] Hence, the concept of "mathematics useful in physics" arguably covers the same area as mathematics *tout court*, and anyone embarking on the publication of a book such as the present one does so in the certainty that no single book can do justice to the intricate interpenetration of mathematics and physics. It is quite possible to write a second, and perhaps a third volume of an encyclopedia such as ours.

Let us quickly mention significant areas that are *not* being covered here. These include combinatorics, deterministic chaos, fractals, nonlinear partial differential

1) The exact relation between physics and biology is still shrouded in mystery, but certainly objects of interest in biology are physical objects. The time for a truly useful *Mathematical Tools for Biologists* has not yet arrived, but is awaited eagerly.

equations, and symplectic geometry. It was also felt that a separate chapter on context-free modeling was not necessary as there is an ample literature on modeling case studies.

What this book offers is an attractive mix of classical areas of applications of mathematics to physics and of areas that have only come to prominence recently. Thus, we have substantive chapters on asymptotic methods, calculus of variations, differential geometry and topology of manifolds, dynamical systems theory, functional analysis, group theory, numerical methods, partial differential equations of mathematical physics, special functions, and transform methods. All these are up-to-date surveys; for example, the chapter on asymptotic methods discusses recent renormalized group-based approaches, while the chapter on variational methods considers examples where no smooth minimizers exist. These chapters appear side by side with a decidedly modern computational take on algebraic topology and in-depth reviews of such increasingly important areas as graph and network theory, Monte Carlo simulations, stochastic differential equations, and algorithms in symbolic computation, an important complement to analytic and numerical problem-solving approaches.

It is hoped that the layout of the text allows for easy cross-referencing between chapters, and that by the end of a chapter, the reader will have a clear view of the area under discussion and will be know where to go to learn more. *Bon voyage!*

Part I
Probability

Mathematical Tools for Physicists, Second Edition. Edited by Michael Grinfeld.

1
Stochastic Processes

James R. Cruise, Ostap O. Hryniv, and Andrew R. Wade

1.1
Introduction

Basic probability theory deals, among other things, with *random variables* and their properties. A random variable is the mathematical abstraction of the following concept: we make a measurement on some physical system, subject to randomness or uncertainty, and observe the value. We can, for example, construct a mathematical model for the system and try to predict the behavior of our random observable, perhaps through its distribution, or at least its average value (mean). Even in the simplest applications, however, we are confronted by systems that change over time. Now we do not have a single random variable, but a family of random variables. The nature of the physical system that we are modeling determines the structure of dependencies of the variables.

A *stochastic* (or random) *process* is the mathematical abstraction of these systems that change randomly over time. Formally, a stochastic process is a family of random variables $(X_t)_{t \in T}$, where T is some index set representing time. The two main examples are $T = \{0, 1, 2, \dots\}$ (*discrete* time) and $T = [0, \infty)$ (*continuous* time); different applications will favor one or other of these. Interesting classes of processes are obtained by imposing additional structure on the family X_t, as we shall see.

The aim of this chapter is to give a tour of some of the highlights of stochastic process theory and its applications in the physical sciences. In line with the intentions of this volume, our emphasis is on *tools*. However, the combination of a powerful tool and an unsteady grip is a hazardous one, so we have attempted to maintain mathematical accuracy. For reasons of space, the presentation is necessarily concise. While we cover several important topics, we omit many more. We include references for further reading on the topics that we do cover throughout the text and in Section 1.8. The tools that we exhibit include *generating functions* and other transforms, and *renewal structure*, including the Markov property, which can be viewed loosely as a notion of statistical self-similarity.

In the next section, we discuss some of the tools that we will use, with some examples. The basic notions of probability

Mathematical Tools for Physicists, Second Edition. Edited by Michael Grinfeld.

theory that we use are summarized in Section 1.A.

1.2 Generating Functions and Integral Transforms

1.2.1 Generating Functions

Given a sequence $(a_k)_{k\geq 0}$ of real numbers, the function

$$G(s) = G_a(s) = \sum_{k\geq 0} a_k s^k \tag{1.1}$$

is called the *generating function* of $(a_k)_{k\geq 0}$. When $G_a(s)$ is finite for some $s \neq 0$, the series (1.1) converges in the disc $\{z \in \mathbb{C} : |z| < |s|\}$ and thus its coefficients a_k can be recovered via differentiation

$$a_k = \frac{1}{k!}\left(\frac{\mathrm{d}}{\mathrm{d}s}\right)^k G_a(s)\bigg|_{s=0}, \tag{1.2}$$

or using the Cauchy integral formula

$$a_k = \frac{1}{2\pi i}\oint_{|z|=r} \frac{G_a(z)}{z^{k+1}}\mathrm{d}z, \quad (i^2 = -1), \tag{1.3}$$

with properly chosen $r > 0$. This observation is often referred to as the *uniqueness property*: if two generating functions, say $G_a(s)$ and $G_b(s)$, are finite and coincide in some open neighborhood of the origin, then $a_k = b_k$ for all $k \geq 0$. In particular, one can identify the sequence from its generating function.

The generating function is one of many transforms very useful in applications. We will discuss some further examples in Section 1.2.3.

Example 1.1 Imagine one needs to pay the sum of n pence using only one pence and two pence coins. In how many ways can this be done?[1)]

(a) If the order matters, that is, when $1 + 2$ and $2 + 1$ are two distinct ways of paying 3 pence, the question is about counting the number a_n of monomer/dimer configurations on the interval of length n. One easily sees that $a_0 = a_1 = 1$, $a_2 = 2$, and in general $a_n = a_{n-1} + a_{n-2}$, $n \geq 2$, because the leftmost monomer can be extended to a full configuration in a_{n-1} ways, whereas the leftmost dimer is compatible with exactly a_{n-2} configurations on the remainder of the interval. A straightforward computation using the recurrence relation above now gives

$$\begin{aligned} G_a(s) &= 1 + s + \sum_{n\geq 2}(a_{n-1} + a_{n-2})s^n \\ &= 1 + sG_a(s) + s^2 G_a(s), \end{aligned}$$

so that $G_a(s) = (1 - s - s^2)^{-1}$. The coefficient a_n can now be recovered using (1.2), (1.3), or partial fractions and then power series expansion.

(b) If the order does not matter, that is, when $1 + 2$ and $2 + 1$ are regarded as identical ways of paying 3 pence, the question above boils down to calculating the number b_n of nonnegative integer solutions to the equation $n_1 + 2n_2 = n$. One easily sees that $b_0 = b_1 = 1$, $b_2 = b_3 = 2$, and a straightforward induction shows that b_n is the coefficient of s^k in the product

1) Early use of generating functions was often in this enumerative vein, going back to de Moivre and exemplified by Laplace in his *Théorie analytique des probabilités* of 1812. The full power of generating functions in the theory of stochastic processes emerged later with work of Pólya, Feller, and others.

$$\left(1 + s + s^2 + s^3 + s^4 + \cdots\right) \times \left(1 + s^2 + s^4 + s^6 + \cdots\right).$$

In other words, $G_b(s) = \sum_{n\geq 0} b_n s^n = (1-s)^{-1}(1-s^2)^{-1}$.

In this chapter, we will make use of generating functions for various sequences with probabilistic meaning. In particular, given a $\mathbb{Z}_+$-valued[2] random variable X, we can consider the corresponding *probability generating function*, which is the generating function of the sequence (p_k), where $p_k = \mathbb{P}[X = k]$, $k \in \mathbb{Z}_+$, describes the probability mass function of X. Thus, the probability generating function of X is given by (writing $\mathbb{E}$ for expectation: see Section 1.A)

$$G(s) = G_X(s) = \sum_{k\geq 0} s^k \mathbb{P}[X = k] = \mathbb{E}[s^X]. \tag{1.4}$$

Example 1.2

(a) If $Y \sim \text{Be}(p)$ and $X \sim \text{Bin}(n, p)$ (see Example 1.22), then $G_Y(s) = (1-p) + ps$ and, by the binomial theorem, $G_X(s) = \sum_{k\geq 0} \binom{n}{k}(sp)^k(1-p)^{n-k} = ((1-p) + ps)^n$.

(b) If $X \sim \text{Po}(\lambda)$ (see Example 1.23), then Taylor's formula implies

$$G_X(s) = \sum_{k\geq 0} s^k \frac{\lambda^k}{k!} e^{-\lambda} = e^{\lambda(s-1)}.$$

Notice that $G_X(1) = \mathbb{P}[X < \infty]$, and thus for $|s| \leq 1$, $|G_X(s)| \leq G_X(1) \leq 1$, implying that the power series $G_X(s)$ can be differentiated in the disk $|s| < 1$ any number of times. As a result,

$$G_X^{(k)}(s) = \left(\frac{d}{ds}\right)^k G_X(s) = \mathbb{E}[(X)_k s^{X-k}], \quad |s| < 1,$$

where $(x)_k = \frac{x!}{(x-k)!} = x(x-1)\cdots(x-k+1)$. Taking the limit $s \nearrow 1$ we obtain the kth *factorial moment* of X, $\mathbb{E}[(X)_k] = G_X^{(k)}(1_-)$, where the last expression denotes the value of the kth left derivative of $G_X(\cdot)$ at 1.

Example 1.3 If $X \sim \text{Po}(\lambda)$, from Example 1.2b, we deduce $\mathbb{E}[(X)_k] = \lambda^k$.

Theorem 1.1 *If X and Y are independent $\mathbb{Z}_+$-valued random variables, then the sum $Z = X + Y$ has generating function $G_Z(s) = G_X(s)G_Y(s)$.*

Proof. As X and Y are independent, so are s^X and s^Y; consequently, the expectation factorizes, $\mathbb{E}\left[s^X s^Y\right] = \mathbb{E}\left[s^X\right]\mathbb{E}\left[s^Y\right]$, by Theorem 1.18.

Example 1.4

(a) If $X \sim \text{Po}(\lambda)$ and $Y \sim \text{Po}(\mu)$ are independent, then $X + Y \sim \text{Po}(\lambda + \mu)$. Indeed, by Theorem 1.1, $G_{X+Y}(s) = e^{(\lambda+\mu)(s-1)}$, which is the generating function of the $\text{Po}(\lambda + \mu)$ distribution; the result now follows by uniqueness. This fact is known as the *additive* property of Poisson distributions.

(b) If $\mathbb{Z}_+$-valued random variables $X_1, \ldots, X_n$ are independent and identically distributed (i.i.d.), then $S_n = X_1 + \cdots + X_n$ has generating function $G_{S_n}(s) = (G_X(s))^n$. If $S_n \sim \text{Bin}(n, p)$, then S_n is a sum of n independent Bernoulli variables with parameter p, and so $G_{S_n}(s) = ((1-p) + ps)^n$ (see Example 1.22).

The following example takes this idea further.

Lemma 1.1 *Let $X_1, X_2, \ldots$ be i.i.d. $\mathbb{Z}_+$-valued random variables, and let*

2) Throughout we use the notation $\mathbb{Z}_+ = \{0, 1, 2, \ldots\}$ and $\mathbb{N} = \{1, 2, \ldots\}$.

N be a $\mathbb{Z}_+$-valued random variable independent of the X_i. Then the random sum $S_N = X_1 + \cdots + X_N$ has generating function $G_{S_N}(s) = G_N\big(G_X(s)\big)$.

Proof. The claim follows from the partition theorem for expectations (see Section 1.A):

$$\mathbb{E}[s^{S_N}] = \sum_{n\geq 0} \mathbb{E}[s^{S_N} \mid N = n]\mathbb{P}[N = n] = \sum_{n\geq 0} (G_X(s))^n \mathbb{P}[N = n].$$

Example 1.5 If $(X_k)_{k\geq 1}$ are independent $\mathrm{Be}(p)$ random variables and if $N \sim \mathrm{Po}(\lambda)$ is independent of $(X_k)_{k\geq 1}$, then

$$G_{S_N}(s) = G_N(G_X(s)) = \mathrm{e}^{\lambda(G_X(s)-1)} = \mathrm{e}^{\lambda p(s-1)},$$

that is, $S_N \sim \mathrm{Po}(\lambda p)$. This result has the following important interpretation: if each of $N \sim \mathrm{Po}(\lambda)$ objects is independently selected with probability p, then the sample contains $S_N \sim \mathrm{Po}(\lambda p)$ objects. This fact is known as the *thinning* property of Poisson distributions.

A further important application of Lemma 1.1 is discussed in Section 1.2.2.

Example 1.6 [Renewals] A diligent janitor replaces a light bulb on the same day as it burns out. Suppose the first bulb is put in on day 0, and let X_i be the lifetime of the ith bulb. Suppose that the X_i are i.i.d. random variables with values in $\mathbb{N}$ and common generating function $G_f(s)$. Define $r_n = \mathbb{P}[\text{a bulb was replaced on day } n]$ and $f_k = \mathbb{P}[\text{the first bulb was replaced on day } k]$. Then $r_0 = 1, f_0 = 0$, and for $n \geq 1$,

$$r_n = f_1 r_{n-1} + f_2 r_{n-2} + \cdots + f_n r_0 = \sum_{k=1}^{n} f_k r_{n-k}. \tag{1.5}$$

Proceeding as in Example 1.1, we deduce $G_r(s) = 1 + G_f(s)\,G_r(s)$ for all $|s| \leq 1$, so that

$$G_r(s) = \frac{1}{1 - G_f(s)}. \tag{1.6}$$

Remark 1.1 *The ideas behind Example 1.6 have a vast area of applicability. For example, the hitting probabilities of discrete-time Markov chains have property similar to the* Ornstein–Zernike relation (1.5)*, see, for example, Lemma 1.3.*

Notice also that by repeatedly expanding each r_i on the right-hand side of (1.5)*, one can rewrite the latter as*

$$r_n = \sum_{\ell\geq 1} \sum_{\{k_1,k_2,\ldots,k_\ell\}} \prod_{j=1}^{\ell} f_{k_j}, \tag{1.7}$$

where the middle sum runs over all decompositions of the integer n into ℓ positive integer parts $k_1, k_2, \ldots, k_\ell$ (cf. Example 1.1a). The decomposition (1.7) *is an example of the so-called* polymer representation*; it arises in many models of statistical mechanics.*

The *convolution* of sequences $(a_n)_{n\geq 0}$ and $(b_n)_{n\geq 0}$ is $(c_n)_{n\geq 0}$ given by

$$c_n = \sum_{k=0}^{n} a_k b_{n-k}, \quad (n \geq 0); \tag{1.8}$$

we write $c = a \star b$. A key property of convolutions is as follows.

Theorem 1.2 (Convolution theorem) *If $c = a \star b$, then the associated generating functions $G_c(s)$, $G_a(s)$, and $G_b(s)$ satisfy $G_c(s) = G_a(s)G_b(s)$.*

Proof. Compare the coefficients of s^n on both sides of the equality.

The convolution appears very often in probability theory because the probability mass function of the sum $X + Y$ of independent variables X and Y is the convolution of their respective probability mass functions. In other words, Theorem 1.1 is a particular case of Theorem 1.2.

Remark 1.2 *The sequence $(b_n)_{n\geq 0}$ in Example 1.1b is a convolution of the sequence $1, 1, 1, 1, \ldots$ and the sequence $1, 0, 1, 0, 1, 0, \ldots$.*

The uniqueness property of generating functions affords them an important role in studying convergence of probability distributions.

Theorem 1.3 *For every fixed $n \geq 1$, let the sequence $a_{n,0}, a_{n,1}, \ldots$ be a probability distribution on $\mathbb{Z}_+$, that is, $a_{n,k} \geq 0$ and $\sum_{k\geq 0} a_{n,k} = 1$. Moreover, let $G_n(s) = \sum_{k\geq 0} a_{n,k} s^k$ be the generating function of the sequence $(a_{n,k})_{k\geq 0}$. Then*

$$\lim_{n\to\infty} a_{n,k} = a_k, \text{ for all } k \geq 0 \iff \lim_{n\to\infty} G_n(s) = G(s), \text{ for all } s \in [0, 1),$$

where $G(s)$ is the generating function of the limiting sequence $(a_k)_{k\geq 0}$.

Example 1.7 [Law of rare events] Let $(X_n)_{n\geq 1}$ be random variables such that $X_n \sim \text{Bin}(n, p_n)$. If $n \cdot p_n \to \lambda$ as $n \to \infty$, then the distribution of X_n converges to that of $X \sim \text{Po}(\lambda)$. Indeed, for every fixed $s \in [0, 1)$, we have, as $n \to \infty$,

$$G_{X_n}(s) = (1 + p_n(s - 1))^n \to \exp\{\lambda(s - 1)\},$$

which is the generating function of a $\text{Po}(\lambda)$ random variable.

1.2.2 Example: Branching Processes

Let $(Z_{n,k})$, $n \in \mathbb{N}$, $k \in \mathbb{N}$, be a family of i.i.d. $\mathbb{Z}_+$-valued random variables with common probability mass function $(p_k)_{k\geq 0}$ and finite mean. The corresponding *branching process* $(Z_n)_{n\geq 0}$ is defined via $Z_0 = 1$, and, for $n \geq 1$,

$$Z_n = Z_{n,1} + Z_{n,2} + \cdots + Z_{n,Z_{n-1}}, \tag{1.9}$$

where an empty sum is interpreted as zero.

The interpretation is that Z_n is the number of individuals in the nth generation of a population that evolves via a sequence of such generations, in which each individual produces offspring according to $(p_k)_{k\geq 0}$, independently of all other individuals in the same generation; generation $n + 1$ consists of all the offspring of individuals in generation n. The process starts (generation 0) with a single individual and the population persists as long as the generations are successful in producing offspring, or until *extinction* occurs. Branching processes appear naturally when modeling chain reactions, growth of bacteria, epidemics, and other similar phenomena[3]: a crucial characteristic of the process is the probability of extinction.

Let $\varphi_n(s) := \mathbb{E}[s^{Z_n}]$ be the generating function of Z_n; for simplicity, we write $\varphi(s)$ instead of $\varphi_1(s) = \mathbb{E}[s^{Z_1}]$. Then $\varphi_0(s) = s$, and a straightforward induction based on Lemma 1.1 implies

$$\varphi_n(s) = \varphi_{n-1}(\varphi(s)), \quad (n > 1). \tag{1.10}$$

Equation (1.10) can be used to determine the distribution of Z_n for any $n \geq 0$. In particular, one easily deduces that $\mathbb{E}[Z_n] = m^n$, where $m = \mathbb{E}[Z_1]$ is the expected number of offspring of a single individual.

3) The simplest branching processes, as discussed here, are known as *Galton–Watson* processes after F. Galton and H. Watson's work on the propagation of human surnames; work on branching processes in the context of nuclear fission, by S. Ulam and others, emerged out of the Manhattan project.

The long-term behavior of a branching process is determined by the expected value m: the process can be *subcritical* ($m < 1$), *critical* ($m = 1$), or *supercritical* ($m > 1$).

Remark 1.3 *By Markov's inequality (see Section 1.A),* $\mathbb{P}[Z_n > 0] = \mathbb{P}[Z_n \geq 1] \leq \mathbb{E}[Z_n] = m^n$. *Hence, in the* subcritical *case,* $\mathbb{P}[Z_n > 0] \to 0$ *as* $n \to \infty$ *(i.e.,* $Z_n \to 0$ *in* probability*). Moreover, the* average *total* population *in this case is finite, because* $\mathbb{E}[\sum_{n\geq 0} Z_n] = \sum_{n\geq 0} m^n = (1-m)^{-1} < \infty$. *It follows that, with probability 1,* $\sum_{n\geq 0} Z_n < \infty$, *which entails* $Z_n \to 0$ almost surely*. This last statement can also be deduced from the fact that* $\sum_{n\geq 0} \mathbb{P}[Z_n > 0] < \infty$ *using the* Borel–Cantelli lemma *(see, e.g.,* [1, 2]*).*

Extinction is the event $\mathcal{E} = \cup_{n=1}^{\infty}\{Z_n = 0\}$. As $\{Z_n = 0\} \subseteq \{Z_{n+1} = 0\}$ for all $n \geq 0$, the *extinction probability* $\rho = \mathbb{P}[\mathcal{E}]$ is well defined (by continuity of probability measure: see Section 1.A) via $\rho = \lim_{n\to\infty} \mathbb{P}[Z_n = 0]$, where $\mathbb{P}[Z_n = 0] = \varphi_n(0)$ is the probability of extinction *before* the $(n+1)$th generation.

Theorem 1.4 *If* $0 < p_0 < 1$, *then the extinction probability* ρ *is given by the smallest positive solution to the equation*

$$s = \varphi(s). \tag{1.11}$$

In particular, if $m = \mathbb{E}[Z] \leq 1$, *then* $\rho = 1$; *otherwise, we have* $0 < \rho < 1$.

Remark 1.4 *The relation* (1.11) *has a clear probabilistic sense. Indeed, by independence,* $\mathbb{P}[\mathcal{E} \mid Z_1 = k] = \rho^k$ *(as extinction only occurs if each of the independent branching processes associated with the* k *individuals dies out). Then, by the law of total probability (see Section 1.A), we get*[4)]

$$\rho = \mathbb{P}[\mathcal{E}] = \sum_{k\geq 0} \mathbb{P}[\mathcal{E} \mid Z_1 = k]\mathbb{P}[Z_1 = k]$$
$$= \sum_{k\geq 0} \rho^k \mathbb{P}[Z_1 = k] = \varphi(\rho).$$

Notice that Theorem 1.4 characterizes the extinction probability without the necessity to compute $\varphi_n(\cdot)$. The excluded values $p_0 \in \{0, 1\}$ are trivial: if $p_0 = 0$, then $Z_n \geq 1$ for all $n \geq 0$ so that $\rho = 0$; if $p_0 = 1$, then $\mathbb{P}[Z_1 = 0] = \rho = 1$.

Proof of Theorem 1.4 Denote $\rho_n = \mathbb{P}[Z_n = 0] = \varphi_n(0)$. By continuity and strict monotonicity of the generating function $\varphi(\cdot)$, we have (recall (1.10))

$$0 < \rho_1 = \varphi(0) < \rho_2 = \varphi(\rho_1) < \cdots < 1 = \varphi(1),$$

so that $\rho_n \nearrow \rho \in (0, 1]$ with $\rho = \varphi(\rho)$.

Now if $\bar{\rho}$ is another fixed point of $\varphi(\cdot)$ in $[0, 1]$, that is, $\bar{\rho} = \varphi(\bar{\rho})$, then, by induction,

$$0 < \rho_1 = \varphi(0) < \rho_2 < \cdots < \varphi(\bar{\rho}) = \bar{\rho},$$

so that $\rho = \lim_{n\to\infty} \rho_n \leq \bar{\rho}$, that is, ρ is the smallest positive solution to (1.11).

Finally, by continuity and convexity of $\varphi(\cdot)$ together with the fact $\varphi(1) = 1$, the condition $m = \varphi'(1) \leq 1$ implies $\rho = 1$ and the condition $m = \varphi'(1) > 1$ implies that ρ is the unique solution in $(0, 1)$ to the fixed point equation (1.11).

We thus see that the branching process exhibits a phase transition: in the subcritical or critical regimes ($m \leq 1$), the process dies out with probability 1, whereas in the supercritical case ($m > 1$) it survives forever with positive probability $1 - \rho$.

4) In Section 1.3, we will see this calculation as exploiting the *Markov property* of the branching process.

1.2.3 Other Transforms

1.2.3.1 Moment Generating Functions

The *moment generating function* of a real-valued random variable X is defined by $M_X(t) = \mathbb{E}[e^{tX}]$. When finite for t in some neighborhood of 0, $M_X(t)$ behaves similarly to the generating function $G_X(s)$ in that it possesses the uniqueness property (identifying the corresponding distribution), maps convolutions (i.e., distributions of sums of independent variables) into products, and can be used to establish convergence in distribution.

If X is $\mathbb{Z}_+$-valued and its generating function $G_X(s)$ is finite for some $s > 1$, then $M_X(t)$ is also finite for some $t \neq 0$, and the two are related by $M_X(t) = G_X(e^t)$. For example, if $X \sim \text{Bin}(n,p)$, then $M_X(t) = (1 + p(e^t - 1))^n$ (see Example 1.2a). The terminology arises from the fact that if $M_X(t)$ is differentiable at 0, then the kth derivative of $M_X(t)$ evaluated at 0 gives $\mathbb{E}[X^k]$, the kth *moment* of X.

Example 1.8 In the case of a continuous distribution X with probability density f, the moment generating function $M_X(t)$ becomes the integral transform $\int e^{tx} f(x)dx$.

(a) Let $\lambda > 0$. If X has density $f(x) = \lambda e^{-\lambda x}$ for $x > 0$, and 0 elsewhere, then X has the *exponential distribution* with parameter λ and

$$M_X(t) = \int_0^\infty \lambda e^{-(\lambda - t)x} dx = \frac{\lambda}{\lambda - t}, \quad (t < \lambda).$$

(b) If X has density $f(x) = e^{-x^2/2}/\sqrt{2\pi}$, $x \in \mathbb{R}$, then X has the *standard normal* $\mathcal{N}(0,1)$ or *Gaussian* distribution and

$$M_X(t) = \frac{1}{\sqrt{2\pi}} \int_{-\infty}^\infty e^{t^2/2} e^{-(x-t)^2/2} dx = e^{t^2/2}, \quad (t \in \mathbb{R}).$$

The *normal distribution* was in part so named[5] for its ubiquity in real data. It is also very common in probability and mathematical statistics, owing to a large extent to results of the following kind.

Theorem 1.5 (de Moivre–Laplace central limit theorem) *Let $X_n \sim \text{Bin}(n,p)$ with fixed $p \in (0,1)$. Denote $X_n^* = (X_n - \mathbb{E}[X_n])/\sqrt{\mathbb{V}\text{ar}[X_n]}$. Then for any $t \in \mathbb{R}$,*

$$M_{X_n^*}(t) \to e^{t^2/2}, \quad \text{as } n \to \infty;$$

in other words, the distribution of X_n^* converges to that of $\mathcal{N}(0,1)$.

Proof. Recall that X_n can be written as a sum $Y_1 + Y_2 + \cdots + Y_n$ of independent $\text{Be}(p)$ random variables. In particular, $\mathbb{E}[X_n] = np$ and $\mathbb{V}\text{ar}[X_n] = np(1-p)$ (see Example 1.25). Using the simple relation $M_{aZ+b}(t) = e^{bt} M_Z(at)$, we deduce that the random variable $\hat{Y} = (Y - p)/\sqrt{n}$ has the moment generating function (with fixed t)

$$e^{-tp/\sqrt{n}}\left(1 + p\left(e^{t/\sqrt{n}} - 1\right)\right) = 1 + \frac{t^2}{2n} p(1-p) + O\left(\frac{t^3}{\sqrt{n}}\right).$$

5) The term *normal* (in the sense of "usual") was apparently attached to the distribution by Francis Galton and others and popularized by Karl Pearson. The distribution arose in the work of Gauss and Laplace on least squares and errors of measurement, and also in Maxwell's work on statistical physics. Perhaps its first tentative appearance, however, is in the work of Abraham de Moivre for whom Theorem 1.5 is named. See [3] for a discussion.

Noticing that $M_{X_n^*}(t) = (M_{\hat{Y}}(t/\sqrt{p(1-p)}))^n \to e^{t^2/2}$ as $n \to \infty$, we deduce the result.

One other application of moment generating functions that we will see is to *large deviations*: see Section 1.4.4.

1.2.3.2 Laplace Transforms

In general, $M_X(t)$ might be infinite for all $t \neq 0$. However, for nonnegative variables $X \geq 0$, we have $M_X(t) = \mathbb{E}[e^{tX}] \leq 1$, for all $t \leq 0$; in particular, $M_X(t)$ is an analytic function at every point of the complex plane with negative real part. In this case, $M_X(t)$ behaves very similarly to generating functions and inherits the main properties described above. In such a situation, the function $M_X(t)$ (or, sometimes $M_X(-t)$) is called the *Laplace transform* of the variable X. See Chapter 15 in the present volume for background on Laplace transforms.

1.2.3.3 Characteristic Functions

Unlike the moment generating function $M_X(t)$, which might be infinite for real $t \neq 0$, the *characteristic function* $\psi_X(t) = \mathbb{E}[e^{itX}]$ (where $i^2 = -1$) always exists and uniquely identifies the distribution, hence the name.[6] The characteristic functions inherit all nice properties of (moment) generating functions, though inverting them is not always straightforward.

Characteristic functions are the standard tool of choice for proving results such as the following generalization of Theorem 1.5. The proof is similar to that of the previous theorem, based on a Taylor-type formula: if $\mathbb{E}[X^{2n}] < \infty$, then

$$\psi_X(t) = \sum_{\ell=0}^{2n} \frac{(it)^\ell}{\ell!} \mathbb{E}[X^\ell] + o(t^{2n}).$$

Theorem 1.6 (Central limit theorem)
Let $X_n = Y_1 + Y_2 + \cdots + Y_n$, where Y_i are i.i.d. random variables with $\mathbb{E}[Y_i^2] < \infty$. Then, as $n \to \infty$, the distribution of $X_n^* = (X_n - \mathbb{E}[X_n])/\sqrt{\mathbb{V}\mathrm{ar}[X_n]}$ converges to the standard normal, $\mathcal{N}(0, 1)$.

1.3 Markov Chains in Discrete Time

1.3.1 What is a Markov Chain?

Our tour begins with stochastic processes in *discrete time*. Here, we will write our process as $X_0, X_1, X_2, \ldots$. A fundamental class is constituted by the *Markov processes* in which, roughly speaking, *given the present, the future is independent of the past*. In this section, we treat the case where the X_n take values in a *discrete* (i.e., finite or countably infinite) state space S. In this case, the general term Markov process is often specialized to a *Markov chain*, although the usage is not universally consistent.

The process $X = (X_n)$ taking values in the discrete set S satisfies the *Markov property* if, for any n and any $i, j, \in S$,[7]

$$\mathbb{P}[X_{n+1} = j \mid X_n = i, X_{n-1} = i_{n-1}, \ldots, X_0 = i_0] = p_{ij}, \tag{1.12}$$

for all previous histories $i_0, \ldots, i_{n-1} \in S$.

The $p_{ij} = \mathbb{P}[X_{n+1} = j \mid X_n = i]$ are the *one-step transition probabilities* for X, and they satisfy the obvious conditions $p_{ij} \geq 0$ for all i, j, and $\sum_{j \in S} p_{ij} = 1$ for all i. It is convenient to arrange the p_{ij} as a matrix $P = (p_{ij})_{i,j \in S}$ with nonnegative entries and

6) The term *characteristic function* is traditional in the probabilistic context for what elsewhere might be called the *Fourier transform*: see Chapter 15 of the present volume.

7) In (1.12) and elsewhere, we indicate intersections of events by commas for readability.

whose rows all sum to 1: these properties define a *stochastic matrix*. The Markov property specifies the step-by-step evolution of the Markov chain. One can imagine a particle moving at random on S, from state i selecting its next location according to distribution $(p_{ij})_{j\in S}$. It is an exercise in conditional probability to deduce from (1.12) that

$$\mathbb{P}[X_1 = i_1, \ldots, X_n = i_n \mid X_0 = i_0] = p_{i_0 i_1} \cdots p_{i_{n-1} i_n}.$$

It may be that X_0 is itself random, in which case, in order to assign a probability to any particular (finite) sequence of moves for the particle, in addition to the Markov property, we also need to know where the particle starts: this is specified by the *initial distribution* $\mathbb{P}[X_0 = i] = w_i$, $i \in S$.

Now, to compute the probability of getting from i to k in *two* steps, we sum over the j-partition to get

$$\mathbb{P}[X_2 = k \mid X_0 = i] = \sum_{j\in S} \mathbb{P}[X_1 = j, X_2 = k \mid X_0 = i] = \sum_{j\in S} p_{ij} p_{jk} = (P^2)_{ik},$$

the (i, k) entry in $P^2 = P \cdot P$; this *matrix multiplication* yields the *two-step transition probability*. More generally, the n-step transition probabilities are

$$p_{ij}^{(n)} := \mathbb{P}[X_n = j \mid X_0 = i] = (P^n)_{ij}.$$

A similar argument shows the following:

$$p_{ij}^{(n+m)} = \sum_{k\in S} p_{ik}^{(n)} p_{kj}^{(m)}. \tag{1.13}$$

Remark 1.5 *We restrict ourselves to the case of* time-homogeneous *Markov chains, by stipulating that (1.12) should hold simultaneously for all n. One may relax this, and allow* p_{ij} *to depend also on n.*

Remark 1.6 *An equivalent definition of a Markov chain is as a* randomized dynamical system: $X_{n+1} = f(X_n; U_{n+1})$ *where* $f : S \times [0,1] \to S$ *is a fixed update function, and* $U_1, U_2, \ldots$ *are independent* $U[0,1]$ *random variables.*[8] Monte Carlo simulation *of a Markov chain uses such a scheme.*

We make a final remark on notation: for p_{ij}, and similar expressions, we sometimes write $p_{i,j}$ if otherwise the subscripts may be ambiguous.

1.3.2 Some Examples

Example 1.9 [The Ehrenfest model] In 1907, Ehrenfest and Ehrenfest [4] introduced this simple model of diffusion. There are N particles in a container that has two chambers separated by a permeable partition. At each step, a particle is chosen uniformly at random and moved across the partition. The state of the Markov chain at each time will be the number of particles in the first chamber, say, so $S = \{0, \ldots, N\}$.

The one-step transition probabilities are, for $i \in \{1, \ldots, N-1\}$,

$$p_{i,i+1} = \frac{N-i}{N}, \quad p_{i,i-1} = \frac{i}{N},$$

and $p_{0,1} = p_{N,N-1} = 1$.

After a long time, what is the distribution of the particles? See Section 1.3.6.

Example 1.10 [One-dimensional simple random walk] A particle moves at random on the state space $S = \mathbb{Z}_+$. From position $i \neq 0$, the particle jumps one step to the left with probability p_i and one step to the right with probability $1 - p_i$. With partial reflection at 0, we can describe this *random*

8) That is, uniform on $[0, 1]$, having density $f(x) = 1$ for $x \in [0, 1]$ and $f(x) = 0$ elsewhere.

walk by a Markov chain with one-step transition probabilities $p_{0,0} = p_0 \in (0,1)$, $p_{0,1} = q_0 := 1 - p_0$, and for $i \geq 1$,

$$p_{i,i-1} = p_i \in (0,1), \quad p_{i,i+1} = q_i := 1 - p_i.$$

1.3.3 Stationary Distribution

We use the compact notation $\mathbb{P}_i$ for the (conditional) probability associated with the Markov chain started from state $i \in S$, that is, $\mathbb{P}_i[\,\cdot\,] = \mathbb{P}[\,\cdot \mid X_0 = i]$. More generally, if $w = (w_i)_{i \in S}$ is a distribution on S (i.e., $w_i \geq 0$, $\sum_i w_i = 1$), then we write $\mathbb{P}_w$ for the Markov chain started from the initial distribution w, that is, $\mathbb{P}_w[\,\cdot\,] = \sum_i w_i \mathbb{P}_i[\,\cdot\,]$.

A distribution $\pi = (\pi_i)_{i \in S}$ is a *stationary distribution* for a Markov chain X if

$$\mathbb{P}_\pi[X_1 = i] = \pi_i, \text{ for all } i \in S. \tag{1.14}$$

Viewing a stationary distribution π as a row vector, (1.14) is equivalent to the matrix-vector equation $\pi P = \pi$, that is, π is a left eigenvector of P corresponding to the eigenvalue 1. The nomenclature arises from the fact that (1.14) implies that $\mathbb{P}_\pi[X_n = i] = \pi_i$ for *all* times n, so the distribution of the Markov chain started according to π is stationary in time.

Example 1.11 [A three-state chain] Consider a Markov chain (X_n) with the state space $\{1,2,3\}$ and transition matrix

$$P = \begin{pmatrix} \frac{1}{2} & \frac{1}{2} & 0 \\ \frac{1}{3} & \frac{1}{3} & \frac{1}{3} \\ 0 & \frac{1}{2} & \frac{1}{2} \end{pmatrix}. \tag{1.15}$$

We look for a stationary distribution $\pi = (\pi_1, \pi_2, \pi_3)$. Now $\pi P = \pi$ with the fact that $\pi_1 + \pi_2 + \pi_3 = 1$ gives a system of equations with unique solution $\pi = (\frac{2}{7}, \frac{3}{7}, \frac{2}{7})$.

A Markov chain with transition probabilities p_{ij} is *reversible* with respect to a distribution $\pi = (\pi_i)_{i \in S}$ if the *detailed balance equations* hold:

$$\pi_i p_{ij} = \pi_j p_{ji}, \text{ for all } i, j \in S. \tag{1.16}$$

Not every Markov chain is reversible. Any distribution π satisfying (1.16) is necessarily a stationary distribution, because then, for all j, $\sum_i \pi_i p_{ij} = \sum_i \pi_j p_{ji} = \pi_j$. If the chain *is* reversible, then the system of equations (1.16) is often simpler to solve than the equations $\pi P = \pi$: see Example 1.12. The "physical" interpretation of reversibility is that, in equilibrium, the Markov chain is statistically indistinguishable from a copy of the chain running backward in time.

Example 1.12 [Random walk] Consider Example 1.10. We seek a solution $\pi = (\pi_i)$ to (1.16), which now reads $\pi_i q_i = \pi_{i+1} p_{i+1}$ for all $i \geq 0$. The solution is $\pi_i = \pi_0 \prod_{j=0}^{i-1} (q_j / p_{j+1})$. This describes a proper distribution if $\sum_i \pi_i = 1$, that is, if

$$\sum_{i=0}^{\infty} \prod_{j=0}^{i-1} \frac{q_j}{p_{j+1}} < \infty. \tag{1.17}$$

If (1.17) holds, then

$$\pi_i = \frac{\prod_{j=0}^{i-1} \frac{q_j}{p_{j+1}}}{\sum_{i=0}^{\infty} \prod_{j=0}^{i-1} \frac{q_j}{p_{j+1}}}.$$

For example, if $p_i = p \in (0,1)$ for all i, then (1.17) holds if and only if $p > 1/2$, in which case $\pi_i = (q/(1-2p))(q/p)^i$, where $q = 1 - p$, an exponentially decaying stationary distribution.

1.3.4 The Strong Markov Property

One way of stating the Markov property is to say that $(X_n)_{n \geq m}$, conditional on

$\{X_m = i\}$, is distributed as the Markov chain $(X_n)_{n\geq 0}$ with initial state $X_0 = i$. It is often desirable to extend such a statement from deterministic times m to *random* times T. An important class of random times are the *first passage times*,[9]

$$T_i := \min\{n \geq 1 : X_n = i\}, \; i \in S. \quad (1.18)$$

The Markov property cannot hold at every random time. For instance, if $T' = T_i - 1$, then the first transition of the process $X_{T'}, X_{T'+1}, \ldots$ is always from $X_{T'}$ to $X_{T'+1} = i$, regardless of the original transition matrix.

The following *strong Markov property* clarifies these issues. A random time $T \in \mathbb{Z}_+ \cup \{\infty\}$ is a *stopping time* with respect to (X_n) if, for any n, the event $\{T \leq n\}$ depends only on $X_0, \ldots, X_n$ (and not on the future evolution of the chain). The passage times T_i are stopping times, but T' described above is not a stopping time.

Lemma 1.2 *Suppose that T is a stopping time for (X_n). Then, given $T < \infty$ and $X_T = i$, $(X_{T+n})_{n\geq 0}$ has the same distribution as $(X_n)_{n\geq 0}$ started from $X_0 = i$.*

Sketch of proof Partition over the possible values of T. Suppose that $T = m$ and $X_T = X_m = i$; this is a condition only on $X_0, \ldots, X_m$, because T is a stopping time. Now apply the usual Markov property at the deterministic time m.

1.3.5 The One-Step Method

In problems involving Markov chains, often quantities of interest are *hitting probabilities* and *expected hitting times*. One approach to computing these is via the powerful *one-step method*, which makes essential use of the Markov property.

9) Here and elsewhere, the convention $\min \emptyset = \infty$ is in force.

Recall the definition of the passage times T_i from (1.18). The *expected hitting time* of state j starting from state i is $\mathbb{E}_i[T_j]$ for $i \neq j$; if $i = j$ this is the *expected return time* to i. Also of interest is $\mathbb{P}_i[T_j < T_k]$, the probability of reaching state j before state k, starting from i. We illustrate the one-step method by some examples.

Example 1.13 [Three-state chain] We return to Example 1.11. We partition over the *first step* of the process to obtain, via the law of total probability (see Section 1.A),

$$\begin{aligned}\mathbb{P}_2[T_1 < T_3] &= \sum_{k=1}^{3} \mathbb{P}_2[\{T_1 < T_3\} \cap \{X_1 = k\}] \\ &= p_{2,1} \cdot 1 + p_{2,2} \cdot \mathbb{P}_2[T_1 < T_3] \\ &\quad + p_{2,3} \cdot 0,\end{aligned}$$

by the Markov property. This gives $\mathbb{P}_2[T_1 < T_3] = 1/2$.

What about $\mathbb{E}_2[T_1]$? Set $z_i = \mathbb{E}_i[T_1]$. Again we condition on the first step, and now use the partition theorem for expectations (see Section 1.A):

$$\begin{aligned}\mathbb{E}_2[T_1] &= 1 + \mathbb{E}_2[T_1 - 1] \\ &= 1 + \sum_{k=1}^{3} p_{2,k} \mathbb{E}_2[T_1 - 1 \mid X_1 = k].\end{aligned}$$

Now applying the Markov property at time 1, we see that $T_1 - 1$, given $X_0 \neq 1$ and $X_1 = k \neq 1$, has the same distribution as T_1 given $X_0 = k \neq 1$ in the original chain, and, in particular, has expected value z_k. On the other hand, if $X_1 = k = 1$, then $T_1 - 1 = 0$. So we get $z_2 = 1 + \frac{1}{3}z_2 + \frac{1}{3}z_3$. A similar argument starting from state 3 gives $z_3 = 1 + \frac{1}{2}z_2 + \frac{1}{2}z_3$. This system of linear equations is easily solved to give $z_2 = 5$ and $z_3 = 7$.

Example 1.14 [Random walk; gambler's ruin] Recall Example 1.10. Fix $n \in \mathbb{N}$ and for $i \in \{1, \dots, n-1\}$, let $u_i = \mathbb{P}_i[T_n < T_0]$. The one-step method gives

$$u_i = p_i u_{i-1} + q_i u_{i+1}, \quad (1 \le i \le n-1),$$

with boundary conditions $u_0 = 0$, $u_n = 1$. The standard method to solve this system of equations is to rewrite it in terms of the differences $\Delta_i = u_{i+1} - u_i$ to get $\Delta_i = \Delta_{i-1}(p_i/q_i)$ for $1 \le i \le n-1$, which yields $\Delta_j = \Delta_0 \prod_{k=1}^{j}(p_k/q_k)$. Then $u_i = \sum_{j=0}^{i-1} \Delta_j$, using the boundary condition at 0. Using the boundary condition at n to fix Δ_0, the solution obtained is

$$u_i = \frac{\sum_{j=0}^{i-1} \prod_{k=1}^{j} \frac{p_k}{q_k}}{\sum_{j=0}^{n-1} \prod_{k=1}^{j} \frac{p_k}{q_k}}. \tag{1.19}$$

In the special case where $p_i = q_i = 1/2$ for all i, we have the elegant formula $u_i = i/n$. If we imagine that the state of the Markov chain is the wealth of a gambler with initial wealth i who plays a sequence of fair games, each time either gaining or losing a unit of wealth, $1 - u_i$ is the *ruin probability* (and u_i is the probability that the gambler makes his fortune).

1.3.6 Further Computational Methods

We present by example some additional techniques.

Example 1.15 [Matrix diagonalization] In many situations, we want to compute the n-step transition probability $p_{ij}^{(n)}$, that is, an entry in the matrix power P^n. To calculate P^n, we try to *diagonalize* P to obtain $P = T\Lambda T^{-1}$ for an invertible matrix T and a diagonal matrix Λ. The usefulness of this representation is that $P^n = T\Lambda^n T^{-1}$ and Λ^n is easy to write down, because Λ is diagonal. A sufficient condition for P to be diagonalizable is that all its eigenvalues be distinct.

Consider again the three-state chain with transition matrix given by (1.15); we have three eigenvalues, $\lambda_1, \lambda_2, \lambda_3$, say. As P is a stochastic matrix, 1 is always an eigenvalue: $\lambda_1 = 1$, say. Then because $\operatorname{tr} P = \lambda_1 + \lambda_2 + \lambda_3$ and $\det P = \lambda_1\lambda_2\lambda_3$, we find $\lambda_2 = 1/2$ and $\lambda_3 = -1/6$, say.

It follows from the diagonalized representation that

$$\begin{aligned} P^n &= \lambda_1^n U_1 + \lambda_2^n U_2 + \lambda_3^n U_3 \\ &= U_1 + \left(\frac{1}{2}\right)^n U_2 + \left(-\frac{1}{6}\right)^n U_3, \end{aligned} \tag{1.20}$$

where U_1, U_2, U_3 are 3×3 matrices to be determined. One can solve the simultaneous matrix equations arising from the cases $n \in \{0, 1, 2\}$ of (1.20) to obtain U_1, U_2, and U_3, and hence,

$$P^n = \begin{pmatrix} \frac{2}{7} & \frac{3}{7} & \frac{2}{7} \\ \frac{2}{7} & \frac{3}{7} & \frac{2}{7} \\ \frac{2}{7} & \frac{3}{7} & \frac{2}{7} \end{pmatrix} + \left(\frac{1}{2}\right)^n \begin{pmatrix} \frac{1}{2} & 0 & -\frac{1}{2} \\ 0 & 0 & 0 \\ -\frac{1}{2} & 0 & \frac{1}{2} \end{pmatrix} + \left(-\frac{1}{6}\right)^n \begin{pmatrix} \frac{3}{14} & -\frac{3}{7} & \frac{3}{14} \\ -\frac{2}{7} & \frac{4}{7} & -\frac{2}{7} \\ \frac{3}{14} & -\frac{3}{7} & \frac{3}{14} \end{pmatrix}.$$

It follows that $\lim_{n\to\infty} p_{ij}^{(n)}$ exists, does not depend on i, and is equal to π_j, the component of the stationary distribution that we calculated in Example 1.11. After a long time, the chain "forgets" its starting state and approaches a stochastic equilibrium described by the stationary distribution. This is an example of a general phenomenon to which we return in Section 1.3.7.

Example 1.16 [Generating functions] We sketch the use of generating functions to evaluate stationary distributions. Consider the Ehrenfest model of Example 1.9. Suppose that $\pi = (\pi_0, \dots, \pi_N)$ is a stationary distribution for the Markov chain, with generating function $\hat{\pi}(s) = \sum_{i=0}^{N} \pi_i s^i$. In this case, the equation $\pi P = \pi$ reads

$$\pi_{j-1}\frac{N-(j-1)}{N} + \pi_{j+1}\frac{j+1}{N} = \pi_j,$$

which is valid for all $j \in \{0, \dots, N\}$, provided we set $\pi_{-1} = \pi_{N+1} = 0$. Now multiply through by s^j and sum from $j = 0$ to N. After some algebra, we obtain

$$\hat{\pi}(s) = \frac{1-s^2}{N}\hat{\pi}'(s) + s\hat{\pi}(s),$$

so that $\mathrm{d}/\mathrm{d}s \log \hat{\pi}(s) = \hat{\pi}'(s)/\hat{\pi}(s) = N/(1+s)$. Integrating with respect to s and using the fact that $\hat{\pi}(1) = 1$, we obtain

$$\hat{\pi}(s) = \left(\frac{1+s}{2}\right)^N.$$

The binomial theorem now enables us to identify $\pi_i = 2^{-N}\binom{N}{i}$.

1.3.7 Long-term Behavior; Irreducibility; Periodicity

We saw in Example 1.15 a Markov chain for which

$$\lim_{n\to\infty} \mathbb{P}_i[X_n = j] = \lim_{n\to\infty} p_{ij}^{(n)} = \pi_j, \qquad (1.21)$$

for all $i, j \in S$, where π_j is from a stationary distribution. For which Markov chains does such a result hold? There are (at least) three obstacles:

(a) There might be no solutions to $\pi P = \pi$, and hence, no right-hand side in (1.21).

(b) There might be *multiple* solutions to $\pi P = \pi$, and so no uniqueness in (1.21). For example, consider the Markov chain on the state space $\{0, 1, 2\}$ with $p_{00} = 1$, $p_{22} = 1$ (0 and 2 are *absorbing states*) and $p_{10} = p_{12} = 1/2$. Then $p_{i2}^{(n)} = i/2$ for all $n \geq 1$, that is, the limit on the left-hand side of (1.21) depends on the starting state i. Note that here $\pi = (\alpha, 0, 1-\alpha)$ is stationary for any $\alpha \in [0, 1]$.

(c) In the Ehrenfest model of Example 1.9, there is a parity effect, because $p_{00}^{(n)} = 0$ for *odd* n, for instance. This phenomenon is an example of *periodicity,* which is another obstacle to (1.21).

Cases (b) and (c) here can be dealt with after some additional concepts are introduced. A state $i \in S$ has *period* d if d is the greatest common divisor of $\{n \geq 1 : p_{ii}^{(n)} > 0\}$. For example, all states in the Ehrenfest model have period 2.

A Markov chain is *irreducible* if, for all $i, j \in S$, there exist finite m and n for which $p_{ij}^{(n)} > 0$ and $p_{ji}^{(m)} > 0$, that is, it is possible to get between any two states in a finite number of steps. For the rest of this section, we will assume that we have an irreducible Markov chain. We do not discuss the case of nonirreducible (*reducible*) chains in a systematic way, but Section 1.3.10 provides an illustrative example.

For an irreducible chain, it can be shown that all states have the same period, in which case one can speak about the period of the chain itself. If all states have period 1, the chain is called *aperiodic*.

Recall the definition of T_i from (1.18): $\mathbb{E}_i[T_i]$ is the *expected return time* to i. The following result answers our question on the limiting behavior of $p_{ij}^{(n)}$.

Theorem 1.7 *For an irreducible Markov chain, the following are equivalent.*

- *There exists a unique stationary distribution* π.
- *For some* $i \in S$, $\mathbb{E}_i[T_i] < \infty$.
- *For all* $i \in S$, $\mathbb{E}_i[T_i] < \infty$.

If these conditions hold, the Markov chain is called positive recurrent. *For a positive-recurrent chain, the following hold.*

- *For all* $i \in S$, $\pi_i = 1/\mathbb{E}_i[T_i]$.
- *If the chain is aperiodic, then* $\mathbb{P}_i[X_n = j] \to \pi_j$ *for all* $i, j \in S$.

In particular, we have the following result.

Theorem 1.8 *An irreducible Markov chain on a finite state space is positive recurrent.*

Proofs of these results can be found in [5, 6], for instance.

1.3.8 Recurrence and Transience

Recall the definition of T_i from (1.18). A state $i \in S$ is called *recurrent* if $\mathbb{P}_i[T_i < \infty] = 1$ or *transient* if $\mathbb{P}_i[T_i = \infty] > 0$. A Markov chain will return infinitely often to a recurrent state, but will visit a transient state only finitely often. If a Markov chain is irreducible (see Section 1.3.7), then either all states are recurrent, or none is, and so we can speak of recurrence or transience of the chain itself.

If an irreducible chain is positive recurrent (see Theorem 1.7), then it is necessarily recurrent. A chain that is recurrent but not positive recurrent is *null recurrent*, in which case, for all i, $\mathbb{P}_i[T_i < \infty] = 1$ but $\mathbb{E}_i[T_i] = \infty$ (equivalently, it is recurrent but no stationary distribution exists). Because of Theorem 1.8, we know that to observe null recurrence or transience we must look at infinite state spaces.

Example 1.17 [One-dimensional random walk] We return to Example 1.10. Consider $\mathbb{P}_0[T_0 = \infty]$. In order for the walk to never return to 0, the first step must be to 1, and then, starting from 1, the walk must reach n before 0 for *every* $n \geq 2$. Thus

$$\begin{aligned}\mathbb{P}_0[T_0 = \infty] &= q_0 \mathbb{P}_1[T_0 = \infty] \\ &= q_0 \mathbb{P}_1[\cap_{n\geq 2}\{T_n < T_0\}].\end{aligned}$$

Note $\{T_{n+1} < T_0\} \subseteq \{T_n < T_0\}$, so the intersection here is over a decreasing sequence of events. Thus by continuity of probability measures (see Section 1.A), $\mathbb{P}_0[T_0 = \infty] = q_0 \lim_{n\to\infty} \mathbb{P}_1[T_n < T_0]$. Here $\mathbb{P}_1[T_n < T_0] = 1 - u_1$ where u_1 is given by (1.19). So we obtain

$$\mathbb{P}_0[T_0 = \infty] > 0 \text{ if and only if } \sum_{j=0}^{\infty} \prod_{k=1}^{j} \left(\frac{p_k}{q_k}\right) < \infty.$$

(For further discussion, see [7, pp. 65–71]) In particular, if $p_k = p \in (0, 1)$ for all k, the walk is transient if $p < 1/2$ and recurrent if $p \geq 1/2$. The phase transition can be probed more precisely by taking $p_k = 1/2 + c/k$; in this case, the walk is transient if and only if $c < -1/4$, a result due to Harris and greatly generalized by Lamperti [8].

We give one criterion for recurrence that we will use in Section 1.4.

Lemma 1.3 *For any* $i \in S$, $\mathbb{P}_i[T_i < \infty] = 1$ *if and only if* $\sum_{n=0}^{\infty} p_{ii}^{(n)} = \infty$.

We give a proof via generating functions. Write $f_i^{(n)} = \mathbb{P}_i[T_i = n]$, the probability that the first return to i occurs at time n; here $f_i^{(0)} = 0$; note that $\sum_n f_i^{(n)}$ may be less than 1. Denote the corresponding generating function by $\phi_i(s) = \sum_{n=0}^{\infty} f_i^{(n)} s^n$. Also define $\psi_i(s) = \sum_{n=0}^{\infty} p_{ii}^{(n)} s^n$, where $p_{ii}^{(n)} = \mathbb{P}_i[X_n = i]$ (so $p_{ii}^{(0)} = 1$). By conditioning on the value of T_i, the strong Markov property gives

$$p_{ii}^{(n)} = \sum_{m=0}^{n} f_i^{(m)} p_{ii}^{(n-m)}, \ (n \geq 1).$$

Treating the case $n = 0$ carefully, it follows that

$$\psi_i(s) = 1 + \sum_{n=0}^{\infty} \sum_{m=0}^{n} f_i^{(m)} p_{ii}^{(n-m)} s^n.$$

The final term here is a discrete convolution of the generating function (cf. Theorem 1.2), so we deduce the important *renewal relation*

$$\psi_i(s) = 1 + \phi_i(s)\psi_i(s). \tag{1.22}$$

Sketch of proof of Lemma 1.3 We have $\mathbb{P}_i[T_i < \infty] = \lim_{s\uparrow 1} \phi_i(s)$, and (1.22) implies that the latter limit is 1 if and only if $\lim_{s\uparrow 1} \psi_i(s) = \infty$.

1.3.9 Remarks on General State Spaces

In the case of discrete state spaces, (1.13) corresponds to the trivial matrix equation $P^{n+m} = P^n \cdot P^m$, which one could describe, rather grandly, as the *semigroup property* of matrix multiplication. More generally, (1.13) is an instance of the fundamental Chapman–Kolmogorov relation, and the connection to semigroup theory runs deep.

In a general state space, the analogue of the transition probability p_{ij} is a transition kernel $p(x; A)$ given by $p(x; A) = \mathbb{P}[X_{n+1} \in A \mid X_n = x]$. This immediately introduces technical issues that can only be addressed in the context of measure theory. We refer to [2, 9], for example.

1.3.10 Example: Bak–Sneppen and Related Models

Bak and Sneppen [10] introduced a simple stochastic model of evolution that initiated a considerable body of research by physicists and mathematicians. In the original model, N sites are arranged in a ring. Each site, corresponding to a species in the evolution model, is initially assigned an independent $U[0, 1]$ random variable representing a "fitness" value for the species. The Bak–Sneppen model is a discrete-time Markov process, where at each step the minimal fitness value and the values at the two neighboring sites are replaced by three independent $U[0, 1]$ random variables.

This process is a Markov process on the continuous state space $[0, 1]^N$, and its behavior is still not fully understood, despite a large physics literature devoted to these models: see the thesis [11] for an overview of the mathematical results.

Here we treat a much simpler model, following [12]. The state space of our process (X_n) will be the "simplex" of ranked sequence of N fitness values

$$\Delta_N := \{(x^{(1)}, \ldots, x^{(N)}) \in [0, 1]^N : \\ x^{(1)} \leq \cdots \leq x^{(n)}\}.$$

Fix a parameter $k \in \{1, \ldots, N\}$. We start with N independent $U[0, 1]$ values: rank these to get X_0. Given X_n, discard the *kth-ranked* value $X_n^{(k)}$ and replace it by a new independent $U[0, 1]$ random variable; rerank to get X_{n+1}.

For example, if $k = 1$, we replace the minimal value at each step. It is natural to anticipate that X_n should approach (as $n \to \infty$) a limiting (stationary) distribution; observe that the value of the second-ranked fitness cannot decrease. A candidate limit is not hard to come by: the distribution of the random vector $(U, 1, 1, 1, \dots, 1)$ (a $U[0,1]$ variable followed by $N-1$ units) is invariant under the evolution of the Markov chain. We show the following result.

Proposition 1.1 *Let $N \in \mathbb{N}$ and $k \in \{1, 2, \dots, N\}$. If at each step we replace the kth-ranked value by an independent $U[0,1]$ value, then, as $n \to \infty$,*

$$(X_n^{(1)}, X_n^{(2)}, \dots, X_n^{(N)}) \to (0, \dots, 0, U, 1, \dots, 1),$$

in distribution,[10] *where the kth coordinate of the limit vector $U \sim U[0,1]$.*

The process X_n lives on a continuous state space, and it might seem that some fairly sophisticated argument would be needed to show that it has a unique stationary distribution. In fact, we can reduce the problem to a simpler problem on a finite state space as follows.

Sketch of proof of Proposition 1.1 We sketch the argument from [12]. For each $s \in [0,1]$, define the *counting function*[11] $C_n(s) := \sum_{i=1}^{N} \mathbf{1}\{X_n^{(i)} \le s\}$, the number of fitnesses of value at most s at time n. Then $C_n(s)$ is a Markov chain on $\{0, 1, 2, \dots, N\}$. The transition probabilities $p_{x,y} = \mathbb{P}[C_{n+1}(s) = y \mid C_n(s) = x]$ are given for $x \in \{0, \dots, k-1\}$ by $p_{x,x} = 1 - s$ and $p_{x,x+1} = s$, and for $x \in \{k, \dots, N\}$ by $p_{x,x} = s$ and $p_{x,x-1} = 1 - s$. For $s \in (0,1)$, the Markov chain is reducible and all states are transient apart from those in the *recurrent class* $S_k = \{k-1, k\}$. The chain will eventually enter S_k and then never exit. So the problem reduces to that of the two-state restricted chain on S_k. It is easy to compute the stationary distribution and for $s \in (0,1)$, analogously to Theorem 1.7,

$$\lim_{n\to\infty} \mathbb{P}[C_n(s) = x] = \begin{cases} 1-s & \text{if } x = k-1 \\ s & \text{if } x = k \\ 0 & \text{if } n \notin \{k-1, k\} \end{cases}.$$

In particular, for $s \in (0,1)$,

$$\lim_{n\to\infty} \mathbb{P}[X_n^{(k)} \le s] = \lim_{n\to\infty} \mathbb{P}[C_n(s) \ge k] = s.$$

That is, $X_n^{(k)}$ converges in distribution to a $U[0,1]$ variable. Moreover, if $k > 1$, for any $s \in (0,1)$, $\mathbb{P}[X_n^{(k-1)} \le s] = \mathbb{P}[C_n(s) \ge k-1] \to 1$, which implies that $X_n^{(k-1)}$ converges in probability to 0. Similarly, if $k < N$, for any $s \in (0,1)$, $\mathbb{P}[X_n^{(k+1)} \le s] = \mathbb{P}[C_n(s) \ge k+1] \to 0$, which implies that $X_n^{(k+1)}$ converges in probability to 1. Combining these marginal results, an additional technical step gives the claimed joint convergence: we refer to [12] for details.

1.4 Random Walks

> *A drunk man will eventually find his way home, but a drunk bird may get lost for ever.*
>
> – *S. Kakutani's rendering of Pólya's theorem [1, p. 191].*

1.4.1 Simple Symmetric Random Walk

The term *random walk* can refer to many different models or classes of models. Although random walks in one dimension had been studied in the context of games of chance, serious study of random

10) That is, for any $x_1, \dots, x_N \in [0,1]$, $\mathbb{P}[X_n^{(1)} \le x_1, \dots, X_n^{(k)} \le x_k, \dots, X_n^{(N)} \le x_N] \to x_k$ if $x_{k+1} \cdots x_N = 1$ and 0 otherwise.

11) "$\mathbf{1}\{\,\cdot\,\}$" is the indicator random variable of the appended event: see Section 1.A.

walks as stochastic processes emerged in pioneering works in several branches of science around 1900: Lord Rayleigh's [13] theory of sound developed from about 1880, Bachelier's [14] 1900 model of stock prices, Pearson and Blakeman's [15] 1906 theory of random migration of species, and Einstein's [16] theory of Brownian motion developed during 1905–1908.

In this section, we restrict attention to *simple symmetric random walk* on the integer lattice $\mathbb{Z}^d$. This model had been considered by Lord Rayleigh, but the preeminent early contribution came from George Pólya [17]: we describe his recurrence theorem in the following text. The phrase "random walk" was first applied by statistical pioneer Pearson [18] to a different model in a 1905 letter to *Nature*. We refer to [19] for an overview of a variety of random walk models.

Let $\mathbf{e}_1, \dots, \mathbf{e}_d$ be the standard orthonormal lattice basis vectors for $\mathbb{Z}^d$. Let $X_1, X_2, \dots$ be i.i.d. random vectors with

$$\mathbb{P}[X_1 = \mathbf{e}_i] = \mathbb{P}[X_1 = -\mathbf{e}_i] = \frac{1}{2d}, \text{ for } i \in \{1, \dots, d\}.$$

Let $S_0 = 0$ and $S_n = \sum_{i=1}^{n} X_i$. Then $(S_n)_{n\in\mathbb{Z}_+}$ is a simple symmetric random walk on $\mathbb{Z}^d$, started from 0; "simple" refers to the fact that the jumps are of size 1.

1.4.2 Pólya's Recurrence Theorem

Clearly (S_n) is a Markov chain; a fundamental question is whether it is recurrent or transient (see Section 1.3.8). Pólya [17] provided the answer in 1921.

Theorem 1.9 (Pólya's theorem) *A simple symmetric random walk on $\mathbb{Z}^d$ is recurrent if $d = 1$ or 2 but transient if $d \geq 3$.*

A basic component in the proof is a combinatorial statement.

Lemma 1.4 *For $d \in \mathbb{N}$ and any $n \in \mathbb{Z}_+$, we have*

$$\mathbb{P}[S_{2n} = 0] = (2d)^{-2n} \binom{2n}{n} \sum_{n_1+\cdots+n_d=n} \left(\frac{n!}{n_1! \cdots n_d!} \right)^2,$$

where the sum is over d-tuples of nonnegative integers $n_1, \dots, n_d$ that sum to n.

Proof. Each path of length $2n$ (i.e., the possible trajectory for $S_0, S_1, \dots, S_{2n}$) has probability $(2d)^{-2n}$. Any such path that finishes at its starting point must, in each coordinate i, take the same number n_i steps in the positive and negative directions. Enumerating all such paths, we obtain

$$\mathbb{P}[S_{2n} = 0] = (2d)^{-2n} \sum_{n_1+\cdots+n_d=n} \frac{(2n)!}{(n_1! \cdots n_d!)^2},$$

from which the given formula follows.

Lemma 1.4 and a careful asymptotic analysis using Stirling's formula for $n!$ yields the following result.

Lemma 1.5 *For $d \in \mathbb{N}$, as $n \to \infty$,*

$$n^{d/2}\mathbb{P}[S_{2n} = 0] \to \left(\frac{d}{4\pi} \right)^{d/2}.$$

Proof of Theorem 1.9 Apply the criterion in Lemma 1.3 with Lemma 1.5.

1.4.3 One-dimensional Case; Reflection Principle

We consider in more detail the case $d = 1$. Let $T_a := \min\{n \geq 1 : S_n = a\}$. Theorem 1.9 says that $\mathbb{P}[T_0 < \infty] = 1$. The next result gives the distribution of T_0.

Theorem 1.10 *(i) For any $n \in \mathbb{Z}_+$, $\mathbb{P}[T_0 = 2n] = (1/(2n-1))\binom{2n}{n}2^{-2n}$.*
(ii) $\mathbb{E}[T_0^\alpha] < \infty$ if and only if $\alpha < 1/2$.

We proceed by counting sample paths, following the classic treatment by Feller [20, chap. 3]. By an *n-path* we mean a sequence of integers $s_0, \ldots, s_n$ where $|s_{i+1} - s_i| = 1$; for an n-path *from a to b* we add the requirement that $s_0 = a$ and $s_n = b$. We view paths as space–time trajectories $(0, s_0), (1, s_1), \ldots, (n, s_n)$.

Let $N_n(a, b)$ denote the number of n-paths from a to b. Let $N_n^0(a, b)$ be the number of such paths that visit 0. An n-path from a to b must take $(n + b - a)/2$ positive steps and $(n + a - b)/2$ negative steps, so

$$N_n(a, b) = \binom{n}{\frac{1}{2}(n + b - a)}, \tag{1.23}$$

where we interpret $\binom{n}{y}$ as 0 if y is not an integer in the range 0 to n.

Lemma 1.6 (Reflection principle) *If $a, b > 0$, then $N_n^0(a, b) = N_n(-a, b)$.*

Proof. Each n-path from $-a$ to b must visit 0 for the first time at some $c \in \{1, \ldots, n-1\}$. Reflect in the horizontal (time) axis the segment of this path over $[0, c]$ to obtain an n-path from a to b which visits 0. This reflection is one-to-one.

Theorem 1.11 (Ballot theorem) *If $b > 0$, then the number of n-paths from* 0 *to b which* do not revisit 0 *is $\frac{b}{n}N_n(0, b)$.*

Proof. The first step of such a path must be 1, so their number is $N_{n-1}(1, b) - N_{n-1}^0(1, b) = N_{n-1}(1, b) - N_{n-1}(-1, b)$, by Lemma 1.6. Now use (1.23).

Theorem 1.12 *If $b \neq 0$ and $n \geq 1$, $\mathbb{P}[T_0 > n, S_n = b] = \frac{|b|}{n}\mathbb{P}[S_n = b]$.*

Proof. Suppose $b > 0$. The event in question occurs if and only if the walk does not visit 0 during $[1, n]$, and $S_n = b$. By the ballot theorem, the number of such paths is $\frac{b}{n}N_n(0, b)$. Similarly for $b < 0$.

At this point, we are ready to prove Theorem 1.10, but first we take a slight detour to illustrate one further variation on "reflection."

Theorem 1.13 *For $a \neq 0$ and $n \geq 1$, $\mathbb{P}[T_a = n] = \frac{|a|}{n}\mathbb{P}[S_n = a]$.*

Proof via time reversal. Fix n. If the trajectory of the original walk up to time n is

$$(S_0, S_1, S_2, \ldots, S_n) = \left(0, X_1, X_1 + X_2, \ldots, \sum_{i=1}^{n} X_i\right),$$

then the trajectory of the *reversed* walk is

$$(R_0, R_1, R_2, \ldots, R_n) = \left(0, X_n, X_n + X_{n-1}, \ldots, \sum_{i=1}^{n} X_i\right),$$

that is, the increments are taken in reverse order. The reversed walk has the same distribution as the original walk, because the X_i are i.i.d.

Suppose $a > 0$. The original walk has $S_n = a$ and $T_0 > n$ if and only if the reversed walk has $R_n = a$ and $R_n - R_{n-i} = X_1 + \cdots + X_i > 0$ for all $i \geq 1$, that is, the first visit of the reversed walk to a happens at time n. So $\mathbb{P}[T_a = n] = \mathbb{P}[T_0 > n, S_n = a]$. Now apply Theorem 1.12.

Proof of Theorem 1.10 If $T_0 = 2n$, then $S_{2n-1} = \pm 1$. Thus

$$\mathbb{P}[T_0 = 2n] = \mathbb{P}[T = 2n, S_{2n-1} = 1] + \mathbb{P}[T = 2n, S_{2n-1} = -1]$$

$$= \frac{1}{2}\mathbb{P}[T_0 > 2n-1,\, S_{2n-1} = 1] + \frac{1}{2}\mathbb{P}[T_0 > 2n-1,\, S_{2n-1} = -1].$$

Now by Theorem 1.12,

$$\mathbb{P}[T_0 = 2n] = \frac{1}{2}\cdot\frac{1}{2n-1}\left(\mathbb{P}[S_{2n-1} = 1] + \mathbb{P}[S_{2n-1} = -1]\right),$$

and part (i) of the theorem follows from (1.23), after simplification. For part (ii), we have that $\mathbb{E}[T_0^\alpha] = \sum_{n=1}^{\infty}(2n)^\alpha \mathbb{P}[T_0 = 2n]$, and Stirling's formula shows that the summand here is asymptotically a constant times $n^{\alpha-(3/2)}$.

Remark 1.7 *(i) An alternative approach to Theorem 1.10 is via the remarkable identity $\mathbb{P}[T_0 > 2n] = \mathbb{P}[S_{2n} = 0]$, which can be verified by a direct but more sophisticated combinatorial argument: see, for example, [21].*

(ii) Yet another approach uses generating functions. For S_n,

$$\psi(s) := \mathbb{E}[s^{S_n}] = \sum_{n=0}^{\infty} s^{2n}\binom{2n}{n}2^{-2n} = \frac{1}{\sqrt{1-s^2}},$$

by (1.23) and then Maclaurin's theorem. Then if ϕ is the generating function for T_0, we can exploit the renewal relation $\psi(t) = 1 + \phi(t)\psi(t)$ (see (1.22)) to obtain $\phi(s) = 1 - \sqrt{1-s^2}$, from which we can deduce Theorem 1.10 once more.

1.4.4 Large Deviations and Maxima of Random Walks

In this section, we consider more general one-dimensional random walks in order to illustrate some further concepts. Again we take $S_n = \sum_{i=1}^{n} X_i$ where the X_i are i.i.d., but now the distribution of X_i will be arbitrary subject to the existence of the mean $\mathbb{E}[X_i] = \mu$. Suppose that $\mu < 0$. The *strong law of large numbers* shows that $n^{-1}S_n \to \mu$, almost surely, as $n \to \infty$. So if $\mu < 0$, then S_n will tend to $-\infty$, and in particular the *maximum* of the walk $M = \max_{n>0} S_n$ is well defined.

There are many applications for the study of M, for example, the modeling of queues (see [22]). What properties does the random variable M possess? We might want to find $\mathbb{P}[M > x]$, for any x, but it is often difficult to obtain exact results; instead we attempt to understand the asymptotic behavior as $x \to \infty$.

Let $\varphi(t) = \mathbb{E}[e^{tX_1}]$ be the moment generating function of the increments. It can be shown that the behavior of $\mathbb{P}[M > x]$ depends on the form of φ. Here we consider only the classical (*light-tailed*) case in which there exists $\gamma > 0$ such that $\varphi(\gamma) = 1$ and $\varphi'(\gamma) < \infty$. For details of the other cases, see [23].

First, Boole's inequality (see Section 1.A) gives

$$\mathbb{P}[M > x] = \mathbb{P}[\cup_{n=1}^{\infty}\{S_n > x\}] \le \sum_{n=1}^{\infty}\mathbb{P}[S_n > x]. \tag{1.24}$$

Now the Chernoff bound (see Section 1.A) implies that, for any $\theta \in [0, \gamma]$,

$$\mathbb{P}[S_n > x] \le e^{-\theta x}\mathbb{E}[e^{\theta S_n}] = e^{-\theta x}(\varphi(\theta))^n;$$

cf. Example 1.4b. We substitute this into (1.24) to obtain

$$\mathbb{P}[M > x] \le e^{-\theta x}\sum_{n=1}^{\infty}(\varphi(\theta))^n = e^{-\theta x}\frac{\varphi(\theta)}{1-\varphi(\theta)},$$

provided $\varphi(\theta) < 1$, which is the case if $\theta \in (0, \gamma)$. For any such θ, we get

$$\limsup_{x\to\infty} \frac{1}{x} \log \mathbb{P}[M > x] \le -\theta - \lim_{x\to\infty} \frac{1}{x} \log(1 - \varphi(\theta)) = -\theta.$$

As $\theta < \gamma$ was arbitrary, we obtain the sharpest bound on letting $\theta \nearrow \gamma$. The matching lower bound can also be proved (see [22]), to conclude that

$$\lim_{x\to\infty} \frac{1}{x} \log \mathbb{P}[M > x] = -\gamma.$$

This is an example of a general class of results referred to as *large deviations*: further details of the general theory can be found in [24], for example. These techniques have found use in many application areas including statistical physics: see, for example, [25].

1.5 Markov Chains in Continuous Time

1.5.1 Markov Property, Transition Function, and Chapman–Kolmogorov Relation

In many applications, it is natural to work in *continuous time* rather than the discrete time of Section 1.3. As before, we assume that we have a discrete state space S, but now our Markov chains $X = (X(t))$ have a continuous- time parameter $t \in [0, \infty)$. Continuous time introduces analytical difficulties, which we will not dwell on in this presentation.

As in the discrete-time case, we concentrate on *time-homogeneous* chains, and we will specify the law of $(X(t))$ in line with the Markovian idea that "given the present, the future is independent of the past."

The process $(X(t))$ satisfies the *Markov property* in continuous time if, for all $t, h \ge 0$, all $i, j \in S$, all $0 \le t_0 < t_1 < \cdots < t_n < t$, and all $i_1, \dots, i_n \in S$,

$$\mathbb{P}[X(t+h) = j \mid X(t) = i, X(t_n) = i_n, \dots, X(t_1) = i_1] = p_{ij}(h).$$

Here $p_{ij}(\cdot) = \mathbb{P}[X(t+\cdot) = j \mid X(t) = i]$ is the *transition function* of the Markov chain. As in the discrete-time case, it is convenient to use matrix notation:

$$P(t) = (p_{ij}(t))_{i,j\in S} \text{ given by } p_{ij}(t) = \mathbb{P}_i[X(t) = j],$$

where again a subscript on $\mathbb{P}$ indicates an *initial state*, that is, $\mathbb{P}_i[\,\cdot\,] = \mathbb{P}[\,\cdot \mid X(0) = i]$. We can obtain full information on the law of the Markov chain, analogously to the discrete-time case. For example, for $0 < t_1 < \cdots < t_n$ and $j_1, \dots, j_n \in S$,

$$\mathbb{P}_i[X(t_1) = j_1, \dots, X(t_n) = j_n] = p_{ij_1}(t_1) p_{j_1 j_2}(t_2 - t_1) \cdots p_{j_{n-1} j_n}(t_n - t_{n-1}).$$

To this we add information about the initial distribution: for instance, $\mathbb{P}[X(t) = j] = \sum_{i\in S} \mathbb{P}[X(0) = i] p_{ij}(t)$. Here we must have

$$p_{ij}(0) = \delta_{ij} := \begin{cases} 1 & \text{if } i = j \\ 0 & \text{if } i \ne j \end{cases}.$$

We also assume that the transition functions satisfy, for each fixed t, $p_{ij}(t) \ge 0$ for all i, j and $\sum_{j\in S} p_{ij}(t) = 1$ for all i.

To describe our Markov chain now seems a formidable task: we must specify the family of functions $P(t)$. However, we will see in the next section that the Markov property enables a *local* (*infinitesimal*) description. First we state a *global* consequence of the Markov property, namely, Chapman–Kolmogorov relation

$$\text{For any } s, t \ge 0,\ P(s+t) = P(s)P(t). \quad (1.25)$$

The fundamental Markovian relation (1.25) is a special case of the relation known to probabilists as the *Chapman–Kolmogorov equation*, and which, in its most general form, is often taken as the starting point of the general theory of Markov processes. Physicists refer to a relation such as (1.25) as a *master equation*. The derivation of (1.25) in our setting is direct from the Markov property:

$$\begin{aligned}\mathbb{P}_i[X(s+t)=j] \\ &= \sum_{k\in S} \mathbb{P}_i[X(s)=k, X(s+t)=j] \\ &= \sum_{k\in S} \mathbb{P}_i[X(s)=k]\mathbb{P}_i[X(s+t)=j \mid X(s)=k] \\ &= \sum_{k\in S} \mathbb{P}_i[X(s)=k]\mathbb{P}_k[X(t)=j],\end{aligned}$$

which is the equality for the (i,j) entry in the matrix equation (1.25).

1.5.2 Infinitesimal Rates and Q-matrices

In continuous time, there is no smallest time step and so no concept of a one-step transition. Often, however, one can encapsulate the information in the functions $p_{ij}(t)$ in a single fundamental matrix associated with the Markov chain, which will serve as an analogue to the P-matrix in the discrete theory. This is the Q-matrix.

To proceed, we need to assume some regularity. We call the chain *standard* if the transition probabilities are continuous at 0, that is, if $p_{ij}(t) \to p_{ij}(0) = \delta_{ij}$ as $t \downarrow 0$.

Lemma 1.7 *Suppose that X is a standard Markov chain with transition functions $p_{ij}(t)$. Then for each i,j, $p_{ij}(t)$ is a continuous and differentiable function of t. The derivatives $p'_{ij}(t)$ evaluated at $t=0$ we denote by $q_{ij} := p'_{ij}(0)$; then $0 \le q_{ij} < \infty$ for $i \ne j$ and $0 \le -q_{ii} \le \infty$.*

The proof of this result relies on the Chapman–Kolmogorov relation, but is somewhat involved: see, for example, [26, Section 14.1]. A Taylor's formula expansion now reads

$$\begin{aligned}p_{ij}(h) &= \mathbb{P}[X(t+h)=j \mid X(t)=i] \\ &= p_{ij}(0) + q_{ij}h + o(h) \qquad (1.26)\\ &= \delta_{ij} + q_{ij}h + o(h), \text{ as } h \downarrow 0. \qquad (1.27)\end{aligned}$$

So, for $i \ne j$, q_{ij} is the (instantaneous) *transition rate* of the process from state i to state j. It is convenient to define $q_i := -q_{ii}$ for all i (so $q_i \ge 0$). Then q_i is the rate of departure from state i.

We further assume that the chain is *conservative*, meaning

$$\sum_{j\ne i} q_{ij} = q_i < \infty, \text{ for all } i. \qquad (1.28)$$

Note that $\sum_{j\ne i} p_{ij}(t) = 1 - p_{ii}(t)$, so, for example, if S is *finite* we can differentiate to immediately get the equality in (1.28), and then $\sum_{j\ne i} q_{ij} < \infty$ by Lemma 1.7, so a finite Markov chain is always conservative. Note that (1.28) implies that $\sum_j q_{ij} = 0$ and $q_i = \sum_{j\ne i} q_{ij}$, so the rows of Q sum to zero.

The matrix $Q = (q_{ij})_{i,j\in S}$ is called the *transition rate matrix*, the *generator matrix*, or simply the *Q-matrix* of the Markov chain; it effectively describes the chain's *dynamics*. In particular, under reasonable conditions (see the following text) the functions $p_{ij}(\cdot)$ are uniquely determined by Q. Thus, in applications, a Markov process is often *defined* via a Q-matrix and an initial distribution.[12)]

Conversely, given a matrix $Q = (q_{ij})_{i,j\in S}$ with nonpositive diagonal entries and nonnegative entries elsewhere for which (1.28) holds, there always exists a Markov process with Q as transition rate matrix. This fact

12) This is also essentially the approach taken in [6].

can be proved by actually *constructing* the paths of such a process: see Section 1.5.4.

Example 1.18 [*Birth-and-death process*]
Here $S = \mathbb{Z}_+$ and $X(t)$ represents a population size at time t. The size of the population increases on a *birth* or decreases on a *death*. The nonzero entries in the Q-matrix are

$$q_{i,i+1} = \lambda_i, \quad i \geq 0 \quad \text{(birth rate in state } i\text{)},$$
$$q_{i,i-1} = \mu_i, \quad i \geq 1 \quad \text{(mortality rate in state } i\text{)},$$
$$q_{0,0} = -\lambda_0 \text{ and } q_{i,i} = -(\lambda_i + \mu_i), \quad i \geq 1.$$

In a *linear* process, $\lambda_i = \lambda i$ and $\mu_i = \mu i$, so λ and μ can be interpreted as *per individual* rates of birth and mortality, respectively.

1.5.3 Kolmogorov Differential Equations

We now consider some differential equations which, given Q, can be used to determine the functions $p_{ij}(\cdot)$. The starting point is the Chapman–Kolmogorov relation $p_{ij}(s+t) = \sum_{k \in S} p_{ik}(s) p_{kj}(t)$. If S is *finite*, say, then it is legitimate to differentiate with respect to s to get

$$p'_{ij}(s+t) = \sum_{k \in S} p'_{ik}(s) p_{kj}(t).$$

Now setting $s = 0$, we obtain

$$p'_{ij}(t) = \sum_{k \in S} q_{ik} p_{kj}(t),$$

which is the *Kolmogorov backward equation*. If instead, we differentiate with respect to t and then put $t = 0$, we obtain (after a change of variable)

$$p'_{ij}(t) = \sum_{k \in S} p_{ik}(t) q_{kj},$$

which is the *Kolmogorov forward equation*. These differential equations are particularly compact in matrix form.

Theorem 1.14 *Given Q satisfying (1.28), we have*

$$P'(t) = P(t)Q, \qquad (\textit{Kolmogorov forward equation});$$
$$P'(t) = QP(t), \qquad (\textit{Kolmogorov backward equation}).$$

We sketched the derivation in the case where S is finite. In the general case, a proof can be found in, for example, [26, Section 14.2].

Remark 1.8 *Suitable versions of the Kolmogorov differential equations also apply to processes on continuous state spaces, such as* diffusions, *where they take the form of partial differential equations. In this context, the forward equation can be framed as a* Fokker–Planck *equation for the evolution of a probability density. The connections among diffusions, boundary-value problems, and potential theory are explored, for example, in [2]; an approach to Fokker–Planck equations from a more physical perspective can be found in [27].*

We give one example of how to use the Kolmogorov equations, together with generating functions, to compute $P(t)$ from Q.

Example 1.19 [*Homogeneous birth process*]
We consider a special case of Example 1.18 with only births, where, for all i, $\lambda_i = \lambda > 0$ and $\mu_i = 0$. So $q_{i,i} = -\lambda$ and $q_{i,i+1} = \lambda$. The Kolmogorov forward equation in this case gives

$$p'_{i,j}(t) = -\lambda p_{i,j}(t) + \lambda p_{i,j-1}(t),$$

where we interpret $p_{i,-1}(t)$ as 0. In particular, if $i = 0$, we have

$$p'_{0,j}(t) = -\lambda p_{0,j}(t) + \lambda p_{0,j-1}(t). \qquad (1.29)$$

The initial conditions are assumed to be $p_{0,0}(0) = 1$ and $p_{0,i}(0) = 0$ for $i \geq 1$, so that the process starts in state 0. Consider the generating function

$$\phi_t(u) = \mathbb{E}[u^{X(t)}] = \sum_{j=0}^{\infty} p_{0,j}(t)u^j, \quad |u| < 1.$$

Multiplying both sides of (1.29) by s^j and summing over j we get

$$\begin{aligned}\frac{\partial}{\partial t}\phi_t(u) &= -\lambda\phi_t(u) + \lambda u\phi_t(u) \\ &= -\lambda(1-u)\phi_t(u).\end{aligned}$$

It follows that $\phi_t(u) = A(u)\mathrm{e}^{-(1-u)\lambda t}$. The initial conditions imply that $\phi_0(u) = 1$, so in fact $A(u) = 1$ here, and $\phi_t(u) = \mathrm{e}^{-(1-u)\lambda t}$, which is the probability generating function of a Poisson distribution with mean λt (see Example 1.2). Hence, $X(t) \sim \mathrm{Po}(\lambda t)$. In fact $X(t)$ is an example of a *Poisson process*: see, for example, [2, 5, 6, 26].

By analogy with the scalar case, under suitable conditions one can define the matrix exponential

$$\exp\{Qt\} = \sum_{k=0}^{\infty} \frac{Q^k t^k}{k!},$$

with $Q^0 = I$ (identity). Then $P(t) = \exp\{Qt\}$ is a formal solution to both the Kolmogorov forward and backward equations. In analytic terminology, Q is the generator of the semigroup P.

1.5.4 Exponential Holding-Time Construction; "Gillespie's Algorithm"

Given the Q-matrix one can construct *sample paths* of a continuous- time Markov chain. The following scheme also tells you how to *simulate* a continuous-time Markov chain.

Suppose the chain starts in a fixed state $X(0) = i$ for $i \in S$. Let $\tau_0 = 0$ and define recursively for $n \geq 0$,

$$\tau_{n+1} = \inf\{t \geq \tau_n : X(t) \neq X(\tau_n)\}.$$

Thus τ_n is the nth *jump time* of X, that is, the nth time at which the process changes its state.

How long does the chain stay in a particular state? We have

$$\begin{aligned}&\mathbb{P}_i[\tau_1 > t + h \mid \tau_1 > t] \\ &\quad = \mathbb{P}_i[\tau_1 > t + h \mid \tau_1 > t, X(t) = i] \\ &\quad = \mathbb{P}_i[\tau_1 > h],\end{aligned}$$

by the Markov property. This *mermoryless* property is indicative of the *exponential distribution* (see Example 1.8a). Recall that $Y \sim \exp(\lambda)$ if $\mathbb{P}[Y > t] = \mathrm{e}^{-\lambda t}$, $t \geq 0$. A calculation shows that

$$\begin{aligned}&\mathbb{P}[Y > t + h \mid Y > t] = \mathbb{P}[Y > h] \\ &\quad = \mathrm{e}^{-\lambda h} = 1 - \lambda h + o(h),\end{aligned} \qquad (1.30)$$

as $h \to 0$.

In fact, the exponential distribution is essentially the only distribution with this property. So it turns out that τ_1 is exponential. A heuristic calculation, which can be justified, suggests that $\mathbb{P}_i[\tau_1 > h] \sim \mathbb{P}_i[X(h) = i] = p_{ii}(h) = 1 - q_i h + o(h)$. A comparison with (1.30) suggests that $\tau_1 \sim \exp(q_i)$.

When the chain does jump, where does it go? Now, for $j \neq i$, $\mathbb{P}[X(t + h) = j$

$\mid X(t) = i] = p_{ij}(h) = q_{ij}h + o(h)$, while $\mathbb{P}[X(t+h) \neq i \mid X(t) = i] = q_i h + o(h)$, so a conditional probability calculations gives

$$\mathbb{P}[X(t+h) = j \mid X(t) = i, X(t+h) \neq i] = \frac{q_{ij}}{q_i} + o(1).$$

Careful argument along these lines (see, e.g., [26, Section 14.3]) gives the next result.[13)]

Theorem 1.15 *Under the law $\mathbb{P}_i$ of the Markov chain started in $X(0) = i$, the random variables τ_1 and $X(\tau_1)$ are independent. The distribution of τ_1 is exponential with rate q_i. Moreover, $\mathbb{P}_i[X(\tau_1) = j] = q_{ij}/q_i$.*

Perhaps the most striking aspect of this result is that the holding time and the jump destination are independent. Theorem 1.15 tells us how to construct the Markov chain, by iterating the following procedure.

- Given τ_n and $X(\tau_n) = i$, generate an $\exp(q_i)$ random variable Y_n (this is easily done via $Y_n = -q_i^{-1} \log U_n$, where $U_n \sim U[0, 1]$). Set $\tau_{n+1} = \tau_n + Y_n$.
- Select the next state $X(\tau_{n+1})$ according to the distribution q_{ij}/q_i.

Although this standard construction of Markov chain sample paths goes back to classical work of Doeblin, Doob, Feller, and others in the 1940s, and was even implemented by Kendall and Bartlett in pioneering computer simulations in the early 1950s, the scheme is known in certain applied circles as *"Gillespie's algorithm"* after Gillespie's 1977 paper that rederived the construction in the context of chemical reaction modeling.

13) Actually Theorem 1.15 assumes that we are working with the *minimal* version of the process: see, for example, [6, 26] for details of this technical point.

Remark 1.9 *Let $X_n^* = X(\tau_n)$. Then $(X_n^*)_{n \in \mathbb{Z}_+}$ defines a discrete-time Markov chain, called the* jump chain *associated with $X(t)$, with one-step transitions*

$$p_{ij}^* = \begin{cases} \frac{q_{ij}}{q_i} & \text{if } i \neq j, \\ 0 & \text{if } i = j. \end{cases},$$

as long as $q_i > 0$. If $q_i = 0$, then $p_{ii}^* = 1$, that is, i is an absorbing state.

1.5.5 Resolvent Computations

Consider a Markov chain with transition functions $p_{ij}(t)$ determined by its generator matrix Q. The Laplace transform of p_{ij} is r_{ij} given by

$$r_{ij}(\lambda) = \int_0^\infty e^{-\lambda t} p_{ij}(t) dt. \tag{1.31}$$

Then $R(\lambda) = (r_{ij}(\lambda))_{i,j \in S}$ is the *resolvent matrix* of the chain. A formal calculation, which can be justified under the conditions in force in this section, shows that $R(\lambda)$ can be expressed as the matrix inverse $R(\lambda) = (\lambda I - Q)^{-1}$, where I is the identity. See Chapter 15 of the present volume for background on Laplace transforms.

Let τ be an $\exp(\lambda)$ random variable, independent of the Markov chain. Then

$$\lambda r_{ij}(\lambda) = \int_0^\infty \lambda e^{-\lambda t} p_{ij}(t) dt = \int_0^\infty \mathbb{P}[\tau \in dt] \mathbb{P}_i[X(t) = j \mid \tau = t],$$

which is just $\mathbb{P}_i[X(\tau) = j]$, the probability that, starting from state i, the chain is in state j at the random time τ.

Resolvents play an important role in the theoretical development of Markov processes, and in particular in the abstract semigroup approach to the theory. Here, however, we view the resolvent as a computational tool, which enables, in principle, calculation of probabilities and hitting-time distributions.

Specifically, (1.31) implies that $p_{ij}(t)$ can be recovered by inverting its Laplace transform $\hat{p}_{ij}(\lambda) = r_{ij}(\lambda)$. Moreover, let $T_i := \inf\{t \geq 0 : X(t) = i\}$, the first hitting time of state i. Write $F_{ij}(t) := \mathbb{P}_i[T_j \leq t]$, and set $f_{ij}(t) = F'_{ij}(t)$ for the density of the hitting-time distribution. We proceed analogously to the discrete argument for the proof of Lemma 1.3. An application of the *strong Markov property* for continuous-time chains (cf Section 1.3.4) gives,

$$p_{ij}(t) = \int_0^t f_{ij}(s) p_{jj}(t-s) \mathrm{d}s.$$

The convolution theorem for Laplace transforms (see Chapter 15) implies that the Laplace transform of f_{ij} is given by

$$\hat{f}_{ij}(\lambda) = \frac{r_{ij}(\lambda)}{r_{jj}(\lambda)}. \tag{1.32}$$

In the next section, we give some examples of using resolvent ideas in computations.

1.5.6 Example: A Model of Deposition, Diffusion, and Adsorption

We describe a continuous-time Markov model of deposition of particles that subsequently perform random walks and interact to form barriers according to an occupation criterion, inspired by models of submonolayer film growth [28, 29]. Particles arrive randomly one by one on a one-dimensional substrate $S_N := \{0, 1, \dots, N+1\}$ and diffuse until $M \geq 2$ particles end up at the same site, when they clump together ("nucleate") to form an "island." Islands form absorbing barriers with respect to the diffusion of other particles. We assume that initially, sites 0 and $N+1$ are occupied by M particles (so are already islands) but all other sites are empty.

The Markov dynamics are as follows.

- At each site $x \in S_N$, new particles arrive independently at rate $\rho > 0$.
- If at any time a site is occupied by M or more particles, all those particles are held in place and are *inactive*. Particles that are not inactive are *active*.
- Each active particle independently performs a symmetric simple random walk at rate 1, that is, from x it jumps to $x+1$ or $x-1$ each at rate $1/2$.

A state ω of the Markov process is a vector of the occupancies of the sites $1, \dots, N$ (it is not necessary to keep track of the occupancies of 0 or $N+1$): $\omega(x)$ is the number of particles at site x. We can simulate the process via the exponential holding-time construction of Section 1.5.4. To do so, we need to keep track of $T(\omega) = \sum_{1 \leq x \leq N} \omega(x) \mathbf{1}\{\omega(x) < M\}$, the total number of active particles in state ω. The waiting time in a state ω is then exponential with parameter $T(\omega) + N\rho$; at the end of this time, with probability $T(\omega)/(T(\omega) + N\rho)$, one of the active particles jumps (chosen uniformly from all active particles, and equally likely to be a jump left or right), else a new particle arrives at a uniform random site in $\{1, \dots, N\}$.

An analysis of the general model just described would be interesting but is beyond the scope of this presentation. We use small examples (in terms of M and N) to illustrate the resolvent methods described in Section 1.5.5. For simplicity, we take $M = 2$ and stop the process the first time that two particles occupy any internal site. Configurations can be viewed as elements of $\{0, 1, 2\}^{\{1,2,\dots,N\}}$, but symmetry can be used to further reduce the state space for our questions of interest.

1.5.6.1 $N = 1$

Take $N = 1$. The state space for our Markov chain $X(t)$ is $\{0, 1, 2\}$, the number of particles in position 1, with $X(0) = 0$, and 2 as the absorbing state. Clearly $X(t) = 2$ eventually; the only question is how long we have to wait for absorption. The answer is not trivial, even in this minimal example.

The generator matrix for the Markov chain is

$$Q = \begin{pmatrix} -\rho & \rho & 0 \\ 1 & -1-\rho & \rho \\ 0 & 0 & 0 \end{pmatrix}, \text{ and so}$$

$$\lambda I - Q = \begin{pmatrix} \lambda + \rho & -\rho & 0 \\ -1 & 1+\lambda+\rho & -\rho \\ 0 & 0 & \lambda \end{pmatrix}.$$

To work out $p_{02}(t)$, we compute $r_{02}(\lambda)$:

$$r_{02}(\lambda) = (\lambda I - Q)^{-1}_{02} = \frac{\det \begin{pmatrix} -\rho & 0 \\ 1+\lambda+\rho & -\rho \end{pmatrix}}{\det(\lambda I - Q)}$$
$$= \frac{\rho^2}{\lambda((\lambda+\rho)^2 + \lambda)}.$$

Inverting the Laplace transform, we obtain

$$1 - p_{02}(t) = e^{-((1+2\rho)/(2))t} \left(\cosh\left(\left(\frac{t}{2}\right) \sqrt{1+4\rho} \right) + \frac{1+2\rho}{\sqrt{1+4\rho}} \sinh\left(\left(\frac{t}{2}\right) \sqrt{1+4\rho} \right) \right)$$

$$\sim \frac{1}{2}\left(1 + \frac{1+2\rho}{\sqrt{1+4\rho}}\right) \exp\left\{ -\frac{t}{2}\left(1 + 2\rho - \sqrt{1+4\rho}\right) \right\}, \quad \text{as } t \to \infty.$$

For example, if $\rho = 3/4$, the exact expression simplifies to

$$p_{02}(t) = 1 - \frac{9}{8}e^{-t/4} + \frac{1}{8}e^{-9t/4}.$$

As in this case, 2 is absorbing, $\mathbb{P}[T_2 \le t] = \mathbb{P}[X(t) = 2] = p_{02}(t)$, so

$$f_{02}(t) = \frac{d}{dt} p_{02}(t) = \frac{9}{32}(e^{-t/4} - e^{-9t/4}),$$

in the $\rho = 3/4$ example. The same answer can be obtained using (1.32).

1.5.6.2 $N = 3$

For the problem of the distribution of the point of the first collision, the first nontrivial case is $N = 3$. Making use of the symmetry, we can now describe the Markov chain with the 8 states

$$(0,0,0),\ (1,0,0),\ (0,1,0),\ (1,1,0),$$
$$(1,0,1),\ (1,1,1),\ (2,*,*),\ (*,2,*),$$

in that order, where each $*$ indicates either a 0 or a 1; so for example $(1, 0, 0)$ stands for $(1, 0, 0)$ or $(0, 0, 1)$. The generator matrix is now

$$Q = \begin{pmatrix} -3\rho & 2\rho & \rho & 0 & 0 & 0 & 0 & 0 \\ \frac{1}{2} & -1-3\rho & \frac{1}{2} & \rho & \rho & 0 & \rho & 0 \\ 0 & 1 & -1-3\rho & 2\rho & 0 & 0 & 0 & \rho \\ 0 & 0 & \frac{1}{2} & -2-3\rho & \frac{1}{2} & \rho & \frac{1}{2}+\rho & \frac{1}{2}+\rho \\ 0 & 1 & 0 & 1 & -2-3\rho & \rho & 2\rho & 0 \\ 0 & 0 & 0 & 1 & 0 & -3-3\rho & 1+2\rho & 1+\rho \\ 0 & 0 & 0 & 0 & 0 & 0 & 0 & 0 \\ 0 & 0 & 0 & 0 & 0 & 0 & 0 & 0 \end{pmatrix}.$$

The probability of the first collision occurring at the midpoint is

$$z(\rho) = \mathbb{P}\left[\lim_{t\to\infty} X(t) = (*, 2, *)\right].$$

According to MAPLE, inverting the appropriate Laplace transform gives

$$z(\rho)=\frac{1}{9}\cdot\frac{486\rho^4+1354\rho^3+1375\rho^2+598\rho+87}{162\rho^4+414\rho^3+392\rho^2+160\rho+23}.$$

In particular, if deposition dominates diffusion,

$$\lim_{\rho\to\infty} z(\rho) = \frac{1}{9}\cdot\frac{486}{162} = \frac{1}{3},$$

which is as it should be!

1.6 Gibbs and Markov Random Fields

We have so far focused on stochastic processes that vary through time. In this section and in Section 1.7, we take a detour into *spatial processes*. The role of space in our models is played by an underlying *graph* structure, describing vertices V and the edges E that connect them. This section is devoted to *random fields*, that is, ensembles of random variables associated with the vertices subject to certain constraints imposed by the edges.

Let $G = (V, E)$ be an undirected finite graph with vertex set V and edge set E consisting of pairs (i, j) where $i, j \in V$; $(i, j) \in E$ indicates an edge between vertices i and j.[14] With this graph, we associate random variables $\{X_i\}$ for $i \in V$; to keep things simple, we take $X_i \in \{-1, 1\}$.

We consider two ways of specifying these random variables – as Gibbs or Markov random fields – and the relationship between the two. Finally, we will consider how to use ideas from Markov chains (Section 1.3) to simulate a Markov random field.

1.6.1 Gibbs Random Field

To define a Gibbs random field, we need the concept of a *clique*. A clique in $G = (V, E)$ is a subset K of V such that E contains all the possible edges between members of K; we include the set of no vertices $\emptyset$ as a clique. Let $\mathcal{K}$ be the set of all cliques of graph G.

Let $\mathbf{x} = (x_1, \dots, x_{|V|}) \in \{-1, 1\}^V$ denote an assignment of a value ± 1 to each vertex of G. For $K \subseteq V$ we write $\mathbf{x}|_K$ for the restriction of $\mathbf{x}$ to K. The random variables $\{X_i\}_{i\in V}$ on G constitute a *Gibbs random field* if, for all $\mathbf{x} \in \{-1, 1\}^V$,

$$\begin{aligned}\mathbb{P}[X_i = x_i \text{ for all } i \in V] \\ = \frac{1}{Z}\exp\left\{\sum_{K\in\mathcal{K}} f_K(\mathbf{x}|_K)\right\},\end{aligned} \quad (1.33)$$

where $f_K : \{-1, 1\}^{|K|} \to \mathbb{R}$ for each clique K, and Z is a normalization:

$$Z = \sum_{\mathbf{x}\in\{-1,1\}^V} \exp\left\{\sum_{K\in\mathcal{K}} f_K(\mathbf{x}|_K)\right\}.$$

To see why this is a natural definition for the law of $\{X_i\}$ in many situations, consider associating an *energy* function $\mathcal{E} : \{-1, 1\}^V \to \mathbb{R}$ to the states. We seek the maximum entropy distribution for $\{X_i\}_{i\in V}$ for a given mean energy. So we want to find the probability mass function, f, on $\{-1, 1\}^V$ that maximizes $-\sum_{\mathbf{x}\in\{-1,1\}^V} f(\mathbf{x}) \log f(\mathbf{x})$ subject to $\sum_{\mathbf{x}\in\{-1,1\}^V} f(\mathbf{x})\mathcal{E}(x) = \text{const.}$ One can show this is achieved by

$$f(\mathbf{x}) \propto \exp\{-\beta\mathcal{E}(x)\},$$

14) Our graphs are undirected, which means that $(i, j) = (j, i)$ is an unordered pair.

where $\beta \in (0, \infty)$ is chosen to obtain the required energy. (The analogy here is between β and the inverse temperature $1/(kT)$ in thermodynamics.) Now if $\mathcal{E}$ can be decomposed as a sum over the cliques of the graph, we recover (1.33).

Example 1.20 [Ising model] Consider the $N \times N$ grid in two dimensions. We label the vertices by members of the set $L_N = \{1, \dots, N\} \times \{1, \dots, N\}$. We put an edge between $i = (i_1, i_2)$ and $j = (j_1, j_2)$ if $|i_1 - j_1| + |i_2 - j_2| = 1$; this adjacency condition we write as $i \sim j$. Here the clique set $\mathcal{K}$ is made up of the empty set, singleton nodes, and pairs of nodes that are distance one apart in the lattice.

Given constants $\beta > 0$, $J > 0$, and $h \in \mathbb{R}$ we consider the Gibbs random field with probability mass function $f(\mathbf{x}) = Z^{-1} \exp\{-\beta\mathcal{E}(x)\}$, where

$$\mathcal{E}(x) = -J \sum_{i,j \in L_N : i \sim j} X_i X_j - h \sum_{i \in L_N} X_i.$$

The sum over pairs of nodes (the *interaction* term) means that neighboring nodes have a propensity to be in the same state. The second (*external field*) term leads to nodes more likely to be in either of the states 1 or -1, depending on the sign of h.

This model has been studied widely as a model for ferromagnetism and was initially proposed by Ising [30] under the guidance of Lenz. The *Potts model* is a generalization with q-valued states and more general interactions: see [31].

1.6.2 Markov Random Field

We now consider a second specification of random field that adapts the Markov property to spatial processes. For a given subset $W \subseteq V$, we define its boundary as

$$\partial W = \{v \in V \setminus W \, : \, (v, w) \in E \\ \text{for some } w \in W\}.$$

The concept of Markov random field extends the temporal Markov property (1.12), which said that, conditional on the previous states of the process, the future depends on the past only through the present, to a spatial (or topological) one. This "Markov property" will say that the state of nodes in some set of vertices W conditioned on the state of all the other vertices only depends on the state of the vertices in ∂W.

The random variables $\{X_i\}_{i \in V}$ on G constitute a *Markov random field* if

- they have a *positive* probability mass function,

$$\mathbb{P}[\{X_i\}_{i \in V} = \mathbf{x}] > 0, \\ \text{for all } \mathbf{x} \in \{-1, 1\}^V,$$

- and obey the global *Markov property*: for all $W \subseteq V$,

$$\mathbb{P}[\{X_i\}_{i \in W} = \mathbf{x}|_W \mid \{X_i\}_{i \in V \setminus W} = \mathbf{x}|_{V \setminus W}] \\ = \mathbb{P}[\{X_i\}_{i \in W} = \mathbf{x}|_W \mid \{X_i\}_{i \in \partial W} = \mathbf{x}|_{\partial W}].$$

Example 1.21 As in Example 1.20, consider a random field taking values ± 1 on the vertices of the $N \times N$ lattice. We specify the (conditional) probability that a vertex, i, is in state 1, given the states of its neighbors to be

$$\frac{e^{\beta(h+Jy_i)}}{e^{-\beta(h+Jy_i)} + e^{\beta(h+Jy_i)}}, \quad \text{where } y_i = \sum_{j \sim i} x_j. \tag{1.34}$$

Here $\beta > 0$, $J > 0$, and $h \in \mathbb{R}$ are parameters. The larger y_i, which is the number of neighbours of i with spin $+1$ minus the number with spin -1, the greater the probability that vertex i will itself have state 1.

1.6.3 Connection Between Gibbs and Markov Random Fields

In Examples 1.20 and 1.21, we have used the same notation for the parameters. In fact, both specifications (one Gibbs, the other Markov) define the same probability measure on $\{-1,1\}^V$. This is an example of the following result.

Theorem 1.16 (Hammersley–Clifford theorem) *The ensemble of random variables $\{X_i\}_{i\in V}$ on G is a Markov random field if and only if it is a Gibbs random field with a positive probability mass function.*

A proof can be found in [31]. From this point forward, we will use the terms Gibbs random field and Markov random field interchangeably.

1.6.4 Simulation Using Markov Chain Monte Carlo

Direct simulation of a Gibbs random field on a graph is computationally difficult because the calculation of the normalizing constant, Z, requires a sum over all the possible configurations. In many situations, this is impractical. Here we consider an alternative way to simulate a Gibbs random field making use of Markov chains.

We saw in Section 1.3.7 that an irreducible, aperiodic Markov chain converges to its stationary distribution. The idea now is to design a Markov chain on the state space $\{-1,1\}^V$ whose stationary distribution coincides with the desired Gibbs random field. We simulate the Markov chain for a long time to obtain what should be a distribution close to stationarity, and hence, a good approximation to a realization of the Gibbs random field.

We initialize the Markov chain with any initial state $\sigma \in \{-1,1\}^V$. To update the state of the chain, we randomly select a vertex uniformly from all vertices in the graph. We will update the state associated with this vertex by randomly selecting a new state using the conditional probabilities given, subject to the neighboring vertices' states, taking advantage of the Markov random field description. For example, in Example 1.21, we set the node state to 1 with probability given by (1.34) and to -1, otherwise.

It is easy to check that this Markov chain is irreducible, aperiodic, and has the required stationary distribution. This is an example of a more general methodology of using a Markov chain with simple update steps to simulate from a distribution that is computationally difficult to evaluate directly, called *Markov chain Monte Carlo*. Specifically, we have used a *Gibbs sampler* here, but there are many other schemes for creating a Markov chain with the correct stationary distribution. Many of these techniques have been developed within the setting of Bayesian statistics but have applications in many other fields, including spin glass models and theoretical chemistry.

1.7 Percolation

Consider the infinite square lattice $\mathbb{Z}^2$. Independently, for each edge in the lattice, the edge is declared *open* with probability p; else (with probability $1-p$) it is *closed*. This model is called *bond percolation* on the square lattice.

For two vertices $\mathbf{x}, \mathbf{y} \in \mathbb{Z}^2$ write $\mathbf{x} \longleftrightarrow \mathbf{y}$ if $\mathbf{x}$ and $\mathbf{y}$ are joined by a path consisting of open edges in the percolation model on $\mathbb{Z}^2$.

The *open cluster* containing vertex $\mathbf{x}$, $C(\mathbf{x})$, is the (random) set of all vertices joined to $\mathbf{x}$ by an open path

$$C(\mathbf{x}) := \{\mathbf{y} \in \mathbb{Z}^2 : \mathbf{y} \longleftrightarrow \mathbf{x}\}.$$

A fundamental question in percolation regards the nature of $C(\mathbf{0})$, the open cluster at $\mathbf{0}$, and how its (statistical) properties depend on the parameter p.

We write $\mathbb{P}_p$ for the probability associated with percolation with parameter p. Write $|C(\mathbf{0})|$ for the number of vertices in the open cluster at $\mathbf{0}$. The *percolation probability* is

$$\theta(p) = \mathbb{P}_p[|C(\mathbf{0})| = \infty].$$

By translation invariance, $\theta(p) \in [0,1]$ is not specific to the origin: $\mathbb{P}_p[|C(\mathbf{x})| = \infty] = \theta(p)$ for any $\mathbf{x}$.

Let H_∞ be the event that $|C(\mathbf{x})| = \infty$ for *some* $\mathbf{x}$. It is not hard to show that

$$\theta(p) = 0 \implies \mathbb{P}_p[H_\infty] = 0$$
$$\theta(p) > 0 \implies \mathbb{P}_p[H_\infty] = 1.$$

We state a fundamental result that may seem obvious; the proof we give, due to Hammersley, demonstrates the effectiveness of another probabilistic tool: *coupling.*

Lemma 1.8 *$\theta(p)$ is nondecreasing as a function of p.*

Proof. List the edges of the lattice $\mathbb{Z}^2$ in some order as $e_1, e_2, \ldots$. Let $U_1, U_2, \ldots$ be independent uniform random variables on $[0,1]$. Assign U_i to edge e_i.

Let $E_p = \{e_i : U_i \le p\}$. Then E_p is the set of open edges in bond percolation with parameter p. This construction *couples* bond percolation models for every $p \in [0,1]$ in a monotone way: if $e_i \in E_p$ then $e_i \in E_q$ for all $q \ge p$.

Let $C_p(\mathbf{0})$ denote the cluster containing $\mathbf{0}$ using edges in E_p. If $p \le q$, then by construction $C_p(\mathbf{0}) \subseteq C_q(\mathbf{0})$. So $\{|C_p(\mathbf{0})| = \infty\} \subseteq \{|C_q(\mathbf{0})| = \infty\}$, and hence, $\theta(p) \le \theta(q)$.

As $\theta(p)$ is nondecreasing, and clearly $\theta(0) = 0$ and $\theta(1) = 1$, there must be some threshold value

$$p_c := \inf\{p \in [0,1] : \theta(p) > 0\}.$$

So for $p < p_c$, $\mathbb{P}_p[H_\infty] = 0$, while for $p > p_c$, $\mathbb{P}_p[H_\infty] = 1$.

The first question is this: is there a nontrivial *phase transition*, that is, is $0 < p_c < 1$? This question was answered by Broadbent and Hammersley in the late 1950s.

Proposition 1.2 $1/3 \le p_c \le 2/3$.

Proof of $p_c \ge 1/3$ Let A_n be the event that there exists a self-avoiding open path starting at $\mathbf{0}$ of length n. Then

$$A_1 \supseteq A_2 \supseteq A_3 \cdots \quad \text{and}$$
$$\bigcap_{n=1}^{\infty} A_n = \{|C(\mathbf{0})| = \infty\}.$$

So $\theta(p) = \mathbb{P}_p[|C(\mathbf{0})| = \infty] = \lim_{n\to\infty} \mathbb{P}_p[A_n]$, by continuity of probability measure (see Section 1.A). Let Γ_n be the set of all possible self-avoiding paths of length n starting at $\mathbf{0}$. Then

$$\mathbb{P}_p[A_n] = \mathbb{P}_p \bigcup_{\gamma\in\Gamma_n} \{\gamma \text{ is open}\}$$
$$\le \sum_{\gamma\in\Gamma_n} \mathbb{P}_p[\gamma \text{ is open}] = |\Gamma_n| p^n$$
$$\le 4 \cdot 3^{n-1} \cdot p^n,$$

which tends to 0 if $p < 1/3$. So $p_c \ge 1/3$.

On the basis of pioneering Monte Carlo simulations, Hammersley conjectured that

p_c was 1/2. Harris proved in 1960 that $\theta(1/2) = 0$, which implies that $p_c \geq 1/2$. It was not until 1980 that a seminal paper of Kesten settled things.

Theorem 1.17 (Harris 1960, Kesten 1980) $p_c = 1/2$.

Harris's result $\theta(1/2) = 0$ thus means that $\theta(p_c) = 0$; this is conjectured to be the case in many percolation models (e.g., on $\mathbb{Z}^d$, it is proved for $d = 2$ and $d \geq 19$, but is conjectured to hold for all $d \geq 2$). In recent years, there has been much interest in the detailed structure of percolation when $p = p_c$. The *Schramm–Loewner evolution* has provided an important new mathematical tool to investigate physical predictions, which often originated in conformal field theory; see [32].

1.8 Further Reading

A wealth of information on stochastic processes and the tools that we have introduced here can be found in [2, 5, 9, 20, 26], for example. All of those books cover Markov chains. The general theory of Markov processes can be found in [2, 9], which also cover the connection to semigroup theory. Feller gives a masterly presentation of random walks and generating functions [20] and Laplace transforms, characteristic functions, and their applications [9]. Branching processes can be found in [5, 20]. A thorough treatment of percolation is presented in [33]. We have said almost nothing here about Brownian motion or diffusions, for which we refer the reader to Chapter 3 of this volume as well as [2, 5, 9, 26]. Physicists and mathematicians alike find it hard not to be struck by the beauty of the connection between random walks and electrical networks, as exposited in [34]. Applications of stochastic processes in physics and related fields are specifically treated in [27, 35]; the array of applications of random walks alone is indicated in [19, 36, 37].

1.A Appendix: Some Results from Probability Theory

There is no other simple mathematical theory that is so badly taught to physicists as probability.

– *R. F. Streater [38, p. 19].*

Essentially, the theory of probability is nothing but good common sense reduced to mathematics.

– *P.-S. de Laplace,* Essai philosophique sur les probabilités, 1813.

Kolmogorov's 1933 axiomatization of probability on the mathematical foundation of measure theory was fundamental to the development of the subject and is essential for understanding the modern theory. Many excellent textbook treatments are available. Here we emphasize a few points directly relevant for the rest of this chapter.

1.A.1 Set Theory Notation

A *set* is a collection of *elements*. The set of no elements is the *empty set* $\emptyset$. Finite nonempty sets can be listed as $S = \{a_1, \dots, a_n\}$. If a set S contains an element a, we write $a \in S$. A set R is a *subset* of a set S, written $R \subseteq S$, if every $a \in R$ also satisfies $a \in S$. For two sets S and T, their *intersection* is $S \cap T$, the set of elements that are in *both* A and B, and their *union* is

$S \cup T$, the set of elements in at least one of S or T. For two sets S and T, "S minus T" is the set $S \setminus T = \{a \in S : a \notin T\}$, the set of elements that are in S but not in T.

Note that $S \cap \emptyset = \emptyset$, $S \cup \emptyset = S$, and $S \setminus \emptyset = S$.

1.A.2 Probability Spaces

Suppose we perform an experiment that gives a random outcome. Let Ω denote the set of all possible outcomes: the *sample space*. To start with, we take Ω to be *discrete*, which means it is *finite* or *countably infinite*. This means that we can write Ω as a (possibly infinite) list:

$$\Omega = \{\omega_1, \omega_2, \omega_3, \dots\},$$

where $\omega_1, \omega_2, \omega_3, \dots$ are the possible outcomes to our experiment.

A set $A \subseteq \Omega$ is called an *event*. Given events $A, B \subseteq \Omega$, we can build new events using the operations of set theory:

- $A \cup B$ ("A or B"), the event that A happens, or B happens, or both.
- $A \cap B$ ("A and B"), the event that A and B both happen.

Two events A and B are called *disjoint* or *mutually exclusive* if $A \cap B = \emptyset$.

We want to assign probabilities to events. Let Ω be a nonempty discrete sample space. A function $\mathbb{P}$ that gives a value $\mathbb{P}[A] \in [0, 1]$ for every subset $A \subseteq \Omega$ is called a discrete *probability measure* on Ω if

(P1) $\mathbb{P}[\emptyset] = 0$ and $\mathbb{P}[\Omega] = 1$;

(P2) For any $A_1, A_2, \dots$, pairwise disjoint subsets of Ω (so $A_i \cap A_j = \emptyset$ for $i \neq j$),

$$\mathbb{P}\left[\bigcup_{i=1}^{\infty} A_i\right] = \sum_{i=1}^{\infty} \mathbb{P}[A_i] \quad (\sigma\text{-additivity}).$$

Given Ω and a probability measure $\mathbb{P}$, we call $(\Omega, \mathbb{P})$ a *discrete probability space*.

Remark 1.10 *For nondiscrete sample spaces, we may not be able to assign probabilities to all subsets of Ω in a sensible way, and so smaller collections of events are required. In this appendix, we treat the discrete case only, as is sufficient for most (but not all) of the discussion in this chapter. For the more general case, which requires a deeper understanding of measure theory, there are many excellent treatments, such as [1, 2, 39, 40].*

For an event $A \subseteq \Omega$, we define its *complement*, denoted A^c and read "not A", to be $A^c := \Omega \setminus A = \{\omega \in \Omega : \omega \notin A\}$. Note that $(A^c)^c = A$, $A \cap A^c = \emptyset$, and $A \cup A^c = \Omega$.

If $(\Omega, \mathbb{P})$ is a discrete probability space, then

- For $A \subseteq \Omega$, $\mathbb{P}[A^c] = 1 - \mathbb{P}[A]$;
- If $A, B \subseteq \Omega$ and $A \subseteq B$, then $\mathbb{P}[A] \leq \mathbb{P}[B]$ (monotonicity);
- If $A, B \subseteq \Omega$, then $\mathbb{P}[A \cup B] = \mathbb{P}[A] + \mathbb{P}[B] - \mathbb{P}[A \cap B]$. In particular, if $A \cap B = \emptyset$, $\mathbb{P}[A \cup B] = \mathbb{P}[A] + \mathbb{P}[B]$.

It follows from this last statement that $\mathbb{P}[A \cup B] \leq \mathbb{P}[A] + \mathbb{P}[B]$; more generally, we have the elementary but useful *Boole's inequality*: $\mathbb{P}[\cup_n A_n] \leq \sum_n \mathbb{P}[A_n]$.

At several points we use the *continuity property of probability measures*:

- If $A_1 \subseteq A_2 \subseteq A_3 \subseteq \cdots$ are events, then $\mathbb{P}[\cup_{i=1}^{\infty} A_i] = \lim_{n\to\infty} \mathbb{P}[A_n]$.
- If $A_1 \supseteq A_2 \supseteq A_3 \supseteq \cdots$ are events, then $\mathbb{P}[\cap_{i=1}^{\infty} A_i] = \lim_{n\to\infty} \mathbb{P}[A_n]$.

To see the first statement, we can write $\cup_{i=1}^{\infty} A_i = \cup_{i=1}^{\infty} (A_i \setminus A_{i-1})$, where we set $A_0 =$

$\emptyset$, and the latter union is over pairwise disjoint events. So, by σ-additivity,

$$\mathbb{P}[\cup_{i=1}^{\infty} A_i] = \sum_{i=1}^{\infty} \mathbb{P}[A_i \setminus A_{i-1}]$$
$$= \lim_{n\to\infty} \sum_{i=1}^{n} \mathbb{P}[A_i \setminus A_{i-1}].$$

But $\sum_{i=1}^{n} \mathbb{P}[A_i \setminus A_{i-1}] = \mathbb{P}[\cup_{i=1}^{n} (A_i \setminus A_{i-1})] = \mathbb{P}[A_n]$, giving the result. The second statement is analogous.

1.A.3 Conditional Probability and Independence of Events

If A and B are events with $\mathbb{P}[B] > 0$ then the *conditional probability* $\mathbb{P}[A \mid B]$ of A given B is defined by

$$\mathbb{P}[A \mid B] := \frac{\mathbb{P}[A \cap B]}{\mathbb{P}[B]}.$$

A countable collection of events $E_1, E_2, \dots$ is called a *partition* of Ω if

(a) for all i, $E_i \subseteq \Omega$ and $E_i \neq \emptyset$;

(b) for $i \neq j$, $E_i \cap E_j = \emptyset$ (the events are disjoint);

(c) $\bigcup_i E_i = \Omega$ (the events fill the sample space).

Let $E_1, E_2, \dots$ be a partition of Ω. Following from the definitions is the basic *law of total probability*, which states that for all $A \subseteq \Omega$,

$$\mathbb{P}[A] = \sum_i \mathbb{P}[E_i]\mathbb{P}[A \mid E_i].$$

A countable collection $(A_i, i \in I)$ of events is called *independent* if, for every *finite* subset $J \subseteq I$,

$$\mathbb{P}\left[\bigcap_{j\in J} A_j\right] = \prod_{j\in J} \mathbb{P}[A_j].$$

In particular, two events A and B are independent if $\mathbb{P}[A \cap B] = \mathbb{P}[A]\mathbb{P}[B]$ (i.e., if $\mathbb{P}[B] > 0$, $\mathbb{P}[A \mid B] = \mathbb{P}[A]$).

1.A.4 Random Variables and Expectation

Let $(\Omega, \mathbb{P})$ be a discrete probability space. A function $X : \Omega \to \mathbb{R}$ is a *random variable*. So each $\omega \in \Omega$ is mapped to a real number $X(\omega)$. The set of possible values for X is $X(\Omega) = \{X(\omega) : \omega \in \Omega\} \subset \mathbb{R}$. Notice that because Ω is discrete, $X(\Omega)$ must be also.

If X and Y are two random variables on $(\Omega, \mathbb{P})$, then $X + Y$, XY, and so on, are also random variables. For example, $(X + Y)(\omega) = X(\omega) + Y(\omega)$. In some cases (such as the expected hitting times defined at (1.18)), we extend the domain of a random variable from $\mathbb{R}$ to $\mathbb{R} \cup \{\infty\}$.

The *probability mass function* of a discrete random variable X is the collection $\mathbb{P}[X = x]$ for all $x \in X(\Omega)$. The *distribution function* of X is $F : \mathbb{R} \to [0, 1]$ given by $F(x) = \mathbb{P}[X \leq x]$.

Example 1.22 [Bernoulli and binomial distributions] Let n be a positive integer and $p \in [0, 1]$. We say X has a *binomial distribution* with parameters (n, p), written $X \sim \text{Bin}(n, p)$, if $\mathbb{P}[X = k] = \binom{n}{k} p^k (1-p)^{n-k}$ for $k \in \{0, 1, \dots, n\}$. The case $Y \sim \text{Bin}(1, p)$ is the *Bernoulli* distribution; here $\mathbb{P}[Y = 1] = p = 1 - \mathbb{P}[Y = 0]$ and we write $Y \sim \text{Be}(p)$. The binomial distribution has the following interpretation: Perform n independent "trials" (e.g., coin tosses) each with probability p of "success" (e.g., "heads"), and count the total number of successes.

Example 1.23 [Poisson distribution] Let $\lambda > 0$ and $p_k := e^{-\lambda}(\lambda^k / k!)$ for $k \in \mathbb{Z}_+$. If $\mathbb{P}[X = k] = p_k$, X is a Poisson random variable with parameter λ.

Let X be a discrete random variable. The *expectation*, *expected value*, or *mean* of X is given by

$$\mathbb{E}[X] = \sum_{x \in X(\Omega)} x\mathbb{P}[X = x],$$

provided the sum is finite. The *variance* of X is $\mathbb{V}\mathrm{ar}[X] = \mathbb{E}[(X - \mathbb{E}[X])^2] = \mathbb{E}[X^2] - (\mathbb{E}[X])^2$.

Example 1.24 [Indicator variables.] Let A be an event. Let $\mathbf{1}_A$ denote the *indicator random variable* of A, that is, $\mathbf{1}_A : \Omega \to \{0, 1\}$ given by

$$\mathbf{1}_A(\omega) := \begin{cases} 1 \text{ if } \omega \in A \\ 0 \text{ if } \omega \notin A \end{cases}$$

So $\mathbf{1}_A$ is 1 if A happens and 0 if not. Then

$$\begin{aligned}\mathbb{E}[\mathbf{1}_A] &= 1 \cdot \mathbb{P}[\mathbf{1}_A = 1] + 0 \cdot \mathbb{P}[\mathbf{1}_A = 0] \\ &= \mathbb{P}[\mathbf{1}_A = 1] = \mathbb{P}[A].\end{aligned}$$

Expectation has the following basic properties. For X and Y random variables with well-defined expectations and $a, b \in \mathbb{R}$,

(a) $\mathbb{E}[aX + bY] = a\mathbb{E}[X] + b\mathbb{E}[Y]$ (linearity).

(b) If $\mathbb{P}[X \le Y] = 1$, then $\mathbb{E}[X] \le \mathbb{E}[Y]$ (monotonicity).

(c) $|\mathbb{E}[X]| \le \mathbb{E}[|X|]$ (triangle inequality).

(d) If $h : X(\Omega) \to \mathbb{R}$, then $\mathbb{E}[h(X)] = \sum_{x \in X(\Omega)} h(x)\mathbb{P}[X = x]$ ("law of the unconscious statistician").

Let X be a nonnegative random variable. Then, for any $x > 0$, $x\mathbf{1}\{X \ge x\} \le X$ holds with probability 1. Taking expectations and using monotonicity yields $\mathbb{P}[X \ge x] \le x^{-1}\mathbb{E}[X]$. This is usually known as *Markov's inequality*, although it is also sometimes referred to as *Chebyshev's inequality*.[15] Applying Markov's inequality to $\mathrm{e}^{\theta X}$, $\theta > 0$, gives

$$\mathbb{P}[X \ge x] = \mathbb{P}[\mathrm{e}^{\theta X} \ge \mathrm{e}^{\theta x}] \le \mathrm{e}^{-\theta x}\mathbb{E}[\mathrm{e}^{\theta X}],$$

which is sometimes known as *Chernoff's inequality*.

Let $(\Omega, \mathbb{P})$ be a discrete probability space. A family $(X_i, i \in I)$ of random variables is called *independent* if for any finite subset $J \subseteq I$ and all $x_j \in X_j(\Omega)$,

$$\mathbb{P}\Big(\bigcap_{j \in J}\{X_j = x_j\}\Big) = \prod_{j \in J} \mathbb{P}(X_j = x_j).$$

In particular, random variables X and Y are independent if $\mathbb{P}[X = x, Y = y] = \mathbb{P}[X = x]\mathbb{P}[Y = y]$ for all x and y.

Theorem 1.18 *If X and Y are independent, then $\mathbb{E}[XY] = \mathbb{E}[X]\mathbb{E}[Y]$.*

A consequence of Theorem 1.18 is that if X and Y are independent, then $\mathbb{V}\mathrm{ar}[X + Y] = \mathbb{V}\mathrm{ar}[X] + \mathbb{V}\mathrm{ar}[Y]$.

Example 1.25

(a) If $Y \sim \mathrm{Be}(p)$ then $\mathbb{E}[Y] = \mathbb{E}[Y^2] = p \cdot 1 + (1 - p) \cdot 0 = p$, so $\mathbb{V}\mathrm{ar}[Y] = p - p^2 = p(1 - p)$.

(b) If $X \sim \mathrm{Bin}(n, p)$ then we can write $X = \sum_{i=1}^n Y_i$ where $Y_i \sim \mathrm{Be}(p)$ are independent. By linearity of expectation, $\mathbb{E}[X] = \sum_{i=1}^n \mathbb{E}[Y_i] = np$. Also, by independence, $\mathbb{V}\mathrm{ar}[X] = \sum_{i=1}^n \mathbb{V}\mathrm{ar}[Y_i] = np(1 - p)$.

1.A.5 Conditional Expectation

On a discrete probability space $(\Omega, \mathbb{P})$, let B be an event with $\mathbb{P}[B] > 0$ and let X be

15) Chebyshev's inequality is the name more commonly associated with Markov's inequality applied to the random variable $(X - \mathbb{E}[X])^2$, to give $\mathbb{P}[|X - \mathbb{E}[X]| \ge x] \le x^{-2}\mathbb{V}\mathrm{ar}[X]$.

a random variable. The *conditional expectation* of X given B is

$$\mathbb{E}[X \mid B] = \sum_{x \in X(\Omega)} x\mathbb{P}[X = x \mid B].$$

So $\mathbb{E}[X \mid B]$ can be thought of as expectation with respect to the conditional probability measure $\mathbb{P}[\,\cdot \mid B]$. An alternative is

$$\mathbb{E}[X \mid B] = \frac{\mathbb{E}[X\mathbf{1}_B]}{\mathbb{P}[B]}, \qquad (1.35)$$

where $\mathbf{1}_B$ is the indicator random variable of B. The proof of (1.35) is an exercise in interchanging summations. First,

$$\begin{aligned}\mathbb{E}[X \mid B] &= \sum_{x \in X(\Omega)} x\mathbb{P}[X = x \mid B] \\ &= \sum_{x \in X(\Omega)} x\frac{\mathbb{P}[\{X = x\} \cap B]}{\mathbb{P}[B]}. \qquad (1.36)\end{aligned}$$

On the other hand, the random variable $\mathbf{1}_B X$ takes values $x \neq 0$ with

$$\begin{aligned}\mathbb{P}[\mathbf{1}_B X = x] &= \sum_{\omega \in \Omega : \omega \in B \cap \{X = x\}} \mathbb{P}[\{\omega\}] \\ &= \mathbb{P}[\{X = x\} \cap B],\end{aligned}$$

so by comparison we see that the final expression in (1.36) is indeed $\mathbb{E}[\mathbf{1}_B X]/\mathbb{P}[B]$.

Let $(E_i, i \in I)$ be a partition of Ω, so $\sum_{i \in I} \mathbf{1}_{E_i} = 1$. Hence,

$$\begin{aligned}\mathbb{E}[X] &= \mathbb{E}\left[X \sum_{i \in I} \mathbf{1}_{E_i}\right] = \mathbb{E}\left[\sum_{i \in I} X\mathbf{1}_{E_i}\right] \\ &= \sum_{i \in I} \mathbb{E}[X\mathbf{1}_{E_i}],\end{aligned}$$

by linearity. By (1.35), $\mathbb{E}[X\mathbf{1}_{E_i}] = \mathbb{E}[X \mid E_i]\mathbb{P}[E_i]$. Thus we verify the *partition theorem* for expectations:

$$\mathbb{E}[X] = \sum_{i \in I} \mathbb{E}[X \mid E_i]\mathbb{P}[E_i].$$

Given two discrete random variables X and Y, the *conditional expectation* of X given Y, denoted $\mathbb{E}[X \mid Y]$, is the *random variable* $\mathbb{E}[X \mid Y](\omega) = \mathbb{E}[X \mid Y = Y(\omega)]$, which takes values $\mathbb{E}[X \mid Y = y]$ with probabilities $\mathbb{P}[Y = y]$.

References

1. Durrett, R. (2010) *Probability: Theory and Examples*, 4th edn, Cambridge University Press, Cambridge.
2. Kallenberg, O. (2002) *Foundations of Modern Probability*, 2nd edn, Springer-Verlag, New York.
3. Fischer, H. (2011) *A History of the Central Limit Theorem*, Springer-Verlag, New York.
4. Ehrenfest, P. and Ehrenfest, T. (1907) Über zwei bekannte Einwände gegen das Boltzmannsche H-Theorem. *Phys. Z.*, **8**, 311–314.
5. Karlin, S. and Taylor, H.M. (1975) *A First Course in Stochastic Processes*, 2nd edn, Academic Press.
6. Norris, J.R. (1997) *Markov Chains*, Cambridge University Press, Cambridge.
7. Chung, K.L. (1967) *Markov Chains with Stationary Transition Probabilities*, 2nd edn, Springer-Verlag, Berlin.
8. Lamperti, J. (1960) Criteria for the recurrence or transience of stochastic process. I. *J. Math. Anal. Appl.*, **1**, 314–330.
9. Feller, W. (1971) *An Introduction to Probability Theory and its Applications*, Vol. II, 2nd edn, John Wiley & Sons, Inc., New York.
10. Bak, P. and Sneppen, K. (1993) Punctuated equilibrium and criticality in a simple model of evolution. *Phys. Rev. Lett.*, **71**, 4083–4086.
11. Gillett, A.J. (2007) Phase Transitions in Bak–Sneppen Avalanches and in a Continuum Percolation Model. PhD dissertation, Vrije Universiteit, Amsterdam.
12. Grinfeld, M., Knight, P.A., and Wade, A.R. (2012) Rank-driven Markov processes. *J. Stat. Phys.*, **146**, 378–407.
13. Rayleigh, L. (1880) On the resultant of a large number of vibrations of the same pitch and of arbitrary phase. *Philos. Mag.*, **10**, 73–78.
14. Bachelier, L. (1900) Théorie de la spéculation. *Ann. Sci. École Norm. Sup.*, **17**, 21–86.

15. Pearson, K. and Blakeman, J. (1906) *A Mathematical Theory of Random Migration, Drapers' Company Research Memoirs Biometric Series*, Dulau and co., London.
16. Einstein, A. (1956) *Investigations on the Theory of the Brownian Movement*, Dover Publications Inc., New York.
17. Pólya, G. (1921) Über eine Aufgabe der Wahrscheinlichkeitsrechnung betreffend die Irrfahrt im Straßennetz. *Math. Ann.*, **84**, 149–160.
18. Pearson, K. (1905) The problem of the random walk. *Nature*, **72**, 342.
19. Hughes, B.D. (1995) *Random Walks and Random Environments*, vol. 1, Oxford University Press, New York.
20. Feller, W. (1968) *An Introduction to Probability Theory and its Applications*, Vol. I, 3rd edn, John Wiley & Sons, Inc., New York.
21. Doherty, M. (1975) *An Amusing Proof in Fluctuation Theory, Combinatorial Mathematics III*, vol. 452, Springer, 101–104.
22. Ganesh, A., O'Connell, N., and Wischik, D. (2004) *Big Queues*, Springer, Berlin.
23. Foss, S., Korshunov, D., and Zachary, S. (2011) *An Introduction to Heavy-Tailed and Subexponential Distributions*, Springer-Verlag.
24. Dembo, A. and Zeitouni, O. (1998) *Large Deviations Techniques and Applications*, Springer, New York.
25. Touchette, H. (2009) The large deviation approach to statistical mechanics. *Phys. Rep.*, **478**, 1–69.
26. Karlin, S. and Taylor, H.M. (1981) *A Second Course in Stochastic Processes*, Academic Press.
27. Lax, M., Cai, W., and Xu, M. (2006) *Random Processes in Physics and Finance*, Oxford University Press.
28. Blackman, J.A. and Mulheran, P.A. (1996) Scaling behaviour in submonolayer film growth: a one-dimensional model. *Phys. Rev. B*, **54**, 11 681.
29. O'Neill, K.P., Grinfeld, M., Lamb, W., and Mulheran, P.A. (2012) Gap-size and capture-zone distributions in one-dimensional point-island nucleation and growth simulations: Asymptotics and models. *Phys. Rev. E*, **85**, 21 601.
30. Ising, E. (1925) Beitrag zur Theorie des Ferromagnetismus. *Z. Phys.*, **31**, 253–258.
31. Grimmett, G. (2010) *Probability on Graphs*, Cambridge University Press.
32. Lawler, G.F. (2005) *Conformally Invariant Processes in the Plane*, American Mathematical Society.
33. Grimmett, G. (1999) *Percolation*, 2nd edn, Springer-Verlag, Berlin.
34. Doyle, P.G. and Snell, J.L. (1984) *Random Walks and Electric Networks*, Mathematical Association of America, Washington, DC.
35. Kac, M. (1959) *Probability and Related Topics in Physical Sciences*, vol. 1957, Interscience Publishers, London and New York.
36. Shlesinger, M.F. and West, B.J. (1984) *Random Walks and Their Applications on the Physical and Biological Sciences*, American Institute of Physics, New York.
37. Weiss, G.H. and Rubin, R.J. (1983) Random walks: theory and selected applications. *Adv. Chem. Phys.*, **52**, 363–505.
38. Streater, R.F. (2007) *Lost Causes in and Beyond Physics*, Springer-Verlag.
39. Billingsley, P. (1995) *Probability and Measure*, 3rd edn, John Wiley & Sons, Inc. New York.
40. Chung, K.L. (2001) *A Course in Probability Theory*, 3rd edn, Academic Press Inc., San Diego, CA.

2
Monte-Carlo Methods

Kurt Binder

2.1
Introduction and Overview

Many problems in science are very complex: for example, statistical thermodynamics considers thermal properties of matter resulting from the interplay of a large number of particles. A deterministic description in terms of the equations of motion of all these particles would make no sense and a probabilistic description is required. A probabilistic description may even be intrinsically implied by the quantum-mechanical nature of the basic processes (e.g., emission of neutrons in radioactive decay) or because the problem is incompletely characterized, only some degrees of freedom being considered explicitly, while the others act as a kind of background causing random noise. While thus the concept of probability distributions is ubiquitous in physics, often it is not possible to compute these probability distribution functions analytically in explicit form, because of the complexity of the problem. For example, interactions between atoms in a fluid produce strong and nontrivial correlations between atomic positions, and, hence, it is not possible to calculate these correlations analytically.

Monte-Carlo methods now aim at a numerical estimation of probability distributions (as well as of averages that can be calculated from them), making use of (pseudo) random numbers. By "pseudorandom numbers" one means a sequence of numbers produced on a computer with a deterministic procedure from a suitable "seed." Hence, this sequence is not truly random: See Section 2.2 for a discussion of this problem.

The outline of the present article is as follows. As all Monte-Carlo methods heavily rely on the use of random numbers, we briefly review random-number generation in Section 2.2. In Section 2.3, we then elaborate on the discussion of "simple sampling," that is, problems where a straightforward generation of probability distributions using random numbers is possible. Section 2.4 briefly mentions some applications to transport problems, such as radiation shielding, and growth

Mathematical Tools for Physicists, Second Edition. Edited by Michael Grinfeld.

phenomena, such as "diffusion-limited aggregation" (DLA).

Section 2.5 then considers the importance-sampling methods of statistical thermodynamics, including the use of different thermodynamic ensembles. This chapter emphasizes applications in statistical mechanics of condensed matter, because this is the field where most activity with Monte-Carlo methods occurs.

Some more practical aspects important for the implementation of algorithms and the judgment of the tractability of simulation approaches to physical problems are then considered in Section 2.6: effects resulting from the finite size of simulation boxes, effects of choosing various boundary conditions, dynamic correlation of errors, and the application to studies of the dynamics of thermal fluctuations. In addition, methods for the sampling of free energies and of free energy barriers in configuration space will be briefly discussed (Section 2.7).

The extension to quantum-mechanical problems is mentioned in Section 2.8 and the application to elementary particle theory (lattice gauge theory) in Section 2.9. Section 2.10 then illustrates some of the general concepts with a variety of applications taken from condensed matter physics, while Section 2.11 contains concluding remarks.

We do not discuss problems of applied mathematics such as applications to the solution of linear operator equations (Fredholm integral equations, the Dirichlet boundary-value problem, eigenvalue problems, etc.); for a concise discussion of such problems, see, for example, Hammersley and Handscomb [1]. Nor do we discuss simulations of chemical kinetics, such as polymerization processes (see, e.g., [2]).

2.2 Random-Number Generation

2.2.1 General Introduction

The precise definition of "randomness" is a problem in itself (see, e.g., [3]) and is outside the scope of this chapter. Truly random numbers are unpredictable in advance and must be produced by an appropriate physical process such as radioactive decay, but are not useful for computer simulation in practice.

Pseudorandom numbers are produced in the computer by one of various algorithms, some of which will be discussed below, and thus are predictable, as their sequence is exactly reproducible. (This reproducibility, of course, is desirable, as it allows detailed checks of Monte-Carlo simulation programs.) They are thus not truly random, but they have statistical properties (nearly uniform distribution, nearly vanishing correlation coefficients, etc.) that are very similar to the statistical properties of truly random numbers. Thus, a given sequence of (pseudo)random numbers appears "random" for many practical purposes. In the following, the prefix "pseudo" will be omitted throughout.

2.2.2 Properties That a Random-Number Generator (RNG) Should Have

What one needs are numbers that are uniformly distributed in the interval [0,1] and that are uncorrelated. By "uncorrelated" we not only mean vanishing pair correlations for arbitrary distances along the random-number sequence, but also vanishing triplet and higher correlations. No algorithm exists that satisfies these desirable requirements fully, of course; the extent to which the remaining correlations between the generated random numbers lead to erroneous results of Monte-Carlo simulations has been a matter of long-standing concern [4–6]; even random-number generators (RNGs) that have passed all common statistical tests and have been used successfully for years may fail for a new application, in particular, if it involves a new type of Monte-Carlo algorithm. Therefore, the testing of RNGs is a field of research in itself (see, e.g., [7–9]).

A limitation due to the finite word length of computers is the finite period: Every generator begins after a long but finite period to produce exactly the same sequence again. For example, simple generators for 32-bit computers have a maximum period of 2^{30} ($\approx 10^9$) numbers only. This is not enough for recent high-quality applications! Of course, one can get around this problem [4, 5], but, at the same time, one likes the code representing the RNG to be "portable" (i.e., in a high-level programming language such as FORTRAN usable for computers from different manufacturers) and "efficient" (i.e., extremely fast so it does not unduly slow down the simulation program as a whole). Thus, inventing new RNGs that are in certain respects a better compromise between these partially conflicting requirements is still of interest (e.g., [10]).

2.2.3 Comments about a Few Frequently Used Generators

The best known among the frequently used RNG is the linear multiplicative algorithm [11], which produces random integers X_i recursively from the formula

$$x_i = aX_{i-1} + c \quad (\text{modulo} \quad m), \tag{2.1}$$

which means that m is added when the result is negative, but multiples of m are subtracted when the result is larger than m. For 32-bit computers, $m = 2^{31} - 1$ (the largest integer that can be used for that computer). The integer constants a, c, and X_0 (the starting value of the recursion, the so-called "seed") need to be appropriately chosen (e.g., $a = 16\,807$, $c = 0$, X_0 odd). Obviously, the "randomness" of the X_i results because, after a few multiplications with a, the result would exceed m and hence is truncated, and so the leading digits of X_i are more or less random. But there are severe correlations: if d-tuples of such numbers are used to represent points in d-dimensional space having a lattice structure, they lie on a certain number of hyperplanes [12].

Equation (2.1) produces random numbers between 0 and m. Converting them into real numbers and dividing by m yields random numbers in the interval [0,1], as desired.

More popular now are shift-register generators [13, 14] based on the formula

$$X_i = X_{i-p} \cdot \text{XOR} \cdot X_{i-q}, \tag{2.2}$$

where ·XOR· is the bitwise "exclusive or" operation, and the "lags" p, q have to be properly chosen (the popular "R250" [14] uses $p = 109$, $q = 250$ and thus needs 250 initializing integers). "Good" generators based on (2.2) have fewer correlations between random numbers than those resulting from (2.1), and much larger period.

A third type of generators is based on the Fibonacci series and is also recommended in the literature [4, 5, 15]. But a general recommendation is that users of random numbers should not rely on their quality blindly and should perform their own tests in the context of the application.

2.3 Simple Sampling of Probability Distributions Using Random Numbers

In this section, we give a few simple examples of the use of Monte-Carlo methods that will be useful for the understanding of later sections. More material on this subject can be found in standard textbooks such as Koonin [16] and Gould and Tobochnik [17].

2.3.1 Numerical Estimation of Known Probability Distributions

A known probability distribution in which a (discrete) state i, $1 \le i \le n$, occurs with probability p_i with $\sum_{i=1}^{n} p_i = 1$, is numerically realized using random numbers uniformly distributed in the interval from zero to unity as follows: defining $P_i = \sum_{j=1}^{i} p_i$, we choose state i if the random number ζ satisfies $P_{i-1} < \zeta < P_i$, with $P_0 = 0$. In the limit of a large number (M) of trials, the generated distribution approximates p_i, with errors of order $1/\sqrt{M}$.

Monte-Carlo methods in statistical mechanics can be viewed as an extension of this simple concept to the situation in which the probability that a point $\mathbf{X}$ in phase space occurs is given by the Boltzmann probability

$$P_{\text{eq}}(\mathbf{X}) = \frac{1}{Z}\exp\left[\frac{-\mathcal{H}(\mathbf{X})}{k_B T}\right],$$

k_B being Boltzmann's constant, T the absolute temperature, and $Z = \sum \exp\left[-\frac{\mathcal{H}(\mathbf{X})}{k_B T}\right]$ the partition function, although in general neither Z nor $P_{\text{eq}}(\mathbf{X})$ can be written explicitly (as function of the variables of interest, such as T, particle number N, volume V, etc.). The term $\mathcal{H}(\mathbf{X})$ denotes the Hamiltonian of the (classical) system.

2.3.2 "Importance Sampling" versus "Simple Sampling"

The sampling of the Boltzmann probability $P_{\text{eq}}(\mathbf{X})$ by Monte-Carlo methods is not completely straightforward: One must not choose the points $\mathbf{X}$ in phase space completely at random, because $P_{\text{eq}}(\mathbf{X})$ is extremely sharply peaked. Thus, one needs "importance-sampling" methods that generate points $\mathbf{X}$ preferably from the "important" region of space where this narrow peak occurs.

Before we treat this problem of statistical mechanics in more detail, we emphasize the more straightforward applications of "simple sampling" techniques. In the following, we list a few problems where simple sampling is useful. Suppose one wishes to generate a configuration of a randomly mixed crystal of a given lattice structure, for example, a binary mixture of composition A_xB_{1-x}. Again, one uses pseudorandom numbers ζ uniformly distributed in [0,1] to choose the occupancy of lattice sites $\{j\}$: If $\zeta_j < x$, the site j is taken by an A atom, else by a B atom. Such configurations now can be used as starting point for a numerical study of the dynamical matrix if one is interested in the phonon spectrum of mixed crystals. One can study the distribution of sizes of "clusters" formed by neighboring A atoms if one is interested in the "site percolation problem" [18], and so on.

If one is interested in simulating transport processes such as diffusion, a basic approach is the generation of simple random walks (RWs). Such RWs, resulting from addition of vectors whose orientation is random, can be generated both on lattices and in the continuum. Such simulations are desirable if one wishes to consider complicated geometries or boundary conditions of the medium where the diffusion takes place. In addition, it is straightforward to include competing processes (e.g., in a reactor, diffusion of neutrons in the moderator competes with loss of neutrons due to nuclear reactions, radiation going to the outside, etc., or gain due to fission events). Actually, this problem of reactor criticality (and related problems for nuclear weapons!) was the starting point for the first large-scale applications of Monte-Carlo methods by Fermi, von Neumann, Ulam, and their coworkers [1].

2.3.3 Monte-Carlo as a Method of Integration

Many Monte-Carlo computations may be viewed as attempts to estimate the value of (multiple) integrals. To give the flavor of this idea, we discuss the one-dimensional integral

$$I = \int_0^1 f(x)\,\mathrm{d}x \equiv \int_0^1\int_0^1 g(x,y)\,\mathrm{d}x\mathrm{d}y;\ \text{with} \tag{2.3}$$

$$g(x,y) = \begin{Bmatrix} 0 & \text{if} & f(x) < y, \\ 1 & \text{if} & f(x) \geq y, \end{Bmatrix}$$

as an example (suppose, for simplicity, that also $0 \leq f(x) \leq 1$ for $0 \leq x \leq 1$). Then I simply is interpreted as the fraction of the unit square $0 \leq x,\ y \leq 1$ lying underneath the curve $y = f(x)$. Now a straightforward (though often not very efficient) Monte-Carlo estimation of (2.3) is the "hit-or-miss" method: We take n points (ζ_x, ζ_y) uniformly distributed in the unit square $0 \leq \zeta x \leq 1$, $0 \leq \zeta y \leq 1$. Then I is estimated by the average $\bar{g}$ of $g(x,y)$,

$$\bar{g} = \frac{1}{n}\sum_{i=1}^{n} g\left(\zeta_{xi}, \zeta_{yi}\right) = \frac{n^*}{n}, \tag{2.4}$$

n^* being the number of points for which $f(\zeta_{xi}) \geq \zeta_{yi}$. Thus, we count the fraction of points that lie underneath the curve $y = f(x)$. Of course, such Monte-Carlo methods are inferior to many other techniques of numerical integration, if the integration space is low dimensional, but the situation for standard integration methods is worse for high-dimensional spaces: For any method using a regular grid of points for which the integrand needs to be evaluated, the number of points sampled along each coordinate is $M^{1/\mathrm{d}}$ in d dimensions, which is small for any reasonable sample size M if d is very large.

2.3.4 Infinite Integration Space

Not always is the integration space limited to a bounded interval in space. For example, the ϕ^4 model of field theory considers a field variable $\phi(x)$, where x is drawn from a d-dimensional space and $\phi(x)$ is a real variable with distribution

$$P(\phi) \propto \exp\left[-\alpha\left(-\frac{1}{2}\phi^2 + \frac{1}{4}\phi^4\right)\right]; \quad \alpha > 0. \tag{2.5a}$$

While $-\infty < \phi < +\infty$, the distribution $P'(y)$

$$P'(y) = \frac{\int_{-\infty}^{y} P(\phi)\mathrm{d}\phi}{\int_{-\infty}^{+\infty} P(\phi)\mathrm{d}\phi} \tag{2.5b}$$

varies in the unit interval, $0 \leq P' \leq 1$. Hence, defining $Y = Y(P')$ as the inverse function of $P'(y)$, we can choose a random number ζ uniformly distributed between zero and one, to obtain $\phi = Y(\zeta)$ distributed according to the chosen distribution $P(\phi)$. Of course, this method works not only for the example chosen in (2.5a) but for any distribution of interest. Often it will not be possible to obtain $Y(P')$ analytically, but then one can compute numerically a table before the start of the sampling [19].

2.3.5 Random Selection of Lattice Sites

A problem that occurs very frequently (e.g., in solid-state physics) is one that considers a large lattice (e.g., a model of a simple cubic crystal with $N = L_x \times L_y \times L_z$ sites), and one wishes to select a lattice site (n_x, n_y, n_z) at random. This is trivially done using the integer arithmetics of standard computers, converting a uniformly distributed random number $\zeta x(0 \leq \zeta x < 1)$ to an integer n_x with $1 \leq n_x \leq L_x$ via the statement $n_x = \text{int}(\zeta_x L_x + 1)$. This is already an example where one must be careful, however, when three successive pseudorandom numbers drawn from an RNG are used for this purpose: If one uses an RNG with bad statistical qualities, the frequency with which individual sites are visited may deviate distinctly from a truly random choice. In unfavorable cases, successive pseudorandom numbers are so strongly correlated that certain lattice sites would never be visited.

2.3.6 The Self-Avoiding Walk Problem

As an example of the straightforward use of simple sampling techniques, we now discuss the study of self-avoiding walks (SAWs) on lattices (which may be considered as a simple model for polymer chains in good solvents; see [20]). Suppose one considers a square or simple cubic lattice with coordination number (number of nearest neighbors) z. Then, for a RW with N steps, we would have $Z_{\text{RW}} = z^N$ configurations, but many of these RWs intersect themselves and thus would not be self-avoiding. For SAWs, one only expects of the order of $Z_{\text{SAW}} = \text{const.} \times N^{\gamma-1} z_{\text{eff}}^N$ configurations, where $\gamma > 1$ is a characteristic exponent (which is not known exactly for $d = 3$ dimensions), and $z_{\text{eff}} \leq z - 1$ is an effective coordination number, which also is not known exactly. But it is already obvious that an exact enumeration of all configurations would be possible for rather small N only, while most questions of interest refer to the behavior for large N; for example, one wishes to study the end-to-end distance of the SAW,

$$\left\langle R^2 \right\rangle_{\text{SAW}} = \frac{1}{Z_{\text{SAW}}} \sum_X [R(X)]^2, \tag{2.6}$$

the sum being extended over all configurations of SAWs which we denote formally as points $\mathbf{X}$ in phase space. One expects that $\left\langle R^2 \right\rangle_{\text{SAW}} \propto N^{2\nu}$, where ν is another characteristic exponent. A Monte-Carlo estimation of $\left\langle R^2 \right\rangle_{\text{SAW}}$ now is based on generating a sample of only $M \ll Z_{\text{SAW}}$ configurations $\mathbf{X}_l$, that is,

$$\overline{R^2} = \frac{1}{M} \sum_{l=1}^{M} \left[R\left(\mathbf{X}_l\right)\right]^2 \approx \left\langle R^2 \right\rangle_{\text{SAW}}. \tag{2.7}$$

If the M configurations are statistically independent, standard error analysis applies, and we expect that the relative error behaves as

$$\frac{\overline{(\delta R^2)^2}}{\left(\overline{R^2}\right)^2} \approx \frac{1}{M-1} \left[\frac{\left\langle R^4 \right\rangle_{\text{SAW}}}{\langle R^2 \rangle_{\text{SAW}}^2} - 1 \right]. \tag{2.8}$$

While the law of large numbers then implies that $\overline{R^2}$ is Gaussian distributed around $\left\langle R^2 \right\rangle_{\text{SAW}}$ with a variance determined by (2.8), one should note that the variance does not decrease with increasing N. Statistical mechanics tells us that fluctuations decrease with increasing the number of degrees of freedom N; that is, one equilibrium configuration differs in its energy $E(x)$ from the average $\langle E \rangle$ only by an amount of order $1/\sqrt{N}$. This property is called *self-averaging*. Obviously, such a property is not true for $\langle R^2 \rangle_{\text{SAW}}$. This "lack of self-averaging" is easy to show for ordinary RWs [21].

2.3.7
Simple Sampling versus Biased Sampling: the Example of SAWs Continued

Apart from this problem, that the accuracy of the estimation of $\langle R^2 \rangle$ does not increase with the number of steps of the walk, it is also not easy to generate a large sample of configurations of SAWs for large N. Suppose we do this at each step by choosing one of $z-1$ neighbors at random (eliminating, from the start, immediate reversals, which would violate the SAW condition). Whenever the chosen lattice site is already taken, the attempted walk must be terminated. Now the fraction of walks that will continue successfully for N steps will only be of the order of $Z_{\text{SAW}}/(z-1)^N \propto [z_{\text{eff}}/(z-1)]^N N^{\gamma-1}$, which decreases to zero exponentially ($\propto \exp(-N\kappa)$ with $\kappa = \ln[(z-1)/z_{\text{eff}}]$ for large N); this failure of success in generating long SAWs is called the *attrition problem*.

The obvious recipe, to select at each step only from among the lattice sites that do not violate the SAW restriction, does not give equal statistical weight for each configuration generated, of course, and so the average would not be the averaging that one needs in (2.6). One finds that this method would create a "bias" toward more compact configurations of the walk. But one can calculate the weights of configurations $w(\mathbf{X})$ that result in this so-called "inversely restricted sampling" [22], and, in this way, correct for the bias and estimate the SAW averages as

$$\overline{R^2} = \left\{ \sum_{l=1}^{M} \left[w\left(\mathbf{X}_l\right)\right]^{-1} \right\}^{-1} \times \sum_{l=1}^{M} \left[w\left(\mathbf{X}_l\right)\right]^{-1} \left[\mathrm{R}\left(\mathbf{X}_l\right)\right]^2 . \qquad (2.9)$$

However, error analysis of this biased sampling is delicate [20].

A popular alternative to overcome the above attrition problem is the "enrichment technique," founded on the principle "Hold fast to that which is good." Namely, whenever a walk attains a length that is a multiple of s steps without intersecting itself, n independent attempts to continue it (rather than a single attempt) are made. The numbers n, s are fixed, and, if we choose $n \approx \exp(\kappa s)$, the numbers of walks of various lengths generated will be approximately equal. Enrichment has the advantage over inversely restricted sampling that all walks of a given length have equal weights, while the weights in (2.9) vary over many orders of magnitude for large N. But the linear dimensions of the walks are highly correlated, because some of them have many steps in common. For these reasons, simple sampling and its extensions are useful only for a few problems in polymer science; "importance sampling" (Section 2.5) is much more used. But we emphasize that related problems are encountered for the sampling of "random surfaces" (this problem arises in the field theory of quantum gravity), in path-integral Monte-Carlo treatments of quantum problems, and in many other contexts.

A variant of such techniques that still is widely used to simulate single polymer chains is the PERM algorithm ("pruned-enriched Rosenbluth method," [23, 24]). In this method, a population of chains growing in their chain length step by step is generated and their statistical weights are considered. From time to time, members of the population with low weight are removed ("pruning") and members with high weight are enriched.

2.4
Survey of Applications to Simulation of Transport Processes

There are many possibilities to simulate the random motion of particles. Therefore, it is difficult to comment about such problems in general. Thus, we only discuss a few examples that illustrate the spirit of such approaches.

2.4.1
The "Shielding Problem"

A thick shield of absorbing material is exposed to γ radiation (energetic photons), of specified distribution of energy and angle of incidence. We want to know the intensity and energy distribution of the radiation that penetrates that shield.

The description is here that one generates a lot of "histories" of those particles traveling through the medium. The paths of these γ particles between scattering events are straight lines and different γ particles do not interact with each other. A particle with energy E, instantaneously at the point $\mathbf{r}$ and traveling in the direction of the unit vector $\mathbf{w}$, continues to travel in the same direction with the same energy, until a scattering event with an atom of the medium occurs. The standard assumption is that these atoms are distributed randomly in space. Then the total probability that the particle collides with an atom while traveling a length δs of its path is $\sigma_c(E)\delta s$, $\sigma_c(E)$ being the cross section. In a region of space where $\sigma_c(E)$ is constant, the probability that a particle travels without collision a distance s is $F_c(s) = 1 - \exp[-\sigma_c(E)s]$. If a collision occurs, it leads to absorption or scattering, and the cross sections for these types of events are assumed to be known.

A Monte-Carlo solution now simply involves the tracking of simulated particles from collision to collision, generating the distances s that the particles travel without collision from the exponential distribution quoted above. Particles leaving a collision point are sampled from the appropriate conditional probabilities as determined from the respective differential cross sections. For increasing sampling efficiency, many obvious tricks are known. For example, one may avoid losing particles by absorption events: If the absorption probability (i.e., the conditional probability that absorption occurs given that a collision has occurred) is α, one may replace $\sigma_c(E)$ by $\sigma_c(E)(1 - \alpha)$, and allow only scattering to take place with the appropriate relative probability. Special methods for the shielding problem have been extensively developed already and have been reviewed by Hammersley and Handscomb [1].

2.4.2
Diffusion-Limited Aggregation (DLA)

DLA is a model for the irreversible formation of random aggregates by diffusion of particles that get stuck at random positions on the already formed object if they hit its surface in the course of their diffusion (see Vicsek [25], Meakin [26], and Herrmann [19] for detailed reviews of this problem and related phenomena). Many structures (shapes of snowflakes, size distribution of asteroids, roughness of crack surfaces, etc.) can be understood as the end product of similar random irreversible growth processes. DLA is just one example of them. It is simulated by iterating the following steps: From a randomly selected position on a spherical surface of radius $R_{\rm m}$ that encloses the aggregate (grown in the previous steps, its center of gravity being in the center of the sphere), a particle of unit mass is launched to start a simple RW trajectory. If it touches the aggregate, it sticks irreversibly on its surface. After the particle has either stuck or moved a distance $R_{\rm f}$ from the center of the aggregate such that it is unlikely that it will hit in the future, a new particle is launched. Ideally, one would like to have $R_{\rm f} \to \infty$ but, in practice, $R_{\rm f} = 2R_{\rm m}$ suffices. By this irreversible aggregation of particles, one forms fractal clusters. This means that the dimension $d_{\rm f}$ characterizing the relation between the mass of the grown object and its (gyration) radius $R, M \propto R^{d_{\rm f}}$, is less than the dimension d of space in which the growth takes place. Again there are some tricks to make such simulations more efficient: For example, one may allow the particles to jump over larger steps when they travel in empty regions. From such studies, researchers have found that $d_{\rm f} = 1.715 \pm 0.004$ for DLA in $d = 2$, while $d_{\rm f} = 2.485 \pm 0.005$ in $d = 3$ [27].

2.5 Monte-Carlo Methods in Statistical Thermodynamics: Importance Sampling

2.5.1 The General Idea of the Metropolis Importance-Sampling Method

In the canonical ensemble, the average of an observable $A(\mathbf{X})$ takes the form

$$\langle A\rangle = \frac{1}{Z}\int_{\Omega} d^k X A(\mathbf{X}) \exp\left[-\frac{\mathcal{H}(\mathbf{X})}{k_{\rm B}T}\right], \quad (2.10)$$

where Z is the partition function,

$$Z = \int_{\Omega} d^k X \exp\left[-\frac{\mathcal{H}(\mathbf{X})}{k_{\rm B}T}\right], \quad (2.11)$$

Ω denoting the (k-dimensional) volume of phase space $\{\mathbf{X}\}$ over which the integration is extended, $\mathcal{H}(\mathbf{X})$ being the (classical) Hamiltonian. For this problem, a simple sampling analogue to Section 2.3 would not work: The probability distribution $p(\mathbf{X}) = (1/Z)\exp[-\mathcal{H}(\mathbf{X})/k_{\rm B}T]$ has a very sharp peak in phase space where all extensive variables $A(\mathbf{X})$ are close to their average values $\langle A\rangle$. This peak may be approximated by a Gaussian centered at $\langle A\rangle$, with a relative half-width of order $1/\sqrt{N}$ only, if we consider a system of N particles. Hence, for a practically useful method, one cannot sample the phase space uniformly, but the points $\mathbf{X}_\nu$ must be chosen preferentially from the important region of phase space, that is, from the vicinity of the peak of $p(\mathbf{X})$. This goal is achieved by the importance-sampling method [28]: Starting from some initial configuration $\mathbf{X}_1$, one constructs a sequence of configurations $\mathbf{X}_\nu$ defined in terms of a transition probability $W(\mathbf{X}_\nu \to \mathbf{X}'_\nu)$ that rules stochastic "moves" from an old state $\mathbf{X}_\nu$ to a new state $\mathbf{X}'_\nu$, and, hence, one creates a "RW through phase space." The idea is to choose $W(\mathbf{X} \to \mathbf{X}')$ such that the probability with which a point $\mathbf{X}$ is chosen converges toward the canonical probability

$$P_{\rm eq}(\mathbf{X}) = \left(\frac{1}{Z}\right)\exp\left[-\frac{\mathcal{H}(\mathbf{X})}{k_{\rm B}T}\right]$$

in the limit where the number M of states $\mathbf{X}$ generated goes to infinity. A condition sufficient to ensure this convergence is the so-called principle of detailed balance,

$$P_{\rm eq}(\mathbf{X})W(\mathbf{X} \to \mathbf{X}') = P_{\rm eq}(\mathbf{X}')W(\mathbf{X}' - \mathbf{X}). \quad (2.12)$$

For a justification that (2.12) actually yields this desired convergence, we refer to Hammersley and Handscomb [1], Binder [29], Binder and Heermann [30], and Kalos and Whitlock [31]. In this importance-sampling technique, the average (2.10) then is estimated in terms of a simple arithmetic average,

$$\overline{A} = \frac{1}{M - M_0} \sum_{\nu=M_0+1}^{M} A(\mathbf{X}_\nu). \tag{2.13}$$

Here it is anticipated that it is advantageous to eliminate the residual influence of the initial configuration $\mathbf{X}_1$ by eliminating a large enough number M_0 of states from the average. (The judgment of what is "large enough" is often difficult; see Binder [29] and Section 2.5.3 below.) Note that this Metropolis method can be used for sampling any distribution $P(\mathbf{X})$: One simply must choose a transition probability $W(\mathbf{X} \to \mathbf{X}')$ that satisfies a detailed balance condition with $P(\mathbf{X})$ rather than with $P_{\text{eq}}(\mathbf{X})$.

2.5.2
Comments on the Formulation of a Monte-Carlo Algorithm

What is now meant in practice by the transition $\mathbf{X} \to \mathbf{X}'$? Again, there is no general answer to this question; the choice of the process may depend both on the model studied and the purpose of the study. As (2.12) implies that $W(\mathbf{X} - \mathbf{X}')/W(\mathbf{X}' \to \mathbf{X}) = \exp(-\delta\mathcal{H}/k_B T)$, $\delta\mathcal{H}$ being the energy change caused by the move from $\mathbf{X} \to \mathbf{X}'$, typically it is necessary to consider small changes of the state X only. Otherwise, $|\delta\mathcal{H}|$ would be rather large, and then either $W(\mathbf{X} \to \mathbf{X}')$ or $W(\mathbf{X}' \to \mathbf{X})$ would be very small, and the procedure would be poorly convergent. For example, in the lattice gas model at constant particle number, a transition $\mathbf{X} \to \mathbf{X}'$ may consist of moving one particle to a randomly chosen neighboring site. In the lattice gas at constant chemical potential, one removes (or adds) just one particle at a time, which is isomorphic to single flips in the Ising model of anisotropic magnets.

Another arbitrariness concerns the order in which the particles are selected for considering a move. Often, one selects them in the order of their labels (in the simulation of a fluid at constant particle number) or to go through the lattice in a regular typewriter-type manner (in the case of spin models, for instance). For lattice systems, it may be convenient to use sublattices (e.g., the "checkerboard algorithm," where the white and black sublattices are updated in alternation, for the sake of an efficient "vectorization" of the program; see [32]). An alternative is to choose the lattice sites (or particle numbers) randomly. The latter procedure is somewhat more time consuming, but it is a more faithful representation of a dynamic time evolution of the model described by a master equation (see below).

It is also helpful to realize that often the transition probability $W(\mathbf{X} \to \mathbf{X}')$ can be written as a product of an "attempt frequency" times an "acceptance frequency." By clever choice of the attempt frequency, one sometimes can attempt large moves and still have a high acceptance, making the computations more efficient.

For spin models on lattices, such as Ising or Potts models, XY, and Heisenberg ferromagnets, and so on, algorithms have been devised where one does not update single spins in the move $\mathbf{X} \to \mathbf{X}'$, but, rather, one updates specially constructed clusters of spins (see [33], for a review). These algorithms have the merit that they reduce critical slowing down, which hampers the efficiency of Monte-Carlo simulations near second-order phase transitions. “Critical slowing down” means a dramatic increase of relaxation times at the critical point of such transitions and these relaxation times also control statistical errors in simulations; see Section 2.6. As these “cluster algorithms” work for rather special models only, they will not be discussed further here.

There is also some arbitrariness in the choice of the transition probability $W(\mathbf{X}\to\mathbf{X}')$ itself. The original choice of Metropolis *et al.* [28] is

$$W(\mathbf{X}\to\mathbf{X}') = \begin{cases} \exp\left(\frac{-\delta\mathcal{H}}{k_\mathrm{B}T}\right) & \text{if } \delta\mathcal{H} > 0, \\ 1 & \text{otherwise.} \end{cases} \tag{2.14}$$

An alternative choice is the so-called “heat-bath method.” There one assigns the new value α_i' of the ith local degree of freedom in the move from $\mathbf{X}$ to $\mathbf{X}'$ irrespective of what the old value α_i was. Considering the local energy $\mathcal{H}_i(\alpha_i')$, one chooses the state α_i' with probability

$$\frac{\exp\left[\frac{-\mathcal{H}_i(\alpha_i')}{k_\mathrm{B}T}\right]}{\sum\limits_{\{\alpha_i''\}}\exp\left[\frac{-\mathcal{H}_i(\alpha_i'')}{k_\mathrm{B}T}\right]}.$$

We now outline the realization of the sequence of states $\mathbf{X}$ with chosen transition probability W. At each step of the procedure, one performs a trial move $\alpha_i\to\alpha_i'$, computes $W(\mathbf{X}\to\mathbf{X}')$ for this trial move, and compares it with a random number κ, uniformly distributed in the interval [0,1]. If $W<\kappa$, the trial move is rejected, and the old state (with α_i) is counted once more in the average, (2.13). Then another trial is made. If $W>\kappa$, on the other hand, the trial move is accepted, and the new configuration thus generated is taken into account in the average, (2.13). It serves then also as a starting point of the next step.

As subsequent states $\mathbf{X}_\nu$ in this Markov chain differ by the coordinate α_i of one particle only (if they differ at all), they are highly correlated. Therefore, it is not straightforward to estimate the error of the average, (2.13). Let us assume for the moment that, after n steps, these correlations have died out. Then we may estimate the statistical error δA of the estimate $\overline{A}$ from the standard formula,

$$\overline{(\delta A)^2} = \frac{1}{m(m-1)}\sum_{\mu=\mu_0}^{m+\mu_0-1}[A(\mathbf{X}_\mu)-\overline{A}]^2, \quad m \gg 1, \tag{2.15}$$

where the integers μ_0, μ, m are defined by $m=(M-M_0)/n$, μ_0 labels the state $\nu=M_0+1$, $\mu=\mu_0+1$ labels the state $\nu=M_0+1+n$, and so on. Then, for consistency, $\overline{A}$ should be taken as

$$\overline{A} = \frac{1}{m}\sum_{\mu=\mu_0}^{m+\mu_0-1}A(\mathbf{X}_\mu). \tag{2.16}$$

If the computational effort of carrying out the “measurement” of $A(\mathbf{X}_\mu)$ in the simulation is rather small, it is advantageous to keep taking measurements every Monte-Carlo step (MCS) per degree of freedom but to construct block averages over n successive measurements, varying n until uncorrelated block averages are obtained. Details on error estimation for Monte-Carlo simulations can be found in Berg [34].

2.5.3 The Dynamic Interpretation of the Monte-Carlo Method

It is not always easy to estimate the number of states M_0 after which the correlations to the initial state $\mathbf{X}_1$, which typically is a state far from equilibrium, have died out, nor is it easy to estimate the number n between steps after which correlations in equilibrium have died out. A formal answer to this problem, in terms of relaxation times of the associated master equation describing the Monte-Carlo process, is discussed in Section 2.5.4. This interpretation of Monte-Carlo sampling in terms of master equations is also the basis for Monte-Carlo studies of the dynamics of fluctuations near thermal equilibrium, and is discussed now. One introduces the probability $P(\mathbf{X},t)$ that a state $\mathbf{X}$ occurs at time t. This probability then decreases by all moves $\mathbf{X} \to \mathbf{X}'$, where the system reaches a neighboring state $\mathbf{X}'$; inverse processes $\mathbf{X}' \to \mathbf{X}$ lead to a gain of probability. Thus, one can write down a rate equation, similar to chemical kinetics, considering the balance of all gain and loss processes:

$$\frac{\mathrm{d}}{\mathrm{d}t}P(\mathbf{X},t) = -\sum_{\mathbf{X}'} W(\mathbf{X} \to \mathbf{X}')P(\mathbf{X},t) + \sum_{\mathbf{X}'} W(\mathbf{X}' \to \mathbf{X})P(\mathbf{X}',t). \quad (2.17)$$

The Monte-Carlo sampling (i.e., the sequence of generated states $\mathbf{X}_1 \to \mathbf{X}_2 \to \cdots \to \mathbf{X}_\nu \to \cdots$) can hence be interpreted as a numerical realization of the master equation, (2.17), and then a "time" t is associated with the index ν of subsequent configurations. In a system with N particles, we may normalize the "time unit" such that N single-particle moves are attempted per unit time. This is often called a *sweep step* or *1 Monte-Carlo step (MCS).*

For the thermal equilibrium distribution $P_{\mathrm{eq}}(\mathbf{X})$, because of the detailed balance principle, (2.12), there is no change of probability with time, $\mathrm{d}P(\mathbf{X},t)/\mathrm{d}t = 0$; thus, thermal equilibrium arises as the stationary solution of the master equation, (2.17). Thus, it is also plausible that Markov processes described by (2.17) describe a relaxation that always leads toward thermal equilibrium, as desired.

Now, for a physical system (whose trajectory in phase space, according to classical statistical mechanics, follows from Newton's laws of motion), it is clear that the stochastic trajectory through phase space that is described by (2.17) in general has nothing to do with the actual dynamics. For example, (2.17) never describes any propagating waves (such as spin waves in a magnet, or sound waves in a fluid, etc.).

In spite of this observation, the dynamics of the Monte-Carlo "trajectory" described by (2.17) sometimes does have physical significance. In many situations, one does not consider the full set of dynamical variables of the system, but rather a subset only: For instance, in modeling the diffusion processes in an interstitial alloy, the diffusion of the interstitials may be described by a stochastic hopping between the available lattice sites. As the mean time between two successive jumps is orders of magnitude larger than the time scale of atomic vibrations in the solid, the phonons can be approximated as a heat bath, as far as the diffusion is concerned.

There are many examples where such a separation of time scales for different degrees of freedom occurs: For example, describing the Brownian motion of polymer chains in polymer melts, the fast bond-angle and bond-length vibrations may be treated as heat bath, and so on. As a rule of thumb, any very slow relaxation phenomena (kinetics of nucleation, decay of remnant magnetization in spin glasses, growth of ordered domains in adsorbed monolayers at surfaces, etc.) can be modeled by Monte-Carlo methods. Of course, one must build in relevant conservation laws into the model properly (e.g., in an interstitial alloy, the overall concentration of interstitials is conserved; in a spin glass, the magnetization is not conserved) and choose microscopically reasonable elementary steps for the move $\mathbf{X} \to \mathbf{X}'$. The great flexibility of the Monte-Carlo method, where one can choose the level of the modeling appropriately for the system at hand and identify the degrees of freedom that one wishes to consider, as well as the type and nature of transitions between them, is a great advantage and thus allows complementary applications to more atomistically realistic simulation approaches such as the molecular dynamics (MD) method where one numerically integrates Newton's equations of motion [35].

2.5.4 Monte-Carlo Study of the Dynamics of Fluctuations Near Equilibrium and of the Approach toward Equilibrium

Accepting (2.17), the average in (2.13) then is interpreted as a time average along the stochastic trajectory in phase space,

$$\overline{A} = \frac{1}{t_M - t_{M_0}} \int_{t_{M_0}}^{t_M} A(t)\mathrm{d}t, \qquad t_M = \frac{M}{N}, \quad t_{M_0} = \frac{M_0}{N}. \tag{2.18}$$

It is thus no surprise that, for importance-sampling Monte-Carlo, one needs to consider carefully the problem of ergodicity: Time averages need not agree with ensemble averages. For example, near first-order phase transitions, there may be long-lived metastable states. Sometimes, the considered moves do not allow one to reach all configurations (e.g., in dynamic Monte-Carlo methods for SAWs; see [20]).

One can also define time-displaced correlation functions: $\langle A(t)B(0)\rangle$, where A, B stand symbolically for any physical observables, is estimated by

$$\overline{A(t)B(0)} = \frac{1}{t_M - t - t_{M_0}} \int_{t_{M_0}}^{t_M - t} A(t+t') B(t')\mathrm{d}t', \quad t_M - t > t_{M_0}. \tag{2.19}$$

Equation (2.19) refers to a case where t_{M_0} is chosen large enough such that the system has relaxed toward equilibrium during the time t_{M_0}; then the pair correlation depends only on t and not the two individual times t', $t' + t$ separately.

However, it is also interesting to study the nonequilibrium process by which equilibrium is approached. In this region, $A(t) - \overline{A}$ depends systematically on the observation time t, and an ensemble average $\langle A(t)\rangle_T - \langle A(\infty)\rangle_T$ ($\lim_{t\to\infty} \overline{A} = \langle A\rangle_T \equiv \langle A(\infty)\rangle_T$ if the system is ergodic, that is, the time average agrees with the ensemble average at the chosen temperature T) is nonzero. One may define

$$\langle A(t)\rangle_T = \sum_{\{\mathbf{X}\}} P(\mathbf{X}, t)A(\mathbf{X}) = \sum_{\{\mathbf{X}\}} P(\mathbf{X}, 0)A(\mathbf{X}(t)). \tag{2.20}$$

In the second step of (2.20), the fact was used that the ensemble average involved is actually an average weighted by $P(\mathbf{X}, 0)$ over an ensemble of initial states $\mathbf{X}(t = 0)$, which then evolve as described by the master equation, (2.17). In practice, (2.20) means an average over a large number $n_{\text{run}} \gg 1$ statistically independent runs,

$$\left[\overline{A(t)}\right]_{av} = \frac{1}{n_{\text{run}}} \sum_{l=1}^{n_{\text{run}}} A(t, l), \tag{2.21}$$

where $A(t, l)$ is the observable A observed at time t in the lth run of this nonequilibrium Monte-Carlo averaging.

Many concepts of nonequilibrium statistical mechanics can immediately be used in such simulations. For instance, one can introduce "fields" that can be switched off to study the dynamic response functions, for both linear and nonlinear responses [36].

2.5.5
The Choice of Statistical Ensembles

While so far the discussion has been (implicitly) restricted to the case of the canonical ensemble (NVT ensemble, for the case of a fluid, particle number N, volume V, and temperature T being held fixed), it is sometimes useful to use other statistical ensembles. Particularly useful is the grand canonical ensemble (μVT), where the chemical potential μ rather than the particle number N is fixed. In addition to moves where the configuration of particles in the box relaxes, one has moves where one attempts to add or remove a particle from the box.

In the case of binary (AB) mixtures, a useful variation is the "semi–grand canonical" ensemble, where $\Delta\mu = \mu_A - \mu_B$ is held fixed and moves where an A particle is converted into a B particle (or vice versa) are considered, in an otherwise identical system configuration.

The isothermal-isobaric (NpT) ensemble, on the other hand, fixes the pressure and then volume changes $V \to V' = V + \Delta V$ need to be considered (rescaling properly the positions of the particles).

It also is possible to define artificial ensembles that are not in the textbooks on statistical mechanics. An example is the Gaussian ensemble (interpolating between the canonical and microcanonical ensemble, useful for the study of first-order phase transitions, such as the so-called "multicanonical ensemble"). Particularly useful is the "Gibbs ensemble," where one considers the equilibrium between two simulation boxes (one containing liquid, the other gas), which can exchange both volume ΔV and particles (ΔN), while the total volume and total particle number contained in the two boxes are held fixed. The Gibbs ensemble is useful for the simulation of gas-fluid coexistence, avoiding interfaces [37, 38].

A simulation at a given state point (NVT) contains information not only on averages at that point but also on neighboring states (NVT'), via suitable reweighting of the energy distribution $P_N(E)$ with a factor $\exp(E/k_B T)\exp(-E/k_B T')$. Such "histogram methods" are particularly useful near critical points [33].

2.6
Accuracy Problems: Finite-Size Problems, Dynamic Correlation of Errors, Boundary Conditions

2.6.1
Finite-Size–Induced Rounding and Shifting of Phase Transitions

A prominent application of Monte-Carlo simulation in statistical thermodynamics and lattice theory is the study of phase transitions. it is well known now in statistical physics that sharp phase transitions can occur in the thermodynamic limit only, $N \to \infty$. This is no practical problem in everyday life – even a small water droplet freezing into a snowflake contains about $N = 10^{18} H_2O$ molecules, and, thus, the rounding and shifting of the freezing are on a relative scale of $\frac{1}{\sqrt{N}} = 10^{-9}$ and thus completely negligible. But the situation differs for simulations, which often consider very small systems (e.g., a d-dimensional box with linear dimensions L, $V = L^d$, and periodic boundary conditions), where only $N \sim 10^3 - 10^6$ particles are involved.

In such small systems, phase transitions are strongly rounded and shifted [39–42]. Thus, care needs to be applied when simulated systems indicate phase changes. It turns out, however, that these finite-size effects can be used as a valuable tool to infer properties of the infinite system from the finite-size behavior. As a typical example, we discuss the phase transition of an Ising ferromagnet (Figure 2.1), which has a second-order phase transition at a critical temperature T_c. For $L \to \alpha$, the spontaneous magnetization M_{spont} vanishes according to a power law, $M_{spont} = B(1 - T/T_c)^\beta$, B being a critical amplitude and β a critical exponent [43], and the susceptibility χ and correlation length ξ diverge,

$$\chi \propto \left|1 - \frac{T}{T_c}\right|^{-\gamma}, \quad \xi \propto \left|1 - \frac{T}{T_c}\right|^{-\nu} \tag{2.22}$$

where γ and ν are critical exponents. In a finite system, ξ cannot exceed L, hence, these singularities are smeared out.

Now finite-size scaling theory [39, 42] implies that these finite-size effects are understood from the principle that "L scales with ξ"; that is, the order-parameter probability distribution $P_L(M)$ can be written [40, 41]

$$P_L = L^{\beta/\nu}\tilde{P}\left(\frac{L}{\xi}, ML^{\frac{\beta}{\nu}}\right), \tag{2.23}$$

where $\tilde{P}$ is a "scaling function." From (2.23), one immediately obtains the finite-size scaling relations for order parameter $\langle|M|\rangle$ and the susceptibility (defined from a fluctuation relation) by taking the moments of the distribution P_L:

$$\langle|M|\rangle = L^{-\beta/\nu}\tilde{M}\left(\frac{L}{\xi}\right), \tag{2.24}$$

$$k_B T\chi' = L^d\left(\langle M^2\rangle - \langle|M|\rangle^2\right) = L^{\gamma/\nu}\tilde{X}\left(\frac{L}{\xi}\right), \tag{2.25}$$

where $\tilde{M}, \tilde{\chi}$ are scaling functions that follow from $\tilde{P}$ in (2.23) At T_c where $\xi \to \infty$, we thus have $\chi' \propto L^{\gamma/\nu}$; from this variation of χ' with L, hence, the exponent γ/ν can be extracted.

The fourth-order cumulant U_L is a function of L/ξ only,

$$U_L \equiv 1 - \frac{\langle M^4\rangle}{\left(3\langle M^2\rangle^2\right)} = \tilde{U}\left(\frac{L}{\xi}\right). \tag{2.26}$$

Here $U_L \to 0$ for a Gaussian centered at $M = 0$, that is, for $T > T_c$; $U_L \to 2/3$ for the double-Gaussian distribution, that is, for $T < T_c$; while $U_L = \tilde{U}(0)$ is a universal nontrivial constant for $T = T_c$. Cumulants for different system sizes hence intersect at T_c, and this can be used to locate T_c precisely [40, 41].

A simple discussion of finite-size effects at first-order transitions is similarly possible [41]: one describes the various phases that coexist at the first-order transition in terms of Gaussians if $L \gg \xi$ (note that ξ stays finite at the first-order transition). In a finite system, these phases coexist not only right at the transition but over a finite parameter region. The weights of the respective peaks are given in terms of the free-energy difference of the various phases. From this description, energy and order-parameter distributions and their moments can be worked out. Of course, this description applies only for long enough runs where the system jumps from one phase to the other many times, while for short runs where the systems stay in a single phase, one observes hysteresis.

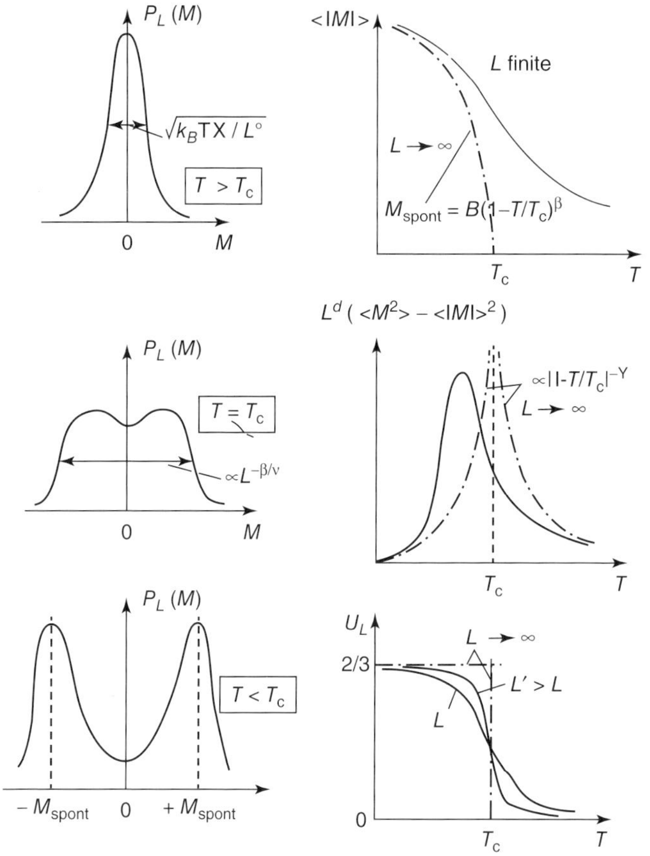

Figure 2.1 (a–f) Schematic evolution of the order-parameter probability distribution $P_L(M)$ from $T > T_c$ to $T < T_c$ (from above to below, left part), for an Ising ferromagnet (where M is the magnetization) in a box of volume $V = L^d$. The right part shows the corresponding temperature dependence of the mean order parameter $\langle |M| \rangle$, the susceptibility $k_B T \chi' = L^d \left(\left\langle M^2 \right\rangle - \langle |M| \rangle^2 \right)$, and the reduced fourth-order cumulant $U_L = 1 - \langle M^4 \rangle / [3 \langle M^2 \rangle^2]$. Dash-dotted curves indicate the singular variation that results in the thermodynamic limit, $L \to \infty$.

2.6.2 Different Boundary Conditions: Simulation of Surfaces and Interfaces

We now briefly discuss the effect of various boundary conditions. Typically, one uses periodic boundary conditions to study bulk properties of systems not obscured by surface effects. However, it also is possible to choose boundary conditions to study surface effects deliberately; for example, one may simulate thin films in an $L \times L \times D$ geometry with two free $L \times L$ surfaces and periodic boundary conditions otherwise. If the film thickness D is large enough, the two surfaces do not influence each other, and one can infer the properties of a semi-infinite system. One may choose special interactions near the free surfaces, apply surface “fields” (even if they cannot be applied in the laboratory, it may nevertheless be useful to study the response to them in the simulation), and so on.

Sometimes, the boundary conditions may stabilize interfaces in the system (e.g., in an Ising model for $T < T_c$ a domain wall between phases with opposite magnetization will be present, if we apply strong enough surface fields of opposite sign). Such interfaces also are often the object of study. It may be desirable to simulate interfaces without having the systems disturbed by free surfaces. In an Ising system, this may simply be done by choosing antiperiodic boundary conditions. Combining antiperiodic and staggered periodic boundary conditions, even tilted interfaces may be stabilized in the system. In all such simulations of systems containing interfaces, one must keep in mind, however, that because of capillary-wave excitations, interfaces usually are very slowly relaxing objects, and often a major computing effort is needed to equilibrate them. A further difficulty (when one is interested in interfacial profiles) is the fact that the center of the interface is typically delocalized.

2.6.3 Estimation of Statistical Errors

We now return to the problem of judging the time needed for having reasonably small errors in Monte-Carlo sampling. If the subsequent configurations used were uncorrelated, we could use (2.15), but in the case of correlations we have rather (here the index μ should not be confused with the chemical potential of Section 2.5.5, of course)

$$\begin{aligned}\langle(\delta A)^2\rangle &= \left\langle\left[\frac{1}{n}\sum_{\mu=1}^{n}A_\mu - \langle A\rangle\right]^2\right\rangle \\ &= \frac{1}{n}\left[\langle A^2\rangle - \langle A\rangle^2 + 2\sum_{\mu=1}^{n}\left(1-\frac{\mu}{n}\right)\right.\\ &\quad\times \left.(\langle A_0A_\mu\rangle - \langle A\rangle^2)\right]. \qquad (2.27)\end{aligned}$$

Now we remember that a time $t_\mu = \mu\delta t$ is associated with the Monte-Carlo process, δt being the time interval between two successive observations A_μ, $A_{\mu+1}$. Transforming the summation to a time integration yields

$$\begin{aligned}\langle(\delta A)^2\rangle &= \frac{1}{n}\left(\langle A^2\rangle - \langle A\rangle^2\right)\\ &\quad\times\left[1+\frac{2}{\delta t}\int_0^{t_n}\left(1-\frac{t}{t_n}\right)\phi_A(t)\mathrm{d}t\right], \qquad (2.28)\end{aligned}$$

where

$$\phi_A(t) \equiv \frac{\langle A(0)A(t)\rangle - \langle A\rangle^2}{\langle A^2\rangle - \langle A\rangle^2}.$$

Defining a relaxation time $\tau_A = \int_0^\infty \mathrm{d}t\phi_A(t)$, one obtains for $\tau_A \ll n\delta t = \tau_{\text{obs}}$ (the observation time)

$$\begin{aligned}\left\langle(\delta A)^2\right\rangle &= \frac{1}{n}[\langle A^2\rangle - \langle A\rangle^2]\left(1+\frac{2\tau_A}{\delta t}\right)\\ &\approx 2\left(\frac{\tau_A}{\tau_{\text{obs}}}\right)[\langle A^2\rangle - \langle A\rangle^2]. \qquad (2.29)\end{aligned}$$

In comparison with (2.15), the dynamic correlations inherent in a Monte-Carlo sampling as described by the master equation, (2.17), lead to an enhancement of the expected statistical error $\left\langle(\delta A)^2\right\rangle$ by a “dynamic factor” $1 + 2\tau_A/\delta t$ (sometimes also called the *statistical inefficiency*).

This dynamic factor is particularly cumbersome near second-order phase transitions (τ_A diverges: critical slowing down) and near first-order phase transitions (τ_A diverges at phase coexistence, because of the large lifetime of metastable states). Thus, even if one is interested only in static quantities in a Monte-Carlo simulation, understanding the dynamics may be useful for estimating errors. In addition, the question of how many configurations (M_0) must be omitted at the start of the averaging for the sake of equilibrium (2.18) can be formally answered in terms of a nonlinear relaxation function

$$\phi^{(nl)}(t) = \frac{\langle A(t)\rangle_T - \langle A(\infty)\rangle_T}{\langle A(0)\rangle_T - \langle A(\infty)\rangle_T}$$

and its associated time $\tau_A^{(nl)} = \int_0^\infty \phi_A^{(nl)}(t)\mathrm{d}t$ by the condition $t_{M_0} \gg \tau_A^{(nl)}$. Here t_{M_0} is the time corresponding to the number of states M_0 omitted.

2.7 Sampling of Free Energies and Free Energy Barriers

2.7.1 Bulk Free Energies

Carrying out the step from (2.10) to (2.13), the explicit knowledge on the partition function Z of the system, and hence its free energy $F = -k_B T \ln Z$ is lost. In many cases, however, knowledge of F would be very useful: for example, at a first-order phase transition the free energies F_1, F_2 of two distinct phases 1, 2 of a system become equal.

There are many ways how this drawback of importance-sampling Monte-Carlo can be overcome (see [30, 36]). We describe here very briefly the most popular method, Wang-Landau sampling [44]. It starts from the concept that Z can be expressed in terms of the energy density of states $g(E)$ as

$$Z = \int \mathrm{d}E g(E) \exp\left(\frac{-E}{k_B T}\right).$$

Suppose $g(E)$ is known and one defines a Markov process via a transition probability $p(E \to E')$ as

$$p\left(E \to E'\right) = \min\left\{\frac{g(E)}{g(E')}, 1\right\}$$

This process would generate a flat histogram $H(E) = \text{const}$. As $g(E)$ is not known beforehand, we can use the process defined by the above equation to construct an iteration that yields an estimate for $g(E)$: when we know $g(E)$, Z and hence F are also known.

In the absence of any a priori knowledge on $g(E)$, the iteration starts with $g(E) \equiv 1$ for all E (for simplicity, we consider here an energy spectrum that is discrete and bounded, $E_{\min} \le E \le E_{\max}$). Now Monte-Carlo moves (e.g., spin flips in an Ising model) are carried out to accumulate a histogram $H(E)$, starting out with $H(E) = 0$ for all E and replacing $H(E)$ by $H(E) + 1$ whenever E is visited. We also replace $g(E)$ by $g(E)f$ where the modification factor f initially is large, for example, $f = e^1$. The sampling of the histogram is continued, until $H(E)$ is reasonably "flat" (e.g., nowhere less than 80% of its maximum value).

Then $H(E)$ is reset to $H(E) = 0$ for all E, f is replaced by $\sqrt{f}$, and the process continues: in the ith step of the iteration $f_i = \sqrt{f_{i-1}}$, so f tends to unity; that is, we have reached the above $p(E \to E')$, as desired, and can use the corresponding $g(E)$ to compute the desired averages.

This algorithm is robust and widely used (see Landau and Binder [36] for examples). We stress, however, that many other schemes for obtaining $g(E)$ exist, such as the so-called "umbrella sampling," "multicanonical Monte-Carlo," "transition matrix Monte-Carlo," and so on. Often the density of states is required as a function of several variable (e.g., $g(E,M)$ where M is the magnetization, for an Ising model). We refer to the more specialized literature [36] for more details on these techniques.

2.7.2
Interfacial Free Energies

When in the volume taken by a system several phases coexist, the interfaces cause excess contribution to the free energy. The simplest case again is provided by the Ising model, considering again the distribution $P_L(M)$ of the magnetization M in a system of finite linear dimension L, cf. Figure 2.1, but now for $T \ll T_c$ (Figure 2.1): A state with $M \approx 0$ actually is characterized by two-phase coexistence in the simulation box: a domain with $M \approx +M_{\text{spont}}$ is separated from a (roughly equally large) domain with $M \approx -M_{\text{spont}}$ by two planar interfaces (of area L^{d-1} if the volume is L^d; note the periodic boundary conditions). As a result, one can argue that $P_L(M \approx 0)/P_L(M = M_{\text{spont}}) \approx \exp(-2L^{d-1} \; f_{\text{int}}/k_B T)$, with f_{int} the interface free-energy density between the coexisting phases.

Accurate sampling of $P_L(M)$ near its minimum (which can be done by umbrella sampling, for instance) in fact is a useful method for estimating f_{int} in various systems. Of course, there exist many alternative techniques, and also methods to estimate the excess free energies due to external walls, and related quantities such as contact angles of droplets, line tensions, and so on. This subject is still a very active area of research [45, 46].

2.7.3
Transition Path Sampling

Suppose we consider a system with a complex free-energy landscape, where the system can stay in some minimum (denoted as A) for a long time, and in a rare event may pass through some saddle point region to another minimum (denoted as B). Being interested in the rates of such process, one must consider the possibility that many trajectories from A to B in configuration space may compete with each other. Transition path sampling [47] is a technique to explore the dominating trajectories and thus the rate of such rare events. Starting out from one trial trajectory that leads from A to B, a sampling of stochastic variations of this trajectory is performed. Again, the technique and its theoretical foundations are rather involved, and still under development.

2.8
Quantum Monte-Carlo Techniques

2.8.1
General Remarks

Development of Monte-Carlo techniques to study ground-state and finite-temperature properties of interacting quantum many-body systems is an active area of research (for reviews, see [48–56]). These methods are of interest for problems such as the structure of nuclei [57] and elementary particles [58], superfluidity of ^{3}He and ^{4}He [53, 54], high-T_c superconductivity (e.g., [59]), magnetism (e.g., [60, 61]), surface physics [62], and so on. Despite this widespread interest, much of this research has the character of "work in progress" and hence cannot feature more prominently in this chapter. Besides, there is not just one quantum Monte-Carlo method, but many variants that exist: variational Monte-Carlo (VMC), Green's-function Monte-Carlo (GFMC), projector Monte-Carlo (PMC), path-integral Monte-Carlo (PIMC), grand canonical quantum Monte-Carlo (GCMC), world-line quantum Monte-Carlo (WLQMC), and so on.

Some of these (such as VMC, GFMC) address ground-state properties, others (such as PIMC) finite temperatures. Here only the PIMC technique will be briefly sketched, following Gillan and Christodoulos [63].

2.8.2 Path-Integral Monte-Carlo Methods

We wish to calculate thermal averages for a quantum system and thus rewrite (2.10) and (2.11) appropriately,

$$\langle A\rangle = \frac{1}{Z}\mathrm{Tr}\exp\left(-\frac{\hat{\mathcal{H}}}{k_B T}\right)\hat{A},$$
$$Z = \mathrm{Tr}\exp\left(-\frac{\hat{\mathcal{H}}}{k_B T}\right), \tag{2.30}$$

using a notation that emphasizes the operator character of the Hamiltonian $\hat{\mathcal{H}}$ and of the quantity $\hat{A}$ associated with the variable A that we consider. For simplicity, we consider first a system of a single particle in one dimension acted on by a potential $V(x)$. Its Hamiltonian is

$$\hat{\mathcal{H}} = -\frac{\hbar^2}{2m}\frac{d^2}{dx^2} + V(x). \tag{2.31}$$

Expressing the trace in the position representation, the partition function becomes

$$Z = \int \mathrm{d}x \left\langle x \middle| \exp\left(\frac{-\hat{\mathcal{H}}}{k_B T}\right) \middle| x \right\rangle, \tag{2.32}$$

where $|x\rangle$ is an eigenvector of the position operator. Writing $\exp(-\hat{\mathcal{H}}/k_B T)$ formally as $[\exp(-\hat{\mathcal{H}}/k_B TP)]^P$, where P is a positive integer, we can insert a complete set of states between the factors:

$$Z = \int \mathrm{d}x_1 \cdots \int \mathrm{d}x_P \langle x_1|\exp\left(\frac{-\hat{\mathcal{H}}}{k_B TP}\right)|x_2\rangle$$
$$\langle x_2|\cdots|x_p\rangle\langle x_P|\exp\left(\frac{-\hat{\mathcal{H}}}{k_B TP}\right)|x_1\rangle. \tag{2.33}$$

For large P, it is a good approximation to ignore the fact that kinetic and potential energy do not commute. Hence, one gets

$$\langle x|\exp\left(\frac{-\hat{\mathcal{H}}}{k_B TP}\right)|x'\rangle \approx \left(\frac{k_B TmP}{2\pi\hbar^2}\right)^{1/2}$$
$$\times \exp\left[\frac{-k_B TmP}{2\pi\hbar^2}(x-x')^2\right]$$
$$\times \exp\left\{\frac{-1}{2k_B TP}\left[V(x) + V(x')\right]\right\} \tag{2.34}$$

and

$$Z \approx \left(\frac{k_B TmP}{2\pi\hbar^2}\right)^{P/2} \int \mathrm{d}x_1 \cdots \int \mathrm{d}x_P$$
$$\times \exp\left\{-\frac{1}{k_B T}\left[\frac{1}{2}\sum_{s=1}^{P}\kappa\left(x_s - x_{s+1}\right)^2\right]\right.$$
$$\left. + P^{-1}\sum_{s=1}^{P} V(x_s)\right\}, \tag{2.35}$$

where

$$\kappa \equiv \left(\frac{k_B T}{\hbar}\right)^2 mP. \tag{2.36}$$

In the limit $P \to \infty$, (2.35) becomes exact. Apart from the prefactor, (2.35) is precisely the configurational partition function of a classical system, a ring polymer consisting of P beads coupled by harmonic springs with spring constant κ. Each bead is in a potential $V(x)/P$.

This approach is straightforwardly generalized to a system of N interacting quantum particles: one gets a system of N classical cyclic "polymer" chains. As a result of this isomorphism, the Monte-Carlo method for simulating classical systems can be carried over to such quantum-mechanical problems too. It is also easy to see that the system always behaves classically at high temperatures: κ gets very large, and then the cyclic chains contract essentially to a point, while at low temperatures, they are spread out, representing zero-point motion. However, PIMC becomes increasingly difficult at low temperatures, because P has to be the larger the lower that T is: If σ is a characteristic distance over which the potential $V(x)$ changes, one must have $\hbar^2/m\sigma^2 \ll k_B TP$ in order that two neighbors along the "polymer chain" are at a distance much smaller than σ. In PIMC simulations, one empirically determines and uses that P beyond which the thermodynamic properties do not effectively change.

This approach can be generalized to the density matrix $\rho(x - x') = \langle x| \exp(-\hat{\mathscr{H}}/k_B T)|x'\rangle$, while there are problems with time-displaced correlation functions $\langle A(t)B(0)\rangle$, where t is now the true time (associated with the time evolution of states following from the Schrödinger equation, rather than the "time" of Section 2.5.3 related to the master equation).

The step leading to (2.34) can be viewed as a special case of the Suzuki–Trotter formula [51]

$$\exp\left(\hat{A} + \hat{B}\right) = \lim_{P\to\infty} \left[\exp\left(\frac{\hat{A}}{P}\right) \exp\left(\frac{\hat{B}}{P}\right)\right]^P, \qquad (2.37)$$

which is also used for mapping d-dimensional quantum problems on lattices to equivalent classical problems (in $d+1$ dimensions, because of the "Trotter direction" corresponding to the imaginary time direction of the path integral).

2.8.3 A Classical Application: the Momentum Distribution of Fluid ^{4}He

We now consider the dynamic structure factor $S(\mathbf{k}, \omega)$, which is the Fourier transform of a time-displaced pair correlation function of the density at a point r_1 at time t_1 and the density at point r_2 at time t_2 ($\hbar\mathbf{k}$ being the momentum transfer and $\hbar\omega$ the energy transfer of an inelastic scattering experiment by which one can measure $S(\mathbf{k}, \omega)$). In the "impulse approximation," $S(\mathbf{k}, \omega)$ can be related to the Fourier transform of the single-particle density matrix $\rho_1(\mathbf{r})$, which for ^{4}He can be written in terms of the wave function $\psi(\mathbf{r})$ as $\rho_1(\mathbf{r}) = \left\langle \psi^+\left(r' + r\right)\psi\left(r\right)\right\rangle$. This relation is

$$S(\mathbf{k}, \omega) \propto J(Y) = \frac{1}{\pi}\int_0^\infty \rho_1(\mathbf{r})\cos(Yr)\mathrm{d}r,$$

where $Y \equiv m(\omega - k^2/2m)/k$ [64]. As $S(\mathbf{k}, \omega)$ has been measured via neutron scattering [65], a comparison between experiment and simulation can be performed without adjustable parameters (Figure 2.2). Thus, the PIMC method yields accurate data in good agreement with experiment.

The studies of ^{4}He have also yielded qualitative evidence for superfluidity [53]. For a quantitative analysis of the λ transition, a careful assessment of finite-size effects (Figure 2.1) is needed because one works with very small particle numbers (of the order of 10^2 ^{4}He atoms only). This has not been possible so far.

2.8.4 A Few Qualitative Comments on Fermion Problems

Particles obeying Fermi–Dirac statistics (such as electrons or ^{3}He, for instance, see Ceperley [54]) pose particular challenges to Monte-Carlo simulation. If one tries to solve the Schrödinger equation of a many-body system by Monte-Carlo methods, one exploits its analogy with a diffusion equation [48, 49]. Diffusion processes correspond to RWs and are hence accessible to Monte-Carlo simulation. However, while the diffusion equation (for one particle)

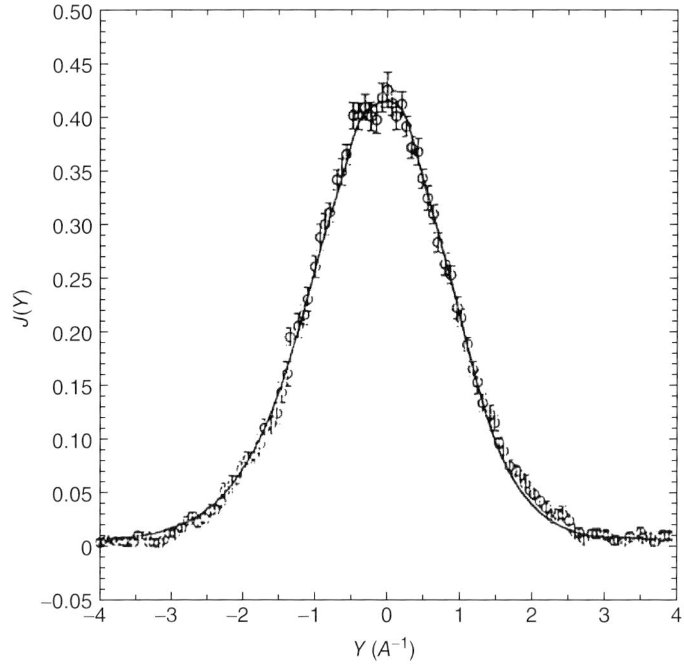

Figure 2.2 The measured momentum distribution $J(Y)$ of ^{4}He at $T = 3.3$ K (circles, from [65]) compared with the PIMC result of Ceperley and Pollock [66] (solid line). (Source: From Schmidt and Ceperley [67].)

considers the probability $P(r, t)$ that a particle starting at time $t = 0$ at the origin has reached the position r at time t, the wave function ψ in the Schrödinger equation is not positive definite. This fact creates severe problems for wave functions of many-fermion systems, because these wave functions must be antisymmetric, and the "nodal surface" in configuration space (where ψ changes sign) is unknown.

Formally, the difficulty of applying importance-sampling techniques to distributions $\rho(\mathbf{r})$ that are not always positive can be overcome by splitting $\rho(\mathbf{r})$ into its sign, $s = \text{sign}(\rho)$, and its absolute value, $\rho = s|\rho|$, and one can use $\tilde{\rho}(\mathbf{r}) = \frac{|\rho(\mathbf{r})|}{\int |\rho(\mathbf{r})| d^3r}$ as a probability density for importance sampling, and absorb the sign of $\rho(r)$ in the quantity to be measured. Symbolically, the average of an observable A is obtained as

$$\langle A \rangle = \frac{\int d^3r A(\mathbf{r}) s(\mathbf{r}) \tilde{\rho}(\mathbf{r})}{\int s(\mathbf{r}) \tilde{\rho}(\mathbf{r}) d^3r} = \frac{\langle As \rangle_{\tilde{\rho}}}{\langle s \rangle_{\tilde{\rho}}},$$

where $\langle \ldots \rangle_{\tilde{\rho}}$ means averaging with $\tilde{\rho}$ as weight function. However, as is not unexpected, using $\tilde{\rho}$ one predominantly samples unimportant regions in phase space; therefore, in sampling the sign $\langle s \rangle_{\tilde{\rho}}$, one has large cancellations from regions where the sign is negative, and, for N degrees of freedom, one gets $\langle s \rangle_{\tilde{\rho}} \propto \exp(-\text{const.} \times N)$. This difficulty is known as the *minus-sign problem* and still hampers applications to fermion problems significantly.

Sometimes it is possible to start with a trial wave function where nodes are a reasonable first approximation to the actual nodes, and, starting with the population of RWs from this fixed-node approximation given by the trial function, one now admits walks that cross this nodal surface and sample the sign as indicated above. In this way, it has been possible to estimate the exchange-correlation energy of the homogeneous electron gas [68] over a wide range of densities very well.

2.9 Lattice Gauge Theory

Monte-Carlo simulation has become the primary tool for nonperturbative quantum chromodynamics, the field theory of quarks and hadrons and other elementary particles (e.g., [58, 69]). In this section, we first stress the basic problem, to make the analogy with the calculations of statistical mechanics clear. Then we very briefly highlight some of the results that have been obtained so far.

2.9.1 Some Basic Ideas of Lattice Gauge Theory

The theory of elementary particles is a field theory of gauge fields and matter fields. Choice of a lattice is useful to provide a cutoff that removes the ultraviolet divergences that would otherwise occur in these quantum field theories. The first step, hence, is the appropriate translation from the four-dimensional continuum (3 space + 1 time dimensions) to the lattice.

The generating functional (analogous to the partition function in statistical mechanics) is

$$Z = \int DAD\overline{\psi}D\psi \exp\left[-S_g(A, \overline{\psi}, \psi)\right], \quad (2.38)$$

where A represents the gauge fields, $\overline{\psi}$ and ψ represent the (fermionic) matter field, S_g is the action of the theory (containing a coupling constant g, which corresponds to inverse temperature in statistical mechanics as const.$/g^2 \rightarrow 1/k_B T$), and the symbols $\int D$ stand for functional integration. The action of the gauge field itself is, using the summation convention that indices that appear twice are summed over,

$$S_G = \frac{1}{4}\int d^4x F^{\alpha}_{\mu\nu}(x)F^{\mu\nu}_{\alpha}(x), \quad (2.39)$$

$F^{\alpha}_{\mu\nu}$ being the fields that derive from the vector potential $A^{\alpha}_{\mu}(x)$. These are

$$F^{\alpha}_{\mu\nu}(x) = \partial_\mu A^{\alpha}_{\nu}(x) - \partial_\nu A^{\alpha}_{\mu}(x) + g f^{\alpha}_{\beta\gamma} A^{\beta}_{\mu}(x) A^{\gamma}_{\nu}(x), \quad (2.40)$$

$f^{\alpha}_{\beta\gamma}$ being the structure constants of the gauge group, and g a coupling constant.

The variables that one then introduces are elements $U_\mu(x)$ of the gauge group G, which are associated with the links of the four-dimensional lattice, connecting x and a nearest-neighbor point $x + \mu$:

$$U_\mu(x) = \exp\left[igaT^{\alpha}A^{\alpha}_{\mu}(x)\right],$$
$$\left[U_m(x+m)\right]^{\dagger} = U_m(x), \quad (2.41)$$

where a is the lattice spacing and T^{α} a group generator. Here U^{+} denotes the Hermitean conjugate of U. Wilson [70] invented a lattice action that reduces in the continuum limit to (2.39), namely,

$$\frac{S_U}{k_B T} = \frac{1}{g^2}\sum_n \sum_{\mu>\nu} \mathrm{Re}\,\mathrm{Tr} U_\mu(n) \times U_\nu(n+\mu)$$
$$U^{\dagger}_{\mu}(n+\nu)U^{\dagger}_{\nu}(n), \quad (2.42)$$

where the links in (2.42) form a closed contour along an elementary plaquette of the lattice.

Using (2.42) in (2.38), which amounts to the study of a "pure" gauge theory (no matter fields), we recognize that the problem is equivalent to a statistical mechanics problem (such as spins on a lattice), the difference being that now the dynamical variables are the gauge group elements $U_\mu(n)$. Thus importance-sampling Monte-Carlo algorithms can be put to work, just as in statistical mechanics.

To include matter fields, one starts from a partition function of the form

$$Z = \int DUD\overline{\psi}D\psi \exp\left\{-\frac{S_U}{k_B T} + \sum_{i=1}^{n_f} \overline{\psi}\underline{M}\psi\right\}$$
$$= \int DU(\det \underline{M})^{n_f} \exp\left(-\frac{S_U}{k_B T}\right), \quad (2.43)$$

where we have assumed fermions with n_f degenerate "flavors." It has also been indicated that the fermion fields can be integrated out analytically, but the price is that one has to deal with the "fermion determinant" of the matrix M. In principle, for any change of the U's this determinant needs to be recomputed; together with the fact that one needs to work on rather large lattices in four dimensions, in order to reproduce the continuum limit, this problem is responsible for the huge requirement of computing resources in this field.

It is clear that lattice gauge theory cannot be explained in depth on two pages – we only intend to give a vague idea of what these calculations are about to a reader who is not familiar with this subject.

2.9.2
A Famous Application

Among the many Monte-Carlo studies of various problems (which include problems in cosmology, such as the phase transition from the quark-gluon plasma to hadronic matter in the early universe), we focus here on the problem of predicting the masses of elementary particles. Butler *et al.* [71] used a new parallel supercomputer with 480 processors ("GF11") exclusively for one year to run lattice sizes ranging from $8^3 \times 32$ to $24^3 \times 32$, $24^3 \times 36$, and $30 \times 32^2 \times 40$. Their program executed at a speed of more than 5 Gflops (Giga floating point operations per second), and the rather good statistics reached allowed a meaningful elimination of finite-size effects by an extrapolation to the infinite-volume limit. This problem is important, because the wave function of a hadron is spread out over many lattice sites.

Even with this impressive effort, several approximations were necessary.

The fermion determinant mentioned above is neglected (this is called *quenched approximation*). One cannot work at the (physically relevant) very small quark mass m_q, but rather data on the hadron masses were taken for a range of quark masses and extrapolated (Figure 2.3). After a double extrapolation ($m_q \to 0$, lattice spacing at fixed volume $\to 0$), one obtained mass ratios that are in satisfactory agreement with experiment. For example, for the nucleon the mass ratio for the finite volume is $m_N/m_æ = 1.285 \pm 0.070$, extrapolated to infinite volume 1.219 ± 0.105, the experimental value being 1.222 (all masses in units of the mass $m_æ$ of a rho meson).

Fifteen years later [72], it became possible to avoid the "quenched approximation" and other limitations of the early work and calculate the light hadron masses, obtaining very good agreement with experiment, using supercomputers that were a million times faster.

2.10
Selected Applications in Classical Statistical Mechanics of Condensed Matter

In this section, we mention a few applications, to give the flavor of the type of work that is done and the kind of questions that

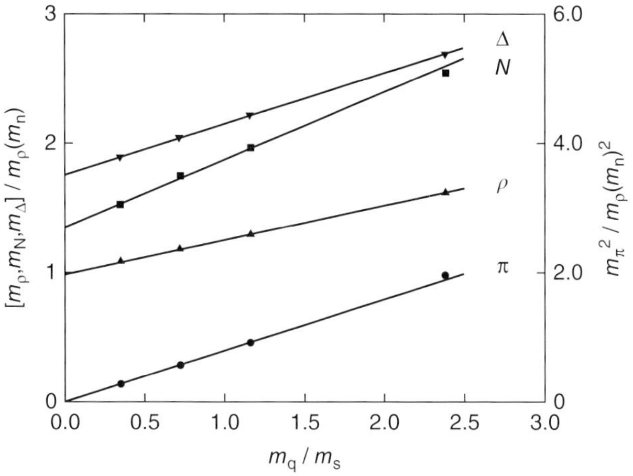

Figure 2.3 For a 30 × 32^2× 40 lattice at $(k_B T)^{-1}$ = 6.17, $m^2_ß$, $m_æ$, m_N, and $m_€$ in units of the physical rho meson mass $m_æ(m_n)$, as functions of the quark mass m_q in units of the strange quark mass m_s. Particles studied are pion, rho meson, nucleon, and delta baryon, respectively. (Source: From Butler *et al.* [71].)

are answered by Monte-Carlo simulations. More extensive reviews can be found in the literature [29, 36, 73, 74].

2.10.1 Metallurgy and Materials Science

A widespread application of Monte-Carlo simulation in this area is the study of order-disorder phenomena in alloys: One tests analytical approximations to calculate phase diagrams, such as the cluster variation (CV) method, and one tests to what extent a simple model can describe the properties of complicated materials.

An example (Figure 2.4) shows the order parameter for long-range order (LRO) and short-range order (SRO, for nearest neighbors) as function of temperature for a model of Cu_3Au alloys on the fcc lattice. An Ising model with antiferromagnetic interactions between nearest neighbors is studied, and the Monte-Carlo data (filled symbols and symbols with crosses) are compared to CV calculations (broken curve) and other analytical calculations (dash-dotted curve) and to experiment (open circles). The simulation shows that the analytical approximations describe the ordering of the model only qualitatively. Of course, there is no perfect agreement with the experiment either; this is to be expected, of course, because in real alloys the interaction range is considerably larger than just extending to nearest neighbors only.

2.10.2 Polymer Science

One can study phase transitions not only for models of magnets or alloys, but also for complex systems such as mixtures of flexible polymers. A question heavily debated in the literature is the dependence of the critical temperature of unmixing of a symmetric polymer mixture (both constituents have the same degree of

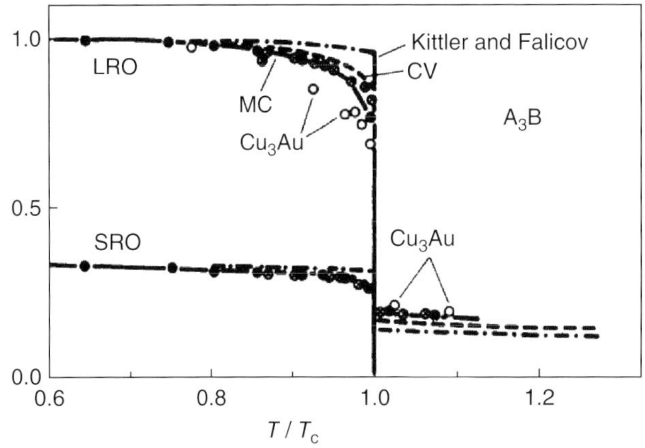

Figure 2.4 Long-range order parameter (LRO) and absolute value of nearest-neighbor short-range order parameter (SRO) plotted versus temperature T (in units of the temperature T_c where the first-order transition occurs) for a nearest-neighbor model of binary alloys on the face-centered cubic lattice with A_3B structure. Open circles: experimental data for Cu_3Au; broken and dash-dotted curves: results of analytical theories. Full dots: Monte-Carlo results obtained in the semi-grand canonical ensemble (chemical potential difference between A and B atoms treated as independent variable); circles with crosses: values obtained from a canonical ensemble simulation (concentration of B atoms fixed at 25%). (Source: From Binder *et al.* [75].)

polymerization $N_A = N_B = N$) on chain length N. The classical Flory–Huggins theory predicted $T_c \propto N$, while an integral equation theory predicted $T_c \propto \sqrt{N}$ [76]. This law would lead in the plot of Figure 2.5 to a straight line through the origin. Obviously, the data seem to rule out this behavior, and are rather qualitatively consistent with Flory–Huggins theory (though the latter significantly overestimates the prefactor in the relation $T_c \propto N$).

Polymer physics provides examples for many questions where simulations could contribute significantly to provide a better understanding. Figure 2.6 provides one more example [78]. The problem is to provide truly microscopic evidence for the reptation concept [79]. This concept implies that, as a result of "entanglements" between chains in a dense melt, each chain moves snakelike along its own contour. This behavior leads to a special behavior of mean square displacements: After a characteristic time τ_e, one should see a crossover from a law $g_1(t) \equiv \langle [r_i(t) - r_i(0)]^2 \rangle \propto t^{1/2}$ (Rouse model) to a law $g_1(t) \propto t^{1/4}$, and, at a still later time (τ_R), one should see another crossover to $g_1(t) \propto t^{1/2}$ again. At the same time, the center of gravity displacement should also show an intermediate regime of anomalously slow diffusion, $g_3(t) \propto t^{1/2}$. Figure 2.6 provides qualitative evidence for these predictions, although the effective exponents indicated do not quite have the expected values.

While this is an example where dynamic Monte-Carlo simulations are used to check theories and pose further theoretical questions, one can also compare to experiment if one uses data in suitably normalized form. In Figure 2.7, the diffusion constant D of the chains is normalized by its value in the Rouse regime (limit for small N) and plotted versus N/N_e where the characteristic "entanglement chain length" N_e is extracted from τ_e shown in Figure 2.6

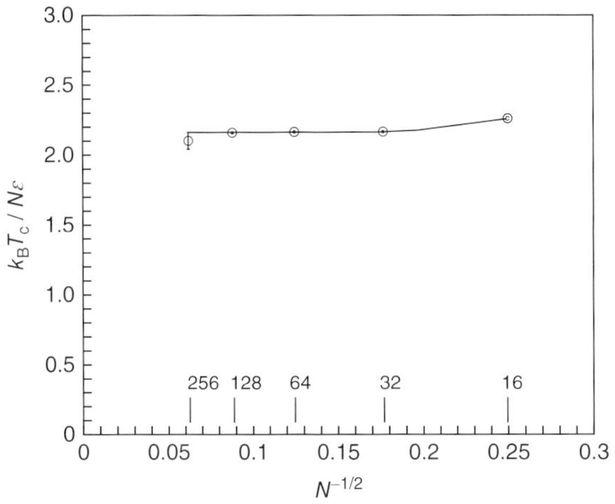

Figure 2.5 Normalized critical temperature $k_BT_c/N\varepsilon$ of a symmetric polymer mixture (N is the chain length, ε is the energy parameter describing the repulsive interaction between A–B pairs of monomers) plotted versus $N^{-1/2}$. Data are results of simulations for the bond-fluctuation model, $16 \leq N \leq 256$. The data are consistent with an asymptotic extrapolation $k_BT_c/\varepsilon \approx 2.15\,N$, while Flory–Huggins theory (in the present units) would yield $k_BT_c/\varepsilon \approx 7N$, and the integral equation theory $\frac{k_BT_c}{\varepsilon} \propto \sqrt{N}$. (Source: From Deutsch and Binder [77].)

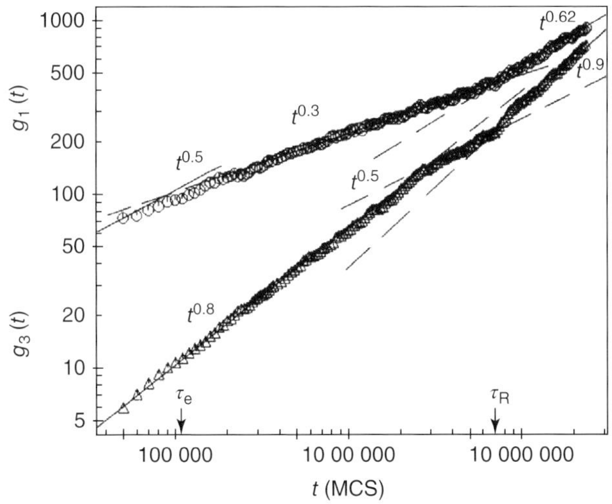

Figure 2.6 Log-log plot of the mean square displacements of inner monomers [$g_1(t)$] and of the center of gravity of the chain [$g_3(t)$] versus time t (measured in units of Monte-Carlo steps, while lengths are measured in units of the lattice spacing). Straight lines show various power laws as indicated; various characteristic times are indicated by arrows (see text). Data refer to the bond-fluctuation model on the simple cubic lattice, for an athermal model of a polymer melt with chain length $N = 200$ and a volume fraction $\phi = 0.5$ of occupied lattice sites. (Source: From Paul *et al.* [78].)

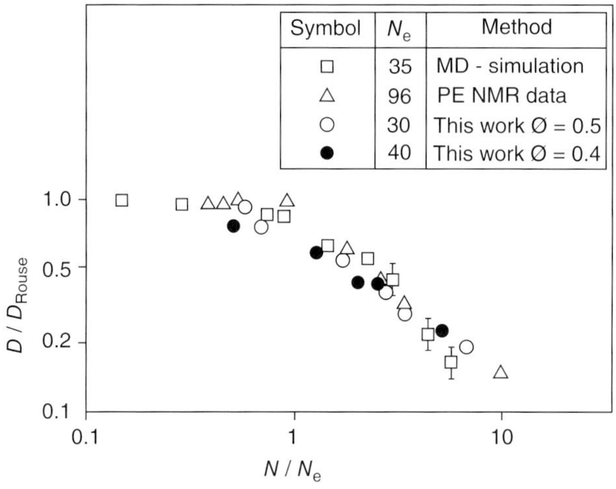

Figure 2.7 Log-log plot of the self-diffusion constant D of polymer chains, normalized by the Rouse diffusivity, versus N/N_e (N_e is the entanglement chain length, estimated independently, and indicated in the inset). Circles, from Monte-Carlo simulations of Paul *et al.* [80]; squares, from molecular dynamics simulations [81]; triangles, experimental data [82]. (Source: From [80].)

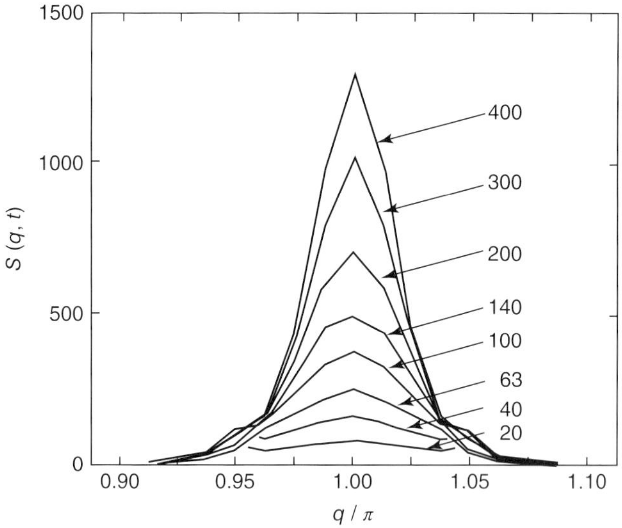

Figure 2.8 Structure factor $S(q, t)$ versus wavevector q at various times t (in units MCS per lattice site), after the system was started in a completely random configuration. Temperature (measured in units of the nearest-neighbor repulsion W_{nn}) is 1.33 ($T_c \approx 2.07$ in these units), and coverage $\theta = 1/2$. (Source: From Sadiq and Binder [83].)

(see [78, 80], for details). The Monte-Carlo data presented in this scaled form agree with results from both a MD simulation [81] and experiment on polyethylene (PE) [82].

This example also shows that, for slow diffusive motions, Monte-Carlo simulation is competitive to MD, although it does not describe the fast atomic motions realistically.

2.10.3 Surface Physics

Our last example considers phenomena far from thermal equilibrium. Studying the ordering behavior of superstructures, we treat the problem where initially the adatoms are adsorbed at random, and one gradually follows the formation of ordered domains out of initially disordered configurations. In a scattering experiment, one sees this by the gradual growth of a peak at the Bragg position q_B. Figure 2.8 shows a simulation for the case of the (2×1) structure on the square lattice, where the Bragg position is at the Brillouin-zone boundary ($q_B = \pi$ if lengths are measured in units of the lattice spacing). Here a lattice gas with repulsive interactions between nearest and next-nearest neighbors (of equal strength) was used [83], using a single-spin-flip kinetics (if the lattice gas is translated to an Ising spin model). This is appropriate for a description of a monolayer in equilibrium with surrounding gas, the random "spin flips" then correspond to random evaporation-condensation events. Figure 2.9 presents evidence that these data on the kinetics of ordering satisfy a scaling hypothesis, namely,

$$S(q,t) = [L(t)]^2 \tilde{S}(|q - q_B|L(t)), \tag{2.44}$$

where $\tilde{S}$ is a scaling function. This hypothesis, (2.44), was first proposed using simulations, and later it was established to describe experimental data as well.

2.11 Concluding Remarks

In this chapter, the basic features of the most widely used numerical techniques that fall into the category of Monte-Carlo calculations were described. There is a vast literature on the subject; the author estimates the number of papers using Monte-Carlo methods in condensed matter physics of the order of 10^5, in lattice gauge theory of the order of 10^4. Thus many important variants of algorithms could not be treated here and interesting applications (e.g., the study of neural-network models) were completely omitted.

There also exist other techniques for the numerical simulation of complex systems, which sometimes are an alternative approach to Monte-Carlo simulation. The MD method (numerical integration of Newton's equations) has already been mentioned in the text, and there exist combinations of both methods ("hybrid Monte-Carlo," "Brownian dynamics," etc.). A combination of MD with the local-density approximation of quantum mechanics is the basis of the Car–Parrinello method.

Problems such as that shown in Figures 2.8 and 2.9 can also be formulated in terms of numerically solving appropriate differential equations, which may in turn even be discretized to cellular automata. When planning a Monte-Carlo simulation, hence, some thought to the question "When which method?" should be given.

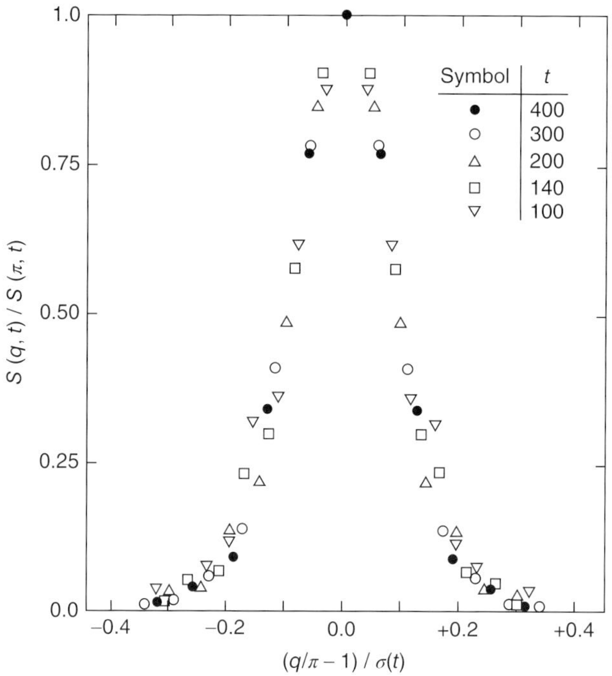

Figure 2.9 Structure factor of Figure 2.8 plotted in scaled form, normalizing $S(q,t)$ by its peak value $S(\pi,t)$ and normalizing $q/\pi - 1$ by the half-width $\sigma(t) = L^{-1}(t)$, where $L(t)$ thus defined is the characteristic domain size. (Source: From Sadiq and Binder [83].)

Glossary

Critical slowing down: Divergence of the relaxation time of dynamic models of statistical mechanics at a second-order phase transition (critical point).

Detailed balance principle: Relation linking the transition probability for a move and the transition probability for the inverse move to the ratio of the probability for the occurrence of these states in thermal equilibrium. This condition is sufficient for a Markov process to tend toward thermal equilibrium.

Ergodicity: Property that ensures equality of statistical ensemble averages (such as the "canonic ensemble") and time averages along the trajectory of the system through phase space.

Finite-size scaling: Theory that describes the finite-size-induced rounding of singularities that would occur at phase transitions in the thermodynamic limit.

Heat-bath method: Choice of transition probability where the probability to "draw" a trial value for a degree of freedom does not depend on its previous value.

Importance-sampling: Monte-Carlo method that chooses the states that are generated according to the desired probability distribution. For example, for statistical mechanics, states are chosen with weights proportional to the Boltzmann factor.

Lattice gauge theory: Field theory of quarks and gluons in which space and time are discretized into a four-dimensional lattice, gauge field variables being associated to the links of the lattice.

Master equation: Rate equation describing the "time" evolution of the probability that a state occurs as a function of a "time" coordinate labeling the sequence of states (in the context of importance-sampling Monte-Carlo methods).

Molecular dynamics method: Simulation method for interacting many-body systems based on numerical integration of the Newtonian equations of motion.

Monte-Carlo step: Unit of (pseudo) time in (dynamically interpreted) importance sampling where, on the average, each degree of freedom in the system gets one chance to be changed (or "updated").

Random-number generator (RNG): Computer subroutine to produce pseudo-random numbers that are approximately not correlated with each other and approximately uniformly distributed in the interval from zero to one. RNGs typically are strictly periodic, but the period is large enough that, for practical applications, this periodicity does not matter.

Simple sampling: Monte-Carlo method that chooses states uniformly and at random from the available phase space.

Transition probability: Probability that controls the move from one state to the next one in a Monte-Carlo process.

References

1. Hammersley, J. M., and Handscomb, D. C. (1964) *Monte-Carlo Methods*, Chapman & Hall, London.
2. Bruns, W., Motoc, I., and O'Driscoll, K. F. (1981) *Monte-Carlo Applications in Polymer Science*, Springer, Berlin.
3. Compagner, A. (1991) *Am. J. Phys.* **59**, 700–705.
4. Knuth, D. (1969) *The Art of Computer Programming*, Vol. **2**, Addison-Wesley, Reading, MA.
5. James, F. (1990) *Comput. Phys. Commun.* **60**, 329–344.
6. Mascagni, M. and Srinivasan, A. (2000) *ACM Trans. Math. Software* **26**, 436–461.
7. Marsaglia, G. A. (1985) in: L. Billard Ed. *Computer Science and Statistics: The Interface*, Chapter 1, Elsevier, Amsterdam.
8. Compagner, A. and Hoogland, A. (1987) *J. Comput. Phys.* **71**, 391–428.
9. Coddington, P. D. (1994) *Int. J. Mod. Phys.* **C5**, 547.
10. Marsaglia, G. A., Narasumhan, B., and Zaman, A. (1990) *Comput. Phys. Commun.* **60**, 345–349.
11. Lehmer, D. H. (1951) *Proceedings of the 2nd Symposium on Large-Scale Digital Computing Machinery*, Harvard University, Cambridge, pp. 142–145.
12. Marsaglia, G. A. (1986), Proc. Natl. Acad. Sci. U.S.A. 61, 25–28.
13. Tausworthe, R. C. (1965) *Math. Comput.* **19**, 201–208.
14. Kirkpatrick, S. and Stoll, E. (1981) *J. Comput. Phys.* **40**, 517–526.
15. Ahrens, J. H. and Dieter, U. (1979) *Pseudo Random Numbers*, John Wiley & Sons, Inc., New York.
16. Koonin, S. E. (1981) *Computational Physics*, Benjamin, Reading, MA.
17. Gould, H. and Tobochnik, J. (1988) *An Introduction to Computer Simulation Methods/Applications to Physical Systems*, Parts 1 and 2, Addison-Wesley, Reading, MA.
18. Stauffer, D. and Aharony, A. (1994) *An Introduction to Percolation Theory*, Taylor & Francis, London.
19. Herrmann, H. J. (1986) *Phys. Rep.* **136**, 153–227.
20. Kremer, K. and Binder, K. (1988) *Comput. Phys. Rep.* **7**, 259–310.
21. Milchev, A., Binder, K., and Heermann, D. W. (1986) *Z. Phys. B* **63**, 521–535.
22. Rosenbluth, M. N. and Rosenbluth, A. W. (1955) *J. Chem. Phys.* **23**, 356–362.
23. Grassberger, P. (1997) *Phys. Rev.* **E56**, 3682–3693.
24. Hsu, H.-P., and Grassberger, P. (2011) *J. Stat. Phys.* **144**, 597–637.
25. Vicsek, T. (1989) *Fractal Growth Phenomena*, World Scientific, Singapore.

26. Meakin, P. (1998) *Fractals, Scaling and Growth Far From Equilibrium.* Cambridge University Press, Cambridge.
27. Tolman, S. and Meakin, P. (1989) *Phys. Rev.* **A40**, 428–437.
28. Metropolis, N., Rosenbluth, A. W., Rosenbluth, M. N., Teller, A.M., and Teller, E. (1953) *J. Chem. Phys.* **21**, 1087–1092.
29. Binder, K. (1976) in: C. Domb, M.S. Green Eds. *Phase Transitions and Critical Phenomena*, Vol. **5b**, p. 1, Academic Press, New York.
30. Binder, K. and Heermann, D. W. (2010) *Monte-Carlo Simulation in Statistical Physics: An Introduction*, 5th edn, Springer, Berlin.
31. Kalos, M. H. and Whitlock, P. A. (1986) *Monte-Carlo Methods,* Vol. **1**, John Wiley & Sons, Inc., New York.
32. Landau, D. P. (1992) in: K. Binder Ed. *The Monte-Carlo Method in Condensed Matter Physics,* Chapter 2, Springer, Berlin.
33. Swendsen, R. H., Wang, J. S., and Ferrenberg, A. M. (1992), in: K. Binder Ed., *The Monte-Carlo Method in Condensed Matter Physics*, Chapter 4, Springer, Berlin.
34. Berg, B. A. (2004) *Markov Chain Monte-Carlo Simulations and Their Statistical Analysis.* World Scientific, Singapore.
35. Rapaport, D. C. (2004) *The Art of Molecular Dynamics Simulation*, 2nd edn, Cambridge University Press, Cambridge.
36. Landau, D. P. and Binder, K. (2009) *A Guide to Monte-Carlo Simulation in Statistical Physics*, 3rd edn., Cambridge University Press Cambridge.
37. Panagiotopoulos, A. Z. (1992) *Mol. Simul.* **9**, 1–23.
38. Frenkel, D. and Smit, B. (2002) *Understanding Molecular Simulation: From Algorithms to Applications,* 2nd edn., Academic Press, San Diego, CA.
39. Barber, M. N. (1983), in: C. Domb, and J. L. Lebowitz Eds, *Phase Transitions and Critical Phenomena*, Chapter 2, Vol. **8**, Academic Press, New York.
40. Binder, K. (1987) *Ferroelectrics* **73**, 43–67.
41. Binder K. (1997) *Rep. Progr. Phys.* **60**, 487–559.
42. Privman, V. Ed. (1990) *Finite Size Scaling and Numerical Simulation of Statistical Systems*, World Scientific, Singapore.
43. Stanley, H. E. (1971) *An Introduction to Phase Transitions and Critical Phenomena*, Oxford University Press, Oxford.
44. Wang, F. and Landau, D. P. (2001) *Phys. Rev. E* **64**, 056101.
45. Binder, K., Block, B., Das, S. K., Virnau, P., and Winter, D. (2011) *J. Stat. Phys.* **144**, 690–729.
46. Grzelak, E. M. and Errington, J. R. (2008) *J. Chem. Phys.* **128**, 014710.
47. Bolhuis, P. G., Chandler, D., Dellago, C., and Geissler, P. L. (2002) *Ann. Rev. Phys. Chem.* **53**, 291.
48. Ceperley, D. M. and Kalos, M. H. (1979) in: K. Binder Ed. *Monte-Carlo Methods in Statistical Physics*, Springer, Berlin, pp. 145–194.
49. Kalos, M. H. Ed. (1984) *Monte-Carlo Methods in Quantum Problems*, Reidel, Dordrecht.
50. Berne, B. J. and Thirumalai, D. (1986) *Annu. Rev. Phys. Chem.* **37**, 401.
51. Suzuki, M. Ed. (1986) *Quantum Monte-Carlo Methods*, Springer, Berlin.
52. Schmidt, K. E. and Ceperley, D. M. (1992) in K. Binder Ed., *The Monte-Carlo Methods in Condensed Matter Physics*, Springer, Berlin, pp. 105–148.
53. Ceperley, D. M. (1995) *Rev. Mod. Phys.* **67**, 279–355.
54. Ceperley, D. M. (1996) in K. Binder and G. Ciccotti Eds., *Monte-Carlo and Molecular Dynamics of Condensed Matter Systems*, Societa Italiana di Fisica, Bologna, pp. 443–482.
55. Rothman, S. ed. (2002) *Recent Advances in Quantum Monte-Carlo Methods II*, World Scientific, Singapore.
56. Trebst, S. and Troyer M. (2006) In: M. Ferrario, G. Ciccotti, and K. Binder eds. *Computer Simulations in Condensed Matter: From Materials to Chemical Biology*, Vol. **1**,Springer, Berlin, pp. 591–640.
57. Carlson, J. (1988) *Phys. Rev. C* **38**, 1879–1885.
58. De Grand, T. (1992) in: H. Gausterer and C. B. Lang Eds., *Computational Methods in Field Theory*, Springer, Berlin, pp. 159–203.
59. Frick, M., Pattnaik, P. C., Morgenstern, I., Newns, D. M., von der Linden, W. (1990) *Phys. Rev. B* **42**, 2665–2668.
60. Reger, J. D. and Young, A. P. (1988) *Phys. Rev. B* **37**, 5978–5981.

61. Sachdev, S. (2000) *Quantum Phase Transitions*, Cambridge University Press, Cambridge.
62. Marx, D., Opitz, O., Nielaba, P., and Binder, K. (1993) *Phys. Rev. Lett.* **70**, 2908–2911.
63. Gillan, M. J. and Christodoulos, F. (1993) *Int. J. Mod. Phys.* **C4**, 287–297.
64. West, G. B. (1975), *Phys. Rep.* **18C**, 263–323.
65. Sokol, P. E., Sosnick, T. R., and Snow, W. M. (1989), in R. E. Silver and P. E. Sokol Eds. *Momentum Distributions*, Plenum Press, New York.
66. Ceperley and Pollock (1987) *Can. J. Phy.* **65**, 1416–1423.
67. Schmidt, K. E. and Ceperley, D. M. (1992), in: K. Binder Ed. *The Monte-Carlo Method in Condensed Matter Physics*, pp. 203–248, Springer, Berlin.
68. Ceperley, D. M., Alder, B. J. (1980) *Phys. Rev. Lett.* **45**, 566–569.
69. Rebbi, C. (1984), in: K. Binder Ed. *Application of the Monte-Carlo Method in Statistical Physics*, Springer, Berlin, p. 277.
70. Wilson, K. (1974) *Phys. Rev.* **D10**, 2445–2453.
71. Butler, F., Chen, H., Sexton, J., Vaccarino, A., and Weingarten, D. (1993) *Phys. Rev. Lett.* **70**, 2849–2852.
72. Dürr, S. *et al.* (2008) *Science* **322**, 1224–1227.
73. Binder, K. Ed. (1979) *Monte-Carlo Methods in Statistical Physics*, Springer, Berlin.
74. Binder, K. Ed. (1995) *The Monte-Carlo Method in Condensed Matter Physics*, Springer, Berlin.
75. Binder, K., Lebowitz, J. L., Phani, M. K., and Kalos, M. H. (1981) *Acta Metall.* **29**, 1655–1665.
76. Schweizer, K. S. and Curro, J. G. (1990) *Chem. Phys.* **149**, 105–127.
77. Deutsch, H. P. and Binder, K. (1992) *Macromolecules* **25**, 6214–6230.
78. Paul, W., Binder, K., Heermann, D. W., and Kremer, K. (1991) *J. Chem. Phys.* **95**, 7726–7740.
79. Doi, M. and Edwards, S. F. (1986) *Theory of Polymer Dynamics*, Clarendon Press, Oxford.
80. Paul, W., Binder, K., Heermann, D. W., and Kremer, K. (1991) *J. Phys. (Paris)* **111**, 37–60.
81. Kremer, K. and Grest, G. S. (1990) *J. Chem. Phys.* **92**, 5057–5086.
82. Pearson, D. S., Verstrate, G., von Meerwall, E., and Schilling, F. C. (1987) *Macromolecules* **20**, 113–1139.
83. Sadiq, A. and Binder, K. (1984) *J. Stat. Phys.* **35**, 517–585.

Further Reading

A textbook describing for the beginner how to learn to write Monte-Carlo programs and to analyze the output generated by them has been written by Binder, K., Heermann, D.W. (2010), Monte-Carlo Simulation in Statistical Physics: An Introduction, 5th edn, Berlin: Springer. This book emphasizes applications of statistical mechanics such as random walks, percolation, and the Ising model.

More extensive and systematic textbooks have been written by D. Frenkel and B. Smit (2002), Understanding Molecular Simulation: From Algorithms to Applications, San Diego, CA: Academic Press, and by D. P. Landau and K. Binder A Guide to Monte-Carlo Simulations in Statistical Physics, Cambridge: Cambridge University Press Practical Considerations are emphasized in B. A. Berg (2004) Markov Chain Monte-Carlo Simulations and Their Statistical Analysis, Singapore: World Scientific.

A useful book that gives much weight to applications outside of statistical mechanics is Kalos, M. H., Whitlock, P. A. (1986), Monte-Carlo Methods, vol. **1**, New York: John Wiley & Sons, Inc.

A more general but pedagogic introduction to computer simulation is presented in Gould, H., Tobochnik, J. (1988), An Introduction to Computer Simulation Methods/Applications to Physical Systems, Parts 1 and 2, Reading, MA: Addison-Wesley.

3
Stochastic Differential Equations

Gabriel J. Lord

3.1
Introduction

The consequence of inclusion of random effects in ordinary differential equations (ODEs) falls in to two broad classes – one in which the solution has a smooth path and one where the path is not smooth (i.e., it is not differentiable). If the random effect is included as a random choice of a parameter or as a random initial condition, u_0, then we have *random differentiable equations*. Each realization (i.e., each pick of the parameter from a random set) can be solved using standard deterministic methods (see Chapter 14). If we need to solve a large number of times to get averaged properties using Monte Carlo techniques (see Chapter 2), then efficient schemes are required.

If the forcing is rough, then the solutions are not differentiable and new techniques are required (in fact, a new calculus is required). Here we concentrate on differential equations that are coupled to some random forcing $\zeta(t)$ that is Gaussian white noise, so the random forces have mean zero and are uncorrelated at different times t. When $\zeta(t)$ is white, it can be characterized as the derivative of Brownian motion $W(t)$, so that "$\zeta(t) = \mathrm{d}W/\mathrm{d}t$." Ordinary differential equations with this sort of forcing are called *stochastic differential equations* (SDEs).

Example 3.1 [Ornstein–Uhlenbeck (OU) process] Consider a particle of unit mass moving with momentum $p(t)$ at time t in a liquid. The dynamics of the particle can be modeled by a dissipative force $-\lambda p(t)$, where $\lambda > 0$ is known as the dissipation constant. In the absence of any external forcing, we have

$$\frac{\mathrm{d}p}{\mathrm{d}t} = -\lambda p, \quad \text{given } p(0) = p_0 \in \mathbb{R} \qquad (3.1)$$

and the particle will converge to the rest state $p = 0$ as $t \to \infty$. This is not physical as the particle is subject to irregular bombardment by molecules in the liquid which can be modeled as a fluctuating force $\sigma\zeta(t)$, where $\zeta(t)$ is white noise and $\sigma > 0$ is called the diffusion constant. Newton's second law of motion gives the acceleration $\mathrm{d}p/\mathrm{d}t$ as the sum of the two forces. We have

Mathematical Tools for Physicists, Second Edition. Edited by Michael Grinfeld.

$$\frac{dp}{dt} = -\lambda p + \sigma\,\zeta(t), \tag{3.2}$$

which is a linear SDE with additive noise $\zeta(t) = dW/dt$. It is also written as

$$dp = -\lambda p\,dt + \sigma\,dW(t). \tag{3.3}$$

The solution $p(t)$ is known as the *Ornstein–Uhlenbeck (OU) process*. We will see that dW/dt does not strictly speaking make sense as a function and we need to interpret the SDE as an integral equation. In Figure 3.1, we show two sample realizations of (3.3) computed numerically and we also plot, for reference, the solution of the corresponding deterministic equation (3.1), all with the same initial data.

Note that (3.2) can also be interpreted as a model for thermal noise in an electrical resistor.

Our aim is to make sense of equations such as (3.3) and to introduce some of the techniques for looking at them. Consider the general ODE $du/dt = f(u)$ with initial data $u(0) = u_0$ and recall that it can be integrated to get an equivalent integral equation

$$u(t) = u(0) + \int_0^t f(s, u(s))ds.$$

When we include the effects of a stochastic forcing term, $\zeta(t)$, we have the SDE

$$\frac{du}{dt} = f(u) + g(u)\zeta(t). \tag{3.4}$$

When "$\zeta(t) = dW/dt$," where W is Brownian motion, we will understand this as the integral equation

$$u(t) = u_0 + \int_0^t f(u(s))ds + \int_0^t g(u(s))dW(s). \tag{3.5}$$

Equation (3.4) and the integral equation above are often written in shorthand as the SDE

$$du = f(u)dt + g(u)dW. \tag{3.6}$$

In (3.6), $f(u)$ is called the *drift term* and $g(u)$ is called the *diffusion term* . When g is independent of u, such as in (3.3) where $g = \sigma$,

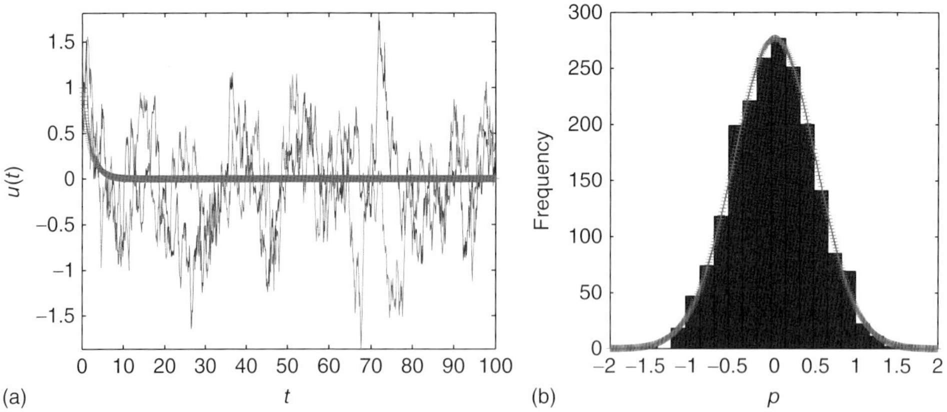

Figure 3.1 (a) Two numerical solutions of the SDE (3.3) on the interval $[0, T]$ with $\lambda = 0.5$ and $\sigma = 0.5$. The initial data is the same in each case, $u(0) = 1$. With each realization of the noise we obtain a different solution path. Also shown is the solution of the deterministic ODE (3.1). In (b), we examine the distribution at $t = 100$ showing a histogram from 2000 different realizations; note that as $t \to \infty$ $p(t) \to N(0, \sigma^2/2\lambda)$.

the SDE is said to have *additive* (or extrinsic) noise. When $g(u)$ varies with u, then the SDE has *multiplicative* (or intrinsic) noise.

Remark 3.1 *Often in the probabilistic and stochastic literature, time dependence of a variable u is denoted by u_t, so that $u(t) \equiv u_t$, $W(t) \equiv W_t$. We have not used that notation here as it can lead to confusion with partial derivatives.*

3.2 Brownian Motion / Wiener Process

Although there are many different types of stochastic processes that could be taken, we look at one type of noise (probably the most common) that arises from Brownian motion. A Scottish botanist, Robert Brown, observed that a small particle such as a seed in a liquid or gas moves about in a seemingly random way. This is because it is being hit continuously by the gas or liquid molecules. Later, Norbert Wiener, an American mathematician, gave a mathematical description of the process and proved its existence. In the literature, both the term Brownian motion and Wiener process are widely used; here we will use Brownian motion and denote it by W.

Definition 3.1 *A scalar standard Brownian motion or standard Wiener process is a collection of random variables $W(t)$ that satisfy*

1. *$W(0) = 0$ with probability 1.*
2. *For $0 \le s \le t$, the random variable given by the increment $W(t) - W(s)$ is normally distributed with mean zero and variance $t - s$. Equivalently, $W(t) - W(s) \sim \sqrt{t-s}\; N(0,1)$.*
3. *For $0 \le s < t \le u < v$ the increments $W(t) - W(s)$ and $W(v) - W(u)$ are independent.*
4. *$W(t)$ is continuous in t.*

Brownian motion is often used to model random effects that are, on a microscale, small, random, independent, and additive, where a random behavior is seen on a larger scale. The central limit theorem (see Chapter 1) then gives that the sum of these small contributions converges to increments that are normally distributed on the large scale. We recall that the *covariance* is a measure of how a random variable X changes against a different random variable Y and

$$\begin{aligned}\mathrm{Cov}\,(X,Y) &:= \mathbf{E}[(X - \mathbf{E}[X])(Y - \mathbf{E}[Y])] \\ &= \mathbf{E}[XY] - \mathbf{E}[X]\mathbf{E}[Y].\end{aligned}$$

Thus the *variance* of a random variable is given by $\mathrm{Var}\,(X) = \mathrm{Cov}\,(X,X)$. To look at the correlation in time of a random variable X and Y, we consider the *covariance function* $c(s,t) := \mathrm{Cov}\,(X(s), Y(t))$, for $s, t \in [0,T]$. If $X = Y$, then this is called the *autocovariance*. When $c(s,t) = c(s-t)$, the covariance is said to be *stationary*.

The following properties of W are straightforward to see.

Lemma 3.1 *Let $W(t)$, $t \ge 0$, be a standard Brownian motion. Then,*

1. *$\mathbf{E}[W(t)] = 0$ and $\mathbf{E}\left[(W(t))^2\right] = t$.*
2. *$\mathrm{Cov}\,(W(s), W(t)) = \min(s,t)$.*
3. *the process $Y(t) = -W(t)$ is also a standard Brownian motion.*

Proof. As $W(0) = 0$ and $W(t) = W(t) - W(0) \sim N(0,t)$, we find that

$$\mathbf{E}[W(t)] = 0 \qquad \mathbf{E}\left[(W(t))^2\right] = t.$$

The covariance $\mathrm{Cov}\,(W(s), W(t)) = \mathbf{E}[W(t)W(s)]$ because $\mathbf{E}[W(t)] = 0$. Let us take $0 \le s \le t$. Then

$$\begin{aligned}\mathbf{E}[W(t)W(s)] = \mathbf{E}[(W(s) + W(t) \\ - W(s))W(s)]\end{aligned}$$

and so

$$\mathbf{E}[W(t)W(s)] = \mathbf{E}\left[(W(s))^2\right] + \mathbf{E}[(W(t) - W(s))W(s)].$$

From Definition 3.1, the increments $W(t) - W(s)$ and $W(s) - W(0)$ are independent, so

$$\mathbf{E}[W(t)W(s)] = s + \mathbf{E}[(W(t) - W(s))] \\ \mathbf{E}[W(s)] = s.$$

Thus

$$\mathbf{E}[W(t)W(s)] = \min(s, t), \qquad \text{for all } s, t \geq 0.$$

To show that $Y(t) = -W(t)$ is also a Brownian motion, it is straightforward to verify the conditions of Definition 3.1.

The increment $W(t) - W(s)$ of a Brownian motion over the interval $[s, t]$ has distribution $N(0, t - s)$ and hence the distribution of $W(t) - W(s)$ is the same as that of $W(t + h) - W(s + h)$ for any $h > 0$; that is, the distribution of the increments is independent of translations and we say that Brownian motion has *stationary increments* (recall the stationary covariance). Another important property is that the Brownian motion is *self-similar*, that is, by a suitable scaling, we obtain another Brownian motion.

Lemma 3.2 (self-similarity) *If $W(t)$ is a standard Brownian motion then for $\alpha > 0$, $Y(t) := \alpha^{1/2} W(t/\alpha)$ is also a Brownian motion.*

Proof. To show that Y is Brownian motion, we need to check the conditions of Definition 3.1.

1. $Y(0) = W(0) = 0$ with probability 1
2. Consider $s < t$ and increment $Y(t) - Y(s) = \alpha^{1/2}(W(t/\alpha) - W(s/\alpha))$. This has distribution $\alpha^{1/2} N(0, \alpha^{-1}(t - s)) = N(0, (t - s))$.
3. Increments are independent. Consider $Y(t) - Y(s) = \alpha^{1/2}(W(t/\alpha) - W(s/\alpha))$ and $Y(v) - Y(u) = \alpha^{1/2}(W(v/\alpha) - W(u/\alpha))$ for $0 \leq s \leq t \leq u \leq v \leq T$, because $(W(t/\alpha) - W(s/\alpha))$ and $(W(v/\alpha) - W(u/\alpha))$ are independent.
4. Continuity of Y follows from the continuity of W.

Hence we have shown $Y(t)$ is also a Brownian motion.

We now examine how rough Brownian motion is. To do this, we will need to consider convergence of random variables. We say that the random variables X_j converge to X in *mean square* if

$$\mathbf{E}\left[\|X_j - X\|^2\right] \to 0, \quad \text{as} \quad j \to \infty$$

and they converge in *root mean square* if

$$\left\{\mathbf{E}\left[\|X_j - X\|^2\right]\right\}^{1/2} \to 0, \quad \text{as} \quad j \to \infty.$$

We need to specify a norm. For vectors, we will take the standard Euclidean norm $\| \, \|_2$ on $\mathbb{R}^d$ with inner product $\langle \boldsymbol{x}, \boldsymbol{y} \rangle$ for $\boldsymbol{x}$, $\boldsymbol{y} \in \mathbb{R}^d$, so $\|x\|_2^2 = \langle \boldsymbol{x}, \boldsymbol{x} \rangle$. For a matrix $X \in \mathbb{R}^{d \times m}$, we will use the Frobenius norm

$$\|X\|_F = \left(\sum_{i=1}^{d} \sum_{j=1}^{m} |x_{ij}|^2 \right)^{1/2} = \sqrt{\text{trace}(X^* X)},$$

where X^* is the conjugate transpose.

We start by examining the *quadratic variation*, $\sum_{j=0}^{N} (W(t_{j+1}) - W(t_j))^2$ as $N \to \infty$. The following lemma will also be useful for developing stochastic integrals later.

Lemma 3.3 (Quadratic variation) *Let $0 = t_0 < t_1 < \cdots < t_N = T$ be a partition of $[0, T]$ where it is understood that $\Delta t := \max_{1\le j\le N} |t_{j+1} - t_j| \to 0$ as $N \to \infty$. Let $\Delta W_j := (W(t_{j+1}) - W(t_j))$. Then $\sum_{j=0}^{N}(\Delta W_j)^2 \to T$ in mean square as $N \to \infty$.*

Proof. Let $\Delta t_j := t_{j+1} - t_j$, $V_N := \sum_{j=0}^{N}(\Delta W_j)^2$ and note that

$$V_N - T = \sum_{j=0}^{N-1} \left[(\Delta W_j)^2 - \Delta t_j\right].$$

The terms $(\Delta W_j)^2 - \Delta t_j$ are independent and have mean 0. Therefore, because the cross terms are 0,

$$\mathbf{E}\left[(V_N - T)^2\right] = \sum_{j=0}^{N-1} \mathbf{E}\left[\left((\Delta W_j)^2 - \Delta t_j\right)^2\right],$$

and

$$\begin{aligned}
&\left((\Delta W_j)^2 - \Delta t_j\right)^2 \\
&= (\Delta W_j)^4 - 2\Delta t_j(\Delta W_j)^2 + \Delta t_j^2 \\
&= \left(\frac{(\Delta W_j)^4}{\Delta t_j^2} - 2\frac{(\Delta W_j)^2}{\Delta t_j} + 1\right)\Delta t_j^2 \\
&= (X_j^2 - 1)^2 \Delta t_j^2,
\end{aligned}$$

where $X_j := \Delta W_j/\sqrt{\Delta t_j} \sim N(0, 1)$. Thus, for some $C > 0$, we have

$$\left[(\Delta W_j)^2 - \Delta t_j\right]^2 \le C \sum_{j=0}^{N-1} \Delta t_j^2 \le CT\Delta t.$$

Hence as $N \to \infty$, $\mathbf{E}\left[(V_N - T)^2\right] \to 0$.

We can now show that, although the quadratic variation is finite, the total variation of Brownian paths is unbounded.

Lemma 3.4 (Total variation) *Let $0 = t_0 < t_1 < \cdots < t_N = T$ be a partition of $[0, T]$ as in Lemma 3.3. Then $\sum_{j=0}^{N} |W(t_{j+1}) - W(t_j)| \to \infty$ as $N \to \infty$.*

Proof. The proof is by contradiction and we outline the idea here. Note that

$$\sum_{j=0}^{N}(W(t_{j+1}) - W(t_j))^2 \le \max_{1\le j\le N} |W(t_{j+1}) - W(t_j)| \sum_{j=0}^{N} |W(t_{j+1}) - W(t_j)|.$$

Now Brownian motion is continuous, so as $N \to \infty$, we have $\max_{1\le j\le N} |W(t_{j+1}) - W(t_j)| \to 0$. If $\sum |W(t_{j+1}) - W(t_j)|$ is finite then $\sum_{j=0}^{N}(W(t_{j+1}) - W(t_j))^2 \to 0$, but this contradicts Lemma 3.3.

The fact that $\sum_{j=0}^{N} |W(t_{j+1}) - W(t_j)| \to \infty$ hints that Brownian motion, even if it is continuous, may in fact be very rough. In Section 3.2.1 we argue that Brownian motion is not differentiable. First, however, we extend the definition of a Brownian motion on $\mathbb{R}$ to $\mathbb{R}^d$. We can easily extend Definition 3.1 to processes $\boldsymbol{W}(t) = (W_1(t), W_2(2), \ldots, W_d(t))^T \in \mathbb{R}^d$ where increments $\boldsymbol{W}(t) - \boldsymbol{W}(s) \sim \sqrt{t-s}\, N(0, I)$. Here $N(0, I)$ is the d-dimensional normal distribution with covariance matrix I, the $\mathbb{R}^{d\times d}$ identity matrix. It is straightforward to show the following lemma.

Lemma 3.5 *The process $\boldsymbol{W}(t) = (W_1(t), W_2(2), \ldots, W_d(t))^T \in \mathbb{R}^d$ is a Brownian motion (or a Wiener process) if and only if each of the components $W_i(t)$ is a Brownian motion.*

3.2.1 White and Colored Noise

Despite the nice properties of Brownian motion, a well-known theorem of Dvoretzky, Erdös, and Kakutani states that sample paths of Brownian motion are *not* differentiable anywhere. To see why this might be the case, note that from Definition 3.1, the increment $W(t + h) - W(t)$, for $h > 0$, is a normal random variable with mean 0 and variance h. Now consider $Y(t) = (W(t+h) - W(t))/h$. This is a normal random variable with mean 0 and variance h^{-1}. For Y to approximate a derivative of W, we need to look at the limit as $h \to 0$. We see that the variance of the random variable Y goes to infinity as $h \to 0$ and, in the limit, Y is not well behaved. The numerical derivative of two different Brownian paths is plotted in Figure 3.2. As $\Delta t \to 0$ these numerical derivatives do not converge to a well-defined function.

Although we have just argued that Brownian motion $W(t)$ is not differentiable, often in applications, one will see that $\mathrm{d}W(t)/\mathrm{d}t$ used. To understand this properly, we will need to develop some stochastic integration theory. Before we do this, let us first relate the term $\mathrm{d}W(t)/\mathrm{d}t$ to "white noise" with covariance (or the autocorrelation function), which is the Dirac delta function, so that

$$\mathbf{E}\left[\frac{\mathrm{d}W(t)}{\mathrm{d}t}\frac{\mathrm{d}W(s)}{\mathrm{d}s}\right] = \delta(s-t). \qquad (3.7)$$

Recall that the Dirac delta satisfies the following properties:

$$\delta(s) = 0, \text{ for } s \neq 0, \qquad \int_{-\infty}^{\infty} \delta(s)\phi(s)\mathrm{d}s = \phi(0)$$

for any continuous function $\phi(s)$. In particular, if $\phi(s) \equiv 1$, then $\int_{-\infty}^{\infty} \delta(s)\mathrm{d}s = 1$.

To see why (3.7) might be true, let us fix s and t and consider

$$d(s) := \mathbf{E}\left[\frac{(W(t+h) - W(t))}{h}\frac{(W(s+h) - W(s))}{h}\right].$$

For small h, this should approximate the Dirac delta. We can simplify $d(s)$ to get

$$\begin{aligned} d(s) = \frac{1}{h^2}\Big(&\mathbf{E}\big[W(t+h)W(s+h)\big] \\ &-\mathbf{E}\big[W(t+h)W(s)\big] \\ &-\mathbf{E}\big[W(t)W(s+h)\big] + \mathbf{E}[W(t)W(s)]\Big) \end{aligned}$$

and use Lemma 3.1 on the Brownian motion to obtain

$$\begin{aligned} d(s) = \frac{1}{h^2}(&\min(t+h, s+h) - \min(t+h, s) \\ &- \min(t, s+h) + \min(t, s)). \end{aligned}$$

We see that d is a piecewise linear function with nodes at $s = t-h, t, t+h$ and

$$d(s) = \begin{cases} 0 & \text{for } s \leq t-h \\ \frac{(s-t+h)}{h^2} & \text{for } t-h < s < t \\ \frac{(t+h-s)}{h^2} & \text{for } t < s < t+h \\ 0 & \text{for } s \geq t+h \end{cases}.$$

Furthermore, $\int_{-\infty}^{\infty} d(s)\mathrm{d}s = 1$. Hence, we see d approximates $\delta(s-t)$ for small h, which suggests (3.7) holds.

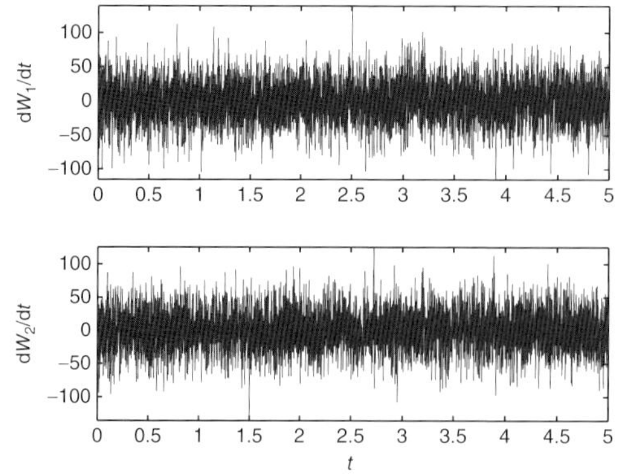

Figure 3.2 Numerical derivatives of $W_1(t)$ and $W_2(t)$ shown in Figure 3.3(a).

We briefly discuss two ways to see that $\mathrm{d}W(t)/\mathrm{d}t$ can be interpreted as white noise and what this means. The first of these is based on the Fourier transform of the covariance function. The *spectral density* of a stochastic process X with stationary covariance function c (i.e., $c(s,t) = c(s-t)$) is defined for $\lambda \in \mathbb{R}$ by

$$f(\lambda) := \frac{1}{2\pi}\int_{-\infty}^{\infty} \mathrm{e}^{-i\lambda t} c(t)\mathrm{d}t.$$

When $c(t) = \delta(t)$ we find for all frequencies λ,

$$f(\lambda) = \frac{1}{2\pi}\int_{-\infty}^{\infty} \mathrm{e}^{-i\lambda t}\delta(t)\mathrm{d}t = \frac{1}{2\pi}.$$

Therefore, by analogy with white light, all frequencies contribute equally.

Remark 3.2 *Note that the covariance function can also be found from the inverse Fourier transform of the spectrum. One practical application of this is to estimate the correlation function from numerical data using the Fourier transform.*

The precise statement is given in the following theorem.

Theorem 3.1 (Wiener–Khintchine) *The following statements are equivalent.*

1. *There exists a mean square continuous stationary process $\{X(t) : t \in \mathbb{R}\}$ with stationary covariance $c(t)$.*
2. *The function $c : \mathbb{R} \to \mathbb{R}$ is such that*

$$c(t) = \int_{\mathbb{R}} \mathrm{e}^{i\nu t} f(\nu)\,\mathrm{d}\nu.$$

A second way to illustrate why $\mathrm{d}W(t)/\mathrm{d}t$ and the covariance (3.7) are called white noise is to consider $t \in [0,1]$ and let $\left\{\phi_j(t)\right\}_{j=1}^{\infty}$ be an orthonormal basis, for example, $\phi_j(t) = \sqrt{2}\sin(j\pi t)$. Now construct $\zeta(t)$ by the random series with $t \in [0,1]$

$$\zeta(t) := \sum_{j=1}^{\infty} \zeta_j \phi_j(t),$$

where each $\zeta_j \sim N(0,1)$ and are independent of each other. Then we have formed a random $\zeta(t)$ that has a homogeneous mix of the different basis functions ϕ_j – just like white light consists of a homogeneous mix of wavelengths. Let us look at the covariance function for $\zeta(t)$,

$$\begin{aligned}\mathrm{Cov}\,(\zeta(s),\zeta(t)) &= \sum_{j,k=1}^{\infty} \mathrm{Cov}\,\left(\zeta_j,\zeta_k\right)\phi_j(s)\phi_k(t)\\ &= \sum_{j}^{\infty} \phi_j(s)\phi_k(t).\end{aligned}$$

Although the Dirac delta δ is not strictly speaking a function, formally we can write

$$\delta(s) = \sum_{j=1}^{\infty}\left\langle \delta, \phi_j(s)\right\rangle \phi_j(s) = \sum_{j=1}^{\infty} \phi_j(0)\phi_j(s)$$

and so the covariance function for $\zeta(t)$ is $\mathrm{Cov}\,(\zeta(s),\zeta(t)) = c(s,t) = \delta(s-t)$; that is, $\zeta(t)$ is a stochastic process with a homogeneous mix of the basis functions that has the Dirac delta as covariance function and the noise is uncorrelated.

Colored noise, as the name suggests, has a heterogeneous mix of the basis functions. For example, consider the stochastic process

$$\nu(t) = \sum_{j=1}^{\infty} \sqrt{\nu_j}\zeta_j\phi_j(t), \tag{3.8}$$

where each $\zeta_j \sim N(0,1)$ and are independent and $\nu_j \geq 0$. As we vary the ν_j with j, the process $\nu(t)$ is said to be colored noise and the random variables $\nu(t)$ and

$\nu(s)$ are correlated. Expansions of the form (3.8) are a useful way to examine many stochastic processes. When the basis functions are chosen as the eigenfunctions of the covariance function, this is termed the *Karhunen–Loève expansion*.

Theorem 3.2 (Karhunen–Loève)
Consider a stochastic process $\{X(t) : t \in [0,T]\}$ and suppose that $\mathbf{E}[\int_0^T (X(s))^2 \mathrm{d}s] < \infty$. Then,

$$X(t) = \mu(t) + \sum_{j=1}^{\infty} \sqrt{\nu_j}\phi_j(t)\xi_j,$$
$$\xi_j := \frac{1}{\sqrt{\nu_j}} \int_0^T (X(s) - \mu(s))\phi_j(s)\mathrm{d}s \qquad (3.9)$$

and $\{\nu_j, \phi_j\}$ denote the eigenvalues and eigenfunctions of the covariance operator so that $\int_0^T c(s,t)\phi_j(s)\mathrm{d}s = \nu_j\phi_j(t)$. The sum in (3.9) converges in a mean square sense. The random variables ξ_j have mean 0, unit variance, and are pairwise uncorrelated. Furthermore, if the process is Gaussian, then $\xi_j \sim N(0,1)$ independent and identically distributed.

Another way to generate colored noise is to solve a stochastic SDE, see Section 3.4, whose solution has a particular frequency distribution. The OU process of Example 3.1 is often used to generate colored noise. For more information on colored noise, see, for example, [1]. The Wiener–Khintchine theorem is covered in a wide range of books, including [2–4]. The Karhunen–Loève expansion is widely used to construct random fields as well as in data analysis and signal processing; see also [4, 5].

3.2.2 Approximation of a Brownian Motion

We can use Definition 3.1 directly to construct a numerical approximation W_n to a Brownian motion $W(t)$ at times t_n, where $0 = t_0 < t_1 < \cdots < t_N$. From (1) of Definition 3.1 we have that $W_0 = W(0) = 0$. We construct the approximation by noting that $W(t_{n+1}) = W(t_n) + (W(t_{n+1}) - W(t_n))$ and from (2) of Definition 3.1 we know how increments are distributed.

Letting $W_n \approx W(t_n)$, we get that

$$W_{n+1} = W_n + \mathrm{d}W_n, \qquad n = 1, 2, \ldots, N,$$

where $\mathrm{d}W_n \sim \sqrt{\Delta t}N(0,1)$. This is described in Algorithm 3.1 and two typical results of this process are shown in Figure 3.3a. In the figure, we use a piecewise linear approximation to $W(t)$ for $t \in [t_n, t_{n+1}]$. To approximate a d-dimensional $\boldsymbol{W}$ by Lemma 3.5, we can simply use Algorithm 3.1 for each component; the result of this is shown in Figure 3.3b.

Algorithm 3.1 Brownian motion in one dimension. We assume $t_0 = 0$.

INPUT : vector $t = [t_0, t_1, \ldots, t_N]$
OUTPUT: vector W such that component $W_n = W(t_n)$.

1: $W_0 = 0$
2: **for** $n = 1 : N$ **do**
3: $\quad \Delta t = t_n - t_{n-1}$
4: $\quad$ find $z \sim N(0,1)$
5: $\quad \mathrm{d}W_n = \sqrt{\Delta t}z$
6: $\quad W_n = W_{n-1} + \mathrm{d}W_n$
7: **end for**

3.3 Stochastic Integrals

Before we consider a stochastic integral, let us briefly recall how the standard deterministic Riemann integral can be defined. Given a bounded function $g : [0,T] \to \mathbb{R}$,

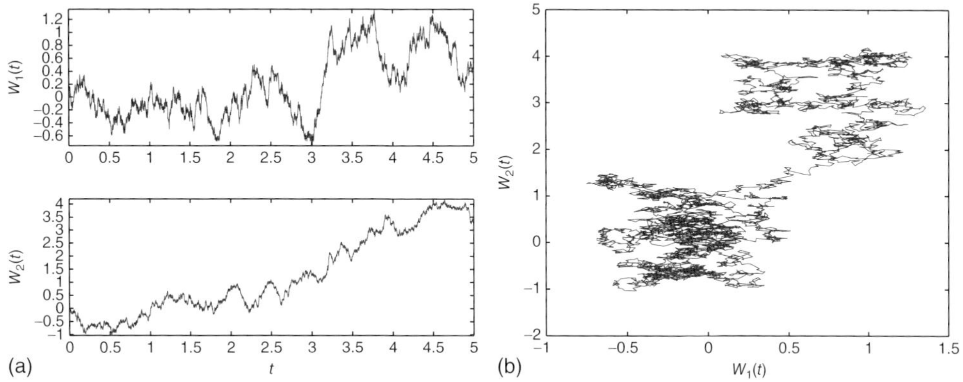

Figure 3.3 (a) Two discretized Brownian motions $W_1(t)$, $W_2(t)$ constructed over $[0, 5]$ with $N = 5000$, so $\Delta t = 0.001$. (b) Brownian motion $W_1(t)$ plotted against $W_2(t)$. The paths start at $(0, 0)$ and the final point at $t = 5$ is marked with a $\star$.

we can define the Riemann integral using the Riemann sum

$$\int_0^T g(t)\mathrm{d}t = \lim_{\Delta t\to 0}\sum_{j=0}^{N-1} g(z_j)(t_{j+1} - t_j) \quad (3.10)$$

where $0 = t_0 < t_1 < \ldots < t_N$ is a partition of the interval $[0, T]$, $\Delta t = \max_{1\leq j\leq N} |t_{j+1} - t_j|$ and we take *any* $z_j \in [t_j, t_{j+1}]$. The key elements are to say for which functions the integral holds, define the partition, and examine the limit as more points are added to the partition (and so that $\Delta t \to 0$).

In the stochastic case, we will need to say what type of functions we can integrate and how we take the limit. It will also turn out that the choice of z_j will also become important when we try and integrate a stochastic process.

It is standard to illustrate the ideas of a stochastic integral by considering

$$I = \int_0^T W(s)\mathrm{d}W(s) := \lim_{N\to\infty} S_N,$$

$$S_N := \sum_{j=0}^{N-1} W(z_j)(W(t_{j+1}) - W(t_j)), \quad (3.11)$$

where $z_j = (1 - \alpha)t_j + \alpha t_{j+1}$ and the limit as $N \to \infty$ refines the partition on $[0, T]$. The idea of defining I in (3.11) follows by analogy with the deterministic integral (3.10). For the stochastic integrals, we examine these limits in a mean square sense. Note that for $\alpha = 0$, $z_j = t_j$ (this will correspond to the Itô integral) and for $\alpha = 1/2$, $z_j = (t_j + t_{j+1})/2$ (which will correspond to a Stratonovich integral).

Lemma 3.6 *Let* $0 = t_0 < t_1 < \cdots < t_N = T$ *be a partition of* $[0, T]$ *where it is understood that* $\Delta t := \max_{1\leq j\leq N} |t_{j+1} - t_j| \to 0$ *as* $N \to \infty$. *Then in mean square we have*

$$I = \lim_{N\to\infty} S_N = \frac{(W(T))^2}{2} + \left(\alpha - \frac{1}{2}\right)T.$$

Proof. Let $\Delta W_j := W(t_{j+1}) - W(t_j)$. Then

$$S_N = \frac{1}{2}\sum_{j=0}^{N-1}(W(t_{j+1})^2 - W(t_j)^2)$$
$$-\frac{1}{2}\sum_{j=0}^{N-1}(\Delta W_j)^2 + \sum_{j=0}^{N-1}(W(z_j) - W(t_j))^2$$

$$+\sum_{j=0}^{N-1}(W(t_{j+1})-W(z_j))(W(z_j)-W(t_j))$$
$$=: A+B+C+D.$$

Now the first sum A is a telescoping sum so that $A = 1/2(W(T))^2$. By Lemma 3.3, the second sum $B \to T/2$ as $N \to \infty$ and a similar lemma gives that $C \to \alpha T$. For D, using the fact that the increments in the sum are independent, it can be shown that $D \to 0$ as $N \to \infty$.

Lemma 3.6 shows that for $\int_0^t W(s)\mathrm{d}W(s)$, the choice of α for z_j is important.

- When $\alpha = 1/2$, an average is taken of W over the time interval $\Delta t_j = t_{j+1} - t_j$ and gives rise to the Stratonovich integral, which is denoted by $\circ$, and we have

$$\int_0^T W(s) \circ \mathrm{d}W(s) = \frac{1}{2}(W(T))^2. \quad (3.12)$$

  This follows directly from Lemma 3.6 with $\alpha = 1/2$. The result is similar to the deterministic integral where $\int_0^T s\mathrm{d}s = s^2/2$.
- When $\alpha = 0$ on the interval, we have the important case of the Itô integral. From Lemma 3.6, with $\alpha = 0$, we find the Itô integral

$$\int_0^T W(s)\mathrm{d}W(s) = \frac{1}{2}(W(T))^2 - \frac{T}{2}. \quad (3.13)$$

  Compared to the deterministic integral, we have an additional term. On the interval Δt_j, we only use information at t_j and no information in the future, and this gives the Itô integral some special and useful properties. Because the Itô integral does not anticipate the future, it is widely used in financial modeling.

We now discuss in detail the Itô version of the stochastic integral.

3.3.1 Itô Integral

Now let us extend the definition of the Itô integral (3.11) (where $\alpha = 0$) to a wider class of functions. We define the Itô integral through the mean square limit as follows

$$\int_0^T X(t)\mathrm{d}W(t) := \lim_{N\to\infty}\sum_{j=0}^{N-1} X(t_j)\left(W(t_{j+1}) - W(t_j)\right),$$

where $0 = t_0 < t_1 < \cdots < t_N = T$ is a partition as in Lemma 3.6. We need to say something about the sort of X that we can integrate, as we have seen X may be a stochastic process. Most importantly, for the Itô integral, we do not want X to see into the future – the ideas required to make this precise are described in the following paragraph.

We have the idea that the stochastic process, $X(t)$, is a model of something that evolves in time and we can therefore ask at time t questions about the past $s < t$ ($X(t)$ is known from observations of the past) and the future $s > t$. A *probability space* $(\Omega, \mathcal{F}, P)$ consists of a sample space Ω, a set of events $\mathcal{F}$, and a probability measure P. A *filtered probability space* consists of $(\Omega, \mathcal{F}, \mathcal{F}_t, P)$ where $\mathcal{F}_t$ is a filtration of $\mathcal{F}$. The *filtration* $\mathcal{F}_t$ is a way of denoting the events that are observable by time t and so, as t increases, $\mathcal{F}_t$ contains more and more events. If $X(t)$, $t \in [0, T]$ is *nonanticipating* (or $\mathcal{F}_t$ *adapted*), then $X(t)$ is $\mathcal{F}_t$ measurable for all $t \in [0, T]$ (roughly $X(t)$ does not see into the future). Finally, $X(t)$ is *predictable* if it is nonanticipating and can be approximated by a sequence $X(s_j) \to X(s)$ if $s_j \to s$ for all $s \in [0, T]$, $s_j < s$.

Theorem 3.3 (Itô integral) *Let $X(t)$, $t \in [0, T]$ be a predictable stochastic* *process such that*

$$\left(\mathbf{E}\left[\int_0^t |X(s)|^2 \mathrm{d}s\right]\right)^{1/2} < \infty.$$

Then

1. *$\int_0^t X(s)\,\mathrm{d}W(s)$ for $t \in [0, T]$ has continuous sample paths.*
2. *The Itô integral has the martingale property . That is, best predictions of $\int_0^t X(s)\,\mathrm{d}W(s)$ based on information up to time r is $\int_0^r X(s)\,\mathrm{d}W(s)$. In particular, the integral has mean 0 :*

$$\mathbf{E}\left[\int_0^t X(s)\,\mathrm{d}W(s)\right] = 0.$$

3. *We have the Itô isometry*

$$\mathbf{E}\left[\left|\int_0^t X(s)\,\mathrm{d}W(s)\right|^2\right] = \int_0^t \mathbf{E}\left[|X(s)|^2\right]\mathrm{d}s, \quad t \in [0, T]. \quad (3.14)$$

Example 3.2 Reconsider the stochastic integral

$$I(t) = \int_0^t W(s)\,\mathrm{d}W(s), \qquad t \geq 0.$$

The martingale property gives $\mathbf{E}[I(t)] = 0$ and the Itô isometry (3.14) gives

$$\mathbf{E}\left[I(t)^2\right] = \int_0^t \mathbf{E}\left[(W(s))^2\right]\mathrm{d}s = \int_0^t s\,\mathrm{d}s = \frac{1}{2}t^2. \quad (3.15)$$

Note that using (3.13), we can write $I(t)$ as $I(t) = \frac{1}{2}(W(t))^2 - \frac{t}{2}$, or rearranged as $\frac{1}{2}(W(t))^2 = \frac{1}{2}t + I(t)$. This is often written in shorthand as

$$\mathrm{d}\left(\frac{1}{2}(W(t)\right)^2 = \frac{1}{2}\,\mathrm{d}t + W(t)\,\mathrm{d}W(t). \quad (3.16)$$

The Itô integral can be extended to be vector-valued.

Definition 3.2 *Let $\boldsymbol{W}(t) = (W_1(t), W_2(2), \ldots, W_m(t))^T \in \mathbb{R}^m$ and let $X \in \mathbb{R}^{d\times m}$ be such that each X_{ij} is a predictable stochastic process such that*

$$\left(\mathbf{E}\left[\int_0^t |X_{ij}(s)|^2 \mathrm{d}s\right]\right)^{1/2} < \infty. \quad (3.17)$$

Then $\int_0^t X(s)\mathrm{d}W(s)$ is the random variable in $\mathbb{R}^d$ with ith component

$$\sum_{j=1}^m \int_0^t X_{ij}(s)\mathrm{d}W_j(s).$$

It can be shown that the following properties hold for the vector-valued Itô integral.

1. The integral $\int_0^t X(s)\,\mathrm{d}\boldsymbol{W}(s)$ for $t \in [0, T]$ is a predictable process.
2. The martingale property holds and the Itô integral has mean 0:

$$\mathbf{E}\left[\int_0^t X(s)\,\mathrm{d}\boldsymbol{W}(s)\right] = 0.$$

3. Given two stochastically integrable processes X and Y, that is, X and Y are predictable and (3.17) holds, then if $\hat{t} := \min(t_1, t_2)$,

$$\mathbf{E}\left[\left\langle \int_0^{t_1} X(s)\,\mathrm{d}\boldsymbol{W}(s), \int_0^{t_2} Y(s)\,\mathrm{d}\boldsymbol{W}(s)\right\rangle\right] = \int_0^{\hat{t}} \sum_{i=1}^m \mathbf{E}\left[\left\langle \boldsymbol{X}_i(s), \boldsymbol{Y}_i(s)\right\rangle\right]\mathrm{d}s,$$

where $\boldsymbol{X}_i, \boldsymbol{Y}_i$ denote the ith column of X and Y and $\langle\cdot,\cdot\rangle$ is the $\mathbb{R}^d$ inner product. From this, the Itô isometry follows

$$\mathbf{E}\left[\|\int_0^t X(s)\,\mathrm{d}\boldsymbol{W}(s)\|_2^2\right] = \int_0^t \mathbf{E}\left[\|X(s)\|_F^2\right]\mathrm{d}s, \quad t \in [0, T]. \quad (3.18)$$

In fact, the class of processes for which the stochastic integral can be developed is wider, see, for example, [5, 6]. With this definition, we can examine systems of Itô SDEs.

3.4 Itô SDEs

We described in the introduction that (3.6) is shorthand for the integral equation (3.5) and in Section 3.2.1 we saw that Brownian motion is not differentiable.

We now consider the Itô SDEs where $\boldsymbol{u} \in \mathbb{R}^d$ and we are given a process with drift $\boldsymbol{f}(\boldsymbol{u}) : \mathbb{R}^d \to \mathbb{R}^d$, diffusion $G(\boldsymbol{u}) : \mathbb{R}^d \to \mathbb{R}^{d\times m}$, and $\boldsymbol{W}(t) = (W_1(t), W_2(2), \dots, W_m(t))^T \in \mathbb{R}^m$. Instead of writing

$$\frac{\mathrm{d}\boldsymbol{u}}{\mathrm{d}t} = \boldsymbol{f}(\boldsymbol{u}) + G(\boldsymbol{u})\frac{\mathrm{d}\boldsymbol{W}(t)}{\mathrm{d}t},$$

we realize that W is not differentiable and hence write

$$\mathrm{d}\boldsymbol{u} = \boldsymbol{f}(\boldsymbol{u})\mathrm{d}t + G(\boldsymbol{u})\mathrm{d}\boldsymbol{W}(t), \tag{3.19}$$

which we interpret as an Itô stochastic integral equation

$$\boldsymbol{u}(t) = \boldsymbol{u}(0) + \int_0^t \boldsymbol{f}(\boldsymbol{u}(s))\mathrm{d}s + \int_0^t G(\boldsymbol{u}(s))\mathrm{d}\boldsymbol{W}(s). \tag{3.20}$$

The last integral in (3.20) is the stochastic integral from Definition 3.2 and $G(\boldsymbol{u})$ is a $d \times m$ matrix.

Example 3.3 Consider the system of SDEs with two independent Brownian motions $W_1(t)$ and $W_2(t)$

$$\mathrm{d}u_1 = f_1(u_1, u_2)\,\mathrm{d}t + g_{11}(u_1, u_2)\mathrm{d}W_1 + g_{12}(u_1, u_2)\mathrm{d}W_2$$

$$\mathrm{d}u_2 = f_2(u_1, u_2)\,\mathrm{d}t + g_{21}(u_1, u_2)\mathrm{d}W_1 + g_{22}(u_1, u_2)\mathrm{d}W_2. \tag{3.21}$$

We can write (3.21) in the form of (3.19) with $d = m = 2$ by taking $\boldsymbol{u} = (u_1, u_2)^T$ and $f : \mathbb{R}^2 \to \mathbb{R}^2$, $G : \mathbb{R}^2 \to \mathbb{R}^{2\times 2}$ given by

$$\boldsymbol{f}(\boldsymbol{u}) = \begin{pmatrix} f_1(u_1, u_2) \\ f_2(u_1, u_2) \end{pmatrix},$$

$$G(\boldsymbol{u}) = \begin{pmatrix} g_{11}(u_1, u_2) & g_{12}(u_1, u_2) \\ g_{21}(u_1, u_2) & g_{22}(u_1, u_2) \end{pmatrix}.$$

If the following conditions on the drift $\boldsymbol{f}$ and diffusion G are satisfied, it can be shown that a solution exists to the SDE (3.19).

Assumption 3.1 (linear growth/Lipschitz condition) *There exists a constant $L > 0$ such that the linear growth condition holds for $\boldsymbol{u} \in \mathbb{R}^d$,*

$$\|\boldsymbol{f}(\boldsymbol{u})\|^2 \le L(1 + \|\boldsymbol{u}\|^2),$$
$$\|G(\boldsymbol{u})\|^2 \le L(1 + \|\boldsymbol{u}\|^2), \qquad \boldsymbol{u} \in \mathbb{R}^d,$$

and the global Lipschitz condition holds for $\boldsymbol{u}_1, \boldsymbol{u}_2 \in \mathbb{R}^d$,

$$\|\boldsymbol{f}(\boldsymbol{u}_1) - \boldsymbol{f}(\boldsymbol{u}_2)\| \le L\,\|\boldsymbol{u}_1 - \boldsymbol{u}_2\|,$$
$$\|G(\boldsymbol{u}_1) - G(\boldsymbol{u}_2)\| \le L\,\|\boldsymbol{u}_1 - \boldsymbol{u}_2\|.$$

Theorem 3.4 (existence and uniqueness for SDEs) *Suppose that Assumption 3.1 holds and that $W(t)$ is a Brownian motion. For each $T > 0$ and $\boldsymbol{u}_0 \in \mathbb{R}^d$, there is a unique $\boldsymbol{u}$ with $\sup_{t\in[0,T]} \left(\mathbf{E}\left[\|\boldsymbol{u}(t)\|^2\right]\right)^{1/2} < \infty$ such that for $t \in [0, T]$*

$$\boldsymbol{u}(t) = \boldsymbol{u}_0 + \int_0^t \boldsymbol{f}(\boldsymbol{u}(s))\,\mathrm{d}s + \int_0^t G(\boldsymbol{u}(s))\,\mathrm{d}\boldsymbol{W}(s). \tag{3.22}$$

The notion of solution that we have taken is a stochastic process that is determined for any Brownian path – this is called a *strong solution*. We could also ask for solutions that are valid only for particular Brownian paths, in which case the Brownian path also becomes part of the solution. This is the idea of a *weak* solution. We do not consider this further. In Example 3.1, we gave a simple example of an SDE for which $f(u) = -\lambda u$ and $G(u) = \sigma$. As G is independent of u, the noise is additive and (3.3) is equivalent to the Itô integral equation

$$p(t) = p(0) - \lambda \int_0^t p(s)\mathrm{d}s + \sigma \int_0^s \mathrm{d}W(s).$$

We now present two further SDEs. The first describes the dynamics of a particle subject to noise. The second models the fluctuation of an asset price on a stock market.

Example 3.4 [Langevin equation] Denote by p the momentum, q the position, and by $H(q,p)$ the system energy of a particle. If the particle in Example 3.1 has potential energy $V(q)$ at position $q \in \mathbb{R}$, its dynamics are described by the following SDE for $(q(t), p(t))^T$

$$\begin{aligned} \mathrm{d}q &= p\,\mathrm{d}t \\ \mathrm{d}p &= -\lambda p\,\mathrm{d}t - V'(q)\,\mathrm{d}t + \sigma\,\mathrm{d}W(t) \end{aligned} \tag{3.23}$$

for parameters $\lambda, \sigma > 0$. Here $d = 2$, $m = 1$, and for $\boldsymbol{u} = (q,p)^T$

$$\boldsymbol{f}(\boldsymbol{u}) = \begin{pmatrix} p \\ -\lambda p - V'(q) \end{pmatrix}, \qquad G(\boldsymbol{u}) = \begin{pmatrix} 0 \\ \sigma \end{pmatrix}.$$

The noise is again additive and acts directly only on the momentum p.

In Figure 3.4, we plot a solution of (3.23) for one sample path of the noise. The potential $V(q) = (q^2 - 1)^2/4$ is a double-well potential and we take $\lambda = 0.1$, $\sigma = 0.2$. In the absence of noise, there are three fixed points $(0,0)^T$ (which is unstable) and $(\pm 1, 0)^T$ (which are stable). With the stochastic forcing we see that the particle no longer comes to rest and can cross the energy barrier between $(-1,0)^T$ and $(1,0)^T$.

The SDEs of Examples 3.1 and 3.4 both have additive noise. The following example has multiplicative noise.

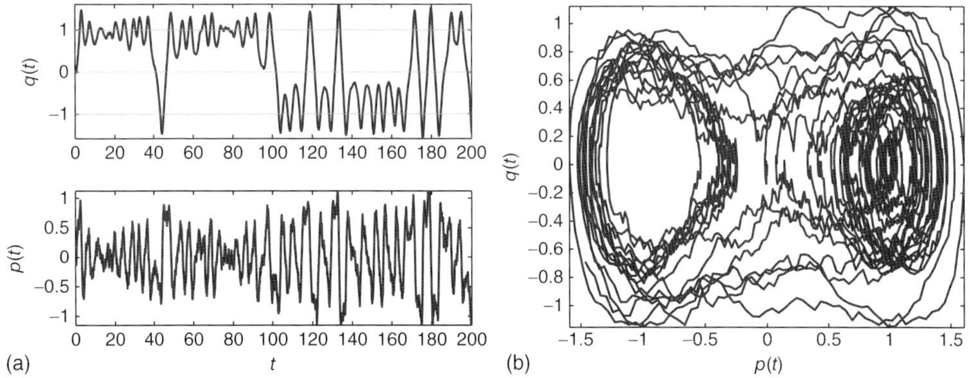

Figure 3.4 Solution of Langevin equation (3.23) with $V(q) = (q^2 - 1)^2/4$, $\lambda = 0.1$, $\sigma = 0.2$. In (a) we show $p(t)$ and $q(t)$ for a sample path of the noise. In (b) we plot the phase portrait.

Example 3.5 [geometric Brownian motion] In financial modeling, the price $u(t)$ at time t of a risk-free asset with interest rate r obeys a differential equation $\mathrm{d}u/\mathrm{d}t = ru$. On the stock market, stock prices fluctuate rapidly and the fluctuations are modeled by replacing the risk-free interest rate r by a stochastic process $r + \sigma dW/\mathrm{d}t$, where σ is the volatility and $\mathrm{d}W/\mathrm{d}t$ is white noise. We obtain the SDE with multiplicative noise

$$\mathrm{d}u = r\,u\,\mathrm{d}t + \sigma\,u\,\mathrm{d}W(t), \qquad u(0) = u_0, \tag{3.24}$$

where $u_0 \geq 0$ is the price at time $t = 0$. Here $d = m = 1$, the drift $f(u) = ru$, and the diffusion $G(u) = \sigma u$. The solution $u(t)$ depends on the interpretation of the integral $\int_0^t u(s)\,\mathrm{d}W(s)$. The Itô SDE is chosen to model stock prices, because the current price $u(t)$ of the stock is supposed to be determined by past events and is independent of the current fluctuations $\mathrm{d}W(t)$. The solution process $u(t)$ of (3.24) is called *geometric Brownian motion*. In Figure 3.5a we plot five sample realizations of (3.24) computed numerically with $r = 0.5$ and $\sigma = 0.5$ and all with initial data $u(0) = 1$. In Figure 3.5a we also plot $\mathbf{E}[u(t)]$ found from the exact solution, see Example 3.7.

Note that although we have described (3.24) from a mathematical finance point of view, it could also be considered as a basic population growth model where exponential population growth is modulated with random fluctuations.

3.4.1 Itô Formula and Exact Solutions

The chain rule of ordinary calculus changes in the Itô calculus to the so-called Itô formula. We now discuss the Itô formula (also called Itô's Lemma) for (3.19) with $d = m = 1$ (so, in one dimension). We apply the Itô formula to geometric Brownian motion (see Example 3.5) and the mean reverting OU process of which Example 3.1 is one case. Finally, we quote a general form of the Itô formula.

Lemma 3.7 (one-dimensional Itô formula) *Let $\Phi : [0, T] \times \mathbb{R} \to \mathbb{R}$ have continuous partial derivatives $\partial\Phi/\partial t$, $\partial\Phi/\partial u$ and $\partial^2\Phi/\partial u^2$. Let u satisfy the Itô SDE* (3.19) *with $d = m = 1$, so*

$$\mathrm{d}u = f(u)\mathrm{d}t + G(u)\mathrm{d}W(t), \qquad u(0) = u_0$$

and suppose Assumption 3.1 holds. Then, almost surely,

$$\mathrm{d}\Phi = \frac{\partial\Phi}{\partial t}\,\mathrm{d}t + \frac{\partial\Phi}{\partial u}\,\mathrm{d}u + \frac{1}{2}\frac{\partial^2\Phi}{\partial u^2}G^2\,\mathrm{d}t \tag{3.25}$$

or written in full

$$\begin{aligned}\Phi(t, u(t)) = \Phi(0, u_0) + \int_0^t & \frac{\partial\Phi}{\partial t}(s, u(s)) \\ &+ \frac{\partial\Phi}{\partial u}(s, u(s))f(u(s)) \\ &+ \frac{1}{2}\frac{\partial^2\Phi}{\partial u^2}(s, u(s))G(u(s))^2\,\mathrm{d}s \\ &+ \int_0^t \frac{\partial\Phi}{\partial u}(s, u(s))G(u(s))\,\mathrm{d}W(s).\end{aligned}$$

To interpret Lemma 3.7, consider the solution $u(t)$ of the deterministic ODE $\mathrm{d}u/\mathrm{d}t = \lambda$ and let $\phi(u) = 1/2u^2$. The standard chain rule implies that

$$\frac{d\phi(u)}{\mathrm{d}t} = u\,\frac{\mathrm{d}u}{\mathrm{d}t} = \lambda u(t).$$

However, if $u(t)$ satisfies $\mathrm{d}u = \lambda\,\mathrm{d}t + \sigma\,\mathrm{d}W(t)$, the Itô formula (3.25) says that

$$\mathrm{d}\phi(u) = u\,\mathrm{d}u + \frac{\sigma^2}{2}\,\mathrm{d}t \tag{3.26}$$

and we pick up an unexpected extra term $\sigma^2/2\mathrm{d}t$. This shows that the Itô calculus is different from the standard deterministic

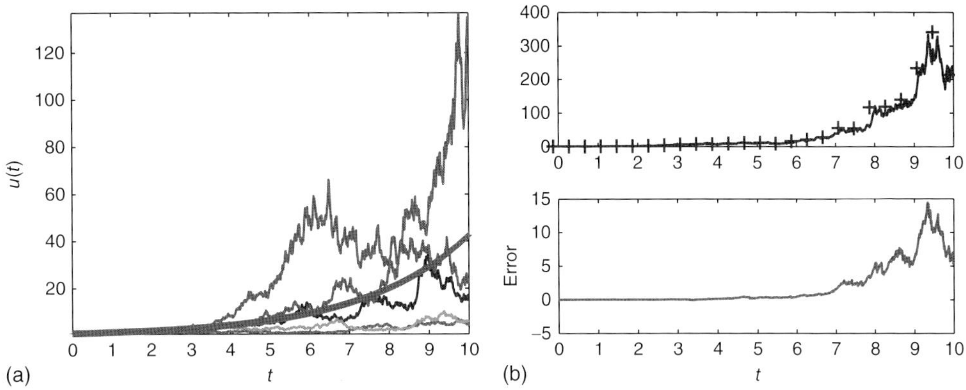

Figure 3.5 (a) Five numerical solutions of the SDE (3.24) for geometric Brownian motion on the interval [0, 10] with $r = 0.5$ and $\sigma = 0.5$. The initial data is the same in each case, $u(0) = 1$. With each realization of the noise we obtain a different solution path. The solid smooth line shows $\mathbf{E}[u(t)]$ from (3.28). In (b) (top) we plot the exact solution and approximate numerical solution + together on the interval [0, 10]. In (b) (bottom) we plot the error.

calculus. We can expand (3.26) using the SDE to get

$$\mathrm{d}\left(\frac{1}{2}u^2\right) = u\,(\lambda u\mathrm{d}t + \sigma \mathrm{d}W) + \frac{\sigma^2}{2}\,\mathrm{d}t$$
$$= \left(\lambda u^2 + \frac{\sigma^2}{2}\right)\mathrm{d}t + \sigma u \mathrm{d}W. \quad (3.27)$$

When $\lambda = 0$ and $\sigma = 1$, the SDE is simply $\mathrm{d}u = \mathrm{d}W$ and $u(t) = W(t)$. In which case, (3.27) becomes

$$\mathrm{d}\left(\frac{1}{2}W^2\right) = \frac{1}{2}\mathrm{d}t + W\mathrm{d}W,$$

and we have recovered (3.16) using the Itô formula.

Example 3.6 [Integration by parts] Reconsider the simple SDE $\mathrm{d}u = \mathrm{d}W$ and let $\phi(t, u) = tu$. Then by the Itô formula, $d\phi = u\mathrm{d}t + t\mathrm{d}u$ and hence $\mathrm{d}(tW) = W\mathrm{d}t + t\mathrm{d}W$. Written in integral form, we have $tW(t) = \int_0^t W(s)\mathrm{d}s + \int_0^t s\mathrm{d}W(s)$ or

$$\int_0^t s\mathrm{d}W(s) = tW(t) - \int_0^t W(s)\mathrm{d}s,$$

which is a formula for integration by parts.

Remark 3.3 *The integral $Y(T) = \int_0^T W(s)\mathrm{d}s$ is the area under the path W and because W is stochastic so is the value of the integral. In fact, for fixed T, $Y \sim N(0, T^3/3)$. This follows from the definition of the Riemann integral. For a partition of $[0, T]$ with $T = N\Delta t$ we have*

$$Y = \lim_{N\to\infty} \sum_{n=1}^{N} \Delta t W(t_n)$$
$$= \lim_{\Delta t\to 0} \big[NW(t_1) + (N-1)(W(t_2) - W(t_1)) + \cdots + (W(t_N) - W(t_{N-1}))\big].$$

On the right, we have a sum of independent normal random variables and clearly $\mathbf{E}[Y] = 0$. For the variance, we have

$$\mathrm{Var}\,(Y) = \lim_{\Delta t\to 0} \big[\Delta t^2\,\mathrm{Var}\,\big(NW(t_1)\big) + \cdots + \Delta t^2\,\mathrm{Var}\,\big(W(t_N) - W(t_{N-1})\big)\big]$$
$$= \lim_{\Delta t\to 0} \big[\Delta t^3 (N^2 + \cdots + 1)\big]$$
$$= \lim_{\Delta t\to 0} \left[\Delta t^3 \frac{1}{6} N(N+1)(2N+1)\right]$$
$$= \frac{T^3}{3}.$$

We now look at two applications of the Itô formula to solve for geometric Brownian motion (Example 3.5) and the OU process (Example 3.1).

Example 3.7 [geometric Brownian motion] We show the solution of the geometric Brownian motion SDE (3.24) is given by

$$u(t) = \exp\left(\left(r - \frac{\sigma^2}{2}\right)t + \sigma W(t)\right)u_0. \quad (3.28)$$

For $u(0) = 0$, $u(t) = 0$ is clearly the solution to (3.24). For $u > 0$, let $\phi(u) = \log u$, so that $\phi'(u) = u^{-1}$ and $\phi''(u) = -u^{-2}$. By the Itô formula with $\Phi(t, u) = \phi(u)$,

$$d\phi(u) = r\,dt + \sigma\,dW(t) - \frac{1}{2}\sigma^2\,dt.$$

Hence,

$$\phi(u(t)) = \phi(u_0) + \int_0^t \left(r - \frac{\sigma^2}{2}\right) ds + \int_0^t \sigma\,dW(s)$$

and $\log u(t) = \log(u_0) + (r - (\sigma^2/2))t + \sigma W(t)$. Taking the exponential, we find (3.28). It is clear that, when $u_0 \geq 0$, the solution $u(t) \geq 0$ for all $t \geq 0$. This is desirable in financial modeling, where stock prices are nonnegative and also for models of populations where populations are positive.

From (3.28), we see that because $\mathbf{E}[W(t)] = 0$, we have $\mathbf{E}[u(t)] = \exp((r - \sigma^2/2)t)u_0$. This is plotted in Figure 3.5a along with five sample paths of the SDE (3.24).

Example 3.8 [mean reverting OU process] We consider the following generalization of (3.3)

$$du = \lambda(\mu - u)dt + \sigma dW(t), \qquad u(0) = u_0,$$

for $\lambda, \mu, \sigma \in \mathbb{R}$. We solve this SDE by using the Itô formula (3.25) with $\Phi(t, u) = e^{\lambda t}u$. Then,

$$d\Phi(t, u) = \lambda\,e^{\lambda t}u\,dt + e^{\lambda t}\big(\lambda(\mu - u)dt + \sigma dW(t)\big) + 0$$

and

$$\Phi(t, u(t)) - \Phi(0, u_0) = e^{\lambda t}u(t) - u_0 = \lambda\mu \int_0^t e^{\lambda s}\,ds + \sigma \int_0^t e^{\lambda s}\,dW(s).$$

After evaluating the deterministic integral, we find

$$u(t) = e^{-\lambda t}u_0 + \mu\left(1 - e^{-\lambda t}\right) + \sigma \int_0^t e^{\lambda(s-t)}dW(s) \quad (3.29)$$

and this is known as the *variation of constants* solution. Notice that $u(t)$ is a Gaussian process and can be specified by its mean $\mu(t) = \mathbf{E}[u(t)]$ and covariance $c(s, t) = \mathrm{Cov}\,(u(s), u(t))$. Using the mean zero property of the Itô integral (Theorem 3.3), the mean is

$$\mu(t) = \mathbf{E}[u(t)] = e^{-\lambda t}u(0) + \mu(1 - e^{-\lambda t}) \quad (3.30)$$

so that $\mu(t) \to \mu$ as $t \to \infty$ and the process is "mean reverting." For the covariance, first note that

$$\begin{aligned} c(s, t) &= \mathrm{Cov}\,(u(t), u(s)) \\ &= \mathbf{E}\big[\big(u(s) - \mathbf{E}[u(s)]\big)\,\big(u(t) - \mathbf{E}[u(t)]\big)\big] \\ &= \mathbf{E}\left[\int_0^s \sigma e^{\lambda(r-s)}dW(r) \int_0^t \sigma e^{\lambda(r-t)}dW(r)\right] \\ &= \sigma^2 e^{-\lambda(s+t)}\mathbf{E}\left[\int_0^s e^{\lambda r}dW(r) \int_0^t e^{\lambda r}dW(r)\right]. \end{aligned}$$

Then, using the Itô isometry, property (3) from Definition 3.2,

$$c(s, t) = \frac{\sigma^2}{2\lambda}e^{-\lambda(s+t)}\left(e^{2\lambda \min(s,t)} - 1\right).$$

In particular, the variance $\text{Var}\,(u(t)) = \sigma^2(1-\mathrm{e}^{-2\lambda t})/2\lambda$ so that $\text{Var}\,(u(t)) \to \sigma^2/2\lambda$ and $u(t) \to N(\mu, \sigma^2/2\lambda)$ in the distribution as $t \to \infty$.

Reconsider Example 3.1, which is the mean reverting OU process with $\mu = 0$.

Example 3.9 For $\mathrm{d}p = -\lambda p \mathrm{d}t + \sigma \mathrm{d}W(t)$ the variation of constants (3.29) gives

$$p(t) = \mathrm{e}^{-\lambda t} p_0 + \sigma \int_0^t \mathrm{e}^{-\lambda(t-s)}\, \mathrm{d}W(s).$$

We can use this to derive an important relationship in statistical physics. By definition, the expected kinetic energy per degree of freedom of a system at temperature T is given by $k_B T/2$, where k_B is the Boltzmann constant. The temperature of a system of particles in thermal equilibrium can be determined from λ and σ. For the OU model, the expected kinetic energy $\mathbf{E}\left[p(t)^2/2\right]$ is easily calculated

$$\begin{aligned}\mathbf{E}\left[p(t)^2\right] &= \mathrm{e}^{-2\lambda t} p_0^2 + \sigma^2 \int_0^t \mathrm{e}^{-2\lambda(t-s)}\, \mathrm{d}s \\ &= \mathrm{e}^{-2\lambda t} p_0^2 + \frac{\sigma^2}{2\lambda}(1-\mathrm{e}^{-2\lambda t}) \\ &\to \frac{\sigma^2}{2\lambda} \quad \text{as } t \to \infty.\end{aligned}$$

We see that the expected kinetic energy $\mathbf{E}\left[1/2p(t)^2\right]$ converges to $\sigma^2/4\lambda$ as $t \to \infty$. Thus, $\sigma^2/4\lambda$ is the equilibrium kinetic energy and the equilibrium temperature is given by $k_B T = \sigma^2/2\lambda$, which is known as the fluctuation-dissipation relation.

We saw in Example 3.8 that $p(t) \to N(0, \sigma^2/2\lambda)$ in distribution. This is illustrated by numerical simulations in Figure 3.1. This limiting distribution is known as the Gibbs canonical distribution, often written $N(0, k_B T)$ with probability density function $p(q,p) = \mathrm{e}^{-\beta H(q,p)}/Z$, where $\beta = 1/k_B T$ is the inverse temperature, Z is a normalization constant, and $H(q,p)$ is the system energy for a particle with position q and momentum p. In this case, $H = p^2/2$.

We now state a more general form of Itô's formula.

Lemma 3.8 (Itô formula) *If $\Phi(t, \boldsymbol{u}_0)$ is a continuously differentiable function of $t \in [0, T]$ and twice continuously differentiable functions of $\boldsymbol{u}_0 \in \mathbb{R}^d$ and $\boldsymbol{u}(t)$ denotes the solution of*

$$\mathrm{d}\boldsymbol{u} = \boldsymbol{f}(\boldsymbol{u})\mathrm{d}t + G(\boldsymbol{u})\mathrm{d}\boldsymbol{W}(t)$$

under Assumption 3.1, then

$$\begin{aligned}\Phi(t, \boldsymbol{u}(t)) &= \Phi(0, \boldsymbol{u}_0) + \int_0^t \left(\frac{\partial}{\partial t} + \mathcal{L}\right) \Phi(s, \boldsymbol{u}(s))\, \mathrm{d}s \\ &\quad + \sum_{k=1}^m \int_0^t \mathcal{L}^k \Phi(s, \boldsymbol{u}(s))\, \mathrm{d}W_k(s), \end{aligned} \tag{3.31}$$

where

$$\begin{aligned}\mathcal{L}\Phi(t, \boldsymbol{u}) &:= \boldsymbol{f}(\boldsymbol{u})^T \nabla \Phi(t, \boldsymbol{u}) \\ &\quad + \frac{1}{2} \sum_{k=1}^m \boldsymbol{g}_k(\boldsymbol{u})^T \nabla^2 \Phi\,(t, \boldsymbol{u}) \boldsymbol{g}_k(\boldsymbol{u}), \\ \mathcal{L}^k \Phi(t, \boldsymbol{u}) &:= \nabla \Phi(t, \boldsymbol{u})^T \boldsymbol{g}_k(\boldsymbol{u}), \end{aligned} \tag{3.32}$$

for $\boldsymbol{u} \in \mathbb{R}^d$ and $t > 0$. Here $\boldsymbol{g}_k$ denotes the kth column of the diffusion matrix G. $\nabla\Phi$ is the gradient and $\nabla^2\Phi$ the Hessian matrix of second partial derivatives of $\Phi(t, \boldsymbol{x})$ with respect to $\boldsymbol{x}$.

Itô's formula can be generalized further, for example, see [6]. With the Itô formula (3.31) it is possible to generalize the variation of constants formula (3.29). Consider the semilinear Itô SDE

$$\mathrm{d}\boldsymbol{u} = \left[-A\boldsymbol{u} + \boldsymbol{f}(\boldsymbol{u})\right] \mathrm{d}t + G(\boldsymbol{u})\, \mathrm{d}\boldsymbol{W}(t),$$
$$\boldsymbol{u}(0) = \boldsymbol{u}_0 \in \mathbb{R}^d$$

Then, by the same techniques as in Example 3.8 we have

$$\boldsymbol{u}(t) = \mathrm{e}^{-tA}\boldsymbol{u}_0 + \int_0^t \mathrm{e}^{-(t-s)A}\boldsymbol{f}(\boldsymbol{u}(s))\,\mathrm{d}s + \int_0^t \mathrm{e}^{-(t-s)A}G(\boldsymbol{u}(s))\,\mathrm{d}\boldsymbol{W}(s).$$

In Section 3.7 the Itô formula is used to obtain deterministic equations from the SDEs.

3.5 Stratonovich Integral and SDEs

Let $W(t)$ be the Brownian motion and let $g : [0,T] \to \mathbb{R}$ be such that $(\mathbf{E}[\int_0^t |g(s)|^2 \mathrm{d}s])^{1/2} < \infty$. Then the one-dimensional Stratonovich integral is defined through the mean square limit as follows

$$\int_0^T g(t)\mathrm{d}W(t) := \lim_{N\to\infty} \sum_{j=0}^{N-1} g\left(\frac{(t_j + t_{j+1})}{2}\right)\left(W(t_{j+1}) - W(t_j)\right), \quad (3.33)$$

where $0 = t_0 < t_1 < \cdots < t_N = T$ is a partition as in Lemma 3.6. The multidimensional Stratonovich integral is defined analogously to Definition 3.2. We saw that the Itô integral has mean 0; however, this is not the case for the Stratonovich one. We can illustrate this for the stochastic integral $I = \int_0^T W(s) \circ \mathrm{d}W(s)$. Lemma 3.6 shows that $I = \frac{1}{2}(W(T))^2$ and

$$\mathbf{E}[I] = \frac{1}{2}\mathbf{E}\left[(W(T))^2\right] = \frac{T}{2}.$$

We can transform between the Stratonovich and Itô integrals by ensuring the mean zero property. This is often called a *drift correction*. Provided g is differentiable, we have

$$\int_0^t g(W(s)) \circ \mathrm{d}W(s) = \int_0^t g(W(s))\,\mathrm{d}W(s) + \frac{1}{2}\int_0^t g'(W(s))\,\mathrm{d}s. \quad (3.34)$$

Example 3.10 Consider the case where $g(W) = W$. Then from (3.34), we have using (3.13)

$$\int_0^t W(s) \circ \mathrm{d}W(s) = \int_0^t W(s)\,\mathrm{d}W(s) + \frac{1}{2}\int_0^t \mathrm{d}s = \frac{(W(T))^2}{2},$$

and we have recovered (3.12) exactly.

We write the Stratonovich SDE with drift f and diffusion g as

$$\mathrm{d}v = f(v)\,\mathrm{d}t + g(v) \circ \mathrm{d}W(t), \quad v(0) = v_0, \quad (3.35)$$

which we interpret as the integral equation

$$v(t) = v_0 + \int_0^t f(v(s))\,\mathrm{d}s + \int_0^t g(v(s)) \circ W(s) \quad (3.36)$$

and the stochastic integral is a Stratonovich integral.

We can replace the Stratonovich integral using the transformation to find that $v(t)$ satisfies the Itô SDE

$$v(t) = v_0 + \int_0^t \left[f(v(s)) + \tfrac{1}{2}\tfrac{\mathrm{d}g(v(s))}{\mathrm{d}v} g(v(s))\right] \mathrm{d}s + \int_0^t g(v(s))\mathrm{d}W(s).$$

Example 3.11 Consider the SDE

$$\mathrm{d}u = ru\mathrm{d}t + \sigma u \circ \mathrm{d}W, \quad \text{given} \quad u(0) = u_0. \quad (3.37)$$

This looks like the geometric Brownian motion SDE, only here the stochastic forcing is interpreted in the Stratonovich sense.

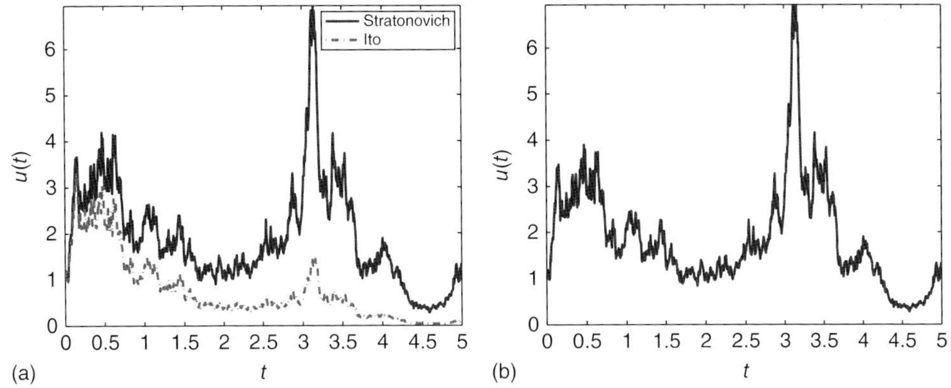

Figure 3.6 In both (a) and (b) the sample path for the Brownian motion is fixed. In (a), we compare the Itô SDE (3.24) and Stratonovich SDEs (3.37), computed using the Heun method. In (b), we plot the solution of the equivalent Itô SDE, (3.38) to (3.37) and see that these agree.

In Figure 3.6a, we fix a sample path for the Brownian motion and plot a sample path of (3.37) and compare it to a sample path of (3.24). It is clear the solutions of the two SDEs do not agree. Equation (3.37) is equivalent to the following Itô SDE

$$du = \left(ru + \frac{1}{2}\sigma^2 u\right) dt + \sigma u dW, \quad \text{given} \quad u(0) = u_0. \tag{3.38}$$

In Figure 3.6b, we plot the solution of this SDE using the same Brownian motion as in Figure 3.6a. We observe that this now agrees with the Stratonovich path in Figure 3.6b.

Example 3.12 Now let us start with an Itô SDE and convert to an equivalent Stratonovich form. Let us consider the geometric Brownian motion of (3.24), this can be rewritten as the equivalent Stratonovich SDE

$$du = \left(ru - \frac{1}{2}\sigma^2 u\right) dt + \sigma u \circ dW, \quad \text{given} \quad u(0) = u_0.$$

For a system of Stratonovich SDEs in dimension $d > 1$ with forcing by m Brownian motions, we have

$$d\boldsymbol{v} = \boldsymbol{f}(\boldsymbol{v})\, dt + G(\boldsymbol{v}) \circ d\boldsymbol{W}(t), \quad \boldsymbol{v}(0) = \boldsymbol{v}_0, \tag{3.39}$$

for $f: \mathbb{R}^d \to \mathbb{R}^d$, $G: \mathbb{R}^d \to \mathbb{R}^{d\times m}$, and $\boldsymbol{W}(t) = (W_1(t), \dots, W_m(t))^T$ for independent Brownian motions $W_i(t)$. Then, $\boldsymbol{v}(t) = (v_1(t), \dots, v_d(t))^T$ is a solution to (3.39) if the components $v_i(t)$ for $i = 1, \dots, d$ satisfy

$$v_i(t) = v_{0,i} + \int_0^t f_i(\boldsymbol{v}(s))\, ds + \sum_{j=1}^{m} \int_0^t g_{ij}(\boldsymbol{v}(s)) \circ dW_j(s), \quad t > 0,$$

where the last term is the Stratonovich integral defined as in (3.33).

Provided G is differentiable, we are able to convert between the Itô and Stratonovich interpretations of the SDEs. We write (3.39) as an Itô SDE. Define $\tilde{f}(\boldsymbol{v}) = (\tilde{f}_1(\boldsymbol{v}), \dots, \tilde{f}_d(\boldsymbol{v}))^T$, where

$$\tilde{f}_i(\boldsymbol{v}) = f_i(\boldsymbol{v}) + \frac{1}{2}\sum_{k=1}^{d}\sum_{j=1}^{m} g_{kj}(\boldsymbol{v})\frac{\partial g_{ij}(\boldsymbol{v})}{\partial v_k},$$
$$i = 1, \dots, d, \quad \boldsymbol{v} \in \mathbb{R}^d. \qquad (3.40)$$

Then the solution $\boldsymbol{v}(t)$ of (3.39) satisfies the equivalent Itô SDE

$$d\boldsymbol{v} = \tilde{\boldsymbol{f}}(\boldsymbol{v})\,dt + G(\boldsymbol{v})\,d\boldsymbol{W}(t), \qquad \boldsymbol{v}(0) = \boldsymbol{v}_0.$$

Note that it is clear that for additive noise the Itô and Stratonovich interpretations coincide as $\partial g_{ij}/\partial v_k = 0$ for all i, j. Finally, we quote the corresponding Stratonovich version of the Itô formula of Lemma 3.8. This shows that the chain rule for Stratonovich calculus resembles the deterministic version of the chain rule.

Lemma 3.9 (Stratonovich formula) *Suppose that Φ is as in Lemma 3.8 and $\boldsymbol{v}(t)$ is the solution of (3.39). Then*

$$\Phi(t, \boldsymbol{v}(t)) = \Phi(0, \boldsymbol{v}_0) + \int_0^t \left(\frac{\partial}{\partial t} + \mathcal{L}_{\text{strat}}\right)\Phi(s, \boldsymbol{v}(s))\,ds + \sum_{k=1}^{m}\int_0^t \mathcal{L}^k\Phi(s, \boldsymbol{v}(s)) \circ dW_k(s), \quad (3.41)$$

where $\mathcal{L}_{\text{strat}}\Phi = \boldsymbol{f}^T\nabla\Phi$ and $\mathcal{L}^k\Phi := \nabla\Phi^T\boldsymbol{g}_k$ is defined by (3.32).

Remark 3.4 *One advantage of the Stratonovich integral is that it follows the standard rules of calculus. The Itô integral is better suited to modeling financial applications and for analysis. Which form of integral is appropriate is essentially a modeling question. However, the integrals are the same if the integrand is not stochastic and we can convert between them provided the diffusion term G is differentiable.*

3.6 SDEs and Numerical Methods

We construct approximations $\boldsymbol{u}_n$ to the solution $\boldsymbol{u}(t_n)$ of a system of SDEs where $t_n = n\Delta t$ and we are given initial data $\boldsymbol{u}(0) = \boldsymbol{u}_0$. As we have seen the solution depends on the sample path of the noise that is taken and so convergence of numerical methods for SDEs has to be approached with care.

If we are worried about approximating the sample path of the solution $\boldsymbol{u}(t)$ for a given sample path of the noise, then we need a strong approximation and we are interested in strong convergence, that is we look at the root mean square error

$$\sup_{0\le t_n\le T}\left(\mathbf{E}\left[\|\boldsymbol{u}(t_n) - \boldsymbol{u}_n\|\right]\right)^{1/2}.$$

Using the Borel–Cantelli lemma, it is possible to obtain from the strong convergence, a pathwise error $\sup_{0\le t_n\le T}\|\boldsymbol{u}(t_n) - \boldsymbol{u}_n\|$, example, see [7–9].

In contrast, in many situations it is average quantities that are of interest and not the individual sample paths. For example, we may be actually interested in $\mathbf{E}[\phi(\boldsymbol{u}(T))]$ for some given function $\phi : \mathbb{R}^d \to \mathbb{R}$, where $\boldsymbol{u}(T)$ is the solution of an SDE at time T. This leads to the notion of weak convergence where we examine

$$\sup_{0\le t_n\le T}\left|\mathbf{E}\left[\phi(\boldsymbol{u}(t_n))\right] - \mathbf{E}\left[\phi(\boldsymbol{u}_n)\right]\right|.$$

We note that if strong convergence can be established, then weak convergence follows, for example, see [10].

In practice, to approximate $\mathbf{E}[\phi(\boldsymbol{u}(T))]$, we need a combination of Monte Carlo and numerical methods for SDEs. We need to generate M independent samples $\boldsymbol{u}_N^j$ for $j = 1, \dots, M$ of an approximation $\boldsymbol{u}_N$ to $\boldsymbol{u}(T)$ for $t_N = T$ and approximate $\mathbf{E}[\phi(\boldsymbol{u}(T))]$ by the sample average

$$\mu_M := \frac{1}{M}\sum_{j=1}^{M} \phi(\boldsymbol{u}_N^j). \tag{3.42}$$

It can be shown that the Monte Carlo error from only taking M samples satisfies

$$P\left(\left|\mathbf{E}\left[\phi(\boldsymbol{u}_N)\right] - \mu_M\right| < \frac{2\sqrt{\operatorname{Var}\left(\phi(\boldsymbol{u}_N)\right)}}{\sqrt{M}}\right) > 0.95 + O(M^{-1/2}),$$

(see Chapter 2). The total error made in approximating $\mathbf{E}\left[\phi(\boldsymbol{u}(t_n))\right]$ can then be divided as the sum of the weak discretization error because of approximating $\boldsymbol{u}(T)$ by $\boldsymbol{u}_N$ and the Monte Carlo error because of taking M samples,

$$\begin{aligned}\mathbf{E}[\phi(\boldsymbol{u}(T))] - \mu_M &= \underbrace{\left[\mathbf{E}[\phi(\boldsymbol{u}(T))] - \mathbf{E}\left[\phi(\boldsymbol{u}_N)\right]\right]}_{\text{weak discretization error}} \\ &+ \underbrace{\left[\mathbf{E}\left[\phi(\boldsymbol{u}_N)\right] - \mu_M\right]}_{\text{Monte Carlo error}}. \end{aligned} \tag{3.43}$$

3.6.1 Numerical Approximation of Itô SDEs

Consider the Itô SDE (3.19) that we can write as the integral equation (3.20). We now want to approximate the solution of the SDE over a fixed time interval $[0, T]$. For convenience, we take a fixed step Δt, so that $N\Delta t = T$. The simplest of such methods is the *Euler–Maruyama method* and we now outline its derivation. We have from (3.20),

$$\boldsymbol{u}(t_{n+1}) = \boldsymbol{u}_0 + \int_0^{t_{n+1}} \boldsymbol{f}(\boldsymbol{u}(s))\,\mathrm{d}s + \int_0^{t_{n+1}} G(\boldsymbol{u}(s))\,\mathrm{d}\boldsymbol{W}(s),$$

and

$$\boldsymbol{u}(t_n) = \boldsymbol{u}_0 + \int_0^{t_n} \boldsymbol{f}(\boldsymbol{u}(s))\,\mathrm{d}s + \int_0^{t_n} G(\boldsymbol{u}(s))\,\mathrm{d}\boldsymbol{W}(s).$$

Subtracting, we obtain

$$\boldsymbol{u}(t_{n+1}) = \boldsymbol{u}(t_n) + \int_{t_n}^{t_{n+1}} \boldsymbol{f}(\boldsymbol{u}(s))\,\mathrm{d}s + \int_{t_n}^{t_{n+1}} G(\boldsymbol{u}(s))\,\mathrm{d}\boldsymbol{W}(s). \tag{3.44}$$

We approximate both the integrands $\boldsymbol{f}$ and G by a constant over $[t_n, t_{n+1})$ on the basis of the value at t_n to obtain the Euler–Maruyama method

$$\begin{aligned}\boldsymbol{u}_{n+1} &= \boldsymbol{u}_n + \boldsymbol{f}(\boldsymbol{u}_n)\,\Delta t + G(\boldsymbol{u}_n)\Delta\boldsymbol{W}_n, \\ \Delta\boldsymbol{W}_n &:= \boldsymbol{W}(t_{n+1}) - \boldsymbol{W}(t_n).\end{aligned} \tag{3.45}$$

As on each subinterval, G is evaluated at the left-hand end point, the approximation is consistent with our definition of the Itô integral. The noise increments $\Delta\boldsymbol{W}_n := \boldsymbol{W}(t_{n+1}) - \boldsymbol{W}(t_n)$ have mean 0 and variance Δt. In implementation, we can usually get standard normal random variables, that is, $\zeta_n \sim N(0, 1)$; thus, to form the increments $\Delta\boldsymbol{W}_n$, we need to use these. To do this, the Euler–Maruyama scheme is rewritten as

$$\boldsymbol{u}_{n+1} = \boldsymbol{u}_n + \boldsymbol{f}(\boldsymbol{u}_n)\,\Delta t + \sqrt{\Delta t}G(\boldsymbol{u}_n)\boldsymbol{\zeta}_n,$$

where $\boldsymbol{\zeta}_n \sim N(0, I)$ (see Algorithm 3.2). This scheme only gives values at the discrete points $t_n = n\Delta t$. If values are required at intermediate points, linear interpolation can be used [5].

Example 3.13 Apply the Euler–Maruyama scheme to the geometric Brownian motion SDE (3.24), to find

Algorithm 3.2 Euler–Maruyama to approximate solution of SDE $d\boldsymbol{u} = \boldsymbol{f}(\boldsymbol{u})dt + G(\boldsymbol{u})d\boldsymbol{W}$.

INPUT : Initial data u_0, Final time T, N number of steps to take.
OUTPUT: Vector t of time and vector u such that $u_n \approx u(t_n)$.

1: $\Delta t = T/N$, $t_0 = 0$
2: **for** $n = 1 : N$ **do**
3: $t_n = n\Delta t$
4: find $z \sim N(0, 1)$
5: $dW_n = \sqrt{\Delta t} z$
6: $u_n = u_{n-1} + \Delta t f(u_{n-1}) + G(u_{n-1}) dW_n$
7: **end for**

$u_{n+1} \approx u(t_{n+1})$ given $u_0 = u(0)$, from

$$u_{n+1} = u_n + r u_n \Delta t + \sigma u_n \Delta W_n, \quad \Delta W_n := W(t_{n+1}) - W(t_n).$$

Figure 3.5b top shows a numerical solution and an explicit solution given from (3.28), and in Figure 3.5b (bottom), we plot the error $u(t_n) - u_n$. We took $r = 0.5$, $\sigma = 0.5$, and $\Delta t = 0.01$ and we see a reasonable agreement between the numerical and exact solution.

For deterministic ODEs numerical methods, we often need to impose a timestep restriction to get the correct long-time dynamics, for example, if $u(t) \to 0$ and $t \to \infty$, we ask that $u_n \to 0$ as $n \to \infty$. In the context of SDEs, such stability questions are treated in a mean square sense, that is if $\mathbf{E}\left[\|\boldsymbol{u}(t)\|^2\right] \to 0$ as $t \to \infty$, we ask $\mathbf{E}\left[\|\boldsymbol{u}_n\|^2\right] \to 0$ as $n \to \infty$. To counter stability constraints on the step size Δt, the drift term can be approximated at $\boldsymbol{f}(\boldsymbol{u}_{n+1})$. This gives an implicit method

$$\boldsymbol{u}_{n+1} = \boldsymbol{u}_n + \boldsymbol{f}(\boldsymbol{u}_{n+1}) \Delta t + \sqrt{\Delta t} G(\boldsymbol{u}_n) \boldsymbol{\zeta}_n.$$

Note that the noise term is still approximated at the left-hand end point and is consistent with the definition of the Itô integral. It can be shown that for stability of the Euler–Maruyama approximation in Example 3.13, we require $0 < \Delta t < -2(r + \sigma^2/2)/r^2$. For stability of numerical methods for SDEs, see [11–14].

Theorem 3.5 (strong convergence) *Let $\boldsymbol{u}(t)$ be the solution of (3.19) and $\boldsymbol{u}_n$ be the Euler–Maruyama approximation to $\boldsymbol{u}(t_n)$. Then the Euler–Maruyama method converges strongly with order $\Delta t^{1/2}$, that is, there is a $C > 0$, independent of Δt such that*

$$\sup_{0 \le t_n \le T} \left(\mathbf{E}\left[\|\boldsymbol{u}(t_n) - \boldsymbol{u}_n\|^2\right]\right)^{1/2} \le C \Delta t^{1/2}.$$

If $G(\boldsymbol{u})$ is independent of $\boldsymbol{u}$, then the rate of convergence is improved and

$$\sup_{0 \le t_n \le T} \left(\mathbf{E}\left[\|\boldsymbol{u}(t_n) - \boldsymbol{u}_n\|^2\right]\right)^{1/2} \le C \Delta t.$$

The $\Delta t^{1/2}$ rate of strong convergence for the general case is very slow, and much slower than that for additive noise. Higher-order numerical schemes can be constructed, and the most popular of these is the Milstein scheme, which is convergent with order Δt (see Figure 3.7a). In the case of additive noise, the Euler–Maruyama and Milstein are equivalent (see (3.46)) and hence we have the improved rate of convergence in Theorem 3.5 in this case.

We examine the Milstein method for the SDE (3.19) with $d = m = 1$ only. In (3.44), instead of approximating G at the left-hand point we now use a Taylor expansion

$$G(u(t)) = G(u(t_n) + u(t) - u(t_n)) \approx G(u(t_n)) + G'(u(t_n))(u(t) - u(t_n)).$$

From the Euler–Maruyama approximation, neglecting the drift, we can approximate $u(t) - u(t_n)$ for $t \in [t_n, t_{n+1}]$ by $G(u(t_n))(W(t) - W(t_n))$. Thus,

$$\begin{aligned}\int_{t_n}^{t_{n+1}} G(u(s))\,\mathrm{d}W(s) &\approx \int_{t_n}^{t_{n+1}} G(u(t_n))\mathrm{d}W(s) \\ &\quad + G'(u(t_n))G(u(t_n))(W(s) \\ &\quad - W(t_n))\,\mathrm{d}W(s).\end{aligned}$$

Now using the integral (3.16), we find

$$\begin{aligned}\int_{t_n}^{t_{n+1}} G(u(s))\,\mathrm{d}W(s) &\approx G(u(t_n))\left(W(t_{n+1}) - W(t_n)\right) \\ &\quad + \frac{1}{2}G'(u(t_n))G(u(t_n)) \\ &\quad \left[\left(W(t_{n+1}) - W(t_n)\right)^2 - \Delta t\right]. \end{aligned} \tag{3.46}$$

Combined with (3.44) and approximating f as before, (3.46) gives the *Milstein method* for SDEs in $\mathbb{R}$

$$\begin{aligned}u_n = u_{n-1} &+ \Delta t f(u_{n-1}) + G(u_{n-1})\mathrm{d}W_n \\ &+ \frac{1}{2}G'(u_{n-1})G(u_{n-1})(\Delta W_n^2 - \Delta t),\end{aligned}$$

(see Algorithm 3.3). The scheme can be extended to general SDEs in $\mathbb{R}^d$ with noise in $\mathbb{R}^m$. In general, however, such higher-order methods require the nontrivial approximation of iterated stochastic integrals (see [5]). Numerically, we observe in Figure 3.7 a strong rate of convergence of order Δt for the Milstein method.

The following theorem states the rate of weak convergence for the Euler–Maruyama and Milstein methods.

Algorithm 3.3 Milstein method to approximate solution of SDE $du = f(u)dt + G(u)dW$.

INPUT : Initial data u_0, Final time T, N number of steps to take.

OUTPUT: Vector t of time and vector u such that $u_n \approx u(t_n)$.

1: $\Delta t = T/N$, $t_0 = 0$
2: **for** $n = 1 : N$ **do**
3: $\quad t_n = n\Delta t$
4: $\quad$ find $z \sim N(0,1)$
5: $\quad \mathrm{d}W_n = \sqrt{\Delta t} z$
6: $\quad u_n = u_{n-1} + \Delta t f(u_{n-1}) + G(u_{n-1})\mathrm{d}W_n + \frac{1}{2}G'(u_{n-1})G(u_{n-1})(\mathrm{d}W_n^2 - \Delta t)$
7: **end for**

Theorem 3.6 (weak convergence)
Suppose that $\boldsymbol{f}$ and G are C^∞ functions and all derivatives of $\boldsymbol{f}$ and G are bounded. Suppose $\phi : \mathbb{R}^d \to \mathbb{R}$ is Lipschitz. Let $\boldsymbol{u}(t)$ be the solution of (3.19) and $\boldsymbol{u}_n$ be either the Euler–Maruyama or Milstein approximation to $\boldsymbol{u}(t_n)$. Then there exists $C > 0$, independent of Δt, such that

$$\left|\mathbf{E}[\phi(\boldsymbol{u}(T))] - \mathbf{E}\left[\phi(\boldsymbol{u}_N)\right]\right| \le C\Delta t.$$

In Figure 3.7, we observe from numerical simulations, the weak convergence rate of $O(\Delta t)$ for the scheme applied to geometric Brownian motion (3.24) with $\phi(u) = u$ and $T = 1$.

3.6.2 Numerical Approximation of Stratonovich SDEs

The numerical solution of the Stratonovich SDE (3.35) can be achieved by transforming the equation to an Itô SDE with a modified drift $\tilde{\boldsymbol{f}}$ and applying, for example, the Euler–Maruyama method. It is also possible to approximate solutions of Stratonovich SDEs directly. Heun's method is the natural equivalent of the Euler–Maruyama method to Stratonovich SDEs. Let us consider the Stratonovich SDE (3.39) and approximate this over a fixed

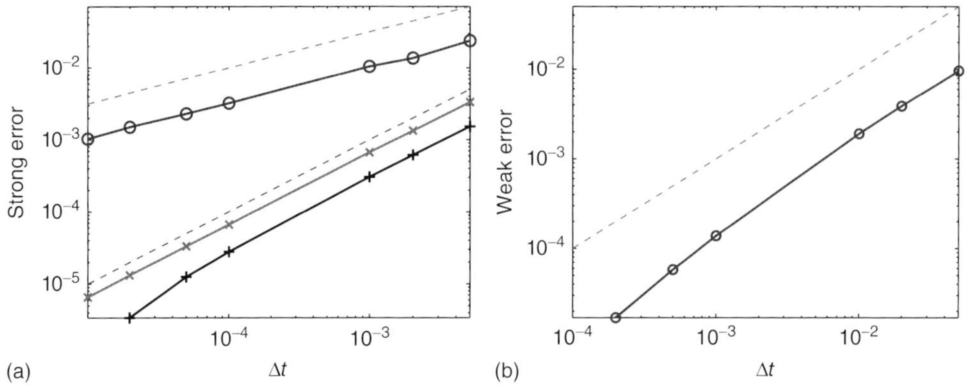

Figure 3.7 (a) a loglog plot illustrating strong convergence of the Euler–Maruyama method for (3.24) marked by *o* and for the Milstein scheme, marked by *x*. The Euler–Maruyama method converges with order $\Delta t^{1/2}$, whereas the Milstein scheme is of order Δt. We also plot the solution of (3.3) with additive noise marked by + solved with the Euler–Maruyama method. Other lines on the plot are reference lines with slopes 0.5 and 1. We took 1000 realizations to estimate the expectations. In (b), we illustrate weak convergence of the Euler–Maruyama method for (3.24) and estimate $\mathbf{E}[u(1)]$. We see an improvement in the rate of convergence over the strong convergence in (a) for (3.24). We took 10 000 realizations to estimate the expectations.

time interval $[0, T]$ with fixed step Δt, so that $N\Delta t = T$. We outline the derivation,

$$\boldsymbol{u}(t_{n+1}) = \boldsymbol{u}_0 + \int_0^{t_{n+1}} \boldsymbol{f}(\boldsymbol{u}(s))\,\mathrm{d}s + \int_0^{t_{n+1}} G(\boldsymbol{u}(s)) \circ \mathrm{d}\boldsymbol{W}(s),$$

and

$$\boldsymbol{u}(t_n) = \boldsymbol{u}_0 + \int_0^{t_n} \boldsymbol{f}(\boldsymbol{u}(s))\,\mathrm{d}s + \int_0^{t_n} G(\boldsymbol{u}(s)) \circ \mathrm{d}\boldsymbol{W}(s).$$

Subtracting, we see that

$$\boldsymbol{u}(t_{n+1}) = \boldsymbol{u}(t_n) + \int_{t_n}^{t_{n+1}} \boldsymbol{f}(\boldsymbol{u}(s))\,\mathrm{d}s + \int_{t_n}^{t_{n+1}} G(\boldsymbol{u}(s)) \circ \mathrm{d}\boldsymbol{W}(s).$$

Taking both $\boldsymbol{f}$ and G constant over $[t_n, t_{n+1})$, we obtain the Heun method. This time, we evaluate G at the midpoint to be consistent with the Stratonovich calculus, that is, we examine

$$\boldsymbol{u}_{n+1} = \boldsymbol{u}_n + \boldsymbol{f}(\boldsymbol{u}_n)\,\Delta t + G(\boldsymbol{u}_{n+1/2})\Delta \boldsymbol{W}_n, \qquad \Delta \boldsymbol{W}_n := \boldsymbol{W}(t_{n+1}) - \boldsymbol{W}(t_n).$$

To implement this, we need, however, an estimate of $G(\boldsymbol{u}_{n+1/2})$. Let us consider this with $d = m = 1$. By Taylor's theorem on G, we have

$$G(u_{n+1/2}) \approx G(u(t_n)) + \frac{\Delta t}{2} G'(u(t_n))$$

and by a finite difference approximation G', we get

$$G(u_{n+1/2}) \approx \frac{1}{2}\Big(G(u(t_n)) + G(u(t_{n+1}))\Big).$$

Using the Euler–Maruyama to approximate $u(t_{n+1})$, we can get a prediction v for

$u(t_{n+1})$:

$$v = u_n + f(u_n)\Delta t + G(u_n)\Delta W_n$$

and then

$$u_{n+1} = u_n + f(u_n)\Delta t + \frac{1}{2}\Big(G(u_n) + G(v)\Big)\Delta W_n.$$

Note that the Brownian increment ΔW_n is the same for the predicting step to get v and the update step to get u_{n+1}. The increments $\Delta W_n \sim N(0, \Delta t)$. We can rewrite the scheme in terms of $\zeta_n \sim N(0, 1)$ as follows:

$$v = u_n + f(u_n)\Delta t + \sqrt{\Delta t} G(u_n)\zeta_n$$

and

$$u_{n+1} = u_n + f(u_n)\Delta t + \frac{1}{2}(G(u_n) + G(v))\sqrt{\Delta t}\zeta_n.$$

This gives the *Heun method* to approximate the Stratonovich SDE, which, in general, becomes

$$\boldsymbol{v} = \boldsymbol{u}_n + \boldsymbol{f}(\boldsymbol{u}_n)\Delta t + \sqrt{\Delta t} G(\boldsymbol{u}_n)\boldsymbol{\zeta}_n$$

and

$$\boldsymbol{u}_{n+1} = \boldsymbol{u}_n + \boldsymbol{f}(\boldsymbol{u}_n)\Delta t + \frac{1}{2}(G(\boldsymbol{u}_n) + G(\boldsymbol{v}))\sqrt{\Delta t}\boldsymbol{\zeta}_n,$$

where $\boldsymbol{\zeta}_n \sim N(0, I)$ and is given in Algorithm 3.4.

Theorem 3.7 (convergence) *Let* $\boldsymbol{u}(t)$ *be the solution of* (3.39) *and* $\boldsymbol{u}_n$ *be the Heun approximation to* $\boldsymbol{u}(t_n)$. *Then, in general, the Heun method converges strongly with order* $\Delta t^{1/2}$, *that is, there is a* $C > 0$, *independent of* Δt, *such that*

$$\sup_{0 \le t_n \le T} \left(\mathbf{E}\left[\|\boldsymbol{u}(t_n) - \boldsymbol{u}_n\|^2\right]\right)^{1/2} \le C\Delta t^{1/2}.$$

Suppose that $\boldsymbol{f}$ *and* G *are* C^∞ *functions and all derivatives of* $\boldsymbol{f}$ *and* G *are bounded. Suppose* $\phi : \mathbb{R}^d \to \mathbb{R}$ *is Lipschitz. Then, there exists* $C > 0$, *independent of* Δt, *such that*

$$\left|\mathbf{E}[\phi(\boldsymbol{u}(T))] - \mathbf{E}\left[\phi(\boldsymbol{u}_N)\right]\right| \le C\Delta t.$$

Algorithm 3.4 Heun method to approximate solution of SDE $du = f(u)dt + G(u) \circ dW$.

Input : Initial data u_0, Final time T, N number of steps to take.

Output: Vector t of time and vector u such that $u_n \approx u(t_n)$.

$\Delta t = T/N$, $t_0 = 0$
for $n = 1 : N$ **do**
  $t_n = n\Delta t$
  find $z \sim N(0, 1)$
  $dW_n = \sqrt{\Delta t} z$
  $v = u_{n-1} + \Delta t f(u_{n-1}) + G(u_{n-1})dW_n$
  $u_n = u_{n-1} + \Delta t f(u_{n-1}) + \frac{1}{2}(G(u_{n-1}) + G(v))dW_n$
end for

The Heun method is the Stratonovich equivalent of the Euler–Maruyama method. As with the Itô case, we can also derive a Stratonovich Milstein and other higher-order methods [5].

3.6.3 Multilevel Monte Carlo

Suppose we need to approximate $\mathbf{E}[\phi(\boldsymbol{u}(T))]$ to a given accuracy ϵ. To achieve this, both the weak discretization error and the Monte Carlo sampling error should be $O(\epsilon)$ in (3.43). We have seen that the weak discretization error for the Heun and Euler–Maruyama methods is $O(\Delta t)$, so we require $\Delta t = O(\epsilon)$. The Monte Carlo error is $O(1/\sqrt{M})$, so we require $M = O(\epsilon^{-2})$.

We can measure the computational cost by counting the total number of steps taken by the numerical method. Finding one sample of $\boldsymbol{u}_N$ requires $T/\Delta t$ steps and finding M

samples requires $MT/\Delta t$ steps. Thus, with $M = O(\epsilon^{-2})$ and $\Delta t = O(\epsilon)$, the total cost to obtain a result with accuracy ϵ is

$$\text{cost}(\mu_M) = \frac{MT}{\Delta t} = O(\epsilon^{-3}). \tag{3.47}$$

A more efficient method, in many cases, is the multilevel Monte Carlo method, which is essentially a form of variance reduction. This method uses a hierarchy of time discretizations Δt_ℓ to compute $\boldsymbol{u}_\ell \approx \boldsymbol{u}(t_n)$ on different levels $\ell = L, L-1, \dots, \ell_0$, so, for example, $\Delta t_\ell = \Delta t_{\ell-1}/2$.

The starting point of the multilevel Monte Carlo method is the telescoping sum

$$\mathbf{E}\left[\phi(\boldsymbol{u}_L)\right] = \mathbf{E}\left[\phi(\boldsymbol{u}_{\ell_0})\right] + \sum_{\ell=\ell_0+1}^{L} \mathbf{E}\left[\phi(\boldsymbol{u}_\ell) - \phi(\boldsymbol{u}_{\ell-1})\right]. \tag{3.48}$$

Thus, $\mathbf{E}\left[\phi(\boldsymbol{u}_L)\right]$, the expected value with the smallest time step Δt_L, is rewritten into $\mathbf{E}\left[\phi(\boldsymbol{u}_{\ell_0})\right]$, computed using the largest step size Δt_0 and a sum of correction terms. The sample average is used to independently estimate the expectations with a different number of samples on each level.

To estimate $\mathbf{E}\left[\phi(\boldsymbol{u}_{\ell_0})\right]$ at level ℓ_0, we use the sample average μ_{ℓ_0} with M_{ℓ_0} independent samples, given by

$$\mu_{\ell_0} := \frac{1}{M_{\ell_0}} \sum_{j_0=1}^{M_{\ell_0}} \phi(\boldsymbol{u}_{\ell_0}^{j_0}). \tag{3.49}$$

Similarly, to estimate the correction at level $\ell = \ell_0 + 1, \dots, L$, we use the sample average μ_ℓ with M_ℓ samples, defined by

$$\mu_\ell := \frac{1}{M_\ell} \sum_{j_\ell=1}^{M_\ell} \left(\phi(\boldsymbol{u}_\ell^{j_\ell}) - \phi\left(\boldsymbol{u}_{\ell-1}^{j_\ell}\right) \right). \tag{3.50}$$

Note that $\phi(\boldsymbol{u}_\ell^k) - \phi(\boldsymbol{u}_{\ell-1}^k)$ is computed using two different time steps Δt_ℓ and $\Delta t_{\ell-1}$ but using the same Brownian path for $\boldsymbol{u}_\ell^k$ and $\boldsymbol{u}_{\ell-1}^k$. Independent sample paths are used for each k and for each level ℓ. Finally, $\mathbf{E}[\phi(\boldsymbol{u}(T))]$ is estimated by

$$\mu_{ML} := \sum_{\ell=\ell_0}^{L} \mu_\ell. \tag{3.51}$$

Because of strong convergence of the numerical method, the correction terms should become small and the variance of these terms decay.

As reappearing in the literature in [15], the multilevel Monte Carlo method has received a great deal of interest. It can readily be extended to Milstein or other methods and also to other noise processes [16, 17]. It has been examined for nonglobal Lipschitz functions ϕ [18] and combined with other variance reduction techniques [19].

3.7 SDEs and PDEs

The solutions of SDEs are closely related to some partial differential equations (PDEs) and we investigate these now. Suppose we have that u satisfies the Itô SDE (3.19). We could estimate statistics on the solution $u(t)$ by direct simulations and using Monte Carlo techniques (see Chapter 2). Here, we examine some alternative approaches. There are two basic types of question we can ask.

1. Given knowledge of the initial data, what is the probability of being in a particular state at time t. Here we are looking forward.
2. What is the expected value of some function of the solution at time t as a

function of the initial data. Here we are looking backward in time.

We can obtain deterministic PDEs that help answer these two questions. The main tool is to use the Itô formula and the mean zero property of the Itô integrals.

The solution of the SDE (3.19) is a *Markov process* (see Chapter 1). This roughly says the future evolution of u given what has happened up to time t is the same as when starting at $u(t)$.

$$P(u(t+s) \in A | u(r) : 0 \le r \le t) = P(u(t+s) \in A | u(t)).$$

The probability $P(u(t) \in A | u(s) = v)$ is called the *transition probability* and the time-dependent probability density function or *transition density* $p(u, t; v, s)$ satisfies

$$P(u(t) \in A | u(s) = v) = \int_A p(u, t; v, s) \mathrm{d}u.$$

To be a Markov process, the transition density needs to satisfy the Chapman–Kolmogorov equation for $s < r < t$,

$$p(u, t; v, s) = \int_{\mathbb{R}} p(z, r; v, s) p(z, r; u, t) \mathrm{d}z.$$

This states that we can get from v at time s to u at time t by passing through z at intermediate times r. The Chapman–Kolmogorov equation gives $p(u, t; v, s)$ by integrating over all possible intermediate z. For further details, see [2, 6, 20, 21].

3.7.1 Fokker–Planck Equation

We start by looking at the forward problem where we know the system at time 0 and we want to have information about the system at time $t > 0$. We develop ideas for an Itô SDE (3.19) with $d = m = 1$ and show that $p(u, t; u_0, 0)$ satisfies the following PDE

$$\frac{\partial}{\partial t} p(u, t; u_0, 0) = -\frac{\partial}{\partial u}\big(f(u(t)) p(u, t; u_0, 0)\big) + \frac{1}{2}\frac{\partial^2}{\partial u^2}\big((G(u(t)))^2 p(u, t; u_0, 0)\big).$$

This is called the (forward) *Fokker–Planck equation* or *forward Chapman–Kolmogorov equation.*

Sketch of derivation. Given any smooth function ϕ we examine $\mathbf{E}^{u_0}[\phi(u(t))] := \mathbf{E}\big[\phi(u(t)) | u(0) = u_0\big]$, that is, the expectation conditional on being at u_0 at time 0, so

$$\mathbf{E}^{u_0}[\phi(u(t))] = \int_{\mathbb{R}} \phi(u(t)) p(u, t; u_0, 0) \mathrm{d}u.$$

From the Itô formula (3.25) and using that Itô integrals have mean zero (see Definition 3.2), we find that

$$\mathbf{E}^{u_0}[\phi(u(t))] = \mathbf{E}^{u_0}[\phi(u(s))] + \mathbf{E}^{u_0}\left[\int_0^t \frac{\partial \phi}{\partial u} f(u(s)) + \frac{1}{2}\frac{\partial^2 \phi}{\partial u^2}(G(u(s)))^2 \mathrm{d}s\right].$$

We differentiate with respect to t to get

$$\frac{\mathrm{d}}{\mathrm{d}t}\mathbf{E}^{u_0}[\phi(u(t))] = \mathbf{E}^{u_0}\left[\frac{\partial \phi}{\partial u} f(u(t)) + \frac{1}{2}\frac{\partial \phi}{\partial u^2}(G(u(t)))^2\right].$$

Now use the definition of the expectation to get

$$\int_{\mathbb{R}} \phi(u) \frac{\partial}{\partial t} p(u, t; u_0, 0) \mathrm{d}u = \int_{\mathbb{R}} \left(\frac{\partial \phi}{\partial u} f(u(t)) + \frac{1}{2}\frac{\partial^2 \phi}{\partial u^2}(G(u(t)))^2\right) p(u, t; u_0, 0) \mathrm{d}u.$$

If we make assumptions on $p \to 0$ as $u \to \infty$, we can use integration by parts once on $\partial\phi/\partial u\, fp$ and twice on $\partial^2\phi/\partial u^2 Gp$ to get

$$\int_{\mathbb{R}} \phi(u)\frac{\partial}{\partial t}p(u,t;u_0,0)\mathrm{d}u$$
$$= \int_{\mathbb{R}} \phi(u)\Big(-\frac{\partial}{\partial u}(f(u(t))p(u,t;u_0,0))$$
$$+\frac{1}{2}\frac{\partial^2}{\partial u^2}\left((G(u(t)))^2 p(u,t;u_0,0)\right)\Big)\,\mathrm{d}u.$$

The above relation is true for any smooth ϕ. Hence, we must have that

$$\frac{\partial}{\partial t}p(u,t;u_0,0)$$
$$= -\frac{\partial}{\partial u}(f(u(t))p(u,t;u_0,0))$$
$$+\frac{1}{2}\frac{\partial^2}{\partial u^2}\left((G(u(t)))^2 p(u,t;u_0,0)\right). \quad (3.52)$$

As initial data, we need $p(u,0;u_0,0) = \delta(u_0)$. The solution of the SDE u takes any value in $\mathbb{R}$ and so the PDE (3.52) is posed on $\mathbb{R}$. Sometimes, (3.52) is rewritten with $p = p(u,0;u_0,0)$ as

$$\frac{\partial p}{\partial t} = \frac{\partial J}{\partial x},$$

where $J := fp + \frac{1}{2}\frac{\partial}{\partial u}(G^2 p)$ is a "probability flux."

Example 3.14 [Brownian motion] For the SDE $\mathrm{d}u = \mathrm{d}W$, $f = 0$ and $G = 1$. Hence the Fokker–Planck equation for the probability density p satisfies the heat equation

$$\frac{\partial}{\partial t}p = \frac{1}{2}\frac{\partial^2}{\partial u^2}p,$$

with $u \in \mathbb{R}$. If the initial condition is $p(u,0) = \delta(u_0)$, the solution is given by $p(u,t) = 1/\sqrt{2\pi t}\,\mathrm{e}^{-u^2/(2t)}$.

We can use the Fokker–Planck equation to examine *moments* defined by

$$M_n := \int_{\mathbb{R}} u^n p(u,t)\mathrm{d}x \quad \text{for} \quad n = 0,1,2,\dots.$$

Example 3.15 [OU process] For the SDE $\mathrm{d}u = -\lambda u + \sigma \mathrm{d}W$, $f = -\lambda u$ and $G = \sigma$. Here the Fokker–Planck equation for the probability density p satisfies the advection diffusion equation

$$\frac{\partial}{\partial t}p = \lambda\frac{\partial}{\partial u}(up) + \frac{\sigma^2}{2}\frac{\partial^2}{\partial u^2}p$$

$u \in \mathbb{R}$. Let us use this to look at the first few moments. For $n = 0$, we integrate the Fokker–Planck equation over $\mathbb{R}$

$$\int_{\mathbb{R}} \frac{\partial}{\partial t}p\mathrm{d}u = \lambda\int_{\mathbb{R}} \frac{\partial}{\partial u}(up)\mathrm{d}u + \int_{\mathbb{R}} \frac{\sigma^2}{2}\frac{\partial^2}{\partial u^2}p\mathrm{d}u.$$

Integration by parts and using the decay of p as $u \to \pm\infty$ gives $\mathrm{d}/\mathrm{d}t\, M_0 = 0$ and so $M_0(t) = M_0(0) = 1$. This is a statement that probability is conserved.

For $n = 1$,

$$\int_{\mathbb{R}} u\frac{\partial}{\partial t}p\mathrm{d}u = \lambda\int_{\mathbb{R}} u\frac{\partial}{\partial u}(up)\mathrm{d}u$$
$$+ \int_{\mathbb{R}} u\frac{\sigma^2}{2}\frac{\partial^2}{\partial u^2}p\mathrm{d}u.$$

Once again, integration by parts and decay of p as $u \to \pm\infty$ gives $\mathrm{d}/\mathrm{d}t\, M_1 = -\lambda M_1$ and so the first moment decays exponentially fast, $M_1(t) = \mathrm{e}^{-\lambda t}M_1(0)$. We have already seen this in (3.30).

For $n \geq 2$, we can proceed in the same way and get a differential equation for M_n,

$$\frac{\mathrm{d}}{\mathrm{d}t}M_n = -\lambda n M_n + \frac{\sigma^2}{2}n(n-1)M_{n-2},$$

which has solution by variation of constants

$$M_n(t) = \mathrm{e}^{-\lambda nt} M_n(0) + \frac{\sigma^2}{2} n(n-1) \times \int_0^t \mathrm{e}^{-\lambda n(t-s)} M_{n-2}(s)\mathrm{d}s.$$

This method of obtaining equations for moments can be applied to SDEs in general. These equations for the moments can then be solved or simulated to obtain statistics on the solutions. To obtain a finite-sized system, the infinite system of equations for the moments needs to be approximated by a finite set. This is often called *moment closure.*

We now state the general form for the Fokker–Planck equation.

Theorem 3.8 (Fokker–Planck) *Consider the Itô SDE* (3.19)

$$\mathrm{d}\boldsymbol{u} = \boldsymbol{f}(\boldsymbol{u})\mathrm{d}t + G(\boldsymbol{u})\mathrm{d}\boldsymbol{W}(t).$$

Then the transition probability density $p = p(u, t; u_0, 0)$ *satisfies*

$$\frac{\partial p}{\partial t} = L^* p,$$

where

$$L^* p(u) = -\sum_{i=1}^{d} \frac{\partial}{\partial u} \left(f_i(u) p(u) \right) + \frac{1}{2} \sum_{i,j=1}^{d} \frac{\partial^2}{\partial u_i \partial u_j} \left((G(u)G(u)^T)_{ij} p(u) \right).$$

If the stochastic process $u(t)$ *has an invariant density* p_0*, then it satisfies* $L^* p_0 = 0$.

Another application of the Fokker–Planck equation is to examine the long-term time behavior of the transition density. The SDE is said to have a *invariant density* if $p_\infty(u) = \lim_{t\to\infty} p(u, t; u_0, 0)$ converges for all initial data u_0. For Brownian motion of Example 3.14, we see that $p_\infty = 0$. For the OU process, we see from Example 3.15 that $p_\infty(u) = C\mathrm{e}^{-\lambda u^2/\sigma^2}$, where C is such that $\int_{\mathbb{R}} p_\infty(u)\mathrm{d}u = 1$, that is, the invariant density of the OU process is a Gaussian.

Example 3.16 Consider the SDE

$$\mathrm{d}u = \frac{\partial}{\partial u} \log(q(u))\mathrm{d}t + \sqrt{2}\mathrm{d}W.$$

This SDE has an invariant density $p_\infty = q$. Let us look at the transition density for this SDE. From (3.52), we find

$$\frac{\partial}{\partial t} p = -\frac{\partial}{\partial u} \left(\frac{\partial}{\partial u} q \frac{1}{q} p \right) + \frac{\partial^2}{\partial u^2} p$$

and thus if $p = q$, we have $\partial p/\partial t = 0$. This idea can be used for Monte Carlo Markov Chain methods and drawing samples from given distributions q for Metropolis Hastings algorithms (see, for example, [22]).

Example 3.17 [Langevin equations] Consider a particle in a potential $V(u)$ subject to some white noise

$$\mathrm{d}u = -\lambda \frac{\partial V}{\partial u} \mathrm{d}t + \sqrt{2\sigma}\mathrm{d}W$$

for $\lambda, \sigma > 0$. Assume that V is twice differentiable. The Fokker–Planck equation is given by

$$\frac{\partial p}{\partial t} = \lambda \frac{\partial}{\partial u} \left(\frac{\partial V}{\partial u} p \right) + \sigma \frac{\partial^2 p}{\partial u^2}.$$

For the steady-state solution, we need the flux $\partial V/\partial u\, p + \sigma\, \partial p/\partial u$ to be constant. As p is assumed to decay to zero at infinity, we can argue that

$$\frac{\partial V}{\partial u} p + \sigma \frac{\partial p}{\partial u} = 0,$$

which is solved by $p_\infty(u) = C\mathrm{e}^{-\lambda V(u)/\sigma^2}$.

Now reconsider the Example 3.4 and the Langevin equation (3.23). It can be shown using the Fokker–Planck equation that the equilibrium distribution has probability density function $p(q,p) = e^{-\beta H(q,p)}/Z$ for $H(q,p) = p^2/2 + V(q)$, normalization constant Z, and inverse temperature $\beta = 1/k_B T = 2\lambda/\sigma^2$.

3.7.2 Backward Fokker–Planck Equation

Now we are interested in $\mathbf{E}[\Psi(u(T))|u(t) = v]$, that is, we want the expected value of some function Ψ of $u(T)$ at $T > t$, given that $u(t) = v$. Let us consider a simple case. Suppose u satisfies the Itô SDE (3.19) with $d = m = 1$, so

$$du = f(u)dt + G(u)dW.$$

We show that $p(u,t) := \mathbf{E}[\Psi(u(T)|u(t)=v]$ satisfies the PDE

$$\frac{\partial}{\partial t}p(u,t) = -f(u)\frac{\partial}{\partial u}p(u,t) - \frac{1}{2}\frac{\partial^2}{\partial u^2}p(u,t), \quad \text{for} \quad t < T, \quad p(u,T) = \Psi(u(T)). \tag{3.53}$$

3.7.2.1 Sketch of Derivation.

The techniques are essentially the same as for the forward Fokker–Planck equation. Let $\phi(u(t),t)$ be the solution of (3.53). We show that $\phi(u(t),t) = \mathbf{E}[\Psi(u(T)|u(t) = v]$.

We apply the Itô formula (3.25) to $\phi(u(t),t)$ to get

$$\begin{aligned} d(\phi(u(s),s)) &= \frac{\partial}{\partial u}\phi du + \frac{1}{2}\frac{\partial^2}{\partial u^2}\phi du + \frac{\partial}{\partial s}\phi ds \\ &= \left(\frac{\partial}{\partial s}\phi + f(u)\frac{\partial}{\partial u}\phi + \frac{1}{2}(G(u))^2\frac{\partial^2}{\partial u^2}\phi\right) dt \\ &\quad + G(u)\frac{\partial}{\partial u}\phi dW. \end{aligned}$$

Thus, because $\phi(u(t),t)$ is solution of (3.53), we have

$$\phi(u(T),T) = \phi(u(t),t) + \int_t^T G(u)\frac{\partial}{\partial u}\phi dW(s).$$

Taking the expectation and using the property that the Itô integral has mean value 0, we get

$$\mathbf{E}[\phi(u(T),T)|u(t) = v] = \phi(u(t),t).$$

As the solution of (3.53) is unique, we have the result.

Example 3.18 [Black–Scholes PDE] Consider the geometric Brownian motion SDE (3.24). Then $p(u,t) = \mathbf{E}[\phi(u(T),T)]$ satisfies the advection diffusion equation

$$\frac{\partial}{\partial t}p = -ru\frac{\partial}{\partial u}(p) + \frac{\sigma^2}{2}u^2\frac{\partial^2}{\partial u^2}(p),$$

$u \in \mathbb{R}$, $t \in [0,T]$, and $p(u,T) = \Phi(u(T))$. This is of interest in finance as it used for pricing of options at $t = 0$ with the so-called payoff Ψ at time T.

In fact, the backward Fokker–Planck can also be written for transition probabilities.

Theorem 3.9 (backward Fokker–Planck) *Consider the Itô SDE*

$$d\boldsymbol{u} = \boldsymbol{f}(\boldsymbol{u})dt + G(\boldsymbol{u})d\boldsymbol{W}(t).$$

The transition probability density $p(u,t;v,s)$ satisfies

$$\frac{\partial p}{\partial t} = Lp,$$

where

$$Lp(u) = -\sum_{i=1}^{d} f_i(u)\frac{\partial}{\partial u}(p(u)) + \frac{1}{2}\sum_{i,j=1}^{d}(G(u)G(u)^T)_{ij}\frac{\partial^2}{\partial u_i \partial u_j}(p(u)).$$

Remark 3.5 *Suppose the SDE*

$$\mathrm{d}\boldsymbol{u} = \boldsymbol{f}(\boldsymbol{u})\mathrm{d}t + G(\boldsymbol{u})\mathrm{d}\boldsymbol{W}(t)$$

has initial data $u(0) = v$. *Let* $\phi : \mathbb{R}^d \to \mathbb{R}$ *be a bounded (measurable) function. We can associate with the SDE a family of linear operators* $S(t)$, $t \geq 0$ *such that*

$$S(t)\phi(u) = \mathbf{E}[\phi(u(t))].$$

The generator of the semigroup S is the operator L such that

$$L\phi(u) = \lim_{t\to 0} \frac{S(t)\phi(u) - \phi(u)}{t}.$$

The operator L^* *in Theorem 3.8 is the adjoint of L.*

3.7.2.2 Boundary Conditions

In our discussion of the Fokker–Planck equations, we assumed that $u \in \mathbb{R}$ and we imposed decay conditions on $p(u, t; u_0, 0)$ as $u \to \pm\infty$. We could also consider imposing some boundaries on u and hence pose the Fokker–Planck equations on a finite (or semi-infinite) domain. Consider, for example, the Langevin equation for the motion of a particle in Example 3.17. We may be interested in the particle arriving at a certain part of phase space and in particular the first time that it arrives there. This gives the idea of first passage time.

If the particle is absorbed as it reaches the boundary, then we get an absorbing boundary condition so that $p(u, t) = 0$ on the boundary. For example, if we have $u \in [a, b] \subset \mathbb{R}$ and the particle gets absorbed when it reaches either a or b, then $p(a, t) = p(b, t) = 0$ (also called a *zero Dirichlet boundary condition*). If the particle is reflected, we have reflecting boundary conditions for the PDE and the outward normal derivative of p is zero at the boundary ($n \cdot \nabla P = 0$). In one dimension, with reflecting boundaries at a and b, we have $\partial p/\partial u|_{u=a} = \partial p/\partial u|_{u=b} = 0$.

The backward Fokker–Planck equation can be used to obtain expressions for the mean first passage times. This is of particular interest for Langevin-type dynamics of a particle moving in a potential V. We can seek the first exit time from some domain $\mathcal{D}$. For the one-dimensional SDE in Example 3.17, it can be shown that the expected first passage is of the order e^2R/σ^2 where R is the minimal potential difference required to leave $\mathcal{D}$. This is known as *Arrhenius' law*. There are a large number of results in this area (see, for example, [2, 20, 21]). An alternative approach is based on *large deviation theory*. Although the sample paths of the solution of an SDE are not differentiable, the idea is that for small noise they should be close to some smooth curves that are often taken from the deterministic system. Suppose we have a smooth curve $\psi : [0 : T] \to \mathbb{R}^d$. A rate function J is introduced that measures the cost of the sample path $u(t)$ being close to the curve ϕ over the interval $[0, T]$. The theory of Wentzell–Freidlin then gives estimates on staying close to the mean path. For a review of large deviation theory, mean first passage times, and Kramer's method see [2, 3, 6, 21, 23, 24].

3.7.3 Filtering

Filtering is a classic problem of trying to estimate a function of the state of a noisy system $X(t)$ at time t where you are only given noisy observations $Y(s)$, $s \leq t$ up to time t.

The *Kalman–Bucy* model assumes that X and Y are determined by linear SDEs so that, in one dimension, we have the signal

X is the solution of

$$dX = \lambda(t)X(t)dt + \sigma(t)dW_1(t), \quad X(0) = X_0,$$

and the observations satisfy

$$dY = \mu(t)X(t)dt + dW_2(t), \quad Y(0) = Y_0.$$

It is assumed that $\lambda(t)$, $\mu(t)$, and $\sigma(t)$ satisfy $\int_0^T |a(t)|dt < \infty$, $a \in \{\lambda, \mu, \sigma\}$. W_1 and W_2 are assumed to be independent Brownian motions. The idea is to estimate the expected value of X given the observations Y, $m(t) := \mathbf{E}[X(t)|Y]$, and the mean square error $z(t) := \mathbf{E}\left[(X(t) - m(t))^2\right]$

Theorem 3.10 (Kalman–Bucy) *The estimate $m(t)$ satisfies the SDE*

$$dm = (\lambda(t) - z(t)(\mu(t))^2)m(t)dt + z(t)\mu(t)dY,$$
$$m(0) = \mathbf{E}\left[X_0 | Y_0\right],$$

where $z(t)$ satisfies the deterministic Ricatti equation

$$\frac{dz}{dt} = 2\lambda(t)z(t) - (z(t))^2(\mu(t))^2 + (\sigma(t))^2,$$
$$z(0) = \mathbf{E}\left[X_0 - m(0)\right].$$

Furthermore, the solution of the equations is unique.

For further examples on SDEs and filtering and related problems, see [6] and reviews [25–28].

Further Reading

SDEs and their mathematical theory are described in many texts, including [6, 29–31] and from a physical perspective in [2, 21, 32, 33]. Numerical methods and their analysis are covered in [5, 34]. The book [4] covers much of the material and also considers both random and stochastic PDEs and their numerical simulation.

In place of a Brownian motion $\boldsymbol{W}(t)$, other types of stochastic process can be used to force the SDE and this is currently an active research topic. For example, we may consider forcing by a fractional Brownian motion $B_H(t)$ in place of $W(t)$. See, for example, [35] and numerical methods for such equations in [36]. More generally, Lyon's theory of rough paths [37, 38] allows very general driving terms to be considered.

This article has only covered some of the basic ideas related to SDEs. It has not discussed, for example, the dynamics related to SDEs. There has been a lot of interest in phenomena such as stochastic resonance where the presence of a small amount of noise in a system can lead to larger scale effects. This is often through the effects of noise when the deterministic system is close to a bifurcation. For more general random dynamical systems, see [39]. A good book on noise-induced phenomena for slow–fast systems, including stochastic resonance, is [3].

There is a growing interest in stochastically forced partial differential equations (SPDEs). The classic theoretical reference is [40]. More recent expositions on the theory include [41, 42].

Glossary

expectation Let $p(x)$ be the pdf of X. Then the expected value of X is given by $\mathbf{E}[X] = \int_{\mathbb{R}} xp(x)$. This can be approximated by the sample average.

Gaussian A random variable X follows the Gaussian or normal distribution ($X \sim N(\mu, \sigma^2)$) if its pdf is $p(x) = 1/\sqrt{2\pi\sigma^2}\exp(-(x-\mu)^2/(2\sigma^2))$.

independent If X and Y are real values with densities p_X and p_Y they are independent if and only if the joint probability density function $p_{X,Y}(x, y) = p_X(x), p_Y(y)$, $x, y \in \mathbb{R}$. Independent random variables are uncorrelated.

multivariate Gaussian $X \in \mathbb{R}^d$ is a multivariate Gaussian if it follows the distribution $N(\boldsymbol{\mu}, C)$ where C is a covariance matrix. Gaussian processes are uniquely determined by their mean and their covariance functions.

probability density function (pdf) is the function such that $P(B) = \int_B p(x)\mathrm{d}x$, so $P(X \in (a, b)) = \int_a^b p(x)\mathrm{d}x$.

probability space $(\Omega, \mathcal{F}, P)$ consists of a sample space Ω, a set of events $\mathcal{F}$ and a probability measure P.

sample path/realisation For a fixed $\omega \in \Omega$ a $X(t, \omega)$ is sample path.

stochastic process is a collection of random variables $X(t)$ that represents the evolution of a system over time that depends on the probability space $(\Omega, \mathcal{F}, P)$. To emphasize dependence on the probability space $X(t)$ is sometimes written as $X(t, \omega)$, $\omega \in \Omega$.

uncorrelated If $\mathrm{Cov}\,(X, Y) = 0$ then the random variables X and Y are uncorrelated.

References

1. Hänggi, P. and Jung, P. (1995) Colored noise in dynamical systems. *Adv. Chem. Phys.*, 239–326.
2. Gardiner, C. (2009) *Stochastic Methods: A Handbook for the Natural and Social Sciences, Springer Series in Synergetics*, 4th edn, Springer, Berlin.
3. Berglund, N. and Gentz, B. (2006) *Noise-Induced Phenomena in Slow-Fast Dynamical Systems. Probability and its Applications (New York)*. Springer-Verlag London Ltd, London, A sample-paths approach.
4. Lord, G.J., Powell, C.E., and Shardlow, T. (2014) *Introduction to Computational Stochastic Partial Differential Equations*, CUP.
5. Peter, E. and Eckhard Platen, K. (1992) *Numerical Solution of Stochastic Differential Equations, Applications of Mathematics*, vol. 23, Springer.
6. Øksendal, B. (2003) *Stochastic Differential Equations*, 6th edn, Universitext Springer, Berlin.
7. Kloeden, P.E. and Neuenkirch, A. (2007) The pathwise convergence of approximation schemes for stochastic differential equations. *LMS J. Comput. Math.*, **10**, 235–253.
8. Jentzen, A., Kloeden, P.E., and Neuenkirch, A. (2008) Pathwise convergence of numerical schemes for random and stochastic differential equations, in *Foundations of Computational Mathematics, London Mathematical Society Lecture Note Series*, vol. 363 (ed. H. Kong), Cambridge University Press, pp. 140–161.
9. Jentzen, A., Kloeden, P.E., and Neuenkirch, A. (2009) Pathwise approximation of stochastic differential equations on domains: higher order convergence rates without global Lipschitz coefficients. *Numer. Math.*, **112**(1), 41–64, doi: 10.1007/s00211-008-0200-8. URL *http://dx.doi.org/10.1007/s00211-008-0200-8*.
10. Williams, D. (1991) *Probability with Martingales*, Cambridge University Press.
11. Buckwar, E. and Sickenberger, T. (2011) A comparative linear mean-square stability analysis of Maruyama- and Milstein-type methods. *Math. Comput. Simulat.*, **81**(6), 1110–1127.
12. Higham, D.J., Mao, X., and Stuart, A.M. (2003) Exponential mean-square stability of numerical solutions to stochastic differential equations. *London Math. Soc. J. Comput. Math.*, **6**, 297–313.
13. Burrage, K., Burrage, P., and Mitsui, T. (2000) Numerical solutions of stochastic differential equations — implementation and stability issues. *J. Comput. Appl. Math.*, **125**(1–2), 171–182, doi: 10.1016/S0377-0427(00)00467-2. URL *http://dx.doi.org/10.1016/S0377-0427(00)00467-2*.
14. Higham, D.J. (2000) Mean-square and asymptotic stability of the stochastic theta method. *SIAM J. Numer. Anal.*, **38**(3), 753–769, doi: 10.1137/S003614299834736X.

URL *http://dx.doi.org/10.1137/S0036142 99834736X.*

15. Giles, M.B. (2008) Multilevel Monte Carlo path simulation. *Oper. Res.*, **56**(3), 607–617, doi: 10.1287/opre.1070.0496. URL *http://dx.doi.org/10.1287/opre.1070.0496.*
16. Giles, M.B. (2008) Improved multilevel Monte Carlo convergence using the Milstein scheme, in *Monte Carlo and Quasi-Monte Carlo Methods 2006*, Springer, Berlin, pp. 343–358, doi: 10.1007/978-3-540-74496-2_20. URL *http://dx.doi.org/10.1007/978-3-540-74496-2_20.*
17. Dereich, S. (2011) Multilevel Monte Carlo algorithms for Lévy-driven SDEs with Gaussian correction. *Ann. Appl. Probab.*, **21**(1), 283–311, doi: 10.1214/10-AAP695. URL *http://dx.doi.org/10.1214/10-AAP695.*
18. Giles, M.B., Higham, D.J., and Mao, X. (2009) Analysing multi-level Monte Carlo for options with non-globally Lipschitz payoff. *Finance Stoch.*, **13**(3), 403–413, doi: 10.1007/s00780-009-0092-1. URL *http://dx.doi.org/10.1007/s00780-009-0092-1.*
19. Giles, M.B. and Waterhouse, B.J. (2009) Multilevel quasi-Monte Carlo path simulation, in *Advanced Financial Modelling, Radon Series on Computational and Applied Mathematics*, vol. 8 (eds H. Abrecher, W.J. Runggaldier, and W. Schachermayer), Walter de Gruyter, Berlin, pp. 165–181, doi: 10.1515/9783110213140.165. URL *http://dx.doi.org/10.1515/9783110213140.165.*
20. Risken, H. (1996) *The Fokker-Planck Equation: Methods of Solution and Applications, Lecture Notes in Mathematics*, Springer-Verlag.
21. van Kampen, N.G. (1997) *Stochastic Processes in Physics and Chemistry*, 2nd edn, North–Holland.
22. Lelièvre, T., Stoltz, G., and Rousset, M. (2010) *Free Energy Computations: A Mathematical Perspective*, Imperial College Press.
23. Freidlin, I. and Wentzell, A.D. (1998) *Random Perturbations of Dynamical Systems, Die Grundlehren der mathematischen Wissenschaften*, Springer-Verlag.
24. Varadhan, S.R.S. (1984) *Large Deviations and Applications, CBMS-NSF Regional Conference Series in Applied Mathematics*, Society for Industrial and Applied Mathematics.
25. Rozovskiĭ, B.L. (1990) *Stochastic Evolution Systems, Mathematics and its Applications (Soviet Series)*, vol. 35, Kluwer Academic Publishers Group, Dordrecht.
26. Bain, A. and Crisan, D. (2009) *Fundamentals of Stochastic Filtering, Stochastic Modelling and Applied Probability*, vol. 60, Springer, New York.
27. Stuart, A.M. (2010) Inverse problems: a Bayesian perspective. *Acta Numer.*, **19**, 451–559, doi: 10.1017/S0962492910000061. URL *http://dx.doi.org/10.1017/S09624 92910000061.*
28. Crisan, D. and Rozovskiĭ, B. (2011) Introduction, in *The Oxford Handbook of Nonlinear Filtering*, Oxford University Press, Oxford, pp. 1–15.
29. Karatzas, I. and Shreve, S.E. (1991) *Brownian Motion and Stochastic Calculus, Graduate Texts in Mathematics*, vol. 113, 2nd edn, Springer, New York.
30. Mao, X. (2008) *Stochastic Differential Equations and Applications*, 2nd edn, Horwood.
31. Protter, P.E. (2005) *Stochastic Integration and Differential Equations, Stochastic Modelling and Applied Probability*, vol. 21, 2nd edn, Springer.
32. Öttinger, H.C. (1996) *Stochastic Processes in Polymeric Fluids*, Springer, Berlin.
33. Nelson, E. (1967) *Dynamical Theories of Brownian Motion*, Princeton University Press.
34. Milstein, G.N. and Tretyakov, M.V. (2004) *Stochastic Numerics for Mathematical Physics*, Springer, Berlin.
35. Mishura, Y.S. (2008) *Stochastic Calculus for Fractional Brownian Motion and Related Processes, Lecture Notes in Mathematics*, vol. 1929, Springer, Berlin, doi: 10.1007/978-3-540-75873-0. URL *http://dx.doi.org/10.1007/978-3-540-75873-0.*
36. Neuenkirch, A. (2008) Optimal pointwise approximation of stochastic differential equations driven by fractional Brownian motion. *Stochastic Process. Appl.*, **118**(12), 2294–2333, doi: 10.1016/j.spa.2008.01.002. URL *http://dx.doi.org/10.1016/j.spa.2008.01.002.*
37. Friz, P.K. and Victoir, N.B. (2010) *Multidimensional Stochastic Processes as Rough Paths, Cambridge Studies in*

Advanced Mathematics, vol. 120, Cambridge University Press.

38. Davie, A.M. (2007) Differential equations driven by rough paths: an approach via discrete approximation. *Appl. Math. Res. Express.*, **2**, 40.
39. Arnold, L. (1998) *Random Dynamical Systems, Monographs in Mathematics*, Springer.
40. Da Prato, G. and Zabczyk, J. (1992) *Stochastic Equations in Infinite Dimensions*, Encyclopedia of Mathematics and its Applications, vol. 44, Cambridge University Press.
41. Prévôt, C. and Röckner, M. (2007) *A Concise Course on Stochastic Partial Differential Equations, Lecture Notes in Mathematics*, vol. 1905, Springer, Berlin.
42. Chow, P.-L. (2007) *Stochastic Partial Differential Equations*, Chapman & Hall/CRC, Boca Raton, FL.

Part II
Discrete Mathematics, Geometry, Topology

Mathematical Tools for Physicists, Second Edition. Edited by Michael Grinfeld.

4
Graph and Network Theory

Ernesto Estrada

4.1
Introduction

Graph Theory was born in 1736 when Leonhard Euler [1] published "*Solutio problematic as geometriam situs pertinentis*" (The solution of a problem relating to the theory of position). This history is well documented [2] and widely available in any textbook of graph or network theory. However, the word graph appeared for the first time in the context of natural sciences in 1878, when the English mathematician James J. Sylvester [3] wrote a paper entitled "Chemistry and Algebra" which was published in *Nature*, where he wrote that "*Every invariant and covariant thus becomes expressible by a* ***graph*** *precisely identical with a Kekulean diagram or chemicograph.*" The use of graph theory in condensed matter physics, pioneered by many chemical and physical graph theorists [4, 5], is today well established; it has become even more popular after the recent discovery of graphene.

There are few, if any, areas of physics in the twenty-first century in which graphs and networks are not involved directly or indirectly. Hence it is impossible to cover all of them in this chapter. Thus I owe the reader an apology for the incompleteness of this chapter and a promise to write a more complete treatise. For instance, quantum graphs are not considered in this chapter and the reader is referred to a recent introductory monograph on this topic for details [6]. In this chapter, we will cover some of the most important areas of applications of graph theory in physics. These include condensed matter physics, statistical physics, quantum electrodynamics, electrical networks, and vibrational problems. In the second part, we summarize some of the most important aspects of the study of complex networks. This is an interdisciplinary area that has emerged with tremendous impetus in the twenty-first century and that studies networks appearing in complex systems. These systems range from molecular and biological to ecological, social, and technological systems. Thus graph theory and network theory have helped to broaden the horizons of physics to embrace the study of new complex systems.

Mathematical Tools for Physicists, Second Edition. Edited by Michael Grinfeld.

We hope this chapter motivates the reader to find more about the connections between graph/network theory and physics, consolidating this discipline as an important part of the curriculum for the physicists of the twenty-first century.

4.2 The Language of Graphs and Networks

The first thing that needs to be clarified is that the terms *graphs* and *networks* are used indistinctly in the literature. In this chapter, we will reserve the term *graph* for the abstract mathematical concept, in general referred to small, artificial formations of nodes and edges. The term *network* is then reserved for the graphs representing real-world objects in which the nodes represent entities of the system and the edges represent the relationships among them. Therefore, it is clear that we will refer to the system of individuals and their interactions as a "social network" and not as a "social graph." However, they should mean exactly the same.

For the basic concepts of graph theory, the reader is recommended to consult the introductory book by Harary [7]. We start by defining a graph formally. Let us consider a finite set $V = \{v_1, v_2, \ldots, v_n\}$ of unspecified elements and let $V \otimes V$ be the set of all ordered pairs $[v_i, v_j]$ of the elements of V. A relation on the set V is any subset $E \subseteq V \otimes V$. The relation E is symmetric if $[v_i, v_j] \in E$ implies $[v_j, v_i] \in E$ and it is reflexive if $\forall v \in V, [v, v] \in E$. The relation E is antireflexive if $[v_i, v_j] \in E$ implies $v_i \neq v_j$. Now we can define a *simple graph* as the pair $G = (V, E)$, where V is a finite set of nodes, vertices, or points and E is a symmetric and antireflexive relation on V, whose elements are known as the *edges* or *links* of the graph. In a *directed graph*, the relation E is nonsymmetric. In many physical applications, the edges of the graphs are required to support weights, that is, real numbers indicating a specific property of the edge. In this case, the following more general definition is convenient. A *weighted graph* is the quadruple $G = (V, E, W, f)$ where V is a finite set of nodes, $E \subseteq V \otimes V = \{e_1, e_2, \ldots, e_m\}$ is a set of edges, $W = \{w_1, w_2, \ldots, w_r\}$ is a set of weights such that $w_i \in \mathbb{R}$, and $f : E \rightarrow W$ is a surjective mapping that assigns a weight to each edge. If the weights are natural numbers, then the resulting graph is a multigraph in which there could be multiple edges between pairs of vertices; that is, if the weight between nodes p and q is $k \in N$, it means that there are k links between the two nodes.

In an undirected graph, we say that two nodes p and q are adjacent if they are joined by an edge $e = \{p, q\}$. In this case, we say that the nodes p and q are incident to the link e, and the link e is incident to the nodes p and q. The two nodes are called the *end nodes* of the edge. Two edges $e_1 = \{p, q\}$ and $e_2 = \{r, s\}$ are adjacent if they are both incident to at least one node. A simple but important characteristic of a node is its degree, which is defined as the number of edges that are incident to it or equivalently as the number of nodes adjacent to it. Slightly different definitions apply to directed graphs. The node p is adjacent to node q if there is a directed link from p to q, $e = (p, q)$. We also say that a link from p to q is incident from p and incident to q; p is incident to e and q is incident from e. Consequently, we have two different kinds of degrees in directed graphs. The in-degree of a node is the number of links incident to it and its out-degree is the number of links incident from it.

4.2.1 Graph Operators

The incidence and adjacency relations in graphs allow us to define the following graph operators. We consider an undirected graph for which we construct its *incidence matrix* with an arbitrary orientation of its entries. This is necessary to consider that the incidence matrix is a discrete analogue of the gradient; that is, for every edge $\{p, q\}$, p is the positive (head) and q the negative (tail) end of the oriented link. Let the links of the graph be labeled as $e_1, e_2, \ldots, e_m$. Hence the *oriented incidence matrix* $\nabla(G)$ is

$$\nabla_{ij}(G) = \begin{cases} +1 & \text{node } v_i \text{ is the head of link } e_j \\ -1 & \text{node } v_i \text{ is the tail of link } e_j \\ 0 & \text{otherwise} \end{cases} .$$

We remark that the results obtained below are independent of the orientation of the links but assume that once the links are oriented, this orientation is not changed. Let the vertex L_V and edge L_E spaces be the vector spaces of all real-valued functions defined on V and E, respectively. The *incidence operator* of the graph is then defined as

$$\nabla(G) : L_V \to L_E, \qquad (4.1)$$

such that for an arbitrary function $f : V \to \mathbb{R}$, $\nabla(G)f : E \to \mathfrak{R}$ is given by

$$\left(\nabla(G)f\right)(e) = f(p) - f(q), \qquad (4.2)$$

where p are the starting (head) and q the ending (tail) points of the oriented link e. Here we consider that f is a real or vector-valued function on the graph with $|f|$ being μ-measurable for certain measure μ on the graph.

On the other hand, let H be a Hilbert space with scalar product $\langle \cdot, \cdot \rangle$ and norm $\|\cdot\|$. Let $G = (V, E)$ be a simple graph. The *adjacency operator* is an operator acting on the Hilbert space $H := l^2(V)$ defined as

$$\left(\mathbf{A}f\right)(p) := \sum_{u,v \in E} f(q), f \in H, i \in V. \qquad (4.3)$$

The adjacency operator of an undirected network is a *self-adjoint operator,* which is bounded on $l^2(V)$. We recall that l^2 is the Hilbert space of square summable sequences with inner product, and that an operator is self-adjoint if its matrix is equal to its own conjugate transpose, that is, it is Hermitian. It is worth pointing out here that the adjacency operator of a directed network might not be self-adjoint. The matrix representation of this operator is the adjacency matrix $\mathbf{A}$, which, for a simple graph, is defined as

$$A_{ij} = \begin{cases} 1 & \text{if } i, j \in E \\ 0 & \text{otherwise.} \end{cases} \qquad (4.4)$$

A third operator related to the previous two, which plays a fundamental role in the applications of graph theory in physics is the *Laplacian operator.* This operator is defined by

$$\mathbf{L}(G)f = -\nabla \cdot \left(\nabla f\right), \qquad (4.5)$$

and it is the graph version of the Laplacian operator

$$\Delta f = \frac{\partial^2 f}{\partial {x_1}^2} + \frac{\partial^2 f}{\partial {x_2}^2} + \cdots + \frac{\partial^2 f}{\partial {x_n}^2}. \qquad (4.6)$$

The negative sign in (4.5) is used by convention. Then the Laplacian operator acting on the function f previously defined is given by

$$\left(\mathbf{L}(G)f\right)(u) = \sum_{\{u,v\} \in E} \left[f(u) - f(v)\right], \qquad (4.7)$$

which in matrix form is given by

$$L_{uv}(G) = \sum_{e \in E} \nabla_{eu} \nabla_{ev} = \begin{cases} -1 & \text{if } uv \in E, \\ k_u & \text{if } u = v, \\ 0 & \text{otherwise.} \end{cases} \tag{4.8}$$

Using the degree matrix **K**, which is a diagonal matrix of the degrees of the nodes in the graph, the Laplacian and adjacency matrices of a graph are related by

$$\mathbf{L} = \mathbf{K} - \mathbf{A}. \tag{4.9}$$

4.2.2 General Graph Concepts

Other important general concepts of graph theory that are fundamental for the study of graphs and networks in physics are the following. Two graphs G_1 and G_2 are *isomorphic* if there is a one-to-one correspondence between the nodes of G_1 and those of G_2, such as, the number of edges joining each pair of nodes in G_1 is equal to that joining the corresponding pair of nodes in G_2. If the graphs are directed, the edges must coincide not only in number but also in direction. The graph $S = \left(V', E'\right)$ is a *subgraph* of a graph $G = (V, E)$ if, and only if, $V' \subseteq V$ and $E' \subseteq E$. The *clique* is a particular kind of subgraph, which is a maximal complete subgraph of a graph. A *complete graph* is one in which every pair of nodes are connected. A (directed) *walk* of length L from v_1 to v_{L+1} is any sequence of (not necessarily different) nodes $v_1, v_2, \ldots, v_L, v_{L+1}$ such that for each $i = 1, 2, \ldots, L$ there is link from v_i to v_{i+1}. A walk is closed (CW) if $v_{L+1} = v_1$. A particular kind of walk is the *path* of length L, which is a walk of length L in which all the nodes (and all the edges) are distinct. A *trial* has all the links different but not necessarily all the nodes. A *cycle* is a closed walk in which all the edges and all the nodes (except the first and last) are distinct. The *girth* of the graph is the size (number of nodes) of the minimum cycle in the graph.

A graph is *connected* if there is a path between any pair of nodes in the graph. Otherwise, it is disconnected. Every connected subgraph is a *connected component* of the graph. The analogous concept in a directed graph is that of a *strongly connected* graph. A directed graph is strongly connected if there is a directed path between each pair of nodes. The strongly connected components of a directed graph are its maximal strongly connected subgraphs.

In an undirected graph, the *shortest-path distance* $d(p, q) = d_{pq}$ is the number of edges in the shortest path between the nodes p and q in the graph. If p and q are in different connected components of the graph the distance between them is set to infinite, $d(p, q) := \infty$. In a directed graph, it is typical to consider the *directed distance* $\vec{d}(p, q)$ between a pair of nodes p and q as the length of the directed shortest path from p to q. However, in general $\vec{d}(p, q) \neq \vec{d}(q, p)$, which violates the symmetry property of a metric, so that $\vec{d}(p, q)$ is not a distance but a *pseudo-distance* or a *pseudo-metric*. The distance between all pairs of nodes in a graph can be arranged in a distance matrix **D**, which, for undirected graphs, is a square symmetric matrix. The maximum entry for a given row/column of the distance matrix of an undirected (strongly connected directed) graph is known as the *eccentricity* $e(p)$ of the node p, $e(p) = \max_{x \in V(G)} \{d(p, x)\}$. The maximum eccentricity among the nodes of a graph is the *diameter* of the graph, which is $diam(G) = \max_{x,y \in V(G)} \left\{d\left(x, y\right)\right\}$. The *average path length* $\bar{l}$ of a graph with n nodes is

$$\bar{l} = \frac{1}{n(n-1)} \sum_{x,y} d\left(x, y\right). \tag{4.10}$$

An important measure for the study of networks was introduced by Watts and Strogatz [8] as a way of quantifying how clustered a node is. For a given node i, the *clustering coefficient* is the number of triangles connected to this node $|C_3(i)|$ divided by the number of triples centered on it

$$C_i = \frac{2\,|C_3(i)|}{k_i\,(k_i - 1)}, \tag{4.11}$$

where k_i is the degree of the node. The average value of the clustering for all nodes in a network $\overline{C}$

$$\overline{C} = \frac{1}{n}\sum_{i=1}^{n} C_i \tag{4.12}$$

has been extensively used in the analysis of complex networks (see Section 4.9 of this chapter).

A second clustering coefficient has been introduced as a global characterization of network cliquishness [9]. This index, which is also known as *network transitivity*, is defined as the ratio of three times the number of triangles divided by the number of connected triples (two paths):

$$C = \frac{3\,|C_3|}{|P_2|}. \tag{4.13}$$

4.2.3 Types of Graphs

The simplest type of graph is the tree. A *tree* of n nodes is a graph that is connected and has no cycles. The simplest tree is the *path* P_n. The path (also known as *linear path* or *chain*) is the tree of n nodes, $n - 2$ of which have degree 2 and two nodes have degree 1. For any kind of graph, we can find a spanning tree, which is a subgraph of this graph that includes every node and is a tree. A *forest* is a disconnected graph in which every connected component is a tree. A *spanning forest* is a subgraph of the graph that includes every node and is a forest.

An *r-regular graph* is a graph with $rn/2$ edges in which all nodes have degree r. A particular case of a regular graph is the complete graph previously defined. Another type of regular graph is the *cycle*, which is a regular graph of degree 2, that is, a 2-regular graph, denoted by C_n. The *complement* of a graph G is the graph $\overline{G}$ with the same set of nodes as G but two nodes in $\overline{G}$ are connected if, and only if, they are not connected in G. An empty or *trivial graph* is a graph with no links. It is denoted as $\overline{K}_n$ as it is the complement of the complete graph.

A graph is *bipartite* if its nodes can be split into two disjoint (nonempty) subsets $V_1 \subset V$ $(V_1 \neq \phi)$ and $V_2 \subset V$ $(V_2 \neq \phi)$ and $V_1 \cup V_2 = V$, such that each edge joins a node in V_1 and a node in V_2. Bipartite graphs do not contain cycles of odd length. If all nodes in V_1 are connected to all nodes in V_2 the graph is known as a *complete bipartite graph*, denoted by K_{n_1,n_2}, where $n_1 = |V_1|$ and $n_2 = |V_2|$ are the number of nodes in V_1 and V_2, respectively. Finally, a graph is *planar* if it can be drawn in a plane in such a way that no two edges intersect except at a node with which they are both incident.

4.3 Graphs in Condensed Matter Physics

4.3.1 Tight-Binding Models

In condensed matter physics, it is usual to describe solid-state and molecular systems by considering the interaction between N electrons whose behavior is determined by a Hamiltonian of the following form:

$$\mathbf{H}=\sum_{n=1}^{N}\left[-\frac{\hbar^2\nabla_n^2}{2m}+U\left(r_n\right)+\frac{1}{2}\sum_{m\neq n}V\left(r_n-r_m\right)\right], \tag{4.14}$$

where $U\left(r_n\right)$ is an external potential and $V\left(r_n-r_m\right)$ is the potential describing the interactions between electrons. Using the second quantization formalism of quantum mechanics, this Hamiltonian can be written as

$$\hat{H}=-\sum_{ij}t_{ij}\hat{c}_i^{\dagger}\hat{c}_j+\frac{1}{2}\sum_{ijkl}V_{ijkl}\hat{c}_i^{\dagger}\hat{c}_k^{\dagger}\hat{c}_l\hat{c}_j, \tag{4.15}$$

where $\hat{c}_i^{\dagger}$ and $\hat{c}_i$ are "ladder operators," t_{ij} and V_{ijkl} are integrals that control the hopping of an electron from one site to another and the interaction between electrons, respectively. They are usually calculated directly from finite basis sets [10].

In the tight-binding approach for studying solids and certain classes of molecules, the interaction between electrons is neglected and $V_{ijkl}=0, \forall i,j,k,l$. This method, which is known as the *Hückel molecular orbital method* in chemistry, can be seen as very drastic in its approximation, but let us think of the physical picture behind it [11, 12]. We concentrate our discussion on alternant conjugated molecules in which single and double bonds alternate. Consider a molecule like benzene in which every carbon atom has an sp^2 hybridization. The frontal overlapping sp^2-sp^2 of adjacent carbon atoms creates very stable œ-bonds, while the lateral overlapping p–p between adjacent carbon atoms creates very labile ß-bonds. Thus it is clear from the reactivity of this molecule that a œ – ß separation is plausible and we can consider that our basis set consists of orbitals centered on the particular carbon atoms in such a way that there is only one orbital per spin state at each site. Then we can write the Hamiltonian of the system as

$$\hat{H}_{tb}=-\sum_{ij}t_{ij}\hat{c}_{i\rho}^{\dagger}\hat{c}_{i\rho}, \tag{4.16}$$

where $\hat{c}_{i\rho}^{(\dagger)}$ creates (annihilates) an electron with spin ρ in a ß (or other) orbital centered at the atom i. We can now separate the in-site energy α_i from the transfer energy β_{ij} and write the Hamiltonian as

$$\hat{H}_{tb}=\sum_{ij}\alpha_i\hat{c}_{i\rho}^{\dagger}\hat{c}_{i\rho}+\sum_{\langle ij\rangle\rho}\beta_{ij}\hat{c}_{i\rho}^{\dagger}\hat{c}_{i\rho}, \tag{4.17}$$

where the second sum is carried out over all pairs of nearest neighbors. Consequently, in a molecule or solid with N atoms the Hamiltonian equation (4.16) is reduced to an $N\times N$ matrix,

$$H_{ij}=\begin{cases}\alpha_i & \text{if} \quad i=j\\ \beta_{ij} & \text{if} \quad i\,\text{isconnectedto}\,j\\ 0 & \text{otherwise.}\end{cases} \tag{4.18}$$

Owing to the homogeneous geometrical and electronic configuration of many systems analyzed by this method, we may take $\alpha_i=\alpha, \forall i$ (Fermi energy) and $\beta_{ij}=\beta\approx -2.70\text{eV}$ for all pairs of connected atoms. Thus,

$$\mathbf{H}=\alpha\mathbf{I}+\beta\mathbf{A}, \tag{4.19}$$

where I is the identity matrix and $\mathbf{A}$ is the adjacency matrix of the graph representing the carbon skeleton of the molecule. The Hamiltonian and the adjacency matrix of the graph have the same eigenfunctions φ_j and their eigenvalues are simply related by

$$\mathbf{H}\varphi_j=E_j\mathbf{A}, \mathbf{A}\varphi_j=\lambda_j\mathbf{H}E_j=\alpha+\beta\lambda_j. \tag{4.20}$$

Hence, everything we have to do in the analysis of the electronic structure of molecules or solids that can be represented by a tight-binding Hamiltonian is to study the spectra of the graphs associated with them. The study of spectral properties of graphs represents an entire area of research in algebraic graph theory. The spectrum of a matrix is the set of eigenvalues of the matrix together with their multiplicities. For the case of the adjacency matrix, let $\lambda_1(\mathbf{A}) \geq \lambda_2(\mathbf{A}) \geq \cdots \geq \lambda_n(\mathbf{A})$ be the distinct eigenvalues of $\mathbf{A}$ and let $m(\lambda_1(\mathbf{A}))$, $m(\lambda_2(\mathbf{A})), \ldots, m(\lambda_n(\mathbf{A}))$ be their algebraic multiplicities, that is, the number of times each of them appears as an eigenvalue of $\mathbf{A}$. Then the spectrum of $\mathbf{A}$ can be written as

$$\mathrm{Sp}\mathbf{A} = \begin{pmatrix} \lambda_1(\mathbf{A}) & \lambda_2(\mathbf{A}) & \cdots & \lambda_n(\mathbf{A}) \\ m(\lambda_1(\mathbf{A})) & m(\lambda_2(\mathbf{A})) & \cdots & m(\lambda_n(\mathbf{A})) \end{pmatrix}. \tag{4.21}$$

The total ß (molecular) energy is given by

$$E = \alpha n_e + \beta \sum_{j=1}^{n} g_j \lambda_j, \tag{4.22}$$

where n_e is the number of ß-electrons in the molecule and g_j is the occupation number of the jth molecular orbital. For neutral conjugated systems in their ground state, we have [13]

$$E = \begin{cases} 2\sum\limits_{j=1}^{n/2} \lambda_j & n \text{ even}, \\ 2\sum\limits_{j=1}^{(n+1)/2} \lambda_j + \lambda_{(j+1)/2} & n \text{ odd}. \end{cases} \tag{4.23}$$

Because an alternant conjugated hydrocarbon has a bipartite molecular graph, $\lambda_j = -\lambda_{n-j+1}$ for all $j = 1, 2, \ldots, n$. In a few molecular systems, the spectrum of the adjacency matrix is known. For instance [11], we have

1. Polyenes C_nH_{n+2}

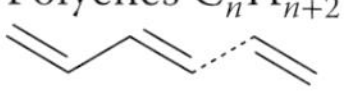

$$\lambda_j(\mathbf{A}) = 2\cos\left(\frac{\pi j}{n+1}\right), \quad j = 1, \ldots, n, \tag{4.24}$$

2. Cyclic polyenes C_nH_n

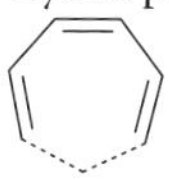

$$\lambda_j(\mathbf{A}) = 2\cos\left(\frac{2\pi j}{n}\right), \quad j = 1, \ldots, n, \lambda_j = \lambda_{n-j} \tag{4.25}$$

3. Polyacenes

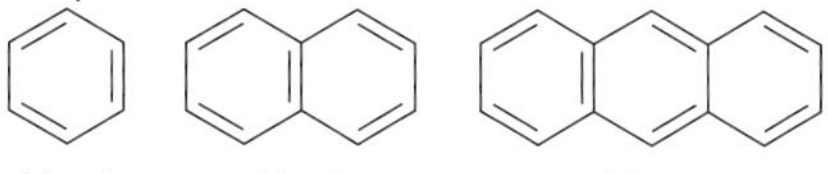

$N = 1$ $N = 2$ $N = 3$

$$\lambda_r(\mathbf{A}) = 1; \lambda_s(\mathbf{A}) = -1;$$
$$\lambda_k(\mathbf{A}) = \pm\frac{1}{2}\left\{1 \pm 9 + 8\cos\frac{k\pi}{N+1}\right\}, \quad k = 1, \ldots, N \tag{4.26}$$

A few bounds exist for the total energy of systems represented by graphs with n vertices and m edges. For instance,

$$\sqrt{2m + n(n-1)(\det \mathbf{A})^{n/2}} \leq E \leq \sqrt{mn} \tag{4.27}$$

and if G is a bipartite graph with n vertices and m edges, then

$$E \leq \frac{4m}{n} + \sqrt{(n-2)\left(2m - \frac{8m^2}{n^2}\right)}. \tag{4.28}$$

4.3.1.1 Nullity and Zero-Energy States

Another characteristic of a graph that is related to an important molecular property is the nullity. The *nullity* of a graph, denoted by $\eta = \eta(G)$, is the algebraic multiplicity of the zero eigenvalue in the spectrum of the adjacency matrix of the graph [14]. This property is very relevant for the stability of alternant unsaturated conjugated hydrocarbons. An alternant unsaturated conjugated hydrocarbon with $\eta = 0$ is predicted to have a closed-shell electron configuration. Otherwise, the respective molecule is predicted to have an open-shell electron configuration, that is, when $\eta > 0$, the molecule has unpaired electrons in the form of radicals that are relevant for several electronic and magnetic properties of materials. In a molecule with an even number of atoms, η is either zero or it is an even positive integer.

The following are a few important facts about the nullity of graphs. Let $M = M(G)$ be the size of the maximum matching of a graph, that is, the maximum number of mutually nonadjacent edges of G. Let T be a tree with $n \geq 1$ vertices. Then,

$$\eta\,(T) = n - 2M. \tag{4.29}$$

If G is a bipartite graph with $n \geq 1$ vertices and no cycle of length $4s$ ($s = 1, 2, \ldots$), then

$$\eta(G) = n - 2M. \tag{4.30}$$

Also for a bipartite graph G with incidence matrix ∇, $\eta(G) = n - 2r\,(\nabla)$, where $r\,(\nabla)$ is the rank of $\nabla = \nabla(G)$. In the particular case of benzenoid graphs Bz, which may contain cycles of length $4s$, the nullity is given by

$$\eta\,(\mathrm{Bz}) = n - 2M. \tag{4.31}$$

The following are some known bounds for the nullity of graphs [15]. Let G be a graph with n vertices and at least one cycle,

$$\eta(G) \leq \begin{cases} n - 2g(G) + 2 & g(G) \equiv 0 \pmod 4, \\ n - 2g(G) & \text{otherwise,} \end{cases} \tag{4.32}$$

where $g(G)$ is the girth of the graph.

If there is a path of length $d\,(p, q)$ between the vertices p and q of G

$$\eta(G) \leq \begin{cases} n - d(p, q) & \text{if } d\,(p, q) \text{ is even,} \\ n - d\,(p, q) - 1 & \text{otherwise.} \end{cases} \tag{4.33}$$

Let G be a simple connected graph of diameter D. Then

$$\eta(G) \leq \begin{cases} n - D & \text{if } D \text{ is even,} \\ n - D - 1 & \text{otherwise.} \end{cases} \tag{4.34}$$

4.3.2 Hubbard Model

Let us now consider one of the most important models in theoretical physics: the Hubbard model. This model accounts for the quantum-mechanical motion of electrons in a solid or conjugated hydrocarbon and includes nonlinear repulsive interactions between electrons. In brief, the interest in this model is due to the fact that it exhibits various interesting phenomena including metal–insulator transition, antiferromagnetism, ferrimagnetism, ferromagnetism, Tomonaga–Luttinger liquid, and superconductivity [16].

The Hubbard model can be seen as an extension of the tight-binding Hamiltonian we have studied in the previous section in which we introduce the electron–electron interactions. To keep things simple, we allow onsite interactions only, that is, we consider one orbital per site and $V_{ijkl} \neq 0$ in (4.15) if, and only if, i, j, k, and l all refer to the same orbital. In this case, the Hamiltonian is written as

$$\mathbf{H} = -t \sum_{i,j,\sigma} A_{ij} \hat{c}^{\dagger}_{i\sigma} \hat{c}_{j\sigma} + U \sum_{i} \hat{c}^{\dagger}_{i\uparrow} \hat{c}_{i\uparrow} \hat{c}^{\dagger}_{i\downarrow} \hat{c}_{i\downarrow}, \tag{4.35}$$

where t is the hopping parameter and $U > 0$ indicates that the electrons repel each other.

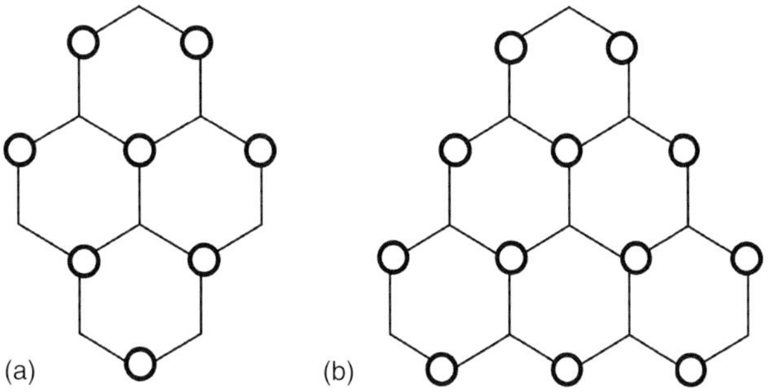

Figure 4.1 Representation of two graphene nanoflakes with closed-shell (a) and open-shell (b) electronic configurations.

Notice that if there is no electron–electron repulsion ($U = 0$), we recover the tight-binding Hamiltonian studied in the previous section. Thus, in that case, all the results given in the previous section are valid for the Hubbard model without interactions. In the case of nonhopping systems, $t = 0$ and the Hamiltonian is reduced to the electron interaction part only. In this case, the remaining Hamiltonian is already in a diagonal form and the eigenstates can be easily obtained. The main difficulty arises when both terms are present in the Hamiltonian. However, in *half-filled systems*, the model has nice properties from a mathematical point of view and a few important results have been proved. These systems have attracted a lot of attention after the discovery of graphene. A system is a half-filled one if the number of electrons is the same as the number of sites, that is, because the total number of electrons can be $2n$, these systems have only a half of the maximum number of electrons allowed. This is particularly the case of graphene and other conjugated aromatic systems. Owing to the œ – ß separation that we have seen in the previous section, these systems can be considered as half-filled, in which each carbon atom provides one ß-electron.

A fundamental result in the theory of half-filled systems is the theorem proved by Lieb [17]. *Lieb's theorem* for repulsive Hubbard model states the following. Let $G = (V, E)$ be a bipartite connected graph representing a Hubbard model, such that $|V| = n$ is even and the nodes of the graph are partitioned into two disjoint subsets V_1 and V_2. We assume that the hopping parameters are nonvanishing and that $U > 0$. Then the ground states of the model are nondegenerate apart from the trivial spin degeneracy, and have total spin $S_{\text{tot}} = ||V_1| - |V_2||/2$.

In order to illustrate the consequences of Lieb's theorem, let us consider two benzenoid systems that can represent graphene nanoflakes. The first of them is realized in the polycyclic aromatic hydrocarbon known as *pyrene* and it is illustrated in Figure 4.1a. The second is a hypothetical graphene nanoflake known as *triangulene* and is illustrated in Figure 4.1b. In both cases, we have divided the bipartite graphs into two subsets, the one marked by empty circles corresponds to the set V_1 and the unmarked nodes form the set V_2. In the structure of pyrene, we can easily check that $|V_1| = |V_2| = 8$ so that the total spin according to Lieb's theorem is $S_{\text{tot}} = 0$. In addition, according to (4.31) given in the previous section pyrene has no zero-energy levels as its nullity is zero, that is, $\eta(\text{Bz}) = 0$. In this case, the mean-field Hubbard model solution for this structure reveals no magnetism.

In the case of triangulene, it can be seen that $|V_1| = 12$ and $|V_2| = 10$, which gives a total spin $S_{\text{tot}} = 1$. In addition, the nullity of this graph is equal to 2, indicating that it has two zero-energy states. The result given by Lieb's theorem indicates that triangulene has a spin-triplet ground state, which means that it has a magnetic moment of $2\mu_B$ per molecule. Thus triangulene and more ß-extended analogs have intramolecular ferromagnetic interactions owing to ß-spin topological structures. Analogs of this molecule have been already obtained in the laboratory [18].

4.4 Graphs in Statistical Physics

The connections between statistical physics and graph theory are extensive and have a long history. A survey on these connections was published in 1971 by Essam [19]; it mainly deals with the Ising model. In the Ising model, we consider a set of particles or "spins," which can be in one of two states. The state of the ith particle is described by the variable σ_i which takes one of the two values ± 1. The connection with graph theory comes from the calculation of the partition function of the model. In this chapter, we consider that the best way of introducing this connection is through a generalization of the Ising model, the Potts model [20, 21].

The Potts model is one of the most important models in statistical physics. In this model, we consider a graph $G = (V, E)$ with each node of which we associate a spin. The spin can have one of q values. The basic physical principle of the model is that the energy between two interacting spins is set to zero for identical spins and it is equal to a constant if they are not. A remarkable property of the Potts model is that for $q = 3, 4$ it exhibits a continuous phase transition between high- and low-temperature phases. In this case, the critical singularities in thermodynamic functions are different from those obtained by using the Ising model. The Potts model has found innumerable applications in statistical physics, for example, in the theory of phase transitions and critical phenomena, but also outside this context in areas such as magnetism, tumor migration, foam behavior, and social sciences.

In the simplest formulation of the Potts model with q states $\{1, 2, \ldots, q\}$, the Hamiltonian of the system can have either of the two following forms:

$$\mathbf{H}_1(\omega) = -J \sum_{(i,j)\in E} \delta\left(\sigma_i, \sigma_j\right), \tag{4.36}$$

$$\mathbf{H}_2(\omega) = J \sum_{(i,j)\in E} \left[1 - \delta\left(\sigma_i, \sigma_j\right)\right], \tag{4.37}$$

where ω is a configuration of the graph, that is, an assignment of a spin to each node of $G = (V, E)$; σ_i is the spin at node i and δ is the Kronecker symbol. The model is called *ferromagnetic* if $J > 0$ and *antiferromagnetic* if $J < 0$. We notice here that the Ising model with zero external field is a special case with $q = 2$, so that the spins are $+1$ and -1.

The probability $p(\omega, \beta)$ of finding the graph in a particular configuration (state) ω at a given temperature is obtained by considering a Boltzmann distribution and it is given by

$$p(\omega, \beta) = \frac{\exp\left(-\beta \mathbf{H}_i(\omega)\right)}{Z_i(G)}, \tag{4.38}$$

where $Z_i(G)$ is the partition function for a given Hamiltonian in the Potts model, that is,

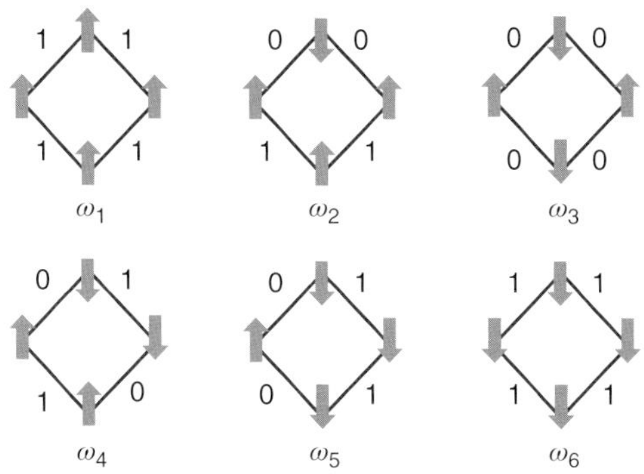

Figure 4.2 Representation of spin configurations in a cycle with four nodes.

$$Z_i(G) = \sum_{\omega} \exp\left(-\beta \mathbf{H}_i(\omega)\right), \tag{4.39}$$

where the sum is over all configurations (states) and $\mathbf{H}_i$ may be either $\mathbf{H}_1$ or $\mathbf{H}_2$. Here $\beta = \left(k_B T\right)^{-1}$, where T is the absolute temperature of the system, and k_B is the Boltzmann constant.

For instance, let us consider all the different spin configurations for a cyclic graph with $n = 4$ as given in Figure 4.2. It may be noted that there are four equivalent configurations for ω_2, ω_4, and ω_5 as well as two equivalent configurations for ω_3. The Hamiltonians $\mathbf{H}_1(\omega)$ for these configurations are

$$\mathbf{H}_1(\omega_1) = -4J; \mathbf{H}_1(\omega_2) = -2J;$$
$$\mathbf{H}_1(\omega_3) = 0; \mathbf{H}_1(\omega_4) = -2J;$$
$$\mathbf{H}_1(\omega_5) = -2J; \mathbf{H}_1(\omega_6) = -4J.$$

Then, the partition function of the Potts model for this graph is

$$Z_1(G) = 12\exp(2\beta J) + 2\exp(4\beta J) + 2. \tag{4.40}$$

It is usual to set $K = \beta J$. The probability of finding the graph in the configuration ω_2 is

$$p\left(\omega_2, \beta\right) = \frac{\exp(2K)}{12\exp(2K) + 2\exp(4K) + 2}. \tag{4.41}$$

The important connection between the Potts model and graph theory comes through the equivalence of this physical model and the graph-theoretic concept of the Tutte polynomial, that is, the partition functions of the Potts model can be obtained in the following form:

$$Z_1(G, q, \beta) = q^{k(G)} v^{n-k(G)} T\left(G; x, y\right), \tag{4.42}$$
$$Z_2(G, q, \beta) = \exp(-mK) Z_1(G, q, \beta), \tag{4.43}$$

where q is the number of spins in the system, $k(G)$ is the number of connected components of the graph, $v = \exp(K) - 1$, n and m are the number of nodes and edges in the graph, respectively, and $T\left(G; x, y\right)$ is the Tutte polynomial, where $x = (q + v)/v$ and $y = \exp(K)$. Proofs of the relationship between the Potts partition function and the Tutte polynomial will not be considered here and the interested reader is directed to the literature to find the details [22].

Let us define the *Tutte polynomial* [23, 24]. First, we define the following graph operations. The *deletion* of an edge e in the graph G, represented by $G - e$, consists of removing the corresponding edge without changing the rest of the graph, that is, the end nodes of the edge remain in the graph. The other operation is the *edge contraction* denoted by G/e, which consists in gluing together the two end nodes of the edge e and then removing e. Both operations, edge deletion and contraction, are commutative, and the operations $G - S$ and G/S, where S is a subset of edges, are well defined. We notice here that the graphs created by these transformations are no longer simple graphs, they are pseudographs that may contain self-loops and multiple edges. Let us also define the following types of edges: a *bridge* is an edge whose removal disconnects the graph. A (self) *loop* is an edge having the two end points incident at the same node. Let us denote by B and L the sets of edges that are bridges or loops in the graph.

Then the Tutte polynomial $T\left(G; x, y\right)$ is defined by the following recursive formulae:

1. $T(G; x, y) = T(G - e; x, y) + T(G/e; x, y)$ if $e \notin B, L$;
2. $T(G; x, y) = x^i y^j$ if $e \in B, L$,

where the exponents i and j represent the number of bridges and self-loops in the subgraph, respectively.

Using this definition, we can obtain the Tutte polynomial for the cyclic graph with four nodes C_4, as illustrated in the Figure 4.3, that is, the Tutte polynomial for C_4 is $T\left(C_4; x, y\right) = x^3 + x^2 + x + y$. We can substitute this expression into (4.42) to obtain the partition function for the Potts model of this graph,

$$Z_1\left(G; 2, \beta\right) = 2^{k(G)} \nu^{n-k(G)} \left[\left(\frac{q+\nu}{\nu}\right)^3 + \left(\frac{q+\nu}{\nu}\right)^2 + \left(\frac{q+\nu}{\nu}\right) + 1 + \nu \right], \tag{4.44}$$

and so we obtain $Z_1(G; 2, \beta) = 12\exp(2K) + 2\exp(4K) + 2$.

The following is an important mathematical result related to the universality of the Tutte polynomial [23, 24]. Let $f(G)$ be a function on graphs having the following properties:

1. $f(G) = 1$ if $|V| = 1$ and $|E| = 0$
2. $f(G) = af(G - e) + bf(G/e)$ if $e \notin B, L$,
3. $f(G \cup H) = f(G)f(H)$;
   $f(G * H) = f(G)f(H)$, where $G * H$

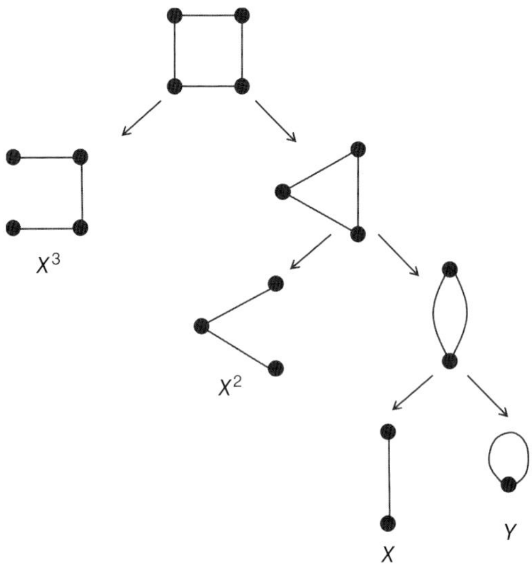

Figure 4.3 Edge deletion and contraction in a cyclic graph with four nodes.

means that G and H shares at most one node.

Then $f(G)$ is an evaluation of the Tutte polynomial, meaning that it is equivalent to the Tutte polynomial with some specific values for the parameters, and takes the form

$$f(G)=a^{m-n+k(G)}b^{n-k(G)}T\left(G;\frac{f(K_2)}{b},\frac{f(L)}{a}\right), \quad (4.45)$$

where L is the graph consisting of a single node with one loop attached, K_2 is the complete graph with two nodes.

More formally, the Tutte polynomial is a *generalized Tutte–Gröthendieck* (T-G for short) invariant. To define the T-G invariant, we need the following concepts. Let S and S' be two disjoint subsets of edges. A minor of G is a graph H that is isomorphic to $(G-S)/S'$. Let Γ be a class of graphs such that if G is in Γ, then any minor of G is also in the class. This class is known as *minor closed*. A graph invariant is a function f on the class of all graphs such that if G and H are isomorphic, then $f(G)=f(H)$. Then, a T-G invariant is a graph invariant f from Γ to a commutative ring $\mathfrak{R}$ with unity, such as the conditions (i)–(iii) above are fulfilled. A graph invariant is a function f on the class of all graphs such that $f(G_1)=f(G_2)$ whenever the graphs G_1 and G_2 are isomorphic. For more details the reader is referred to the specialized literature on this topic.

Some interesting evaluations of the Tutte polynomial are the following:

$T(G;1,1)$	Number of spanning trees of the graph G
$T(G;2,1)$	Number of spanning forests
$T(G;1,2)$	Number of spanning connected subgraphs
$T(G;2,2)$	$2^{\lvert E\rvert}$

Let us now consider a proper coloring of a graph G, which is an assignment of a color to each node of G such that any two adjacent nodes have different colors. The *chromatic polynomial* $\chi(G;q)$ of the graph G is the number of ways in which q colors can be assigned to the nodes of G such that no two adjacent nodes have the same color. The following are two interesting characteristics of the chromatic polynomial:

1. $\chi(G;q)=\chi(G-e;q)-\chi(G/e;q)$,
2. $\chi(G;q)=q^n$ for the trivial graph on n nodes.

Thus, the chromatic polynomial fulfills the same contraction/deletion rules as the Tutte polynomial. Indeed, the chromatic polynomial is an evaluation of the Tutte polynomial,

$$\chi(G;q)=q^{k(G)}(-1)^{n-k(G)}T(G;1-q,0). \quad (4.46)$$

To see the connection between the Potts model and the chromatic polynomial, we have to consider the Hamiltonian $H_1(G;\omega)$ in the zero temperature limit, that is, $T\to 0$ $(\beta\to\infty)$. When $\beta\to\infty$, the only spin configurations that contribute to the partition function are the ones in which adjacent spins have different values. Then we have that $Z_1(G;q,\beta)\to 1$ in the antiferromagnetic model ($J<0$). Thus, $Z_1(G;q,\beta\to\infty)$ counts the number of proper colorings of the graph using q colors. The partition function in the $T=0$ limit of the Potts model is given by the chromatic polynomial

$$\begin{aligned}Z_1(G;q,-1)&=\chi(G)\\&=(-1)^{k(G)}(-1)^nT(G;1-q,0).\end{aligned} \quad (4.47)$$

4.5 Feynman Graphs

When studying elementary-particle physics, the calculation of higher-order corrections in perturbative quantum field theory naturally leads to the evaluation of Feynman integrals. Feynman integrals are associated to *Feynman graphs*, which are graphs $G = (V, E)$ with n nodes and m edges and some special characteristics [25–27]. For instance, the edges play a fundamental role in the Feynman graphs as they represent the different particles, such as fermions (edges with arrows), photons (wavy lines), gluons (curly lines). Scalar particles are represented by simple lines. Let us assign a D-dimensional momentum vector q_j and a number representing the mass m_j to the jth edge representing the jth particle, where D is the dimension of the space-time. In the theory of Feynman graphs, the nodes with degree one are not represented, leaving the edge without the end node. This edge is named an *external edge* (they are sometimes called *legs*). The rest of edges are called *internal*. Also, nodes of degree two are omitted as they represent mass insertions. Thus, Feynman graphs contain only nodes of degree $k \geq 3$, which represent the interaction of k particles. At each of these nodes, the sum of all momenta flowing into the node equals that of the momenta flowing out of it. As usual the number of basic cycles, here termed *loops*, is given by the cyclomatic number $l = m - n + C$, where C is the number of connected components of the graph.

Here we will only consider Feynman graphs with scalar propagators and we refer to them as *scalar theories*. In scalar theories, the D-dimensional Feynman integral has the form

$$I_G = \left(\mu^2\right)^{\nu - lD/2} \int \prod_{r=1}^{l} \frac{d^D k_r}{i\pi^{D/2}} \prod_{j=1}^{n} \frac{1}{\left(-q_j^2 + m_j^2\right)^{\nu_j}}, \tag{4.48}$$

where l is the number of loops (basic cycles) in the Feynman diagram, μ is an arbitrary scale parameter used to make the expressions dimensionless, ν_j is a positive integer number that gives the power to which the propagator occurs, $\nu = \nu_1 + \cdots + \nu_m$, k_r is the independent loop momentum, m_j is the mass of the jth particle, and

$$q_j = \sum_{j=1}^{l} \rho_{ij} k_j + \sum_{j=1}^{m} \sigma_{ij} p_j, \quad \rho_{ij}, \sigma_{ij} \in \{-1, 0, 1\}, \tag{4.49}$$

represents the momenta flowing through the internal lines.

The correspondence between the Feynman integral and the Feynman graph is as follows. An internal edge represents a propagator of the form

$$\frac{i}{q_j^2 - m_j^2}, \tag{4.50}$$

whereby abusing of the notation q_j^2 represents the inner product of the momentum vector with itself, that is, $q_j^2 = q_j \cdot q_j^T$. Notice that this is a relativistic propagator that represents a Greens function for integrations over space and time.

Nodes and external edges have weights equal to one. For each internal momentum not constrained by momentum conservation there is also an integration associated.

Now, in order to compute the integral equation (4.48), we need to assign a (real or complex) variable x_j to each internal edge, which are known as the *Feynman parameters*. Then, we need to use the Feynman parameter trick for each propagator and evaluate the integrals over the loop momenta $k_1, \dots, k_l$. As a consequence, we obtain

$$I_G = \frac{\Gamma(\nu - lD/2)}{\prod_{j=1}^{m}\Gamma(\nu_j)} \int_{x_j \geq 0} \left(\prod_{j=1}^{m} dx_j x_j^{\nu_j - 1} \right) \times \delta\left(1 - \sum_{i=1}^{m} x_i\right) \frac{U^{\nu-(l+1)D/2}}{F^{\nu - lD/2}}. \quad (4.51)$$

The real connection with the theory of graphs comes from the two terms U and F, which are graph polynomials, known as the *first and second Symanzik polynomials* (sometimes called *Kirchhoff–Symanzik polynomials*). We will now specify some methods for obtaining these polynomials in Feynman graphs.

4.5.1 Symanzik Polynomials and Spanning Trees

The first *Symanzik polynomial* can be obtained by considering all spanning trees in the Feynman graph. Let τ_1 be the set of spanning trees in the Feynman graph G. Then,

$$U = \sum_{T \in \tau_1} \prod_{e_j \notin T} x_j, \quad (4.52)$$

where T is a spanning tree and x_j is the Feynman parameter associated with edge e_j.

In order to obtain the second Symanzik polynomial F, we have to consider the set of spanning 2-forest τ_2 in the Feynman graph. A *spanning 2-forest* is a spanning forest formed by only two trees. Then, the elements of τ_2 are denoted by (T_i, T_j). The second Symanzik polynomial is given by

$$F = F_0 + U \sum_{i=1}^{m} x_i \frac{m_i^2}{\mu^2}. \quad (4.53)$$

The term F_0 is a polynomial obtained from the sets of spanning 2-forests of G in the following way: let P_{T_i} be the set of external momenta attached to the spanning tree T_i, which is part of the spanning 2-forest (T_i, T_j). Let $p_k \cdot p_r$ be the Minkowski scalar product of the two momenta vectors associated with the edges e_k and e_r, respectively. Then

$$F_0 = \sum_{(T_i,T_j)\in\tau_2} \left(\prod_{e_k \notin (T_i,T_j)} x_k \right) \left(\sum_{p_k \in P_{T_i}} \sum_{p_r \in P_{T_j}} \frac{p_k \cdot p_r}{\mu^2} \right). \quad (4.54)$$

Let us now show how to obtain the Symanzik polynomials for the simple Feynman graph illustrated in the Figure 4.4. For the sake of simplicity, we take all internal masses to be zero.

We first obtain all spanning trees of this graph, which are given in Figure 4.5.

Hence, the first Symanzik polynomial is obtained as follows:

$$\begin{aligned} U &= x_1x_2 + x_1x_3 + x_1x_5 + x_2x_4 + x_2x_5 \\ &\quad + x_3x_4 + x_3x_5 + x_4x_5 \\ &= (x_1 + x_4)(x_2 + x_3) \\ &\quad + (x_1 + x_2 + x_3 + x_4)x_5. \end{aligned} \quad (4.55)$$

Now, for the second Symanzik polynomial, we obtain all the spanning 2-forests of the graph, which are given in Figure 4.6.

We should notice that the terms $x_1x_2x_5$ and $x_3x_4x_5$ do not contribute to F_0 because the momentum sum flowing through all cut edges is zero. Thus, we can obtain F_0 as follows:

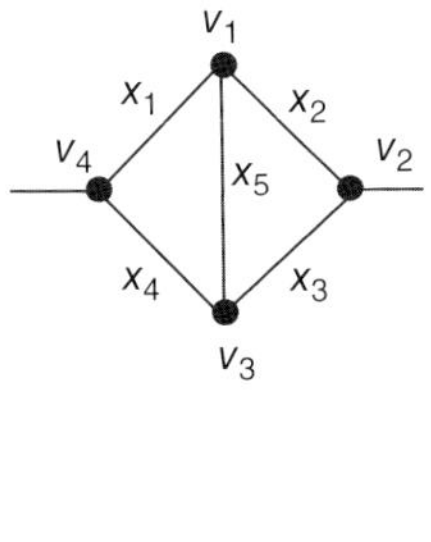

Figure 4.4 Illustration of a Feynman graph with four nodes, five internal, and two external edges. The Feynman parameters are represented by x_i on each internal edge.

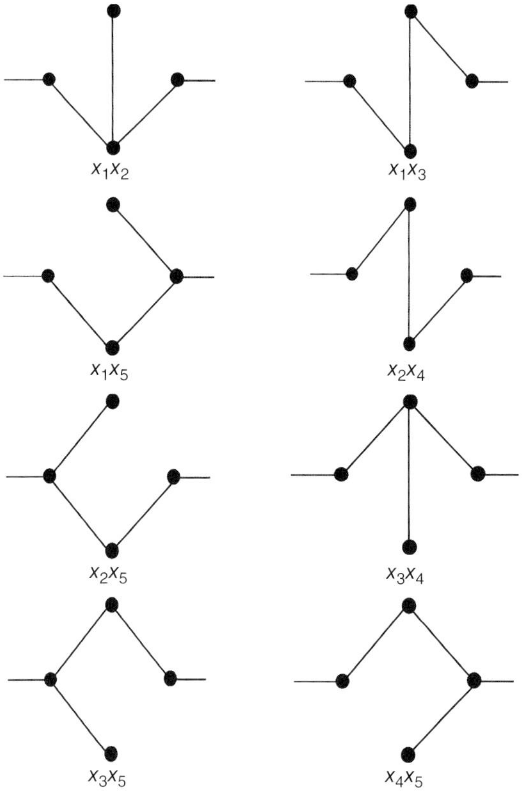

Figure 4.5 Spanning trees of the Feynman graphs represented in Figure 4.4.

$$F = F_0 = (x_1x_2x_3 + x_1x_2x_4 + x_1x_3x_4 + x_1x_3x_5 + x_1x_4x_5 + x_2x_3x_4 + x_2x_3x_5 + x_2x_4x_5)\left(\frac{-p^2}{\mu^2}\right)$$
$$= [(x_1+x_2)(x_3 + x_4)x_5 + x_1x_4(x_2 + x_3) + x_2x_3(x_1 + x_4)]\left(\frac{-p^2}{\mu^2}\right). \tag{4.56}$$

4.5.2 Symanzik Polynomials and the Laplacian Matrix

Another graph-theoretic way of obtaining the Symanzik polynomials is through the use of the Laplacian matrix. The Laplacian matrix for the Feynman graphs is defined as usual for any weighted graph. For instance, for the Feynman graph given in Figure 4.4, the Laplacian matrix is

$$\mathbf{L} = \begin{pmatrix} x_1 + x_2 + x_5 & -x_2 & -x_5 & -x_1 \\ -x_2 & x_2 + x_3 & -x_3 & 0 \\ -x_5 & -x_3 & x_3 + x_4 + x_5 & -x_4 \\ -x_1 & 0 & -x_4 & x_1 + x_4 \end{pmatrix}. \tag{4.57}$$

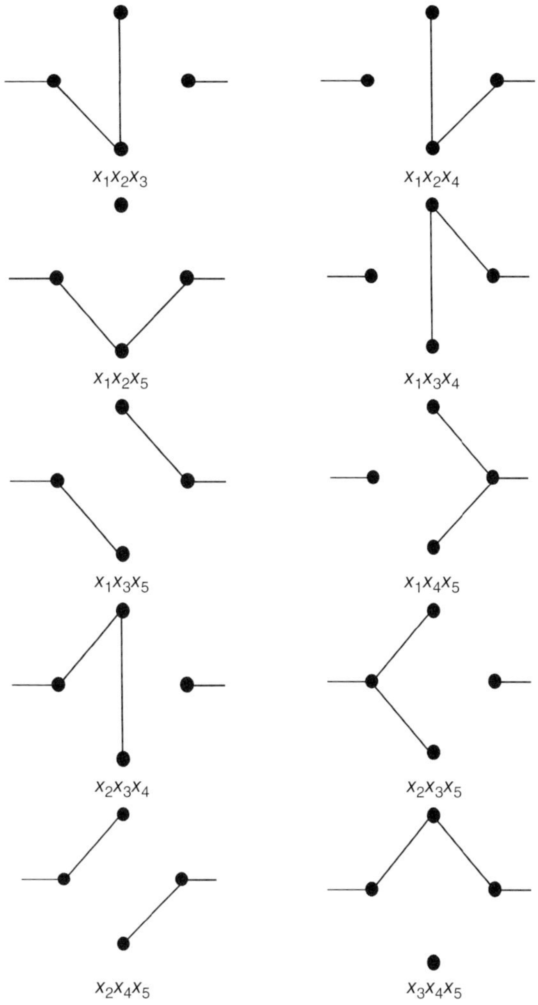

Figure 4.6 Spanning 2-forest of the Feynman graph represented in Figure 4.4.

Then, we can define the auxiliary polynomial $K = \det \mathbf{L}[i]$, where $\mathbf{L}[i]$ denotes the minor of the Laplacian matrix obtained by removing the ith row and column of $\mathbf{L}$. This polynomial is known as the *Kirchhoff polynomial* of the graph and it is easy to see that it can be defined by

$$K = \sum_{T \in \tau_1} \prod_{e_j \in T} x_j. \tag{4.58}$$

For instance,

$$\begin{aligned} K &= \det \mathbf{L}[1] \\ &= \begin{vmatrix} x_2 + x_3 & -x_3 & 0 \\ -x_3 & x_3 + x_4 + x_5 & -x_4 \\ 0 & -x_4 & x_1 + x_4 \end{vmatrix} \\ &= x_1x_2x_3 + x_1x_2x_4 + x_1x_2x_5 + x_1x_3x_4 \\ &\quad + x_1x_3x_5 + x_2x_3x_4 + x_2x_4x_5 \\ &\quad + x_3x_4x_5. \end{aligned} \tag{4.59}$$

We transform the Kirchhoff polynomial into the first Symanzik polynomial by setting $U = x_1 \cdots x_m K\left(x_1^{-1}, \ldots, x_m^{-1}\right)$, that is,

$$U = \frac{x_1 x_2 x_3 x_4 x_5}{x_1 x_2 x_3} + \frac{x_1 x_2 x_3 x_4 x_5}{x_1 x_2 x_4} + \cdots$$
$$+ \frac{x_1 x_2 x_3 x_4 x_5}{x_3 x_4 x_5}$$
$$= x_1 x_2 + x_1 x_3 + x_1 x_5 + x_2 x_4 + x_2 x_5$$
$$+ x_3 x_4 + x_3 x_5 + x_4 x_5$$
$$= (x_1 + x_4)(x_2 + x_3) + (x_1 + x_2 + x_3 + x_4)x_5. \tag{4.60}$$

$$\tilde{\mathbf{L}} = \begin{pmatrix} x_1 + x_2 + x_5 & -x_2 & -x_5 & -x_1 \\ -x_2 & x_2 + x_3 + z_1 & -x_3 & 0 \\ -x_5 & -x_3 & x_3 + x_4 + x_5 & -x_4 \\ -x_1 & 0 & -x_4 & x_1 + x_4 + z_2 \end{pmatrix}, \tag{4.64}$$

To calculate the second Symanzik polynomial using the Laplacian matrix, we have to introduce some modifications. First assign a new parameter z_j to each of the external edges of the Feynman graph. Now, build a diagonal matrix whose diagonal are $D_{ii} = \sum_{j \to i} z_j$, that is the ith diagonal entry of $\mathbf{D}$ represents the sum of the parameters z_j for all the external edges incident with the node i. Modify the Laplacian matrix as follows: $\tilde{\mathbf{L}} = \mathbf{L} + \mathbf{D}$. The modified Laplacian matrix $\tilde{\mathbf{L}}$ is the minor of a Laplacian matrix constructed for a modification of the Feynman graph in which all rows and columns corresponding to the external edges are removed [26, 27]. The determinant of the modified Laplacian matrix is

$$W = \det \tilde{\mathbf{L}}, \tag{4.61}$$

and let us expand it in a series of polynomials homogeneous in the variables z_j, such that

$$W = W^{(0)} + W^{(1)} + W^{(2)} + \cdots + W^{(t)}, \tag{4.62}$$

where t is the number of external edges. Then the Symanzik polynomials are
$$U = x_1 \cdots x_m W_j^{(1)}(x_1^{-1}, \ldots, x_m^{-1}) \text{ for any } j,$$
and
$$F_0 = x_1 \cdots x_m \sum_{(j,k)} \left(\frac{p_j \cdot p_k}{\mu^2} \right) W_{(j,k)}^{(2)} \left(x_1^{-1}, \ldots, x_m^{-1} \right). \tag{4.63}$$

For the Feynman graph given in the previously analyzed example, we have

and $W = \det \tilde{\mathbf{L}} = W^{(1)} + W^{(2)}$, where

$$W^{(1)} = (z_1 + z_2)(x_1 x_2 x_3 + x_1 x_2 x_4 + x_1 x_3 x_4$$
$$+ x_2 x_3 x_4 + x_1 x_2 x_5 + x_1 x_3 x_5$$
$$+ x_2 x_4 x_5 + x_3 x_4 x_5), \tag{4.65}$$
$$W^{(2)} = z_1 z_2 (x_1 x_3 + x_2 x_3 + x_1 x_4 + x_2 x_4$$
$$+ x_1 x_5 + x_2 x_5 + x_3 x_5 + x_4 x_5). \tag{4.66}$$

With this information, the first and second Symanzik polynomials can be easily obtained.

4.5.3
Symanzik Polynomials and Edge Deletion/Contraction

The Symanzik polynomials can also be obtained through the graph transformations used to define the Tutte polynomial, that is, the Symanzik polynomials obey the rules for edge deletion and contraction operations that we encountered in the previous section. Recall that the deletion of an edge e in the graph G is represented by $G - e$, and the edge contraction denoted by G/e, and that B and L are the sets of edges that are bridges or loops in the graph (see Section 4.4). Then

$$U(G) = U\left(\frac{G}{e_j}\right) + x_j U\left(G - e_j\right), \qquad (4.67)$$

$$F_0(G) = F_0\left(\frac{G}{e_j}\right) + x_j F_0\left(G - e_j\right), \qquad (4.68)$$

for any $e_j \notin B, L$.

Finally, let us mention that there exist factorization theorems for the Symanzik polynomials that are based on a beautiful theorem due to Dodgson [28]. The reader is reminded that Charles L. Dodgson is better known as *Lewis Carroll* who has delighted many generations with his *Alice in Wonderland.* These factorization theorems are not given here and the reader is directed to the excellent reviews of Bogner and Weinzierl for details [26, 27].

4.6 Graphs and Electrical Networks

The relation between electrical networks and graphs is very natural and is documented in many introductory texts on graph theory. The idea is that a simple *electrical network* can be represented as a graph $G = (V, E)$ in which we place a fixed electrical resistor at each edge of the graph. Therefore, these networks can also be called *resistor networks.* Let us suppose that we connect a battery across the nodes u and v. There are several parameters of an electrical network that can be considered in terms of graph-theoretic concepts but we concentrate here on one that has important connections with other parameters of relevance in physics, namely, the effective resistance [29]. Let us calculate the effective resistance $\Omega(u, v)$ between two nodes by using the Kirchhoff and Ohm laws. For the sake of simplicity, we always consider here resistors of 1 ohm. In the simple case of a tree, the effective resistance is simply the sum of the resistances along the path connecting u and v, that is, for a tree, $\Omega(u, v) = d(u, v)$, where $d(u, v)$ is the shortest-path distance between the corresponding nodes (number of links in the shortest path connecting both nodes). However, in the case of two nodes connected by multiple routes, the effective resistance $\Omega(u, v)$ can be obtained by using Kirchhoff's laws. A characteristic of the effective resistance $\Omega(u, v)$ is that it decreases with the increase of the number of routes connecting u and v. Thus, in general, $\Omega(u, v) \le d(u, v)$.

An important result about the effective resistance was obtained by Klein and Randić [30]: the effective resistance is a proper distance between the pairs of nodes of a graph, that is,

1. $\Omega(u, v) \ge 0$ for all $u \in V(G), v \in V(G)$.
2. $\Omega(u, v) = 0$ if and only if $u = v$.
3. $\Omega(u, v) = \Omega(v, u)$ for all $u \in V(G), v \in V(G)$.
4. $\Omega(u, w) \le \Omega(u, v) + \Omega(v, w)$ for all $u \in V(G), v \in V(G), w \in V(G)$.

The *resistance distance* $\Omega(u, v)$ between a pair of nodes u and v in a connected component of a network can be calculated by using the Moore–Penrose generalized inverse $\mathbf{L}^+$ of the graph Laplacian $\mathbf{L}$:

$$\Omega(u, v) = \mathbf{L}^+(u, u) + \mathbf{L}^+(v, v) - 2\mathbf{L}^+(u, v), \qquad (4.69)$$

for $u \ne v$.

Another way of computing the resistance distance for a pair of nodes in a network is as follows. Let $\mathbf{L}(G - u)$ be the matrix resulting from removing the uth row and column of the Laplacian and let $\mathbf{L}(G - u - v)$ be the matrix resulting from removing both the uth and vth rows and columns of **L**. Then it has been proved [31] that

$$\Omega(u,v) = \frac{\det \mathbf{L}(G-u-v)}{\det \mathbf{L}(G-u)}, \quad (4.70)$$

Notice that $\det \mathbf{L}(G-u)$ is the Kirchhoff (Symanzik) polynomial we discussed in the previous section. Yet another way for computing the resistance distance between a pair of nodes in the network is given on the basis of the Laplacian spectra [32]

$$\Omega(u,v) = \sum_{k=2}^{n} \frac{1}{\mu_k} \left[U_k(u) - U_k(v)\right]^2, \quad (4.71)$$

where $U_k(u)$ is the uth entry of the kth orthonormal eigenvector associated to the Laplacian eigenvalue μ_k, written in the ordering $0 = \mu_1 < \mu_2 \leq \cdots \leq \mu_n$.

The resistance distance between all pairs of nodes in the network can be represented in the resistance matrix $\boldsymbol{\Omega}$ of the network. This matrix can be written as

$$\boldsymbol{\Omega} = |\mathbf{1}\rangle \operatorname{diag}\left\{\left[\mathbf{L} + \left(\frac{1}{n}\right)\mathbf{J}\right]^{-1}\right\}^{\mathrm{T}} + \operatorname{diag}\left[\mathbf{L} + \left(\frac{1}{n}\right)\mathbf{J}\right]^{-1} \langle\mathbf{1}| - 2\left(\mathbf{L} + \left(\frac{1}{n}\right)\mathbf{J}\right)^{-1}, \quad (4.72)$$

where $\mathbf{J} = |\mathbf{1}\rangle\langle\mathbf{1}|$ is a matrix having all entries equal to 1.

For the case of connected networks, the resistance distance matrix can be related to the Moore–Penrose inverse of the Laplacian as shown by Gutman and Xiao [33]:

$$\mathbf{L}^{+} = -\frac{1}{2}\left[\boldsymbol{\Omega} - \frac{1}{n}(\boldsymbol{\Omega}\mathbf{J} + \mathbf{J}\boldsymbol{\Omega}) + \frac{1}{n^2}\mathbf{J}\boldsymbol{\Omega}\mathbf{J}\right], \quad (4.73)$$

where **J** is as above.

The resistance distance matrix is a matrix of squared *Euclidean distances.* A matrix $\mathbf{M} \in \mathbb{R}^{n \times n}$ is said to be *Euclidean* if there is a set of vectors $x_1, \ldots, x_n$ such that $M_{ij} = \|x_i - x_j\|^2$. Because it is easy to construct vectors such that $\Omega_{ij} = \|x_i - x_j\|^2$ the resistance distance matrix is squared Euclidean and the resistance distance satisfies the weak triangle inequality

$$\Omega_{ik}^{1/2} \leq \Omega_{ij}^{1/2} + \Omega_{jk}^{1/2}, \quad (4.74)$$

for every pair of nodes in the network.

As we noted in the introduction to this section, effective resistance has connections with other concepts that are of relevance in the applications of mathematics in physics. One of these connections is between the resistance distance and Markov chains. In particular, the resistance distance is proportional to the expected commute time between two nodes for a Markov chain defined by a weighted graph [29, 34]. If w_{uv} be the weight of the edge $\{u,v\}$, the probability of transition between u and v in the Markov chain defined on the graph is

$$P_{uv} = \frac{w_{uv}}{\sum\limits_{u,v \in E} w_{uv}}. \quad (4.75)$$

The *commuting time* is the time taken by "information" starting at node u to return to it after passing through node v. The expected commuting time $\hat{C}_{uv}$ is related to the resistance distance [29, 34] by

$$\hat{C}_{uv} = 2\left(\mathbf{1}^T \mathbf{w}\right)\Omega(u,v), \quad (4.76)$$

where **1** is vector of 1's and **w** is the vector of link weights. Note that if the network is unweighted $\hat{C}_{uv} = 2m\Omega(u,v)$.

4.7 Graphs and Vibrations

In this section, we develop some connections between vibrational analysis, which is important in many areas of physics ranging from classical to quantum mechanics, and the spectral theory of graphs. Here we consider the one-dimensional case with a graph $G = (V, E)$ in which every node represents a ball of mass m and every edge represents a spring with the spring constant $m\omega^2$ connecting two balls. The ball–spring network is assumed to be submerged in a thermal bath at temperature T. The balls in the graph oscillate under thermal excitation. For the sake of simplicity, we assume that there is no damping and no external forces are applied to the system. Let x_i, $i = 1, 2, \ldots, n$ be the coordinates of each node; this measures the displacement of the ball i from its equilibrium state $x_i = 0$. For a complete guide to the results to be presented here, the reader is directed to Estrada *et al.* [35].

4.7.1 Graph Vibrational Hamiltonians

Let us start with a Hamiltonian of the oscillator network in the form

$$\mathbf{H}_A = \sum_i \left[\frac{p_i^2}{2m} + (K - k_i) \frac{m\omega^2 x_i^2}{2} \right] + \frac{m\omega^2}{2} \sum_{\substack{i,j \\ (i<j)}} A_{ij} (x_i - x_j)^2, \quad (4.77)$$

where k_i is the degree of the node i and K is a constant satisfying $K \geq \max_i k_i$. The second term in the right-hand side is the potential energy of the springs connecting the balls, because $x_i - x_j$ is the extension or the contraction of the spring connecting the nodes i and j. The first term in the first set of square parentheses is the kinetic energy of the ball i, whereas the second term in the first set of square parentheses is a term that avoids the movement of the network as a whole by tying the network to the ground. We add this term because we are only interested in small oscillations around the equilibrium; this will be explained below again.

The Hamiltonian equation (4.77) can be rewritten as

$$\mathbf{H}_A = \sum_i \left(\frac{p_i^2}{2m} + \frac{Km\omega^2}{2} x_i^2 \right) - \frac{m\omega^2}{2} \sum_{i,j} x_i A_{ij} x_j. \quad (4.78)$$

Let us next consider the Hamiltonian of the oscillator network in the form

$$\mathbf{H}_L = \sum_i \frac{p_i^2}{2m} + \frac{m\omega^2}{2} A_{ij} (x_i - x_j)^2 \quad (4.79)$$

instead of the Hamiltonian $\mathbf{H}_A$ in (4.78). Because the Hamiltonian $\mathbf{H}_L$ lacks the springs that tie the whole network to the ground (the second term in the first set of parentheses in the right-hand side of (4.78), this network can undesirably move as a whole. We will deal with this motion shortly.

The expansion of the Hamiltonian equation (4.79) as in (4.77) and (4.78) now gives

$$\mathbf{H}_L = \sum_i \frac{p_i^2}{2m} + \frac{m\omega^2}{2} \sum_{i,j} x_i L_{ij} x_j, \quad (4.80)$$

where L_{ij} denotes an element of the network Laplacian $\mathbf{L}$.

4.7.2 Network of Classical Oscillators

We start by considering a network of classical harmonic oscillators with the Hamiltonian H_A. Here the momenta p_i and the coordinates x_i are independent variables, so that the integration of the factor

$$\prod \exp \left[-\beta \left(\frac{p_i^2}{2m} \right) \right] \quad (4.81)$$

over the momenta $\{p_i\}$ reduces to a constant term, which does not affect the integration over $\{x_i\}$. As a consequence, we do not have to consider the kinetic energy and we can write the Hamiltonian in the form

$$\mathbf{H}_A = \frac{m\omega^2}{2} x^T (K\mathbf{I} - \mathbf{A}) x, \tag{4.82}$$

where $x = (x_1, x_2, \ldots, x_n)^T$ and I is the $n \times n$ identity matrix.

The partition function is given by

$$\begin{aligned} Z &= \int \mathrm{e}^{-\beta \mathbf{H}_A} \prod_i \mathrm{d}x_i \\ &= \int \mathrm{d}x \exp\left(-\frac{\beta m\omega^2}{2} x^T (K\mathbf{I} - \mathbf{A}) x \right), \end{aligned} \tag{4.83}$$

where the integral is an n-fold one and can be evaluated by diagonalizing the matrix $\mathbf{A}$. The adjacency matrix can be diagonalized by means of an orthogonal matrix $\mathbf{O}$ as in

$$\mathbf{\Lambda} = \mathbf{O} (K\mathbf{I} - \mathbf{A}) \mathbf{O}^T, \tag{4.84}$$

where $\mathbf{\Lambda}$ is the diagonal matrix with eigenvalues λ_μ of $(K\mathbf{I} - \mathbf{A})$ on the diagonal. Let us consider that K is sufficiently large, so that we can make all eigenvalues λ_μ positive. By defining a new set of variables y_μ by $y = \mathbf{O}x$ and $x = \mathbf{O}^T y$, we can transform the Hamiltonian equation (4.82) to the form

$$\begin{aligned} \mathbf{H}_A &= \frac{m\omega^2}{2} y^T \mathbf{\Lambda} y \\ &= \frac{m\omega_0^2}{2} \sum_\mu y_\mu^2 + \frac{m\omega^2}{2} \sum_\mu \lambda_\mu y_\mu^2. \end{aligned} \tag{4.85}$$

Then the integration measure of the n-fold integration in (4.83) is transformed as $\prod_i \mathrm{d}x_i = \prod_\mu \mathrm{d}y_\mu$, because the Jacobian of the orthogonal matrix $\mathbf{O}$ is unity. Therefore, the multifold integration in the partition function (4.83) is decoupled to give

$$Z = \prod_\mu \sqrt{\frac{2\pi}{\beta m\omega^2 \lambda_\mu}}, \tag{4.86}$$

which can be rewritten in terms of the adjacency matrix as

$$Z = \left(\frac{2\pi}{\beta m} \omega^2 \right)^{n/2} \frac{1}{\sqrt{\det (K\mathbf{I} - \mathbf{A})}}. \tag{4.87}$$

As we have made all the eigenvalues of $(K\mathbf{I} - \mathbf{A})$ positive, its determinant is positive.

Now we define an important quantity, the mean displacement of a node from its equilibrium position. It is given by

$$\left\langle x_p^2 \right\rangle = \frac{1}{Z} \int x_p^2 \mathrm{e}^{-\beta H_A} \prod_i \mathrm{d}x_i, \tag{4.88}$$

which, by using the spectral decomposition of $\mathbf{A}$, yields

$$\left\langle x_p^2 \right\rangle = \frac{1}{Z} \int \left[\sum_\sigma \left(\mathbf{O}^T\right)_{p\sigma} y_\sigma \right]^2 \mathrm{e}^{-\beta \mathbf{H}_A} \prod_\mu \mathrm{d}y_\mu \tag{4.89}$$

In the integrand, the odd functions with respect to y_μ vanish. Therefore, only the terms of y_σ^2 survive after integration in the expansion of the square parentheses in the integrand. This gives

$$\begin{aligned} \left\langle x_p^2 \right\rangle &= \frac{1}{Z} \sum_\sigma O_{\sigma p}^2 \int y_\sigma^2 \exp\left(-\frac{\beta m\omega^2}{2} \lambda_\sigma y_\sigma^2 \right) \mathrm{d}y_\sigma \\ &\times \prod_{\mu (\neq \sigma)} \left[\int \exp\left(-\frac{\beta m\omega^2}{2} \lambda_\mu y_\mu^2 \right) \mathrm{d}y_\mu \right]. \end{aligned} \tag{4.90}$$

Comparing this expression with (4.86), we have

$$\left\langle x_p^2 \right\rangle = \frac{1}{\beta m K\omega^2} \left[\left(\mathbf{I} - \frac{\mathbf{A}}{K} \right)^{-1} \right]_{pp}. \tag{4.91}$$

The mean node displacement may be given by the thermal Green's function in the framework of classical mechanics by

$$\left\langle x_p^2 \right\rangle = \frac{1}{\beta K m\omega^2}\left[\left(\mathbf{I}-\frac{\mathbf{A}}{K}\right)^{-1}\right]_{pq}. \quad (4.92)$$

This represents a correlation between the node displacements in a network due to small thermal fluctuations.

The same calculation using the Hamiltonian equation (4.80) gives

$$\left\langle x_p^2 \right\rangle' = \frac{1}{\beta m\omega^2}\left(\mathbf{L}^{+}\right)_{pq}, \quad (4.93)$$

where $\mathbf{L}^{+}$ is the Moore–Penrose generalized inverse of the Laplacian.

4.7.3 Network of Quantum Oscillators

Here we consider the quantum-mechanical version of the Hamiltonian $\mathbf{H}_A$ in (4.78) by considering that the momenta p_j and the coordinates x_i are not independent variables. In this case, they are operators that satisfy the commutation relation,

$$\left[x_i, p_j\right] = i\hbar\delta_{ij}. \quad (4.94)$$

We use the boson creation and annihilation operators $a_i^{\dagger}$ and a_i, which allow us to write the coordinates and momenta as

$$x_i = \sqrt{\frac{\hbar}{2m\Omega}}\left(a_i^{\dagger} + a_i\right), \quad (4.95)$$

$$p_i = \sqrt{\frac{\hbar}{2m\Omega}}\left(a_i^{\dagger} - a_i\right), \quad (4.96)$$

where $\Omega = \sqrt{K/m\omega}$. The commutation relation (4.94) yields

$$\left[a_i, a_j^{\dagger}\right] = \delta_{ij}. \quad (4.97)$$

With the use of these operators, we can recast the Hamiltonian equation (4.78) into the form

$$\mathbf{H}_A = \sum_i \hbar\Omega\left(a_i^{\dagger}a_i + \frac{1}{2}\right) - \frac{\hbar\omega^2}{4\Omega}\sum_{i,j}\left(a_i^{\dagger}+a_i\right)A_{ij}\left(a_j^{\dagger}+a_j\right). \quad (4.98)$$

Using the spectral decomposition of the adjacency matrix, we generate a new set of boson creation and annihilation operators given by

$$b_\mu = \sum_i O_{\mu i}a_i = \sum_i a_i\left(\mathbf{O}^T\right)_{i\mu}, \quad (4.99)$$

$$b_\mu^{\dagger} = \sum_i O_{\mu i}a_i^{\dagger} = \sum_i a_i^{\dagger}\left(\mathbf{O}^T\right)_{i\mu}, \quad (4.100)$$

Applying the transformations ((4.99) and (4.100)) to the Hamiltonian equation (4.98), we can decouple it as

$$\mathbf{H}_A = \sum_\mu \mathbf{H}_\mu, \quad (4.101)$$

with

$$\mathbf{H}_\mu = \hbar\Omega\left[1 + \frac{\omega^2}{2\Omega^2}\left(\lambda_\mu - K\right)\right]\left(b_\mu^{\dagger}b_\mu + \frac{1}{2}\right) + \frac{\hbar\omega^2}{4\Omega}\left(\lambda_\mu - K\right)\left[\left(b_\mu^{\dagger}\right)^2 + \left(b_\mu\right)^2\right]. \quad (4.102)$$

In order to go further, we now introduce an approximation in which each mode of oscillation does not get excited beyond the first excited state. In other words, we restrict ourselves to the space spanned by the ground state (the vacuum) $|\mathrm{vac}\rangle$ and the first excited states $b_\mu^{\dagger}|\mathrm{vac}\rangle$. Then the second term of the Hamiltonian equation (4.102) does not contribute and we therefore have

$$\mathbf{H}_\mu = \hbar\Omega\left[1 + \frac{\omega^2}{2\Omega^2}\left(\lambda_\mu - K\right)\right]\left(b_\mu^{\dagger}b_\mu + \frac{1}{2}\right) \quad (4.103)$$

within this approximation. This approximation is justified when the energy level spacing $\hbar\Omega$ is much greater than the energy scale of external disturbances, (specifically the temperature fluctuation $k_B T = 1/\beta$, in assuming the physical metaphor that the system is submerged in a thermal bath at the temperature T), as well as than the energy of the network springs $\hbar\omega$, that is, $\beta\hbar\Omega \gg 1$ and $\Omega \gg \omega$. This happens when the mass of each oscillator is small, when the springs connecting to the ground $m\Omega^2$ are strong, and when the network springs $m\omega^2$ are weak. Then an oscillation of tiny amplitude propagates over the network. We are going to work in this limit hereafter.

We are now in a position to compute the partition function as well as the thermal Green's function quantum-mechanically. As stated above, we consider only the ground state and one excitation from it. Therefore we have the quantum-mechanical partition function in the form

$$Z^A = \langle \mathrm{vac}| \, \mathrm{e}^{-\beta \mathbf{H}_A} \, |\mathrm{vac}\rangle = \prod_{\mu} \exp\left\{ -\frac{\beta\hbar\Omega}{2} \left[1 + \frac{\omega^2}{2\Omega^2} \left(\lambda_\mu - K \right) \right] \right\}. \quad (4.104)$$

The diagonal thermal Green's function giving the mean node displacement in the quantum-mechanical framework is given by

$$\left\langle x_p^2 \right\rangle = \frac{1}{Z} \langle \mathrm{vac}| \, a_p \mathrm{e}^{-\beta \mathbf{H}_A} a_p^\dagger \, |\mathrm{vac}\rangle , \quad (4.105)$$

which indicates how much an excitation at the node p propagates throughout the graph before coming back to the same node and being annihilated. Let us compute the quantity equation (4.105) by

$$\left\langle x_p^2 \right\rangle = \mathrm{e}^{-\beta\hbar\Omega} \left(\exp\left[\frac{\beta\hbar\omega^2}{2\Omega} \mathbf{A} \right] \right)_{pp} , \quad (4.106)$$

where we have used (4.84). Similarly, we can compute the off-diagonal thermal Green's function as

$$\left\langle x_p, x_q \right\rangle = \mathrm{e}^{-\beta\hbar\Omega} \left(\exp\left[\frac{\beta\hbar\omega^2}{2\Omega} \mathbf{A} \right] \right)_{pq} . \quad (4.107)$$

The same quantum-mechanical calculation by using the Hamiltonian H_L in (4.79) gives

$$\left\langle x_p, x_q \right\rangle = 1 + \lim_{\Omega \to 0} O_{2p} O_{2q} \exp\left[-\frac{\beta\hbar\omega^2}{2\Omega} \mu_2 \right], \quad (4.108)$$

where μ_2 is the second eigenvalue of the Laplacian matrix.

4.8 Random Graphs

The study of random graphs is one of the most important areas of theoretical graph theory. Random graphs have found multiple applications in physics and they are used today as a standard null model in simulating many physical processes on graphs and networks. There are several ways of defining a random graph, that is, a graph in which, given a set of nodes, the edges connecting them are selected in a random way. The simplest model of random graph was introduced by Erdös and Rényi [36]. The construction of a random graph in this model starts by considering n isolated nodes. Then, with probability $p > 0$, a pair of nodes is connected by an edge. Consequently, the graph is determined only by the number of nodes and edges such that it can be written as $G(n, m)$ or $G(n, p)$. In Figure 4.7, we illustrate some examples of *Erdös–Rényi (ER) random graphs* with the same number of nodes and different linking probabilities.

A few properties of ER random graphs are summarized as follows.

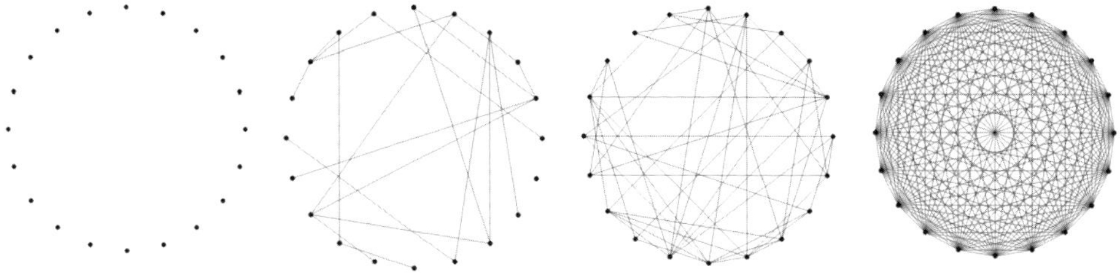

Figure 4.7 Illustration of the changes of an Erdös–Rényi random network with 20 nodes and probabilities that increases from zero (left) to one (right).

1. The expected number of edges per node:

$$\overline{m} = \frac{n(n-1)p}{2}. \tag{4.109}$$

2. The expected node degree:

$$\overline{k} = (n-1)p.$$

3. The average path length for large n:

$$\bar{l}(H) = \frac{\ln n - \gamma}{\ln(pn)} + \frac{1}{2}, \tag{4.110}$$

where $\gamma \approx 0.577$ is the Euler–Mascheroni constant.

4. The average clustering coefficient (see (4.12)):

$$\overline{C} = p = \delta(G). \tag{4.111}$$

5. When increasing p, most nodes tends to be clustered in one giant component, while the rest of nodes are isolated in very small components (see Figure 4.8).
6. The structure of $G_{ER}(n,p)$ changes as a function of $p = \overline{k}/(n-1)$ giving rise to the following three stages (see Figure 4.9):
   a *Subcritical* $\overline{k} < 1$, where all components are simple and very small. The size of the largest component is $S = O(\ln n)$.
   b *Critical* $\overline{k} = 1$, where the size of the largest component is $S = \Theta\left(n^{2/3}\right)$.
   c *Supercritical* $\overline{k} > 1$, where the probability that $\left(f - \varepsilon\right)n < S < \left(f + \varepsilon\right)n$ is 1 when $n \to \infty \quad \varepsilon > 0$, where $f = f(\overline{k})$ is the positive solution of the equation: $\mathrm{e}^{-\overline{k}f} = 1 - f$. The rest of the components are very small, with the second largest having size about $\ln n$.
7. The largest eigenvalue of the adjacency matrix in an ER network grows proportionally to n [37]: $\lim_{n\to\infty}\left(\lambda_1(\mathbf{A})/n\right) = p$.
8. The second largest eigenvalue grows more slowly than λ_1: $\lim_{n\to\infty}\left(\lambda_2(\mathbf{A})/n^{\varepsilon}\right) = 0$ for every $\varepsilon > 0.5$.
9. The smallest eigenvalue also grows with a similar relation to $\lambda_2(\mathbf{A})$: $\lim_{n\to\infty}\left(\lambda_n(\mathbf{A})/n^{\varepsilon}\right) = 0$ for every $\varepsilon > 0.5$.
10. The spectral density of an ER random network follows *Wigner's semicircle law* [38], which is simply written as (see Figure 4.10):

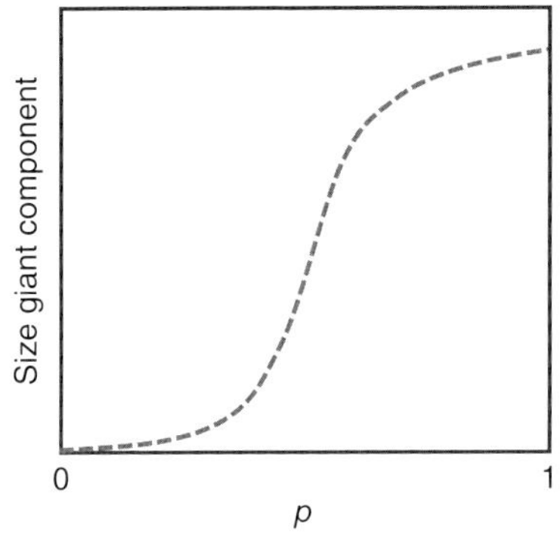

Figure 4.8 Change of the size of the giant connected component in an ER random graph as probability is increased.

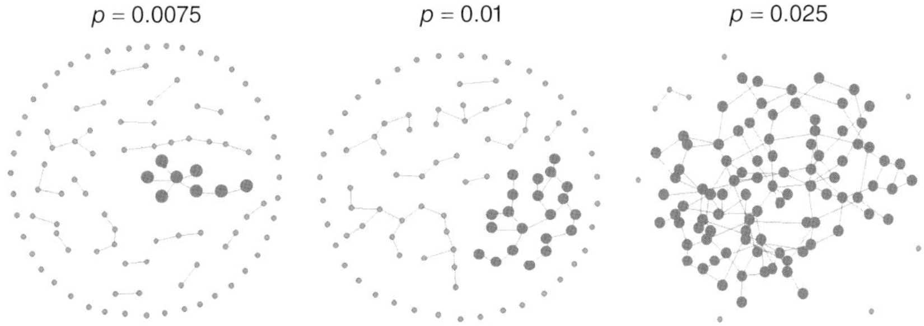

Figure 4.9 Examples of the different stages of the change of an ER random graph with the increase in probability: subcritical (a), critical (b), and supercritical (c).

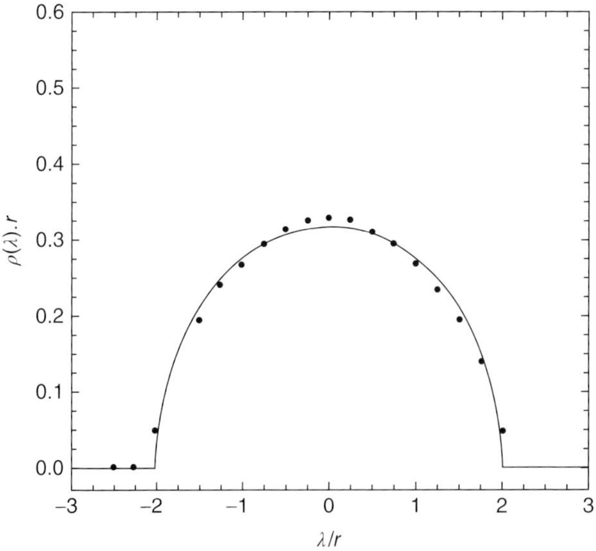

Figure 4.10 Illustration of the Wigner semicircle law for the spectral density of an ER random graph.

$$\rho(\lambda) = \begin{cases} \frac{\sqrt{4-\lambda^2}}{2\pi} & -2 \leq \frac{\lambda}{r} \leq 2, \; r = \sqrt{np(1-p)} \\ 0 & \text{otherwise.} \end{cases} \quad (4.112)$$

4.9 Introducing Complex Networks

In the rest of this chapter,we are going to study the so-called complex networks. Complex networks can be considered as the skeleton of complex systems in a variety of scenarios ranging from social and ecological to biological and technological systems. Their study has become a major field of interdisciplinary research in the twenty-first century with an important participation of physicists who have contributed significantly by creating new models and adapting others known in physics to the study of the topological and dynamical properties of these networks. A number of universal topological properties that explain some of the dynamical and functional properties of networks have been introduced, such as "small-world" and "scale-free" phenomena; these will be analyzed briefly in the next sections.

There is much confusion about what a complex network is. To start with we should attempt a clarification about what a complex system is. There is no clear-cut definition of a complex system. First, it must be clear that the concept of complexity is a twofold one: it may refer to a quality of the system or to a quantitative concept. In the first case, complexity is what makes the system complex. In the second, it is a continuum embracing both the simple and the complex according to a given measure of complexity. Standish [39] has stressed that as a quality "*complexity of a system refers to the presence of emergence in the system, or the exhibition of behaviour not specified in the system specification.*" In other words, complex systems "*display organization without any external organizing principle being applied*" [40]. When we speak of complexity as a quantity, it "*refers to the amount of information needed to specify the system.*"

Then, what is a complex network? Before attempting to answer this question let us try to make a classification of some of the systems represented by networks (see [41]) by considering the nature of the links they represent. The following are some examples of these classes.

- ***Physical linking.*** Pairs of nodes are physically connected by a *tangible link,* such as a cable, a road, a vein, and so on. Examples are the Internet, urban street networks, road networks, vascular networks, and so on.
- ***Physical interactions.*** The links between pairs of nodes represent *interactions* that are determined by a *physical force.* Examples are protein residue networks, protein–protein interaction networks, and so on.
- ***"Ethereal" connections.*** The links between pairs of nodes are *intangible,* such that information sent from one node is received at another irrespective of the "physical" trajectory. Examples are WWW, airports network.
- ***Geographic closeness.*** Nodes represent regions of a surface and their connections are determined by their *geographic proximity.* Examples arecountries in a map, landscape networks, and so on.
- ***Mass/energy exchange.*** The links connecting pairs of nodes indicate that some *energy or mass* has been *transferred* from one node to another.

Examples are reaction networks, metabolic networks, food webs, trade networks, and so on.

- ***Social connections.*** The links represent any kind of *social relationship* between nodes. Examples are friendship, collaboration, and so on.
- ***Conceptual linking.*** The links indicate *conceptual relationships* between pairs of nodes. Examples are dictionaries, citation networks, and so on.

Now, let us try to characterize the complexity of these networks by giving the minimum amount of information needed to describe them. For the sake of comparison, let us also consider a regular and a random graph of the same size as that of the real-world networks we want to describe. For the case of a regular graph, we only need to specify the number of nodes and the degree of the nodes (recall that every node has the same degree). With this information, many non-isomorphic graphs can be constructed, but many of their topological and combinatorial properties are determined by the information provided. In the case of the random network, we need to specify the number of nodes and the probability for joining pairs of nodes. As we have seen in the previous section, most of the structural properties of these networks are determined by this information. In contrast, to describe the structure of one of the networks representing a real-world system, we need an awful amount of information, such as number of nodes and links, degree distribution, degree–degree correlation, diameter, clustering, presence of communities, patterns of communicability, and other properties that we will study in this section. However, even in this case, a complete description of the system is still far away. Thus, the network representation of these systems deserves the title of *complex networks* because their topological structures cannot be trivially described as in the cases of random or regular graphs. In closing, when referring to complex networks, we are making an implicit allusion to the topological or structural complexity of the graphs representing a complex system. We will consider some general topological and dynamical properties of these networks in the following sections and the reader is recommended to consult the Further Reading section at the end of this chapter for more details and examples of applications.

4.10 Small-World Networks

One of the most popular concepts in network theory is that of the "small-world." In practically every language and culture, we have a phrase saying that the world is small enough so that a randomly chose person has a connection with some of our friends. The empirical grounds for this "concept" come from an experiment carried out by Stanley Milgram in 1967 [42]. Milgram asked some randomly selected people in the US cities of Omaha (Nebraska) and Wichita (Kansas) to send a letter to a target person who lives in Boston (Massachusetts) on the East Coast. The rules stipulate that the letter should be sent to somebody the sender knows personally. Despite the senders and the target being separated by about 2000 km, the results obtained by Milgram were surprising for the following reasons:

1. The average number of steps needed for the letters to arrive to its target was around six.
2. There was a large group inbreeding, which resulted in acquaintances of one individual feedback into his/her own

circle, thus usually eliminating new contacts.

The assumption that the underlying social network is a random one with characteristics such as the ER network fails to explain these findings. We already know that an ER random network displays a very small average path length, but it fails in reproducing the large group inbreeding observed because the number of triangles and the clustering coefficient in the ER network are very small. In 1998, Watts and Strogatz [8] proposed a model that reproduces the two properties mentioned in a simple way. Let n be the number of nodes and k be an even number, the Watt–Strogatz model starts by using the following construction. Place all nodes in a circle and connect every node to its first $k/2$ clockwise nearest neighbors as well as to its $k/2$ counterclockwise nearest neighbors (see Figure 4.11). This will create a ring, which, for $k > 2$, is full of triangles and consequently has a large clustering coefficient. The average clustering coefficient for these networks is given by Barrat and Weigt [43]:

$$\overline{C} = \frac{3(k-2)}{4(k-1)}, \tag{4.113}$$

which means that $\overline{C} = 0.75$ for very large values of k.

As can be seen in Figure 4.11 (top left), the shortest path distance between any pair of nodes that are opposite to each other in the network is relatively large. This distance is, in fact, equal to $\left\lceil \frac{n}{k} \right\rceil$. Then

$$\bar{l} \approx \frac{(n-1)(n+k-1)}{2kn}. \tag{4.114}$$

This relatively large average path length is far from that of the Milgram experiment. In order to produce a model with small average path length and still having relatively large clustering, Watts and Strogatz consider a probability for rewiring the links in that ring. This rewiring makes the average path length decrease very fast, while the clustering coefficient still remains high. In Figure 4.12, we illustrate what happens to the clustering and average path length as the rewiring probability changes from 0 to 1 in a network.

4.11 Degree Distributions

One of the network characteristics that has received much attention in the literature is the statistical distribution of the node degrees. Let $p(k) = n(k)/n$, where $n(k)$ is the number of nodes having degree k in a network of size n. That is, $p(k)$ represents the probability that a node selected

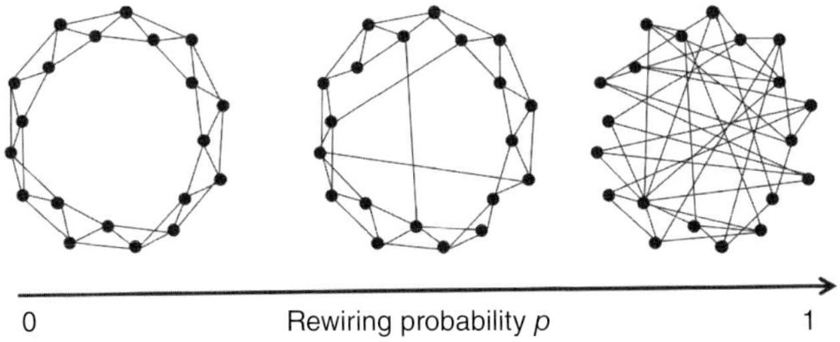

Figure 4.11 Schematic representation of the evolution of the rewiring process in the Watts–Strogatz model.

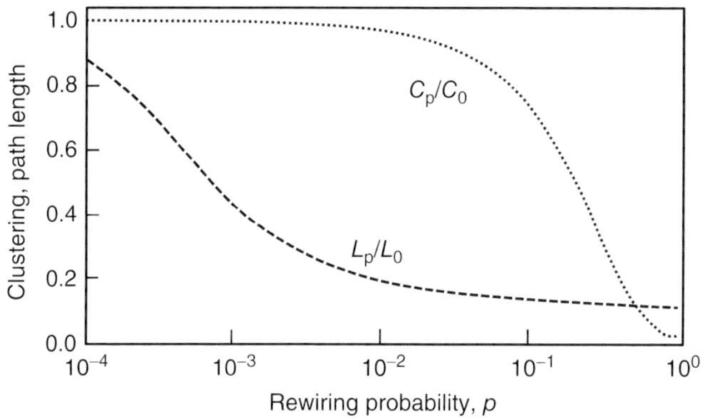

Figure 4.12 Schematic representation of the variation in the average path length and clustering coefficient with the change of the rewiring probability in the Watts–Strogatz model.

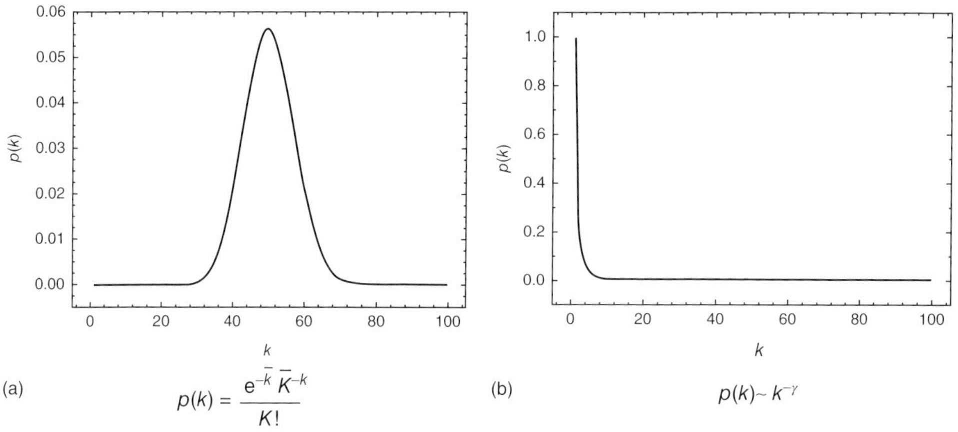

Figure 4.13 (a,b) Illustration of the Poisson and power-law degree distributions found in complex networks.

uniformly at random has degree k. The histogram of $p(k)$ versus k represents the degree distribution for the network. There are hundreds of statistical distributions in which the node degrees of a network can fit. A typical distribution that is expected for a random network of the type of ER is the Poisson distribution. However, a remarkable characteristic of complex networks is that many of them display some kind of "fat-tailed" degree distributions. In these distributions, a few nodes appear with very large degree, while most of the nodes have relatively small degrees. The prototypical example of these distributions is the power-law one, which is illustrated in the Figure 4.13, but others such as log normal, Burr, logGamma, Pareto, and so on [44] fall in the same category.

In the case of power-law distributions (see Figure 4.13a), the probability of finding a node with degree k decays as a negative power of the degree: $p(k) \sim k^{-\gamma}$. This means that the probability of finding a high-degree node is relatively small in comparison with the high probability of finding low-degree nodes. These networks are usually referred to as *scale-free networks*. The term *scaling* describes the existence of a power-law relationship between the probability and the node degree: $p(k) = Ak^{-\gamma}$. Scaling the degree by a constant factor c only produces a proportionate scaling of the probability:

$$p(k, c) = A(ck)^{-\gamma} = Ac^{-\gamma} \cdot p(k). \quad (4.115)$$

Power-law relations are usually represented in a logarithmic scale, leading to a straight line, $\ln p(k) = -\gamma \ln k + \ln A$, where $-\gamma$ is the slope and $\ln A$ the intercept of the function. Scaling by a constant factor c means that only the intercept of the straight line changes but the slope is exactly the same as before: $\ln p(k, c) = -\gamma \ln k - \gamma Ac$.

Determining the degree distribution of a network is a complicated task. Among the difficulties, we can mention the fact that sometimes the number of data points used to fit the distribution is too small and sometimes the data are very noisy. For instance, in fitting power-law distributions, the tail of the distribution, the part which corresponds to high degrees, is usually very noisy. There are two main approaches in use for reducing this noise effect in the tail of probability distributions. One is the binning procedure, which consists in building a histogram using bin sizes that increase exponentially with degree. The other approach is to consider the cumulative distribution function (CDF) [45]. The cumulative distributions of power-law and Poisson distributions are given as follows:

$$P(k) = \sum_{k'=k}^{\infty} p(k'), \quad (4.116)$$

$$P(k) = e^{-\bar{k}} \sum_{i=1}^{\lfloor k \rfloor} \frac{\left(\bar{k}\right)^i}{i!}, \quad (4.117)$$

which represent the probability of choosing at random a node with degree greater than or equal to k. In the case of power-law degree distributions, $P(k)$ also shows a power-law decay with degree

$$P(k) \sim \sum_{k'=k}^{\infty} k'^{-\gamma} \sim k^{-(\gamma-1)}, \quad (4.118)$$

which means that we will also obtain a straight line for the logarithmic plot of $P(k)$ versus k in scale-free networks.

4.11.1 "Scale-Free" Networks

Among the many possible degree distributions existing for a given network the "scale-free" one is one of the most ubiquitously found distributions. Consequently, it is important to study a model that is able to produce random networks with such kind of degree distribution, that is, a model in which the probability of finding a node with degree k decreases as a power-law of its degree. The most popular of these models is the one introduced by Barabási and Albert [46], which is described below.

In the Barabási–Albert (BA) model, a network is created by using the following procedure. Start from a small number m_0 of nodes. At each step, add a new node u to the network and connect it to $m \le m_0$ of the existing nodes $v \in V$ with probability

$$p_u = \frac{k_v}{\sum_w k_w}. \quad (4.119)$$

We can assume that we start from a connected random network of the ER type with m_0 nodes, $G_{ER} = (V, E)$. In this case, the BA process can be understood as a process in which small inhomogeneities in the degree distribution of the ER network grow in time. Another option is the one developed by Bollobás and Riordan [47] in which it is first assumed that $d = 1$ and that the ith node is attached to the jth one with probability

$$p_i = \begin{cases} \dfrac{k_j}{1+\sum_{j=0}^{i-1} k_j} \text{if} j < i \\ \dfrac{1}{1+\sum_{j=0}^{i-1} k_j} \text{if} j = i \end{cases}. \quad (4.120)$$

Then, for $d > 1$, the network grows as if $d = 1$ until nd nodes have been created and the size is reduced to n by contracting groups of d consecutive nodes into one. The network is now specified by two parameters and we denote this by $BA(n, d)$. Multiple links and self-loops are created during this process and they can be simply eliminated if we need a simple network.

A characteristic of BA networks is that the probability that a node has degree $k \geq d$ is given by

$$p(k) = \frac{2d(d-1)}{k(k+1)(k+2)} \sim k^{-3}, \quad (4.121)$$

which immediately implies that the cumulative degree distribution is given by

$$P(k) \sim k^{-2}. \quad (4.122)$$

For fixed values $d \geq 1$, Bollobás *et al.* [48] have proved that the expected value for the clustering coefficient $\overline{C}$ is given by

$$\overline{C} \sim \frac{d-1}{8} \frac{\log^2 n}{n}, \quad (4.123)$$

for $n \to \infty$, which is very different from the value $\overline{C} \sim n^{-0.75}$ reported by Barabási and Albert [46] for $d = 2$.

On the other hand, the average path length has been estimated for the BA networks to be as follows [47]:

$$\bar{l} = \frac{\ln n - \ln(d/2) - 1 - \gamma}{\ln \ln n + \ln(d/2)} + \frac{3}{2}, \quad (4.124)$$

where γ is the Euler–Mascheroni constant. This means that for the same number of nodes and average degree, BA networks have smaller average path length than their ER analogs. Other alternative models for obtaining power-law degree distributions with different exponents γ can be found in the literature [49]. In closing, using this preferential attachment algorithm, we can generate random networks that are different from those obtained by using the ER method in many important aspects including their degree distributions, average clustering, and average path length.

4.12 Network Motifs

The concept of *network motifs* was introduced by Milo *et al.* [50] in order to characterize recurring, significant patterns in real-world networks [50, 51]. A network motif is a subgraph that appears more frequently in a real network than could be expected if the network were built by a random process. In order to measure the statistical significance of a given subgraph, the *Z-score* is used, which is defined as follows for a given subgraph i:

$$Z_i = \frac{N_i^{\text{real}} - \langle N_i^{\text{random}} \rangle}{\sigma_i^{\text{random}}}, \quad (4.125)$$

where N_i^{real} is the number of times the subgraph i appears in the real network, $\langle N_i^{\text{random}} \rangle$ and σ_i^{random} are the average and standard deviation of the number of times that i appears in the ensemble of random networks, respectively. Some of the motifs found by Milo *et al.* [50] in different real-world directed networks are illustrated in the Figure 4.14.

4.13 Centrality Measures

Node centrality in a network is one of the many concepts that have been created in the analysis of social networks and then imported to the study of any kind of networked system. Measures of centrality try to capture the notion of "importance" of nodes in networks by quantifying the ability of a node to communicate directly with other nodes, or its closeness to many other nodes, or the number of pairs of nodes that need a specific node as intermediary in their communications. Here we describe some of the most relevant centrality measures currently in use for studying complex networks.

The *degree* of a node was defined in the first section. It was first considered as a centrality measure for nodes in a network by Freeman [52] as a way to account for immediate effects taking place in a network. The degree centrality can be written as

$$k_i = \sum_{j=1}^{n} A_{ij}. \tag{4.126}$$

In directed networks, we have two different kinds of degree centrality, namely, the *in-* and *out-degree* of a node:

$$k_i^{\text{in}} = \sum_{j=1}^{n} A_{ij}, \tag{4.127}$$

$$k_j^{\text{out}} = \sum_{i=1}^{n} A_{ij}. \tag{4.128}$$

Another type of centrality is the *closeness centrality*, which measures how close a node is from the rest of the nodes in the network. The closeness centrality [52] is expressed mathematically as follows:

$$\text{CC}(u) = \frac{n-1}{s(u)}, \tag{4.129}$$

where the distance sum $s(u)$ is

$$s(u) = \sum_{v \in V(G)} d(u,v). \tag{4.130}$$

The *betweenness centrality* quantifying the importance of a node is in the communication between other pairs of nodes in the network [52]. It measures the proportion of information that passes through a given node in the communications between other pairs of nodes in the network and it is defined as follows:

$$\text{BC}(k) = \sum_i \sum_j \frac{\rho(i,k,j)}{\rho(i,j)}, \quad i \neq j \neq k, \tag{4.131}$$

where $\rho(i,j)$ is the number of shortest paths from node i to node j, and $\rho(i,k,j)$ is the number of these shortest paths that pass through node k in the network.

The *Katz centrality index* for a node in a network is defined as [53]

$$K_i = \left\{ \left[\left(\mathbf{I} - \eta^{-1}\mathbf{A} \right)^{-1} - I \right] \mathbf{1} \right\}_i, \tag{4.132}$$

where I is the identity matrix, $\eta \neq \lambda_1$ is an attenuation factor (λ_1 is the principal eigenvector of the adjacency matrix), and $\mathbf{1}$ is column vector of 1's. This centrality index can be considered as an extension of the degree in order to consider the influence

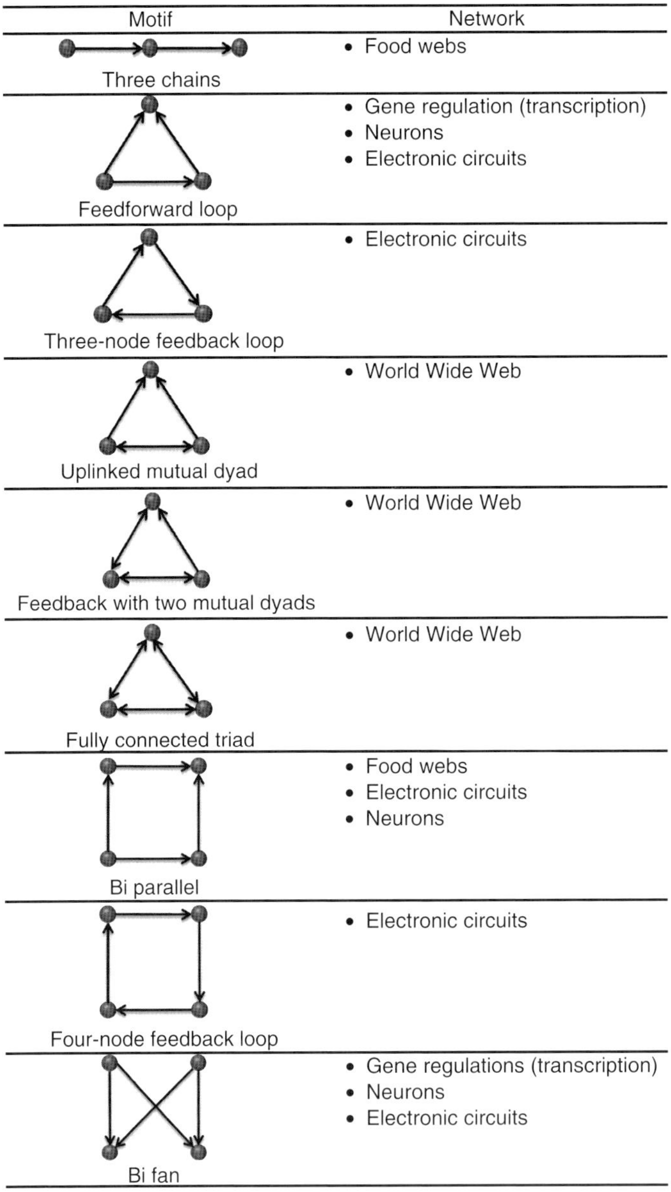

Figure 4.14 Illustration of some of the motifs found in real-world networks.

not only of the nearest neighbors but also of the most distant ones.

The Katz centrality index can be defined for directed networks:

$$K_i^{\text{out}} = \left\{ \left[\left(\mathbf{I} - \eta^{-1}\mathbf{A} \right)^{-1} - I \right] \mathbf{1} \right\}_i, \qquad (4.133)$$

$$K_i^{\text{in}} = \left\{ \mathbf{1}^T \left[\left(\mathbf{I} - \eta^{-1}\mathbf{A} \right)^{-1} - \mathbf{I} \right] \right\}_i. \qquad (4.134)$$

The K_i^{in} is a measure of the "prestige" of a node as it accounts for the importance that a node has due to other nodes that point to it.

Another type of centrality that captures the influence not only of nearest neighbors but also of more distant nodes in a network is the *eigenvector centrality*. This index was introduced by Bonacich [54, 55] and is the ith entry of the principal eigenvector of the adjacency matrix

$$\boldsymbol{\varphi}_1(i) = \left(\frac{1}{\lambda_1}\mathbf{A}\boldsymbol{\varphi}_1\right)_i. \quad (4.135)$$

In directed networks, there are two types of eigenvector centralities that can be defined by using the principal right and left eigenvectors of the adjacency matrix:

$$\boldsymbol{\varphi}_1^R(i) = \left(\frac{1}{\lambda_1}\mathbf{A}\boldsymbol{\varphi}_1^R\right)_i, \quad (4.136)$$

$$\boldsymbol{\varphi}_1^L(i) = \left(\frac{1}{\lambda_1}\mathbf{A}^T\boldsymbol{\varphi}_1^L\right)_i. \quad (4.137)$$

Right eigenvector centrality accounts for the "importance" of a node by taking into account the "importance" of nodes to which it points on, that is, it is an extension of the out-degree concept by taking into account not only nearest neighbors. On the other hand, the left-eigenvector centrality measures the importance of a node by considering those nodes pointing toward the corresponding node and it is an extension of the in-degree centrality. This is frequently referred to as *prestige* in social sciences contexts.

There is an important difficulty when we try to apply right- and left-eigenvector centralities to networks where there are nodes having out-degree or in-degree equal to zero, respectively. In the first case, the nodes pointing to a given node do not receive any score for pointing to it. When the in-degree is zero, the left-eigenvector centrality or prestige of this node is equal to zero as no node points to it, even though it can be pointing to some important nodes. A solution for this problem is obtained by the following centrality measure.

The *PageRank centrality* measure is the tool used by Google in order to rank citations of web pages in the WWW [56]. Its main idea is that the importance of a web page should be proportional to the importance of other web pages pointing to it. In other words, the PageRank of a page is the sum of the PageRanks of all pages pointing into it. Mathematically, this intuition is captured by the following definition. The PageRank is obtained by the vector

$$\left(\boldsymbol{\pi}^{k+1}\right)^T = \left(\boldsymbol{\pi}^{k}\right)^T \mathbf{G}. \quad (4.138)$$

The matrix $\mathbf{G}$ is defined by

$$\mathbf{G} = \alpha\mathbf{S} + \left(\frac{1-\alpha}{n}\right)\mathbf{1}\mathbf{1}^T, \quad (4.139)$$

where $0 \le \alpha \le 1$ is a "teleportation" parameter, which captures the effect in which a web surfer abandons his random approach of bouncing from one page to another and initiates a new search simply by typing a new destination in the browser's URL command line. The matrix $\mathbf{S}$ solves the problem of dead-end nodes in ranking web pages, and it is defined as

$$\mathbf{S} = \mathbf{H} + \mathbf{a}\left[\left(\frac{1}{n}\right)\mathbf{1}^T\right], \quad (4.140)$$

where the entries of the dangling vector $\mathbf{a}$ are given by

$$a_i = \begin{cases} 1 & \text{if} k_i^{\text{out}} = 0 \\ 0 & \text{otherwise.} \end{cases} \quad (4.141)$$

Finally, the matrix $\mathbf{H}$ is defined as a modified adjacency matrix for the network

$$H_{ij} = \begin{cases} \frac{1}{k_i^{\text{out}}} & \text{if thre is a link from } i \text{ to } j \\ 0 & \text{otherwise.} \end{cases}$$

The matrix $\mathbf{G}$ is row stochastic, which implies that its largest eigenvalue is equal to one and the *principal left-hand eigenvector* of $\mathbf{G}$ is given by

$$\boldsymbol{\pi}^T = \boldsymbol{\pi}^T \mathbf{G}, \tag{4.142}$$

where $\boldsymbol{\pi}^T \mathbf{1} = 1$ [56].

Another type of node centrality that is based on the spectral properties of the adjacency matrix of a graph is the *subgraph centrality*. The subgraph centrality counts the number of closed walks starting and ending at a given node, which are mathematically given by the diagonal entries of $\mathbf{A}^k$. In general terms, the subgraph centrality is a family of centrality measures defined on the basis of the following mathematical expression:

$$f_i(\mathbf{A}) = \left(\sum_{l=0}^{\infty} c_l \mathbf{A}^l \right)_{ii}, \tag{4.143}$$

where coefficients c_l are selected such that the infinite series converges. One particularly useful weighting scheme is the following, which eventually converges to the exponential of the adjacency matrix [57]:

$$\mathrm{EE}(i) = \left(\sum_{l=0}^{\infty} \frac{\mathbf{A}^l}{l!} \right)_{ii} = \left(\mathrm{e}^{\mathbf{A}} \right)_{ii}. \tag{4.144}$$

We can also define subgraph centralities that take into account only contributions from odd or even closed walks in the network:

$$\mathrm{EE}_{\mathrm{odd}}(i) = (\sinh \mathbf{A})_{ii}, \tag{4.145}$$

$$\mathrm{EE}_{\mathrm{even}}(i) = (\cosh \mathbf{A})_{ii}. \tag{4.146}$$

A characteristic of the subgraph centrality in directed networks is that it accounts for the participation of a node in directed walks. This means that the subgraph centrality of a node in a directed network is $\mathrm{EE}(i) > 1$ only if there is at least one closed walk that starts and returns to this node. In other cases, $\mathrm{EE}(i) = 1$, that is, the subgraph centrality in a directed network measures the returnability of "information" to a given node.

4.14 Statistical Mechanics of Networks

Let us consider that every link of a network is weighted by a parameter β. Evidently, the case $\beta = 1$ corresponds to the simple network. Let $\mathbf{W}$ be the adjacency matrix of this homogeneously weighted network. It is obvious that $\mathbf{W} = \beta \mathbf{A}$ and the spectral moments of the adjacency matrix are $M_r(\mathbf{W}) = \mathrm{Tr}\mathbf{W}^r = \beta^r \mathrm{Tr}\mathbf{A}^r = \beta^r M_r$. Let us now count the total number of closed walks in this weighted network. It is straightforward to realize [58] that this is given by

$$Z(G;\beta) = \mathrm{Tr} \sum_{r=0}^{\infty} \frac{\beta^r \mathbf{A}^r}{r!} = \mathrm{Tr}\mathrm{e}^{\beta \mathbf{A}} = \sum_{j=1}^{n} \mathrm{e}^{\beta \lambda_j}. \tag{4.147}$$

Let us now consider that the parameter $\beta = \left(k_{\mathrm{B}} T \right)^{-1}$ is the inverse temperature of a thermal bath in which the whole network is submerged. Here the temperature is a physical analogy for the external "stresses" that a network is continuously exposed to. For instance, let us consider the network in which nodes represent corporations and the links represent their business relationships. In this case, the external stress can represent the economical situation of the world at the moment in which the network is analyzed. In "normal" economical situations we are in the presence of a low level of external stress. In situations of economical crisis, the level of external stress is elevated.

We can consider the probability that the network is in a configuration (state) with an energy given by the eigenvalue λ_j. The configuration or state of the network can be considered here as provided by the corresponding eigenvector of the adjacency matrix associated with λ_j. This probability is then given by

$$p_j = \frac{e^{\beta\lambda_j}}{\sum_j e^{\beta\lambda_j}} = \frac{e^{\beta\lambda_j}}{Z(G;\beta)}, \qquad (4.148)$$

which identifies the normalization factor as the partition function of the network. This index, introduced by Estrada [59], is known in the graph theory literature as the *Estrada index* of the graph/network and usually denoted by EE(G).

We can define the *entropy* for the network,

$$S(G;\beta) = -k_B \sum \left[p_j\left(\beta\lambda_j - \ln Z\right)\right], \qquad (4.149)$$

where we wrote $Z(G;\beta) = Z$ for the sake of economy.

The *total energy* $H(G)$ and Helmholtz free energy $F(G)$ of the network, respectively [58], are given by

$$\begin{aligned} H(G,\beta) &= -\frac{1}{\mathrm{EE}}\sum_{j=1}^{n}\left(\lambda_j e^{\beta\lambda_j}\right) \\ &= -\frac{1}{\mathrm{EE}}\mathrm{Tr}\left(\mathbf{A}e^{\beta\mathbf{A}}\right) \\ &= -\sum_{j=1}^{n}\lambda_j p_j, \end{aligned} \qquad (4.150)$$

$$F(G,\beta) = -\beta^{-1}\ln \mathrm{EE}. \qquad (4.151)$$

Known bounds for the physical parameters defined above, are the following:

$$0 \le S(G,\beta) \le \beta \ln n, \qquad (4.152)$$

$$-\beta(n-1) \le H(G,\beta) \le 0, \qquad (4.153)$$

$$-\beta(n-1) \le F(G,) \le -\beta \ln n, \qquad (4.154)$$

where the lower bounds are obtained for the complete graph as $n \to \infty$ and the upper bounds are reached for the null graph with n nodes [58].

Next, let us analyze the thermodynamic functions of networks for extreme values of the temperature. At very low temperatures, the total energy and Helmholtz free energy are reduced to the interaction energy of the network [58]: $H(G,\beta \to \infty) = F(G,\beta \to \infty) = -\lambda_1$. At very high temperatures, $\beta \to 0$, the entropy of the system is completely determined by the partition function of the network, $S(G;\beta \to 0) = k_B \ln Z$ and $F(G,\beta \to 0) \to -\infty$.

The introduction of these statistical mechanics parameters allows the study of interesting topological and combinatorial properties of networks by using a well-understood physical paradigm. For finding examples of these applications, the reader is referred to the specialized literature [41].

4.14.1 Communicability in Networks

The concept of network communicability is a very recent one. However, it has found applications in many different areas of network theory [35]. This concept captures the idea of correlation in a physical system and translates it into the context of network theory. We define here that the *communicability* between a pair of nodes in a network depends on all routes that connect these two nodes [35]. Among all these routes, the shortest path is the one making the most important contribution as it is the most "economic" way of connecting two nodes in a network. Thus we can use the weighted sum of all walks of different lengths between a pair of nodes as a measure of their communicability, that is,

$$G_{pq} = \sum_{k=0}^{\infty} \frac{\left(\mathbf{A}^k\right)_{pq}}{k!} = \left(\mathrm{e}^{\mathbf{A}}\right)_{pq}, \quad (4.155)$$

where $\mathrm{e}^{\mathbf{A}}$ is the matrix exponential function. We can express the communicability function for a pair of nodes in a network by using the eigenvalues and eigenvectors of the adjacency matrix:

$$G_{rs} = \sum_{j=1}^{n} \varphi_j(r)\,\varphi_j(s)\,\mathrm{e}^{\lambda_j}. \quad (4.156)$$

By using the concept of inverse temperature introduced above, we can also express the communicability function in terms of this parameter [60]

$$G_{rs}(\beta) = \sum_{k=0}^{\infty} \frac{\left(\beta\mathbf{A}^k\right)_{rs}}{k!} = \left(\mathrm{e}^{\beta\mathbf{A}}\right)_{rs}. \quad (4.157)$$

Intuitively, the communicability between the two nodes connected by a path should tend to zero as the length of the path tends to infinity. In order to show that this is exactly the case, we can write the expression for $G_{rs}(\beta)$ for the path P_n:

$$G_{rs} = \frac{1}{n+1}\left(\sum_j \cos\frac{j\pi(r-s)}{n+1} - \cos\frac{j\pi(r+s)}{n+1}\right)\mathrm{e}^{2\cos\left(\frac{j\pi}{(n+1)}\right)}, \quad (4.158)$$

where we have used $\beta \equiv 1$ without any loss of generality. Then, it is straightforward to realize by simple substitution in (4.158) that $G_{rs} \to 0$ for the nodes at the end of a linear path as $n \to \infty$. At the other extreme, we find the complete network K_n, for which

$$G_{rs} = \frac{\mathrm{e}^{n+1}}{n} + \mathrm{e}^{-1}\sum_{j=2}^{n}\phi_j(r)\phi_j(s)$$
$$= \frac{\mathrm{e}^{n+1}}{n} - \frac{1}{n\mathrm{e}} = \frac{1}{n\mathrm{e}}\left(\mathrm{e}^n - 1\right), \quad (4.159)$$

which means that $G_{rs} \to \infty$ as $n \to \infty$. In closing, the communicability measure quantifies very well our intuition that communication decays to zero when only one route exists for connecting two nodes at an infinite distance (the end nodes of a path) and it tends to infinity when there are many possible routes of very short distance (any pair of nodes in a complete graph).

4.15 Communities in Networks

The study of *communities* in complex networks is a large area of research with many existing methods and algorithms. The aim of all of them is to identify subsets of nodes in a network the density of whose connections is significantly larger than the density of connections between them and the rest of the nodes. It is impossible to give a complete survey of all the methods for detecting communities in networks in this short section. Thus we are going to describe some of the main characteristics of a group of methods currently used for detecting communities in networks. An excellent review with details on the many methods available can be found in [61].

The first group of methods used for detecting communities in networks is that of partitioning methods. Their aim is to obtain a partition of the network into p disjoint sets of nodes such that

(i) $\bigcup_{i=1}^{p} V_i = V$ and $V_i \bigcap V_j = \phi$ for $i \neq j$,
(ii) the number of edges crossing between subsets (cut size or *boundary*) is minimized,

(iii) $|V_i| \approx n/p$ for all $i = 1, 2, \ldots, p$, where the vertical bars indicates the cardinality of the set.

When condition (iii) is fulfilled, the corresponding partition is called *balanced*. There are several algorithms that have been tested in the literature for the purpose of network partitioning that include local improvement methods and spectral partitioning. The last family of partition methods is based on the adjacency, Laplacian, or normalized Laplacian matrices of graphs. In general, their goal is to find a separation between the nodes of the network based on the eigenvectors of these matrices. This separation is carried out *grosso modo* by considering that two nodes v_1, v_2 are in the same partition if $\operatorname{sgn} \varphi_2^M(v_1) = \operatorname{sgn} \varphi_2^M(v_2)$ for $M = A, L, \tilde{L}$. Otherwise, they are considered to be in two different partitions $\{V_1, V_2\}$. Sophisticated versions of these methods exist and the reader is referred to the specialized literature for details [61].

The second group of methods is based on edge centralities. We have defined a number of centrality measures for nodes in a previous section of this chapter; these can be extended to edges of a network in a straightforward way. In these methods, the aim is to identify edges that connect different communities. The best known technique is based on edge betweenness centrality, defined for edges in a similar way as for nodes. This method, known as the *Girvan–Newman algorithm* [62], can be summarized in the following steps:

1. Calculate the edge betweenness centrality for all links in the network.
2. Remove the link with the largest edge betweenness or any of them if more than one exists.
3. Recalculate the edge betweenness for the remaining links.
4. Repeat until all links have been removed.
5. Use a dendrogram for analyzing the community structure of the network.

Using this dendrogram, a hierarchy of different communities is identified, which can be discriminated by using different quality criteria. The most popular among these quality criteria is the so-called modularity index. In a network consisting of n_V partitions, $V_1, V_2, \ldots, V_{n_C}$, the modularity is the sum over all partitions of the difference between the fraction of links inside each partition and the expected fraction by considering a random network with the same degree for each node [63]:

$$Q = \sum_{k=1}^{n_C} \left[\frac{|E_k|}{m} - \left(\frac{\sum_{j \in V_k} k_j}{2m} \right)^2 \right], \tag{4.160}$$

where $|E_k|$ is the number of links between nodes in the kth partition of the network. Modularity is interpreted in the following way. If $Q = 0$, the number of intracluster links is not bigger than the expected value for a random network. Otherwise, $Q = 1$ means that there is a strong community structure in the network given by the partition analyzed.

The third group of community detection methods is based on *similarity measures* for the nodes in a network. Such similarity measures for the nodes of a network can be based on either rows or columns of the adjacency matrix of the network. For instance, we can consider as a measure of similarity between two nodes the angle between the corresponding rows or columns of these two nodes in the adjacency matrix of the graph. This angle is defined as

$$\sigma_{ij} = \cos \vartheta_{ij} = \frac{\mathbf{x}^T \mathbf{y}}{\|\mathbf{x}\| \cdot \|\mathbf{y}\|}, \tag{4.161}$$

which can be seen to equal

$$\sigma_{ij} = \frac{\eta_{ij}}{\sqrt{k_i k_j}}, \tag{4.162}$$

where η_{ij} is the number of common neighbors of nodes i and j.

Other similarity measures between rows or columns of the adjacency matrices are the Pearson correlation coefficient, different types of norms and distances (Manhattan, Euclidean, infinite), and so on. Once a similarity measure has been chosen, any of the variety of similarity-based methods for detecting communities in a network can be used.

4.16 Dynamical Processes on Networks

There are many dynamical processes that can be defined on graphs and networks. The reader should be aware that this is a vast area of multidisciplinary research with a huge number of publications in different fields.

4.16.1 Consensus

We will start here with a simple model for analyzing consensus among the nodes in a network. We consider a graph $G = (V, E)$ whose nodes represent agents in a complex system and the edges represent interactions between such agents. In such multiagent system, *consensus* means an agreement regarding a certain quantity of interest. An example is a collection of autonomous vehicles engaged in cooperative teamwork in civilian and military applications. Such coordinated activity allows them to perform missions with greater efficacy than if they perform solo missions [64].

Let $n = |V|$ be the number of agents forming a network. The collective dynamics of the group of agents is represented by the following equations for the continuous-time case:

$$\dot{\boldsymbol{\varphi}} = -\mathbf{L}\boldsymbol{\varphi}, \boldsymbol{\varphi}(0) = \boldsymbol{\varphi}_0, \tag{4.163}$$

where $\boldsymbol{\varphi}_0$ is the original distribution, which may represent opinions, positions in space, or other quantities with respect to which the agents should reach a consensus. The reader surely already has recognized that (4.163) is identical to the heat equation,

$$\frac{\partial u}{\partial t} = h\Delta u, \tag{4.164}$$

where h is a positive constant and $\nabla^2 = -\mathbf{L}$ is the Laplace operator. In general, this equation is used to model the diffusion of "information" in a physical system, where, by information we can understand heat, a chemical substance, or opinions in a social network.

A consensus is reached if, for all $\varphi_i(0)$ and all $i, j = 1, \ldots, n$, $\left|\varphi_i(t) - \varphi_j(t)\right| \to 0$ as $t \to 0$. The discrete-time version of the model has the form

$$\boldsymbol{\varphi}_i(t+1) = \boldsymbol{\varphi}_i(t) + \varepsilon \sum_{j \sim i} \mathbf{A}_{ij} \left[\boldsymbol{\varphi}_j(t) - \boldsymbol{\varphi}_i(t)\right],$$
$$\boldsymbol{\varphi}(0) = \boldsymbol{\varphi}_0, \tag{4.165}$$

where $\boldsymbol{\varphi}_i(t)$ is the value of a quantitative measure on node i, $\varepsilon > 0$ is the step-size, and $j \sim i$ indicates that node j is connected to node i. It has been proved that the consensus is asymptotically reached in a connected graph for all initial states if $0 < \varepsilon < 1/\delta_{\max}$, where $\delta_{\max}$ is the maximum degree of the graph. The discrete-time collective

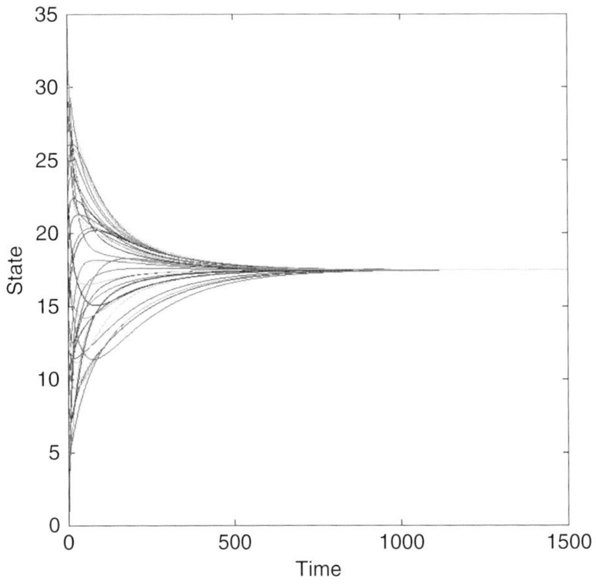

Figure 4.15 Time evolution of consensus dynamics in a real-world social network with random initial states for the nodes.

dynamics of the network can be written in matrix form as [64] as

$$\boldsymbol{\varphi}(t+1) = \mathbf{P}\boldsymbol{\varphi}(t), \boldsymbol{\varphi}(0) = \boldsymbol{\varphi}_0, \quad (4.166)$$

where $\mathbf{P} = \mathbf{I} - \varepsilon\mathbf{L}$, and $\mathbf{I}$ is the $n \times n$ identity matrix. The matrix $\mathbf{P}$ is the Perron matrix of the network with parameter $0 < \varepsilon < 1/\delta_{\max}$. For any connected undirected graph, the matrix $\mathbf{P}$ is an irreducible, doubly stochastic matrix with all eigenvalues μ_j in the interval $[-1, 1]$ and a trivial eigenvalue of 1. The reader can find the previously mentioned concepts in any book on elementary linear algebra. The relation between the Laplacian and Perron eigenvalues is given by $\mu_j = 1 - \varepsilon\lambda_j$.

In Figure 4.15, we illustrate the consensus process in a real-world social network having 34 nodes and 78 edges.

4.16.2 Synchronization in Networks

A problem closely related to that of consensus in networks is one of *synchronization* [65, 66]. The phenomenon of synchronization appears in many natural systems consisting of a collection of oscillators coupled to each other. These systems include animal and social behavior, neurons, cardiac pacemaker cells, among others. We can start by considering a network $G = (V, E)$ with $|V| = n$ nodes representing coupled identical oscillators. Each node is an N-dimensional dynamical system that is described by the following equation:

$$\dot{x}_i = f(x_i) + c\sum_{j=1}^{n} L_{ij} H(t) x_j, \quad i = 1, \dots, n, \quad (4.167)$$

where $x_i = (x_{i1}, x_{i2}, \dots, x_{iN}) \in \mathbb{R}^N$ is the state vector of the node i, $f(\cdot) : \mathbb{R}^N \to \mathbb{R}^N$ is a smooth vector-valued function that defines the dynamics, c is a constant representing the coupling strength, $H(\cdot) : \mathbb{R}^N \to \mathbb{R}^N$ is a fixed output function also known as the *outer coupling matrix*, t is the time, and L_{ij} are the elements of the Laplacian matrix of the network (sometimes the negative of the $H(x_i) \approx H(s) + \xi_i H'(s)$ Laplacian matrix is taken here). The network is said to achieve synchronization if

$$x_1(t)=x_2(t)=\cdots=x_n(t) \to s(t), \text{as} t \to \infty. \tag{4.168}$$

Let us now consider a small perturbation ξ_i such that $x_i = s + \xi_i$ ($\xi_i \ll s$) and let us analyze the stability of the synchronized manifold $x_1 = x_2 = \cdots = x_n$. First, we expand the terms in (4.167) as

$$f(x_i) \approx f(s) + \xi_i f'(s), \tag{4.169}$$

$$H(x_i) \approx H(s) + \xi_i H'(s), \tag{4.170}$$

where the primes refers to the derivatives respect to s. Thus, the evolution of the perturbations is determined by the following equation:

$$\dot{\xi}_i = f'(s)\xi_i + c\sum_j \left[L_{ij}H'(s)\right]\xi_j. \tag{4.171}$$

It is known that the system of equations for the perturbations can be decoupled by using the set of eigenvectors of the Laplacian matrix, which are an appropriate set of linear combinations of the perturbations. Let ϕ_j be an eigenvector of the Laplacian matrix of the network associated with the eigenvalue μ_j. Recall that the Laplacian is positive semidefinite, that is, $0 = \mu_1 \le \mu_2 \le \cdots \le \mu_n \equiv \mu_{max}$. Then

$$\dot{\phi}_i = \left[f'(s) + c\mu_i H'(s)\right]\phi_i. \tag{4.172}$$

Let us now assume that at short times the variations of s are small enough to allow us to solve these decoupled equations, with the solutions being

$$\dot{\phi}_i(t) = \phi_i^0 \exp\left\{\left[f'(s) + c\mu_i H'(s)\right]t\right\}, \tag{4.173}$$

where ϕ_i^0 is the initially imposed perturbation.

We now consider the term in the exponential of (4.173), $\Lambda_i = f'(s) + c\mu_i H'(s)$. If $f'(s) > c\mu_i H'(s)$, the perturbations will increase exponentially, while if $f'(s) < c\mu_i H'(s)$, they will decrease exponentially. So, the behavior of the perturbations in time is controlled by the magnitude of μ_i. Then, the stability of the synchronized state is determined by the master stability function:

$$\Lambda(\alpha) \equiv \max_s \left[f'(s) + \alpha H'(s)\right], \tag{4.174}$$

which corresponds to a large number of functions f, and H is represented in the Figure 4.16.

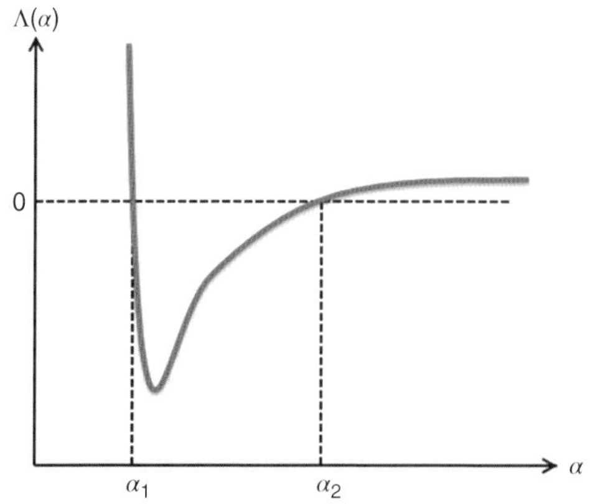

Figure 4.16 Schematic representation of the typical behavior of the master stability function.

Figure 4.17 Diagrammatic representation of an SIR model.

As can be seen, the necessary condition for stability of the synchronous state is that $c\mu_i$ is between α_1 and α_2, which is the region where $\Lambda(\alpha) < 0$. Then, the condition for synchronization is [67]:

$$Q := \frac{\mu_N}{\mu_2} < \frac{\alpha_2}{\alpha_1}, \tag{4.175}$$

that is, synchronizability of a network is favored by a small eigenratio Q, which indeed depends only on the topology of the network. There are many studies on the synchronizability of networks using different types of oscillators and the reader is referred to the specialized literature for the details (See Further Reading, [68]).

4.16.3 Epidemics on Networks

Another area in which the dynamical processes on networks play a fundamental role is the study of the spread of epidemics. These models are extensions of the classical models used in epidemiology that consider the influence of the topology of a network on the propagation of an epidemic [69]. The simplest model assumes that an individual who is susceptible (S) to an infection could become infected (I). In a second model, the infected individual can also recover (R) from infection. The first model is known as an *SI model*, while the second is known as a *susceptible–infected–recovered* (*SIR*) model. In a third model, known as the *susceptible–infected–susceptible* (*SIS*), an individual can be reinfected, so that infections do not confer immunity on an infected individual. Finally, a model known as *SIRS* allows for recovery and reinfection as an attempt to model the temporal immunity conferred by certain infections. Here we briefly consider only the SIR and SIS models on networks.

In the SIR model there are three compartments as sketched in the Figure 4.17, that is, in a network $G = (V, E)$, a group of nodes $S \subseteq V$ are considered susceptible and they can be infected by directed contact with infected individuals. Let s_i, x_i, and r_i be the probabilities that the node i is susceptible, infected, or has recovered. The evolution of these probabilities in time is governed by the following equations that define the *SIR model*:

$$\dot{s}_i = -\beta s_i \sum_j A_{ij} x_j, \tag{4.176}$$

$$\dot{x}_i = \beta s_i \sum_j A_{ij} x_j - \gamma x_i, \tag{4.177}$$

$$\dot{r}_i = \gamma x_i, \tag{4.178}$$

where β is the spreading rate of the pathogen, A_{ij} is an entry of the adjacency matrix of the network, and γ is the probability that a node recovers or dies, that is, the recovery rate.

In the SIS model, the general flow chart of the infection can be represented as in Figure 4.18:

The equations governing the evolution of the probabilities of susceptible and infected individuals are given as follows:

$$\dot{s}_i = -\beta s_i \sum_j A_{ij} x_j + \gamma x_i, \tag{4.179}$$

$$\dot{x}_i = \beta s_i \sum_j A_{ij} x_j - \gamma x_i. \tag{4.180}$$

The analysis of epidemics in networks is of tremendous importance in modern life. Today, there is a large mobility of people across cities, countries, and the entire world and an epidemic can propagate

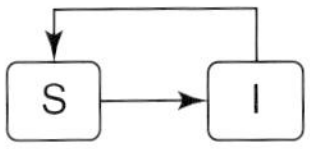

Figure 4.18 Diagrammatic representation of an SI model.

through the social networks at very high rates. The reader can find a few examples in the specialized literature [69].

Glossary

Adjacency matrix of a simple graph: a binary symmetric matrix whose row and columns represent the vertices of the graph, where the i, j entry is one if the corresponding vertices i and j are connected.
Betweenness centrality a centrality measure for a node that characterizes how central a node is in passing information from other nodes.
Bipartite graph a graph with two sets of vertices, the nodes of each set being connected only to nodes of the other set.
Bridge an edge whose deletion increases the number of connected components of the graph.
Centrality measure an index for a node or edge of a graph/network that characterizes its topological or structural importance.
Closeness centrality a centrality measure for a node that characterizes how close the node is with respect to the rest in terms of the shortest-path distance.
Clustering coefficient the ratio of the number of triangles incident to a node in a graph to the maximum possible number of such triangles.
Communicability a measure of how well-communicated a pair of nodes is by considering all possible routes of communication in a graph/network.
Complete graph a graph in which every pair of vertices are connected to each other.
Connected graph a graph in which there is a path connecting every pair of nodes.
Cycle a path in which the initial and end vertices coincide.
Cycle graph a graph in which every node has degree two.
Degree a centrality measure for a node that counts the number of edges incident to a node.
Degree distribution the statistical distribution of the degrees of the nodes of a graph.
Edge contraction a graph operation in which an edge of the graph is removed and the two end nodes are merged together.
Edge deletion a graph operation in which an edge of the graph is removed leaving the end nodes in the graph.
Erdös–Rényi graph a random graph formed from a given set of nodes and a probability of create edges among them.
Forest a graph formed by several components all of which are trees.
Girth the size of the minimum cycle in a graph.
Graph a pair formed by a set of vertices or nodes and a set of edges.
Graph diameter the length of the largest shortest-path distance in a graph.
Graph diameter the maximum shortest-path distance in a graph.
Graph invariant a characterization of a graph that does not depend on the labeling of vertices or edges.
Graph nullity the multiplicity of the zero eigenvalue of the adjacency matrix, that is, the number of times eigenvalue zero occurs in the spectrum of the adjacency matrix.
Hydrocarbon a molecule formed only by carbon and hydrogen.
Incidence matrix of a graph a matrix whose rows correspond to vertices and whose columns correspond to edges of the

graph and the i, j entry is one or zero if the ith vertex is incident with the jth edge or not, respectively.

Laplacian matrix a square symmetric matrix with diagonal entries equal to the degree of the corresponding vertex and off-diagonal entries equal to -1 or zero depending on whether the corresponding vertices are connected or not, respectively.

Loop an edge that is doubly incident to the same node.

Matching of a graph the number of mutually nonadjacent edges in the graph.

Mean displacement of an atom (vertex): refers to the oscillations of an atoms from its equilibrium position due to thermal fluctuations.

Molecular Hamiltonian the operator representing the energy of the electrons and atomic nuclei in a molecule.

Network community a subset of nodes in a graph/network that are better connected among themselves than with the rest of the nodes.

Network motif a subgraph in a graph that is overrepresented in relation to a random graph of the same size.

Path a sequence of different consecutive vertices and edges in a graph.

Path graph a tree in which all nodes have degree two except two nodes, which has degree one.

Regular graph a graph in which every node has the same degree.

Resistance distance the distance between any pair of vertices of the graph, determined by the Kirchhoff rules for electrical sets.

Scale-free network a network/graph with a power-law degree distribution.

Shortest path between two nodes a path having the least number of edges among all paths connecting two vertices.

Simple graph a graph without multiple edges, self-loops, and weights.

Spanning forest a subgraph of a graph that contains all the nodes of the graph and is a forest.

Spanning tree a subgraph of a graph that contains all the nodes of the graph and is also a tree.

Star graph a tree consisting of a node with degree $n-1$ and $n-1$ nodes of degree one.

Tree a graph that does not have any cycle.

Vertex degree the number of vertices adjacent to a given vertex.

Walk a sequence of (not necessarily) different consecutive vertices and edges in a graph.

References

1. Euler L. (1736) *Comm. Acad. Sci. Imp. Petrop.*, **8**, 128–140.
2. Biggs, N.L., Lloyd, E. K. and Wilson, L. (1976) *Graph Theory 1736–1936*. Clarendon Press, Oxford.
3. Sylvester, J. J. (1877–1878) *Nature* **17**, 284–285.
4. Harary, F. Ed. (1968) *Graph Theory and Theoretical Physics*. Academic Press.
5. Trinajstić, N. (1992) *Chemical Graph Theory*. CRC Press, Boca Raton, FL.
6. Berkolaiko, G., Kuchment, P. (2013) *Introduction to Quantum Graphs*, vol. 186, American Mathematical Society, Providence, RI.
7. Harary, F. (1969) *Graph Theory*. Addison-Wesley, Reading, MA.
8. Watts, D. J., Strogatz, S. H. (1998) *Nature* **393**, 440–442.
9. Newman, M. E. J., Strogatz, S. H., Watts, D. J. (2001) *Phys. Rev. E* **64**, 026118.
10. Canadell, E., Doublet, M.-L., Iung, C. (2012) *Orbital Approach to the Electronic Structure of Solids*. Oxford University Press, Oxford.
11. Kutzelnigg, W. (2006) *J. Comput. Chem.* **28**, 25–34.
12. Powell, B.J. (2009) An Introduction to Effective Low-Energy Hamiltonians in Condensed Matter Physics and Chemistry. arXiv preprint arXiv:0906.1640.
13. Gutman, I. (2005). *J. Serbian Chem. Soc.* **70**, 441–456.

14. Borovićanin, B., Gutman, I. (2009) In *Applications of Graph Spectra*, D. Cvetković and I. Gutman Eds. Mathematical Institute SANU, pp. 107–122.
15. Cheng, B., Liu, B. (2007) *Electron. J. Linear Algebra* **16** (2007), 60–67.
16. Tasaki, H. (1999) *J. Phys.: Cond. Mat.*, **10**, 4353.
17. Lieb, E. H. (1989) *Phys. Rev. Lett.* **62**, 1201 (Erratum **62**, 1927, (1989)).
18. Morita, Y., Suzuki, S., Sato, K., Takui, T. (2011) *Nat. Chem.* **3**, 197–204.
19. Essam, J. W. (1971) *Discrerte Math.* **1**, 83–112.
20. Beaudin, L., Ellis-Monaghan, J., Pangborn, G., Shrock, R. (2010) *Discrete Math.* **310**, 2037–2053.
21. Welsh, D. J. A., Merino, C. (2000) *J. Math. Phys.* **41**, 1127–1152.
22. Bollobás, B., (1998) *Modern Graph Theory*. Springer-Verlag, New York.
23. Ellis-Monaghan, J.A. and Merino, C. (2011) In *Structural Analysis of Complex Networks*, ed. M. Dehmer, Birkhauser, Boston, MA, pp. 219–255.
24. Welsh, D. (1999) *Random Struct. Alg.*, **15**, 210–228.
25. Bogner, C. (2010) *Nucl. Phys. B Proc. Suppl.* **205**, 116–121.
26. Bogner, C. and Weinzierl S. (2010) *Int. J. Mod. Phys. A* **25**, 2585.
27. Weinzierl, S. (2010) Introduction to Feynman Integrals. arXiv preprint arXiv:1005.1855.
28. Dodgson, C. L. (1866) *Proc. R. Soc. London* **15**, 150–155.
29. Doyle, P. and Snell, J. (1984) *Random Walks and Electric Networks*. Carus Mathematical Monographs, vol. 22, The Mathematical Association of America, Washington, DC.
30. Klein, D. J., Randić, M. (1993) *J. Math. Chem.* **12**, 81–95.
31. Bapat, R. B., Gutman, I., Xiao, W. (2003) *Z. Naturforsch.* **58a**, 494 – 498.
32. Xiao, W., Gutman, I., (2003) *Theor. Chem. Acc.* **110**, 284–289.
33. Gutman, I., Xiao, O. (2004) *Bull. Acad. Serb. Sci. Arts.* **29**, 15–23.
34. Ghosh, A., Boyd, S., Saberi, A. (2008) *SIAM Rev.* **50**, 37–66.
35. Estrada, E., Hatano, N., Benzi, M., (2012) *Phys. Rep.* **514**, 89–119.
36. Erdös, P., Rényi, A. (1959) *Publ. Math. Debrecen* **5**, 290–297.
37. Janson, S. (2005) *J. Combin. Prob. Comput.* **14**, 815–828.
38. Wigner, E. P. (1955) *Ann. Math.* **62**, 548–564.
39. Standish, R. K. (2008) In *Intelligent Complex Adaptive Systems*, Yang, A. and Shan, Y. eds, IGI Global: Hershey, PA, pp. 105–124, arXiv:0805.0685.
40. Ottino, J. M. (2003) *AIChE J.*, **49**, 292–299.
41. Estrada, E. (2011) *The Structure of Complex Networks. Theory and Applications*. Oxford University Press, Oxford.
42. Milgram, S. (1967) *Psychol. Today* **2**, 60–67.
43. Barrat, A., Weigt, M. (2000) *Eur. Phys. J. B* **13**, 547–560.
44. Foss, S., Korshunov, D., Zachary, S. (2011) *An Introduction to Heavy-Tailed and Subexponential Distributions*. Springer, Berlin.
45. Clauset, A., Rohilla Shalizi, C., Newman, M. E. J. (2010) *SIAM Rev.* **51**, 661–703.
46. Barabási, A.-L. Albert, R. (1999) *Science* **286**, 509–512.
47. Bollobás, B., Riordan, O. (2004) *Combinatorica* **24**, 5–34.
48. Bollobás, B. (2003) In *Handbook of Graph and Networks: From the Genome to the Internet*, Bornholdt, S., Schuster, H. G. Eds. Wiley-VCH Verlag GmbH, Weinheim, pp. 1–32.
49. Dorogovtsev, S. N. and Mendes, J. F. F. (2003) *Evolution of Networks: From Biological Nets to the Internet and WWW*. Oxford University Press, Oxford.
50. Milo, R., Shen-Orr, S., Itzkovitz, S. Kashtan, N. Chklovskii, D. Alon, U. (2002) *Science* **298**, 824–827.
51. Milo, R., Itzkovitz, S., Kashtan, N., Levitt, R., Shen-Orr, S., Ayzenshtat, I., Sheffer, M., Alon, U. (2004) *Science* **303**, 1538–1542.
52. Freeman, L. C. (1979) *Social Networks* **1**, 215–239.
53. Katz, L. (1953) *Psychometrica* **18**, 39–43.
54. Bonacich, P. (1972) *J. Math. Sociol.* **2**, 113–120.
55. Bonacich, P., (1987) *Am. J. Soc.* **92**, 1170–1182.
56. Langville, A. N., Meyer, C. D. (2006). *Google's PageRank and Beyond. The Science of Search Engine Rankings*. Princeton University Press, Princeton, NJ.
57. Estrada, E., Rodríguez-Velázquez, J. A. (2005) *Phys. Rev. E* **71**, 056103.

58. Estrada, E., Hatano, N. (2007) *Chem. Phys. Lett.* **439**, 247–251.
59. Estrada, E. (2000) *Chem. Phys. Lett.* **319**, 713–718.
60. Estrada, E., Hatano, N. (2008) *Phys. Rev. E* **77**, 036111.
61. Fortunato, S. (2010) *Phys. Rep.* **486**, 75–174.
62. Girvan, M., Newman, E. J. (2002) *Proc. Natl. Acad. Sci. U.S.A.* **99**, 7821–7826.
63. Newman, M. E. J. (2006) *Proc. Natl. Acad. Sci. U.S.A.* **103**, 8577–8582.
64. Olfati-Saber, R., Fax, J. A., Murray, R. M. (2007) *Proc. IEEE* **95**, 215–233.
65. Arenas, A., Diaz-Guilera, A., Pérez-Vicente, C. J. (2006). *Physica D* **224**, 27–34.
66. Chen, G., Wang, X., Li, X., Lü , J. (2009) In *Recent Advances in Nonlinear Dynamics and Synchronization,* K. Kyamakya Ed., Springer-Verlag, Berlin, pp. 3–16.
67. Barahona, M., Pecora, L. M. (2002) *Phys. Rev. Lett.* **89**, 054101.
68. Barrat, A., Barthélemy, M., Vespignani, A. (2008) *Dynamical Processes on Complex Networks*. Cambridge University Press, Cambridge.
69. Keeling, M. J., Eames, K. T. (2005) *J. R. Soc. Interface* **2**, 295–307.

Further Reading

Bollobás, B. (1998) *Modern Graph Theory*. Springer, Berlin.

Caldarelli, G. (2007) *Scale-Free Networks. Complex Webs in Nature and Technology*. Oxford University Press, Oxford.

Cvetković, D., Rowlinson, P., Simić, S. (2010) *An Introduction to the Theory of Graph Spectra*. Cambridge University Press, Cambridge.

Nakanishi, N. (1971) *Graph Theory and Feynman Integrals*. Gordon and Breach.

5
Group Theory

Robert Gilmore

5.1
Introduction

Symmetry has sung its siren song to physicists since the beginning of time, or since even before there were physicists. Today, the ideas of symmetry are incorporated into a subject with the less imaginative and suggestive name of *group theory*. This chapter introduces many of the ideas of group theory that are important in the natural sciences.

Natural philosophers in the past have come up with many imaginative arguments for estimating physical quantities. They have often used out of the box methods that were proprietary, to pull rabbits out of hats. When these ideas were made available to a wider audience, they were often improved upon in unexpected and previously unimaginable ways. A number of these methods are precursors of group theory. These are dimensional analysis, scaling theory, and dynamical similarity. We review these three methods in Section 5.2.

In Section 5.3, we get down to the business at hand, introducing the definition of a group and giving a small set of important definitions (e.g., isomorphism and homomorphism). Others will be introduced later in a context in which they make immediate sense. In Sections 5.4–5.6, we present some useful examples of groups ranging from finite and infinite discrete groups through matrix groups to Lie groups. These examples include transformation groups, which played an important if under-recognized rôle in the development of classical physics, in particular, the theories of special and general relativity. The relation between these theories and group theory is indicated in Section 5.9.

The study of Lie groups is greatly simplified when carried out on their "infinitesimal" versions. These are Lie algebras, which are introduced in Section 5.7. In this section, we introduce many of the important concepts and provide examples to illustrate all of them. One simple consequence of these beautiful developments is the possibility of studying and classifying Riemannian symmetric spaces. These are Riemannian spaces with a particular symmetry that is, effectively, time reversal invariance at each point. These spaces

Mathematical Tools for Physicists, Second Edition. Edited by Michael Grinfeld.

are cosets (quotients) of one Lie group by another. They are introduced in Section 5.8.

Despite this important rôle in the development of physics, groups existed at the fringe of the physics of the early twentieth century. It was not until the theory of the linear matrix representations of groups was invented that the theory of groups migrated from the outer fringes to play a more central rôle in physics. Important points in the theory of representations are introduced in Section 5.10. Representations were used in an increasingly imaginative number of ways in physics throughout the twentieth century. Early on, they were used to label states in quantum systems with a symmetry group: for example, the rotation group $SO(3)$. Once states were named, degeneracies could be predicted and computations simplified. Such applications are indicated in Section 5.11. Later, they were used when symmetry was not present, or just the remnant of a broken symmetry was present. When used in this sense, they are often called "dynamical groups." This type of use greatly extended the importance of group theory in physics. Some such applications of group theory are presented in Section 5.12.

As a latest tour de force in the development of physics, groups play a central rôle in the formulation of gauge theories. These theories describe the interactions between fermions and the bosons and lie at the heart of the standard model. We provide the simplest example of a gauge theory, based on the simplest compact one-parameter Lie group $U(1)$, in Section 5.13.

For an encore, in Section 5.14, we show how the theory of the special functions of mathematical physics (Legendre and associated Legendre functions, Laguerre and associated Laguerre functions, Gegenbauer, Chebyshev, Hermite, Bessel functions, and others) are subsumed under the theory of representations of some low-dimensional Lie groups. The classical theory of special functions came to fruition in the mid nineteenth century, long before Lie groups and their representations were even invented.

5.2 Precursors to Group Theory

The axioms used to define a group were formulated in the second half of the nineteenth century. Long before then, the important ideas underlying these axioms were used to derive classical results (for example, Pythagoras' theorem: see the following text) in alternative, simpler, and/or more elegant ways, to obtain new results, or to consolidate different results under a single elegant argument. In this section, we survey some of these imaginative lines of thought. We begin with a simple argument due to Barenblatt that has been used to derive Pythagoras' theorem. We continue with a discussion of the central features of dimensional analysis and illustrate how this tool can be used to estimate the size of a hydrogen atom. We continue in the same vein, using scaling arguments to estimate the sizes of other "atom-like" structures based on the known size of the hydrogen atom. We conclude this section with a brief description of dynamical similarity and how the arguments intrinsic to this line of thinking can be used to estimate one of Kepler's laws and to place four classical mechanics laws (Kepler, Newton, Galileo, Hooke) in a common framework.

We emphasize that group theory is not used explicitly in any of these arguments but its fingerprints are everywhere. These digressions should serve as appetizers to indicate the power of the tool called group theory in modern physical theories.

5.2.1 Classical Geometry

Barenblatt [1] has given a beautiful derivation of Pythagoras' theorem that is out of the box and suggests some of the ideas behind dimensional analysis. The area of the right triangle $\Delta(a, b, c)$ is $1/2ab$ (Figure 5.1). Dimensionally, the area is proportional to square of any of the sides, multiplied by some factor. We make a unique choice of side by choosing the hypotenuse, so that $\Delta(a, b, c) = c^2 \times f(\theta)$, θ is one of the two acute angles, and $f(\theta) \neq 0$ unless $\theta = 0$ or $\pi/2$. Equating the two expressions

$$f(\theta) = \frac{1}{2}\left(\frac{a}{c}\right)\left(\frac{b}{c}\right) = \frac{1}{2}\left(\frac{b}{c}\right)\left(\frac{a}{c}\right) \overset{\text{symmetry}}{=} f\left(\frac{\pi}{2} - \theta\right). \quad (5.1)$$

This shows (a) that the same function $f(\theta)$ applies for all similar triangles and (b) $f(\theta) = f(\pi/2 - \theta)$. The latter result is due to reflection "symmetry" of the triangle about the bisector of the right angle: the triangle changes but its area does not. We need (a) alone to prove Pythagoras' theorem. The proof is in the figure caption.

5.2.2 Dimensional Analysis

How big is a hydrogen atom?

The size of the electron "orbit" around the proton in the hydrogen atom ought to depend on the electron mass m_e, or more precisely the electron–proton reduced mass $\mu = m_e M_P/(m_e + M_P)$. It should also depend on the value of Planck's constant h or reduced Planck's constant $\hbar = h/2\pi$. Since the interaction between the proton with charge e and the electron with charge $-e$ is electromagnetic, of the form $V(r) = -e^2/r$ (Gaussian units), it should depend on e^2.

Mass is measured in grams. The dimensions of the charge coupling e^2 are determined by recognizing that e^2/r is a (potential) energy, with dimensions $M^1L^2T^{-2}$. We will use capital letters M (mass), L (length), and T (time) to characterize the three independent dimensional "directions." As a result, the charge coupling strength e^2 has dimensions ML^3T^{-2} and is measured in $g(cm)^3\ s^{-2}$. The quantum of action $\hbar$ has dimensions $[\hbar] = ML^2T^{-1}$. Here and below we use the standard convention that $[*]$ is to be read "the dimensions of $*$ are."

Constant	Dimensions	Value	Units
μ	M	9.10442×10^{-28}	g
$\hbar$	ML^2T^{-1}	1.05443×10^{-27}	$g\ cm^2\ s^{-1}$
e^2	ML^3T^{-2}	2.30655×10^{-19}	$g\ cm^3\ s^{-2}$
a_0	L	?	cm

Can we construct something (e.g., Bohr orbit a_B) with the dimensions of length from m, e^2, and $\hbar$? To do this, we introduce three unknown exponents a, b, and c and

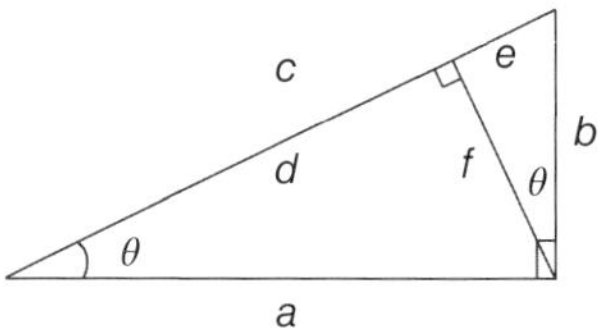

Figure 5.1 The area of the large right triangle is the sum of the areas of the two similar smaller right triangles: $\Delta(a, b, c) = \Delta(d, f, a) + \Delta(f, e, b)$, so that $c^2 f(\theta) = a^2 f(\theta) + b^2 f(\theta)$. Since $f(\theta) \neq 0$ for a nondegenerate right triangle, $a^2 + b^2 = c^2$.

write

$$\begin{aligned} a_B &\simeq m^a\, (e^2)^b\, \hbar^c \\ &= (M)^a\, (ML^3T^{-2})^b\, (ML^2T^{-1})^c \\ &= (M)^{a+b+c}\, L^{0a+3b+2c}\, T^{0a-2b-c} \end{aligned} \tag{5.2}$$

and set this result equal to the dimensions of whatever we would like to compute; in this case, the Bohr orbit a_B (characteristic atomic length), with $[a_B] = L$. This results in a matrix equation

$$\begin{bmatrix} 1 & 1 & 1 \\ 0 & 3 & 2 \\ 0 & -2 & -1 \end{bmatrix} \begin{bmatrix} a \\ b \\ c \end{bmatrix} = \begin{bmatrix} 0 \\ 1 \\ 0 \end{bmatrix}. \tag{5.3}$$

We can invert this matrix to find

$$\begin{bmatrix} 1 & 1 & 1 \\ 0 & 3 & 2 \\ 0 & -2 & -1 \end{bmatrix}^{-1} = \begin{bmatrix} 1 & -1 & -1 \\ 0 & -1 & -2 \\ 0 & 2 & 3 \end{bmatrix}. \tag{5.4}$$

This allows us to determine the values of the exponents that provide the appropriate combinations of important physical parameters to construct the characteristic atomic length:

$$\begin{bmatrix} a \\ b \\ c \end{bmatrix} = \begin{bmatrix} 1 & -1 & -1 \\ 0 & -1 & -2 \\ 0 & 2 & 3 \end{bmatrix} \begin{bmatrix} 0 \\ 1 \\ 0 \end{bmatrix} = \begin{bmatrix} -1 \\ -1 \\ 2 \end{bmatrix}. \tag{5.5}$$

This result tells us that

$$a_0 \sim m^{-1}(e^2)^{-1}(\hbar)^2 = \frac{\hbar^2}{me^2} \sim 10^{-8}\ \text{cm}. \tag{5.6}$$

To construct a characteristic atomic time, we can replace the vector col[0, 1, 0] in (5.5) by the vector col[0, 0, 1], giving us the result $\tau_0 \sim \hbar^3/m(e^2)^2$. Finally, to get a characteristic energy, we can form the combination $\mathcal{E} \sim ML^2T^{-2} = m(\hbar^2/me^2)^2 (\hbar^3/me^4)^{-2} = me^4/\hbar^2$. Another, and more systematic, way to get this result is to substitute the vector col$[1, 2, -2]^t$ for $[0, 1, 0]^t$ in (5.5).

Note that our estimate would be somewhat different if we had used h instead of $\hbar = h/2\pi$ in these arguments. We point out that this method is *very* useful for estimating the order of magnitude of physical parameters and in practised hands usually gets the prefactor within a factor of 10. The most critical feature of dimensional analysis is to identify the parameters that are most important in governing the science of the problem, and then to construct a result depending on only those parameters.

5.2.3 Scaling

Positronium is a bound state of an electron e with a positron $\bar{e}$, its antiparticle with mass m_e and charge $+e$. How big is positronium?

To address this question, we could work very hard and solve the Schrödinger equation for the positronium. This is identical to the Schrödinger equation for the hydrogen atom, except for replacing the hydrogen atom reduced mass $m_eM_p/(m_e + M_p) \simeq m_e$ by the positronium reduced mass $m_em_e/(m_e + m_e) = \frac{1}{2}m_e$. Or we could be lazy and observe that the hydrogen atom radius is inversely proportional to the reduced electron–proton mass, so the positronium radius should be inversely proportional to the reduced electron–positron mass $m_e/2$. Since the reduced electron–proton mass is effectively the electron mass, the positronium atom is approximately twice as large as the hydrogen atom.

In a semiconductor, it is possible to excite an electron (charge $-e$) from an almost filled (valence) band into an almost empty (conduction) band. This leaves a "hole" of charge $+e$ behind in the valence band. The positively charged hole in the valence band interacts with the excited electron in the conduction band through a reduced Coulomb interaction: $V(r) = -e^2/\epsilon r$. The strength of the interaction is reduced by screening effects that are swept into a phenomenological dielectric constant ϵ. In addition, the effective masses m_e^* of the excited electron and the left-behind hole m_h^* are modified from the free-space electron mass values by many-particle effects.

How big is an exciton in gallium arsenide (GaAs)? For this semiconductor, the phenomenological parameters are $\epsilon = 12.5$, $m_e^* = 0.07m_e$, $m_h^* = 0.4m_e$.

We extend the scaling argument above by computing the reduced mass of the electron–hole pair: $\mu_{e-h} = (0.07m_e)(0.4m_e)/(0.07 + 0.4)m_e = 0.06m_e$ and replacing e^2 in the expression (5.4) for the Bohr radius a_0 by e^2/ϵ. The effect is to multiply a_0 by $12.5/0.06 = 208$. The ground-state radius of the exciton formed in GaAs is about 10^{-6} cm. The ground-state binding energy is lower than the hydrogen atom binding energy of 13.6 eV by a factor of $0.06/12.5^2 = 3.8 \times 10^{-4}$ so it is 5.2 meV.

Scaling arguments such as these are closely related to renormalization group arguments as presented in Chapter 12.

5.2.4 Dynamical Similarity

Jupiter is about five times further (5.2 AU) from our Sun than the Earth. How many earth years does it take for Jupiter to orbit the Sun?

Landau and Lifshitz [2] provide an elegant solution to this simple question using similarity (scaling) arguments. The equation of motion for the Earth around the Sun is

$$m_E \frac{d^2\mathbf{x}_E}{dt_E^2} = -Gm_E M_S \frac{\hat{\mathbf{x}}_E}{|\mathbf{x}_E|^2}, \tag{5.7}$$

where $\mathbf{x}_E$ is a vector from the sun to the earth and $\hat{\mathbf{x}}_E$ the unit vector in this direction. If Jupiter is in a geometrically similar orbit, then $\mathbf{x}_J = \alpha\mathbf{x}_E$, with $\alpha = 5.2$. Similarly, time will evolve along the Jupiter trajectory in a scaled version of its evolution along the Earth's trajectory: $t_J = \beta t_E$. Substituting these scaled expressions into the equation of motion for Jupiter, and canceling out m_J from both sides, we find

$$\frac{\alpha}{\beta^2} \frac{d^2\mathbf{x}_E}{dt_E^2} = -\frac{1}{\alpha^2} GM_S \frac{\hat{\mathbf{x}}_E}{|\mathbf{x}_E|^2}. \tag{5.8}$$

This scaled equation for Jupiter's orbit can only be equated to the equation for the Earth's trajectory (the orbits are similar) provided $\alpha^3/\beta^2 = 1$. That is, $\beta = \alpha^{3/2}$, so that the time-scaling factor is $5.2^{3/2} = 12.5$.

We have derived Kepler's third law without even solving the equations of motion! Landau and Lifshitz point out that you can do even better than that. You don't even need to know the equations of motion to construct scaling relations when motion is described by a potential $V(\mathbf{x})$ that is homogeneous of degree k. This means that $V(\alpha\mathbf{x}) = \alpha^k V(\mathbf{x})$. When the equations of motion are derivable from a variational principle $\delta I = 0$, where

$$I = \int \left(m \left(\frac{d\mathbf{x}}{dt} \right)^2 - V(\mathbf{x}) \right) dt, \tag{5.9}$$

then the scaling relations $\mathbf{x} \to \mathbf{x}' = \alpha\mathbf{x}$, $t \to t' = \beta t$ lead to a modified action

$$I' = \frac{\alpha^2}{\beta} \int \left(m \left(\frac{d\mathbf{x}}{dt} \right)^2 - \alpha^{k-2}\beta^2 V(\mathbf{x}) \right) dt. \tag{5.10}$$

The Action I' is proportional to the original Action I, and therefore leads to the same equations of motion, only when $\alpha^{k-2}\beta^2 = 1$;

that is, the time elapsed, T, is proportional to the distance traveled, D, according to $T \simeq D^{(1-k/2)}$. Four cases are of interest.

$k = -1$ (Coulomb/gravitational potential) The period of a planetary orbit scales as the 3/2 power of the distance from the Sun (Kepler's third law).

$k = 0$ (No forces) The distance traveled is proportional to the time elapsed (essentially Newton's first law). To recover Newton's first law completely, it is only necessary to carry out the variation in (5.10), which leads to $\mathrm{d}/\mathrm{d}t(\mathrm{d}\mathbf{x}/\mathrm{d}t) = 0$.

$k = +1$ (Free fall in a homogeneous gravitational field) The potential $V(z) = mgz$ describes free fall in a homogeneous gravitational field. Galileo is reputed to have dropped rocks off the Leaning Tower of Pisa to determine that the distance fallen was proportional to the square of the time elapsed. The story is apocryphal: in fact, he rolled stones down an inclined plane to arrive at the result $\Delta z \simeq \Delta t^2$.

$k = +2$ (Harmonic oscillator potential) The period is independent of displacement: $\beta = 1$ independent of α. Hooke's law, $F = -kx$, $V(x) = 1/2kx^2$ leads to oscillatory motion whose frequency is independent of the amplitude of motion. This was particularly useful for constructing robust clocks.

These four historical results in the development of early science are summarized in Table 5.1 and Figure 5.2.

Table 5.1 Four important results in the historical development of science are consequences of scaling arguments.

k	Scaling	Law
-1	$T^2 \simeq D^3$	Kepler #3
0	$D \simeq T$	Newton #1
$+1$	$\Delta z \simeq \Delta t^2$	Galileo: rolling stones
$+2$	$T \simeq D^0$	Hooke

5.3 Groups: Definitions

In this section, we finally get to the point of defining what a group is by stating the group axioms (see [3–8]). These are illustrated in the following sections with a number of examples: finite groups, including the two-element group, the group of transformations that leaves the equilateral triangle invariant, the permutation group, point groups, and discrete groups with a countable infinite number of group elements, such as space groups. Then we introduce groups of transformations in space as matrix groups. Lie groups are introduced and examples of matrix Lie groups are presented. Lie groups are linearized to form their Lie algebras, and groups are recovered from their algebras by reversing the linearization procedure using the exponential mapping. Many of the important properties of Lie algebras are introduced, including isomorphisms among different representations of a Lie algebra. A powerful disentangling theorem is presented and illustrated in a very simple case that plays a prominent rôle in the field of quantum optics. We will use this result in Section 5.14.

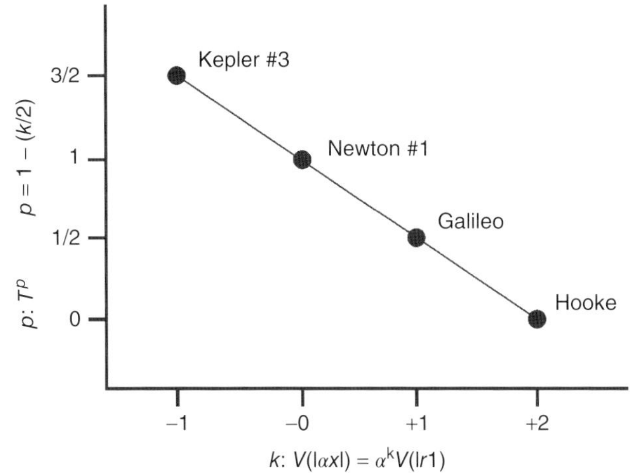

Figure 5.2 Four substantial advances in the development of early physics are summarized. Each is a consequence of using a homogeneous potential with a different degree k in a variational description of the dynamics. Scaling relates the size scale of the trajectory α to the time scale $\beta = \alpha^p$, $p = 1 - \frac{1}{2}k$ of the motion.

5.3.1 Group Axioms

A group G consists of

- a set of group elements: $g_0, g_1, g_2, g_3, \ldots \in G$
- a group operation, $\circ$, called group multiplication

that satisfy the following four axioms:

Closure: $g_i \in G,\ g_j \in G \Rightarrow g_i \circ g_j \in G$
Associativity: $(g_i \circ g_j) \circ g_k = g_i \circ (g_j \circ g_k)$
Identity: $g_0 \circ g_i = g_i = g_i \circ g_0$
Unique Inverse: $g_k \circ g_l = g_0 = g_l \circ g_k$.

Group multiplication $\circ$ has two inputs and one output. The two inputs must be members of the set. The first axiom (*Closure*) requires that the output must also be a member of the set.

The composition rule $\circ$ does not allow us to multiply three input arguments. Rather, two can be combined to one, and that output can be combined with the third. This can be done in two different ways that preserves the order (i, j, k). The second axiom (*Associativity*) requires that these two different ways give the same final output.

The third axiom (*Identity*) requires that a special group element exists. This, combined with any other group element, gives back exactly that group element.

The fourth axiom (*Unique Inverse*) guarantees that for each group element g_l, there is another uniquely defined group element g_k, with the property that the product of the two is the unique identity element $g_k \circ g_l = g_0$. It is a simple matter to prove that $g_l \circ g_k = g_0$.

Remark 5.1 – Indexes: The notation (subscripts $i, j, k, \ldots$) may suggest that the indices are integers. This is not generally true: for continuous groups the indices are points in some subspace of a Euclidean space or a more complicated manifold.

Remark 5.2 – Commutativity: In general, the output of the group multiplication depends on the order of the inputs: $g_i \circ g_j \neq g_j \circ g_i$. If the result is independent of the order the group is said to be *commutative*.

It is not entirely obvious that the *Unique Inverse* axiom is needed. It is included among the axioms because many of our uses involve relating measurements made by two observers. For example, if Allyson on the Earth can predict something about the length of a year on Jupiter, then Bob on Jupiter should just as well be able to predict the length of Allyson's year on Earth. Basically, this axiom is an implementation of Galileo's principle of relativity.

5.3.2 Isomorphisms and Homomorphisms

It is often possible to compare two different groups. When it is possible, it is very useful. Suppose we have two groups G with group elements $g_0, g_1, g_2, \ldots$ and group composition law $g_i \circ g_j = g_k$ and H with group elements $h_0, h_1, h_2, \ldots$ and group composition law $h_i \diamond h_j = k_k$. A mapping f from G to H is a *homomorphism* if it preserves the group operation:

$$f(g_i \circ g_j) = f(g_i) \diamond f(g_j). \tag{5.11}$$

In this expression $f(g_*) \in H$, so the two group elements $f(g_i)$ and $f(g_j)$ can only be combined using the combinatorial operation $\diamond$. If (5.11) is true for all pairs of group elements in G, the mapping f is a homomorphism.

If G has four elements I, C_4, C_4^2, C_4^3 and H has two Id, C_2, the mapping $f(I) = f(C_4^2) = \text{Id}$, $\quad f(C_4) = f(C_4^3) = C_2$, the mapping f is a homomorphism. If the mapping f is a homomorphism *and* is also 1 : 1, it is called an *isomorphism*. Under these conditions, the inverse mapping also is an isomorphism:

$$f^{-1}(h_p \diamond h_q) = f^{-1}(h_p) \circ f^{-1}(h_q). \tag{5.12}$$

As an example, an isomorphism exists between the four group elements I, C_4, C_4^2, C_4^3 and the 2×2 matrices with $f(C_4) = \begin{bmatrix} 0 & 1 \\ -1 & 0 \end{bmatrix}$.

5.4 Examples of Discrete Groups

5.4.1 Finite Groups

5.4.1.1 The Two-Element Group Z_2

The simplest nontrivial group has one additional element beyond the identity e: $G = \{e, g\}$ with $g \circ g = e$. This group can act in our three-dimensional space R^3 in several different ways:

Reflection: $(x, y, z) \stackrel{g=\sigma_Z}{\rightarrow} (+x, +y, -z)$

Rotation: $(x, y, z) \stackrel{g=R_Z(\pi)}{\rightarrow} (-x, -y, +z)$

Inversion: $(x, y, z) \stackrel{g=\mathcal{P}}{\rightarrow} (-x, -y, -z)$

These three different actions of the order-two group on R^3 describe: reflections in the x-y plane, σ_Z; rotations around the Z-axis through π radians, $R_Z(\pi)$; and inversion in the origin, the parity element, $\mathcal{P}$. They can be distinguished by their *matrix representations,* which are

$$\overset{\sigma_Z}{\begin{bmatrix} +1 & 0 & 0 \\ 0 & +1 & 0 \\ 0 & 0 & -1 \end{bmatrix}} \quad \overset{R_Z(\pi)}{\begin{bmatrix} -1 & 0 & 0 \\ 0 & -1 & 0 \\ 0 & 0 & +1 \end{bmatrix}}$$

$$\overset{\mathcal{P}}{\begin{bmatrix} -1 & 0 & 0 \\ 0 & -1 & 0 \\ 0 & 0 & -1 \end{bmatrix}}. \qquad (5.13)$$

5.4.1.2 Group of Equilateral Triangle C_{3v}

The six elements that map the equilateral triangle to itself constitute the group C_{3v} (cf. Figure 5.3). There are three distinct types of elements:

Identity. The element e does nothing: it maps each vertex into itself.

Rotations. Two rotations $C_3^{\pm}$ about the center of the triangle through $\pm 2\pi/3$ radians.

Reflections. There are three reflections σ_i, each in a straight line through the center of the triangle and the vertex i ($i = 1, 2, 3$, cf. Figure 5.3).

These elements can be defined by their action on the vertices of the triangle. For example

$$C_3^+ \begin{pmatrix} 1 & 2 & 3 \\ 2 & 3 & 1 \end{pmatrix} \begin{bmatrix} 0 & 1 & 0 \\ 0 & 0 & 1 \\ 1 & 0 & 0 \end{bmatrix}. \qquad (5.14)$$

The first description (( )) says that the rotation C_3^+ maps vertex 1 to vertex 2, $2 \to 3$ and $3 \to 1$. The second description ([ ]) can be understood as follows:

$$\begin{bmatrix} 2 \\ 3 \\ 1 \end{bmatrix} = \begin{bmatrix} 0 & 1 & 0 \\ 0 & 0 & 1 \\ 1 & 0 & 0 \end{bmatrix} \begin{bmatrix} 1 \\ 2 \\ 3 \end{bmatrix}. \qquad (5.15)$$

The group multiplication law can be represented through a 6×6 matrix (Cayley multiplication table) that describes the output of $g_i \circ g_j$, with g_i listed by column and g_j by row:

g_i \ g_j	e	C_3^+	C_3^-	σ_1	σ_2	σ_3
e	e	C_3^+	C_3^-	σ_1	σ_2	σ_3
C_3^+	C_3^+	C_3^-	e	σ_2	σ_3	σ_1
C_3^-	C_3^-	e	C_3^+	σ_3	σ_1	σ_2
σ_1	σ_1	σ_3	σ_2	e	C_3^-	C_3^+
σ_2	σ_2	σ_1	σ_3	C_3^+	e	C_3^-
σ_3	σ_3	σ_2	σ_1	C_3^-	C_3^+	e.

(5.16)

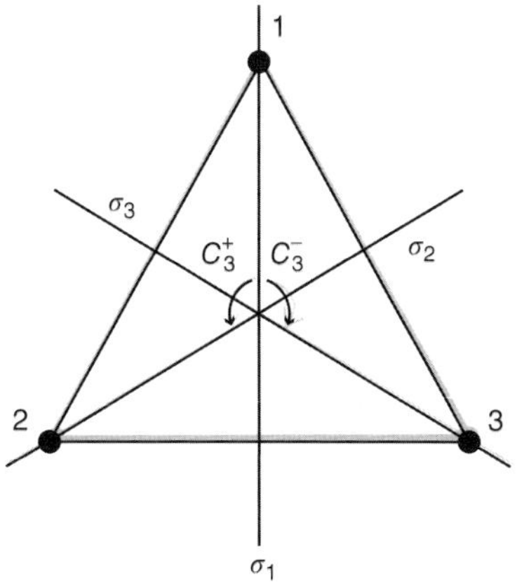

Figure 5.3 The group of the equilateral triangles consists of (a) the identity group element e; (b) two rotations $C_3^{\pm}$ by $\pm 2\pi/3$ about the centroid of the triangle; and (c) three reflections σ_i in straight lines between the centroid and each of the vertices i.

This table makes clear that the group is not commutative: $C_3^- = \sigma_1 \circ \sigma_2 \neq \sigma_2 \circ \sigma_1 = C_3^+$.

The partition of the six elements in this group into three subsets of geometrically equivalent transformations is typical of any group. These subsets are called *classes.* Classes are defined by the condition

$$\textbf{Class}: \; \{h_1, h_2, \ldots\} \; g \circ h_i \circ g^{-1} = h_j \text{ all } g \in G. \tag{5.17}$$

All elements in the same class have essentially the same properties. They are *equivalent* under a group transformation. The three classes for the finite group C_{3v} are $\{e\}, \{C_3^+, C_3^-\}, \{\sigma_1, \sigma_2, \sigma_3\}$.

It is clear from the group multiplication table (5.16) that C_{3v} has a number of (proper) *subgroups*: three subgroups of order two $\{e, \sigma_1\}, \{e, \sigma_2\}, \{e, \sigma_3\}$ and one of order three $\{e, C_3^+, C_3^-\}$. For technical reasons, the single element $\{e\}$ and the entire group C_{3v} are also considered to be subgroups of C_{3v} (they are not proper subgroups). Whenever a group G has a subgroup H, it is always possible to write each group element in G as the product of an element in H with "something else": $g_i = h_j C_k$. For example, if H is the subgroup of order three we can choose the two elements C_1, C_2 $(2 = 6/3)$ as $\{e, \sigma_1\}$. Then from (5.16)

$$\begin{aligned} \{e, C_3^+, C_3^-\} \circ e &= \{e, C_3^+, C_3^-\} \\ \{e, C_3^+, C_3^-\} \circ \sigma_1 &= \{\sigma_1\, \sigma_2, \sigma_3\}. \end{aligned} \tag{5.18}$$

Since in some sense G is a product of group elements in the subgroup H with group elements in C $(G = H \circ C)$, we can formally write C as the "quotient" of G by the subgroup H: $C = H\backslash G$. If we composed in the reversed order: $G = CH$, then we could write $C = G/H$.

The set C is called a *coset.* It is not unique, but for finite groups, its *order* (number of elements in the set) is unique: the quotient of the order G by the order of H. A coset may or may not be a group, depending whether the subgroup H is invariant in G ($gHg^{-1} \subset H$ for all $g \in G$) or not.

Remarks.

When G and H are Lie groups of dimensions d_G and d_H, G/H is a manifold of dimension d_G/d_H, and under a broad set of conditions this manifold has a geometric structure imparted by a Riemannian metric derived from the geometric properties of the two groups G and H [3, 4].

5.4.1.3 Cyclic Groups C_n

The cyclic group consists of all rotations of the circle into itself through the angle $2\pi/n$ radians, and integer multiples of this angle. There are n such elements. The rotation through $2\pi k/n$ radians is obtained by applying the "smallest" rotation (also called C_n or C_n^1) k times. This smallest rotation is called a *generator* of the group. The group is commutative. There are therefore as many classes as group elements. The group elements can be put in 1 : 1 correspondence with the complex numbers and also with real 2×2 matrices:

$$\left[e^{i2\pi k/n}\right] \overset{1\times1}{\leftarrow} C_n^k \overset{2\times2}{\rightarrow} \begin{bmatrix} \cos\frac{2\pi k}{n} & \sin\frac{2\pi k}{n} \\ -\sin\frac{2\pi k}{n} & \cos\frac{2\pi k}{n} \end{bmatrix}. \tag{5.19}$$

with $k = 0, 1, 2, \ldots, n-1$ or $k = 1, 2, \ldots, n$. Every element in the group can be obtained by multiplying C_n^1 by itself. In the same way, the 1×1 complex matrix with $k = 1$ is the generator for the 1×1 matrix representation of this group and the 2×2 real matrix with $k = 1$ is the generator for the 2×2 matrix representation of the group.

5.4.1.4 Permutation Groups S_n

Permutation groups act to interchange things. For example, if we have n numbers $1, 2, 3, \ldots, n$, each permutation group element will act to scramble the order of the integers differently. Two useful ways to describe elements in the permutation group are shown in (5.14) for the permutation group on three vertices of an equilateral triangle. In the first case, the extension of this notation for individual group elements consists of a matrix with two rows, the top showing the ordering before the element is applied, the bottom showing the ordering after the group element has been applied. In the second case shown in (5.14) the extension consists of $n \times n$ matrices with exactly one +1 in each row and each column. The order of S_n is $n!$. Permutation groups are noncommutative for $n > 2$. $S_3 = C_{3v}$.

The permutation group plays a fundamental role in both mathematics and physics. In mathematics, it is used to label the irreducible tensor representations of all Lie groups of interest. In physics, it is required to distinguish among different states that many identical particles (either bosons or fermions) can assume.

5.4.1.5 Generators and Relations

If G is a discrete group, with either a finite or a countable number of group elements, it is useful to introduce a small set of *generators* $\{\sigma_1, \sigma_2, \dots, \sigma_k\}$ to describe the group. Every element in the group can be represented as a product of these generators and/or their inverses in some order.

For example, if there is only one generator $\{\sigma\}$ and every group element can be written in the form $g_n = \sigma^n$, $n = \dots, -2, -1, 0, 1, 2, \dots$ then G has a countable number of group elements. It is called a *free group* with one generator. If there are two generators $\{\sigma_1, \sigma_2\}$, the two generators commute $\sigma_1\sigma_2 = \sigma_2\sigma_1$, and every group element can be expressed in the form $g_{m,n} = \sigma_1^m \sigma_2^n$ (m, n integers), the group is the free group with two commuting generators. Free groups with $k > 2$ generators are defined similarly. Free groups with $1, 2, 3, \dots$ generators are isomorphic to groups that act on periodic lattices in $1, 2, 3, \dots$ dimensions.

Often the generators satisfy *relations*. For example, a single generator σ may satisfy the relation $\sigma^p = I$. Then there are exactly p distinct group elements $I = \sigma^0, \sigma^1, \sigma^2, \dots, \sigma^{p-1}$. The group with one generator and one relation is the cyclic group C_p. Generators $\{\sigma_1, \sigma_2, \dots, \sigma_k\}$ and relations $f_l(\{\sigma_1, \sigma_2, \dots, \sigma_k\}) = I$, $l = 1, 2, \dots$ have been used to define many different groups. In fact, every discrete group is either defined by a set of generators and relations, or else a subgroup of such a group. The symmetric group S_n is defined by $n-1$ generators $\sigma_i, i = 1, 2, \dots, n-1$ and the relations $\sigma_i^2 = I$, $\sigma_i\sigma_j = \sigma_j\sigma_i$ if $j \neq i \pm 1$, and $\sigma_i\sigma_{i+1}\sigma_i = \sigma_{i+1}\sigma_i\sigma_{i+1}$. The tetrahedral ($T$), octahedral ($O$), and icosahedral ($I$) point groups are defined by two generators and three relations: $\sigma_1^2 = I, \sigma_2^3 = I, (\sigma_1\sigma_2)^p = I$ with $p = 3, 4, 5$, respectively. The quaternion group Q_8 can be defined with two generators and two relations $\sigma_1\sigma_2\sigma_1 = \sigma_2, \sigma_2\sigma_1\sigma_2 = \sigma_1$ or in terms of three generators and four relations $\sigma_1^4 = \sigma_2^4 = \sigma_3^4 = I, \sigma_1\sigma_2\sigma_3 = I$. In the latter case, the three generators can be chosen as 2×2 matrices that are the three Pauli spin matrices, multiplied by $i = \sqrt{-1}$.

The study of discrete groups defined by generators and relations has a long and very rich history [9].

5.4.2 Infinite Discrete Groups

5.4.2.1 Translation Groups: One Dimension

Imagine a series of points at locations na along the straight line, where a is a physical parameter with dimensions of length ($[a] = L$) and n is an integer. The group that leaves this set invariant consists of rigid displacements through integer multiples of the fundamental length. The element T_{ka} displaces the point at na to position $(n+k)a$. This group has a single generator T_a, and $T_{ka} = T_a \circ T_a \circ \cdots \circ T_a = T_a^k$. It is convenient to represent these group elements by 2×2 matrices

$$T_{ka} \to \begin{bmatrix} 1 & ka \\ 0 & 1 \end{bmatrix}. \tag{5.20}$$

In this representation, group composition is equivalent to matrix multiplication. The group is commutative. The generator for the group and this matrix representation is obtained by setting $k = 1$. There is also an entire set of 1×1 complex matrix representations indexed by a real parameter p with generator $T_a \to [e^{ipa}]$. The representations with $p' = p + 2\pi/a$ are equivalent, so all the inequivalent complex representations can be parameterized by real values of p in the range $0 \leq p < 2\pi/a$ or, more symmetrically $-\pi/a \leq p \leq \pi/a$, with the end points identified. The real parameter p is in the dual space to the lattice, called the *first Brillouin zone*.

5.4.2.2 Translation Groups: Two Dimensions

Now imagine a series of lattice points in a plane at positions $\mathbf{x} = i_1\mathbf{f}_1 + i_2\mathbf{f}_2$. Here i_1, i_2 are integers and the vectors $\mathbf{f}_1, \mathbf{f}_2$ are not colinear but otherwise arbitrary. Then the set of rigid displacements (j_1, j_2) move lattice points $\mathbf{x}$ to new locations as per

$$T_{j_1\mathbf{f}_1+j_2\mathbf{f}_2}\left(i_1\mathbf{f}_1 + i_2\mathbf{f}_2\right) = (i_1+j_1)\mathbf{f}_1 + (i_2+j_2)\mathbf{f}_2. \tag{5.21}$$

Generalizing (5.20), there is a simple 1 : 1 (or *faithful*) matrix representation for this group of rigid translations:

$$T_{j_1\mathbf{f}_1+j_2\mathbf{f}_2} \to \begin{bmatrix} 1 & 0 & j_1|\mathbf{f}_1| \\ 0 & 1 & j_2|\mathbf{f}_2| \\ 0 & 0 & 1 \end{bmatrix}. \tag{5.22}$$

Extension to groups of rigid displacements of lattices in higher dimensions is straightforward.

5.4.2.3 Space Groups

When $|\mathbf{f}_1| = |\mathbf{f}_2|$ and the two vectors are orthogonal, rotations through $k\pi/2$ $(k = 1, 2, 3)$ radians about any lattice point map the lattice into itself. So also do reflections in lines perpendicular to $|\mathbf{f}_1|$ and $|\mathbf{f}_2|$ as well as lines perpendicular to $\pm|\mathbf{f}_1| \pm |\mathbf{f}_2|$. This set of group elements contains displacements, rotations, and reflections. It is an example of a two-dimensional *space group*. There are many other space groups in two dimensions and very many more in three dimensions. These groups were first used to enumerate the types of regular lattices that nature allows in two and three dimensions [7]. After the development of quantum mechanics, they were used in another way (depending on the theory of representations): to give names to wavefunctions that describe electrons (and also phonons) in these crystal lattices.

5.5 Examples of Matrix Groups

5.5.1 Translation Groups

The group of rigid translations of points in R^3 through distances a_1 in the x-direction, a_2 in the y-direction, and a_3 in the z-direction can be described by simple block 4×4 $(4 = 3 + 1)$ matrices:

$$T_{a_1,a_2,a_3} \rightarrow \left[\begin{array}{ccc|c} 1 & 0 & 0 & a_1 \\ 0 & 1 & 0 & a_2 \\ 0 & 0 & 1 & a_3 \\ \hline 0 & 0 & 0 & 1 \end{array}\right]. \qquad (5.23)$$

If the **a** belong to a lattice, the group is discrete. If they are continuous, the group is continuous and has dimension three.

5.5.2 Heisenberg Group H_3

The Heisenberg group H_3 plays a fundamental role in quantum mechanics. As it appears in the quantum theory it is described by "infinite-dimensional" matrices. However, the group itself is three dimensional. In fact, it has a simple faithful description in terms of 3×3 matrices depending on three parameters:

$$h(a,b,c) = \left[\begin{array}{c|cc} 1 & a & c \\ 0 & 1 & b \\ \hline 0 & 0 & 1 \end{array}\right]. \qquad (5.24)$$

The matrix representation is faithful because any matrix of the form (5.24) uniquely defines the abstract group element $h(a,b,c)$. The group is not commutative. The group multiplication law can be easily seen via matrix multiplication:

$$h_1 h_2 = h_3 = h(a_3, b_3, c_3) = \left[\begin{array}{c|cc} 1 & a_1 + a_2 & c_1 + c_2 + a_1 b_2 \\ 0 & 1 & b_1 + b_2 \\ \hline 0 & 0 & 1 \end{array}\right]. \qquad (5.25)$$

The group composition law given in (5.25) defines the Heisenberg group. The result $c_3 = c_1 + c_2 + a_1 b_2$ leads to remarkable noncommutativity properties among canonically conjugate variables in the quantum theory: $[p, x] = \hbar / i$.

5.5.3 Rotation Group $SO(3)$

The set of rigid rotations of R^3 forms a group. It is conveniently represented by a faithful 3×3 matrix. The 3×3 matrix describing rotations about an axis of unit length $\hat{\mathbf{n}}$ through an angle θ, $0 \leq \theta \leq \pi$ is

$$(\hat{\mathbf{n}}, \theta) \rightarrow I_3 \cos\theta + \hat{\mathbf{n}} \cdot \mathbf{L} \sin\theta + \begin{bmatrix} \hat{n}_1 \\ \hat{n}_2 \\ \hat{n}_3 \end{bmatrix} \times \begin{bmatrix} \hat{n}_1 & \hat{n}_2 & \hat{n}_3 \end{bmatrix} (1 - \cos\theta). \qquad (5.26)$$

Here $\mathbf{L}$ is a set of three 3×3 angular momentum matrices

$$L_x = \begin{bmatrix} 0 & 0 & 0 \\ 0 & 0 & 1 \\ 0 & -1 & 0 \end{bmatrix} \quad L_y = \begin{bmatrix} 0 & 0 & -1 \\ 0 & 0 & 0 \\ 1 & 0 & 0 \end{bmatrix}$$

$$L_z = \begin{bmatrix} 0 & 1 & 0 \\ -1 & 0 & 0 \\ 0 & 0 & 0 \end{bmatrix}. \qquad (5.27)$$

The matrix multiplying $(1 - \cos\theta)$ in (5.26) is a 3×3 matrix: it is the product of a 3×1 column matrix with a 1×3 row matrix. We will show later how this marvelous expression has been derived (cf. (5.51)).

There is a 1:1 correspondence between points in the interior of a ball of radius π and rotations through an angle in the range $0 \leq \theta < \pi$. Two points on the surface $(\hat{\mathbf{n}}, \pi)$ and $(-\hat{\mathbf{n}}, \pi)$ describe the same rotation. The parameter space describing this group is not a simply connected submanifold of R^3: it is a doubly connected manifold. The relation between continuous groups and their underlying parameter space involves some fascinating topology.

5.5.4 Lorentz Group $SO(3,1)$

The Lorentz group is the group of linear transformations that leave invariant the square of the distance between two nearby points in space–time: (cdt, dx, dy, dz) and (cdt', dx', dy', dz'). The distance can be written in matrix form:

$$(cd\tau)^2 = (cdt)^2 - (dx^2 + dy^2 + dz^2)$$
$$= \begin{bmatrix} cdt & dx & dy & dz \end{bmatrix} \begin{bmatrix} +1 & 0 & 0 & 0 \\ 0 & -1 & 0 & 0 \\ 0 & 0 & -1 & 0 \\ 0 & 0 & 0 & -1 \end{bmatrix} \begin{bmatrix} cdt \\ dx \\ dy \\ dz \end{bmatrix}. \tag{5.28}$$

If the infinitesimals in the primed coordinate system are related to those in the unprimed coordinate system by a linear transformation – $dx'^{\mu} = M^{\mu}_{\nu} dx^{\nu}$ – then the matrices M must satisfy the constraint (^{t} means matrix transpose and the summation convention has been used: doubled indices are summed over.)

$$M^t I_{1,3} M = I_{1,3}, \tag{5.29}$$

where $I_{1,3}$ is the diagonal matrix diag$(+1, -1, -1, -1)$. The matrices M belong to the orthogonal group $O(1,3)$. This is a six-parameter group. Clearly the rotations (three dimensions worth) form a subgroup, represented by matrices of the form $\left[\begin{array}{c|c} \pm 1 & 0 \\ \hline 0 & \pm R(\hat{\mathbf{n}}, \theta) \end{array}\right]$, where $R(\hat{\mathbf{n}}, \theta)$ is given in (5.26). This group has four disconnected components, each connected to a 4×4 matrix of the form $\begin{bmatrix} 1 & 0 \\ 0 & I_3 \end{bmatrix}$, $\begin{bmatrix} 1 & 0 \\ 0 & -I_3 \end{bmatrix}$, $\begin{bmatrix} -1 & 0 \\ 0 & I_3 \end{bmatrix}$, or $\begin{bmatrix} -1 & 0 \\ 0 & -I_3 \end{bmatrix}$. We choose the component connected to the identity I_4. This is the special Lorentz group $SO(1,3)$. A general matrix in this group can be written in the form

$$SO(1,3) = B(\boldsymbol{\beta}) R(\boldsymbol{\theta}), \tag{5.30}$$

where the matrices $B(\boldsymbol{\beta})$ describe *boost* transformations and $R(\boldsymbol{\theta}) = R(\hat{\mathbf{n}}, \theta)$. A boost transformation maps a coordinate system at rest to a coordinate moving with velocity $\mathbf{v} = c\boldsymbol{\beta}$ and with axes parallel to those in the stationary coordinate system.

Since every group element in $SO(1,3)$ can be expressed as the product of a rotation element with a boost, we can formally write boost transformations as elements in a coset: $B(\boldsymbol{\beta}) = SO(1,3)/SO(3)$.

A general boost transformation can be written in the form

$$B(\boldsymbol{\beta}) = \left[\begin{array}{c|c} \gamma & \gamma\boldsymbol{\beta} \\ \hline \gamma\boldsymbol{\beta} & I_3 + (\gamma - 1)\frac{\beta_i \beta_j}{\boldsymbol{\beta}\cdot\boldsymbol{\beta}} \end{array}\right]. \tag{5.31}$$

For example, a boost in the x-direction with $\mathbf{v}/c = (\beta, 0, 0)$ has the following effect on coordinates ($\beta = |\boldsymbol{\beta}|$):

$$\begin{bmatrix} ct \\ x \\ y \\ z \end{bmatrix}' = \left[\begin{array}{c|ccc} \gamma & \gamma\beta & 0 & 0 \\ \hline \gamma\beta & \gamma & 0 & 0 \\ 0 & 0 & 1 & 0 \\ 0 & 0 & 0 & 1 \end{array}\right] \begin{bmatrix} ct \\ x \\ y \\ z \end{bmatrix}$$
$$= \begin{bmatrix} \gamma(ct + \beta x) \\ \gamma(x + \beta ct) \\ y \\ z \end{bmatrix}. \tag{5.32}$$

Here $\gamma^2 - (\beta\gamma)^2 = 1$ so $\gamma = 1/\sqrt{1-\beta^2}$. In the nonrelativistic limit, $x' = \gamma(x + \beta ct) \to x + vt$, so β has an interpretation of $\boldsymbol{\beta} = \mathbf{v}/c$.

The product of two boosts in the same direction is obtained by matrix multiplication. This can be carried out on a 2×2 submatrix of that given in (5.32):

$$B(\beta_1)B(\beta_2) = \begin{bmatrix} \gamma_1 & \beta_1\gamma_1 \\ \beta_1\gamma_1 & \gamma_1 \end{bmatrix} \begin{bmatrix} \gamma_2 & \beta_2\gamma_2 \\ \beta_2\gamma_2 & \gamma_2 \end{bmatrix} = \begin{bmatrix} \gamma_{\text{tot}} & \beta_{\text{tot}}\gamma_{\text{tot}} \\ \beta_{\text{tot}}\gamma_{\text{tot}} & \gamma_{\text{tot}} \end{bmatrix}. \quad (5.33)$$

Simple matrix multiplication shows $\beta_{\text{tot}} = (\beta_1 + \beta_2)/(1 + \beta_1\beta_2)$, which is the relativistic velocity addition formula for parallel velocity transformations.

When the boosts are not parallel, their product is a transformation in $SO(1,3)$ that can be written as the product of a boost with a rotation:

$$B(\boldsymbol{\beta}_1)B(\boldsymbol{\beta}_2) = B(\boldsymbol{\beta}_{\text{tot}})R(\boldsymbol{\theta}). \quad (5.34)$$

Multiplying two boost matrices of the form given in (5.31) leads to a simple expression for γ_{tot} and a more complicated expression for $\boldsymbol{\beta}_{\text{tot}}$

$$\begin{aligned} \gamma_{\text{tot}} &= \gamma_1\gamma_2(1 + \boldsymbol{\beta}_1 \cdot \boldsymbol{\beta}_2) \\ \gamma_{\text{tot}}\boldsymbol{\beta}_{\text{tot}} &= \left[\gamma_1\gamma_2 + (\gamma_1 - 1)\gamma_2 \frac{\boldsymbol{\beta}_1 \cdot \boldsymbol{\beta}_2}{\boldsymbol{\beta}_1 \cdot \boldsymbol{\beta}_1}\right] \boldsymbol{\beta}_1 + \gamma_2\boldsymbol{\beta}_2. \end{aligned} \quad (5.35)$$

This shows what is intuitively obvious: the boost direction is in the plane of the two boosts. Less obvious is the rotation required by noncollinear boosts. It is around an axis parallel to the cross product of the two boosts. When the two boosts are perpendicular, the result is

$$\hat{\mathbf{n}} \sin(\theta) = -\boldsymbol{\beta}_1 \times \boldsymbol{\beta}_2 \cdot \frac{\gamma_1\gamma_2}{1 + \gamma_1\gamma_2}. \quad (5.36)$$

When one of the boosts is infinitesimal, we find

$$B(\boldsymbol{\beta})B(\delta\boldsymbol{\beta}) = B(\boldsymbol{\beta} + \mathrm{d}\boldsymbol{\beta})R(\hat{\mathbf{n}}\mathrm{d}\theta). \quad (5.37)$$

Multiplying out these matrices and comparing the two sides gives

$$\begin{aligned} \mathrm{d}\boldsymbol{\beta} &= \gamma^{-1}\delta\boldsymbol{\beta} + \left(\frac{\gamma^{-1} - 1}{\gamma\beta^2}\right)(\boldsymbol{\beta} \cdot \delta\boldsymbol{\beta})\boldsymbol{\beta} \\ \hat{\mathbf{n}}\mathrm{d}\theta &= \left(\frac{1 - \gamma^{-1}}{\beta^2}\right)\delta\beta \times \beta. \end{aligned} \quad (5.38)$$

In the nonrelativistic limit, when $\boldsymbol{\beta}$ is also small, $1 - \gamma^{-1}/\beta^2 \to 1/2$. This (in)famous factor of 1/2 is known as the "Thomas factor" in atomic physics.

5.6 Lie Groups

The group elements g in a Lie group are parameterized by points $\mathbf{x}$ in a manifold $\mathcal{M}^n$ of dimension n: $g = g(\mathbf{x})$, $\mathbf{x} \in \mathcal{M}^n$. The product of two group elements $g(\mathbf{x})$ and $g(\mathbf{y})$ is parameterized by a point $\mathbf{z}$ in the manifold: $g(\mathbf{x}) \circ g(\mathbf{y}) = g(\mathbf{z})$, where $\mathbf{z} = \mathbf{z}(\mathbf{x}, \mathbf{y})$. This composition law can be very complicated. It is necessarily nonlinear (cf. (5.25) for H_3) unless the group is commutative. For example, the parameter space for the group $SO(3)$ consists of points in R^3: $\boldsymbol{\theta} = (\hat{\mathbf{n}}, \theta)$. Only a compact subspace consisting of a sphere of radius π is needed to parameterize this Lie group.

Almost all of the Lie groups of use to physicists exist as matrix groups. For this reason, it is possible for us to skip over the fundamental details of whether the composition law must be analytic and the elegant details of their definition and derivations. A composition law can be determined as follows:

1. Construct a useful way to parameterize each group element as a matrix depending on a suitable number of parameters: $\mathbf{x} \rightarrow g(\mathbf{x})$.
2. Perform matrix multiplication $g(\mathbf{x}) \circ g(\mathbf{y})$ of the matrices representing the two group elements.
3. Find the parameters $\mathbf{z} = \mathbf{z}(\mathbf{x}, \mathbf{y})$ of the group element that corresponds to the product of the two matrices given in Step 2.

We list several types of matrix groups below.

- *GL(n;R), GL(n;C), GL(n;Q).* These *general linear groups* (general means $\det \neq 0$) consist of $n \times n$ invertible matrices, each of whose n^2 matrix elements are real numbers, complex numbers, or quaternions. The group composition law is matrix multiplication. The numbers of real parameters required to specify an element in these groups are $n^2, 2 \times n^2, 4 \times n^2$, respectively.
- *SL(n;R), SL(n;C).* The *special linear groups* ("S" for special means det = +1) are subgroups of $GL(n; R)$ and $GL(n; C)$ containing the subgroup of matrices with determinant +1. The real dimensions of these groups are $(n^2 - 1) \times \dim(F)$ where $\dim(F) = (1, 2)$ for $F = (R, C)$.
- *O(n), U(n), Sp(n).* Three important classes of groups are defined by placing quadratic constraints on matrices. The orthogonal group $O(n)$ is the subgroup of $GL(n; R)$ containing only matrices M that satisfy $M^t I_n M = I_n$. Here we use previously introduced notation: I_n is the unit $n \times n$ matrix and ^{t} signifies the transpose of the matrix. This constraint arises in a natural way when requiring that linear transformations in a real n-dimensional linear vector space preserve a positive-definite inner product. The unitary group $U(n)$ is the subgroup of $GL(n; C)$ for which the matrices M satisfy $M^\dagger I_n M = I_n$, where † signifies the adjoint, or complex conjugate transpose matrix. The symplectic group $Sp(n)$ is defined similarly for the quaternions. In this case, † signifies quaternion conjugate transpose. The real dimensions of these groups are $n(n-1)/2, n^2$, and $n(2n+1)$, respectively.
- *SO(n), SU(n).* For the group $O(n)$, the determinant of any group element is a real number whose modulus is +1: that is , ± 1. Placing the special constraint on the group of orthogonal transformations reduces the "number" of elements in the group by one half (in a measure theoretic sense) but does not reduce the dimension of the space required to parameterize the elements in this group. For the group $U(n)$, the determinant of any group element is a complex number whose modulus is +1: that is, $e^{i\phi}$. Placing the special constraint on $U(n)$ reduces the dimension by one: $\dim SU(n) = \dim U(n) - 1 = n^2 - 1$. The symplectic group $Sp(n)$ has determinant +1.
- *O(p,q), U(p,q), Sp(p,q).* These groups are defined by replacing I_n in the definitions for $O(n), U(n), Sp(n)$ by the matrix $I_{p,q} = \left[\begin{array}{c|c} +I_p & 0 \\ \hline 0 & -I_q \end{array}\right]$. These groups preserve an *indefinite* nonsingular metric in linear vector spaces of dimension $(p+q)$. The groups $O(n), U(n), Sp(n)$ are *compact* (a useful topological concept) and so are relatively easy to deal with. This means effectively that only a finite volume of parameter space is required to parameterize every element in the group. The groups $O(p,q), U(p,q), Sp(p,q)$ are not compact

if both p and q are nonzero. Further, $O(p,q) \simeq O(q,p)$ by a simple similarity transformation, and similarly for the others.

5.7 Lie Algebras

A Lie algebra is a linear vector space on which an additional composition law [,] is defined. If X, Y, Z are elements in a Lie algebra $\mathcal{L}$, then linear combinations are in the Lie algebra: $\alpha X + \beta Y \in \mathcal{L}$ (this is the linear vector space property), the commutator of two elements $[X, Y]$ is in $\mathcal{L}$, and, in addition, the new composition law satisfies the following axioms:

$$\begin{aligned} &[\alpha X + \beta Y, Z] = \alpha [X, Z] + \beta [Y, Z], \\ &[X, Y] + [Y, X] = 0, \\ &[X, [Y, Z]] + [Y, [Z, X]] + [Z, [X, Y]] = 0. \end{aligned} \tag{5.39}$$

The first of these conditions preserves the linear vector space property of $\mathcal{L}$. The second condition defines the commutator bracket [,] as an antisymmetric composition law: $[X, Y] = -[Y, X]$, and the third imposes an integrability constraint called the *Jacobi identity*.

In principle, commutators are defined by the properties presented in (5.39), whether or not composition of the operators X and Y is defined. If this composition is defined, then $[X, Y] = XY - YX$ and the commutator can be computed by applying the operators X and Y in different orders and subtracting the difference. For example, if $X = y\partial_z - z\partial_y$ (here $\partial_z = \partial/\partial z$) and if $Y = z\partial_x - x\partial_z$, then the operator XY can be applied to a general function whose second partial derivatives are continuous to give $XYf(x,y,z) = yf_x + yzf_{xz} - z^2 f_{xy} - xyf_{zz} + zxf_{yz}$. The value of $YXf(x,y,z)$ is computed similarly, and the difference is $(XY - YX)f = yf_x - xf_y = (y\partial_x - x\partial_y)f$. Since this holds for an arbitrary function $f(x,y,z)$ for which all second partial derivatives exist and are independent of the order taken, we find $[X, Y] = (XY - YX) = (y\partial_x - x\partial_y)$.

5.7.1 Structure Constants

When the underlying linear vector space for $\mathcal{L}$ has dimension n, it is possible to choose a set of n basis vectors (matrices, operators) X_i. The commutation relations are encapsulated by a set of *structure constants* C_{ij}^{k} that are defined by

$$\left[X_i, X_j\right] = C_{ij}^{k} X_k. \tag{5.40}$$

A Lie algebra is defined by its structure constants.

5.7.2 Constructing Lie Algebras by Linearization

The Lie algebra for a Lie group is constructed by linearizing the constraints that define the Lie group in the neighborhood of the identity I. Matrix Lie algebras are obtained for $n \times n$ matrix Lie groups by linearizing the matrix group in the neighborhood of the unit matrix I_n. A Lie group and its Lie algebra have the same dimension.

In the neighborhood of the identity the groups, $GL(n;R), GL(n;C),$ *and* $GL(n,Q)$ have the form

$$GL(n;F) \rightarrow I_n + \delta M. \tag{5.41}$$

where δM is an $n \times n$ matrix, all of whose matrix elements are small. Over the real, complex, and quaternion fields the matrix elements are small real or complex numbers or small quaternions. Quaternions q

can be expressed as 2×2 complex matrices using the Pauli spin matrices σ_μ:

$$q \to (c_0, c_1) = (r_0, r_1, r_2, r_3) = \sum_{\mu=0}^{3} r_\mu \sigma_\mu$$
$$= \begin{bmatrix} r_0 + ir_3 & r_1 - ir_2 \\ r_1 + ir_2 & r_0 - ir_3 \end{bmatrix}. \quad (5.42)$$

The Lie algebras $\mathfrak{gl}(n; F)$ of $GL(n; F)$ have dimensions $\dim(F) \times n^2$, with $\dim(F) = 1, 2, 4 = 2^2$ for $F = R, C, Q$.

For the special linear groups, the determinant of a group element near the identity is

$$\det(I_n + \delta M) = 1 + \mathrm{tr}\delta M + \text{h.o.t.} \quad (5.43)$$

In order to ensure the unimodular condition, the Lie algebras of the special linear groups consist of traceless matrices. The Lie algebra $\mathfrak{sl}(n; R)$ of $SL(n; R)$ consists of real traceless $n \times n$ matrices. It has dimension $n^2 - 1$. The Lie algebra $\mathfrak{sl}(n; C)$ of $SL(n; C)$ consists of traceless complex $n \times n$ matrices. It has real dimension $2n^2 - 2$.

Many Lie groups are defined by a metric-preserving condition: $M^\dagger \mathbf{G} M = \mathbf{G}$, where $\mathbf{G}$ is some suitable metric matrix (see the discussion of the Lorentz group $SO(3, 1) \simeq SO(1, 3)$ that preserves the metric $(+1, +1, +1, -1)$ in Section 5.5.4 and the groups *O(p,q)*, *U(p,q)*, *Sp(p,q)* in Section 5.6). The linearization of this condition is

$$M^\dagger \mathbf{G} M = (I_n + \delta M)^\dagger \mathbf{G} (I_n + \delta M) = \mathbf{G}$$

so that from $M^\dagger \mathbf{G} M = \mathbf{G}$, it follows that, neglecting small terms of order two

$$\delta M^\dagger \mathbf{G} + \mathbf{G} \delta M = 0. \quad (5.44)$$

Thus the Lie algebras $\mathfrak{so}(n; R)$, $\mathfrak{su}(n; C)$, $\mathfrak{sp}(n; Q)$, which correspond to the case $\mathbf{G} = I_n$, consist of real antisymmetric matrices $M^t = -M$, complex traceless antihermitian matrices $M^\dagger = -M$, and quaternion antihermitian matrices $M^\dagger = -M$, respectively.

The Lie algebra $\mathfrak{so}(3)$ of the rotation group $SO(3) = SO(3; R)$ consists of real 3×3 antisymmetric matrices. This group and its algebra are three dimensional. The Lie algebra (it is a linear vector space) is spanned by three "basis vectors." These are 3×3 antisymmetric matrices. A standard choice for these basis vectors is given in (5.27). Their commutation relations are given by

$$[L_i, L_j] = -\epsilon_{ijk} L_k. \quad (5.45)$$

The structure constants for $\mathfrak{so}(3)$ are $C_{ij}^k = -\epsilon_{ijk}$, $1 \le i, j, k \le 3$. Here ϵ_{ijk} is the "sign symbol" (antisymmetric Levi–Civita 3-tensor), which is zero if any two symbols are the same, +1 for a cyclic permutation of integers (123), and −1 for a cyclic permutation of (321).

Two Lie algebras with the same set of structure constants are isomorphic [3–6]. The Lie algebra of 2×2 matrices obtained from $\mathfrak{su}(2)$ is spanned by three operators that can be chosen as $(i/2)$ times the Pauli spin matrices (cf. (5.42)):

$$S_1 = \frac{i}{2}\begin{bmatrix} 0 & 1 \\ 1 & 0 \end{bmatrix} \quad S_2 = \frac{i}{2}\begin{bmatrix} 0 & -i \\ +i & 0 \end{bmatrix}$$
$$S_3 = \frac{i}{2}\begin{bmatrix} 1 & 0 \\ 0 & -1 \end{bmatrix}. \quad (5.46)$$

These three operators satisfy the commutation relations

$$[S_i, S_j] = -\epsilon_{ijk} S_k. \quad (5.47)$$

As a result, the Lie algebra for the group $\mathfrak{so}(3)$ of rotations in R^3 is isomorphic to the Lie algebra $\mathfrak{su}(2)$ for the group of unimodular metric-preserving rotations in a complex two-dimensional space, $SU(2)$.

Spin and orbital rotations are intimately connected.

5.7.3 Constructing Lie Groups by Exponentiation

The mapping of a Lie group, with a complicated nonlinear composition, down to a Lie algebra with a simple linear combinatorial structure plus a commutator, would not be so useful if it were not possible to undo this mapping. In effect, the linearization is "undone" by the exponential map. For an operator X the exponential is defined in the usual way:

$$\exp(X) = \mathrm{e}^X = I + X + \frac{X^2}{2!} + \frac{X^3}{3!} + \cdots = \sum_{k=0}^{\infty} \frac{X^k}{k!}. \tag{5.48}$$

The radius of convergence of the exponential function is infinite. This means that we can map a Lie algebra back to its parent Lie group in an algorithmic way.

We illustrate with two important examples. For the first, we construct a simple parameterization of the group $SU(2)$ by exponentiating its Lie algebra. The Lie algebra is given in (5.46). Define $M = i/2\hat{\mathbf{n}} \cdot \boldsymbol{\sigma}\theta$. Then $M^2 = -(\theta/2)^2 I_2$ is a diagonal matrix. The exponential expansion can be rearranged to contain even powers in one sum and odd powers in another:

$$\mathrm{e}^M = I_2\left(1 - \frac{(\theta/2)^2}{2!} + \frac{(\theta/2)^4}{4!} - \cdots\right) + M\left(1 - \frac{(\theta/2)^2}{3!} + \frac{(\theta/2)^4}{5!} - \cdots\right). \tag{5.49}$$

The even terms sum to $\cos(\theta/2)$ and the odd terms sum to $\sin(\theta/2)/(\theta/2)$. The result is

$$\exp\left(\frac{i}{2}\hat{\mathbf{n}} \cdot \boldsymbol{\sigma}\theta\right) = \cos\frac{\theta}{2} I_2 + i\hat{\mathbf{n}} \cdot \boldsymbol{\sigma} \sin\frac{\theta}{2}. \tag{5.50}$$

A similar power series expansion involving the angular momentum matrices in (5.27) leads to the parameterization of the rotation group elements given in (5.26). Specifically, $\exp(\hat{\mathbf{n}} \cdot \mathbf{L}\theta) =$

$$I_3 \cos\theta + \hat{\mathbf{n}} \cdot \mathbf{L} \sin\theta + \begin{bmatrix} \hat{n}_1 & \hat{n}_2 & \hat{n}_3 \end{bmatrix} \begin{bmatrix} \hat{n}_1 \\ \hat{n}_2 \\ \hat{n}_3 \end{bmatrix} (1 - \cos\theta). \tag{5.51}$$

The Lie groups $SO(3)$ and $SU(2)$ possess isomorphic Lie algebras. The Lie algebra is three dimensional. The basis vectors in $\mathfrak{so}(3)$ can be chosen as the angular momentum matrices given in (5.27) and the basis vectors for $\mathfrak{su}(2)$ as $i/2$ times the Pauli spin matrices, as in (5.46). A point in the Lie algebra (e.g., R^3) can be identified by a unit vector $\hat{\mathbf{n}}$ and a radial distance from the origin θ. Under exponentiation, the point $(\hat{\mathbf{n}}, \theta)$ maps to the group element given in (5.51) for $SO(3)$ and in (5.50) for $SU(2)$.

The simplest way to explore how the Lie algebra parameterizes the two groups is to look at how points along a straight line through the origin of the Lie algebra map to elements in the two groups. For simplicity, we choose the z-axis. Then $(\hat{\mathbf{z}}, \theta)$ maps to

$$\begin{bmatrix} \mathrm{e}^{i\theta/2} & 0 \\ 0 & \mathrm{e}^{-i\theta/2} \end{bmatrix} \in SU(2), \quad \begin{bmatrix} \cos\theta & \sin\theta & 0 \\ -\sin\theta & \cos\theta & 0 \\ 0 & 0 & 1 \end{bmatrix} \in SO(3). \tag{5.52}$$

As θ increases from 0 to 2π, the $SU(2)$ group element varies from $+I_2$ to $-I_2$, while the $SO(3)$ group element starts at I_3 and returns to $+I_3$. The $SU(2)$ group element returns to the identity $+I_2$ only after θ increases from 2π to 4π. The rotations by θ and $\theta + 2\pi$ give the same group element in $SO(3)$ but they describe group elements in $SU(2)$ that differ by sign: $(\hat{\mathbf{z}}, 2\pi + \theta) = -I_2 \times (\hat{\mathbf{z}}, \theta)$. Two 2×2 matrices M and $-M$ in $SU(2)$ map to the same group element in $SO(3)$. In words, $SU(2)$ is a *double cover* of $SO(3)$.

For $SU(2)$, all points inside a sphere of radius 2π in the Lie algebra map to different group elements, and all points on the sphere surface map to one group element $-I_2$. The group $SU(2)$ is *simply connected.* Any path starting and ending at the same point (for example, the identity) can be continuously contracted to the identity.

By contrast, for $SO(3)$, all points inside a sphere of radius π in the Lie algebra map to different group elements, and two points $(\hat{\mathbf{n}}, \pi)$ and $-(\hat{\mathbf{n}}, \pi)$ at opposite ends of a straight line through the origin map to the same group element. The group $SO(3)$ is *not* simply connected. Any closed path from the origin that cuts the surface $\theta = \pi$ once (or an odd number of times) cannot be continuously deformed to the identity. The group $SO(3)$ is *doubly connected.*

This is the simplest example of a strong theorem by Cartan. There is a $1:1$ relation between Lie algebras and simply connected Lie groups. Every Lie group with the same (isomorphic) Lie algebra is either simply connected or else the quotient (coset) of the simply connected Lie group by a discrete invariant subgroup.

For matrix Lie groups, discrete invariant subgroups consist of scalar multiples of the unit matrix. For the isomorphic Lie algebras $\mathfrak{su}(2) = \mathfrak{so}(3)$, the Lie group $SU(2)$ is simply connected. Its discrete invariant subgroup consists of multiples of the identity matrix: $\{I_2, -I_2\}$. Cartan's theorem states $SO(3) = SU(2)/\{I_2, -I_2\}$. This makes explicit the $2 \downarrow 1$ nature of the relation between $SU(2)$ and $SO(3)$.

The group $SU(3)$ is simply connected. It discrete invariant subgroup consists of $\{I_3, \omega I_3, \omega^2 I_3\}$, with $\omega^3 = 1$. The only other Lie group with the Lie algebra $\mathfrak{su}(3)$ is the $3 \downarrow 1$ image $SU(3)/\{I_3, \omega I_3, \omega^2 I_3\}$. This group has a description in terms of real eight-dimensional matrices ("the eightfold way").

5.7.4 Cartan Metric

The notation for the structure constants C_{ij}^{k} for a Lie algebra gives the them appearance of being components of a tensor. In fact, they are: the tensor is first-order contravariant (in k) and second-order covariant, and antisymmetric, in i, j. It is possible to form a second-order symmetric covariant tensor (Cartan–Killing metric) from the components of the structure constant by double cross contraction:

$$g_{ij} = \sum_{rs} C_{ir}^{s} C_{js}^{r} = g_{ji}. \tag{5.53}$$

This real symmetric tensor "looks like" a metric tensor. In fact, it has very powerful properties. If g_{**} is nonsingular, the Lie algebra, and its Lie group, is "semisimple" or "simple" (these are technical terms meaning that the matrices describing the Lie algebras are either fully reducible or irreducible). If g_{**} is negative definite, the group is compact. It is quite remarkable that an algebraic structure gives such powerful topological information.

As an example, for $SO(3)$ and $SU(2)$ the Cartan–Killing metric equation (5.53) is

$$g_{ij} = \sum_{r,s} (-\epsilon_{irs})(-\epsilon_{jsr}) = -2\delta_{ij}. \tag{5.54}$$

For the real forms $SO(2,1)$ of $SO(3)$ and $SU(1,1)$ of $SU(2)$, the Cartan–Killing metric tensor is

$$g(\mathfrak{so}(2,1)) = g(\mathfrak{su}(1,1)) = 2\begin{bmatrix} +1 & 0 & 0 \\ 0 & +1 & 0 \\ 0 & 0 & -1 \end{bmatrix}. \quad (5.55)$$

The structure of this metric tensor (two positive diagonal elements or eigenvalues, and one negative) tells us about the topology of the groups: they have two noncompact directions and one compact direction. The compact direction describes the compact subgroups $SO(2)$ and $U(1)$, respectively.

5.7.5 Operator Realizations of Lie Algebras

Each Lie algebra has three useful operator realizations. They are given in terms of boson operators, fermion operators, and differential operators.

Boson annihilation operators b_i and creation operators $b_j^\dagger$ for independent modes $i,j = 1,2,\ldots,$ their fermion counterparts $f_i, f_j^\dagger$, and the operators ∂_i, x_j satisfy the following commutation or anticommutation relations

$$\begin{aligned} \left[b_i, b_j^\dagger\right] &= b_i b_j^\dagger - b_j^\dagger b_i &= \delta_{ij}, \\ \left\{f_i, f_j^\dagger\right\} &= f_i f_j^\dagger + f_j^\dagger f_i &= \delta_{ij}, \\ \left[\partial_i, x_j\right] &= \partial_i x_j - x_j \partial_i &= \delta_{ij}. \end{aligned} \quad (5.56)$$

In spite of the fact that bosons and differential operators satisfy commutation relations and fermion operators satisfy anticommutation (see the + sign in (5.56)) relations, bilinear combinations $Z_{ij} = b_i^\dagger b_j, f_i^\dagger f_j, x_i \partial_j$ of these operators satisfy commutation relations:

$$\left[Z_{ij}, Z_{rs}\right] = Z_{is}\delta_{jr} - Z_{rj}\delta_{si}. \quad (5.57)$$

These commutation relations can be used to associate operator algebras to matrix Lie algebras. The procedure is simple. We illustrate this for boson operators. Assume $A, B, C = [A, B]$ are $n \times n$ matrices in a matrix Lie algebra. Associate operator $\mathcal{A}$ to matrix A by means of

$$A \to \mathcal{A} = b_i^\dagger A_{ij} b_j \quad (5.58)$$

and similarly for other matrices. Then

$$\begin{aligned} [\mathcal{A}, \mathcal{B}] &= \left[b_i^\dagger A_{ij} b_j, b_r^\dagger B_{rs} b_s\right] \\ &= b_i^\dagger \, [A, B]_{is} \, b_s = b_i^\dagger C_{is} b_s = \mathcal{C}. \end{aligned} \quad (5.59)$$

This result holds if the bilinear combinations of boson creation and annihilation operators are replaced by bilinear combinations of fermion creation and annihilation operators or products of multiplication (by x_i) and differentiation (by ∂_j) operators.

One consequence of this matrix Lie algebra to operator algebra isomorphism is that any Hamiltonian that can be expressed in terms of bilinear products of creation and annihilation operators for either bosons or fermions can be studied in a simpler matrix form.

The operator algebra constructed from the spin operators in (5.46) has been used by Schwinger for an elegant construction of all the irreducible representations of the Lie group $SU(2)$ (cf. Section 5.10.5 and Figure 5.4).

We use the matrix-to-operator mapping now to construct a differential operator realization of the Heisenberg group, given in (5.24). Linearizing about the identify gives a three-dimensional Lie algebra of the form

$$\begin{bmatrix} 0 & l & d \\ 0 & 0 & r \\ 0 & 0 & 0 \end{bmatrix} = lL + rR + dD. \quad (5.60)$$

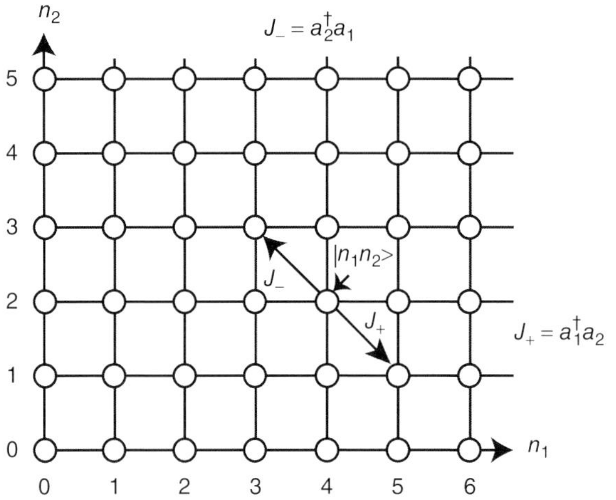

Figure 5.4 Angular momentum operators **J** have isomorphic commutation relations with specific bilinear combinations $b_i^\dagger b_j$ of boson creation and annihilation operators for two modes. The occupation number for the first mode is plotted along the x-axis and that for the second mode is plotted along the y-axis. The number-conserving operators act along diagonals similarly to the operators J_+, J_-, J_z to easily provide states and matrix elements for the $\mathfrak{su}(2)$ operators.

Here L, R, D are 3×3 matrices. The only nonzero commutator is $[L, R] = D$. The corresponding differential operator algebra is

$$\begin{bmatrix} x & y & z \end{bmatrix} \begin{bmatrix} 0 & l & d \\ 0 & 0 & r \\ 0 & 0 & 0 \end{bmatrix} \begin{bmatrix} \partial_x \\ \partial_y \\ \partial_z \end{bmatrix} = l\mathcal{L} + r\mathcal{R} + d\mathcal{D}. \tag{5.61}$$

The three differential operators are

$$\mathcal{L} = x\partial_y \quad \mathcal{R} = y\partial_z \quad \mathcal{D} = x\partial_z. \tag{5.62}$$

Among these operators, none depends on z (so ∂_z has nothing to operate on) and none contains ∂_x, so that in essence x is an irrelevant variable. A more economical representation of this algebra is obtained by zeroing out the cyclic variables z, ∂_x and replacing their duals by ∂_z, x by $+1$ (duality is under the commutator $[\partial_i, x_j] = \delta_{ij}$)

$$\begin{bmatrix} 1 & y & 0 \end{bmatrix} \begin{bmatrix} 0 & l & d \\ 0 & 0 & r \\ 0 & 0 & 0 \end{bmatrix} \begin{bmatrix} 0 \\ \partial_y \\ 1 \end{bmatrix} = l\mathcal{L}' + r\mathcal{R}' + d\mathcal{D}', \tag{5.63}$$

$$\mathcal{L}' = \partial_y, \quad \mathcal{R}' = y, \quad \mathcal{D}' = 1. \tag{5.64}$$

In essence, we have zeroed out the operators coupled to the vanishing rows and columns of the matrix Lie algebra and replaced their dual variables by 1. The representation given is essentially the Heisenberg representation of the position (y) and conjugate momentum ($p_y \simeq \partial_y$) operators in quantum mechanics.

5.7.6 Disentangling Results

It happens surprisingly often in distantly related fields of physics that expressions of the form $e^{x+\partial_x}$ are encountered. Needless to say, these are not necessarily endearing to work with. One approach to simplifying computations involving such operators is to rewrite the operator in such a way that all differential operators ∂_x act first, and all multiplications by x act last. One way to effect this decomposition is to cross one's fingers and write this operator as $e^{ax+b\partial_x} \simeq e^{ax}e^{b\partial_x}$ and hope for the best. Of course, this does not work, since the operators x and ∂_x do not commute.

Exponential operator rearrangements are called *disentangling theorems*. Since the exponential mapping is involved, powerful methods are available when the operators in the exponential belong to a finite-dimensional Lie algebra. Here is the algorithm:

1. Determine the Lie algebra.
2. Find a faithful finite-dimensional matrix representation of this Lie algebra.
3. Identify how you want the operators ordered in the final product of exponentials.
4. Compute this result in the faithful matrix representation.
5. Lift this result back to the operator form.

Here is how this algorithm works. The operators x and ∂_x have one nonzero commutator $[\partial_x, x] = 1$. These three operators close under commutation. They therefore form a Lie algebra. This is the algebra $\mathfrak{h}_3$ of the Heisenberg group, (5.64). We also have a faithful matrix representation of this Lie algebra, given in (5.63). We make the identification

$$e^{ax+b\partial_x} \to \exp\begin{bmatrix} 0 & b & 0 \\ 0 & 0 & a \\ 0 & 0 & 0 \end{bmatrix} = \begin{bmatrix} 1 & b & \frac{ab}{2} \\ 0 & 1 & a \\ 0 & 0 & 1 \end{bmatrix}. \tag{5.65}$$

Now we identify this matrix with

$$e^{rx}e^{dI}e^{l\partial_x} \to \begin{bmatrix} 1 & 0 & 0 \\ 0 & 1 & r \\ 0 & 0 & 1 \end{bmatrix} \begin{bmatrix} 1 & 0 & d \\ 0 & 1 & 0 \\ 0 & 0 & 1 \end{bmatrix} \begin{bmatrix} 1 & l & 0 \\ 0 & 1 & 0 \\ 0 & 0 & 1 \end{bmatrix}. \tag{5.66}$$

By multiplying out the three matrices in (5.66) and comparing with the matrix elements of the 3×3 matrix in (5.65), we learn that $l = b, r = a, d = ab/2$. Porting the results of this matrix calculation back to the land of operator algebras, we find

$$e^{ax+b\partial_x} = e^{ax}e^{ab/2}e^{b\partial_x}. \tag{5.67}$$

We will use this expression in Section 5.14.7 to construct a generating function for the Hermite polynomials.

5.8 Riemannian Symmetric Spaces

A Riemannian symmetric space is a manifold on which a Positive-definite metric can be defined everywhere. In addition, at each point p there is an isometry (transformation that leaves distances between points unchanged) that (i) leaves p fixed; (ii) is not the identity; and (iii) whose square is the identity. It was discovered by Cartan that Riemannian symmetric spaces are very closely related to Lie groups. Specifically, they are quotients of Lie groups by certain Lie subgroups [3, 4, 10]. We illustrate with some examples.

The Lie algebra for the Lorentz group consists of 4×4 matrices:

$$\mathfrak{so}(1,3) = \left[\begin{array}{c|ccc} 0 & \alpha_1 & \alpha_2 & \alpha_3 \\ \hline \alpha_1 & 0 & \theta_3 & -\theta_2 \\ \alpha_2 & -\theta_3 & 0 & \theta_1 \\ \alpha_3 & \theta_2 & -\theta_1 & 0 \end{array}\right]. \tag{5.68}$$

The Cartan–Killing metric for $SO(1,3)$ is given by the trace of the product of this matrix with itself:

$$g(\mathfrak{so}(1,3)) = 2(\alpha_1^2 + \alpha_2^2 + \alpha_3^2 - \theta_1^2 - \theta_2^2 - \theta_3^2). \tag{5.69}$$

The subalgebra of rotations $\mathfrak{so}(3)$ describes the compact subgroup $SO(3)$. The remaining three infinitesimal generators parameterized by α_i span the noncompact part of this group, the coset $SO(1,3)/SO(3)$, and exponentiate to boost elements.

Cartan has pointed out that it is often possible to find a linear transformation, T, of a Lie algebra $\mathfrak{g}$ to itself whose square is the identity: $T \neq I, T^2 = I$. Such a T has two eigenspaces, $\mathfrak{k}$ and $\mathfrak{p}$, with $T\mathfrak{g} = T(\mathfrak{k} \oplus \mathfrak{p}) = \mathfrak{k} \ominus \mathfrak{p}$. The two subspaces are orthogonal under the Cartan metric and satisfy the commutation relations:

$$\left[\mathfrak{k},\mathfrak{k}\right] \subseteq \mathfrak{k} \quad \left[\mathfrak{k},\mathfrak{p}\right] \subseteq \mathfrak{p} \quad \left[\mathfrak{p},\mathfrak{p}\right] \subseteq \mathfrak{k}. \tag{5.70}$$

When this is possible, $\exp(\mathfrak{k}) = K$ is a subgroup of G and $\exp(\mathfrak{p}) = P = G/K$. Further, if the Cartan–Killing metric is negative definite on $\mathfrak{k}$, and positive definite on $\mathfrak{p}$, then K is a maximal compact subgroup of G and the coset $P = G/K$ is a Riemannian symmetric space. A Riemannian symmetric space is *homogeoeous*: every point looks like every other point. It is not necessarily *isotropic*: every direction looks like every other direction. Spheres are homogeneous *and* isotropic.

For the Lorentz group $SO(1,3)$, by Cartan's criterion, $SO(3)$ is the maximal compact subgroup and the coset of boost transformations $B(\boldsymbol{\beta}) = SO(1,3)/SO(3)$ is a three-dimensional Riemannian space with positive-definite metric. In this case, the space is a 3-hyperboloid $(ct)^2 - x^2 - y^2 - z^2 = cst.$ embedded in R^4. The metric on this space is obtained by moving the metric $(1,1,1)$ at the origin $(x,y,z) = (0,0,0)$ over the embedded space using the set of Lorenz group transformations in the quotient space $SO(1,3)/SO(3)$. Cartan also showed that all Riemannian symmetric spaces arise as quotients of (simple) Lie groups by maximal compact subgroups.

5.9 Applications in Classical Physics

Group theory's first important role in physics came even before quantum mechanics was discovered. The two pillars of classical deterministic physics are mechanics and electrodynamics. Group theory played a fundamental role in rectifying the difficulties in describing the interactions between these two fields. The principle tool used, besides group theory, was Galileo's principle of relativity and an assumption about the underlying elegance of physical theories.

5.9.1 Principle of Relativity

The principle of relativity posits as follows: Two observers, S and S', observe the same physical system. Each knows how his coordinate system is related to the other's – that is, the transformation of coordinates that maps one coordinate system into the other. Assume both observers collect data on the same physical system. Given the data that S takes, and the coordinate transformation from S to S', S can predict the data that S' has recorded. *And he will be correct.*

Essentially, without this ability to communicate data among observers, there would be little point in pursuing the scientific method.

A second assumption that is used is usually not stated explicitly. This assumption is as follows: the quantitative formulation of physical laws is simple and elegant (whatever that means).

5.9.2 Making Mechanics and Electrodynamics Compatible

The quantitative formulation of mechanics for a single particle in an inertial frame is

$$\frac{\mathrm{d}\mathbf{p}}{\mathrm{d}t} = \mathbf{F}(\mathbf{x}), \tag{5.71}$$

where $\mathbf{p}$ is defined by $\mathbf{p} = md\mathbf{x}/dt$ for a particle with fixed mass m. The transformations from one inertial coordinate system to another consist of displacements in space $\mathbf{d}$, displacements in time d, rigid rotations R, and boosts with constant velocity $\mathbf{v}$ that keep the axis parallel. The space and time coordinates in S' are related to those in S by

$$\begin{aligned} \mathbf{x}' &= R\mathbf{x} + \mathbf{v}t + \mathbf{d} \\ t' &= t + d. \end{aligned} \tag{5.72}$$

In the inertial coordinate system S', Newton's equations are

$$\frac{\mathrm{d}\mathbf{p}'}{\mathrm{d}t} = \mathbf{F}'(\mathbf{x}') \qquad \begin{aligned} \mathbf{p}' &= R\mathbf{p} \\ \mathbf{F}' &= R\mathbf{F}. \end{aligned} \tag{5.73}$$

The equations of motion have the same vectorial form in both inertial coordinate systems (the simple and elegant assumption).

Newton's laws were incredibly successful in describing planetary motion in our solar system, so when Maxwell developed his laws for electrodynamics, it was natural to assume that they also retained their form under the set of inertial coordinate transformations given in (5.72). Applying these transformations to Maxwell's equations creates a big mess. But if one looks only at signals propagating along or opposite the direction of the velocity $\mathbf{v}$, these assumptions predict

$$(c\,\mathrm{d}t)' \to c'\,\mathrm{d}t, \quad c' = c \pm |\mathbf{v}|. \tag{5.74}$$

The round trip time for a light signal traveling in a cavity of length L as seen by S is $2L/c$, while in S' the time lapse was predicted to be $L/(c+v) + L/(c-v) = (2L/c)/(1-\beta^2)$, with $\beta = |\mathbf{v}|/c$.

This predicted difference in elapsed round trip times in "rest" and "moving" frames was thought to enable us to determine how fast the earth was moving in the Universe. As ever more precise measurements in the late nineteenth and early twentieth century led to greater disappointment, more and more bizarre explanations were created to "explain" this null result. Finally, Einstein and Poincaré returned to the culprit (5.74) and asserted what the experiments showed: $c' = c$, so that

$$\begin{aligned} &(c\,\mathrm{d}t)' \to c\,\mathrm{d}t', \\ &\mathrm{d}t' = \text{linear comb. } \mathrm{d}x, \mathrm{d}y, \mathrm{d}z, \mathrm{d}t. \end{aligned} \tag{5.75}$$

The condition that the distance function

$$\begin{aligned} (c\,\mathrm{d}\tau)^2 &= (c\,\mathrm{d}t)^2 - (\mathrm{d}x^2 + \mathrm{d}y^2 + \mathrm{d}z^2) \\ &= (c\,\mathrm{d}t')^2 - (\mathrm{d}x'^2 + \mathrm{d}y'^2 + \mathrm{d}z'^2) \end{aligned} \tag{5.76}$$

is invariant leads directly to the transformation law for infinitesimals $\mathrm{d}x'^{\mu} = \Lambda^{\mu}_{\nu}\,\mathrm{d}x^{\nu}$, where the 4×4 matrices belong to the Lorentz group $\Lambda \in SO(3,1)$, $\Lambda^{\mu}_{\nu} = \partial x'^{\mu}/\partial x^{\nu}$. The transformation laws taking inertial frames S to inertial frames S' involves inhomogeneous coordinate transformations

$$\begin{bmatrix} x' \\ 1 \end{bmatrix} = \left[\begin{array}{c|c} SO(3,1) & d \\ \hline 0 & 1 \end{array}\right] \begin{bmatrix} x \\ 1 \end{bmatrix}. \tag{5.77}$$

While Maxwell's equations remain unchanged in form under this set of coordinate transformations (inhomogeneous Lorentz group), Newton's force law no longer preserves its form.

In order to find the proper form for the laws of classical mechanics under this new set of transformations the following two-step process was adopted:

1. Find an equation that has the proper transformation properties under the *inhomogeneous Lorentz group.*

2. If the equation reduces to Newton's equation of motion in the nonrelativistic limit $\beta \to 0$, it is the proper generalization of Newton's equation of motion.

Application of the procedure leads to the relativistic equation for particle motion

$$\frac{dp^{\mu}}{d\tau} = f^{\mu}, \tag{5.78}$$

where p^{μ} is defined by $p^{\mu} = m(dx^{\mu}/d\tau)$. The components of the relativistic four-vector f^{μ} are related to the three-vector force $\mathbf{F}$ by

$$\begin{aligned} \mathbf{f} &= \mathbf{F} + (\gamma)\frac{\boldsymbol{\beta}\cdot\mathbf{F}}{\boldsymbol{\beta}\cdot\boldsymbol{\beta}}\boldsymbol{\beta} \\ f^{0} &= \gamma\boldsymbol{\beta}\cdot\mathbf{F} \end{aligned} \tag{5.79}$$

(cf. (5.31)).

5.9.3 Gravitation

Einstein wondered how it could be possible to determine if you were in an inertial frame. He decided that the algorithm for responding to this question, Newton's first law (*In an inertial frame,* an object at rest remains at rest and an object in motion remains in motion with the same velocity unless acted upon by external forces.) was circularly defined (How do you know there are no forces? When you are sufficiently far away from the fixed stars. How do you know you are sufficiently far away? When there are no forces.)

He therefore set out to formulate the laws of mechanics in such a way that they were invariant in form under an arbitrary coordinate transformation. While the Lorentz group is six dimensional, general coordinate transformations form an "infinite-dimensional" group. The transformation properties at any point are defined by a Jacobian matrix $[(\partial x'^{\mu}/\partial x^{\nu})(x)]$. While for the Lorentz group, this matrix is constant throughout space, for general coordinate transformations this 4×4 matrix is position dependent.

Nevertheless, he was able to modify the algorithm described above to formulate laws that are invariant under *all* coordinate transformations. This two-step process is a powerful formulation of the equivalence principle. It is called the principle of general covariance [11] It states that a law of physics holds in the presence of a gravitational field provided

1. The equation is invariant in form under an *arbitrary coordinate transformation* $x \to x'(x)$.
2. In a locally free-falling coordinate system, or the absence of a gravitational field, the equation assumes the form of a law within the special theory of relativity.

Using these arguments, he was able to show that the equation describing the trajectory of a particle is

$$\frac{d^2x^{\mu}}{d\tau^2} = -\Gamma^{\mu}_{\nu,\kappa}\frac{dx^{\nu}}{d\tau}\frac{dx^{\kappa}}{d\tau}. \tag{5.80}$$

The Christoffel symbols are defined in terms of the metric tensor $g_{\mu,\nu}(x)$ and its inverse $g^{\nu,\rho}(x)$ by

$$\Gamma^{\mu}_{\nu,\kappa} = \frac{1}{2}g^{\mu,\rho}\left(\frac{\partial g_{\nu,\rho}}{\partial x^{\kappa}} + \frac{\partial g_{\rho,\kappa}}{\partial x^{\nu}} - \frac{\partial g_{\nu,\kappa}}{\partial x^{\rho}}\right). \tag{5.81}$$

They are not components of a tensor, as coordinate systems can be found (freely falling, as in an elevator) in which they vanish and the metric tensor reduces to its form in special relativity: $g = \text{diag}(1,-1,-1,-1)$.

Neither the left-hand side nor the right-hand side of (5.80) is invariant under arbitrary coordinate changes (extra terms creep in), but the following transformation law is valid [11]

$$\frac{\mathrm{d}^2 x'^{\mu}}{\mathrm{d}\tau^2} + \Gamma'^{\mu}_{\nu,\kappa} \frac{\mathrm{d}x'^{\nu}}{\mathrm{d}\tau} \frac{\mathrm{d}x'^{\kappa}}{\mathrm{d}\tau} = \left(\frac{\partial x'^{\mu}}{\partial x^{\lambda}}\right) \left(\frac{\mathrm{d}^2 x^{\lambda}}{\mathrm{d}\tau^2} + \Gamma^{\lambda}_{\nu,\kappa} \frac{\mathrm{d}x^{\nu}}{\mathrm{d}\tau} \frac{\mathrm{d}x^{\kappa}}{\mathrm{d}\tau}\right). \quad (5.82)$$

This means that the set of terms on the left, or those within the brackets on the right, have the simple transformation properties of a four-vector. In a freely falling coordinate system, the Christoffel symbols vanish and what remains is $\mathrm{d}^2 x^{\lambda}/\mathrm{d}\tau^2$. This special relativity expression is zero in the absence of forces, so the equation that describes the trajectory of a particle in a gravitational field is

$$\frac{\mathrm{d}^2 x^{\lambda}}{\mathrm{d}\tau^2} + \Gamma^{\lambda}_{\nu,\kappa} \frac{\mathrm{d}x^{\nu}}{\mathrm{d}\tau} \frac{\mathrm{d}x^{\kappa}}{\mathrm{d}\tau} = 0. \quad (5.83)$$

5.9.4 Reflections

Two lines of reasoning have entered the reconciliation of the two pillars of classical deterministic physics and the creation of a theory of gravitation. One is group theory and is motivated by Galileo's principle of relativity. The other is more vague. It is a principle of elegance: there is the mysterious assumption that the structure of the "real" equations of physics are simple, elegant, and invariant under a certain class of coordinate transformations. The groups are the 10-parameter inhomogeneous Lorentz group in the case of the special theory of relativity and the much larger group of general coordinate transformations in the case of the general theory of relativity. There is every likelihood that intergalactic travelers will recognize the principle of relativity but no guarantee that their sense of simplicity and elegance will be anything like our own.

5.10 Linear Representations

The theory of representations of groups – more precisely the linear representations of groups by matrices – was actively studied by mathematicians, while physicists actively ignored these results. This picture changed dramatically with the development of the quantum theory, the understanding that the appropriate "phase space" was the Hilbert space describing a quantum system, and that group elements acted in these spaces through their linear matrix representations [8].

A linear matrix representation is a mapping of group elements g to matrices $g \to \Gamma(g)$ that preserves the group operation:

$$g_i \circ g_j = g_k \Rightarrow \Gamma(g_i) \times \Gamma(g_j) = \Gamma(g_k). \quad (5.84)$$

Here, $\circ$ is the composition in the group and $\times$ indicates matrix multiplication. Often the mapping is one way: many different group elements can map to the same matrix (homomorphism). If the mapping is 1 : 1 the mapping is an isomorphism and the representation is called *faithful*.

5.10.1 Maps to Matrices

We have already seen many matrix representations. We have seen representations of the two-element group Z_2 as reflection, rotation, and inversion matrices acting in R^3 (cf. (5.13)).

So, how many representations does a group have? It is clear from the example of Z_2 that we can create an infinite number of representations. However, if we look carefully at the three representations presented in (5.13), we see that all these representations are diagonal: direct sums of essentially two distinct one-dimensional matrix representations:

$$\begin{array}{c|cc} Z_2 & e & f \\ \hline \Gamma^1 & [1] & [1] \\ \Gamma^2 & [1] & [-1] \end{array} \quad . \tag{5.85}$$

Each of the three matrix representations of Z_2 in (5.13) is a direct sum of these two *irreducible representations*:

$$\begin{array}{rcccccc} \sigma_Z & = & \Gamma^1 & \oplus & \Gamma^1 & \oplus & \Gamma^2, \\ R_Z(\pi) & = & \Gamma^1 & \oplus & \Gamma^2 & \oplus & \Gamma^2, \\ \mathcal{P} & = & \Gamma^2 & \oplus & \Gamma^2 & \oplus & \Gamma^2. \end{array} \tag{5.86}$$

A basic result of representation theory is that for large classes of groups (finite, discrete, compact Lie groups), every representation can be written as a direct sum of irreducible representations. The construction of this direct sum proceeds by matrix diagonalization. Irreducible representations are those that cannot be further diagonalized. In particular, one-dimensional representations cannot be further diagonalized. Rather than enumerating all possible representations of a group, it is sufficient to enumerate only the much smaller set of irreducible representations.

5.10.2 Group Element–Matrix Element Duality

The members of a group can be treated as a set of points. It then becomes possible to define a set of functions on this set of points. How many independent functions are needed to span this function space? A not too particularly convenient choice of basis functions are the delta functions $f_i(g) = \delta(g, g_i)$. For example, for C_{3v} there are six group elements and therefore six basis functions for the linear vector space of functions defined on this group.

Each matrix element in any representation is a function defined on the members of a group. It would seem reasonable that the number of matrix elements in all the irreducible representations of a group provide a set of basis functions for the function space defined on the set of group elements. This is true: it is a powerful theorem. There is a far-reaching duality between the elements in a group and the set of matrix elements in its set of irreducible representations. Therefore, if $\Gamma^\alpha(g)$, $\alpha = 1, 2, \ldots$ are the irreducible representations of a group G and the dimension of $\Gamma^\alpha(g)$ is d_α (i.e., $\Gamma^\alpha(g)$ consists of $d_\alpha \times d_\alpha$ matrices), then the total number of matrix elements is the order of the group G:

$$\sum_{\alpha}^{\text{all irreps}} d_\alpha^2 = |G|. \tag{5.87}$$

Further, the set of functions $\sqrt{d_\alpha/|G|}\Gamma^\alpha_{rs}(g)$ form a complete orthonormal set of functions on the group space. The *orthogonality relation* is

$$\sum_{g\in G} \sqrt{\frac{d_{\alpha'}}{|G|}}\Gamma^{\alpha'\,*}_{r's'}(g)\sqrt{\frac{d_\alpha}{|G|}}\Gamma^\alpha_{rs}(g) = \delta(\alpha', \alpha)\delta(r', r)\delta(s', s), \tag{5.88}$$

and the *completeness relation* is

$$\sum_\alpha \sum_r \sum_s \sqrt{\frac{d_\alpha}{|G|}}\Gamma^{\alpha\,*}_{rs}(g')\sqrt{\frac{d_\alpha}{|G|}}\Gamma^\alpha_{rs}(g) = \delta(g', g). \tag{5.89}$$

These complicated expressions can be considerably simplified when written in the Dirac notation. Define

$$\left\langle g \middle| \begin{matrix} \alpha \\ rs \end{matrix} \right\rangle = \sqrt{\frac{d_\alpha}{|G|}}\Gamma^\alpha_{rs}(g),$$

$$\left\langle \begin{smallmatrix} \alpha \\ rs \end{smallmatrix} | g \right\rangle = \sqrt{\frac{d_\alpha}{|G|}} \Gamma^{\alpha\,*}_{rs}(g). \tag{5.90}$$

For convenience, we have assumed that the irreducible representations are unitary: $\Gamma^\dagger(g) = \Gamma(g^{-1})$ and $^\dagger = ^{t\,*}$.

In Dirac notation, the orthogonality and completeness relations are

$$\text{Orthogonality :}$$
$$\left\langle \begin{smallmatrix} \alpha' \\ r's' \end{smallmatrix} | g \right\rangle \left\langle g | \begin{smallmatrix} \alpha \\ rs \end{smallmatrix} \right\rangle = \left\langle \begin{smallmatrix} \alpha' \\ r's' \end{smallmatrix} | \begin{smallmatrix} \alpha \\ rs \end{smallmatrix} \right\rangle,$$
$$\text{Completeness :}$$
$$\left\langle g' | \begin{smallmatrix} \alpha \\ rs \end{smallmatrix} \right\rangle \left\langle \begin{smallmatrix} \alpha \\ rs \end{smallmatrix} | g \right\rangle = \langle g' | g \rangle. \tag{5.91}$$

As usual, doubled dummy indices are summed over.

5.10.3 Classes and Characters

The group element–matrix element duality is elegant and powerful. It leads to yet another duality, somewhat less elegant but, in compensation, even more powerful. This is the *character–class duality*.

We have already encountered classes in (5.17). Two elements c_1, c_2 are in the same class if there is a group element, g, for which $gc_1g^{-1} = c_2$. The *character* of a matrix is its trace. All elements in the same class have the same character in any representation, for

$$\begin{aligned} \mathrm{Tr}\,\Gamma(c_2) &= \mathrm{Tr}\,\Gamma(gc_1g^{-1}) = \mathrm{Tr}\,\Gamma(g)\Gamma(c_1)\Gamma(g^{-1}) \\ &= \mathrm{Tr}\,\Gamma(c_1). \end{aligned} \tag{5.92}$$

The last result comes from invariance of the trace under cyclic permutation of the argument matrices.

With relatively little work, the powerful orthogonality and completeness relations for the group elements–matrix elements can be transformed to corresponding orthogonality and completeness relations for classes and characters. If $\chi^\alpha(i)$ is the character for elements in class i in irreducible representation α and n_i is the number of group elements in that class, the character–class duality is described by the following relations:

$$\text{Orthogonality :}$$
$$\sum_i n_i \chi^{\alpha'*}(i)\chi^\alpha(i) = |G|\delta(\alpha', \alpha), \tag{5.93}$$
$$\text{Completeness :}$$
$$\sum_\alpha n_i \chi^{\alpha*}(i)\chi^\alpha(i') = |G|\delta(i', i). \tag{5.94}$$

5.10.4 Fourier Analysis on Groups

The group C_{3v} has six elements. Its set of irreducible representations has a total of six matrix elements. Therefore $d_1^2 + d_2^2 + \cdots = 6$. This group has three classes. By the character–class duality, it has three irreducible representations. As a result, $d_1 = d_2 = 1$ and $d_3 = 2$. The matrices of the six group elements in the three irreducible representations are:

	Γ^1	Γ^2	Γ^3
e	[1]	[1]	$\begin{bmatrix} 1 & 0 \\ 0 & 1 \end{bmatrix}$
C_3^+	[1]	[1]	$\begin{bmatrix} -a & b \\ -b & -a \end{bmatrix}$
C_3^-	[1]	[1]	$\begin{bmatrix} -a & -b \\ b & -a \end{bmatrix}$
σ_1	[1]	[−1]	$\begin{bmatrix} -1 & 0 \\ 0 & 1 \end{bmatrix}$
σ_2	[1]	[−1]	$\begin{bmatrix} a & b \\ b & -a \end{bmatrix}$
σ_3	[1]	[−1]	$\begin{bmatrix} a & -b \\ -b & -a \end{bmatrix}$

$$a = \tfrac{1}{2} \quad b = \tfrac{\sqrt{3}}{2}. \tag{5.95}$$

The character table for this group is

$$\begin{array}{cccc} & \{e\} & \{C_3^+, C_3^-\} & \{\sigma_1,\sigma_2,\sigma_3\} \\ & 1 & 2 & 3 \\ \hline \chi^1 & 1 & 1 & 1 \\ \chi^2 & 1 & 1 & -1 \\ \chi^3 & 2 & -1 & 0. \end{array} \tag{5.96}$$

The first line shows how the group elements are apportioned to the three classes. The second shows the number of group elements in each class. The remaining lines show the trace of the matrix representatives of the elements in each class in each representation. For example, the -1 in the middle of the last line is $-1 = -1/2 - 1/2$. The character of the identity group element e is the dimension of the matrix representation, d_α. Observe that the rows of the table in (5.96) satisfy the orthogonality relations in (5.93) and the columns of the table in (5.96) satisfy the completeness relations in (5.94).

We use this character table to perform a Fourier analysis on representations of this group. For example, the representation of $C_{3v} = S_3$ in terms of 3×3 permutation matrices is not irreducible (cf. (5.15)). For various reasons, we might like to know which irreducible representations of C_{3v} are contained in this reducible representation. The characters of the matrices describing each class are

$$\begin{array}{cccc} & \{e\} & \{C_3^+, C_3^-\} & \{\sigma_1,\sigma_2,\sigma_3\} \\ \chi^{3\times 3} & 3 & 0 & 1. \end{array} \tag{5.97}$$

To determine the irreducible content of this representation we take the inner product of (5.97) with the rows of (5.96) using (5.93) with the results

$$\begin{aligned} \langle \chi^{3\times 3} | \chi^1 \rangle &= 1 \times 3 \times 1 + 2 \times 0 \times 1 \\ &\quad + 3 \times 1 \times 1 = 6, \\ \langle \chi^{3\times 3} | \chi^2 \rangle &= 1 \times 3 \times 1 + 2 \times 0 \times 1 \\ &\quad + 3 \times 1 \times -1 = 0, \\ \langle \chi^{3\times 3} | \chi^3 \rangle &= 1 \times 3 \times 2 + 2 \times 0 \times -1 \\ &\quad + 3 \times 1 \times 0 = 6. \end{aligned} \tag{5.98}$$

As a result, the permutation representation is reducible and $\chi^{3\times 3} \simeq \chi^1 \oplus \chi^3$.

5.10.4.1 Remark on Terminology

The cyclic group C_n has n group elements g_k, $k = 0, 1, 2, \dots, n-1$ that can be identified with rotations through an angle $\theta_k = 2\pi k/n$. This group is abelian. It therefore has n one-dimensional irreducible matrix representations Γ^m, $m = 0, 1, 2, \dots, n-1$ whose matrix elements are $\Gamma^m(\theta_k) = \left[e^{2\pi i k m/n}\right]$. Any function defined at the n equally spaced points at angles θ_k around the circle can be expressed in terms of the matrix elements of the unitary irreducible representations of C_n. The study of such functions, and their transforms, is the study of Fourier series. This analysis method can be applied to functions defined along the real line R^1 using the unitary irreducible representations (UIR) $\Gamma^k(x) = \left[e^{ikx}\right]$ of the commutative group of translations T_x along the real line through the distance x. This is Fourier analysis on the real line. This idea generalizes to groups and their complete set of unitary irreducible representations.

5.10.5 Irreps of $SU(2)$

The UIR (or "irreps") of Lie groups can be constructed following two routes. One route begins with the group. The second begins with its Lie algebra. The second method is simpler to implement, so we use it here to construct the hermitian irreps of $\mathfrak{su}(2)$ and then exponentiate them to the unitary irreps of $SU(2)$.

The first step is to construct shift operators from the basis vectors in $\mathfrak{su}(2)$:

$$
\begin{aligned}
S_+ &= S_x + iS_y = \begin{bmatrix} 0 & 1 \\ 0 & 0 \end{bmatrix} \\
S_- &= S_x - iS_y = \begin{bmatrix} 0 & 0 \\ 1 & 0 \end{bmatrix} \\
S_z &= \tfrac{1}{2}\begin{bmatrix} 1 & 0 \\ 0 & -1 \end{bmatrix}
\end{aligned}
\qquad
\begin{aligned}
[S_z, S_\pm] &= \pm S_\pm \\
[S_+, S_-] &= 2S_z
\end{aligned} . \tag{5.99}
$$

Next, we use the matrix algebra to operator algebra mapping (cf. Section 5.7.5) to construct a useful boson operator realization of this Lie algebra:

$$
\begin{aligned}
S_+ \to \mathcal{S}_+ &= b_1^\dagger b_2 \\
S_- \to \mathcal{S}_- &= b_2^\dagger b_1
\end{aligned}
\qquad
S_z \to \mathcal{S}_z = \frac{1}{2}\left(b_1^\dagger b_1 - b_2^\dagger b_2\right). \tag{5.100}
$$

The next step introduces representations (of a Lie algebra). Introduce a state space on which the boson operators $b_1, b_1^\dagger$ act, with basis vectors $|n_1\rangle$, $n_1 = 0, 1, 2, \dots$ with the action given as usual by

$$
\begin{aligned}
b_1^\dagger |n_1\rangle &= |n_1 + 1\rangle \sqrt{n_1 + 1}, \\
b_1 |n_1\rangle &= |n_1 - 1\rangle \sqrt{n_1}.
\end{aligned} \tag{5.101}
$$

Introduce a second state space for the operators $b_2, b_2^\dagger$ and basis vectors $|n_2\rangle$, $n_2 = 0, 1, 2, \dots$. In order to construct the irreducible representations of $\mathfrak{su}(2)$, we introduce a grid, or lattice, of states $|n_1, n_2\rangle = |n_1\rangle \otimes |n_2\rangle$. The operators $\mathcal{S}_\pm, \mathcal{S}_z$ are number conserving and move along the diagonal $n_1 + n_2 =$ cnst. (cf. Figure 5.4). It is very useful to relabel the basis vectors in this lattice by two integers. One (j) identifies the diagonal, the other (m) specifies position along a diagonal:

$$
\begin{aligned}
2j &= n_1 + n_2 \\
2m &= n_1 - n_2
\end{aligned}
\quad
\begin{aligned}
n_1 &= j + m \\
n_2 &= j - m
\end{aligned}
\quad
|n_1, n_2\rangle \leftrightarrow \left| \begin{matrix} j \\ m \end{matrix} \right\rangle . \tag{5.102}
$$

The spectrum of allowed values of the quantum number j is $2j = 0, 1, 2, \dots$ and $m = -j, -j+1, \dots, +j$.

The matrix elements of the operators $\mathcal{S}$ with respect to the basis $\left| \begin{matrix} j \\ m \end{matrix} \right\rangle$ are constructed from the matrix elements of the operators $b_i^\dagger b_j$ on the basis vectors $|n_1, n_2\rangle$. For $\mathcal{S}_z$, we find

$$
\begin{aligned}
\mathcal{S}_z \left| \begin{matrix} j \\ m \end{matrix} \right\rangle &= \frac{1}{2}(b_1^\dagger b_1 - b_2^\dagger b_2) |n_1, n_2\rangle \\
&= |n_1, n_2\rangle \frac{1}{2}(n_1 - n_2) = \left| \begin{matrix} j \\ m \end{matrix} \right\rangle m.
\end{aligned} \tag{5.103}
$$

For the shift-up operator

$$
\begin{aligned}
\mathcal{S}_+ \left| \begin{matrix} j \\ m \end{matrix} \right\rangle &= b_1^\dagger b_2 |n_1, n_2\rangle \\
&= |n_1 + 1, n_2 - 1\rangle \sqrt{n_1 + 1}\sqrt{n_2} \\
&= \left| \begin{matrix} j \\ m+1 \end{matrix} \right\rangle \sqrt{(j + m + 1)(j - m)}.
\end{aligned} \tag{5.104}
$$

and similarly for the shift-down operator

$$
\mathcal{S}_- \left| \begin{matrix} j \\ m \end{matrix} \right\rangle = \left| \begin{matrix} j \\ m-1 \end{matrix} \right\rangle \sqrt{(j - m + 1)(j + m)}. \tag{5.105}
$$

In this representation of the (spin) angular momentum algebra $\mathfrak{su}(2)$, $\mathcal{S}_z = J_z$ is diagonal and $\mathcal{S}_\pm = J_\pm$ have one nonzero diagonal row just above (below) the main diagonal. The hermitian irreducible representations of $\mathfrak{su}(2)$ with

$j = 0, \frac{1}{2}, 1, \frac{3}{2}, 2, \frac{5}{2}, \ldots$ form a complete set of irreducible hermitian representations for this Lie algebra.

The UIR of $SU(2)$ are obtained by exponentiating i times the hermitian representations of $\mathfrak{su}(2)$:

$$\mathcal{D}^j\,[SU(2)] = \exp\left(i\hat{\mathbf{n}}\cdot\mathbf{J}\theta\right), \tag{5.106}$$

with $J_x = (J_+ + J_-)/2$ and $J_y = (J_+ - J_-)/2i$, and J_* are the $(2j+1)\times(2j+1)$ matrices whose matrix elements are given in (5.103–5.105). The $(2j+1)\times(2j+1)$ matrices $\mathcal{D}^j$ are traditionally called *Wigner matrices* [8]. For many purposes, only the character of an irreducible representation is needed. The character depends only on the class and the class is uniquely determined by the rotation angle θ since rotations by angle θ about any axis $\hat{\mathbf{n}}$ are geometrically equivalent. It is sufficient to compute the trace of any rotation, for example, the rotation about the z-axis. This matrix is diagonal: $\left(\mathrm{e}^{iJ_z\theta}\right)_{m',m} = \mathrm{e}^{im\theta}\delta_{m',m}$ and its trace is

$$\chi^j(\theta) = \sum_{m=-j}^{+j} \mathrm{e}^{im\theta} = \frac{\sin(j+\frac{1}{2})\theta}{\sin\frac{1}{2}\theta}. \tag{5.107}$$

These characters are orthonormal with respect to the weight $w(\theta) = 1/\pi\sin^2(\theta/2)$.

5.10.6
Crystal Field Theory

The type of Fourier analysis outlined above has found a useful role in crystal (or ligand) field theory [5, 7]. This theory was created to describe the behavior of charged particles (electrons, ions, atoms) in the presence of an electric field that has some symmetry, usually the symmetry of a host crystal. We illustrate this with a simple example.

A many-electron atom with total angular momentum L is placed in a crystal field with cubic symmetry. How do the $2L+1$-fold degenerate levels split?

Before immersion in the crystal field, the atom has spherical symmetry. Its symmetry group is the rotation group, the irreducible representations $\mathcal{D}^L$ have dimension $2L+1$, the classes are rotations through angle θ, and the character for the class θ in representation $\mathcal{D}^L$ is given in (5.107) with $j \to L$ (integer). When the atom is placed in an electric field with cubic symmetry O^h, the irreducible representations of $SO(3)$ become reducible. The irreps of O^h are A_1, A_2, E, T_1, T_2. The irreducible content is obtained through a character analysis. of $\mathcal{D}^L\,[SO(3)]$.

The group O^h has 24 elements partitioned into five classes. These include the identity E, eight rotations C_3 by $2\pi/3$ radians about the diagonals through the opposite vertices of the cube, six rotations C_4 by $2\pi/4$ radians about the midpoints of opposite faces, three rotations C_4^2 by $2\pi/2$ radians about the same midpoints of opposite faces, and six rotations C_2 about the midpoints of opposite edges. The characters for these five classes in the five irreducible representations are collected in the *character table* for O^h. This is shown at the top in Table 5.2. At the bottom left of the table are the characters of the irreducible representations of the rotation group $SO(3)$ in the irreducible representations of dimension $2L+1$. These are obtained from (5.107). A *character analysis* (cf. (5.98)) leads to the O^h irreducible content of each of the lowest six irreducible representations of $SO(3)$, presented at the bottom right of the table.

5.11
Symmetry Groups

Groups first appeared in the quantum theory as a tool for labeling eigenstates of

Table 5.2 (top) Character table for the cubic group O^h. The functions in the right-hand column are some of the basis vectors that "carry" the corresponding representation. (bottom) Characters for rotations through the indicated angle in the irreducible representations of the rotation group.

O^h	E	$8C_3$	$3C_4^2$	$6C_2$	$6C_4$	Basis
A_1	1	1	1	1	1	$r^2 = x^2 + y^2 + z^2$
A_2	1	1	1	−1	−1	
E	2	−1	2	0	0	$(x^2 - y^2, 3z^2 - r^2)$
T_1	3	0	−1	−1	1	$(x, y, z), (L_x, L_y, L_z)$
T_2	3	0	−1	1	−1	(yz, zx, xy)
$L : \theta$	0	$\frac{2\pi}{3}$	$\frac{2\pi}{2}$	$\frac{2\pi}{2}$	$\frac{2\pi}{4}$	Reduction
0 S	1	1	1	1	1	A_1
1 P	3	0	−1	−1	1	T_1
2 D	5	−1	1	1	−1	$E \oplus T_2$
3 F	7	1	−1	−1	−1	$A_2 \oplus T_1 \oplus T_2$
4 G	9	0	1	1	1	$A_1 \oplus E \oplus T_1 \oplus T_2$
5 H	11	−1	−1	−1	1	$E \oplus 2T_1 \oplus T_2$

a Hamiltonian with useful quantum numbers. If a Hamiltonian $\mathcal{H}$ is invariant under the action of a group G, then $g\mathcal{H}g^{-1} = \mathcal{H}$, $g \in G$. If $|\psi^\alpha_\mu\rangle$ satisfies Schrödinger's time-independent equation $\mathcal{H}|\psi^\alpha_\mu\rangle - E|\psi^\alpha_\mu\rangle = 0$, so that

$$\begin{aligned} g(\mathcal{H} - E)|\psi^\alpha_\mu\rangle &= \left\{g(\mathcal{H} - E)g^{-1}\right\} g|\psi^\alpha_\mu\rangle \\ &= (\mathcal{H} - E)|\psi^\alpha_\nu\rangle\langle\psi^\alpha_\nu|g|\psi^\alpha_\mu\rangle \\ &= (\mathcal{H} - E)|\psi^\alpha_\nu\rangle D^\alpha_{\nu,\mu}(g) \\ &= 0. \end{aligned} \quad (5.108)$$

All states $|\psi^\alpha_\nu\rangle$ related to each other by a group transformation $g \in G$ (more precisely, a group representation $\mathcal{D}^\alpha(g)$) have the same energy eigenvalue. The existence of a symmetry group G for a Hamiltonian $\mathcal{H}$ provides representation labels for the quantum states *and also* describes the degeneracy patterns that can be observed. If the symmetry group G is a Lie group, so that $g = e^X$, then $e^X\mathcal{H}e^{-X} = \mathcal{H} \Rightarrow [X, \mathcal{H}] = 0$. The existence of operators X that commute with the Hamiltonian $\mathcal{H}$ is a clear signal that the physics described by the Hamiltonian is invariant under a Lie group.

For example, for a particle in a spherically symmetric potential, $V(r)$, Schrödinger's time-independent equation is

$$\left(\frac{\mathbf{p}\cdot\mathbf{p}}{2m} + V(r)\right)\psi = E\psi \quad (5.109)$$

with $\mathbf{p} = (\hbar/i)\nabla$. The Hamiltonian operator is invariant under rotations. Equivalently, it commutes with the angular momentum operators $\mathbf{L} = \mathbf{r} \times \mathbf{p}$: $[\mathbf{L}, \mathcal{H}] = 0$. The wavefunctions can be partly labeled by rotation group quantum numbers, l and m: $\psi \to \psi^l_m(r, \theta, \phi)$. In fact, by standard separation of variables, arguments in this description can be made more precise: $\psi(r, \theta, \phi) = 1/rR_{nl}(r)Y^l_m(\theta, \phi)$. Here $R_{nl}(r)$ are radial wavefunctions that depend on the potential $V(r)$ but the angular function $Y^l_m(\theta,\phi)$ is "a piece of geometry": it depends only on the existence of rotation symmetry. It is the same no matter what the potential is. In fact, these functions can be constructed from the matrix representations of the group $SO(3)$. The action of a rotation group element g on the angular functions is

$$gY^l_m(\theta,\phi) = Y^l_{m'}(\theta,\phi)\mathcal{D}^l_{m'm}(g), \quad (5.110)$$

where the construction of the Wigner $\mathcal{D}$ matrices has been described in Section 5.10.5.

If the symmetry group is reduced, as in the case of $SO(3) \downarrow O^h$ described in Section 5.10.6, the eigenstates are identified by the labels of the irreducible representations of O^h: A_1, A_2, E, T_1, T_2.

Once the states have been labeled, computations must be done. At this point, the power of group theory becomes apparent. Matrices must be computed – for example, matrix elements of a Hamiltonian. Typically, most matrix elements vanish (by group-theoretic selection rules). Of the small number that do not vanish, many are simply related to a small number of the others. In short, using group theory as a guide, only a small number of computations must actually be done.

This feature of group theory is illustrated by computing the eigenstates and their energy eigenvalues for an electron in the $N = 4$ multiplet of the hydrogen atom under the influence of a constant external field $\mathcal{E}$. The Hamiltonian to be diagonalized is

$$\left\langle N' \begin{array}{c} L' \\ M' \end{array} \left| \frac{\mathbf{p}\cdot\mathbf{p}}{2m} - \frac{e^2}{r} + e\mathcal{E}\cdot\mathbf{r} \right| N \begin{array}{c} L \\ M. \end{array} \right\rangle. \quad (5.111)$$

The first two terms in the Hamiltonian describe the electron in a Coulomb potential, the last is the Stark perturbation, which describes the interaction of a dipole $\mathbf{d} = -e\mathbf{r}$ with a constant external electric field: $\mathcal{H}_{\text{St.}} = -\mathbf{d}\cdot\mathcal{E}$. In the $N = 4$ multiplet, we set $N' = N = 4$, so that $L', L = 0, 1, 2, 3$ and M ranges from $-L$ to $+L$ and $-L' \le M' \le +L'$. The matrix elements of the Coulomb Hamiltonian are $E_N\delta_{N'N}\delta_{L'L}\delta_{M'M}$, with $E_4 = -13.6/4^2$ eV.

There are $\sum_{L=0}^{3=4-1}(2L+1) = 16$ states in the $N = 4$ multiplet, so 16^2 matrix elements of the 16×16 matrix must be computed. We simplify the computation by choosing the z-axis in the direction of the applied uniform electric field, so that $e\mathcal{E}\cdot\mathbf{r} \to e\mathcal{E}z$ ($\mathcal{E} = |\mathcal{E}|$). In addition, we write $z = \sqrt{4\pi/3}rY^1_0(\theta,\phi)$. The matrix elements factor (separation of variables) into a radial part and an angular part, as follows:

$$\langle 4L'M'|e\mathcal{E}z|4LM\rangle \to e\mathcal{E} \times \text{Radial} \times \text{Angular}$$

$$\text{Radial} = \int_0^\infty R_{4L'}(r)r^1R_{4L}(r)\mathrm{d}r$$

$$\text{Angular} = \sqrt{\frac{4\pi}{3}}\int Y^{L'*}_{M'}(\Omega)Y^1_0(\Omega)Y^L_M(\Omega)\mathrm{d}\Omega, \quad (5.112)$$

where $\Omega = (\theta,\phi)$ and $d\Omega = \sin\theta\mathrm{d}\theta\mathrm{d}\phi$.

Selection rules derived from $SO(3)$ simplify the angular integral. First, the integral vanishes unless $\Delta M = M' - M = 0$. It also vanishes unless $\Delta L = \pm 1, 0$. By parity, it vanishes if $\Delta L = 0$, and by time reversal, its value for M and $-M$ are the same. The nonzero angular integrals are

$$\begin{aligned} \mathcal{A}(L,M) &= \sqrt{\frac{4\pi}{3}}\int_\Omega Y^{L*}_{M'}(\Omega)Y^1_0(\Omega)Y^{L-1}_M(\Omega)\mathrm{d}\Omega \\ &= \delta_{M'M}\sqrt{\frac{(L+M)(L-M)}{(2L+1)(2L-1)}}, \end{aligned} \quad (5.113)$$

The useful radial integrals, those with $\Delta L = \pm 1$, are all related:

$$\mathcal{R}(N,L) = \int_0^\infty R_{N,L}(r) r R_{N,L-1} \mathrm{d}r$$
$$= \frac{N\sqrt{N^2 - L^2}}{2\sqrt{3}} \times \mathcal{R}(2,1) \quad (5.114)$$

with $1 \leq L \leq N - 1$. All integrals are proportional to the single integral $\mathcal{R}(2,1)$. This comes from yet another symmetry that the Coulomb potential exhibits (cf. Section 5.12), not shared by other spherically symmetric potentials. The single integral to be evaluated is

$$\mathcal{R}(2,1) = -3\sqrt{3}a_0. \quad (5.115)$$

This integral is proportional to the Bohr radius a_0 of the hydrogen atom, whose value was estimated in (5.6).

The arguments above show drastic simplifications in the computational load for computing the energy eigenfunctions and eigenvalues of a many-electron atom in a uniform external electric field (Stark problem).

Of the $256 = 16^2$ matrix elements to compute, only 18 are nonzero. All are real. Since the Hamiltonian is hermitian (symmetric if real), there are in fact only 9 nonzero matrix elements to construct. Each is a product of two factors, so only 6 (angular) plus 1 (radial) quantities need be computed. These numbers must be stuffed into a 16×16 matrix to be diagonalized. But there are no nonzero matrix elements between states with $M' \neq M$. This means that by organizing the row and columns appropriately the matrix can be written in block diagonal form. The block diagonal form consists of a 1×1 matrix for $M = 3$, a 2×2 matrix for $M = 2$, a 3×3 matrix for $M = 1$, a 4×4 matrix for $M = 0$, a 3×3 matrix for $M = -1$, and so on. The 1×1 matrices are already diagonal. The 2×2 matrices are identical, so only one needs to be diagonalized. Similarly for the two 3×3 matrices. There is only one 4×4 matrix. The computational load for diagonalizing this matrix has been reduced from $T \simeq 16^2 \log 16$ to $T \simeq 2^2 \log 2 + 3^2 \log 3 + 4^2 \log 4$, a factor of 20 (assuming the effort required for diagonalizing an $n \times n$ matrix goes like $n^2 \log n$)!

It gets even better. For the $N = 5$ multiplet the $1 \times 1, 2 \times 2, 3 \times 3, 4 \times 4$ matrices are all proportional to the matrices of the corresponding size for $N = 4$. The proportionality factor is $5/4$. Only *one* new matrix needs to be constructed – the 5×5 matrix. This symmetry extends to all values of N.

This is a rather simple example that can be carried out by hand. This was done when quantum mechanics was first developed, when the fastest computer was a greased abacus. Today, time savings of a factor of 20 on such a simple problem would hardly be noticed. But calculations have also inflated in size. Reducing a $10^6 \times 10^6$ matrix to about 1000 $10^3 \times 10^3$ matrices reduces the computational effort by a factor of 2000. For example, a computation that would take 6 years without such methods could be done in a day with these methods.

Symmetry groups play several roles in quantum mechanics.

- They provide group representation labels to identify the energy eigenstates of a Hamiltonian with symmetry.
- They provide selection rules that save us the effort of computing matrix elements whose values are zero (by symmetry!).
- And they allow transformation of a Hamiltonian matrix to block diagonal form, so that the computational load can be drastically reduced.

5.12 Dynamical Groups

A widely accepted bit of wisdom among physicists is that symmetry implies degeneracy, and the larger the symmetry, the larger the degeneracy. What works forward ought to work backward ("Newton's third law"): if the degeneracy is greater than expected, the symmetry is greater than apparent.

5.12.1 Conformal Symmetry

The hydrogen atom has rotational symmetry $SO(3)$, and this requires $(2L+1)$-fold degeneracy. But the states with the same principal quantum number N are all degenerate in the absence of spin and other relativistic effects, and nearly degenerate in the presence of these effects. It would make sense to look for a larger than apparent symmetry. It exists in the form of the *Runge–Lenz vector* $\mathbf{M} = 1/2m(\mathbf{p}\times\mathbf{L} - \mathbf{L}\times\mathbf{p}) - e^2\mathbf{r}/r$, where $\mathbf{r}, \mathbf{p}, \mathbf{L} = \mathbf{r}\times\mathbf{p}$ are the position, momentum, and orbital angular momentum operators for the electron. The three orbital angular momentum operators L_i and three components of the Runge–Lenz vector close under commutation to form a Lie algebra. The six operators commute with the Hamiltonian, so the "hidden" symmetry group is larger than the obvious symmetry group $SO(3)$. On the bound states, this Lie algebra describes the Lie group $SO(4)$. The irreducible representation labels for the quantum states are N, $N = 1, 2, 3, \ldots, \infty$. The three nested groups $SO(2) \subset SO(3) \subset SO(4)$ and their representation labels and branching rules are

Group	Rep.label	Degeneracy	Branching rules
$SO(4)$	N	N^2	
$SO(3)$	L	$2L+1$	$0, 1, 2, \ldots, N-1$
$SO(2)$	M	1	$-L \leq M \leq +L$.

(5.116)

Branching rules identify the irreducible representations of a subgroup that any representation of a larger group splits into under group–subgroup reduction. We have seen branching rules in Table 5.2.

One advantage of using the larger group is that there are more shift operators in the Lie algebra. The shift operators, acting on one state, moves it to another (cf. $|LM\rangle \xrightarrow{L_+} |L, M+1\rangle$). This means that there are well-defined algebraic relations among states that belong to the same N multiplet. This means that more of any computation can be pushed from the physical domain to the geometric domain, and simplifications accrete.

Why stop there? In the hydrogen atom, the energy difference between the most tightly bound state, the ground state, and the most weakly bound state ($N \to \infty$) is 13.6 eV. When this difference is compared with the electron rest energy of 511 KeV, the *symmetry-breaking* is about $13.6/511\,000 \simeq 0.000\,027$ or $2.7 \times 10^{-3}\%$. This suggests that there is a yet larger group that accounts for this near degeneracy. Searches eventually lead to the noncompact conformal group $SO(4,2) \supset SO(4) \cdots$ as the all-inclusive "symmetry group" of the hydrogen atom. The virtue of using this larger group is that states in different multiplets $N, N \pm 1$ can be connected by shift operators within the algebra $\mathfrak{so}(4,2)$, and ultimately there is only one number to compute [12]. Including this larger group in (5.116) would include inserting it in the row above $SO(4)$, showing there is only one representation label for bound states, indicating its degeneracy is "∞", and adding branching rules $N = 1, 2, \ldots, \infty$ to the $SO(4)$ row.

5.12.2 Atomic Shell Structure

Broken symmetry beautifully accounts for the systematics of the chemical elements. It accounts for the filling scheme as electrons enter a screened Coulomb potential around a nuclear charge $+Ze$ as the nuclear charge increases from $Z = 1$ to $Z > 92$. The screening is caused by "inner electrons." The filling scheme accounts for the "magic numbers" among the chemical elements: these are the nuclear charges of exceptionally stable chemical elements He, Ne, Ar, Kr, Xe, Rn with atomic numbers $2, 10, 18, 36, 54, 86$.

When more than one electron is present around a nuclear charge $+Ze$, then the outer electrons "see" a screened central charge and the $SO(4)$ symmetry arising from the Coulomb nature of the potential is lost. There is a reduction in symmetry, a "broken symmetry": $SO(4) \downarrow SO(3)$. The quantum numbers (N, L) can be used to label states and energies, $E_{N,L}$, and these energy levels are $(2L + 1)$-fold degenerate. The $SO(4)$ multiplet with quantum number N splits into orbital angular momentum multiplets with L values ranging from $L = 0$ to a maximum of $L = N - 1$.

Each additional electron must enter an orbital that is not already occupied by the Pauli exclusion principle. This principle is enforced by the requirement that the total electron wavefunction transforms under the unique antisymmetric representation $\Gamma^{\text{anti.}}(S_k)$ on the permutation group S_k for k electrons.

Generally, the larger the L value the further the outer electron is from the central charge, on average. And the further it is, the larger is the negative charge density contributed by inner electrons that reduces the strength of the central nuclear attraction. As a result, $E_{N,0} < E_{N,1} < \cdots < E_{N,L=N-1}$. There is mixing among levels with different values of N and L. The following energy ordering scheme, ultimately justified by detailed calculations, accounts for the systematics of the chemical elements, including the magic numbers:

$$1S \mid 2S\ 2P \mid 3S\ 3P \mid 4S\ 3D\ 4P \mid 5S\ 4D\ 5P \mid 6S\ 4F\ 5D\ 6P \mid 7S. \tag{5.117}$$

Each level can hold $2(2L + 1)$ electrons. The first factor of $2 = (2s + 1)$ with $s = 1/2$ is due to electron spin. The vertical bar $\mid$ indicates a large energy gap. The cumulative occupancy reproduces the magic numbers of the chemical elements: $2, 10, 18, 36, 54, 86$. The filling order is shown in Figure 5.5. Broken symmetry is consistent with Mendeleev's periodic table of the chemical elements.

5.12.3 Nuclear Shell Structure

Magic numbers among nuclei suggested that, here also, one could possibly describe many different nuclei with a single simple organizational structure. The magic numbers are: $2, 8, 20, 28, 40, 50, 82, 126$, both for protons and for neutrons. The following model was used to organize this information.

Assume that the effective nuclear potential for protons (or neutrons) is that of a three-dimensional isotropic harmonic oscillator. The algebraic properties of the three-dimensional isotropic harmonic oscillator are described by the unitary group $U(3)$ and its Lie algebra $\mathfrak{u}(3)$. The basis states for excitations can

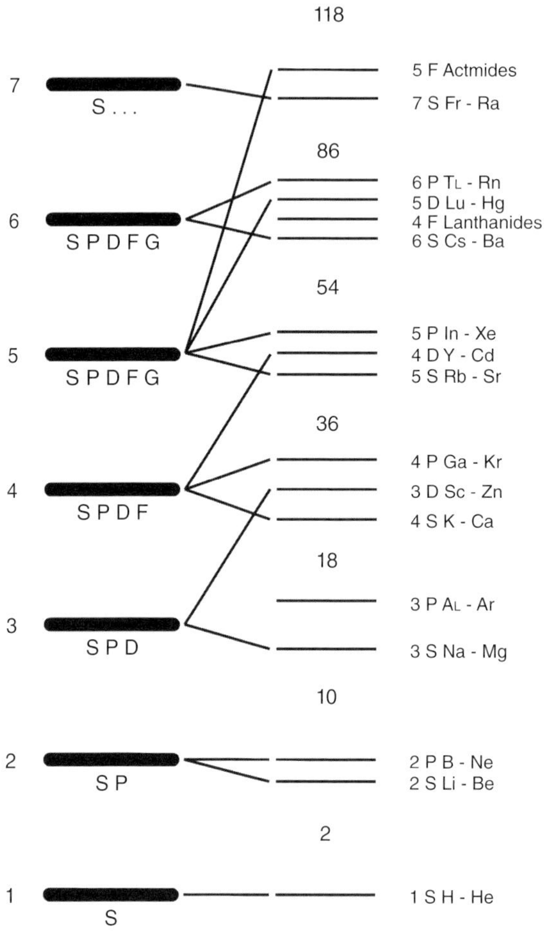

Figure 5.5 Broken $SO(4)$ dynamical symmetry due to screening of the central Coulomb potential by inner electrons successfully accounts for the known properties of the chemical elements, as reflected in Mendeleev's periodic table of the chemical elements.

be described by $|n_1, n_2, n_3\rangle$. One excitation would be threefold degenerate: $|1,0,0\rangle, |0,1,0\rangle, |0,0,1\rangle$, two excitations would be sixfold degenerate, and states with N excitations would have a degeneracy $(N+2)(N+1)/2$. Each integer $N \geq 0$ describes an irrep of $U(3)$. Under a spherically symmetric perturbation, these highly degenerate N multiplets would split into multiplets identified by an angular momentum index L. A character analysis gives this branching result

$U(3)$ N	$SO(3)$ L values	Spectroscopic
0	0	S
1	1	P
2	2, 0	D, S
3	3, 1	F, P
4	4, 2, 0	$G, D, S.$

(5.118)

For example, the $N = 4$ harmonic oscillator multiplet splits into an $L = 4$ multiplet, an $L = 2$ multiplet, and an $L = 0$ multiplet. The larger the angular momentum, the lower the energy. After this splitting, the spin of the proton (or neutron) is coupled to the orbital angular momentum to give values of the total angular momentum $J = L \pm 1/2$, except that for S states only the $J = 1/2$ state occurs. Again, the larger angular momentum occurs at a lower energy than the smaller angular momentum. The resulting filling order, analogous to (5.117), is

$$\begin{aligned} &0S_{1/2}|1P_{3/2}\ 1P_{1/2}|2D_{5/2}\ 2S_{1/2}\ 2D_{3/2}|3F_{7/2}| \\ &3P_{3/2}\ 3F_{5/2}\ 3P_{1/2}\ 4G_{9/2}| \\ &4D_{5/2}\ 4G_{7/2}\ 4S_{1/2}\ 4D_{3/2}\ 5H_{11/2}| \\ &5H_{9/2}\ 5F_{7/2}\ 5F_{5/2}\ 5P_{3/2}\ 5P_{1/2}\ 6I_{13/2}|. \end{aligned} \tag{5.119}$$

Each shell with angular momentum j can hold up to $2j + 1$ nucleons. Broken symmetry is also consistent with the "periodic table" associated with nuclear shell models. The filling order is shown in Figure 5.6 [13, 14].

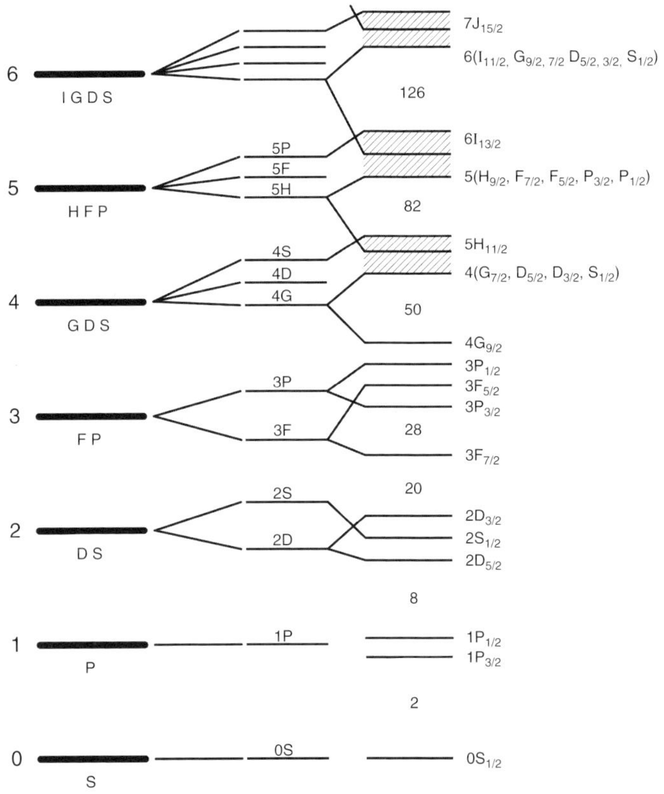

Figure 5.6 The filling order describing very many properties of nuclear ground states is described by the levels of an isotropic harmonic oscillator potential with multiplets having N excitations and degeneracy $(N+1)(N+2)/2$. The degeneracy is broken by a spherically symmetric perturbation and broken further by spin-orbit coupling. For both perturbations, energy increases as angular momentum decreases. The filling order shown successfully accounts for the known properties of the ground states of most even–even nuclei, including the magic numbers. In the higher levels, the largest spin angular momentum state (e.g., $5H_{11/2}$) is pushed down into the next lower multiplet, containing all the remaining $N = 4$ states, with the exception of the $4G_{9/2}$.

At a group theoretical level, our starting point has been the Lie algebra $\mathfrak{u}(3)$ with basis vectors $b_i^\dagger b_j$ $(1 \leq i, j \leq 3)$ whose representations are labeled by an integer index N, the number of excitations present. This algebra can be embedded in a larger Lie algebra containing in addition shift-up operators $b_i^\dagger$, their counterpart annihilation operators b_j, and the identity operator I. The Lie algebra is $9 + 2 \cdot 3 + 1 = 16 = 4^2$ dimensional, and is closely related to the noncompact Lie algebra $\mathfrak{u}(3, 1)$. The embedding $\mathfrak{u}(3) \subset \mathfrak{u}(3, 1)$ is analogous to the inclusion $SO(4) \subset SO(4, 2)$ for the hydrogen atom.

5.12.4
Dynamical Models

In this section, so far we have described the hydrogen atom using a very large group $SO(4, 2)$ and breaking down the symmetry to $SO(4)$ and further to $SO(3)$ when there are Coulomb-breaking perturbations that maintain their spherical symmetry. We have also introduced a sequence of groups and subgroups $U(3, 1) \downarrow U(3) \downarrow SO(3)$ to provide a basis for the nuclear shell model.

Nuclear computations are very difficult because there is "no nuclear force." The force acting between nucleons is a residual force from the quark–quark interaction. This is analogous to the absence of a "molecular force." There is none – the force that binds together atoms in molecules is the residual electromagnetic force after exchange and other interactions have been taken into account.

For this reason, it would be very useful to develop a systematic way for making nuclear models and carrying out calculations within the context of these models. Group theory comes to the rescue!

The first step in creating a simple environment for quantitative nuclear models is to assume that pairs of nucleons bind tightly into boson-like excitations. The leading assumption is that of all the nuclear-pair degrees of freedom, the most important are those with scalar $(S, L = 0)$ and quadrupole $(D, L = 2)$ transformation properties under the rotation group $SO(3)$. States in a Hilbert space describing 2 protons (neutrons, nucleons) can be produced by creation operators $s^\dagger, d_m^\dagger$ acting on the vacuum $|0; 0, 0, 0, 0, 0\rangle$. For n pairs of nucleons, n creation operators act to produce states $|n_s; n_{-2}, n_{-1}, n_0, n_1, n_2\rangle$ with $n_s + \sum_m n_m = n$. There are $(n + 6 - 1)!/n!(6 - 1)!$ states in this Hilbert space. For computational convenience, they can be arranged by their transformation properties under rotations $SO(3)$. For example, the two-boson Hilbert space has 21 states consisting of an $L = 0$ state from $s^\dagger s^\dagger$, an $L = 2$ multiplet from $s^\dagger d_{m'}^\dagger$, and multiplets with $L = 0, 2, 4$ from $d_{m'}^\dagger d_m^\dagger$.

The Hamiltonian acts within the space with a fixed number of bosons. It must therefore be constructed from number-conserving operators: $b_i^\dagger b_j$, where the boson operators include the s and d excitations. These operators must be rotationally invariant. At the linear level, only two such operators exist: $s^\dagger s$ and $d_m^\dagger d_m$. At the quadratic level, there are a small number of additional rotationally invariant operators. The n-boson Hamiltonian can therefore be systematically parameterized by a relatively small number of terms. The parameters can be varied in attempts to fit models to nuclear spectra and transition rates. In the two-boson example with 21 states, it is sufficient to diagonalize this Hamiltonian in the two-dimensional subspace of $L = 0$ multiplets, in another two-dimensional subspace with the two

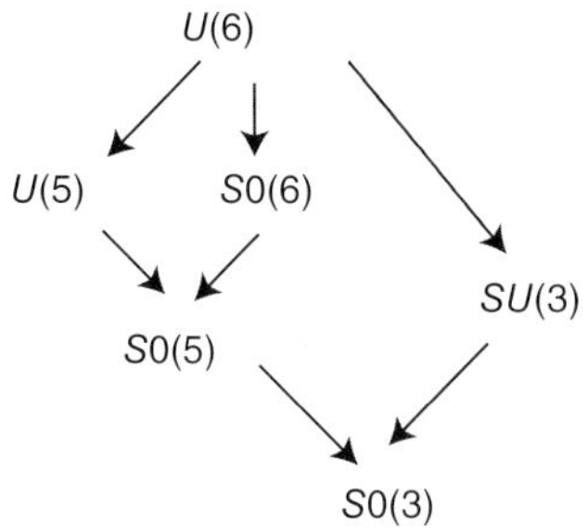

Figure 5.7 States with $2N$ nucleons outside a closed shell are described by N bosons in the interacting boson model. The basis states carry a symmetric representation of the Lie group $U(6)$. Various limiting Hamiltonians that exhibit a group–subgroup symmetry can be diagonalized by hand. The three group–subgroup chains for which this is possible are shown here.

states with $L = 2$ and $M_L = 2$ (all other M_L values will give the same result), and the one-dimensional subspace with $L = 4, M_L = 4$.

The interacting boson model (IBM) outlined above has deeply extended our understanding of nuclear physics [15]. In fact, some Hamiltonians can be solved "by hand." These involve a group–subgroup chain. The chain of groups is shown in Figure 5.7. This model incorporates in a magnificent way the use of groups in their capacity as symmetry groups, implying degeneracy, and dynamical groups, implying relations among multiplets of different energies.

5.13 Gauge Theory

Gauge transformations were introduced by Weyl following Einstein's development (1916) of the theory of general relativity. In crude terms, Weyl's original idea was to introduce a ruler (the "gauge" of gauge theory) whose length was an arbitrary function of position. His original objective was to unify the two then-known forces of nature: gravitation and electromagnetism. His theory is quite beautiful but Einstein raised serious objections, and Weyl eventually relinquished it. Einstein's objection was that if Weyl's theory were correct then the results of laboratory experiments would depend on the history of the material being investigated.

Weyl came back to this general idea following Schrödinger's development (1926) of wave mechanics. In this case, a modified objective was achieved: he succeeded in describing how light interacts with charged matter.

The original theory (GR) involved a *real* scaling transformation that was space–time dependent. As a result, it is in the same spirit as the discussion about scaling in Section 5.2.3, but more general. His modified theory (QM) involved a *complex* phase transformation. In some sense this would be an analytic continuation of the scaling arguments, but the spirit of the discussion given in Section 5.2.3 does not in any sense suggest phase changes.

The starting point of this work is the observation that if $\psi(x,t)$ satisfies Schrödinger's time-dependent equation, so also does $e^{i\phi}\psi(x,t)$, for

$$\left(\mathcal{H} - i\hbar\frac{\partial}{\partial t}\right) e^{i\phi}\psi(x,t) = e^{i\phi}\left(\mathcal{H} - i\hbar\frac{\partial}{\partial t}\right)\psi(x,t) = 0. \quad (5.120)$$

This fails to be true if the phase ϕ depends on space–time coordinates, for then the derivative terms act on this phase when we try to pull it through the Hamiltonian and time-derivative operators:

$$\left(\left(\frac{\mathbf{p}\cdot\mathbf{p}}{2m}\right)+q\Phi(x,t)-\mathrm{i}\hbar\frac{\partial}{\partial t}\right)\mathrm{e}^{\mathrm{i}\phi(x,t)}\psi(x,t)$$
$$=\mathrm{e}^{\mathrm{i}\phi(x,t)}\left(\frac{(\mathbf{p}+\hbar\nabla\phi)^2}{2m}+q\Phi(x,t)\right.$$
$$\left.+\hbar\frac{\partial\phi}{\partial t}-\mathrm{i}\hbar\frac{\partial}{\partial t}\right)\psi(x,t). \tag{5.121}$$

Symmetry is not preserved. What can be done?

It had long been known that the electric and magnetic fields **E**, **B** could be represented by "fictitious" potentials that served to simplify Maxwell's equations but were otherwise "not real." The vector potential **A** and scalar potential Φ are related to the "real" fields by

$$\begin{aligned}\mathbf{B} &= \boldsymbol{\nabla}\times\mathbf{A}\\ \mathbf{E} &= -\boldsymbol{\nabla}\Phi-\frac{1}{c}\frac{\partial\mathbf{A}}{\partial t}.\end{aligned} \tag{5.122}$$

This simplification is not unique. The vector potential can be changed by the addition of the gradient of a scalar field $\chi(x,t)$, and the scalar potential correspondingly changed:

$$\begin{aligned}\mathbf{A}\to\mathbf{A}' &= \mathbf{A}+\nabla\chi &\Rightarrow\quad& \mathbf{B}'=\mathbf{B}\\ \Phi\to\Phi' &= \Phi-\frac{1}{c}\frac{\partial\chi}{\partial t} &\Rightarrow\quad& \mathbf{E}'=\mathbf{E}.\end{aligned} \tag{5.123}$$

The resolution of the difficulty is to assume that the electrostatic part of the interaction is described by the term $q\Phi(x,t)$ in the Hamiltonian and the magnetic part is represented by replacing **p** by $\mathbf{p}-\frac{q}{c}\mathbf{A}(x,t)$ wherever it appears in the Hamiltonian. Under these conditions,

$$\left(\mathbf{p}-\frac{q}{c}\mathbf{A}(x,t)\right)\mathrm{e}^{\mathrm{i}\phi(x,t)}$$
$$=\mathrm{e}^{\mathrm{i}\phi(x,t)}\left(\mathbf{p}-\frac{q}{c}\mathbf{A}(x,t)+\hbar\nabla\phi(x,t)\right) \tag{5.124}$$

and

$$\left(q\Phi-\mathrm{i}\hbar\frac{\partial}{\partial t}\right)\mathrm{e}^{\mathrm{i}\phi(x,t)}$$
$$=\mathrm{e}^{\mathrm{i}\phi(x,t)}\left(q\Phi+\hbar\frac{\partial\phi}{\partial t}-\mathrm{i}\hbar\frac{\partial}{\partial t}\right). \tag{5.125}$$

If we choose $\phi(x,t)=-q/\hbar c\chi(x,t)$, then the added terms on the right in (5.124) are

$$\mathbf{p}-\frac{q}{c}\mathbf{A}(x,t)-\frac{q}{c}\nabla\chi(x,t)=\mathbf{p}-\frac{q}{c}\mathbf{A}'(x,t) \tag{5.126}$$

and those on the right in (5.125) are

$$q\Phi(x,t)-\frac{q}{c}\frac{\partial\chi(x,t)}{\partial t}-\mathrm{i}\hbar\frac{\partial}{\partial t}$$
$$=q\Phi'(x,t)-\mathrm{i}\hbar\frac{\partial}{\partial t}. \tag{5.127}$$

The result is that the structure of the interaction between the electromagnetic fields and charged particles is *invariant* provided the interaction is given in terms of the "fictitious" fields **A**, Φ by

$$\begin{aligned}&\mathbf{p}\to\mathbf{p}-\frac{q}{c}\mathbf{A}(x,t)\\ &-\mathrm{i}\hbar\frac{\partial}{\partial t}\to-\mathrm{i}\hbar\frac{\partial}{\partial t}+q\Phi(x,t).\end{aligned} \tag{5.128}$$

There are several other ways to couple the electromagnetic field with charged particles that are allowed by symmetry [16]. But the structure of the interaction described by (5.128) is sufficient to account for all known measurements. It turns out that Maxwell's equations are also a consequence of the structure of this interaction.

This principle is called the principle of minimal electromagnetic coupling.

The phase transformation introduced in (5.120) belongs to the Lie group $U(1)$. Its generalization to position-dependent phase $\mathrm{e}^{\mathrm{i}\phi(x,t)}$ does not belong to a Lie group.

Questions soon surfaced whether the same process could be used to describe the interaction between more complicated "charged" particles and the fields that cause interactions among them. It seemed that the proton–neutron pair was a good candidate for such a treatment. These two particles seemed to be essentially the same, except that one was charged and the other not. Neglecting charge, these two particles could be treated as an *iso*spin doublet. The nucleon wavefunction ϕ could be treated as a two-state system, $|\phi\rangle = |\begin{matrix}\psi_p\\ \psi_n\end{matrix}\rangle$, and the Hamiltonian describing nuclear interactions should be invariant under a *global* $SU(2)$ transformation, analogous to a global $U(1)$ transformation $e^{i\phi}$ in (5.120). If the $SU(2)$ rotation were allowed to vary with position, perhaps it would be possible to determine the nature of the interaction between the nucleons (fermions) and the bosons ($\pi^{\pm}, \pi^0$, analogous to photons that carry the electromagnetic interaction) responsible for the interaction among the fermions.

This program was carried out by Yang and Mills. They succeeded in determining the nature of the interaction. But we now understand that nuclear interactions are residual forces left over from the strong interactions among the quarks.

Nevertheless, the program persisted. The gauge program can be phrased as follows.

1. Suppose there is a set of n fermion fields that are invariant under a g-parameter Lie group.
2. Assume that the Hamiltonian (Lagrangian, action integral) for these fields, without any interaction, is known.
3. Now assume that the Lie group parameters are allowed to be functions on space–time. What additional terms occur in the Hamiltonian (cf. (5.121) above)?
4. How many boson fields must be introduced in order to leave the structure of the Hamiltonian invariant?
5. How must they be introduced into the Hamiltonian? That is, what is the structure of the "minimal coupling" in terms of the Lie algebra parameters (its structure constants)?
6. How do these new fields transform under the Lie group and its space–time extension?
7. What field equations do the new fields satisfy?

These questions have all been answered [17]. The number of new fields required is exactly the number of generators of the Lie group (i.e., its dimension). Each field is a four-component field. Their dynamical equations are a consequence of this theory. All new fields are massless.

This theory has been applied to describe the electroweak interaction $U(2) \simeq U(1) \times SU(2)$ to predict the massless electromagnetic field and three boson fields called $W^{\pm}, Z^0$ that transmit the weak interaction. This theory was also applied to describe three quarks. The Lie group used was $SU(3)$ and the theory predicted the existence of eight (that is the dimension of the Lie group $SU(3)$) gluon fields, all massless. The gluon fields transmit the strong interaction. In the case of the gluons, the mass seems to be small enough to be consistent with "zero" but that is definitely not the case of the very massive weak gauge bosons $W^{\pm}, Z^0$. A new mechanism was called for, and proposed, to describe how these "massless" particles acquire such a heavy mass. This mechanism was proposed by Higgs, among others, and is called the Higgs mechanism. The discovery of the Higgs boson was announced in 2012.

5.14 Group Theory and Special Functions

5.14.1 Summary of Some Properties

The classical special functions of mathematical physics were developed in the nineteenth century in response to a variety of specific physical problems. They include the Legendre and associated Legendre functions, the Laguerre and associated Laguerre functions, the Gegenbauer, Chebyshev, Hermite, and Bessel functions. They are for the most part orthogonal polynomials. They are constructed by choosing a basis set $f_0, f_1, f_2, \ldots$ that are monomials in the position representation (Dirac notation): $\langle x|f_0\rangle = x^0$, $\langle x|f_1\rangle = x^1$, $\langle x|f_2\rangle = x^2, \ldots$ and then creating an orthogonal set by successive Gram–Schmidt orthogonalization by means of an inner product $\langle f|g\rangle = \int_a^b f^*(x)g(x)w(x)\mathrm{d}x$ with various weights $w(x)$ for the different functions:

$$|\phi_0\rangle = |f_0\rangle, \quad |\phi_1\rangle = |f_1\rangle - \frac{|\phi_0\rangle\langle\phi_0|}{\langle\phi_0|\phi_0\rangle}|f_0\rangle,$$
$$|\phi_j\rangle = |f_j\rangle - \sum_{k=0}^{j-1}\frac{|\phi_k\rangle\langle\phi_k|}{\langle\phi_k|\phi_k\rangle}|f_j\rangle. \tag{5.129}$$

The Bessel functions are the exception to this rule, as they are not polynomials.

These functions obey a common variety of properties

Differential equation

$$g_2(x)y'' + g_1(x)y' + g_0(x)y = 0. \tag{5.130a}$$

Recurrence relations

$$a_{1n}f_{n+1}(x) = (a_{2n} + a_{3n}x)f_n(x) - a_{4n}f_{n-1}(x). \tag{5.130b}$$

Differential relations

$$g_2(x)\frac{\mathrm{d}f_n(x)}{\mathrm{d}x} = g_1(x)f_n(x) + g_0(x)f_{n-1}(x). \tag{5.130c}$$

Generating functions

$$g(x,z) = \sum_{n=0}^{\infty} a_n f_n(x) z^n. \tag{5.130d}$$

Rodrigues' formula

$$f_n(x) = \frac{1}{a_n\rho(x)}\frac{\mathrm{d}^n}{\mathrm{d}x^n}\left\{\rho(x)(g(x))^n\right\}. \tag{5.130e}$$

The coefficients and functions can be found in standard tabulations (e.g., Abramowitz and Stegun [18]). The Bessel functions have similar properties.

5.14.2 Relation with Lie Groups

A Lie group lives on a manifold $\mathcal{M}^n$ of dimension n. Each group element is a function of position in the manifold: $g = g(\mathbf{x}), \mathbf{x} \in \mathcal{M}$. The product of two group elements is defined by an analytic composition law on the manifold:

$$g(\mathbf{x}) \circ g(\mathbf{y}) = g(\mathbf{z}), \quad \mathbf{z} = \mathbf{z}(\mathbf{x}, \mathbf{y}). \tag{5.131}$$

It is not until we construct representations for the group, or on top of the manifold, that really interesting things begin to happen. Representations

$$g(\mathbf{x}) \to \Gamma^{\alpha}_{ij}(g(\mathbf{x})) \tag{5.132}$$

are functions defined on the manifold. Suitably normalized, the set of matrix elements for the complete set of UIR form a complete orthonormal set of functions on the manifold $\mathcal{M}^n$. By duality (the miracles of Hilbert space theory), the triplet of indices α, i, j is described by as many integers as the dimension of $\mathcal{M}^n$. For example, for three-dimensional Lie groups, such as $SO(3), SU(2), SO(2,1), ISO(2), H_3$ the matrix elements are indexed by three integers and can be represented in the form $\Gamma^{\alpha}_{ij}(g(\mathbf{x})) = \langle {\alpha \atop i} | g(\mathbf{x}) | {\alpha \atop j} \rangle$. Including the appropriate normalization factor, they can be expressed as

$$\sqrt{\frac{\dim(\alpha)}{\mathrm{Vol}(G)}}\,\Gamma^{\alpha}_{ij}(g(\mathbf{x})) = \langle g(\mathbf{x}) | {\alpha \atop i,j} \rangle. \quad (5.133)$$

For noncompact groups Vol(G) is not finite, but dim(α) is also not finite, so the ratio under the radical needs to be taken with care.

Representations are powerful because they lie in two worlds: geometric and algebraic. They have one foot in the manifold ($\langle g(\mathbf{x})| \simeq \langle \mathbf{x}|$ above) and the other in algebra ($| {\alpha \atop ij} \rangle \simeq |\mathbf{n}\rangle$ above).

All classical special functions are specific matrix elements, evaluated on specific submanifolds, of specific irreducible representations of some Lie group.

We illustrate these ideas with a few examples without pretending we have even scratched the surface of this vast and fascinating field [19–21]. See also Chapter 7.

5.14.3 Spherical Harmonics and $SO(3)$

For the group $SU(2)$, the underlying manifold is a solid three-dimensional sphere. There are many ways to parameterize an element in this group. We use an Euler-angle-like parameterization introduced by Wigner:

$$\begin{aligned}\mathcal{D}^{j}_{mk}(\phi,\theta,\psi) &= \left\langle {j \atop m} \middle| \mathrm{e}^{-\mathrm{i}\phi J_z}\mathrm{e}^{-\mathrm{i}\theta J_y}\mathrm{e}^{-\mathrm{i}\psi J_z} \middle| {j \atop k} \right\rangle \\ &= \mathrm{e}^{-\mathrm{i}m\phi} d^{j}_{mk}(\theta)\mathrm{e}^{-\mathrm{i}k\psi}. \end{aligned} \quad (5.134)$$

The orthogonality properties of the matrix elements are

$$\begin{aligned}&\int_0^{2\pi} \mathrm{d}\phi \int_0^{\pi} \sin\theta \mathrm{d}\theta \\ &\int_0^{2\pi} \mathrm{d}\psi\, \mathcal{D}^{j'*}_{m'k'}(\phi,\theta,\psi)\mathcal{D}^{j}_{mk}(\phi,\theta,\psi) \\ &\quad = \frac{8\pi^2}{2j+1}\delta^{j'j}\delta_{m'm}\delta_{k'k}. \end{aligned} \quad (5.135)$$

The volume of the group in this parameterization is $8\pi^2 = (2\pi)(2)(2\pi) = (\int_0^{2\pi}\mathrm{d}\phi)(\int_0^{\pi}\sin\theta\mathrm{d}\theta)(\int_0^{2\pi}\mathrm{d}\psi)$. The normalization factor, converting the matrix elements to a complete orthonormal set, is $\sqrt{(2j+1)/8\pi^2}$.

In order to find a complete set of functions on the sphere (θ,ϕ), we search for those matrix elements above that are independent of the angle ψ. These only occur for $k = 0$, which occurs only among the subset of irreducible representations with $j = l$ (integer). Integrating out the dψ dependence in (5.134) leads to a definition of the spherical harmonics in terms of some Wigner $\mathcal{D}$ matrix elements (cf. (5.135):

$$Y^{l}_{m}(\theta,\phi) = \sqrt{\frac{2l+1}{4\pi}}\mathcal{D}^{l*}_{m0}(\phi,\theta,-). \quad (5.136)$$

These functions on the two-dimensional unit sphere surface (θ,ϕ) inherit their orthogonality and completeness properties from the corresponding properties of the UIR matrix elements $\mathcal{D}^{j}_{mk}$ on the three-dimensional solid ball of radius 2π.

Other special functions are similarly related to these matrix elements. The associated Legendre polynomials are

$$P^{m}_{l}(\cos\theta) = \sqrt{\frac{(l+m)!}{(l-m)!}}\, d^{l}_{0,0}(\theta) \quad (5.137)$$

and the Legendre polynomials are

$$P_l(\cos\theta) = \mathcal{D}^{l}_{0,0}(-,\theta,-) = d^{l}_{0,0}(\theta). \quad (5.138)$$

These functions inherit the their measure $w(\theta)$ from the measure on $SU(2)$ and their orthogonality and completeness properties from those of the Wigner rotation matrix elements $\mathcal{D}^j_{mk}[SU(2)]$.

We emphasize again that these functions are specific matrix elements $\mathcal{D}^j_{mk}$, evaluated on specific submanifolds (sphere, line), of specific irreducible representations ($j = l$) of $SU(2)$.

5.14.4 Differential and Recursion Relations

We can understand the wide variety of relations that exist among the special functions (e.g., recursion relations, etc.) in terms of group theory/representation theory as follows. It is possible to compute the matrix elements of an operator $\mathcal{O}$ in either the continuous basis $\langle \mathbf{x}'|\mathcal{O}|\mathbf{x}\rangle$ or the discrete basis $\langle \mathbf{n}'|\mathcal{O}|\mathbf{n}\rangle$. In the first basis, the coordinates $\mathbf{x}$ describe points in a submanifold in the group manifold $\mathcal{M}^n$, and the operator is a differential operator. In the second basis, the indices $\mathbf{n}$ are an appropriate subset of the group representation α and row/column (i, j) index set and the operator is a matrix with entries in the real or complex field.

It is also possible to compute the matrix elements in a *mixed* basis $\langle \mathbf{x}|\mathcal{O}|\mathbf{n}\rangle$. It is in this basis that really exciting things happen, for

$$\begin{array}{c} \langle \mathbf{x}|\mathcal{O}|\mathbf{n}\rangle \\ \swarrow \qquad \searrow \\ \langle \mathbf{x}|\mathcal{O}|\mathbf{x}'\rangle\langle \mathbf{x}'|\mathbf{n}\rangle = \langle \mathbf{x}|\mathbf{n}'\rangle\langle \mathbf{n}'|\mathcal{O}|\mathbf{n}\rangle. \end{array} \tag{5.139}$$

On the left-hand side a differential operator $\langle \mathbf{x}|\mathcal{O}|\mathbf{x}'\rangle$ acts on the special function $\langle \mathbf{x}'|\mathbf{n}\rangle$, while on the right-hand side, a matrix $\langle \mathbf{n}'|\mathcal{O}|\mathbf{n}\rangle$ multiplies the special functions $\langle \mathbf{x}|\mathbf{n}'\rangle$.

For the rotation group acting on the sphere surface (θ, ϕ) and the choice $\mathcal{O} = L_\pm$, we find for $\langle \theta\phi|L_\pm|\begin{smallmatrix} l \\ m \end{smallmatrix}\rangle$ computed as on the left in (5.139),

$$\begin{aligned} &\mathrm{e}^{\pm \mathrm{i}\phi}\left(\pm\frac{\partial}{\partial\theta} + \mathrm{i}\frac{\cos\theta}{\sin\theta}\frac{\partial}{\partial\phi}\right) \\ &\quad \delta(\cos\theta' - \cos\theta)\delta(\phi' - \phi)Y^l_m(\theta', \phi') \\ &\quad = \mathrm{e}^{\pm \mathrm{i}\phi}\left(\pm\frac{\partial}{\partial\theta} + \mathrm{i}\frac{\cos\theta}{\sin\theta}\frac{\partial}{\partial\phi}\right)Y^l_m(\theta, \phi) \end{aligned} \tag{5.140}$$

and as computed on the right

$$\begin{aligned} &\left\langle \theta\phi \left| \begin{matrix} l' \\ m' \end{matrix} \right.\right\rangle \left\langle \begin{matrix} l' \\ m' \end{matrix} \left| L_\pm \right| \begin{matrix} l \\ m \end{matrix} \right\rangle \\ &\quad = Y^l_{m\pm 1}(\theta, \phi)\sqrt{(l \pm m+1)(l \mp m)}. \end{aligned} \tag{5.141}$$

There are a number of Lie groups that can be defined to act on a one-dimensional space. In such cases, the infinitesimal generators take the form of functions of the coordinate x and the derivative $\mathrm{d}/\mathrm{d}x$. We illustrate the ideas behind differential and recursion relations in the context of the Heisenberg group H_3. Its algebra $\mathfrak{h}_3$ is spanned by three operators, universally identified as $a, a^\dagger, I$ with commutation relations $\left[a, a^\dagger\right] = I, [a, I] = \left[a^\dagger, I\right] = 0$. These operators have matrix elements as follows in the continuous basis (geometric) representation:

$$\begin{aligned} \langle x'|a|x\rangle &= \delta(x' - x)\frac{1}{\sqrt{2}}(x + D) \\ \langle x'|a^\dagger|x\rangle &= \delta(x' - x)\frac{1}{\sqrt{2}}(x - D) \\ \langle x'|I|x\rangle &= \delta(x' - x) \end{aligned} \tag{5.142}$$

and discrete basis (algebraic) representation:

$$\begin{aligned} \langle n'|a|n\rangle &= \delta_{n',n-1}\sqrt{n} \\ \langle n'|a^\dagger|n\rangle &= \delta_{n',n+1}\sqrt{n'} \\ \langle n'|I|n\rangle &= \delta_{n',n}. \end{aligned} \tag{5.143}$$

Here $D = \mathrm{d}/\mathrm{d}x$.

The special functions are the mixed basis matrix elements $\langle x|n\rangle$. We can compute these starting with the ground, or lowest, state $|0\rangle$.

$$
\begin{array}{ccc}
 & \langle x|a|0\rangle & \\
 & \swarrow \quad \searrow & \\
\langle x|a|x'\rangle\langle x'|0\rangle & = & \langle x|n\rangle\langle n|a|0\rangle \\
\frac{1}{\sqrt{2}}(x+D)\langle x|0\rangle & = & 0.
\end{array}
\tag{5.144}
$$

This equation has a unique solution $N\langle x|0\rangle = \mathrm{e}^{-x^2/2}$ up to scale factor, $N = 1/\sqrt[4]{\pi}$.

The remaining normalized basis states are constructed by applying the raising operator:

$$
\begin{aligned}
\langle x|n\rangle &= \langle x|\frac{(a^\dagger)^n}{n!}|x'\rangle\langle x'|0\rangle \\
&= \frac{(x-D)^n}{\sqrt{2^n n!\sqrt{\pi}}}\,\mathrm{e}^{-x^2/2} \\
&= \frac{H_n(x)\mathrm{e}^{-x^2/2}}{\sqrt{2^n n!\sqrt{\pi}}}.
\end{aligned}
\tag{5.145}
$$

The Hermite polynomials in (5.145) are defined by

$$H_n(x) = \mathrm{e}^{+x^2/2}(x-D)^n\mathrm{e}^{-x^2/2}. \tag{5.146}$$

The states $\langle x|n\rangle$ are normalized to $+1$.

In order to construct the recursion relations for the Hermite polynomials, choose $\mathcal{O} = x = (a + a^\dagger)/\sqrt{2}$ in (5.139). Then

$$
\begin{aligned}
\langle x|\mathcal{O}|n\rangle &= x\frac{H_n(x)\mathrm{e}^{-x^2/2}}{\sqrt{2^n n!\sqrt{\pi}}} \\
&= \frac{1}{\sqrt{2}}\langle x|n'\rangle\langle n'|(a+a^\dagger)|n\rangle.
\end{aligned}
\tag{5.147}
$$

The two nonzero matrix elements on the right are given in (5.143). They couple $xH_n(x)$ on the left with $H_{n\pm1}(x)$ on the right. When the expression is cleaned up, the standard recursion relation is obtained:

$$2x\,H_n(x) = H_{n+1}(x) + 2n\,H_{n-1}(x). \tag{5.148}$$

The differential relation is obtained in the same way, replacing $x = (a + a^\dagger)/\sqrt{2}$ by $D = (a - a^\dagger)/\sqrt{2}$ in (5.147). On the left-hand side, we find the derivative of $H_n(x)$ as well as the derivative of $\mathrm{e}^{-x^2/2}$, and on the right-hand side a linear combination of $H_{n\pm1}(x)$. When the expression is cleaned up, there results the standard differential relation

$$H_n'(x) = 2n\,H_{n-1}(x). \tag{5.149}$$

5.14.5 Differential Equation

It happens often that an operator can be formed that is quadratic in the basis vectors of the Lie algebra and it also commutes with every element in the Lie algebra. Such operators can always be constructed for semisimple Lie algebras where the Cartan metric g_{ij} (cf. (5.53)) is nonsingular. The operator $g^{ij}X_iX_j$ has this property. The construction of nontrivial quadratic operators with this property is even possible for many Lie algebras that are not semisimple. When it is possible, the left-hand side of (5.139) is a second-order differential operator and the right-hand side is a constant. This constant is the eigenvalue in the differential equation (first property listed above).

For the three-dimensional nonsemisimple group $ISO(2)$ of length-preserving translations and rotations of the plane to itself, the three infinitesimal generators are L^3, which generates rotations around the z-axis, and T_1, T_2, which generate displacements in the x and y directions. The operators T_1 and T_2 commute. The operators $L^3, T^\pm = T_1 \pm iT_2$ satisfy commutation relations

$$[L^3, T^{\pm}] = \pm T^{\pm} \qquad [T^+, T^-] = 0. \quad (5.150)$$

When acting on the plane, the three can be expressed in terms of a radial (r) and angular (ϕ) variable.

$$L^3 = \frac{1}{i}\frac{\partial}{\partial \phi} \qquad T^{\pm} = \mathrm{e}^{\pm i\phi}\left(\pm\frac{\partial}{\partial r} + \frac{i}{r}\frac{\partial}{\partial \phi}\right). \quad (5.151)$$

Basis vectors $|m\rangle$ are introduced that satisfy the condition

$$L^3|m\rangle = m|m\rangle \Rightarrow \langle r\phi|m\rangle = g_m(r)\mathrm{e}^{im\phi}. \quad (5.152)$$

Single-valuedness requires m is an integer. Adjacent basis vectors are defined by

$$\begin{aligned} T^{\pm}|m\rangle = -|m\pm 1\rangle \Rightarrow \left(\pm\frac{\mathrm{d}}{\mathrm{d}r} - \frac{m}{r}\right) g_m(r) \\ = -g_{m\pm 1}(r). \end{aligned} \quad (5.153)$$

Finally, the identity $T^+T^-|m\rangle = |m\rangle$ gives Bessel's equation

$$\left(\frac{1}{r}\frac{\mathrm{d}}{\mathrm{d}r}r\frac{\mathrm{d}}{\mathrm{d}r} + 1 - \frac{m^2}{r^2}\right) g_m(r) = 0. \quad (5.154)$$

5.14.6 Addition Theorems

Addition theorems reflect the group composition property through the matrix multiplication property of representations

$$\begin{array}{c} \langle \mathbf{n}|g(\mathbf{x})g(\mathbf{y})|\mathbf{n}'\rangle \\ \swarrow \qquad \searrow \\ \sum_{\mathbf{k}}\langle \mathbf{n}|g(\mathbf{x})|\mathbf{k}\rangle\langle \mathbf{k}|g(\mathbf{y})|\mathbf{n}'\rangle = \langle \mathbf{n}|g\left[\mathbf{z}(\mathbf{x},\mathbf{y})\right]|\mathbf{n}'\rangle. \end{array} \quad (5.155)$$

The special function at argument $\mathbf{z}$ is expressed as a pairwise product of special functions evaluated at the group elements $g(\mathbf{x})$ and $g(\mathbf{y})$ for which $\mathbf{x}\circ\mathbf{y} = \mathbf{z}$. The best known of these addition results is

$$\begin{aligned} \mathcal{D}^l_{00}(\Theta) &= \mathcal{D}^l_{0m}(g_1^{-1})\mathcal{D}^l_{m0}(g_2) \\ \mathcal{D}^l_{00}(\Theta) &= \mathcal{D}^{l*}_{m0}(g_1)\mathcal{D}^l_{m0}(g_2) \\ \frac{2l+1}{4\pi}P_l(\cos\Theta) &= \sum_m Y^l_m(\theta_1,\phi_1)Y^{l*}_m(\theta_2,\phi_2). \end{aligned} \quad (5.156)$$

Here we have taken $g_1 = (\theta_1, \phi_1, -)$ and $g_2 = (\theta_2, \phi_2, -)$, and Θ is the angle between these two points on the sphere surface, defined by

$$\cos\Theta = \cos\theta_1\cos\theta_2 + \sin\theta_1\sin\theta_2\cos(\phi_2 - \phi_1). \quad (5.157)$$

5.14.7 Generating Functions

Generating functions are constructed by computing the exponential of an operator $\mathcal{O}$ in the Lie algebra in two different ways and then equating the results. We illustrate this for H_3 by computing $\langle x|\mathrm{e}^{\sqrt{2}ta^{\dagger}}|0\rangle$. We first compute the brute-strength Taylor series expansion of the exponential:

$$\langle x|\mathrm{e}^{t(x-D)}|x'\rangle\langle x'|0\rangle = \mathrm{e}^{-x^2/2}\frac{1}{\sqrt[4]{\pi}}\sum_{n=0}^{\infty}\frac{t^n\, H_n(x)}{n!}. \quad (5.158)$$

Here $\langle x|0\rangle = \mathrm{e}^{-x^2/2}/\sqrt[4]{\pi}$. Equation (5.146) was used to obtain this result.

Next, we observe that exponentials of differential operators are closely related to Taylor series expansions, for instance $\mathrm{e}^{-t\mathrm{d}/\mathrm{d}x}f(x) = f(x-t)$. To exploit this, we use the result of the disentangling theorem (5.67) to write

$$\mathrm{e}^{t(x-D)} = \mathrm{e}^{tx}\mathrm{e}^{-t^2/2}\mathrm{e}^{-tD}. \quad (5.159)$$

Then

$$\begin{aligned} \langle x|\mathrm{e}^{t(x-D)}|x'\rangle\langle x'|0\rangle &= \mathrm{e}^{tx}\mathrm{e}^{-t^2/2}\mathrm{e}^{-tD}\langle x|0\rangle \\ &= \frac{1}{\sqrt[4]{\pi}}\mathrm{e}^{tx}\mathrm{e}^{-t^2/2}\mathrm{e}^{-(x-t)^2/2}. \end{aligned} \quad (5.160)$$

By comparing the two calculations, (5.158) with (5.160), we find the standard generating function for the Hermite polynomials.

$$e^{2xt-t^2} = \sum_{n=0}^{\infty} \frac{t^n H_n(x)}{n!}. \tag{5.161}$$

5.15 Summary

The study of symmetry has had a profound influence on the development of the natural sciences. Group theory has been used in constructive ways before groups even existed. We have given a flavor of what can be done with symmetry and related arguments in Section 5.2, which describes three types of arguments that live in the same ballpark as group theory. Groups were formally introduced in Section 5.3 and a number of examples given in the following three sections, ranging from finite groups to Lie groups. Lie algebras were introduced in Sect. 5.7 and a number of their properties discussed in that section. In Sect. 5.8, we introduced the idea of a Riemannian symmetric space and showed the close connection between these spaces and Lie groups, specifically that they are quotients (cosets) of one Lie group by another, subject to stringent conditions.

Transformation groups played a big role in the development of classical physics – mechanics and electrodynamics. In fact, it was the need to formulate these two theories so that their structure remained unchanged under transformations from the same group that led to the theory of special relativity. The group at hand was the inhomogeneous Lorentz group, the 10-parameter Lie group of Lorentz transformations and translations acting on fields defined over space–time. Section 5.9 describes how group theory played a role in the development of special relativity. The next step beyond requiring invariance under the *same* Lorentz transformation at every space–time point involved, allowing the Lorentz transformation to vary from point to point in a continuous way and still requiring some kind of invariance (cf. gauge theories as discussed in Section 5.13). This extended aesthetic led to the theory of general relativity.

Up to this point, physicists could have done without all the particular intricacies of group theory. The next step in the growth of this subject was the intensive study of the linear representations of groups. The growth was indispensible when quantum theory was developed, because groups acted in Hilbert spaces through their linear matrix representations. We provided an overview of representation theory in Section 5.10. At first, groups were applied in the quantum theory as symmetry groups (cf. Section 5.11. In this Capacity, they were used to label energy eigenstates and describe the degeneracies in energy levels that were required by symmetry. Shortly afterward, they were used in a more daring way to describe nondegenerate levels related to each other either by a broken symmetry or simply by operators that had little to do with symmetry but had the good sense to close under commutation with each other. In a very accurate sense, Mendeleev's table of the chemical elements and the table of nuclear isotopes are manifestations of broken symmetry applied to the conformal group $SO(4,2)$ that describes all bound states of the hydrogen atom in a single UIR, and the group $U(3,1)$ that describes all bound states of the three-dimensional harmonic oscillator in one representation. These and other applications of dynamical groups are described in Section 5.12.

Gauge theories were briefly treated in Section 5.13. In such theories, one begins with a symmetry group and requires that a Hamiltonian, Lagrangian, or action remain "invariant" under the transformation when the parameters of the transformation group are allowed to be functions over space–time. It is remarkable that this requirement leads to the prediction of new fields, the nature of the interaction of the new fields with the original fields, the structure of the equations of the new fields, and the mass spectrum of these new fields: all new masses are zero. This problem was overcome by proposing that a new particle, now called the Higgs boson, exists. Its discovery was announced in 2012.

As a closing tribute to the theory of groups and their linear matrix representations, we hint how the entire theory of the special functions of mathematical physics, which was created long before the Lie groups were invented, is a study of the properties of specific matrix elements of specific matrix representations of particular Lie groups acting over special submanifolds of the differentiable manifold that parameterizes the Lie group. These ideas are sketched by simple examples in Section 5.14.

Group theory has progressed from the outer fringes of theoretical physics in 1928, when it was referred to as the *Gruppenpest* (1928 Weyl to Dirac at Princeton), through the mainstream of modern physics, to wind up playing the central role in the development of physical theory. Theoretical physicists now believe that if a theory of fundamental interactions is not a gauge theory it does not have the right to be considered a theory of interactions at all. Gauge theory is the new version of "simple" and "elegant."

We learn that Nature was not tamed until Adam was able to give names to all the animals. Similarly, we cannot even give names to particles and their states without knowing at least a little bit about group theory. Group theory has migrated from the outer fringes of physics (Gruppenpest, 1928) to the central player, even the lingua franca, of modern physics.

Glossary

Group: A group is a set $\{g_0, g_1, g_2, \dots\}$, called group elements, together with a combinatorial operation, $\circ$, called group multiplication, with the property that four axioms are obeyed: Closure, Associativity, Existence of Identity, and Unique Inverse.

Lie Group: A Lie group is a group whose elements $g(\mathbf{x})$ are parameterized by points $\mathbf{x}$ in an n-dimensional manifold. Group multiplication corresponds to mappings of pairs of points in the manifold to points in the manifold: $g(\mathbf{x}) \circ g(\mathbf{y}) = g(\mathbf{z})$, where $\mathbf{z} = \mathbf{z}(\mathbf{x}, \mathbf{y})$.

Lie Algebra: A Lie algebra is a linear vector space on which one additional composition law, the commutator [,], is defined. The commutator satisfies three conditions: it preserves linear vector space properties, it is antisymmetric, and the Jacobi identity is satisfied.

Homomorphism: A homomorphism is a mapping of one set with an algebraic structure (e.g., group, linear vector space, algebra) onto another algebraic structure of the same type that preserves all combinatorial operations.

Isomorphism: An isomorphism is a homomorphism that is 1 : 1.

Structure Constants $C_{ij}{}^{k}$**:** These are expansion coefficients for the commutator of basis vectors in a Lie algebra. If X_i, $i = 1, 2, \dots, n$ are basis vectors for a Lie algebra, then the commutator $[X_i, X_j]$ can be expanded in terms of these basis vectors: $\left[X_i, X_j\right] = \sum_k C_{ij}^{\ k} X_k$.

Cartan Metric: A metric tensor that can be defined on a Lie algebra by double cross contraction on the indices of the structure constants: $g_{ij} = \sum_{kl} C_{ik}^{\ l} C_{jl}^{\ k}$. This metric tensor has remarkable properties.

Symmetry Group: In quantum mechanics, this is a group G that leaves the Hamiltonian $\mathcal{H}$ of a physical system invariant: $G\mathcal{H}G^{-1} = \mathcal{H}$. Through its matrix representations, it maps states of energy E_i into other, often linearly independent states, with the same energy E_i.

Dynamical Group: In quantum mechanics, this is a group H that leaves invariant the time-dependent Schrödinger equation: $(\mathcal{H} - i\hbar\partial/\partial t)\,\psi(x,t) = 0$. Through its matrix representations, it maps states of energy E_i into other, linearly independent states with different energies E_j.

Subgroup: A subgroup H of a group G consists of a subset of group elements of G that obey the group axioms under the same group multiplication operation that is defined on G. For example $C_3 \subset C_{3v}$. Both the full group G and only the identity element e are (improper) subgroups of G.

Coset C: If H is a subgroup of G, then every group element in G can be written as a product of an element in H and a group element in G: $g_i = h_j c_k$. If n_G and n_H are the orders of G and H, then $g_i \in G$, $1 \leq i \leq n_G$, $h_j \in H$, $1 \leq j \leq n_H$ and $c_k \in C \subset G$, $1 \leq k \leq n_C$, where $n_C = n_G/n_H$. The group elements C_k are called coset representatives; the set $\left\{c_1, c_2, \dots, c_{n_C}\right\}$ is called a coset. A coset is not generally a group. If H is an invariant subgroup of G the coset representatives can be chosen so that C is a group.

Representation: A representation is a homomorphism of a set with an algebraic structure (e.g., group, linear vector space, algebra) into matrices. If the mapping is an isomorphism the representation is called faithful or 1 : 1.

Class: This is a set of elements in a group that are related to each other by a similarity transformation: if group elements h_i and h_j are related by $h_i = gh_jg^{-1}$ for some $g \in G$, then they are in the same class. Group elements in the same class are geometrically similar. For example, in the group of transformations that map the cube to itself, there are eight rotations by $2\pi/3$ radians about axes through opposite corners. These are geometrically similar and belong to the same class.

Characters: In the theory of group representations, characters are traces of the representation matrices for the classes of the group.

Gauge Theory: Schrödinger's equation is unchanged if the wavefunction is multiplied by a phase factor that is constant throughout space and time. If the wavefunction has two components, Schrödinger's equation is unchanged if the wavefunction is multiplied (rotated) by a group element from a 2×2 matrix from the group $U(2)$ or $SU(2)$ $(2 \rightarrow 3, SU(2) \rightarrow SU(3)$, etc.). If the rotation depends on space–time coordinates, Schrödinger's equation is not invariant. Gauge theory attempts to show how Schrödinger's equation can remain unchanged in form under space–time-dependent rotations. This requires the inclusion of N extra fields to the physical problem, where N is the dimension of the rotation group, $U(2)$, $SU(2)$, $SU(3)$, … These extra fields describe particles (bosons) that govern the interactions among the original fields described by Schrödinger's equation. For example, for an original charged field with one component, invariance under space–time- dependent phase factor in the group $U(1)$ leads to the prediction that zero-mass photons are responsible for charge-charge interactions. Gauge theories also predict the form of the interaction between the original (fermion) fields and the newly introduced (gauge boson) fields.

References

1. Barenblatt, G. (2003) *Scaling*, Cambridge University Press, Cambridge.
2. Landau, L.D. and Lifshitz, E.M. (1960) *Mechanics*, Addison-Wesley, Reading, MA.
3. Gilmore, R. (1974) *Lie Groups, Lie Alegras, and Some of their Applications*, John Wiley & Sons, Inc. (reprinted in 2005 by Dover in New York), New York.
4. Gilmore, R. (2008) *Lie Groups, Physics, and Geometry, An Introduction for Physicists, Engineers, and Chemists*, Cambridge University Press, Cambridge.
5. Hamermesh, M. (1962) *Group Theory and its Application to Physical Problems*, Addison-Wesley, Reading, MA; reprint (1989) Dover, New York.
6. Sternberg, S. (1994) *Group Theory and Physics*, University Press, Cambridge.
7. Tinkham, M. (1964) *Group Theory and Quantum Mechanics*, McGraw Hill, New York.
8. Wigner, E.P. (1959) *Group Theory, and its Application to the Quantum Mechanics of Atomic Spectra* (ed. J.J. Griffin, translator), Academic Press, New York.
9. Coxeter, H.S.M. and Moser, W.O.J. (1980) *Generators and Relations for Discrete Groups*, 4th edn, Springer-Verlag, Berlin.
10. Helgason, S. (1962) *Differential Geometry and Symmetric spaces*, Academic Press, New York.
11. Weinberg, S. (1972) *Gravitation and Cosmology, Principles and Applications of the General Theory of Relativity*, John Wiley & Sons, Inc., New York.
12. Barut, A.O. and Raczka, R. (1986) *Theory of Group Representations and Applications*, World Scientific, Singapore.
13. Haxel, O., Jensen, J.H.D., and Suess, H.E. (1949), On the "Magic Numbers" in nuclear structure. *Phys. Rev.*, **75**, 1766–1766.
14. Mayer, M.G. (1949) On closed shells in Nuclei. II. *Phys. Rev*, **75**, 1969–1970.
15. Arima, A. and Iachello, F. (1976) Interacting Boson Model of collective states, Part I (the vibrational limit). *Ann. Phys. (New York)*, **99**, 253–317.
16. Bethe, H.A. and Salpeter, E.E. (1957) *Quantum Mechanics of One- and Two-Electron Atoms*, Academic Press, New York.
17. Utiyama, R. (1956) Invariant theoretical interpretation of interaction. *Phys. Rev.*, **101**, 1597–1607.
18. Abramowitz, M. and Stegun, I.A. (1964) *Handbook of Mathematical Functions*, National Bureau of Standards (reprinted in 1972 by Dover in New York), Washington, DC.
19. Miller, W. Jr. (1968) *Lie Theory and Special Functions*, Academic Press, New York.
20. Talman, J.D. (1968) *Special Functions: A Group Theoretic Approach (Based on Lectures by Eugene P. Wigner)*, Benjamin, New York.
21. Vilenkin, N.Ja. (1968) *Special Functions and the Theory of Group Representations*, American Mathematical Society, Providence, RI.

6
Algebraic Topology

Vanessa Robins

6.1
Introduction

Topology is the study of those aspects of shape and structure that do not depend on precise knowledge of an object's geometry. Accurate measurements are central to physics, so physicists like to joke that a topologist is someone who cannot tell the difference between a coffee cup and a doughnut. However, the qualitative nature of topology and its ties to global analysis mean that many results are relevant to physical applications. One of the most notable areas of overlap comes from the study of dynamical systems. Some of the earliest work in algebraic topology was by Henri Poincaré in the 1890s, who pioneered a qualitative approach to the study of celestial mechanics by using topological results to prove the existence of periodic orbits [1]. Topology continued to play an important role in dynamical systems with significant results pertinent to both areas from Smale in the 1960s [2]. More recently, in the 1990s, computer analysis of chaotic dynamics was one of the drivers for innovation in computational topology [3–6].

As with any established subject, there are several branches to topology: *General topology* defines the notion of "closeness" (the neighborhood of a point), limits, continuity of functions and so on, in the absence of a metric. These concepts are absolutely fundamental to modern functional analysis; a standard introductory reference is [7]. *Algebraic topology* derives algebraic objects (typically groups) from topological spaces to help determine when two spaces are alike. It also allows us to compute quantities such as the number of pieces the space has, and the number and type of "holes." *Differential topology* builds on the above and on the differential geometry of manifolds to study the restrictions on functions that arise as the result of the structure of their domain. This chapter is primarily concerned with algebraic topology; it covers the elementary tools and concepts from this field. It draws on definitions and material from Chapter 5 on group theory and Chapters 9 and 10 on differential geometry.

A central question in topology is to decide when two objects are the same in some sense. In general topology, two spaces, A and B, are considered to be

Mathematical Tools for Physicists, Second Edition. Edited by Michael Grinfeld.

the same if there is a *homeomorphism*, f, between them: $f : A \to B$ is a continuous function with a continuous inverse. This captures an *intrinsic* type of equivalence that allows arbitrary stretching and squeezing of a shape and permits changes in the way an object sits in a larger space (its embedding), but excludes any cutting or gluing. So for example, a circle ($x^2 + y^2 = 1$) is homeomorphic to the perimeter of a square and to the trefoil knot, but not to a line segment, and a sphere with a single point removed is homeomorphic to the plane. One of the ultimate goals in topology is to find a set of quantities (called *invariants*) that characterize spaces up to homeomorphism. For arbitrary topological spaces, this is known to be impossible [1] but for closed, compact 2-manifolds this problem is solved by finding the *Euler characteristic* (see p. 222) and orientability of the surface [8, 9].

What is the essential difference between a line segment and a circle? Intuitively, it is the ability to trace your finger round and round the circle as many times as you like without stopping or turning back. Algebraic topology is the mathematical machinery that lets us quantify and detect this. The idea behind algebraic topology is to map topological spaces into groups (or other algebraic structures) in such a way that continuous functions between topological spaces map to homomorphisms between their associated groups.[1)]

In Sections 6.2–6.4, this chapter covers the three basic constructions of algebraic topology: homotopy, homology, and cohomology theories. Each has a different method for defining a group from a topological space, and although there are close links between the three, they capture different qualities of a space. Many of the more advanced topics in algebraic topology involve studying functions on a space, so we introduce the fundamental link between critical points of a function and the topology of its domain in Section 6.5 on Morse theory. The computability of invariants, both analytically and numerically, is vital to physical applications, so the recent literature on computational topology is reviewed in Section 6.6. Finally, we give a brief guide to further reading on applications of topology to physics.

1) A *homomorphism* between two groups, $\phi : G \to H$ is a function that respects the group operation. That is, $\phi(a \cdot b) = \phi(a) * \phi(b)$, for $a, b \in G$, where $\cdot$ is the group operation in G and $*$ is the group operation in H.

6.2 Homotopy Theory

A homotopy equivalence is a weaker form of equivalence between topological spaces than a homeomorphism that allows us to collapse a space onto a lower-dimensional subset of itself (as we will explain in Section 6.2.3), but it captures many essential aspects of shape and structure. When applied to paths in a space, homotopy equivalence allows us to define an algebraic operation on loops, and provides our first bridge between topology and groups.

6.2.1 Homotopy of Paths

We begin by defining a *homotopy* between two continuous functions $f, g : X \to Y$. These maps will be homotopic if their images $f(X)$, $g(X)$, can be continuously morphed from one to the other within Y, that is, there is a parameterized set of images that starts with one and ends with the other. Formally, this deformation is achieved by defining a continuous function

$F : X \times [0,1] \to Y$ with $F(x,0) = f(x)$ and $F(x,1) = g(x)$.

For example, consider two maps from the unit circle into the unit sphere, $f, g : S^1 \to S^2$. We use the angle $\theta \in [0, 2\pi)$ to parameterize S^1 and the embedding $x^2 + y^2 + z^2 = 1$ in $\mathbb{R}^3$ to define points in S^2. Define $f(\theta) = (\cos\theta, \sin\theta, 0)$ to be a map from the circle to the equator and $g(\theta) = (0,0,1)$, a constant map from the circle to the north pole. A homotopy between f and g is given by $F(\theta, t) = (\cos(\pi t/2)\cos\theta, \cos(\pi t/2)\sin\theta, \sin(\pi t/2))$ and illustrated in Figure 6.1. Any function that is homotopic to a constant function, as in this example, is called *null homotopic*.

When the domain is the unit interval, $X = [0,1]$, and Y is an arbitrary topological space, the continuous functions f and g are referred to as *paths* in Y. A space in which every pair of points may be joined by a path is called *path connected*. It is often useful to consider homotopies between paths that fix their end points, y_0 and y_1, say, so we have the additional conditions on F that for all $t \in [0,1]$, $F(0,t) = f(0) = g(0) = y_0$, and $F(1,t) = f(1) = g(1) = y_1$. If a path starts and ends at the same point, $y_0 = y_1$, it is called a *loop*. A loop that is homotopic to a single point, that is, a null-homotopic loop is also said to be *contractible* or *trivial*. A path-connected space in which every loop is contractible is called *simply connected*. So the real line and the surface of the sphere are simply connected, but the circle and the surface of a doughnut (the torus) are not.

6.2.2
The Fundamental Group

We are now in a position to define our first algebraic object, the *fundamental group*. The first step is to choose a *base point* y_0 in the space Y and consider all possible loops in Y that start and end at y_0. Two loops belong to the same *equivalence class* if they are homotopic: given a loop $f : [0,1] \to Y$ with $f(0) = f(1) = y_0$, we write $[f]$ to represent the set of all loops that are homotopic to f. The appropriate group operation $[f] * [g]$ on these equivalence classes is a concatenation of loops defined by tracing each at twice the speed. Specifically, choose f and g to be representatives of their respective equivalence classes and define $f * g(x) = f(2x)$ when $x \in [0, \frac{1}{2}]$ and $f * g(x) = g(2x-1)$ when $x \in [\frac{1}{2}, 1]$. As all loops have the same base point, this product is another loop based at y_0. We then simply set $[f] * [g] = [f * g]$. The equivalence class of the product $[f * g]$ is independent of the choice of f and g because we can re-parameterize the

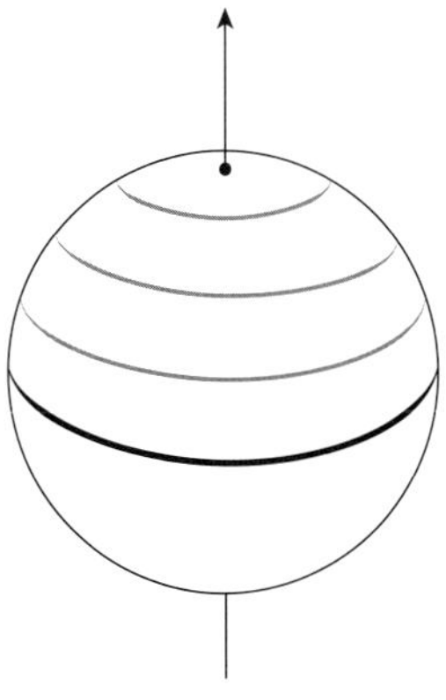

Figure 6.1 The function $f : S^1 \to S^2$ that maps the circle onto the equator of the sphere is homotopic to the function $g : S^1 \to S^2$ that maps the circle to the north pole. Three sections of the homotopy $F : S^1 \times [0,1] \to S^2$ are shown in gray.

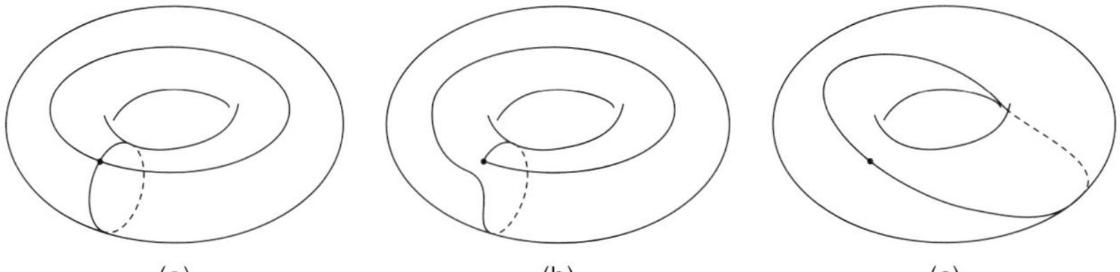

Figure 6.2 (a) Two nonhomotopic loops on the torus with the same base point. (b) A loop homotopic to the concatenation the loops depicted in (a). (c) Another loop in the same homotopy class.

homotopies in the same way as we concatenated the loops. Note, though, that the equivalence class $[f * g]$ consists of more than just concatenated loops; Figure 6.2 depicts an example on the torus.

The set of all homotopy-equivalence classes of loops based at y_0 with the operation $*$ forms a group with the identity element being the class of null-homotopic loops $[e]$ where $e(x) = y_0$, and the inverse of a loop defined to be the same loop traced backwards: $[f]^{-1} = [h]$ where $h(x) = f(1 - x)$. This group is the *fundamental group* of Y with base point y_0: $\pi_1(Y, y_0)$.

The operation taking a topological space to its fundamental group is an example of a *functor*. This word expresses the property alluded to in the introduction that continuous maps between topological spaces transform to homomorphisms between their associated groups. The functorial nature of the fundamental group is manifest in the following fashion. Suppose we have a continuous function $f : X \to Y$ with $f(x_0) = y_0$. Then given a loop in X with base point x_0, we can use simple composition of the loop with f to obtain a loop in Y with base point y_0. Composition also respects the concatenation of loops and homotopy equivalences, so it induces a homomorphism between the fundamental groups: $\pi_1(f) : \pi_1(X, x_0) \to \pi_1(Y, y_0)$. When the function $f : X \to Y$ is a *homeomorphism*, it follows that the induced map $\pi_1(f)$ is an *isomorphism* of their fundamental groups.

The following are some further elementary properties of the fundamental group:

- A simply connected space has a trivial fundamental group, that is, only the identity element.
- If Y is path connected, the fundamental group is independent of the base point, and we write $\pi_1(Y)$.
- The fundamental group respects products[2)] of path-connected spaces: $\pi_1(X \times Y) = \pi_1(X) \times \pi_1(Y)$.
- The wedge product of two path-connected spaces (obtained by gluing the spaces together at a single point) gives a free product[3)] on their fundamental groups: $\pi_1(X \vee Y) = \pi_1(X) * \pi_1(Y)$.
- The *van Kampen theorem* shows how to compute the fundamental group of a space $X = U \cup V$ when U, V, and $U \cap V$ are open, path-connected subspaces of X via a *free product with amalgamation*: $\pi_1(X) = \pi_1(U) * \pi_1(V)/N$, where N is a

2) The (Cartesian or direct) *product* of two spaces (or two groups) $X \times Y$ is defined by ordered pairs (x, y) where $x \in X$ and $y \in Y$.

3) The *free product* of two groups $G * H$ is an infinite group that contains both G and H as subgroups and whose elements are words of the form $g_1h_1g_2h_2 \cdot \cdots$

normal subgroup generated by elements of the form $i_U(\gamma)i_V(\gamma)^{-1}$ and γ is a loop in $\pi_1(U \cap V)$, i_U and i_V are inclusion-induced maps from $\pi_1(U \cap V)$ to $\pi_1(U)$ and $\pi_1(V)$ respectively. See [10] for details.

6.2.3 Homotopy of Spaces

Now we look at what it means for two spaces X and Y to be *homotopy equivalent* or to have the same *homotopy type*: there must be continuous functions $f : X \to Y$ and $g : Y \to X$ such that $fg : Y \to Y$ is homotopic to the identity on Y and $gf : X \to X$ is homotopic to the identity on X. We can show that the unit circle S^1 and the annulus A have the same homotopy type as follows. Let

$$S^1 = \{(r, \theta) \mid r = 1, \theta \in [0, 2\pi)\} \quad \text{and}$$
$$A = \{(r, \theta) \mid 1 \le r \le 2, \theta \in [0, 2\pi)\}$$

be subsets of the plane parameterized by polar coordinates. Define $f : S^1 \to A$ to be the inclusion map $f(1, \theta) = (1, \theta)$ and let $g : A \to S^1$ map all points with the same angle to the corresponding point on the unit circle: $g(r, \theta) = (1, \theta)$. Then $gf : S^1 \to S^1$ is given by $gf(1, \theta) = (1, \theta)$, which is exactly the identity map. The other composition is $fg : A \to A$ is $fg(r, \theta) = (1, \theta)$. This is homotopic to the identity $i_A = (r, \theta)$ via the homotopy $F(r, \theta, t) = (1 + t(r - 1), \theta)$. This example is an illustration of a *deformation retraction*: a homotopy equivalence between a space (e.g., the annulus) to a subset (the circle) that leaves the subset fixed throughout.

Spaces that are homotopy equivalent have isomorphic fundamental groups. A space that has the homotopy type of a point is said to be *contractible* and has trivial fundamental group. This is much stronger than being simply connected: for example, the sphere S^2 is simply connected because every loop can be shrunk to a point, but it is not a contractible space.

6.2.4 Examples

Real space $\mathbb{R}^m$, $m \ge 1$, all spheres S^n with $n \ge 2$, any Hilbert space, and any connected tree (cf. Chapter 4) have trivial fundamental groups.

The fundamental group of the circle is isomorphic to the integers under addition: $\pi_1(S^1) = \mathbb{Z}$. This can be seen by noting that the homotopy class of a loop is determined by how many times it wraps around the circle. A formal proof of this result is quite involved – see [10] for details. Any space that is homotopy equivalent to the circle will have the same fundamental group – this holds for the annulus, the Möbius band, a cylinder, and the "punctured plane" $\mathbb{R}^2 \setminus (0, 0)$.

The projective plane $\mathbb{R}P^2$ is a nonorientable surface defined by identifying antipodal points on the boundary of the unit disk (or equivalently, antipodal points on the sphere). Its fundamental group is isomorphic to $\mathbb{Z}_2$ (the group with two elements, the identity and r which is its own inverse $r^2 = \text{id}$). To see this, consider a loop that starts at the center of the unit disk $(0, 0)$, goes straight up to $(0, 1)$ which is identified with $(0, -1)$ then continues straight back up to the origin. This loop is in a distinct homotopy class to the null-homotopic loop but it is in the same homotopy class as its inverse (to see this, imagine fixing the loop at $(0, 0)$ and rotate it by 180° as illustrated in Figure 6.3).

The fundamental group of a connected graph with v vertices and e edges (cf. Chapter 4) is a free group with n

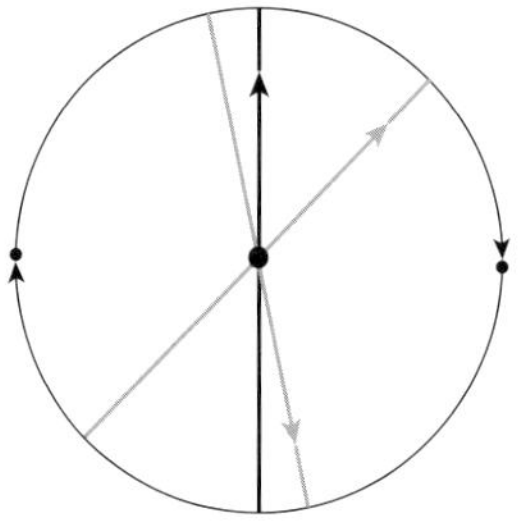

Figure 6.3 The projective plane, $\mathbb{R}P^2$ is modeled by the unit disk with opposite points on the boundary identified. The black loop starting at $(0, 0)$ is homotopic to its inverse with the equivalence suggested by the gray loops.

generators[4] where $n = e - (v - 1)$ is the number of edges in excess of a spanning tree (the *cyclomatic number*). This demonstrates that the fundamental group need not be Abelian (products do not necessarily commute).

The torus $\mathbb{T} = S^1 \times S^1$, so $\pi_1(\mathbb{T}) = \mathbb{Z} \times \mathbb{Z}$. More generally, the fundamental group of an orientable genus-g surface[5] ($g \geq 2$) is isomorphic to a hyperbolic translation group with $2g$ generators.

If a space has a finite *cell structure*, then the fundamental group can be computed as a free group with relations in an algorithmic manner; this is discussed in Section 6.6.

6.2.5 Covering Spaces

The result about the fundamental group of a genus-g surface comes from analyzing the relationship between loops on a surface and paths in its *universal covering space* (the hyperbolic plane when $g \geq 2$). Covering spaces are useful in many other contexts (from harmonic analysis to differential topology to computer simulation), so we describe them briefly here. They are simply a more general formulation of the standard procedure of identifying a real-valued periodic function with a function on the circle.

Given a topological space X, a *covering space* for X is a pair (C, p), where C is another topological space and $p : C \to X$ is a continuous function onto X. The *covering map* p must satisfy the following condition: for every point $x \in X$, there is a neighborhood U of x such that $p^{-1}(U)$ is a disjoint union of open sets each of which is mapped homeomorphically onto U by p. The discrete set of points $p^{-1}(x)$ is called the *fiber* of x. A *universal* covering space is one in which C is simply connected. The reason for the name comes from the fact that a universal covering of X will cover any other connected covering of X. For example, the circle is a covering space of itself with $C = S^1 = \{z \in \mathbb{C} : |z| = 1\}$ and $p_k(z) = z^k$ for all nonzero integers k, while the universal cover of the circle is the real line with $p_{U} : \mathbb{R} \to S^1$ given by $p_{U}(t) = \exp(i2\pi t)$. The point $z = (1, 0) \in S^1$ is then covered by the fiber $p^{-1}(z) = \mathbb{Z} \subset \mathbb{R}$. We illustrate a covering of the torus in Figure 6.4.

When X and C are suitably nice spaces (connected and locally path connected), loops in X based at x_0 *lift* to paths in C between elements of the fiber of x_0. So in the example of S^1, a path in $\mathbb{R}$ between two integers $i < j$ maps under p_{U} to a loop that wraps $j - i$ times around the circle.

Now consider homeomorphisms of the covering space $h : C \to C$ that respect the

4) A *free group* with one generator, a say, is the infinite cyclic group with elements $\ldots, a^{-1}, 1, a, a^2, \ldots$ A free group with two generators a, b, contains all elements of the form $a^{i_1} b^{j_1} a^{i_2} b^{j_2} \cdots$ for integers i_k, j_l. A free group with n generators is the natural generalization of this.

5) Starting with a sphere, you can obtain all closed oriented 2-manifolds by attaching some number of handles (cylinders) to the sphere. The number of handles is the *genus*.

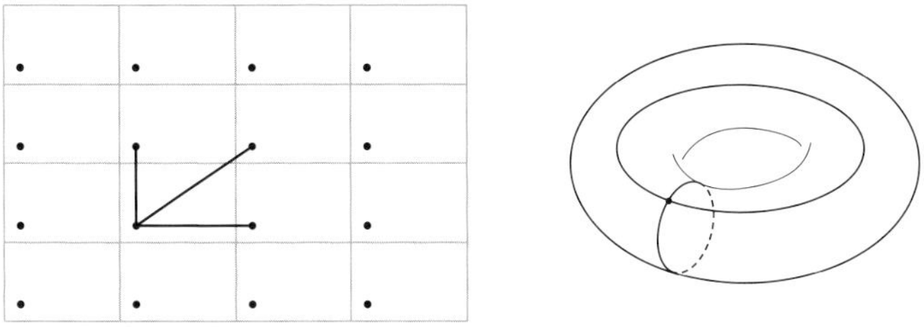

Figure 6.4 The universal covering space of the torus is the Euclidean plane projected onto the closed surface by identifying opposite edges of each rectangle with parallel orientations. The fiber of the base point on the torus is a lattice of points in the cover. The two loops on the torus lift to the vertical and horizontal paths shown in the cover. The lift of the concatenation of these two loops is homotopic to the diagonal path in the cover (see Figure 6.2c). The deck transformation group for this cover is simply the group of translations that preserve the rectangles; it is isomorphic to $\mathbb{Z} \times \mathbb{Z} = \pi_1(\mathbb{T})$.

covering map, $p(h(c)) = p(c)$. The set of all such homeomorphisms forms a group under composition called the *deck transformation group*. When (C, p) is a universal covering space for X, it is possible to show that the deck transformation group must be isomorphic to the fundamental group of X. This gives a technique for determining the fundamental group of a space in some situations; see [10] for details and examples.

6.2.6 Extensions and Applications

As we saw in the examples of Section 6.2.4, the fundamental group of an n-dimensional sphere is trivial for $n \geq 2$, so the question naturally arises how we might capture the different topological structures of S^n. To generalize the fundamental group, we examine maps from an n-dimensional unit cube I^n into the space X where the entire boundary of the cube is mapped to a fixed base point in X, that is, $f : I^n \to X$, with $f(\partial I^n) = x_0$. Elements of the *higher homotopy groups* $\pi_n(X, x_0)$ are then homotopy-equivalence classes of these maps. The group operation is concatenation in the first coordinate just as we defined for one-dimensional closed paths above. The main difference in higher dimensions is that this operation now commutes.

It is perhaps not too difficult to see that $\pi_2(S^2) = \mathbb{Z}$, although the multiple wrapping of the sphere by a piece of paper cannot be physically realized in $\mathbb{R}^3$ in the same way a piece of string wraps many times around a circle. The surprise comes with the result that $\pi_k(S^n)$ is nontrivial for most (but certainly not all) $k \geq n \geq 2$, and in fact mathematicians have not yet determined all the homotopy groups of spheres for arbitrary k and n [10]. Higher-order homotopy groups are a rich and fascinating set of topological invariants that are the subject of active research in mathematics.

One application of homotopy theory arises in the study of *topological defects* in condensed matter physics. A classic example is that of nematic liquid crystals, which are fluids comprised of molecules with an elongated ellipsoidal shape. The order parameter for this system is the (time-averaged) direction of the major axis of the ellipsoidal molecule: $\mathbf{n}$. For identical and symmetrical molecules, the sign and the magnitude of the vector is irrelevant, and so the parameter space for $\mathbf{n}$ is the surface of the sphere with antipodal points

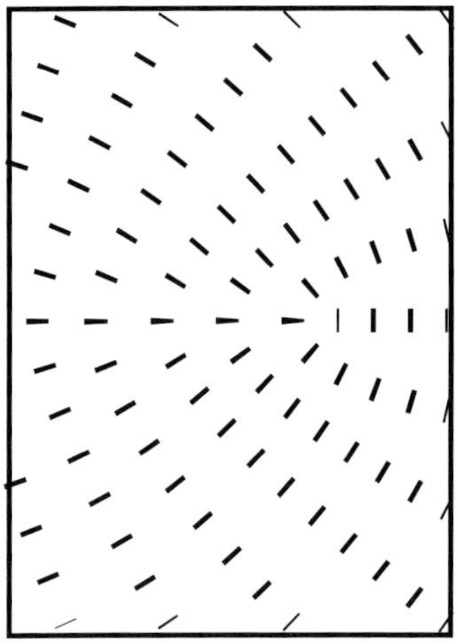

Figure 6.5 A cross section through a nematic fluid with a line defect that runs perpendicular to the page. Each line segment represents the averaged direction of a single molecule.

identified – topologically $\mathbb{R}P^2$. The existence of noncontractible loops in $\mathbb{R}P^2$ is associated with the existence of topological line defects in configurations of molecules in the nematic liquid crystal; see Figure 6.5. The fact that $\pi_1(\mathbb{R}P^2) = \mathbb{Z}_2$ is manifest in the fact that two defects of the *same* type can smoothly cancel one another. The second homotopy group is $\pi_2(\mathbb{R}P^2) = \mathbb{Z}$, and this is manifest in the existence of point defects ("hedgehogs") where the director field points radially away from a central point. See Mermin's original article [11] or [12] for further details.

6.3 Homology

The fundamental group is a useful invariant but it captures only the one-dimensional structure of equivalent loops in a space and cannot distinguish between spheres of dimensions greater than two, for example. The higher homotopy groups do capture this structure but are difficult to compute. The homology groups provide a way to describe structure in all relevant dimensions, but require a bit more machinery to define. This can seem abstract at first, but in fact the methods are quite combinatorial and there has been much recent activity devising efficient algorithms to compute homological quantities from large data sets (see Section 6.6).

There are a number of different formulations of homology theory that give essentially the same results for "nice" spaces (such as differentiable manifolds). The two key ingredients are a discrete cell complex that captures the way a space is put together, and a boundary map that describes incidences between cells of adjacent dimensions. The algebraic structure comes from defining the addition and subtraction of cells.

The earliest formulation of homology theory is *simplicial homology*, based on triangulations of topological spaces called *simplicial complexes.* This theory has some shortcomings when dealing with very general topological spaces and successive improvements over the past century have culminated in the current form based on singular homology and general cell complexes. Hatcher [10] provides an excellent introduction to homology from this modern perspective. We focus on simplicial homology here because it is the most concrete and easy to adapt for implementation on a computer. The notation used in this section is based on that of Munkres [13].

6.3.1
Simplicial Complexes

The basic building block is an *oriented k-simplex*, σ^k, the convex hull of $k+1$ geometrically independent points, $\{x_0, x_1, \dots, x_k\} \subset \mathbb{R}^m$, with $k \le m$. For example, a 0-simplex is just a point, a 1-simplex is a line segment, a 2-simplex a triangle, and a 3-simplex is a tetrahedron. We write $\sigma^k = \langle x_0, x_1, \dots, x_k\rangle$ to denote a k-simplex and its vertices. The ordering of the vertices defines an *orientation* of the simplex. This orientation is chosen arbitrarily but is fixed and coincides with the usual notion of orientation of line segments, triangles, and tetrahedra. Any even permutation of the vertices in a simplex gives another simplex with the same orientation, while an odd permutation gives a simplex with negative orientation.

Given a set V, an abstract *simplicial complex*, C, is a collection of finite subsets of V with the property that if $\sigma^k = \{v_0, \dots, v_k\} \in C$ then all subsets of σ^k (its faces) are also in C. If the simplicial complex is finite, then it can always be embedded in $\mathbb{R}^m$ for some m (certain complexes with infinitely many simplices can also be embedded in finite-dimensional space). An embedded complex is called a *geometric realization* of C. The subset of $\mathbb{R}^m$ occupied by the geometric complex is denoted by $|C|$ and is called a *polytope* or *polyhedron*. When a topological space X is homeomorphic to a polytope, $|C|$, it is called a *triangulated space* and the simplicial complex C is a *triangulation* of X. For example, a circle is homeomorphic to the boundary of a triangle, so the three vertices a, b, c and three 1-simplices, $\langle ab\rangle, \langle bc\rangle, \langle ca\rangle$ are a triangulation of the circle (see Figure 6.6). All *differentiable* manifolds have triangulations, but a complete characterization of the class of topological spaces that have a triangulation is not known. Every topological 2- or 3-manifold has a triangulation, but there is a (nonsmooth) 4-manifold that cannot have a triangulation (it is related to the Lie group E_8 [14]). The situation for nondifferentiable manifolds in higher dimensions remains uncertain.

6.3.2
Simplicial Homology Groups

We now define the group structures associated with a space X that is triangulated by a finite simplicial complex C. Although the triangulation of a space is not unique, the homology groups for any triangulation of the same space are isomorphic (see [13]); this makes simplicial homology well defined.

The set of all k-simplices from C form the basis of a free group called the kth *chain group*, $C_k(X)$. The group operation is an additive one; recall that $-\sigma^k$ is just σ^k with the opposite orientation, so this defines the

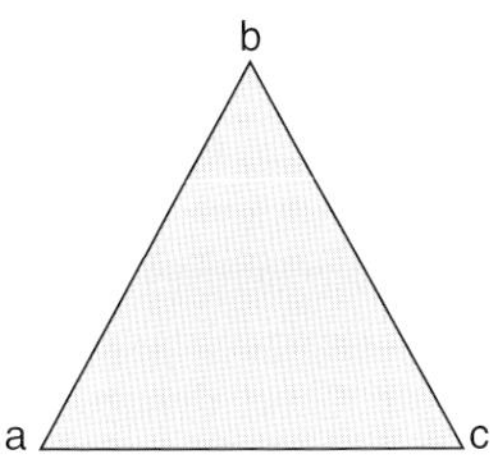

Figure 6.6 The simplicial complex of a triangle consists of one 2-simplex, three 1-simplices, and three vertices (0-simplices).

inverse elements. In general, a k-chain is the formal sum of a finite number of oriented k-simplices: $c_k = \sum_i a_i \sigma_i^k$. The coefficients, a_i, are elements of another group, called the *coefficient group* that is typically the integers $\mathbb{Z}$, but can be any Abelian group such as the integers mod 2 $\mathbb{Z}_2$, the rationals $\mathbb{Q}$, or real numbers $\mathbb{R}$. If the coefficient group G needs to be emphasized, we write $C_k(X; G)$.

When $k \geq 1$, the *boundary operator* $\partial_k : C_k \rightarrow C_{k-1}$ maps a k-simplex onto the sum of the $(k-1)$-simplices in its boundary. If $\sigma^k = \langle x_0, x_1, \ldots, x_k \rangle$ is a k-simplex, we have

$$\partial_k(\sigma^k) = \sum_{i=0}^{k} (-1)^i \langle x_0, \ldots, \hat{x}_i, \ldots, x_k \rangle,$$

where $\langle x_0, \ldots, \hat{x}_i, \ldots, x_k \rangle$ represents the $(k-1)$-simplex obtained by deleting the vertex x_i. The action of the boundary operator on general k-chains is obtained by linear extension from its action on the k-simplices: $\partial_k(\sum_i a_i \sigma_i^k) = \sum_i a_i \partial_k(\sigma_i^k)$. For $k = 0$, the boundary operator is defined to be null: $\partial_0(c_0) = 0$. We drop the subscript from the boundary operator when the dimension is understood.

As an example, consider the simplicial complex consisting of a triangle and all its edges and vertices, as shown in Figure 6.6. The boundary of the 2-simplex $\langle a, b, c \rangle$ is

$$\partial(\langle a, b, c \rangle) = \langle b, c \rangle - \langle a, c \rangle + \langle a, b \rangle,$$

and the boundary of this 1-chain is

$$\begin{aligned} &\partial(\langle b, c \rangle - \langle a, c \rangle + \langle a, b \rangle) \\ &\quad = (c - b) - (c - a) + (b - a) = 0. \end{aligned}$$

This illustrates the fundamental property of the boundary operator, namely,

$$\partial_k \, \partial_{k+1} = 0.$$

We now consider two subgroups of C_k that have important geometric interpretations. The first subgroup consists of k-chains that map to zero under the boundary operator. This group is the *group of cycles* denoted Z_k, it is the kernel (or null space) of ∂_k and its elements are called k-cycles. From the definition of ∂_0 we see that all 0-chains are cycles so $Z_0 = C_0$. The second subgroup of C_k is the group of k-chains that bound a $(k+1)$-chain. This is the *group of boundaries* B_k, it is the image of ∂_{k+1}. It follows from the above equation that every boundary is a cycle, that is, the image of ∂_{k+1} is mapped to zero by ∂_k, so B_k is a subgroup of Z_k. In our example of the triangle simplicial complex, we find no 2-cycles and a 1-cycle, $\langle b, c \rangle - \langle a, c \rangle + \langle a, b \rangle$, such that all other 1-cycles are integer multiples of this. This 1-cycle is also the only 1-boundary, so $Z_1 = B_1$, and the 0-boundaries B_0 are generated by the two 0-chains: $c - b$ and $c - a$ (the third boundary $a - b = (c - b) - (c - a)$).

As $B_k \subset Z_k$, we can form a quotient group, $H_k = Z_k / B_k$; this is precisely the kth *homology group*. The elements of H_k are equivalence classes of k-cycles that do not bound any $k+1$ chain, so this is how homology characterizes k-dimensional holes. Formally, two k-cycles $w, z \in Z_k$ are in the same equivalence class if $w - z \in B_k$; such cycles are said to be homologous. We write $[z] \in H_k$ for the equivalence class of cycles homologous to z. For the simple example of the triangle simplicial complex, we have already seen that $Z_1 = B_1$ so that $H_1 = \{0\}$. The 0-cycles are generated by $\{a, b, c\}$ and the boundaries by $\{(c - b), (c - a)\}$, so H_0 has a single equivalence class, $[c]$, and is isomorphic to $\mathbb{Z}$.

The following are homology groups for some familiar spaces

- Real space $\mathbb{R}^n$ has $H_0 = \mathbb{Z}$ and $H_k = \{0\}$ for $k \geq 1$.

- The spheres have $H_0(S^n) = \mathbb{Z}$, $H_n(S^n) = \mathbb{Z}$, and $H_k(S^n) = \{0\}$ for all other values of k.
- The torus has $H_0 = \mathbb{Z}, H_1 = \mathbb{Z} \oplus \mathbb{Z}$, $H_2 = \mathbb{Z}, H_k = \{0\}$ for all other k. The 2-cycle that generates H_2 is the entire surface. This is in contrast to the second homotopy group for the torus, which is trivial.
- The real projective plane, $\mathbb{R}P^2$ has $H_0 = \mathbb{Z}, H_1 = \mathbb{Z}_2$, and $H_k = \{0\}$ for $k \geq 2$. The fact that H_2 is trivial is a result of the surface being *nonorientable*; even though $\mathbb{R}P^2$ is *closed* as a manifold, the 2-chain covering the surface is not a 2-cycle.

The combinatorial nature of simplicial homology makes it readily computable. We give the classical algorithm and review recent work on fast and efficient algorithms for data in Section 6.6.

6.3.3 Basic Properties of Homology Groups

In general, the homology groups of a finite simplicial complex are finitely generated Abelian groups, so the following theorem tells us about their general structure (see Theorem 4.3 of [13]).

Theorem 6.1 *If G is a finitely generated Abelian group then it is isomorphic to the following direct sum:*

$$G \simeq (\mathbb{Z} \oplus \cdots \oplus \mathbb{Z}) \oplus \mathbb{Z}_{t_1} \oplus \cdots \oplus \mathbb{Z}_{t_m}.$$

The number of copies of the integer group $\mathbb{Z}$ is called the *Betti number* β. The cyclic groups $\mathbb{Z}_{t_i}$ are called the *torsion subgroups* and the t_i are the *torsion coefficients*; they are defined so that $t_i > 1$ and t_1 divides t_2, which divides t_3, and so on. The torsion coefficients of the homology group $H_k(C)$ measure the twistedness of the space in some sense. For example, the real projective plane has $H_1 = \mathbb{Z}_2$, because the 2-chain that represents the whole of the surface has a boundary that is twice the generating 1-cycle. The Betti number β_k of the kth homology group counts the number of nonequivalent nonbounding k-cycles and this can be loosely interpreted as the number of k-dimensional holes. The 0th Betti number, β_0, counts the number of path-connected components of the space.

Some other fundamental properties of the homology groups are as follows:

- If the simplicial complex has N connected components, $X = X_1 \cup \cdots \cup X_N$, then H_0 is isomorphic to the direct sum of N copies of the coefficient group, and $H_k(X) = H_k(X_1) \oplus \cdots \oplus H_k(X_N)$.
- Homology is a functor. If $f : X \to Y$ is a continuous function from one simplicial complex into another, it induces natural maps on the chain groups $f_\sharp : C_k(X) \to C_k(Y)$ for each k, which commute with the boundary operator: $\partial f_\sharp = f_\sharp \partial$. This commutativity implies that cycles map to cycles and boundaries to boundaries, so that the $f_\sharp$ induce homomorphisms on the homology groups $f_* : H_k(X) \to H_k(Y)$.
- If two spaces are homotopy equivalent, they have isomorphic homology groups (this is shown using the above functorial property).
- The first homology group is the *Abelianization* of the fundamental group. When X is a path-connected space, the connection between $H_1(X)$ and $\pi_1(X)$ is made by noticing that two 1-cycles are equivalent in homology if their difference is the boundary of a 2-chain; if we parameterize the 1-cycles as loops then this 2-chain forms a region

through which one can define a homotopy. See [10] for a formal proof.
- The higher-dimensional homology groups have the comforting property that if all simplices in a complex have dimensions $\leq m$ then $H_k = \{0\}$ for $k > m$. This is in stark contrast to the higher-dimensional homotopy groups.

A particularly pleasing result in homology relates the Betti numbers to another topological invariant called the *Euler characteristic*. For a finite simplicial complex, C, define n_k to be the number of simplices of dimension k, then the Euler characteristic is defined to be $\chi(C) = n_0 - n_1 + n_2 + \cdots$. The *Euler–Poincaré theorem* states that the alternating sum of Betti numbers is the same as the Euler characteristic [13]: $\chi = \beta_0 - \beta_1 + \beta_2 + \cdots$. This is one of many results that connect the Euler characteristic with other properties of manifolds. For example, if M is a compact 2-manifold with a Riemannian metric, then the *Gauss–Bonnet theorem* states that $2\pi\chi$ is equal to the integral of Gaussian curvature over the surface plus the integral of geodesic curvature over the boundary of M [15]. Further, if M is orientable and has no boundary, then it must be homeomorphic to a sphere with g handles and $\chi = 2 - 2g$ where g is the genus of the surface. When M is nonorientable without boundary, then it is homeomorphic to a sphere with r cross-caps and $\chi = 2 - r$.

The Euler characteristic is a topological invariant with the property of *inclusion–exclusion*: If a triangulated space $X = A \cup B$ where A and B are both subcomplexes, then

$$\chi(X) = \chi(A) + \chi(B) - \chi(A \cap B).$$

This means the value of χ is a localizable one and can be computed by cutting up a larger space into smaller chunks. This property makes it a popular topological invariant to use in applications [16]. A recent application that exploits the local additivity of the Euler characteristic to great effect is target enumeration in localized sensor networks [17, 18]. The Euler characteristic has also been shown to be an important parameter in the physics of porous materials [19, 20].

The simple inclusion–exclusion property above does not hold for the Betti numbers because they capture global aspects of the topology of a space. Relating the homology of two spaces to their union requires more sophisticated algebraic machinery that we review below.

6.3.4 Homological Algebra

Many results and tools in homology are independent of the details about the way the chains and boundary operators are defined for a topological space; they depend only on the chain groups and the fact that $\partial\partial = 0$. The study of such abstract chain complexes and transformations between them is called *homological algebra* and is one of the original examples in *category theory* [13].

An abstract *chain complex* is a sequence of Abelian groups and homomorphisms

$$\cdots \xrightarrow{d_{k+2}} C_{k+1} \xrightarrow{d_{k+1}} C_k \xrightarrow{d_k} \cdots \xrightarrow{d_1} C_0 \longrightarrow \{0\},$$

with $d_k d_{k+1} = 0$.

The homology of this chain complex is $H_k(C) = \ker d_k / \mathrm{im}\, d_{k+1}$. In certain cases (such as the simplicial chain complex of a contractible space), we find that $\mathrm{im}\, d_{k+1} = \ker d_k$ for $k \geq 1$, so the homology groups are trivial. Such a sequence is said to be *exact*.

This property of exactness has many nice consequences. For example, with a *short exact sequence* of groups,

$$0 \to A \xrightarrow{f} B \xrightarrow{g} C \to 0.$$

The exactness means that f is a monomorphism (one-to-one) and g is an epimorphism (onto). In fact, g induces an isomorphism of groups $C \approx B/f(A)$, and if these groups are finitely generated Abelian, then the Betti numbers satisfy $\beta(B) = \beta(C) + \beta(A)$ (where we have replaced $f(A)$ by A because f is one-to-one).

Now imagine there is a short exact sequence of *chain complexes*, that is,

$$0 \to A_k \xrightarrow{f_k} B_k \xrightarrow{g_k} C_k \to 0.$$

is exact for all k and the maps f_k, g_k commute with the boundary operators in each complex (i.e., $d_B f_k = f_{k-1} d_A$, etc.). Typically, the f_k will be induced by an inclusion map (and so be monomorphisms), and g_k by a quotient map (making them epimorphisms) on some underlying topological spaces. The *zigzag lemma* shows that these short exact sequences can be joined together into a long exact sequence on the homology groups of A, B, and C:

$$\cdots \to H_k(A) \xrightarrow{f_*} H_k(B) \xrightarrow{g_*} H_k(C) \xrightarrow{\Delta}$$
$$H_{k-1}(A) \xrightarrow{f_*} H_{k-1}(B) \to \cdots$$

The maps f_* and g_* are those induced by the chain maps f and g and the boundary maps Δ are defined directly on the homology classes in $H_k(C)$ and $H_{k-1}(A)$ by showing that cycles in C_k map to cycles in A_{k-1} via g_k, the boundary ∂_B, and f_{k-1}. Details are given in Hatcher [10].

One of the most useful applications of this long exact sequence in homology is the *Mayer–Vietoris exact sequence*, a result that describes the relationship between the homology groups of two simplicial complexes, X, Y, their union, and their intersection. This gives us a way to deduce homology groups of a larger space from smaller spaces.

$$\begin{aligned} \cdots &\xrightarrow{j_k} H_k(X) \oplus H_k(Y) \xrightarrow{s_k} H_k(X \cup Y) \\ &\xrightarrow{\nu_k} H_{k-1}(X \cap Y) \xrightarrow{j_{k-1}} H_{k-1}(X) \oplus H_{k-1}(Y) \\ &\xrightarrow{s_{k-1}} \cdots \end{aligned} \tag{6.1}$$

The homomorphisms are defined as follows:

$$\begin{aligned} j_k([u]) &= ([u], -[u]) \\ s_k([w], [w']) &= [w + w'] \\ \nu_k([z]) &= [\partial z'], \end{aligned}$$

Where, in the last equation, z is a cycle in $X \cup Y$ and we can write $z = z' + z''$ where z' and z'' are chains (not necessarily cycles) in X and Y respectively. These homomorphisms are well defined (see, for example, Theorem 33.1 of [13]). Exactness implies that the image of each homomorphism is equal to the kernel of the following one: $\text{im}\, j_k = \ker s_k$, $\text{im}\, s_k = \ker \nu_k$, and $\text{im}\, \nu_k = \ker j_{k-1}$.

The Mayer–Vietoris sequence has an interesting interpretation in terms of Betti numbers. First we define $N_k = \ker j_k$ to be the subgroup of $H_k(X \cap Y)$ defined by the k-cycles that bound in both X and in Y. Then by focusing on the exact sequence around $H_k(X \cup Y)$, it follows that [21, 22]

$$\begin{aligned} \beta_k(X \cup Y) = \beta_k(X) + \beta_k(Y) - \beta_k(X \cap Y) \\ + \text{rank}\, N_k + \text{rank}\, N_{k-1}. \end{aligned}$$

This is where we see the nonlocalizable property of homology and the Betti numbers most clearly.

The Mayer–Vietoris sequence holds in a more general setting than simplicial homology. It is an example of a result that can be derived from the Eilenberg–Steenrod axioms. Any theory for which these five axioms hold is a type of homology theory, see [10] for further details.

6.3.5 Other Homology Theories

Chain complexes that capture topological information may be defined in a number of ways. We have defined simplicial chain complexes above, and will briefly describe some other techniques here.

Cubical homology is directly analogous to simplicial homology using k-dimensional cubes as building elements rather than k-dimensional simplices. This theory is developed in full in [6] and arose from applications in digital image analysis and numerical analysis of dynamical systems.

Singular homology is built from singular k-simplices that are continuous functions from the standard k-simplex into a topological space X, $\sigma : \langle v_0, \dots, v_k \rangle \to X$. Singular chains and the boundary operator are defined as they are in simplicial homology. A greater degree of flexibility is found in singular homology because the maps σ are allowed to collapse the simplices, for example, the standard k-simplex, $k > 0$, may have boundary points mapping to the same point in X, or the entire simplex may be mapped to a single point [10].

An even more general formulation of *cellular homology* is made by considering general cell complexes. A cell complex is built incrementally by starting with a collection of points $X^{(0)} \subset X$, then attaching 1-cells via maps of the unit interval into X so that end points map into $X^{(0)}$ to form the 1-skeleton $X^{(1)}$. This process continues by attaching k-cells to the $(k-1)$-skeleton by continuous maps of the closed unit k-ball, $\phi : B_k(0,1) \to X$ that are homeomorphic on the interior and satisfy $\phi : \partial B_k(0,1) \to X^{(k-1)}$. The definition of the boundary operator for a cell complex requires the concept of *degree* of a map of the k-sphere (i.e., the boundary of a $(k+1)$-dimensional ball). For details see Hatcher [10].

We will see in the section on Morse Theory that it is also possible to define a chain complex from the critical points of a smooth function on a manifold.

6.4 Cohomology

The cohomology groups are derived by a simple dualization procedure on the chain groups (similar to the construction of dual function spaces in analysis). We will again give definitions in the simplicial setting but the concepts carry over to other contexts. A *cochain* ϕ^k is a function from the simplicial chain group into the coefficient group, $\phi^k : C_k(X;G) \to G$ (recall that G is usually the integers, $\mathbb{Z}$, but can be any Abelian group). The space of all k-cochains forms a group called the kth *cochain group* $C^k(X;G)$. The simplicial boundary operators $\partial_k : C_k \to C_{k-1}$ induce *coboundary* operators $\delta^{k-1} : C^{k-1} \to C^k$ on the cochain groups via $\delta(\phi) = \phi\partial$. In other words, the cochain $\delta(\phi)$ is defined via the action of ϕ on the boundary of each k-simplex $\sigma = \langle x_0, x_1, \dots, x_k \rangle$:

$$\delta(\phi)(\sigma) = \sum_i (-1)^i \phi(\langle x_0, \dots, \hat{x}_i, \dots, x_k \rangle).$$

The key property from homology that $\partial_k \partial_{k+1} = 0$ also holds true for the coboundary: $\delta^k \delta^{k-1} = 0$ (coboundaries are mapped to zero), so we define the kth *cohomology group* as $H^k(X) = \ker \delta^k / \mathrm{im}\, \delta^{k-1}$. Note that cochains $\phi \in \ker \delta$ are functions that vanish on the k-boundaries (not the larger group of k-cycles), and a coboundary $\eta^k \in \operatorname{im}\delta$ is one that can be defined via the action of some cochain ϕ^{k-1} on the $(k-1)$-boundaries.

The coboundary operator acts in the direction of increasing dimension and this can be a more natural action in some situations (such as de Rham cohomology of differential forms discussed below) and also has some interesting algebraic consequences (it leads to the definition of the cup product).

$$\cdots \longleftarrow C^{k+1} \overset{\delta^k}{\longleftarrow} C^k \longleftarrow \cdots \longleftarrow C^0 \longleftarrow \{0\}.$$

This action of the coboundary makes cohomology *contravariant* (induced maps act in the opposite direction) where homology is *covariant* (induced maps act in the same direction). If $f : X \to Y$ is a continuous function between two topological spaces then the group homomorphism induced on the cohomology groups acts as $f^* : H^k(Y) \to H^k(X)$.

In simplicial homology, the simplices form bases for the chain groups, and we can similarly use them as bases for the cochain groups by defining an *elementary cochain* $\dot{\sigma}$ as the function that takes the value one on σ and zero on all other simplices. For a finite simplicial complex, it is possible to represent the boundary operator ∂ as a matrix with respect to the bases of oriented k- and $(k-1)$-simplices. If we then use the elementary cochains as bases for the cochain groups, the matrix representation for the coboundary operator is just the transpose of that for the boundary operator. This shows that for finite simplicial complexes, the functional and geometric meanings of duality are the same.

Another type of duality is that between homology and cohomology groups on compact closed oriented manifolds (i.e., without boundary). *Poincaré duality* states that $H^k(M) = H_{m-k}(M)$ for $k \in \{0, \ldots, m\}$ where m is the dimension of the manifold, M; see [10] for further details.

Despite this close relationship between homology and cohomology on manifolds, the cohomology groups have a naturally defined product combining two cochains and this additional structure can help distinguish between some spaces that homology does not. We start with $\phi \in C^k(X; G)$ and $\psi \in C^l(X; G)$ where the coefficient group should now be a ring R (i.e., R should have both addition and multiplication operations; $\mathbb{Z}$, $\mathbb{Z}_p$, and $\mathbb{Q}$ are rings.) The *cup product* is the cochain $\phi \smile \psi \in C^{k+l}(X; R)$ defined by its action on a $(k+l)$-simplex $\sigma = \langle v_0, \ldots, v_{k+l}\rangle$ as follows:

$$(\phi \smile \psi)(\sigma) = \phi(\langle v_0, \ldots, v_k\rangle)\psi(\langle v_k, \ldots, v_{k+l}\rangle).$$

The relation between this product and the coboundary is

$$\delta(\phi \smile \psi) = \delta\phi \smile \psi + (-1)^k \phi \smile \delta\psi.$$

From this, we see that the product of two cocycles is another cocycle, and if the product is between a cocycle and a coboundary, then the result is a coboundary. Thus, the cup product is a well-defined product on the cohomology groups that is *anticommutative*: $[\phi] \smile [\psi] = (-1)^{kl}[\psi] \smile [\phi]$ (provided the coefficient ring, G, is commutative). These rules for products of cocycles should look suspiciously familiar to those who have read Chapter 9. They are similar to those for exterior products of differential forms and this relationship is formalized in the next section when we define de Rham cohomology.

6.4.1 De Rham Cohomology

One interpretation of cohomology that is of particular interest in physics comes from the study of *differential forms* on smooth manifolds; cf. Chapter 9. Recall that a differential form of degree k, ω, defines for each point $p \in M$, an alternating multilinear map on k copies of the tangent space to M at p:

$$\omega_p : T_pM \times \cdots \times T_pM \to \mathbb{R}.$$

The set of all differential k-forms on a manifold M is a vector space, $\Omega^k(M)$, and the *exterior derivative* is a linear operator that takes a k-form to a $(k+1)$-form, $d_k : \Omega^k(M) \to \Omega^{k+1}(M)$ as defined in Chapter 9.

The crucial property $dd = 0$ holds for the exterior derivative. In this context, k-forms in the image of d are called *exact*, that is, $\omega = \mathrm{d}\sigma$ for some $(k-1)$-form σ; and those for which $\mathrm{d}\omega = 0$ are called *closed*. We therefore have a cochain complex of differential forms and can form quotient groups of closed forms modulo the exact forms to obtain the de Rham cohomology groups:

$$H^k_{\mathrm{dR}}(M, \mathbb{R}) = \frac{\ker d_k}{\operatorname{im} d_{k-1}}.$$

The cup product in de Rham cohomology is exactly the exterior (or wedge) product on differential forms.

De Rham's theorem states that the above groups are isomorphic to those derived via simplicial or singular cohomology [23]. And so we see that the topology of a manifold has a direct influence on the properties of differential forms that have it as their domain. For example, the Poincaré lemma states that if M is a contractible open subset of $\mathbb{R}^n$, then all smooth closed k-forms on M are exact (the cohomology groups are trivial). In the language of multivariable calculus, this becomes Helmholtz' theorem that a vector field, $\mathbf{V}$, with $\mathrm{curl}\mathbf{V} = 0$ in a simply connected open subset of $\mathbb{R}^3$ can be expressed as the gradient of a potential function: $\mathbf{V} = \mathrm{grad}f$ in the appropriate domain [24]. These considerations play a key role in the study of electrodynamics via Maxwell's equations [25].

6.5 Morse Theory

We now turn to a primary topic in differential topology: to examine the relationship between the topology of a manifold M and real-valued functions defined on M. The basic approach of Morse theory is to use the *level cuts* of a function $f : M \to \mathbb{R}$ and study how these subsets $M_a = f^{-1}(-\infty, a]$ change as a is varied. For "nice" functions the level cuts change their topology in a well-defined way only at the critical points. This leads to a number of powerful theorems that relate the homology of a manifold to the critical points of a function defined on it.

6.5.1 Basic Results

A *Morse function* $f : M \to \mathbb{R}$ is a smooth real-valued function defined on a differentiable manifold M such that each critical point of f is isolated and the matrix of second derivatives (the *Hessian*) is nondegenerate at each critical point. An example is illustrated in Figure 6.7. The details on how to define these derivatives with respect to a coordinate chart on M are given in Chapter 9. This may seem like a restrictive class of functions but in fact Morse functions are dense in the space

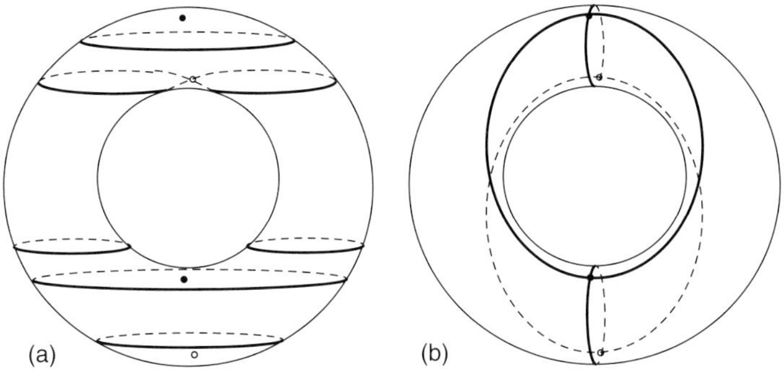

Figure 6.7 Imagine a torus sitting with one point in contact with a plane and tilted slightly into the page as depicted. Define a Morse function by mapping each point on the torus to its height above the plane. This function has four critical points: a minimum, two saddles, and a maximum. (a) Five level cuts of the height function showing how the topology of a level cut changes when passing through a critical point. (b) Gradient flow lines between the maximum and the two saddle points, and from each saddle point to the minimum.

of all smooth functions, so any smooth function can be smoothly perturbed to obtain a Morse function [26]. Now suppose $x \in M$ is a critical point of f, that is, $\mathrm{d}f(x) = 0$. The *index* of this critical point is the number of negative eigenvalues of the Hessian matrix. Intuitively, this is the number of directions in which f is decreasing: a minimum has index 0, and a maximum has index m where m is the dimension of the manifold M. Critical points of intermediate index are called *saddles* because they have some increasing and some decreasing directions.

The two main results about level cuts M_a of a Morse function f are the following:

- When $[a, b]$ is an interval for which there are no critical values of f (i.e., there is no $x \in f^{-1}([a, b])$ for which $\mathrm{d}f(x) = 0$), then M_a and M_b are homotopy equivalent.
- Let x be a nondegenerate critical point of f with index i, let $f(x) = c$, and let $\epsilon > 0$ be such that $f^{-1}[c - \epsilon, c + \epsilon]$ is compact and contains no other critical points. Then $M_{c+\epsilon}$ is homotopy equivalent to $M_{c-\epsilon}$ with an i-cell attached.

(Recall that an i-cell is an i-dimensional unit ball and the attaching map glues the whole boundary of the i-cell continuously into the prior space). The proofs of these theorems rely on homotopies defined via the *negative gradient flow* of f [26].

Gradient flow lines are another key ingredient of Morse theory and allow us to define a chain complex and to compute the homology of M. Each point $x \in M$ has a unique *flow line* or *integral path* $\gamma_x : \mathbb{R} \to M$ such that

$$\gamma_x(0) = x \text{ and } \frac{\partial \gamma_x(t)}{\partial t} = -\nabla f(\gamma_x(t)).$$

Taking the limit as $t \to \pm\infty$, each flow line converges to a *destination* and an *origin* critical point. The *unstable manifold* of a critical point p with index i is the set of all $x \in M$ that have p as their origin; this set is homeomorphic to an open ball of dimension i. Correspondingly, the *stable manifold* is the set of all x that have p as their destination. For suitably "nice" functions, the collection of unstable manifolds form a cell complex for the manifold M [27].

We can also define a more abstract chain complex, sometimes referred to as the *Morse–Smale–Witten complex*, to reflect the history of its development. Let C_i be the chain group derived from formal sums of critical points of index i. A boundary operator $\partial : C_i \to C_{i-1}$ is then defined by mapping $p \in C_i$ to a sum of critical points $\sum \alpha_j q_j \in C_{i-1}$ for which there is a flow line with p as its origin and q as its destination. The coefficients α_j of the q_j in this boundary chain are the number of geometrically distinct flow lines that join p and q_j (one can either count mod 2 or keep track of orientations in a suitable manner). It requires some effort to show that $\partial\partial = 0$ in this setting; see [27] for details.[6] *Morse homology* is the homology computed via this chain complex.

For *finite-dimensional compact manifolds*, Morse homology is isomorphic to singular homology, and we obtain the *Morse inequalities* relating numbers of critical points of $f : M \to \mathbb{R}$ to the Betti numbers of M:

$$\begin{aligned} c_0 &\geq \beta_0 \\ c_1 - c_0 &\geq \beta_1 - \beta_0 \\ c_2 - c_1 + c_0 &\geq \beta_2 - \beta_1 + \beta_0 \\ &\vdots \\ \sum_{0 \leq i \leq m} (-1)^{m-i} c_i &= \sum_{0 \leq i \leq m} (-1)^{m-i} \beta_i = \chi(M), \end{aligned}$$

where c_i is the number of critical points of f of index i and β_i is the ith Betti number of M. Notice that the final relationship is an *equality*; the alternating sum of numbers of critical points is the same as the Euler characteristic $\chi(M)$. It also follows from the above that $c_i \geq \beta_i$ for each i.

6) In 2D, think of the flow lines that join a single maximum, minimum pair. In general, such a region is bounded by flow lines from the maximum to two saddles and from these saddles to the minimum. The boundary of the maximum contains these two saddles and their boundaries contain the minimum in oppositely induced orientations.

6.5.2 Extensions and Applications

Morse theory is primarily used as a powerful tool to prove results in other settings. For example, Morse obtained his results in order to prove the existence of closed geodesics on a Riemannian manifold [28]; most famously, Morse theory forms the foundation of a proof due to Smale of the higher-dimensional Poincaré conjecture [29]. Morse theory has been extended in many ways that relax conditions on the manifold or the function being studied [30]. We mention a few of the main generalizations here.

A *Morse–Bott function* is one for which the critical points may now not be isolated and instead form a critical set that is a closed submanifold. At the very simplest level for example, this lets us study the height function of a torus sitting flat on a table because the circle of points touching the table is critical [31].

The *Conley index* from dynamical systems is a generalization of Morse theory to flows in a more general class than those generated by the gradient of a Morse function. For general flows, invariant sets are no longer single fixed points but may be periodic cycles or even fractal "strange attractors." In the Morse setting, the index is simply the dimension of the unstable manifold of the fixed point, but for general flows a more subtle construction is required. Conley's insight was that an isolated invariant set can be characterized by the flow near the boundary of a neighborhood of the set. The Conley index is then (roughly speaking) the homotopy type of such a neighborhood relative to its boundary. For details see [32–34].

Building on Conley's work and the Morse complex of critical points and connecting orbits, Floer created an infinite-dimensional version of Morse homology now called *Floer homology* [27]. This has various formulations that have been used to study problems in symplectic geometry (the geometry of Hamiltonian dynamical systems) and also the topology of three- and four-dimensional manifolds [35].

There are a number of approaches adapting Morse theory to a discrete setting, of increasing importance in geometric modeling, image and data analysis, and quantum field theory. The approach due to Forman is summarized in the following section.

6.5.3 Forman's Discrete Morse Theory

Discrete Morse theory is a combinatorial analogue of Morse theory for functions defined on cell complexes. Discrete Morse functions are not intended to be approximations to smooth Morse functions, but the theory developed in [36, 37] keeps much of the style and flavor of the standard results from smooth Morse theory.

In keeping with earlier parts of this chapter, we will give definitions for a simplicial complex C, but the theory holds for general cell complexes with little modification. First recall that a simplex α is a *face* of another simplex β if $\alpha \subset \beta$, in which case, β is called a *coface* of α. A function $f : C \to \mathbb{R}$ that assigns a real number to each simplex in C is a *discrete Morse function* if for every $\alpha^{(p)} \in C$, f takes a value less than or equal to $f(\alpha)$ on at most one coface of α and takes a value greater than or equal to $f(\alpha)$ on at most one face of α. In other words,

$$\#\{\beta^{(p+1)} > \alpha \mid f(\beta) \leq f(\alpha)\} \leq 1,$$

and

$$\#\{\gamma^{(p-1)} < \alpha \mid f(\gamma) \geq f(\alpha)\} \leq 1,$$

where # denotes the number of elements in the set. A simplex $\alpha^{(p)}$ is *critical* if all cofaces take strictly greater values and all faces are strictly lower.

A cell α can fail to be critical in two possible ways. There can exist $\gamma < \alpha$ such that $f(\gamma) \geq f(\alpha)$, or there can exist $\beta > \alpha$ such that $f(\beta) \leq f(\alpha)$. Lemma 2.5 of [36] shows that these two possibilities are exclusive: they cannot be true simultaneously for a given cell α. Thus each noncritical cell α may be paired either with a noncritical cell that is a coface of α, or with a noncritical cell that is a face of α.

As noted by Forman (Section 3 of [37]), it is usually simpler to work with pairings of cells with faces than to construct a discrete Morse function on a given complex. So we define a *discrete vector field* V as a collection of pairs $(\alpha^{(p)}, \beta^{(p+1)})$ of cells $\alpha < \beta \in C$ such that each cell of C is in at most one pair of V. A discrete Morse function defines a discrete vector field by pairing $\alpha^{(p)} < \beta^{(p+1)}$ whenever $f(\beta) \leq f(\alpha)$. The critical cells are precisely those that do not appear in any pair. Discrete vector fields that arise from Morse functions are called *gradient* vector fields. See Figure 6.8 for an example.

It is natural to consider the flow associated with a vector field and in the discrete setting the analogy of a flow line is a V-path. A *V-path* is a sequence of cells:

$$\alpha_0^{(p)}, \beta_0^{(p+1)}, \alpha_1^{(p)}, \beta_1^{(p+1)}, \alpha_2^{(p)}, \ldots, \beta_{r-1}^{(p+1)}, \alpha_r^{(p)},$$

where $(\alpha_i, \beta_i) \in V$, $\beta_i > \alpha_{i+1}$, and $\alpha_i \neq \alpha_{i+1}$ for all $i = 0, \ldots, r-1$. A V-path is a *nontrival closed V-path* if $\alpha_r = \alpha_0$ for $r > 1$.

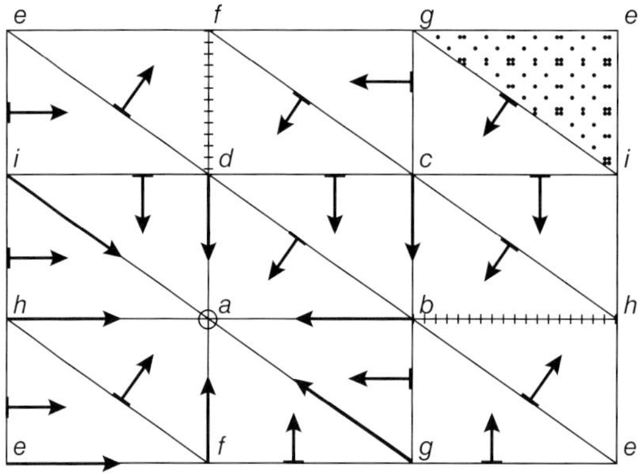

Figure 6.8 A simplicial complex with the topology of the torus (opposite edges of the rectangle are identified according to the vertex labels). The arrows show how to pair simplices in a gradient vector field. A compatible discrete Morse function has a critical 0-cell (a minimum) at a, two critical 1-cells (saddles) at edges $\langle b, h\rangle$ and $\langle d, f\rangle$, and a critical 2-cell (a maximum) at $\langle e, i, g\rangle$.

Forman shows that a discrete vector field is the gradient vector field of a discrete Morse function if, and only if, there are no nontrivial closed V-paths (Theorem 9.3 of [36]).

The four results about Morse functions that we gave earlier all carry over into the discrete setting: the homotopy equivalence of level sets away from a critical point, adding a critical i-cell is homotopy equivalent to attaching an i-cell, the existence of and homology of the Morse chain complex, and the Morse inequalities. One of the notable differences between the discrete and continuous theories is that flow lines for a smooth Morse function on a manifold are uniquely determined at each point, whereas V-paths can merge and split.

6.6 Computational Topology

An algorithmic and combinatorial approach to topology has led to significant results in low-dimensional topology over the past twenty years. There are two main aspects to computational topology: first, research into methods for making topological concepts algorithmic, culminating for example, in the beginnings of an algorithmic classification of (Haken) 3-manifolds [38] (a result analogous to the classification of closed compact 2-manifolds by Euler characteristic and orientability). And second, the challenge to find efficient and useful techniques for extracting topological invariants from data; see [39] for example. We start this section by describing simple algorithms that demonstrate the computability of the fundamental group and homology groups of a simplicial complex, and then survey some recent advances in building cell complexes and computing homology from data.

6.6.1 The Fundamental Group of a Simplicial Complex

In Section 6.2, we saw that the fundamental group of a topological space could be determined from unions and products of smaller spaces or by using a covering space. When the space has a triangulation (i.e., it is homeomorphic to a polyhedron) there is a more systematic and algorithmic approach to finding the fundamental group as the quotient of a free group by a set of relations that we summarize below. See [40] for a complete treatment of this *edge-path group*.

Let C be a connected finite simplicial complex. Any path in $|C|$ is homotopic to one that follows only edges in C, and any homotopy between edge-paths can be restricted to the 2-simplices of C. This means the fundamental group depends only on the topology of the 2-skeleton of C. The algorithm for finding a presentation of $\pi_1(C)$ proceeds as follows.

First find a *spanning tree* $T \subset C^{(1)}$ that is, a connected, contractible subgraph of the 1-skeleton that contains every vertex of C; see Figure 6.9 for an example. One algorithm for doing this simply grows from an initial vertex v (*the root*) by adding adjacent (edge, vertex) pairs only if the other vertex is not already in T. A nontrivial closed edge-path in C (a loop) must include edges that are not in T and in fact *every* edge in $C - T$ generates a *distinct* closed path in $C^{(1)}$. Specifically, for each edge $\langle x_i, x_j \rangle \in C - T$ there is a closed path starting and ending at the root v and lying wholly in T except for the generating edge; we label this closed path g_{ij}. Moreover, any closed path based at v can be written as a concatenation of such generating paths where inverses are simply followed in the opposite direction: $g_{ji} = g_{ij}^{-1}$. The g_{ij} are therefore generators for a free group with coefficients in $\mathbb{Z}$.

Next we use the 2-skeleton $C^{(2)}$ to obtain the homotopy equivalences of closed edge-paths. Each triangle $\langle x_i, x_j, x_k \rangle \in C$ defines a relation in the group via $g_{ij}g_{jk}g_{ki} = \text{id}$ (the identity) where we also set $g_{ij} = \text{id}$ if $\langle x_i, x_j \rangle \in T$. Let $G(C, T)$ be the finitely presented group defined by the above generators and relations. Then it is possible to show that we get isomorphic groups for different choices of T and that $G(C, T)$ is isomorphic to the fundamental group $\pi_1(|C|)$ [40].

If C has many vertices, then the presentation of its fundamental group as $G(C, T)$ may not be a very efficient description. It is possible to reduce the number of generating edges and relations by using any connected and contractible subcomplex that contains all the vertices $C^{(0)} \subset K \subset C$. Generators for the edge-path group are labeled by edges in $C - K$, and the homotopy relations are again defined by triangles in $C^{(2)}$, but we can now ignore all triangles in K. For the example of the torus in Figure 6.9, we could take K to be the eight triangles in the 2×2 lower left corner of the rectangular grid.

6.6.2 Smith Normal form for Homology

There is also a well-defined algorithm for computing the homology groups from a simplicial complex C. This algorithm is based on finding the *Smith normal form* (SNF) of a matrix representation of the boundary operator as outlined below.

Recall that the oriented k-simplices form a basis for the kth chain group, C_k. This means it is possible to represent the boundary operator, $\partial_k : C_k \to C_{k-1}$, by a (non-square) matrix A_k with entries in $\{-1, 0, 1\}$. The matrix A_k has m_k columns and m_{k-1} rows where m_k is the number of k-simplices in C. The entry a_{ij} is 1 if $\sigma_i \in C_{k-1}$ is a face of $\sigma_j \in C_k$ with consistent orientation, -1 if σ_i appears in $\partial\sigma_j$ with opposite orientation, and 0 if σ_i is not a face of σ_j. Thus each column of A_k is a boundary chain in C_{k-1} with respect to a basis of simplices.

The algorithm to reduce an integer matrix to SNF uses row and column operations as in standard Gaussian elimination, but at all stages the entries must remain integers. The row and column operations correspond to changing bases for C_{k-1} and C_k respectively and the resulting matrix

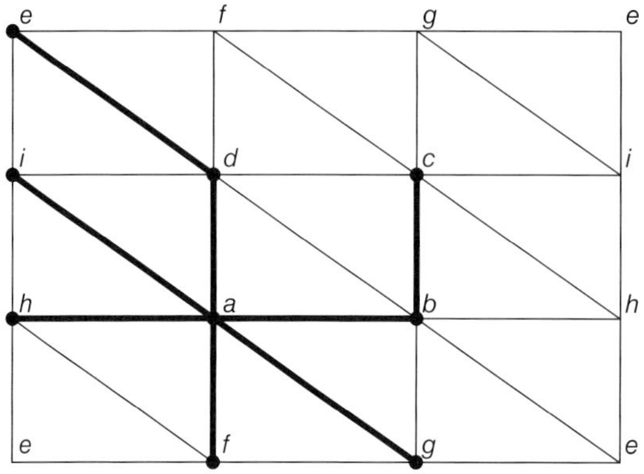

Figure 6.9 A simplicial complex with the topology of a torus (opposite edges of the rectangle are identified according to the vertex labels). A spanning tree T with root vertex a is shown in bold. Any closed path that starts and ends at a can be decomposed into a sum of loops that lie in T except for a single edge.

has the form

$$D_k = \begin{bmatrix} B_k & \mathbf{0} \\ \mathbf{0} & \mathbf{0} \end{bmatrix}, \quad \text{where } B_k = \begin{bmatrix} b_1 & & 0 \\ & \ddots & \\ 0 & & b_{l_k} \end{bmatrix}.$$

B_k is a square matrix with l_k nonzero diagonal entries that satisfy $b_i \geq 1$ and b_1 divides b_2, divides b_3, and so on. For a full description of the basic algorithm see Munkres [13].

The SNF matrices for ∂_{k+1} and ∂_k give a complete characterization of the kth homology group H_k. The rank of the boundary group B_k (im A_{k+1}) is the number of nonzero rows of D_{k+1}, that is, l_{k+1}. The rank of the cycle group Z_k (ker A_k) is the number of zero columns of D_k, that is, $m_k - l_k$. The torsion coefficients of H_k are the diagonal entries b_i of D_{k+1} that are greater than one. The kth Betti number is therefore

$$\beta_k = \text{rank}(Z_k) - \text{rank}(B_k) = m_k - l_k - l_{k+1}.$$

Bases for Z_k and B_k (and hence H_k) are determined by the row and column operations used in the SNF reduction but the cycles found in this way typically have poor geometric properties.

There are two practical problems with the algorithm for reducing a matrix to SNF as it is described in Munkres [13]. First, the time-cost of the algorithm is of a high polynomial degree in the number of simplices; second, the entries of the intermediate matrices can become extremely large and create numerical problems, even when the initial matrix and final normal form have small integer entries. When only the Betti numbers are required, it is possible to do better. In fact, if we construct the homology groups over the rationals, rather than the integers, then we need only apply Gaussian elimination to diagonalize the boundary operator matrices; doing this, however, means we lose all information about the torsion. Devising algorithms that overcome these problems and are fast enough to be effective on large complexes is an area of active research.

6.6.3 Persistent Homology

The concept of persistent homology arose in the late 1990s from attempts to extract meaningful topological information from data [41–43]. To give a finite set of points some interesting topological structure requires the introduction of a parameter to define which points are connected. The key lesson learnt from examining data was that rather than attempting to choose a single best parameter value, it is much more valuable to investigate a range of parameter values and describe how the topology changes with this parameter. So persistent homology tracks the topological properties of a sequence of nested spaces called a *filtration* $\cdots \subset C_a \subset C_b \subset \cdots$ where $a < b \in \mathcal{I}$ is an index parameter. In a continuous setting, the nested spaces might be the level cuts of a Morse function on a manifold, so that $\mathcal{I}$ is a real interval. In a discrete setting, this becomes a sequence of subcomplexes indexed by a finite set of integers. In either case, as the filtration grows, topological features appear and may later disappear. The *persistent homology group*, $H_k(a, b)$ measures the topological features from C_a that are still present in C_b. Formally, $H_k(a, b)$ is the image of the map induced on homology by the simple inclusion of C_a into C_b. Algebraically, it is defined by considering cycles in C_a to be equivalent with respect to the boundaries in C_b:

$$H_k(a, b) = Z_k(a)/\big(B_k(b) \cap Z_k(a)\big).$$

Computationally, persistent homology tracks the birth and death of every equivalence class of cycle and provides a complete picture of the topological structure present at all stages of the filtration. The initial algorithm for doing this, due to Edelsbrunner *et al.* [43], is surprisingly simple and rests on the observation that if we build a cell complex by adding a single cell at each step, then (because all its faces must already be present) this cell either creates a new cycle and is flagged as positive, or "fills in" a cycle that already existed and is labeled negative. If σ is a negative $(k+1)$-cell, its boundary $\partial\sigma$ is a k-cycle and its cells are already flagged as either positive or negative. The new cell σ is then paired with the most recently added (i.e., youngest) unpaired positive cell in $\partial\sigma$. If there are no unpaired positive cells available, we must grow $\partial\sigma$ to successively larger homologous cycles until an unpaired positive cell is found. By doing this carefully, we can guarantee that σ is paired with the positive k-cell that created the homology class of $\partial\sigma$. Determining whether a cell is positive or negative a priori is computationally nontrivial in general, but there is a more recent version of the persistence pairing algorithm due to Zomorodian and Carlsson [44, 45] that avoids doing this as a separate step, and also finds a representative k-cycle for each homology class.

The result of computing persistent homology from a finite filtration is a list of pairs of simplices $(\sigma^{(k)}, \tau^{(k+1)})$ that represent the birth and death of each homology class in the filtration. The persistence interval for each homology class is then given by the indices at which the creator σ and destroyer τ entered the filtration. Some nontrivial homology classes may be present at the final step of the filtration, these *essential classes* have an empty partner and are assigned "infinite" persistence. There are a number of ways to represent this persistence information graphically: the two most popular techniques are the *barcode* [46] and the *persistence diagram* [43, 47]. The barcode has a horizontal axis representing the filtration index; for each homology class a solid line spanning the persistence interval is drawn in a stack above the axis. The persistence diagram plots the (birth, death) index pair for each cycle. These points lie above the diagonal, and points close to the diagonal are homology classes that have low persistence. It is possible to show that persistence diagrams are stable with respect to small perturbations in the data. Specifically, if

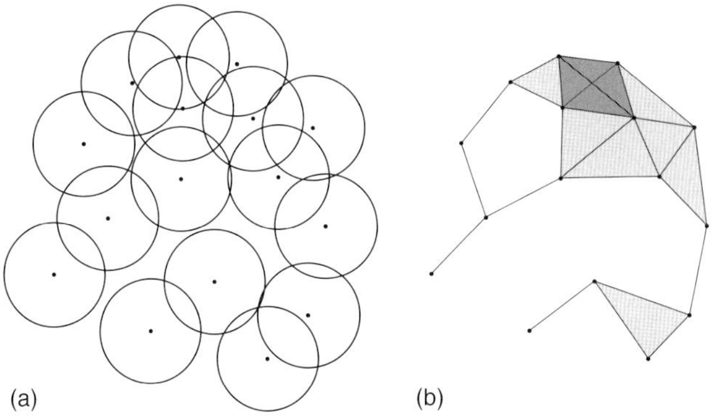
(a) (b)

Figure 6.10 (a) Balls of radius a centered on 16 data points. (b) The nerve of the cover by balls of radius a gives the Čech complex. In this example, the complex consists of points, edges, triangles, and a single tetrahedron (shaded dark gray). As the radius of the balls increases, there are more intersections between them and higher-dimensional simplices are created.

the filtration is defined by the level cuts of a Morse function on a manifold, then a small perturbation to this function will produce a persistence diagram that is close to that of the original one [47].

6.6.4 Cell Complexes from Data

We now address how to build a cell complex and a filtration for use in persistent homology computations. Naturally, the techniques differ depending on the type of data being investigated; we discuss some common scenarios below.

The first construction is based on a general technique from topology called the *nerve of a cover*. Suppose we have a collection of "good" sets (the sets and their intersections should be contractible) $\mathcal{U} = \{U_1, \dots, U_N\}$ whose union $\bigcup U_i$ is the space we are interested in. An abstract simplicial complex $\mathcal{N}(\mathcal{U})$ is defined by making each U_i a vertex and adding a k-simplex whenever the intersection $U_{i0} \cap \cdots \cap U_{ik} \neq \emptyset$. The nerve lemma states that $\mathcal{N}(\mathcal{U})$ has the same homotopy type as $\bigcup U_i$ [10].

If the data set, X, is not too large, and the points are fairly evenly distributed over the object they approximate, it makes sense to choose the U_i to be balls of radius a centered on each data point: $\mathcal{U}_a = \{B(x_i, a), x_i \in X\}$. This is often called the *Čech complex*; see Figure 6.10. If $a < b$, we see that $\mathcal{N}(\mathcal{U}_a) \subset \mathcal{N}(\mathcal{U}_b)$, and we have a filtration of simplicial complexes that captures the topology of the data as they are inflated from isolated points ($a = 0$) to filling all of space ($a \to \infty$).

A similar construction to the Čech complex that is much simpler to compute is the *Vietoris–Rips* or *clique* complex. Rather than checking for higher-order intersections of balls, we build a 1-skeleton from all pairwise intersections and then add a k-simplex when all its edges are present. This construction is not necessarily homotopy equivalent to the union of balls, but is useful when the data set comes from a high-dimensional space, perhaps with only an approximate metric.

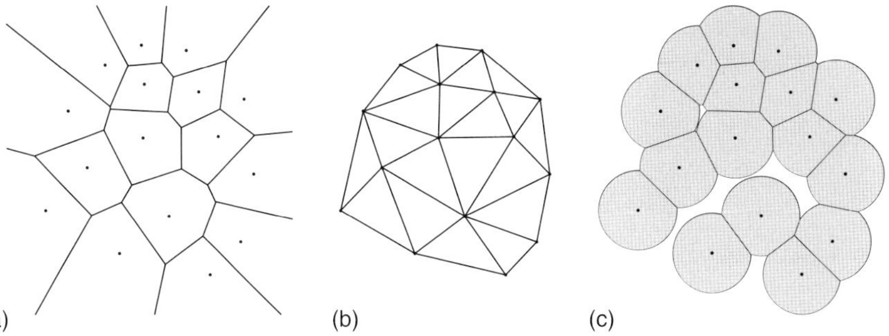

Figure 6.11 (a) The Voronoi diagram of a data set with 16 points. (b) The corresponding Delaunay triangulation. (c) The union of balls of radius a centered on the data points and partitioned by the Voronoi cells. The corresponding triangulation is almost the same as that shown in Figure 6.10: instead of the tetrahedron there are just two acute triangles.

A drawback of the Čech and Vietoris–Rips complexes is that many unnecessary high-dimensional simplices may be constructed. One way to avoid this is to build the Delaunay triangulation. There are many equivalent definitions of this widely used geometric data structure [48]. We start by defining the *Voronoi partition* of space for a data set $\{x_1, \ldots, x_N\} \subset \mathbb{R}^m$, via the closed cells

$$V(x_i) = \{p \text{ such that } d(p, x_i) \leq d(p, x_j) \text{ for } j \neq i\}.$$

That is, the Voronoi cell of a data point is the region of space closer to it than to any other data point. The boundary faces of Voronoi cells are pieces of the $(m-1)$-dimensional bisecting hyperplanes between pairs of data points. The *Delaunay complex* is the geometric dual to the Voronoi complex: when $k+1$ Voronoi cells share an $(m-k)$-dimensional face there is a k-simplex in the Delaunay complex that spans the corresponding $k+1$ data points.[7] See Figure 6.11 for an example in the plane ($m = 2$). The geometry of the Voronoi partition guarantees that there are no simplices of dimension greater than m in the Delaunay complex.

Now consider what happens when we take the intersection of each Voronoi cell with a ball centered on the data point, $B(x_i, a)$. The Voronoi cells partition the union of balls $\bigcup B(x_i, a)$ and the geometric dual is a subset of the Delaunay complex that is commonly referred to as an *alpha complex* or alpha shape (where alpha refers to the radius of the ball [49, 50]). By increasing the ball radius from zero to some large enough value, we obtain a filtration of the Delaunay complex that starts with the finite set of data points and ends with the entire convex hull. The topology and geometry of alpha complexes has been used, for example, in characterizing the

7) This is true for points in general position. Degenerate configurations of points occur in the plane for example, when four Voronoi cells meet at a point. In this case the Delaunay complex may be assigned to contain either a 3-simplex, a quadrilateral cell, or one of two choices of triangle pairs.

shape of and interactions between proteins [51]. The Betti numbers of alpha shapes are also a useful tool for characterizing structural patterns of spatial data [52] such as the distribution of galaxies in the cosmic web [53].

When the data set is very large, a dramatic reduction in the number of simplices used to build a complex is achieved by defining *landmarks* and the *witness complex*. This construction generalizes the Voronoi and Delaunay method, so that only a subset of data points (the landmarks) are used as vertices for the complex, while still maintaining topological accuracy. A further advantage is that only the distances between data points are required to determine whether to include a simplex in the witness complex. See [54] for details and [55] for an extensive review of applications in data analysis.

Another important class of data is *digital images* which can be binary (voxels are black or white), grayscale (voxels take a range of discrete values), or colored (voxels are assigned a multidimensional value). In this setting, the structures of interest arise from level cuts of functions defined on a regular grid. Morse theory is the natural tool to apply here, although in this application, the structures of interest are the level cuts of the function, while the domain (a rectangular box) is simple. There are a number of different approaches to computing homology from such data and this is an area of active research. The works [6, 56, 57] present solutions motivated by applications in the physical sciences.

Further Reading

We give a brief precis of a few standard texts on algebraic topology from mathematical and physical perspectives.

Allen Hatcher's *Algebraic Topology* [10] is one of the most widely used texts in mathematics courses today and has a strong geometric emphasis. Munkres [13] is an older text that remains popular. Spanier [40] is a dense mathematical reference and has one of the most complete treatments of the fundamentals of algebraic topology. A readable introduction to Morse theory is given by Matsumoto [26] and Forman's [37] review article is an excellent introduction to his discrete Morse theory.

Textbooks written for physicists that cover algebraic topology include Nakahara's [12] comprehensive book *Geometry, Topology and Physics*, Schwarz [58] *Topology for Physicists*, and Naber [59] *Topology, Geometry and Gauge Theory*. Each book goes well beyond algebraic topology to study its interactions with differential geometry and functional analysis. A celebrated example of this is the Atiyah–Singer index theorem, which relates the analytic index of an elliptic differential operator on a compact manifold to a topological index of that manifold, a result that has been useful in the theoretical physics of fundamental particles.

References

1. Stillwell, J. (2010) *Mathematics and Its History*, 3rd edn, Springer-Verlag, New York.
2. Smale, S. (1967) Differentiable dynamical systems. *Bull. Am. Math. Soc.*, **73**, 747–817.
3. Mindlin, G., Solari, H., Natiello, M., Gilmore, R., and Hou, X.J. (1991) Topological analysis of chaotic time series data from the Belousov-Zhabotinskii reaction. *J. Nonlinear Sci.*, **1**, 147–173.
4. Muldoon, M., MacKay, R., Huke, J., and Broomhead, D. (1993) Topology from a time series. *Physica D*, **65**, 1–16.
5. Robins, V., Meiss, J., and Bradley, E. (1998) Computing connectedness: An exercise in computational topology. *Nonlinearity*, **11**, 913–922.

6. Kaczynski, T., Mischaikow, K., and Mrozek, M. (2004) *Computational Homology*, Springer-Verlag, New York.
7. Armstrong, M. (1983) *Basic Topology*, Springer-Verlag, New York.
8. Seifert, H. and Threlfall, W. (1980) *A Textbook of Topology*, Academic Press. Published in German 1934. Translated by M.A. Goldman.
9. Francis, G. and Weeks, J. (1999) *Conway's ZIP Proof*, The American Mathematical Monthly, 393–399.
10. Hatcher, A. (2002) *Algebraic Topology*, Cambridge University Press, Cambridge.
11. Mermin, N. (1979) The topological theory of defects in ordered media. *Rev. Mod. Phys.*, **51**, 591–648.
12. Nakahara, M. (2003) *Geometry, Topology and Physics*, 2nd edn, Taylor and Francis.
13. Munkres, J.R. (1984) *Elements of Algebraic Topology*, Addison-Wesley, Reading, MA.
14. Scorpan, A. (2005) *The Wild World of 4-Manifolds*, American Mathematical Society, Providence, RI.
15. Hyde, S., Blum, Z., Landh, T., Lidin, S., Ninham, B., and Andersson, S. (1996) in *The Language of Shape: The Role of Curvature in Condensed Matter* (ed. K. Larsson), Elsevier, New York.
16. Mecke, K. (1998) Integral geometry in statistical physics. *Int. J. Mod. Phys. B*, **12**(9), 861.
17. Baryshnikov, Y. and Ghrist, R. (2009) Target enumeration via Euler characteristic integrals. *SIAM J. Appl. Math.*, **70**(9), 825–844.
18. Curry, J., Ghrist, R., and Robinson, M. (2012) Euler calculus and its applications to signals and sensing. *Proc. Symp. Appl. Math.*, **70**, 75–146.
19. Arns, C., Knackstedt, M., and Mecke, K. (2003) Reconstructing complex materials via effective grain shapes. *Phys. Rev. Lett.*, **91**, 215 506.
20. Scholz, C., Wirner, F., Götz, J., Rüe, U., Schröder-Turk, G., Mecke, K., and Bechinger, C. (2012) Permeability of porous materials determined from the Euler characteristic. *Phys. Rev. Lett.*, **109**, 264 504.
21. Delfinado, C.J.A. and Edelsbrunner, H. (1993) An incremental algorithm for Betti numbers of simplicial complexes, in *SCG '93: Proceedings of the 9th Annual Symposium on Computational Geometry*, ACM Press, New York, pp. 232–239.
22. Alexandroff, P. (1935) in *Topologie* (ed. H. Hopf), Springer-Verlag, Berlin.
23. Bott, R. (1982) in *Differential Forms in Algebraic Topology* (ed. L. Tu), Springer-Verlag, New York.
24. Nash, C. (1983) in *Topology and Geometry for Physicists* (ed. S. Sen), Academic Press, London.
25. Gross, P. (2004) in *Electromagnetic Theory and Computation: A Topological Approach* (ed. P. Kotiuga), Cambridge University Press, Cambridge.
26. Matsumoto, Y. (2002) *An Introduction to Morse Theory*, AMS Bookstore.
27. Banyaga, A. and Hurtubise, D. (2004) *Lectures on Morse Homology*, Kluwer Academic Publishers, The Netherlands.
28. Morse, M. (1934) *The Calculus of Variations in the Large, Colloquium Publications*, vol. 18, American Mathematical Society, Providence, RI.
29. Smale, S. (1961) Generalized Poincaré's conjecture in dimensions greater than four. *Ann. Math.*, **74**(2), 391–406.
30. Bott, R. (1988) Morse theory indomitable. *Publ. Math. de I.H.E.S.*, **68**, 99–114.
31. Bott, R. (1954) Nondegenerate critical manifolds. *Ann. Math.*, **60**(2), 248–261.
32. Conley, C.C. and Easton, R. (1971) Isolated invariant sets and isolating blocks. *Trans. Am. Math. Soc.*, **158**, 35–61.
33. Conley, C.C. (1978) *Isolated Invariant Sets and the Morse Index*, AMS, Providence, RI.
34. Mischaikow, K. (2002) Conley index, in *Handbook of Dynamical Systems*, vol. **2**, Chapter 9 (eds M. Mrozek and B. Fiedler), Elsevier, New York, pp. 393–460.
35. McDuff, D. (2005) Floer theory and low dimensional topology. *Bull. Am. Math. Soc.*, **43**, 25–42.
36. Forman, R. (1998) Morse theory for cell complexes. *Adv. Math.*, **134**, 90–145.
37. Forman, R. (2002) A user's guide to discrete Morse theory. *Séminaire Lotharingien de Combinatoire*, **48**, B48c.
38. Matveev, S. (2003) *Algorithmic Topology and Classification of 3-manifolds*, Springer-Verlag, Berlin.
39. Edelsbrunner, H. and Harer, J. (2010) *Computational Topology: An Introduction*, American Mathematical Society, Providence, RI.

40. Spanier, E. (1994) *Algebraic Topology*, Springer, New York. First published 1966 by McGraw-Hill.
41. Robins, V. (1999) Towards computing homology from finite approximations. *Topol. Proc.*, **24**, 503–532.
42. Frosini, P. and Landi, C. (1999) Size theory as a topological tool for computer vision. *Pattern Recogn. Image Anal.*, **9**, 596–603.
43. Edelsbrunner, H., Letscher, D., and Zomorodian, A. (2002) Topological persistence and simplification. *Discrete Comput. Geom.*, **28**, 511–533.
44. Carlsson, G. and Zomorodian, A. (2005) Computing persistent homology. *Discrete Comput. Geom.*, **33**, 249–274.
45. Zomorodian, A. (2009) Computational topology, in *Algorithms and Theory of Computation Handbook, Special Topics and Techniques*, vol. **2**, 2nd edn, Chapter 3, (eds M. Atallah and M. Blanton), Chapman & Hall/CRC Press, Boca Raton, FL.
46. Carlsson, G., Zomorodian, A., Collins, A., and Guibas, L. (2005) Persistence barcodes for shapes. *Int. J. Shape Modell.*, **11**, 149–187.
47. Cohen-Steiner, D., Edelsbrunner, H., and Harer, J. (2007) Stability of persistence diagrams. *Discrete Comput. Geom.*, **37**, 103–120.
48. Okabe, A., Boots, B., Sugihara, K., and Chiu, S. (2000) *Spatial Tessellations: Concepts and Applications of Voronoi Diagrams*, 2nd edn, John Wiley & Sons, Ltd, Chichester.
49. Edelsbrunner, H., Kirkpatrick, D., and Seidel, R. (1983) On the shape of a set of points in the plane. *IEEE Trans. Inform. Theory*, **29**(4), 551–559.
50. Edelsbrunner, H. and Mücke, E. (1994) Three-dimensional alpha shapes. *ACM Trans. Graphics*, **13**, 43–72.
51. Edelsbrunner, H. (2004) Biological applications of computational topology, in *Handbook of Discrete and Computational Geometry*, Chapter 63 (eds J. Goodman and J.O. Rourke), CRC Press, Boca Raton, FL, pp. 1395–1412.
52. Robins, V. (2006) Betti number signatures of homogeneous Poisson point processes. *Phys. Rev. E*, **74**, 061107.
53. van de Weygaert, R., Vegter, G., Edelsbrunner, H., Jones, B.J.T., Pranav, P., Park, C., Hellwing, W.A., Eldering, B., Kruithof, N., Bos, E.G.P., Hidding, J., Feldbrugge, J., ten Have, E., van Engelen, M., Caroli, M., and Teillaud, M. (2011) Alpha, Betti and the megaparsec universe: on the topology of the cosmic web. *Trans. Comput. Sci.*, **XIV**, 60–101.
54. Carlsson, G. (2004) Topological estimation using witness complexes, in *Eurographics Symposium on Point-Based Graphics* (eds V. de Silva, M. Alexa, and S. Rusinkiewicz), ETH, Zürich, Switzerland.
55. Carlsson, G. (2009) Topology and data. *Bull. Am. Math. Soc.*, **46**(2), 255–308.
56. Robins, V., Wood, P.J., and Sheppard, A.P. (2011) Theory and algorithms for constructing discrete Morse complexes from grayscale digital images. *IEEE Trans. Pattern Anal. Mach. Intell.*, **33**(8), 1646–1658.
57. Bendich, P., Edelsbrunner, H., and Kerber, M. (2010) Computing robustness and persistence for images. *IEEE Trans. Visual. Comput. Graphics*, **16**, 1251–1260.
58. Schwarz, A. (2002) *Topology for Physicists*, Springer-Verlag, Berlin.
59. Naber, G. (2011) *Topology, Geometry and Gauge fields: Foundations*, 2nd edn, Springer-Verlag, Berlin.

7
Special Functions

Chris Athorne

7.1
Introduction

What is so special about a *special* function?

Within the universe of functions, a zoo is set aside for classes of functions that are ubiquitous in applications, particularly well studied, subject to a general algebraic theory, or judged to be important in some other manner. There is no definition of the collection comprising such species beyond writing down a long and never complete list of the specimens in it.

There have been perhaps three attempts at constructing large theories of special functions each of which has been very satisfactory but none entirely comprehensive.

The most fundamental is that of Liouville, later developed by Ritt, in which functions are built from the ground up by extending given function fields via differential equations defined over those fields. Thus, if we allow ourselves to start with rational functions in x over a field of constants, say $\mathbb{C}$, we can extend this field by solutions to ordinary differential equations (ODEs) such as

$$y'(x) + a(x) = 0$$

and

$$y'(x) + a(x)y(x) = 0,$$

$a(x)$ being a rational function. Most simply, we may introduce the "elementary" functions log and exp by choosing $a(x) = 1/x$ and $a(x) = 1$, respectively. The process is then repeated utilizing the newly created functions in place of the rationals. Even then we may not regard *all* such functions as special but only those arising in the most natural manner.

A second way is via the theory of transformation groups due to Lie. Here a function inherits particular properties owing to its being a solution of a differential equation (with boundary conditions) invariant under some underlying geometrical symmetry. The special functions retain a memory of this symmetry by arranging themselves in a highly connected (and beautiful) manner into representations of the symmetry group. For instance, the rotational symmetries of the plane and of Euclidean three-space are responsible

Mathematical Tools for Physicists, Second Edition. Edited by Michael Grinfeld.

for the properties of Bessel functions and spherical harmonics.

A third approach is to start from the point to which the symmetry theory delivers us and to call those functions special that satisfy simple, linear recurrence relations of the form

$$\frac{\partial F(x;n)}{\partial x} = \alpha F(x;n+1).$$

Here x is a continuous variable, n a discrete (integer) variable, and α a constant. Many special functions can be characterized this way.

Interestingly, this last approach dovetails best with modern developments in special function theory. There has been a great flowering of the theory of discrete systems in the last decades, connected with discrete geometries, discrete integrable systems, and "quantum" ($q-$) deformations of classical objects giving rise to q-analogues of all special functions.

In the end, it may be that special functions are simply the flotsam and jetsam thrown up by the sea of mathematical history. There will always be a need for new piles into which as yet unclassified debris may be sorted.

The philosophy behind this article is that since long lists of formulae are to be found in larger treatises on special functions (to which the reader will be directed), we concentrate on a survey of the landscape and the mathematical motivation for describing the properties of such functions. We will nod in the direction of applications now and then but we feel that a tome entitled *Tools for Physicists* really ought to leave the question of application as open as possible.

For the most part, we treat special functions from the point of view of symmetries, summarizing uncomprehensively but hopefully comprehensibly, classes of function associated with discrete and continuous symmetry, referring to the underlying group theory. We use the Laplace operator as a source of many of the paradigms of the theory. We also present the factorization approach that ties in with representation theory, as well as an aside on factorization theory for linear, partial differential operators. Somewhat briefly, we cover cases where symmetries are less apparent: the cases of irregular singular points; and the general theories of Liouville and Picard–Vessiot. Much research effort has been expended in developing applications of functions satisfying nonlinear ODEs and accordingly we introduce the elliptic and theta functions as well as the Painlevé transcendents. Finally, we take a step into the more recently explored world of discrete systems, difference equations, and q-functions, which are related to representations of quantum groups in the way that classical special functions are related to representations of continuous symmetry groups. A consistent theme throughout the treatment is the rôle of finite reflection groups.

For none of these ideas is there room for detailed development and many issues are left untouched. Our treatment throughout is biased toward the algebraic rather than the analytic theory. Serious omissions include the following: asymptotic methods; completeness of function spaces; integral representations, and so on. We direct the reader to cited works as sources of further study. References are representative rather than inclusive and we have tried to cite what is formative but space, and the author's ignorance, have militated against an attempt at completeness.

7.2 Discrete Symmetry

7.2.1 Symmetries

Symmetric objects are distinguished by the property of invariance under a group of transformations or redescriptions [1] . The simplifying consequences of the existence of symmetric objects for all the sciences cannot be overstated and the theory of special functions is a notable example. We must start therefore with a brief heuristic discussion of symmetries.

Let X be a set, assumed finite for the moment. We ignore any mathematical structure this set may possess, so it is really just a set of labels: say, $X = \{1, 2, \dots, n\}$. Functions from X to itself are then maps from the set of labels to itself. For example, we might map every element to the label 1: $f(x) = 1, \forall x \in X$. Such a map is not bijective. The requirement that f *be* bijective is natural. In that case, there exists an inverse function, f^{-1} and given any $x \in X$ there is a unique preimage, $f^{-1}(x)$, mapping to x under f. The composition ($\circ$) of maps is associative,

$$f_1 \circ (f_2 \circ f_3) = (f_1 \circ f_2) \circ f_3,$$

and there is, in particular, a special map $id : x \mapsto x$ that maps each label to itself. So the set of all such maps (which is clearly finite) has a group structure (Chapter 5). This is the symmetric group on n labels, S_n. Its elements are *permutations* of the label set. S_n is the largest group acting on X. Any subgroup, T, of S_n will also act on X and the characteristic properties of any such (sub)group of maps will be that:

$$id \in T;$$

$$f_1, f_2 \in T \Rightarrow f_1 \circ f_2 \in T;$$

$$f \in T \Rightarrow f^{-1} \in T;$$

and

$$id(x) = x,$$

$$f_1(f_2(x)) = (f_1 \circ f_2)(x)$$

$$f \circ f^{-1}(x) = f^{-1} \circ f(x) = x, \quad \forall x.$$

These properties are taken to define the idea of the *group action* of any group on any set (finite or not).

Now consider the set, denoted by Y^X, of maps from X to some other (usually simpler) set Y. Again, we ignore structure on Y. If

$$\phi : X \mapsto Y$$

is such a map, and g an element of the group G acting on X, we define a corresponding action, g; on ϕ by

$$\phi \mapsto g \cdot \phi$$

where

$$(g \cdot \phi)(x) = \phi(g^{-1}(x)) = (\phi \circ g^{-1})(x),$$

the inverse on the right-hand side being necessary to ensure the correct ordering in

$$(g_1 g_2) \cdot \phi = g_1 \cdot (g_2 \cdot \phi),$$

that is, to ensure that the action of G on the set of maps satisfies our prior requirements for a group action.

Already we can see the primitive elements of symmetry creeping in. Let G act on X (finite) and let Y^X be the set of maps from X to Y. A map $\phi \in Y^X$ is said to be *G-invariant* if

$$g \cdot \phi = \phi, \forall g \in G.$$

This means that $\phi(g(x)) = \phi(x)$ for each x and all g. So ϕ is constant on the orbits of G:

$$O_G(x) = \{g(x)|g \in G\} \subseteq X.$$

For example, let X be the set of n elements as before and let G be the subgroup $S_{n-1} \subset S_n$ that fixes only one element, say 1. Then X is a union of two orbits

$$X = \{1\} \cup \{2, 3, \dots, n\}$$

and any invariant map, ϕ, takes at most two distinct values: $\phi(1)$ and $\phi(2) = \phi(3) \cdots = \phi(n)$.

The group S_n has a geometrical interpretation. Consider, for instance, an equilateral triangle with vertices labeled $\{1, 2, 3\}$. The six distinct permutations of these labels correspond to rotations and reflections of the triangle to itself. Likewise, the regular tetrahedron with vertices $\{1, 2, 3, 4\}$ has a set of geometrical symmetries corresponding to the elements of S_4. These are finite subgroups of the group $O(3)$ of isometries (distance preserving maps–rotations and reflections), fixing the origin of $\mathbb{R}^3$.

Apart from generalizing the triangle to regular polygonal prisms, the only regular polyhedra are the cube, octahedron, dodecahedron and icosahedron. Their symmetries complete the list of all finite subgroups of $O(3)$. These are not groups of the type S_n. The symmetry group of the cube and octahedron is generated by S_4 and an inversion,

$$x \mapsto -x.$$

and that of the dodecahedron and icosahedron, is generated by S_5 and the inversion.

It is when we pay attention to extra structure the underlying sets may possess that we start to make a connection with the theory of special functions. Here are two illustrations that we will revisit.

Let K be a field, say $\mathbb{Q}$ or $\mathbb{C}$, and let

$$R = K[x_1, \dots, x_n]$$

be the ring of polynomials in indeterminates x_i, $i = 1, \dots, n$. S_n acts on this ring in the natural way:

$$(\sigma \cdot p)(x_1, \dots, x_n) = p(x_{\sigma^{-1}(1)}, \dots, x_{\sigma^{-1}(n)}),$$

p being an arbitrary polynomial (element of R). Consider

$$G(p) = \sum_{\sigma \in S_n} \sigma \cdot p.$$

This is an invariant element of the ring. For example,

$$G(x_1) = (n-1)!(x_1 + x_2 + \cdots + x_n).$$

Such invariant elements are called *symmetric polynomials* (Section 7.2.3) and although there is clearly an infinite collection of such, all can, in fact, be generated as a polynomial ring over a finite number of basic symmetric polynomials for which a number of canonical choices exist.

As a second example, consider rotations of the plane about a fixed point (the origin of coordinates). Let $R(\theta)$ denote the clockwise rotation through angle θ, a linear map on coordinates, $x, y \in \mathbb{R}^2$:

$$x \mapsto x\cos\theta + y\sin\theta,$$
$$y \mapsto -x\sin\theta + y\cos\theta.$$

It is easily verified that

$$R(\theta)R(\phi) = R(\theta + \phi),$$
$$R(0) = id$$

and that the set of all $R(\theta)$ comprises an infinite group. Real-valued functions on $\mathbb{R}^2$

inherit the group action

$$R(\theta) \cdot f(x, y) = f(R(-\theta)(x, y))).$$

It is natural to exchange rectangular coordinates for polar coordinates,

$$x = \rho \cos \psi,$$
$$y = \rho \sin \psi,$$

which are adapted to the group: rotations act on the coordinate function ψ by translation,

$$\psi \mapsto \psi - \theta,$$

and the function ρ is invariant. All invariant functions are functions of ρ alone. Such functions are constant valued on the orbits of the group, which are the geometrical circles centered on the origin and the origin itself.

Because the group is smooth (Chapter 9), we can also consider a local description of invariance. In the case of small $\theta \sim \epsilon$:

$$x \mapsto x + y\epsilon + O(\epsilon^2),$$
$$y \mapsto y - x\epsilon + O(\epsilon^2).$$

The condition of invariance of a function f is expressed as

$$f(x + \epsilon y, y - \epsilon x) - f(x, y) = O(\epsilon^2)$$

and by using a Taylor expansion in ϵ,

$$(x\partial_y - y\partial_x)f = 0.$$

In polar coordinates, this is simply

$$\partial_\theta f = 0.$$

7.2.2 Coxeter Groups

Although finite, we shall see in later sections that groups such as S_n play a central role in modern aspects of special function Theory, for example, Dunkl operators and Painlevé equations. One ubiquitous class is that of reflection (Coxeter) groups [2–4].

In $\mathbb{R}^n$, imagine a set of hyperplanes ("mirrors") arranged in such a way that the reflection of any one hyperplane in another is again a hyperplane of the set. An example would be three infinite systems of orthogonal mirrors, regularly spaced along the x, y, and z directions in $\mathbb{R}^3$ to form an infinitely extended "milk crate." Another would be mirrors arranged along the edges of a regular triangular lattice in $\mathbb{R}^2$.

If the set of hyperplanes is finite (unlike these two examples), it will associate with each hyperplane a generating element of a finite reflection group and the hyperplanes will all have a common point of intersection.

Suppose we have a finite reflection group with the hyperplanes in $\mathbb{R}^n$ all passing through the origin. To each hyperplane, H, is associated a nonzero normal vector $\mathbf{h}$, defined up to length. The group element representing a reflection in this hyperplane is

$$s_{\mathbf{h}}(\mathbf{x}) = \mathbf{x} - 2\frac{(\mathbf{x}, \mathbf{h})}{(\mathbf{h}, \mathbf{h})}\mathbf{h}$$

where the brackets $(\cdot, \cdot)$ denote the standard inner product on $\mathbb{R}^n$. Clearly $s_{\mathbf{h}} = s_{c\mathbf{h}}$ for any $c \in \mathbb{R} \setminus \{0\}$.

In this way, the set of hyperplanes is associated with a classical *root system* of the kind exemplified in the theory of semisimple Lie algebras (Section 7.3.1). More generally, they are root systems of finite

Coxeter groups generated by reflections,

$$W = \langle r_1, \ldots, r_n | (r_i r_j)^{m_{ij}} = id \rangle,$$

the m_{ij} being the positive integer orders of pair products and, $\forall i$, $m_{ii} = 1$, so that $r_i^2 = e$.

It is the case that W is a finite Coxeter group if and only if it is isomorphic to a finite reflection group. Such groups have been fully classified.

Generalizations of the reflection groups discussed so far are the *affine reflection groups* (affine Weyl groups). In this case, the reflection hyperplanes are not restricted to passing through the origin: one has reflections in translated planes. Such groups are semi-direct products of a reflection group and a translation group associated with a lattice (as in the examples given at the beginning of this section). They also have an important place throughout the modern theory of special functions and we will touch on this in later sections.

7.2.3
Symmetric Functions

Symmetric functions are a class of special functions arising from the most universal finite symmetry groups (permutations of labels) with ubiquitous applications in mathematics and physics [5, 6]. These include combinatorial problems, partitioning questions, labeling of irreducible representations and the parameterization of geometrical objects such as Grassmann varieties.

Let S_n act on polynomials in $x_1, \ldots, x_n$ by permutation of indices. Symmetric polynomials are those invariant under this action and the most general examples are the *Schur* polynomials.

Let Y_μ be the Young diagram of a partition of some positive integer m, that is a list

$$\mu = (m_1, m_2, \ldots)$$

with the properties

$$m_{i-1} \geq m_i \geq 0$$

and

$$\sum_i m_i = m.$$

Thus the partitions of 5 are, writing only the (finitely many) nonzero entries: (5), (4, 1), (3, 2), (3, 1, 1), (2, 2, 1), (2, 1, 1, 1), (1, 1, 1, 1, 1). The corresponding Young diagrams consist of rows of boxes of lengths m_i, left justified and ordered downward. Thus:

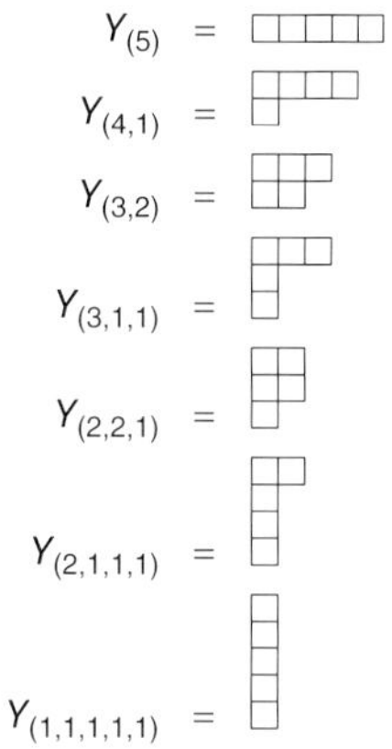

The partition of n 1's is abbreviated to $(1)^n$.

From Young diagrams with m boxes, we create Young *tableaux* on n labels by filling the boxes with choices from the set $\{1, \ldots, n\}$ subject to the rules that the numerical values of entries increase weakly along rows but strictly down columns. Thus two possible tableaux arising from $Y_{(3,2)}$ on four labels are

1	2	2
2	4	

and

3	3	3
4	4	

.

With any Young tableau we associate a monomial by including a factor of x_i for each of the occurrences of i in a box of the tableau. Thus, from the above we get

$$x_1 x_2^3 x_4$$

and

$$x_3^3 x_4^2.$$

The Schur function, S_μ, of a partition μ is the sum (with unit coefficients) over the monomials associated with all possible tableaux coming from the Young diagram of the partition.

Two simple cases correspond to the elementary symmetric functions

$$e_m(x_1, \ldots, x_n) = \sum_{1 \le i_1 < i_2 < \ldots < i_m \le n} x_{i_1} x_{i_2} \ldots x_{i_m}$$

and the complete symmetric functions

$$h_m(x_1, \ldots, x_n) = \sum_{1 \le i_1 \le i_2 \le \ldots \le i_m \le n} x_{i_1} x_{i_2} \ldots x_{i_m}.$$

Note that $e_i = 0$ for $i > n$.

These polynomials have generating functions:

$$E^{(n)}(t) = \sum_0^n e_k t^t = \prod_{i=1}^n (1 + x_i t)$$

$$H^{(n)}(t) = \sum_0^\infty h_k t^t = \prod_{i=1}^n (1 - x_i t)^{-1}$$

The elementary and complete symmetric polynomials are the special cases of Schur polynomials corresponding to Young diagrams of a single column or a single row of length $k: e_k = S_{(1)^k}; h_k = S_{(k)}$.

More generally, a determinantal formula for S_μ exists:

$$S_\mu = \frac{\det\left(x_j^{\mu_i + n - i}\right)_{i,j}}{\prod_{1 \le i < j \le n}(x_i - x_j)}$$

and the Schur polynomials can also be expressed by the *Jacobi–Trudi* formulae in terms of either the e_k's or the h_k's:

$$S_\mu = \det(h_{\mu_i + j - i})_{i,j}$$
$$= \det(e_{\mu_i' + j - i})_{i,j}$$

where μ' denotes the partition of the Young diagram conjugate to μ (transposed about the diagonal). For example, if

$$\mu = (3, 2)$$

then

$Y_{(3,2)} =$ 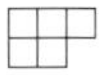

so

$$\mu' = (2, 2, 1)$$

because

$Y_{(2,2,1)} =$ 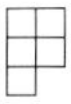

is the transpose of $Y_{(3,2)}$.

Schur polynomials are rich in identities and applications. Perhaps the most important elementary identity is due to Giambelli. This is simply expressed using the *Frobenius* notation for a partition: we count in each row the number (α_i) of boxes to the right of the box in the ith place on the diagonal and in each column the number (β_i) of boxes below that diagonal place. Then we write

$$(\alpha_1, \alpha_2, \ldots | \beta_1, \beta_2, \ldots).$$

For example, the partition $(4, 3, 2, 1, 1)$ associated with the Young diagram

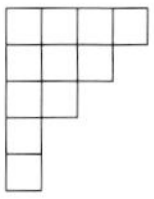

has $\alpha_1 = 3, \alpha_2 = 1, \beta_1 = 4$ and $\beta_2 = 1$, its Frobenius notation being $(3, 1|4, 1)$.

The *Giambelli formula* expresses the Schur polynomial of a partition in terms of simpler Schur polynomials of *hook diagrams*:

$$S_{(\alpha_1,\alpha_2,\ldots|\beta_1,\beta_2,\ldots)} = \det(S_{(\alpha_i|\beta_j)})_{i,j}.$$

Another crucial property is a multiplication formula for Schur polynomials, the *Littlewood–Richardson rule*: Given Young diagrams μ and λ with $\mu_i \leq \lambda_i, \forall i$, a *skew diagram*, λ/μ is the *shape* obtained by removing the leftmost μ_i boxes from the ith row of λ. For example, if $\lambda = (3, 2, 1)$ and $\mu = (1, 1)$, then λ/μ is the skew diagram,

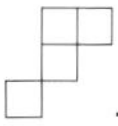 .

A filling of the boxes in the skew diagram with integers according to the same rules as for Young tableaux yields a skew tableau, for example,

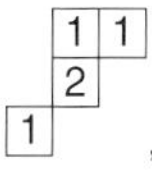

,
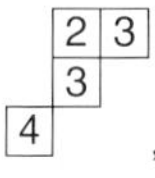

, . . .

and summing over the associated monomials in the same way yields the *skew Schur polynomial* associated with that skew diagram.

According to the RSK (*Robinson–Schensted–Knuth*) correspondence, a skew tableau can be *rectified* to a Young tableau by a simple algorithmic (Shensted) process. The Littlewood–Richardson numbers, $c^{\lambda}_{\mu,\nu}$ count the number of skew tableaux, λ/μ that rectify to a Young tableau with diagram ν. Then

$$S_{\lambda/\mu} = \sum_{\nu} c^{\lambda}_{\mu\nu} S_\nu,$$

and

$$S_\mu \bullet S_\nu = \sum_{\lambda} c^{\lambda}_{\mu\nu} S_\lambda$$

where the $\bullet$ product is again the Schensted algorithm applied to the tableaux and extended linearly to the terms in the Schur polynomials. It is an associative but *not* a commutative multiplication.

Quite generally, any partition λ corresponds to an irreducible representation (*Specht module*) of S_n and, in turn, to an irreducible representation (*Schur* or *Weyl module*) of the general linear group. The Schur polynomials and Littlewood–Richardson formulae encode the properties of tensor products and quotients of these modules.

7.2.4 Invariants of Coxeter Groups

Root systems are associated with finite reflection groups according to a standard classification and we give here some examples with the corresponding invariant polynomials [2, 4].

The simplest example is the group of a single reflection in the hyperplane $x_1 +$

$x_2 = 0$ acting on $\mathbb{R}^2$. Polynomial generators of the ring of invariants are

$$y_1 = x_1^2$$
$$y_2 = x_1 x_2$$
$$y_3 = x_2^2$$

subject to a single relation: $y_1 y_3 - y_2^2$.

For the dihedral group D_{2n} generated by

$$r_1 : x_1 \mapsto x_2, \quad x_2 \mapsto x_1$$
$$r_2 : x_1 \mapsto \zeta x_1, \quad x_2 \mapsto \zeta^{-1} x_2$$

ζ being a primitive nth root of unity, the ring of invariants is generated by

$$y_1 = x_1 x_2$$
$$y_2 = x_1^n + x_2^n$$

without relations.

The Weyl group, $W = S_{n+1}$, associated with the series of semi-simple Lie algebras A_n (or $\mathfrak{sl}_n$) has a root system consisting of $n(n+1)$ vectors:

$$e_i - e_j, \quad 1 \leq i \neq j \leq n+1.$$

The e_i are the standard, unit coordinate vectors. A basis of simple, positive roots is $\Delta = \{e_i - e_{i+1} | i = 1, \ldots n\}$. W acts by permuting coordinate functions x_i, $i = 1, \ldots n+1$ subject to the hyperplane condition

$$x_1 + x_2 + \cdots + x_{n+1} = 0,$$

and the basis of invariant polynomials is given by the first n *power sums*

$$f_i = x_1^i + x_2^i + \cdots + x_{n+1}^i$$

for $i = 1, \ldots, n$, which can be written in terms of elementary symmetric functions using the *Newton identities*.

7.2.5 Fuchsian Equations

Many partial differential equations (PDEs) arising in physics can, in the presence of geometrical symmetry, be reduced to a type of linear ODE having only singular points in $\mathbb{C}$ of a well-controlled type known as *regular singular* points (rsp) described in the following text [7–11].

As the simplest type of singular point, the accompanying theory has been thoroughly developed. We summarize some fundamental results in this section. Further generalization of these ideas will prove important later Section 7.6.

Solutions near an rsp, located at $z = 0$, are of the form $z^\gamma f(z)$ or $z^\gamma (\ln z) g(z)$ where f or g are holomorphic in the local, complex parameter z and nonvanishing at the singular point. The equations of Bessel, Legendre, Hermite, and so on that we shall meet shortly are of this kind. If, in addition, the point at infinity is regular singular, then the ODE is called *Fuchsian*. To determine whether the point at ∞ has this property for a given equation, we transform ∞ in the z-plane using $w = 1/z$ to 0 in the w-plane. We then analyze the singularity at $w = 0$. The general family of Fuchsian equations having three rsps only (at 0, 1, and ∞) is the hypergeometric family.

A second-order ODE in the complex plane,

$$w'' + p(z)w' + q(z)w = 0$$

is said to have a regular singular point at $z = z_0$ if p and q are meromorphic functions on $\mathbb{C}$ with local form,

$$p(z) = (z - z_0)^{-1} P(z), \quad P(z_0) \neq 0,$$
$$q(z) = (z - z_0)^{-2} Q(z), \quad Q(z_0) \neq 0,$$

P and Q both being analytic near $z = z_0$. If we shift the independent variable to make $z_0 = 0$, then a local basis of solutions near 0 is given by

$$w_1 = z^{\nu_1} W_1(z), \quad w_2 = z^{\nu_2} W_2(z)$$

where ν_1 and ν_2 satisfy the quadratic indicial equation

$$\nu(\nu - 1) + P(0)\nu + Q(0) = 0$$

provided the roots are neither equal nor differ by an integer. When the roots are equal or differ by an integer, $\nu_1 - \nu_2 \in \mathbb{Z}$, the solution basis may include logarithms

$$w_1 = z^{\nu_1} W_1(z),$$
$$w_2 = \ln z \quad w_1 + z^{\nu_1} W_2(z).$$

Convergent power series solutions for the W's are obtained by substitution into the ODE to derive second-order recurrence relations on the coefficients in their analytic expansion. This is the *Frobenius method.*

In general, solutions of an ODE near its rsp's are multivalued in the sense that analytic continuation of a solution along a closed path about such a point returns a different solution at the initial point. Given that one has a local basis of analytic solutions at each nonsingular point, any closed path can be associated with a matrix expressing the continued solutions as linear combinations of this local basis. Thus a circuit around the singular point affects the first kind of basis above diagonally,

$$w_1 \mapsto e^{2\pi i\nu_1} w_1, \quad w_2 \mapsto e^{2\pi i\nu_2} w_2,$$

and the second kind of basis triangularly,

$$w_1 \mapsto e^{2\pi i\nu_1} w_1,$$
$$w_2 \mapsto e^{2\pi i\nu_1} w_2 + 2\pi i e^{2\pi i\nu_1} w_1.$$

The matrices formed this way clearly constitute a group: the *monodromy group* of the ODE. This group will be finitely generated if the number of rsp's is finite but the order of any generator need not be finite because the multivaluedness is determined by the exponents ν_i, which may not be rational, and by the possible presence of the logarithmic function. In general, a solution is infinitely ramified (sheeted) over $\mathbb{C}$.

But it is possible to ask, as did Klein, under what circumstances there exist finitely sheeted solutions. It follows from the general theory of Riemann surfaces that such solutions are algebraic: they satisfy a polynomial equation in themselves with coefficients polynomial in $z \in \mathbb{C}$ [12].

It is useful to reduce the ODE to a canonical form by eliminating the coefficient $p(z)$ of w'. A revealing way of doing this is to attempt to "complete the square" in the differential operator part by writing it as

$$\left(\frac{d}{dz} + \frac{p(z)}{2}\right)^2 w + I(z)w = 0$$

where the function

$$I(z) = q - \frac{p'}{2} - \frac{p^2}{4}$$

is a differential *invariant* in the sense that it is unaltered by any z-dependent scaling of w:

$$w(z) \mapsto \lambda(z)w(z).$$

We choose linearly independent solutions w_1 and w_2 to the ODE, introduce a new independent variable

$$s = \frac{w_1}{w_2}$$

and consider z as *dependent* variable. Then using the notation

$$\{Z, s\} = -\frac{1}{Z'^2}\left(\frac{Z'''}{Z'} - \frac{3}{2}\frac{Z''^2}{Z'^2}\right),$$

for the *Schwarzian derivative* of Z with respect to s, we obtain

$$\{z, s\} = 2I(z).$$

The Schwarzian derivative is a projective invariant in the sense that, for α, β, γ, and δ constant with $\alpha\delta - \beta\gamma \neq 0$,

$$\left\{\frac{\alpha Z + \beta}{\gamma Z + \delta}, s\right\} = \{Z, s\}$$

and it also has the property

$$\{s, z\} = \{s, Z\}\left(\frac{dZ}{dz}\right)^2 + \{Z, z\}.$$

Restricting attention to second-order Fuchsian ODEs and to the situation where there are only three rsp's we may, by a linear fractional transformation in z, assume these rsp's lie at the points $z = 0$, $z = 1$, and $z = \infty$. Consider a pair of linearly independent solutions ψ_1 and ψ_2 to the ODE and analytically continue along a closed path around an rsp. The two solutions transform linearly on return to the initial place and so their ratio, $\phi = \psi_1/\psi_2$ transforms by the Möbius map (parameters as above):

$$\phi \mapsto \frac{\alpha\phi + \beta}{\gamma\phi + \delta}.$$

The group of all Möbius transformations is $PSL_2(\mathbb{C})$. We are looking then for ODEs whose monodromy group is a finite subgroup of $PSL_2(\mathbb{C})$. This group is isomorphic to $SO_3(\mathbb{R})$, the group of orientation-preserving isometries of $\mathbb{R}^3$. So we have the answer to the question in the form of the finite subgroups of the rotation group: the symmetry groups of the regular solids and prisms discussed earlier (Section 7.2.1).

Klein used this classification to show that in order for the ODE to have algebraic solutions the invariant $I(z)$ must be one of five possible forms up to an arbitrary rational function, $Z(z)$, namely,

$$I_i(z) = \frac{1}{4}J_i(Z)\left(\frac{dZ}{dz}\right)^2 + \frac{1}{2}\{Z, z\},$$

where $J_i(Z)$ is given by

$$J_1(Z) = \frac{1 - N^{-2}}{Z^2}$$

$$J_i(Z) = \frac{1 - \nu_2^{-2}}{Z^2} + \frac{1 - \nu_1^{-2}}{(Z-1)^2} + \frac{\nu_1^{-2} + \nu_2^{-2} - \nu_3^{-2} - 1}{Z(Z-1)},$$

$$i = 2, 3, 4, 5.$$

The values of $\nu_2, \ldots, \nu_5$ are tabulated thus:

i	ν_1	ν_2	ν_3	Symmetry
1	N	N		cyclic
2	2	2	n	dihedral
3	2	3	3	tetrahedral
4	2	3	4	cubic
5	2	3	5	icoshedral

The solutions are

$$w_1 = s\left(\frac{ds}{dz}\right)^{-1/2}, \quad w_2 = \left(\frac{ds}{dz}\right)^{-1/2},$$

where s is related to $Z(z)$ in each case:

Case 1.

$$Z = \left(\frac{s - s_1}{s - s_2}\right)^N$$

Case 2.

$$Z = -\frac{1}{2}\frac{s^n - 1}{s^n}$$

Case 3.

$$Z = \frac{(s^4 + 2\sqrt{3}s^2 - 1)^3}{(s^4 - 2\sqrt{3}s^2 - 1)^3}$$

Case 4.

$$Z = \frac{(s^8 + 14s^4 + 1)^3}{108s^4(s^4 - 1)^2}$$

Case 5.

$$Z = -\frac{(s^{20} - 228s^{15} + 494s^{10} + 228s^5 + 1)^3}{1728s^5(s^{10} + 11s^5 - 1)^5}.$$

Because the ODEs of the special functions that we shall study below are not always Fuchsian (∞ is not an rsp), we can think of the algebraic solutions above as simple precursors of the (transcendental) special functions.

7.3 Continuous Symmetry

Between 1888 and 1893, Sophus Lie [13] published the three volumes of his *Theorie der Transformationsgruppen*, where he laid the foundations and initiated the systematic applications of continuous groups of symmetries acting on geometric spaces.

The construction of the classical special functions of applied mathematics can be approached using continuous symmetries from two end points.

From one end, we may start with a partial differential operator whose eigenfunctions we wish to understand and use a geometrical symmetry to reduce the PDE to an ODE by "separation of variables." We may solve this equation either by the Frobenius expansion or by a factorization method and study the properties of the resulting families of functions.

Starting from the other end, we may look for representations of symmetry groups in function spaces. When these are constructed, the corresponding representation of the Lie algebra gives rise to differential expressions for raising and lowering operators which present to us the differential equations with which the first approach commenced.

As examples, we will discuss spherical harmonics from the first point of view and Bessel functions from the second. We will also list some results appertaining to other (but by no means *all*) classical special functions.

7.3.1 Lie Groups and Lie Algebras

A group G is said to be a (finite-dimensional) Lie group if it is a finite-dimensional differentiable manifold endowed with a multiplication map,

$$\mu : G \times G \to G$$

and an inversion map,

$$\iota : G \to G,$$

which are smooth with respect to the differentiable structure on G and satisfy the group axioms for some identity point, e :

$$\mu(e, x) = \mu(x, e) = x$$
$$\mu(x, \iota(x)) = \mu(\iota(x), x) = e$$
$$\mu(\mu(x, y), z) = \mu(x, \mu(y, z)).$$

They often arise as symmetries of geometrical spaces and their own, intrinsic geometry allows the methodology of differential geometry (Chapter 9) to be effectively applied [14–17].

In particular, G has, at each point $g \in G$, a tangent space: T_gG. If $L_g : G \to G$ is the map $L_g(h) = gh$, h being an arbitrary point in G, then the derivative of this map, $\mathrm{d}L_g$ takes T_hG to $T_{gh}G$ and is an isomorphism of finite-dimensional vector spaces.

The set of all vector fields on G is a module over the ring of functions on G and an infinite-dimensional vector space over the field of constants.

On the other hand, a vector field, v, on G is *left-invariant* if

$$v(L_g h) = \mathrm{d}L_g v(h), \forall g, h.$$

The set of such left-invariant fields is a finite-dimensional vector space and can be identified with the tangent space at the identity, T_eG. This is the *Lie algebra*, $\mathfrak{g}$, of G. In many practical situations in applied mathematics and theoretical physics, it is through $\mathfrak{g}$ that the action of a smooth symmetry group is recognized. We have already seen an example: the rotation group $SO(2)$ has a Lie algebra $so(2)$, which acts on $\mathbb{R}^2$ via a one-dimensional vector field: $x\partial_y - y\partial_x$.

A Lie algebra, $\mathfrak{g}$, is endowed with a (nonassociative) bilinear product, $\star$, having the following properties: every $g \in \mathfrak{g}$ is nilpotent,

$$g \star g = 0;$$

and every triple $g, h, k \in \mathfrak{g}$ satisfies the *Jacobi* identity,

$$g \star (h \star k) + h \star (k \star g) + k \star (g \star h) = 0.$$

Applying the nilpotence condition to $g + h$,

$$(g + h) \star (g + h) = 0$$

and using bilinearity, gives the skewness of $\star$:

$$g \star h = -h \star g.$$

The *Universal Enveloping Algebra* construction replaces $\mathfrak{g}$ by the full tensor algebra (Chapter 9) $T(\mathfrak{g})$ (noncommutative polynomial algebra or sums of ordered products of elements of $\mathfrak{g}$) quotiented by the ideal generated by elements of the form

$$g \otimes h - h \otimes g - g \star h.$$

It allows us to represent the $\star$ operation by the commutator $[\cdot, \cdot]$.

The structure of a semi-simple (see next section), finite-dimensional Lie algebra is determined by a commutative subalgebra, the Cartan subalgebra, $\mathfrak{h}$, by which $\mathfrak{g}$ can be decomposed as $\mathfrak{h}$-eigenspaces, labeled by *roots*. These roots are elements $\alpha \in \mathfrak{h}^*$, the vector space *dual* of $\mathfrak{h}$, the elements of eigenspaces, $\mathfrak{g}_\alpha$, satisfying $[h, e] = \alpha(h)e$ for every $h \in \mathfrak{h}$, $e \in \mathfrak{g}_\alpha$. The roots can be divided into two types: positive roots $\alpha \in \Delta^+$; and negative $\alpha \in \Delta^-$; the entire algebra decomposing thus:

$$\mathfrak{g} = \mathfrak{h} \oplus \bigoplus_{\alpha\in\Delta} \mathfrak{g}_\alpha \oplus \bigoplus_{\alpha\in\Delta^-} \mathfrak{g}_\alpha.$$

7.3.2 Representations

Very many special functions arise as elements of finite- or infinite-dimensional vector spaces carrying representations of a Lie group [14, 18]. In fact, the language of Lie group and algebra representations is endemic in modern physics.

If G is a Lie group and V a vector space, which may be infinite dimensional, over a field, say $\mathbb{C}$, then a representation of G is a ring homomorphism $\rho : G \to End(V)$, (Chapter 5) from the group into $\mathbb{C}$-linear maps, *endomorphisms*, from V to itself, that is, ρ satisfies the conditions

$$\rho(\alpha)\rho(\beta) = \rho(\alpha\beta)$$
$$\rho(e) = Id_V$$

where e is the group identity and Id_V the identity endomorphism on V. If $\{e_i | i = 1, \ldots, n\}$ is a basis of V (and in a formal sense, we may take n to be infinity) we define a matrix representation for G via the identity

$$\rho(\alpha)e_i = \sum_{j=1}^{n} R(\alpha)_{ij} e_j$$

and the homomorphism condition becomes

$$R_{ij}(\alpha\beta) = \sum_{k=1}^{n} R_{ik}(\alpha)R_{kj}(\beta)$$
$$R_{ij}(e) = \delta_{ij}.$$

Representations ρ on V and ρ' on V' are *equivalent* if there exists an invertible map $S : V \to V'$ satisfying

$$\rho'(\alpha)S = S\rho(\alpha), \forall \alpha \in G.$$

If V has an inner product, $<,>: V \times V \to \mathbb{C}$, then ρ is a *unitary* representation of G if for all $\alpha \in G$ and all $u, v \in V$

$$< \rho(\alpha)u, \rho(\alpha)v >=< u, v > .$$

Any representation of a finite group or of a compact Lie Group is equivalent to such a unitary representation.

A vector subspace $U \subset V$ is *invariant* if for all $\alpha \in G$ and all $u \in U$,

$$\rho(\alpha)u \in U$$

and the representation ρ on V is said to be *reducible* if it contains a nontrivial, proper invariant subspace. If $V = U \oplus W$ where U and W are both invariant, then the representation ρ on V is *completely reducible* and its matrix representation will decompose into a block diagonal form.

For unitary representations, reducibility implies complete reducibility.

These ideas carry over to the local situation via the Lie algebra. λ is a representation on V of the Lie algebra, $\mathfrak{g}$, of the Lie group G, if for $g, h \in \mathfrak{g}$ and $a, b \in \mathbb{C}$,

$$\lambda(ag + bh) = a\lambda(g) + b\lambda(h)$$
$$\lambda(g \star h) = [\lambda(g), \lambda(h)],$$

$[^*, ^*]$ denoting the commutator of matrices

Equivalence, invariance, reducibility, and so on, are defined analogously to the group case.

A Lie algebra $\mathfrak{g}$ is *solvable* if its derived series

$$\mathfrak{g}^{(0)} \supset \mathfrak{g}^{(1)} \supset \mathfrak{g}^{(2)} \cdots$$

defined by

$$\mathfrak{g}^{(i)} = \mathfrak{g}^{(i-1)} \star \mathfrak{g}^{(i-1)}, \quad \mathfrak{g}^{(0)} = \mathfrak{g}$$

is finite. A simple example is the set of upper triangular matrices. Solvability of an ideal of $\mathfrak{g}$ is defined likewise. $\mathfrak{g}$ is *simple* if it is non-abelian and contains no proper, nontrivial ideal. It is *semi-simple* if it has no nontrivial solvable ideal.

If ρ is a representation of $\mathfrak{g}$ on V then V decomposes into *weight spaces*

$$V_\lambda = \{v \in V | h(v) = \lambda(h)v, \forall h \in \mathfrak{h}\}$$

for each $\lambda \in \mathfrak{h}^*$. The action of $\mathfrak{g}_\alpha$ on the weight spaces is

$$\mathfrak{g}_\alpha : V_\lambda \to V_{\lambda+\alpha}.$$

The intersections of distinct weight spaces are trivial, their sum is direct, and in the case of finite-dimensional V,

$$V = \bigoplus_{\lambda \in \mathfrak{h}^*} V_\lambda.$$

An element $v \in V$ is said to be *maximal* of *highest weight* λ if $v \in V_\lambda$ and $\mathfrak{g}_\alpha(v) = 0$ for all $\alpha \in \Delta^+$. The action of the whole of $\mathfrak{g}$ on v then generates a *cyclic representation* of highest weight λ. Two cyclic representations of the same highest weight are isomorphic, hence *standard*. The weights form a lattice in $\mathfrak{h}^*$ and such a standard representation exists for any element in the weight lattice.

Finite-dimensional cyclic, irreducible representations of $\mathfrak{g}$ necessarily have highest weights λ such that, for a standard basis, $\{h_i | i = 1, \ldots, rank(\mathfrak{h})\}$, of $\mathfrak{h}$, the numbers $\lambda(h_i)$ are positive integers.

In the spaces of special functions arising from the actions of symmetry groups on differential operators, these $\lambda(h_i)$ appear as special parameter values and so where the representations are finite dimensional, it is because of integrality conditions on these parameters.

7.3.3 The Laplace Operator

A commonly occurring situation for the classical special functions is the need to solve the Laplace equation in a specific geometry [19, 20]. Of course, one may write down integral formulae for the solutions in terms of a Green's function for almost *any* (asymmetric) geometry, but where there is some symmetry, the solution space is more explicitly manifested.

The Laplace operator on three-dimensional space, $\mathbb{R}^3$,

$$\Delta = \partial_x^2 + \partial_y^2 + \partial_z^2$$

has a number of symmetries. In particular, translations in the x, y, and z variables, described by the Lie group $\mathbb{R}^3$ and rigid rotations, $SO(3)$ are symmetries.

The Lie algebra of translations is generated by the set $\{\partial_x, \partial_y, \partial_z\}$, which all commute with $\Delta : [\Delta, \partial_x] = 0$, and so on.

The Lie algebra $so(3)$ of $SO(3)$ is the real, three-dimensional algebra with basis

$$\{x\partial_y - y\partial_x, y\partial_z - z\partial_x, z\partial_x - x\partial_z\},$$

which acts on functions of x, y and z, and which commutes with Δ.

The commutation property means that if ϕ is a solution of the Laplace equation, then so is $g(\phi)$, g being any of the generators described above.

If ϕ solves the Laplace equation on $\mathbb{R}^3$,

$$\Delta\phi(x, y, z) = 0,$$

then so does the function $a\phi_x + b\phi_y + c\phi_z$, a, b, and c being constants. So the solution space is a vector space of functions closed under this differential algebraic operation. In particular, it suffices to choose functions satisfying

$$\phi_x = \lambda\phi,\ \phi_y = \mu\phi,\ \phi_z = \nu\phi$$

for constants λ, μ and ν. This is just the space of functions such as $\exp(\lambda x + \mu y + \nu z)$. (Note that we include complex-valued functions here because their linear combinations may be real.)

If the geometry of our problem requires conditions of the form

$$\phi(x + L, y + M, z + N) = \phi(x, y, z)$$

for finite, positive L, M and N, then these are reflected in the choice of sine and cosine functions as a basis of the solution space.

An important consequence of these considerations is that these functions obey addition laws: if $\sin x$ is periodic of period 2π, then so is $\sin(x+x')$ and, by linearity, must be a linear combination of $\sin x$ and $\cos x$. Determining the multiplicative coefficients from special values of x results in the addition law:

$$\sin(x+x') = \cos x' \sin x + \sin x' \cos x.$$

This is a common theme in special function theory, including nonlinear situations.

Other solution spaces will be appropriate to other geometries. A semi-infinite, finite width strip, for example, will require a combination of periodic and decaying exponential solutions.

The Laplacian operator is a geometric object that can be defined in a coordinate-free manner [20]. On any smooth, Riemannian manifold with covariant metric g_{ij}, the Laplacian operator is

$$\Delta\phi = \sqrt{g}^{-1}\partial_i(\sqrt{g}g^{ij}\partial_j\phi)$$

g being the determinant of g_{ij}, regarded as a symmetric matrix, and where the Einstein convention for summation of repeated (contra/covariant) index pairs is in operation. Coordinates may be chosen adapted to the geometric symmetry of a problem and the Laplacian written accordingly. Examples are listed in [21].

In the case of cylindrical symmetry about, say, the z-axis, $SO(2)$, it is convenient to change variables to polar coordinates (ρ, ψ, z). For full rotational symmetry, $SO(3)$, the spherical polar coordinates (r, θ, ψ) are appropriate. (To be clear, we use θ here as the azimuthal angle, constant on lines of lattitude; ψ the equatorial angle, constant on lines of longitude.)

In such coordinates, the symmetry transformations are essentially trivial, that is, they amount to simple translations in the variables.

$$SO(2) : (\rho, \psi, z) \mapsto (\rho, \psi + \epsilon, z + \eta)$$
$$SO(3) : (r, \theta, \psi) \mapsto (r, \theta + \zeta, \psi + \epsilon)$$

In the cases of the Laplacian on $\mathbb{R}^3$, we obtain in each case:

$$\Delta\phi = \frac{1}{\rho}\partial_\rho(\rho\partial_\rho\phi) + \frac{1}{\rho^2}\partial_\psi^2\phi + \partial_z^2\phi$$

and

$$\Delta\phi = \frac{1}{r^2}\partial_r(r^2\partial_r\phi) + \frac{1}{r^2\sin\theta}\partial_\theta(\sin\theta\partial_\theta\phi) + \frac{1}{r^2\sin^2\theta}\partial_\psi^2\phi.$$

Because the group transformations are translations in θ and ψ, rotational periodicity in θ requires that ϕ be expanded in terms of the complete system

$$\{\exp in\psi \,|\, n \in \mathbb{Z}\}.$$

Thus

$$\phi(r, \theta, \psi) = \sum_{n\in\mathbb{Z}} \Phi_n e^{in\psi}$$

where the coefficients $\Phi_n(\rho, z)$ or $\Phi_n(r, \theta)$ satisfy in each case the differential equations

$$\frac{1}{\rho}\partial_\rho(\rho\partial_\rho\Phi_n) - \frac{n^2}{\rho^2}\Phi_n + \partial_z^2\Phi_n = 0$$

and

$$\frac{1}{r^2}\partial_r(r^2\partial_r\Phi_n) + \frac{1}{r^2\sin\theta}\partial_\theta(\sin\theta\partial_\theta\Phi_n) - \frac{n^2}{r^2\sin^2\theta}\Phi_n = 0.$$

For cylindrical symmetry, we may have either decaying or periodic boundary conditions on z and we obtain an equation of

Bessel type by expanding in an exponential, e^{mz} series or integral,

$$\frac{1}{\rho}\partial_\rho(\rho\partial_\rho\Phi) + \left(m^2 - \frac{n^2}{\rho^2}\right)\Phi = 0$$

where Φ depends on parameters n and m.

In the case of spherical symmetry, we may expand $\Phi_n(r,\theta)$ in powers of r,

$$\sum_{l=0}^{\infty} \Phi_{nl}(\theta) r^l$$

to obtain the *associated Legendre* equation:

$$\frac{1}{\sin\theta}\partial_\theta(\sin\theta\partial_\theta\Phi) + \left(l(l+1) - \frac{n^2}{\sin^2\theta}\right)\Phi = 0$$

where Φ depends on parameters n and l. This is more usually written with the independent variable $x = \cos\theta$:

$$(1-x^2)y'' - 2xy' + \left(l(l+1) - \frac{n^2}{1-x^2}\right)y = 0.$$

Linearly independent solutions are denoted $P_l^n(x)$ and $Q_l^n(x)$, associated Legendre functions of the first and second kind. The second kind are singular at $x = 1$ and $x = -1$.

When $n = 0$ and $l \in \mathbb{Z}$, the first kind functions are polynomial in x and called *Legendre polynomials*, $P_l(x)$, if the normalization is chosen such that $P_l(1) = 1$. l may be taken to be a positive integer because the defining equation

$$y'' - 2xy' + l(l+1)y = 0$$

is unchanged under the replacement of l by $-(l+1)$.

As follows from later considerations (Section 7.4), the Legendre polynomials are given by the *Rodrigues formula*:

$$P_l(x) = \frac{1}{2^l n!}\frac{d^l}{dx^l}(x^2-1)^l.$$

The associated functions can be derived from the Legendre functions using the *Ferrers formulae*:

$$P_l^n(x) = (1-x^2)^{n/2}\frac{d^n}{dx^n}P_l(x);$$

$$Q_l^n(x) = (1-x^2)^{n/2}\frac{d^n}{dx^n}Q_l(x).$$

7.3.4 Spherical Harmonics

In many applications, the associated Legendre functions appear under the guise of (*surface*) *spherical harmonics*,

$$Y_l^n(\theta,\psi) = \sqrt{\frac{(2l+1)(l-n)!}{4\pi(l+m)!}}P_l^n(\cos\theta)e^{in\psi},$$

for integer values of m, $-l \le m \le l$, or with a monomial factor in r,

$$r^l Y_l^n(\theta,\psi)$$

as (*solid*) *spherical harmonics* [22].

These in turn arise as moments in an integral expansion in the following manner. Consider a complex variable ζ and an arbitrary meromorphic function f. The contour integral

$$\oint_C d\zeta f\left(X\zeta + 2iz + \frac{\overline{X}}{\zeta}, \zeta\right),$$

$X = x + iy$, is easily seen to satisfy the Laplace equation by virtue of

$$(\partial_x^2 + \partial_y^2 + \partial_z^2)f = \left(\left(\zeta + \frac{1}{\zeta}\right)^2 - \left(\zeta - \frac{1}{\zeta}\right)^2 - 4\right)f'' = 0.$$

If we write $\zeta = e^{i\phi}$ and choose the contour to be a unit circle about 0, we obtain as an instance of the above integral an expansion in associated Legendre functions:

$$\int_0^{2\pi} \mathrm{d}\phi\,(z + ix\cos\phi + iy\sin\phi)^n \cos(m\phi)$$
$$= \frac{2\pi i^m n!}{(n+m)!} r^n P_n^m(\cos\theta)\cos(m\psi),$$

θ and ψ being the usual azimuthal and longitudinal coordinates of spherical polar coordinates.

The integral representation above is also of interest as an instance of the (mini-) twistor transform that has proved fundamental to obtaining monopole solutions to Yang–Mills–Higgs gauge theories [23].

7.3.5 Separation of Variables

In a general coordinate system, $\{x_1, x_2, x_3\}$, it is customary to solve the Laplace equation by choosing an ansatz,

$$\Phi(x_1, x_2, x_3) = X_1(x_1)X_2(x_2)X_3(x_3),$$

and decoupling the dependence of each variable to obtain three ODE for X_1, X_2, and X_3 coupled only through some eigenvalue-like parameters, the possible values of these parameters being determined by boundary or initial conditions. This method of *separation of variables* is illustrated now for the case of parabolic cylinder coordinates [24].

The pair of equations

$$y^2 + 2\lambda x - \lambda^2 = 0,$$
$$y^2 - 2\mu x - \mu^2 = 0,$$

represent two mutually orthogonal families of parabola. λ and μ are constant on each parabola and, with z, constitute *parabolic cylinder coordinates* on $\mathbb{R}^3$. The standard metric in these coordinates is,

$$\mathrm{d}s^2 = \frac{\lambda+\mu}{4}\left(\frac{\mathrm{d}\lambda^2}{\lambda} + \frac{\mathrm{d}\mu^2}{\mu}\right) + \mathrm{d}z^2,$$

and using the general definition of the Laplacian given earlier (Section 7.3.3) we obtain,

$$\Delta\Phi = \frac{4}{\lambda+\mu}\left(\sqrt{\lambda}\partial_\lambda(\sqrt{\lambda}\partial_\lambda\Phi) + \sqrt{\mu}\partial_\mu(\sqrt{\mu}\partial_\mu\Phi)\right) + \partial_z^2\Phi.$$

Separation of variables,

$$\Phi(\lambda, \mu, z) = \Lambda(\lambda)M(\mu)Z(z),$$

and decoupling yields equations for Λ and M of the form

$$\frac{\mathrm{d}^2 w}{\mathrm{d}u^2} \pm (\frac{1}{4}u^2 \mp a)w = 0,$$

which are solved by *Whittaker functions* [22].

7.3.6 Bessel Functions

We illustrate in this section how one starts from representations of symmetries on function spaces to arrive at the Bessel differential equation [20, 21, 25–27]. The rigid symmetries (isometries) $\mathcal{E}^2$ of the plane, $\mathbb{R}^2$, are generated by translations, $\mathbf{x} \mapsto \mathbf{x} + \mathbf{a}$ and rotations, $\mathbf{x} \mapsto R(\alpha)\mathbf{x}$. The general transformation we will denote $g_{\mathbf{a},\alpha} : \mathbf{x} \mapsto R(\alpha)\mathbf{x} + \mathbf{a}$. It is easy to see that the group law is

$$g_{\mathbf{a},\alpha} g_{\mathbf{b},\beta} = g_{\mathbf{a}+R(\alpha)\mathbf{b},\alpha+\beta}.$$

This product of groups is a semi-direct product, $\mathbb{R}^2 \rtimes SO(2)$, of translation and rotation subgroups, the former a normal subgroup of the full group.

The corresponding Lie algebra is spanned by $a_1 = \partial_x$, $a_2 = \partial_y$, and $a_3 = y\partial_x - x\partial_y$, x and y being coordinates on $\mathbb{R}^2$. Commutation relations are:

$$[a_1, a_2] = 0$$
$$[a_2, a_3] = a_1$$
$$[a_3, a_1] = a_2.$$

By tensoring with $\mathbb{C}$, we may as well consider the complex Lie algebra.

It is easily checked that for any $z \in \mathbb{C}$, there is a natural representation of $\mathcal{E}^2$ on smooth functions on the unit circle, $\mathbf{x} \cdot \mathbf{x} = 1$, namely,

$$g_{\mathbf{a},\alpha} \cdot f(\mathbf{x}) = \mathrm{e}^{z\mathbf{a}\cdot\mathbf{x}} f(R(-\alpha)\mathbf{x}).$$

Since a $\mathbb{C}$-vector space basis of such functions is given by the set

$$\{v_n := \mathrm{e}^{in\phi} | n \in \mathbb{Z}\},$$

we have a representation of the Lie algebra, obtained by differentiating the group representation with respect to its three parameters:

$$a_1 \cdot v_n = \frac{1}{2} z(v_{n+1} + v_{n-1})$$
$$a_2 \cdot v_n = \frac{1}{2i} z(v_{n+1} - v_{n-1})$$
$$a_3 \cdot v_n = -in\, v_n$$

These relations show that for $z \neq 0$, there is no invariant subspace and so the function space is an irreducible representation of $\mathcal{E}^2$.

Special functions are entries in the matrix representation of this action. On the function space, we have an inner product,

$$(f_1, f_2) = \frac{1}{2\pi} \int_0^{2\pi} \bar{f}_1 f_2 \,\mathrm{d}\theta$$

and we consider the functions

$$(g_{\mathbf{a},\alpha} \cdot v_n, v_m)$$

as functions of the group parameters $\mathbf{a} \in \mathbb{R}^2$ and α. Noting that these depend only on the differences $n - m$ and incorporating a conventional multiplicative factor, we define the Bessel functions in the following way:

$$J_n(x) = \frac{1}{2\pi} \int_0^{2\pi} \mathrm{e}^{ix\sin\theta - in\theta} \,\mathrm{d}\theta.$$

(We have exchanged the ρ of Section 7.3.3 for x.)

Note that

$$J_{-n}(x) = (-1)^n J_n(x).$$

From this definition follow, by differentiation with respect to x and by integration by parts, the *raising* (D_n^+) and *lowering* (D_n^-) operators:

$$J_{n+1}(x) = D_n^+ J_n(x)$$
$$J_{n-1}(x) = D_n^- J_n(x)$$

where

$$D_n^+ = \frac{n}{x} - \frac{\mathrm{d}}{\mathrm{d}x}$$
$$D_n^- = \frac{n}{x} + \frac{\mathrm{d}}{\mathrm{d}x}$$

These operators commute, up to unit shift in the index, in the sense that

$$D_{n+1}^- D_n^+ = D_{n-1}^+ D_n^-$$

and give a representation of the translation algebra on the Bessel functions.

Further one sees that

$$D_{n+1}^- D_n^+ J_n = D_{n-1}^+ D_n^- J_n = J_n,$$

that is, the Bessel function J_n is an eigenfunction, with eigenvalue n^2 of the operator

$$x^2\frac{d^2}{dx^2}+x\frac{d}{dx}+x^2.$$

We return to this "factorization" property in Section 7.4.

7.3.7 Addition Laws

Addition laws between functions defined on groups arise because of the underlying group operation [26, 27]. Thus if g_1 and g_2 are group elements of G, the matrix elements of a representation satisfy the homomorphism property

$$(g_1g_2\cdot v_n, v_m)=\sum_l (g_1\cdot v_n, v_l)(g_2\cdot v_l, v_m).$$

We may take a simple choice for the group elements. In this case, take g_1 to be the translation

$$g_{\mathbf{a},0}:\quad \mathbf{a}=(x_1,0)$$

and g_2 the translation

$$g_{\mathbf{a},0}:\quad \mathbf{a}=(x_2\cos\theta_2, x_2\sin\theta_2).$$

Then the product element is a translation $\mathbf{a}=(x\cos\theta, x\sin\theta)$ with

$$x^2=x_1^2+x_2^2+2x_1x_2\cos\theta_2$$

and

$$e^{i\theta}=\frac{x_1+x_2e^{i\theta_2}}{x}.$$

The definition of Bessel functions as matrix elements then gives

$$e^{in\theta}J_n(x)=\sum_{k=-\infty}^{\infty}e^{ik\theta_2}J_{n-k}(x_1)J_k(x_2).$$

From this general addition formula, simpler ones follow by choices of parameter. For example, if $\theta_2=0$ then we obtain,

$$J_n(x_1+x_2)=\sum_{k=-\infty}^{\infty}J_{n-k}(x_1)J_k(x_2)$$

and for $\theta_2=\pi/2$, we obtain

$$\left(\frac{x_1+ix_2}{x_1-ix_2}\right)^{n/2}J_n\left(\sqrt{x_1^2+x_2^2}\right)=\sum_{k=-\infty}^{\infty}i^kJ_{n-k}(x_1)J_k(x_2).$$

Finally, we cite the *Jansen* formula,

$$\sum_{k=-\infty}^{\infty}i^kJ_{n+k}(x)J_k(x)=\delta_{n,0}.$$

7.3.8 The Hypergeometric Equation

The hypergeometric family of functions, $F(\alpha,\beta,\gamma;z)$, is universal for special functions with three regular singular points in the extended complex plane, that is, for Fuchsian equations (Section 4.2.5)[21, 22, 25]. Many special functions appear as special cases of this family.

Suppose we have a second-order Fuchsian ODE, in z and $w(z)$, with distinct regular singular points at $z=a,b$, and c, carrying exponent pairs (roots of the indicial equation in the Frobenius method), (α_1,α_2), (β_1,β_2) and (γ_1,γ_2). The Riemann P-symbol denotes the set of solutions to this ODE with these local properties:

$$P\left\{\begin{matrix} a & b & c & \\ \alpha_1 & \beta_1 & \gamma_1 & ;z \\ \alpha_2 & \beta_2 & \gamma_2 & \end{matrix}\right\}$$

The exponents of the solutions of the Fuchsian equation, and hence the P-symbol, are unaltered by Möbius maps on the independent z-variable, except for the locations of the singularities:

$$z \mapsto \frac{\lambda z + \mu}{\nu z + \rho},$$
$$a \mapsto a',\ b \mapsto b',\ c \mapsto c',$$

$$P\left\{\begin{matrix} a & b & c & \\ \alpha_1 & \beta_1 & \gamma_1 & ;z \\ \alpha_2 & \beta_2 & \gamma_2 & \end{matrix}\right\} =$$

$$P\left\{\begin{matrix} a' & b' & c' & \\ \alpha_1 & \beta_1 & \gamma_1 & ;z \\ \alpha_2 & \beta_2 & \gamma_2 & \end{matrix}\right\};$$

but under maps on the dependent variable

$$\tilde{w}(z) = \frac{(z-a)^k (z-b)^l}{(z-c)^{k+l}} w(z)$$

the symbol transforms as

$$\frac{(z-a)^k (z-b)^l}{(z-c)^{k+l}} P\left\{\begin{matrix} a & b & c & \\ \alpha_1 & \beta_1 & \gamma_1 & ;z \\ \alpha_2 & \beta_2 & \gamma_2 & \end{matrix}\right\}$$

$$= P\left\{\begin{matrix} a & b & c & \\ \alpha_1 + k & \beta_1 + l & \gamma_1 - k - l & ;z \\ \alpha_2 + k & \beta_2 + k & \gamma_2 - k - l & \end{matrix}\right\}$$

The hypergeometric differential equation is

$$z(1-z)w'' + (\gamma - (\alpha + \beta + 1)z)w' - \alpha\beta w = 0$$

which, it should be noted, is symmetric in α and β, and the hypergeometric function solution obtained by the Frobenius method,

$$F(\alpha, \beta, \gamma; z) = \sum_0^\infty \frac{(\alpha)_n (\beta)_n}{n!(\gamma)_n} z^n,$$

is valid for $|z| < 1$. The *Pochhammer symbols* $(\cdot)_n$ are defined by

$$(\alpha)_n = \frac{\Gamma(\alpha + n)}{\Gamma(\alpha)},$$
$$(\alpha)_0 = 1,$$

the gamma function being the analytic extension of the factorial function

$$\Gamma(z) = \int_0^\infty dt e^{-t} t^{z-1}, \quad \mathrm{Re}(z) > 0;$$
$$\Gamma(z+1) = z!, \quad z \in \mathbb{N}.$$

When $\gamma \notin \mathbb{Z}$, the second solution inside the open disk $|z| < 1$ is given by

$$z^{1-\gamma} F(\alpha - \gamma + 1, \beta - \gamma + 1, 2 - \gamma; z).$$

The solution corresponds to the Riemann P-symbol

$$P\left\{\begin{matrix} 0 & 1 & \infty & \\ 0 & 0 & \alpha & ;z \\ 1-\gamma & \gamma - \alpha - \beta & \beta & \end{matrix}\right\}.$$

Many elementary and special functions are associated with specific choices of the parameters. When $\gamma = \beta$, we recover (rational) binomials. Also, for example,

$$\arcsin z = zF\left(\frac{1}{2}, \frac{1}{2}, \frac{3}{2}; z^2\right);$$
$$\ln \frac{1+z}{1-z} = 2zF\left(\frac{1}{2}, 1, \frac{3}{2}, z^2\right);$$

and the Legendre polynomials,

$$P_n(z) = \frac{(2n)!}{2^n (n!)^2} z^n F\left(\frac{n}{2}, \frac{1-n}{2}, \frac{1}{2} - n, \frac{1}{z^2}\right).$$

A related class of function (including Bessel functions) have *irregular* singular points at infinity that can be controlled by allowing two of the regular singular points in the hypergeometric function to coalesce. These are the *confluent hypergeometric functions*.

The hypergeometric functions satisfy three-term recurrence relations:

$$(\gamma-1)F(\alpha,\beta,\gamma-1,z)-\alpha F(\alpha+1,\beta,\gamma,z)$$
$$-(\gamma-\alpha-1)F(\alpha,\beta,\gamma,z)=0$$

and

$$\gamma F(\alpha,\beta,\gamma,z)-\beta zF(\alpha,\beta+1,\gamma+1,z)$$
$$-\gamma F(\alpha-1,\beta,\gamma,z)=0.$$

A further class of generalization is the hypergeometric series in two sets of parameters,

$${}_rF_s\left(\begin{matrix}a_1\ \dots\ a_r\\ b_1,\dots\ b_s\end{matrix};z\right)=\sum_{n=0}^{\infty}\frac{(a_1,\dots,a_r)_n}{(b_1,\dots,b_s)_n}\frac{z^n}{n!}.$$

Here the symbols $(a_1,\dots,a_r)$, and so on, are defined as products of the Pochhammer symbols:

$$(a_1,\dots,a_r)_n=\prod_{i=1}^{r}(a_i)_n;$$

and the previously defined hypergeometric function corresponds to

$$F(\alpha,\beta,\gamma;z)={}_2F_1\left(\begin{matrix}\alpha\ ,\ \beta\\ \gamma\end{matrix};z\right).$$

7.3.9
Orthogonality

The generality of the classical special functions satisfy the Sturm–Liouville equations:

$$L(y)\equiv\frac{\mathrm{d}}{\mathrm{d}x}\left(k(x)\frac{\mathrm{d}y}{\mathrm{d}x}\right)+(\lambda g(x)-l(x))y=0.$$

$k(x)>0$, $l(x)$ and $g(x)>0$ are real continuous functions on an interval $[a,b]$, λ an eigenvalue, and we assume boundary conditions,

$$\alpha_1y(a)+\alpha_2y'(a)=0,$$
$$\beta_1y(b)+\beta_2y'(b)=0.$$

It can be shown that there is an unbounded, infinite set of eigenvalues $\{\lambda_i\}_0^\infty$ with eigenfunctions $\{y_n(x)\}_0^\infty$ such that $y_n(x)$ has exactly n zeros in the interval (a,b) [9, 22].

The differential operator is self-adjoint:

$$\int_a^b \mathrm{d}x\, uLv=\int_a^b \mathrm{d}x\, vLu.$$

From this follows an integral identity for eigenfunctions,

$$(\lambda_i-\lambda_j)\int_a^b \mathrm{d}x\, g(x)y_iy_j=0,$$

so that eigenfunctions of distinct eigenvalues are orthogonal with respect to the inner product

$$(u,v)=\int_a^b \mathrm{d}x\, g(x)u(x)v(x).$$

In this way, one obtains, for instance, for the associated Legendre functions, the orthogonality relations

$$\int_{-1}^{1}\mathrm{d}x\, P_n^m(x)P_r^m(x)=\frac{2}{2n+1}\frac{(n+m)!}{(n-m)!}\delta_{r,n}.$$

Provided $(y_i,y_i)>0$, we can normalize the eigenfunctions by positive constants to obtain an *orthonormal* set $\{y_n(x)\}_0^\infty$,

$$(y_i,y_j)=\delta_{i,j}.$$

7.3.10
Orthogonal Polynomials

Suppose we have a positive, Borel measure, $\mathrm{d}\mu(x)=g(x)\mathrm{d}x$, on $\mathbb{R}$ [21, 28–30]. We can always construct from it a sequence $\{\phi_n(x)\}_0^\infty$, of monic, *orthogonal polynomials* with respect to this measure in the sense that

$$\int_{\mathbb{R}} \mathrm{d}\mu\, \phi_n(x)\phi_m(x) = \zeta_n \delta_{m,n},$$

where the numbers $\{\zeta_n\}_0^\infty$ are all positive and $\zeta_0 = 1$.

We achieve this by defining *Hankel* determinants,

$$H_n = \det(\mu_{i+j-2})_{1\le i,j\le n}$$

the μ_j being moments of monomials with respect to $\mathrm{d}\mu(x)$:

$$\mu_j = \int_{\mathbb{R}} \mathrm{d}\mu(x)\, x^j.$$

Then define

$$\phi_n(x) = \frac{1}{\Delta_{n-1}} \begin{vmatrix} \mu_0 & \mu_1 & \cdots & \mu_n \\ \mu_1 & \mu_2 & \cdots & \mu_{n+1} \\ \vdots & & & \vdots \\ \mu_{n-1} & \mu_n & \cdots & \mu_{2n-1} \\ 1 & x & \cdots & x^n \end{vmatrix}$$

where Δ_n is the *Hankel determinant*,

$$\Delta_n = \begin{vmatrix} \mu_0 & \mu_1 & \cdots & \mu_n \\ \mu_1 & \mu_2 & \cdots & \mu_{n+1} \\ \vdots & & & \vdots \\ \mu_n & \mu_{n+1} & \cdots & \mu_{2n} \end{vmatrix}.$$

These polynomials constitute such a sequence with

$$\zeta_n = \frac{\Delta_n}{\Delta_{n-1}}.$$

The sequence also satisfies a three-term recurrence relation of the form

$$\phi_{n+1} + (\alpha_n - x)\phi_n + \beta_n \phi_{n-1} = 0,$$

the constants $\alpha_n \in \mathbb{R}$ and $\beta_n > 0$ for $n > 0$, depending on the measure.

Further, the converse holds (*the spectral theorem*): given such a sequence of monic, orthogonal polynomials, satisfying a recurrence relation of this form, there exists a measure such that $\{\phi_n(x)\}_0^\infty$, can be constructed as above with

$$\zeta_n = \beta_1\beta_2\cdots\beta_n.$$

Such orthogonal polynomials also satisfy second-order, linear ODE and differential-difference recurrence relations.

One example, amongst many topical in the current literature because of their q-deformed cousins is the sequence of Jacobi polynomials on $[-1, 1]$:

$$g(x) = \frac{(1-x)^\alpha(1+x)^\beta\Gamma(\alpha+\beta+2)}{2^{\alpha+\beta+1}\Gamma(\alpha+1)\Gamma(\beta+1)},$$

$$p_n(x) = \frac{P_n^{(\alpha,\beta)}(x)}{\sqrt{h_n^{(\alpha,\beta)}}},$$

$$h_n^{(\alpha,\beta)} = \frac{(\alpha+\beta+1)(\alpha+1)_n(\beta+1)_n}{(2n+\alpha+\beta+1)n!(\alpha+\beta+1)_n},$$

and

$$P_n^{(\alpha,\beta)}(x) = \frac{(\alpha+1)_n}{n!}\,{}_2F_1\left(\begin{matrix}-n, \alpha+\beta+n+1\\ \alpha+1\end{matrix}; x\right)$$

7.4 Factorization

The linear differential equations satisfied by special functions of the type described are all factorizable and this property provides a neat way to obtain functional properties of the special functions in question. Originally driven by the analogy with factorizing polynomial equations, this method is developed and its applications in quantum mechanics explored in [31].

7.4.1 The Bessel Equation

We have seen already that the Bessel operator factorizes up to a constant

$$\partial_x^2 + \frac{1}{x}\partial_x + (1 - \frac{m^2}{x^2}) =$$
$$(\partial_x - \frac{m-1}{x})(\partial_x + \frac{m}{x}) - 1.$$

Denoting the factors, as before (Section 7.3.6) by

$$D_m^{\pm} = \mp\partial_x + \frac{m}{x}$$

and the Bessel function, J_m, we have

$$D_{m-1}^{+}D_m^{-}J_m = D_{m+1}^{-}D_m^{+}J_m = J_m.$$

Because this holds for all m, it is the case that

$$D_m^{+}J_m = J_{m+1}$$
$$D_m^{-}J_m = J_{m-1},$$

which are recurrence relations for the Bessel functions. It is easy to obtain a generating function for them. Define

$$J(z) = \sum_{m\in\mathbb{Z}} z^m J_m(x).$$

Then, summing the recurrence formulae above over z^m,

$$\partial_x J(z) = \frac{1}{2}(z - \frac{1}{z})J(x)$$

and so

$$J(z) = \exp\left((z - \frac{1}{z})\frac{x}{2}\right).$$

7.4.2 Hermite

The wave function, ψ, of the quantum harmonic oscillator of a specified energy level (λ_l) obeys the equation,

$$(-\partial_x^2 + x^2)\psi_l = \lambda_l\psi_l.$$

We can write this in either of two factorized forms:

$$-(\partial_x - x)(\partial_x + x)\psi_l = (\lambda_l - 1)\psi_l;$$

and

$$-(\partial_x + x)(\partial_x - x)\psi_l = (\lambda_l + 1)\psi_l.$$

Writing $D^+ = \partial_x - x$ and $D^- = \partial_x + x$, and noting that

$$[D^-, D^+] = 2,$$

one has the system

$$D^-\psi_l = (\lambda_l - 1)\psi_{l-1}$$
$$D^+\psi_{l-1} = (\lambda_l + 1)\psi_l.$$

If consequently, for each value of l, ψ_l satisfies the differential equation with eigenvalue λ_l, these eigenfunctions are related by raising (*creation*) and lowering (*annihilation*) operators, D^- and D^+ and by the eigenvalue relation $\lambda_{l+1} = \lambda_l + 2$.

Assume there is a highest weight vector, ψ_0, in the kernel of D^-. Then $D^-\psi_0 = 0$ implies $\psi_0 = e^{-1/2x^2}$ and $\lambda_0 = 1$. Application of D^+ l times to ψ_0 will yield a product of a polynomial $H_l(x)$ and the Gaussian exponential:

$$\psi_l = (D^+)^l\psi_0 = H_l(x)e^{-x^2/2},$$

an eigenfunction of eigenvalue $2l + 1$.

The $H_l(x)$ are the *Hermite polynomials*, which satisfy the ODE

$$H_l'' - 2xH_l' + 2lH_l = 0$$

obtained from the oscillator via the above substitution for ψ_l.

Note that as operators we may write

$$D^+ = e^{x^2/2}\partial_x e^{-x^2/2},$$

so that

$$\begin{aligned}(D^+)^l &= (e^{x^2/2}\partial_x e^{-x^2/2})\cdots(e^{x^2/2}\partial_x e^{-x^2/2})\\ &= e^{x^2/2}\partial_x^l e^{-x^2/2}.\end{aligned}$$

Combining these, we get a formula for the lth Hermite polynomial:

$$H_l(x) = e^{x^2}\partial_x^l(e^{-x^2}).$$

Using ψ_{l-1}, ψ_l, and ψ_{l+1} to eliminate derivatives in the system for ψ_l, one replaces the second-order differential equation by a three-term recurrence relation for the $H_l(x)$:

$$H_{l+1} - 2xH_l + 2lH_{l-1} = 0.$$

Such a relation is the basis for a *continued fraction* expansion. Put $h_l = \frac{H_l}{H_{l-1}}$ so that the recurrence relation becomes

$$h_{l+1} = 2x - \frac{2l}{h_l}.$$

Then

$$h_l = 2x - \frac{2(l-1)}{2x - \frac{2(l-2)}{2x - \cdots}}.$$

Finally, we may obtain a generating function. Start with the system form for the H_l,

$$\begin{aligned}H_l' &= 2lH_{l-1},\\ H_l' - 2xH_l &= -H_{l+1},\end{aligned}$$

which corresponds to the factorizations, up to constants, of the H-equation,

$$\partial_x(\partial_x - 2x)H_l = -2(l+1)H_l$$

and

$$(\partial_x - 2x)\partial_x H_l = -2lH_l.$$

The generating function

$$H(x,t) = \sum_{l=0}^{\infty}\frac{t^l}{l!}H_l(x)$$

is seen to satisfy the compatible pair of PDEs

$$\begin{aligned}H_x &= 2tH,\\ H_t &= 2(x-t)H,\end{aligned}$$

whose solution is

$$H(x,t) = e^{-t^2+2xt}.$$

7.4.3 Legendre

In this case, introduce the linear differential operators [32]

$$\begin{aligned}L &= e^{i\psi}\left(\partial_\theta + i\cot\theta\partial_\psi)\right)\\ L^* &= -e^{-i\psi}\left(\partial_\theta - i\cot\theta\partial_\psi)\right)\\ M &= \frac{1}{i}\partial_\psi.\end{aligned}$$

Then

$$\begin{aligned}LL^* &= -\Delta - M(M-1)\\ L^*L &= -\Delta - M(M+1),\end{aligned}$$

where

$$\Delta = \partial_\theta^2 + \cot\theta\partial_\theta + \frac{1}{\sin^2\theta}\partial_\psi^2$$

is the Laplacian in spherical coordinates on the sphere (constant r).

The operators $\{L, L^*, M\}$ form a representation of the Lie algebra $su(2)$ of the Lie group $SU(2)$ of unitary, 2×2 matrices of unit determinant, as is seen from the commutation relations:

$$[M, L] = L,$$
$$[M, L^*] = -L^*,$$
$$[L, L^*] = 2M.$$

The equivalence is given by the correspondence

$$L \sim \begin{pmatrix} 0 & i \\ 0 & 0 \end{pmatrix}$$
$$L^* \sim \begin{pmatrix} 0 & 0 \\ -i & 0 \end{pmatrix}$$
$$M \sim \frac{1}{2} \begin{pmatrix} 1 & 0 \\ 0 & -1 \end{pmatrix}$$

One checks that the generators commute with the Laplacian:

$$[L, \Delta] = [L^*, \Delta] = 0,$$
$$[M, \Delta] = 0.$$

In fact, in this sense, the Δ operator is a *Casimir element* of the universal enveloping algebra of $su(2)$: it commutes with every generator and its eigenvalues, $l(l+1)$, indicate the dimension d of a (finite-dimensional) representation via the formula,

$$l(l+1) = \frac{1}{4}(d^2 - 1).$$

Because Δ and M commute, one may form a pair of mutually consistent differential equations

$$MY_l^m = mY_l^m,$$

and

$$\Delta Y_l^m = -\lambda(l) Y_l^m,$$

defining functions Y_l^m where the dependence of λ on l is to be determined.

It is easy to show, using the commutation relations, that the function LY_l^m satisfies

$$M(LY_l^m) = (m+1)LY_l^m,$$
$$\Delta(LY_l^m) = -\lambda(l)(LY_l^m),$$

meaning that L plays the rôle of a *raising* operator:

$$LY_l^m = Y_l^{m+1}.$$

In a similar manner, L^* is a *lowering* operator:

$$L^* Y_l^m = Y_l^{m-1}.$$

If we assume that the set $\{Y_l^m | l' \leq m \leq l\}$ forms the basis of a finite-dimensional representation of $su(2)$, then we will have a highest weight element Y_l^l satisfying

$$LY_l^l = 0.$$

Using the identity for L^*L yields

$$\lambda(l) = l(l+1).$$

From the theory of finite-dimensional representations, we conclude that the set $\{Y_l^m | -l \leq m \leq l\}$ is the basis for an irreducible representation of $su(2)$ of (odd) dimension $2l+1$. These are called integer spin representations. Representations of even dimension (half integer spin) require *spinors* [33].

Solving the defining equations for the $Y_l^m(\theta, \psi)$ gives (up to normalization) the *spherical harmonics*

$$Y_l^m(\theta, \psi) = e^{im\psi} P_l^m(\theta).$$

The $P_l^m(\theta)$ are *associated Legendre functions* and satisfy the associated Legendre equation:

$$\frac{1}{\sin\theta}\partial_\theta(\sin\theta\partial_\theta P_l^m(\theta)) +$$
$$(l(l+1) - \frac{m^2}{\sin^2\theta})P_l^m(\theta) = 0$$

7.4.4 "Factorization" of PDEs

The idea of factorization finds a general context in the theory of the *linear* PDEs that effect the inverse scattering transform for integrable *nonlinear* PDEs [34–36]. The special character of the *soliton* solutions to such integrable equations allows us to think of them as nonlinear special functions.

Given a general second-order PDE of hyperbolic type in two independent variables and in canonical form,

$$L\phi \equiv (\partial_x\partial_y + a\partial_x + b\partial_y + c)\phi = 0,$$

a and b being functions of x and y, we can attempt a natural factorization in two ways:

$$((\partial_x + b)(\partial_y + a) - h)\phi = 0,$$
$$((\partial_y + a)(\partial_x + b) - k)\phi = 0.$$

The functions

$$h = ab + a_x - c$$
$$k = ab + b_y - c$$

are invariants under linear scalings of ϕ :

$$\phi \mapsto \lambda(x,y)\phi$$

and they measure the obstruction to factorizing the linear differential operator. Vanishing of either allows the reduction of the second-order equation to triangular form and the general integration problem to a pair of quadratures.

The invariants classify operators up to scaling maps: two distinct operators sharing the same invariants are necessarily related by such a scaling map.

From each of the forms, we can define new independent variables

$$\phi^\sigma = (\partial_y + a)\phi$$

and

$$\phi^\Sigma = (\partial_x + b)\phi$$

(the *Laplace transforms* of ϕ) satisfying

$$(\partial_x + b)\phi^\sigma = h\phi$$

and

$$(\partial_y + a)\phi^\Sigma = k\phi.$$

Eliminating ϕ for ϕ^σ or ϕ^Σ yields in each case a new equation of the original type but with redefined parameters:

$$a^\sigma = a - (\ln h)_y, \quad b^\sigma = b$$
$$h^\sigma = 2h - k - (\ln h)_{xy}, \quad k^\sigma = h$$

and

$$a^\Sigma = a, \quad b^\Sigma = b - (\ln k)_x$$
$$h^\Sigma = k, \quad k^\Sigma = 2k - h - (\ln k)_{xy}.$$

One checks straightforwardly that, for the invariants,

$$(k^\sigma)^\Sigma = (k^\Sigma)^\sigma,$$
$$(h^\sigma)^\Sigma = (h^\Sigma)^\sigma.$$

However the (noninvariant) coefficients are *not* well behaved: $(a^\sigma)^\Sigma \neq a$, and so on. This is to be expected as scaling transformations cannot alter the factorization properties of the second-order operator even though they alter coefficients.

Repetitions of Laplace maps generate sequences of pairs of invariants. So we label them

$$h_n = h^{\sigma^n}, \quad k_n = k^{\sigma^n}$$

subject to the understanding that

$$h^{\sigma^{-1}} = h^{\Sigma}$$

and so on. Then we may write three-term recurrence relations for the h_n and k_n :

$$h_{n+1} - 2h_n + h_{n-1} = (\ln h_n)_{xy}$$
$$k_{n+1} - 2k_n + k_{n-1} = (\ln k_n)_{xy}$$

and these are a particular form of the two-dimensional Toda field equations studied in the theory of integrable systems [37, 38].

The family of invariants is associated with a rank-one lattice, which may be infinite, semi-infinite or finite depending on whether there are vanishing invariants for some value(s) of n.

Suppose we consider the formal adjoint of the given second-order differential operator, L:

$$L^{\dagger} = \partial_x\partial_y - a\partial_x - b\partial_y + c - a_x - b_y.$$

Its invariants are

$$h^{\dagger} = k, \quad k^{\dagger} = h.$$

and it is easy to check that

$$h^{\dagger\sigma} = h^{\Sigma\dagger}, \quad k^{\dagger\sigma} = k^{\Sigma\dagger}$$

so that the operations $\sigma = \Sigma^{-1}$ and $\dagger$ generate the *infinite dihedral* group:

$$D_{\infty} = < \dagger, \sigma | \sigma \dagger \sigma = \dagger, \dagger^2 = id > .$$

If we start with a second-order hyperbolic operator that is self-adjoint, then the sequence, or one-dimensional lattice, of invariants generated by Laplace maps is symmetric in the sense that,

$$h_n = h_{-n}, \quad k_n = k_{-n}.$$

If at some point an invariant vanishes, then the lattice is finite with an odd number of sites, the central site corresponding to the self-adjoint operator – a situation reminiscent of the Legendre equation with eigenvalue $l(l+1)$.

In such situations (and others where a modified symmetry obtains), there is a natural map between lattices of different lengths. If L is self-adjoint then $h = k$ and it can, by a scaling transformation, be written in the form

$$L = \partial_x\partial_y + h = \partial_x\partial_y + k.$$

Assume ϕ_0 is a function in the kernel of this operator and consider the coupled system

$$(\partial_x + \phi_0^{-1}\phi_{0x})\phi^{\mu} = -(\partial_x - \phi_0^{-1}\phi_{0x})\phi$$
$$(\partial_y + \phi_0^{-1}\phi_{0y})\phi^{\mu} = (\partial_y - \phi_0^{-1}\phi_{0y})\phi$$

for a pair of functions ϕ and ϕ^{μ}. Elimination of ϕ^{μ} gives the original equation

$$(\partial_x\partial_y + h)\phi = 0,$$

but elimination of ϕ gives

$$(\partial_x\partial_y + h + 2(\ln\phi_0)_{xy})\phi^{\mu} = 0,$$

a new, self-adjoint equation with invariant

$$h^{\mu} = k^{\mu} = h + 2(\ln\phi_0)_{xy},$$

ϕ_0 being a solution of the original self-adjoint equation. This so-called *Moutard* [35] transformation is different in character to the Laplace transformation because it takes us outside the differential algebra generated by the invariants alone. In general, the lattice associated with h^μ will have a different character to that associated with h.

The Moutard transformation can be reduced to a one-dimensional situation by assuming a symmetry of the form $h(x, y) = h(x + y)$ and correspondingly $\phi(x, y) = e^{\lambda(x-y)}\psi(x + y)$. Then ψ satisfies

$$\psi'' + h\psi = \lambda^2\psi.$$

Following the previous prescription, set

$$(\partial - \alpha)\psi = \psi^\delta,$$
$$(\partial + \alpha)\psi^\delta = \lambda^2\psi$$

with

$$-\alpha' - \alpha^2 = h - \mu^2 + \lambda^2.$$

It follows that $\alpha = \psi_0'/\psi_0$ where

$$\psi_0'' + h\psi_0 = \mu^2\psi_0$$

and ψ^δ satisfies

$$\psi^{\delta\prime\prime} + h^\delta\psi^\delta = \lambda^2\psi^\delta$$

with

$$h^\delta = h + 2(\ln\psi_0)''.$$

This is the *Darboux* transformation, a map of enormous interest in the theory of soliton equations [39].

If we regard soliton solutions as special functions, parameterized by positions, momenta, and soliton number, for nonlinear equations then Darboux transformations are analogous to the raising operators in the classical theory.

7.4.5 Dunkl Operators

Dunkl operators provide an approach to multivariable generalizations of the classical special functions via finite reflection groups [40–42]. Given such a group, W, with positive roots $\Delta_+ \subset \mathbb{R}^N$, so

$$W =< \sigma_\alpha | \alpha \in \Delta_+ >,$$

the Dunkl operators T_i are differential-difference operators acting on functions, $f : \mathbb{R}^N \to \mathbb{R}$:

$$T_i f(x) = \partial_{x_i} f(x) + \sum_{\alpha\in\Delta_+} k_\alpha \frac{f(x) - f(\sigma_\alpha(x))}{< x, \alpha >}\alpha_i,$$

where $x \in \mathbb{R}^N$, α_i denotes the ith component of α, k_α is a *multiplicity* function (the cardinality of the conjugacy class of σ_α in W), and $< \cdot, \cdot >$ denotes the inner product on $\mathbb{R}^N$.

For example, in the case of the root system A_{N-1}, presented earlier (Section 7.2.4),

$$T_i = \partial_{x_i} + k\sum_{i\neq j}\frac{1 - \sigma_{(ij)}}{x_i - x_j},$$

$\sigma_{(ij)}$ acting on f by interchange of the ith and j^{th} arguments.

A crucial property of the Dunkl operators is that they commute:

$$[T_i, T_j] = 0.$$

The Dunkl Laplacian is defined to be

$$\Delta_k = \sum_{i=1}^{N} T_i^2$$

and the generalized special functions in mind are harmonic with respect to this Laplacian.

Thus for the example of A_{N-1},

$$\Delta_k = \Delta + 2k \sum_{i<j} \frac{1}{x_i - x_j} \left(\partial_{x_i} - \partial_{x_j} - \frac{1 - \sigma_{(ij)}}{x_i - x_j} \right).$$

For A_1 we have the expression,

$$\partial^2 + \frac{2k}{x} \partial,$$

and so Δ_k generalizes to many variables the classical (spherical) Bessel function.

Important players in the theory are the intertwining operators, V_k :

$$T_i V_k = V_k \partial_{x_i}$$

which, like the T_i themselves are W-equivariant, and the Dunkl kernel, $E(\cdot, y)$, the unique solution to the set of PDE,

$$T_i f = y_i f, \quad i = 1, \ldots, N.$$

In terms of E, the generalized Bessel function is defined, for $x, y \in \mathbb{R}^N$, as

$$J_k(x, y) = \frac{1}{|W|} \sum_{g \in W} E_k(gx, y).$$

7.5 Special Functions Without Symmetry

In this section, we present theories of a more general character of which differential Galois theory is arguably the most inclusive. It would be invidious to claim that there is no debt to symmetry but its rôle is less obviously geometric.

7.5.1 Airy Functions

If a singular point of an ODE is not regular, then it is called *irregular* and *if* a (complex) solution of the form

$$y(x) = \mathrm{e}^{Q(x)} w(x)$$

exists near the singular point (assumed to be $x = 0$), $Q(x)$ being polynomial in $1/x$ and $w(x)$ being regular, then it is called a *normal solution* [21, 43–45]. Such solutions need not exist. Note that even so simple an equation as a linear ODE with constant coefficients falls into this class, the point at infinity being an irregular singular point and $Q(x)$ linear in x. The normal solution has an essential singularity.

If the equation has a normal solution in the variable $x^{1/k}$ for some positive integer $k > 1$ (*Puiseux expansion*), then the solution is called *subnormal.*

An example of such a situation is the Airy equation,

$$y''(x) - xy(x) = 0,$$

which has an irregular singular point at infinity. Linearly independent solutions are denoted

$$Ai(x), \quad Bi(x).$$

They have Puiseux expansions expressed via fractional Bessel functions:

$$Ai(-x) = 1/3\sqrt{x}\left(J_{1/3}(2/3x^{3/2}) + J_{-1/3}(2/3x^{3/2})\right);$$
$$Bi(-x) = 1/3\sqrt{x}\left(J_{-1/3}(2/3x^{3/2}) - J_{1/3}(2/3x^{3/2})\right).$$

There are also integral representations:

$$(3a)^{-1/3} \pi Ai(\pm(3a)^{-1/3} x) = \int_0^\infty \mathrm{d}t \, \cos(at^3 \pm xt);$$

and

$$(3a)^{-1/3} \pi Bi(\pm(3a)^{-1/3} x) = = \int_0^\infty \mathrm{d}t \left(\sin(at^3 \pm xt) + \mathrm{e}^{-at^3 \pm xt} \right).$$

7.5.1.1 Stokes Phenomenon

This is an important feature of functions with essential singularities. The coefficients in the Airy equation being entire on $\mathbb{C}$ and the equation being linear, it follows that the general solution is likewise entire on $\mathbb{C}$. However, asymptotic representations of the solutions,

$$y_{\pm}(x) \approx x^{-1/4} e^{\pm 2/3x^{3/2}},$$
$$y''_{\pm} - xy \approx x^{-9/4} e^{\pm 2/3x^{3/2}},$$

are multivalued,

$$y_{\pm} \to i y_{\mp},$$

as x traverses a closed loop about $x = 0$ (equivalently, $x = \infty$.)

Note that unlike the Fuchsian case where the multivaluedness is a genuine property of the solution near a regular singular point, *this* behavior is an artifact of using multivalued approximations near an *essential* singularity. One avoids the confusion created by this representation dependence by allocating representations to specific sectors of $\mathbb{C}$ separated by *Stokes lines.*

In the case of the Airy functions, the asymptotic forms are

$$Ai(x) \equiv \frac{1}{2}\pi^{-1/2} x^{-1/4} e^{2/3x^{3/2}},$$

for $|\arg x| < \pi$, and

$$Ai(-x) \equiv \pi^{-1/2} x^{-1/4} \sin\left(\frac{2}{3}x^{3/2} + \frac{\pi}{4}\right)$$

for $|\arg x| < 2/3\pi$.

7.5.2 Liouville Theory

Just as in number theory there are hierarchies of number system,

$$\mathbb{N} \subset \mathbb{Z} \subset \mathbb{Q} \subset \cdots \subset \mathbb{R},$$

so in function theory too it is helpful to distinguish *elementary* functions from *higher* functions so that we can answer the question as to whether a given differential equation has solutions that cannot be defined by more elementary means [46, 47].

In the case of number theory, the positive integers, $\mathbb{N}$, taken as a starting point, allow us to develop the full set of integers $\mathbb{Z}$ and the rationals $\mathbb{Q}$ as extensions that provide solutions to equations

$$z + n = 0, \quad n \in \mathbb{N},$$

and

$$az + b = 0, \quad a, b \in \mathbb{Z}, \ a \neq 0.$$

Algebraic numbers are defined as real solutions to polynomial equations over $\mathbb{Q}$ in one variable,

$$f(z) = 0, \quad f \in \mathbb{Q}[z].$$

Beyond this point, completions with respect to some norm construct the real, $\mathbb{R}$, and p-adic, Ω_p, numbers. We further extend to complex versions of any of these systems by allowing solutions to the quadratic equation,

$$z^2 + 1 = 0.$$

For functions, we may proceed in an algebraically similar manner. Certain operations are taken as elementary: the construction of an algebraic function, that is, a function $y(x)$ satisfying an equation polynomial in x and y; the exponentiation of a function, that is, solution of the first-order ODE, $y' + f'y = 0$; the taking of a logarithm, that is, solution of an equation $e^y = f(x)$.

In his theory of functions, Liouville defines an *elementary function* as one constructed by a finite sequence of such operations and the *order* of such a function as the minimal number of exponentiations or taking of logarithms in that process. Algebraic operations are neutral: they do not affect the order.

We now give some applications of Liouville's approach to special functions.

A theorem of Abel [48] states that if the integral, along some path in $\mathbb{C}$, of an algebraic function,

$$\int \mathrm{d}x\, y(x),$$

is itself algebraic, then it is necessarily rational in x and y. We can use this to show that the Jacobi elliptic functions (Section 7.6.2) are not elementary. Assume that y is defined by

$$y^2 = (1 - x^2)(1 - k^2x^2).$$

If the integral of y is rational, then according to Abel's result, it must be of the form,

$$\int \frac{\mathrm{d}x}{y(x)} = A(x) + B(x)y,$$

A and B being rational in x. Differentiation with respect to x allows us to write yA' as a function rational in x : a contradiction, unless $A = 0$. Further analysis of the possible poles in the relation leads to a similar contradiction. Hence the Jacobi elliptic functions are not elementary.

Consider secondly the Gaussian integral,

$$\operatorname{erf}(x) = \int_{-\infty}^{x} \mathrm{d}t\, \mathrm{e}^{-t^2}.$$

On the basis of Liouville's theory, it must be of the form

$$w(x)\mathrm{e}^{-x^2} + \text{const},$$

for a rational $w(x)$, if it is to be elementary at all. But then w must satisfy the ODE,

$$1 = w' - 2x,$$

and it is straightforward to show that such a w has no poles either in the finite part of $\mathbb{C}$ or at ∞. Hence, by the Liouville theorem of complex analysis, it is constant and we have a contradiction: the erf function is not elementary.

Finally, we will show that Bessel functions are not in general elementary in an extended sense (Liouville). This requires a result concerning Riccati equations:

$$y' + y^2 = P(x),$$

$P(x)$ being algebraic. We generalize the notion of elementary functions by allowing integrals. Since the logarithm of an algebraic function is also the integral of an algebraic function, we loosen the previous definition by allowing more general integrals of algebraic functions and call the class of extended elementary functions *Liouville functions*. The order is defined as before but replacing the word "logarithm" by "integral."

The result we require is that *if* the given Riccati equation has a particular solution that is Liouville, *then* it has a particular solution that is algebraic.

In the case of Bessel's equation (Section 7.3.6), the transformation $y = u/\sqrt{x}, x = iz$ yields

$$\frac{\mathrm{d}^2u}{\mathrm{d}z^2} = \left(1 + \frac{l(l+1)}{z^2}\right) u,$$

for $l = n - 1/2$ and this is reduced to a Riccati equation,

$$\frac{\mathrm{d}v}{\mathrm{d}z} + v^2 = 1 + \frac{l(l+1)}{z^2},$$

by the differential substitution $v = u'/u$. (In fact, any second-order, linear, homogeneous ODE may be reduced to a Riccati equation by a similar substitution.)

To settle the question of whether the Bessel equation has Liouvillian solutions, we need to see whether it has algebraic solutions or not for given values of l. Using expansions of v, one firstly determines that any algebraic solution is rational in z. Secondly, it is shown that a rational solution exists if and only if l is integral. Consequently, the Bessel equation admits a nontrivial Liouvillian solution (and hence all solutions are Liouville) if and only if $2n$ is an odd integer.

7.5.3 Differential Galois Theory

Strictly speaking, the differential Galois theory *is* concerned with questions of symmetry but not in the geometrical sense described earlier. It is also a development of the Liouville theory [11, 49–52].

It was from the question of the solvability of polynomial equations in a single variable that group theory originally arose. *Solvability* here means obtaining expressions for solutions in terms of rational operations augmented by the extraction of simple algebraic roots. One associates with an irreducible polynomial with coefficients in a field $\mathfrak{F}$, $p(x) \in \mathfrak{F}[x]$, the group of automorphisms, G, of the field $\mathfrak{L} = \mathfrak{F}[x]/(p)$, where (p) is the ideal generated by $p(x)$, which fix $\mathfrak{F}$ elementwise. If $\mathfrak{F}$ is *exactly* the set of points fixed by G, then $\mathfrak{L}$ is called a *normal extension* of $\mathfrak{F}$ and G is the *Galois group*, $Gal(\mathfrak{L}/\mathfrak{F})$ [53].

A group is said to have a *normal series* if it has a finite sequence of subgroups, G_i,

$$G = G_0 \supset G_1 \supset G_2 \supset \cdots \supset G_n = \{e\}$$

such that each G_i is normal in G_{i-1}. G is *solvable* if it has a normal series where each quotient G_{i+1}/G_i is abelian.

The fundamental result of Galois theory is that the equation $p(x) = 0$ is solvable by radicals over $\mathfrak{F}$ if and only if the Galois group $Gal(\mathfrak{L}/\mathfrak{F})$ is solvable. The fact that the symmetric groups S_n for $n \geq 5$ are not solvable accounts for the fact that there is no generic formula for extracting the roots of polynomial equations of degree five or more.

One thinks of the Galois group as a symmetry group "permuting" the elements of a set while fixing those of a subset. It was the link with solvability that inspired Lie [13] to develop a group theory for continuous symmetries (already discussed) and Picard and Vessiot to develop a differential theory closer in spirit to the Galois theory.

In *differential* Galois theory, $\mathfrak{F}$ is taken to be a differential field, that is, it is a field with *derivation*

$$D : \mathfrak{F} \to \mathfrak{F},$$
$$D(a + b) = D(a) + D(b),$$
$$D(ab) = aD(b) + D(a)b.$$

Homomorphisms, in particular automorphisms, between differential fields,

$$\phi : \mathfrak{F} \to \mathfrak{K},$$

are required to satisfy the property,

$$\phi D_{\mathfrak{F}} = D_{\mathfrak{K}} \phi.$$

The analogue of the polynomial equation in Galois theory is a linear differential equation and the analogue of $\mathfrak{L}$, is the corresponding *Picard–Vessiot* extension. This is the differential field generated by a full set of linearly independent (over constants) solutions of the differential equation, say,

$\{y_1, y_2, \ldots, y_n\}$ for a differential equation of order n. Such a set of solutions must have nonvanishing Wronskian:

$$\begin{vmatrix} y_1 & y_2 & \cdots & y_n \\ D(y_1) & D(y_2) & \cdots & D(y_n) \\ \vdots & & & \vdots \\ D^{n-1}(y_1) & D^{n-1}(y_2) & \cdots & D^{n-1}(y_n) \end{vmatrix} \neq 0.$$

We write the Picard–Vessiot extension as

$$\mathfrak{F}\langle y_1, y_2, \ldots, y_n\rangle.$$

It consists of rational, differential functions of the y_i with coefficients in the ground field, $\mathfrak{F}$, subject to the constraint of the linear differential equation. It is important that this field extension of $\mathfrak{F}$ contain *no new constants* and that the field of constants ($ker(D)$) is algebraically closed. For such extensions, the Galois group of differential automorphisms fixing $\mathfrak{F}$ is defined in analogy to the polynomial case. If the differential Galois group has a certain "solvability" property, then the Picard–Vessiot extension can be decomposed into a sequence of subfields, $\mathfrak{L}_{i-1} \supset \mathfrak{L}_i$, each of which is a simple extension of the subsequent one in the sense that, à la Liouville theory, it can be constructed by adjoining solutions of either $D(y) + ay = 0$ or $D(y) + a = 0$, where $a \in \mathfrak{L}_i$.

As an example, consider the (Fuchsian) *Cauchy equation*,

$$x^2y'' + axy' + by = 0,$$

a and b belonging to the field of constants $C \subset \mathfrak{F}$.

Let m_1 and m_2 be the roots of

$$m^2 + (a-1)m + b = 0.$$

If we define functions y_1 and y_2 to be linearly independent solutions of

$$xy_i' = m_iy_i, \quad 1 = 1, 2,$$

then each satisfies the Cauchy equation.

In the case that $m_1 \neq m_2$, the Picard–Vessiot extension is the differential field, $\mathfrak{L} = \mathfrak{F}(y_1, y_2)$, C being the field of constants, the differential Galois group will be $C \times C$ acting multiplicatively and diagonally,

$$y_1 \mapsto c_1y_1, \quad y_2 \mapsto c_2y_2,$$

unless y_1 and y_2 are algebraically dependent (satisfy a polynomial equation $f(y_1, y_2) = 0$ over $\mathfrak{F}$), in which case the Galois group is a proper subgroup of $C \times C$.

In the case that $m_1 = m_2 = m$, $\mathfrak{L} = \mathfrak{F}(y, u)$ where y and u are solutions to

$$xy' = my, \quad xu' = 1,$$

and the differential Galois group is $C \times C'$, the second factor acting additively,

$$y \mapsto cy, \quad u \mapsto u + d.$$

Obviously, these are all solvable cases. If $C = \mathbb{C}$, then the differential Galois group will be isomorphic to the monodromy group (Section 7.2.5).

7.6 Nonlinear Special Functions

7.6.1 Weierstraß Elliptic Functions

One encounters the Weierstraß $\wp$-function as soon as one departs from the linear oscillator model, that is, as soon as the amplitude of oscillations can no longer be regarded as "small." Historically, the $\wp$-function arose from exact analytic treatments of mechanical systems [21, 54–56].

The second-order linear equation

$$y'' + y = 0$$

is the equation of simple harmonic motion, which governs small oscillations. Integrating it once, we can think of $\sin\theta$ and $\cos\theta$ functions as solutions of the ODE $y'^2 = 1 - y^2$ under appropriate boundary conditions. The most natural form of solution is

$$\theta = \int_0^{\sin\theta} \frac{dy}{\sqrt{1-y^2}},$$

that is, we define by the integral the function inverse to sin. The integral is taken around a path in the complex plane avoiding a cut from -1 to 1 . A closed path around the cut returns to its starting point but θ is augmented by the amount

$$2\int_0^1 \frac{dy}{\sqrt{1-y^2}} = 2\pi,$$

the periodicity of the circular function.

An obvious simple nonlinear generalization of this situation is

$$y'' = 6y^2 - \frac{1}{2}g_2.$$

(The prefactor of 6 reflects a choice of canonical form and g_2 is a constant.) Integrating yields

$$y'^2 = 4y^3 - g_2 y - g_3,$$

g_3 another constant, which is an equation defining, in the complex plane, a doubly periodic function with a double pole that can be chosen to lie at the origin. This is the Weierstraß $\wp$-function. As with circular functions, we can present an inverse definition:

$$u = \int_\infty^{\wp(u)} \frac{dy}{\sqrt{4y^3 - g_2 y - g_3}},$$

or, equivalently,

$$\wp'^2 = 4\wp^3 - g_2\wp - g_3.$$

This time, there are cuts between a pair of roots of the cubic in y and between the third root and the point at infinity. Integrating around closed loops in the plane avoiding these cuts augments the variable u by points in a lattice $\Lambda = 2\omega_1\mathbb{Z} + 2\omega_2\mathbb{Z}$, the ω_i being two fundamental complex half-periods (linearly independent over $\mathbb{R}$) of the $\wp$-function.

The $\wp$-function can also be constructed as an infinite sum,

$$\wp(u) = \frac{1}{z^2} + \sum_{m,n}{}' \left(\frac{1}{(z-\omega_{m,n})^2} - \frac{1}{\omega_{m,n}^2} \right),$$

$\omega_{m,n}$ being defined as

$$\omega_{m,n} = 2m\omega_1 + 2n\omega_2$$

and the primed sum denoting omission of the term $\omega_{0,0}$.

The parameters in the differential equation above, satisfied by the $\wp$-function, are related to the lattice parameters via infinite sums:

$$g_2 = 60 \sum_{m,n}{}' \frac{1}{\omega_{m,n}^4};$$

$$g_3 = 140 \sum_{m,n}{}' \frac{1}{\omega_{m,n}^6};$$

As for circular functions, there is a (nonlinear) addition law:

$$\wp(u) + \wp(v) + \wp(u+v) = \frac{1}{4}\left(\frac{\wp'(u) - \wp'(v)}{\wp(u) - \wp(v)}\right)^2$$

In more symmetric form,

$$\begin{vmatrix} 1 & \wp(u) & \wp'(u) \\ 1 & \wp(v) & \wp'(v) \\ 1 & \wp(u+v) & -\wp'(u+v) \end{vmatrix} = 0$$

Geometrically, the addition law expresses the fact that the genus one, plane cubic curve

$$z^2 = 4y^3 - g_2 y - g_3$$

is parameterized by the choice $(y, z) = (\wp, \wp')$ and the point corresponding to parameter $u + v$ is the reflection in the x-axis of the third intersection point with the cubic of a straight line through those points corresponding to u and v.

Thus let (y_1, z_1) and (y_2, z_2) be a pair of points on the cubic curve. The straight line passing through these points is

$$z = \alpha + \beta y,$$

where

$$\alpha = \frac{y_1 z_2 - y_2 z_1}{y_1 - y_2}$$

and

$$\beta = \frac{z_1 - z_2}{y_1 - y_2}.$$

The third point of intersection (y_3, z_3) arises from solving

$$4y^3 - g_2 y - g_3 - (\alpha + \beta y)^2 = 4(y - y_1)(y - y_2)(y - y_3) = 0,$$

so that

$$y_1 + y_2 + y_3 = \frac{1}{4}\beta^2.$$

The construction is that the "sum" of $(y_1, z_1$ and (y_2, z_2) is $(y_3, \bar{z}_3)$ where $\bar{z} = -z$ is the reflection of z in the y-axis. So

$$y_3 = -y_1 - y_2 + \frac{1}{4}\left(\frac{z_1 - z_2}{y_1 - y_2}\right)^2,$$

and the expression

$$\bar{z}_3 = -\alpha - \beta y_3$$

simplifies to the determinantal form given above under the identifications of y with $\wp$ and z with $\wp'$.

Any doubly periodic function on $\mathbb{C}$ must have poles. Otherwise, by Liouville's theorem, it would be constant. All such functions form a differential field generated by $\wp(u)$: any such function can be written as

$$A(\wp) + B(\wp)\wp'$$

where A and B are rational functions over $\mathbb{C}$.

The $\wp$-function is the primary example of an *abelian function*: it is meromorphic, single-valued, and (doubly) periodic.

Two useful, but nonperiodic, functions are the ζ and the Weierstraß σ functions:

$$\wp(u) = -\frac{d}{du}\zeta(u),$$
$$\zeta(u) - \frac{1}{u} \to 0, \ u \to 0;$$
$$\zeta(u) = \frac{d}{du}\ln\sigma(u),$$
$$\frac{\sigma(u)}{u} \to 1, \ u \to 0.$$

Thus,

$$\wp(u) = -\frac{d^2}{du^2}\ln\sigma(u).$$

The ζ-function fails to be periodic:

$$\zeta(u + 2\omega_i) = \zeta(u) + 2\eta_i, \quad i = 1, 2$$

where $\eta_i = \zeta(\omega_i)$ and

$$\eta_1\omega_2 - \eta_2\omega_1 = \frac{1}{2}i\pi.$$

The σ function is likewise not periodic,

$$\sigma(u + 2\omega_i) = -e^{2\eta_i(u+\omega_i)}\sigma(u),\ i = 1, 2,$$

but has the additional, beneficial property of being entire (everywhere holomorphic) on $\mathbb{C}$. Knowing it has a zero only at the origin, one may use it to construct general abelian functions on $\mathbb{C}$. Thus consider

$$f(u) = \prod_{i=1}^{n} \frac{\sigma(u - a_i)}{\sigma(u - b_i)}.$$

This function is holomorphic apart from zeros at the a_i's and poles at the b_i's. It is further periodic with half-periods ω_1 and ω_2, provided

$$\sum_{i=1}^{n} a_i + \sum_{i=1}^{n} b_i \in \Lambda,$$

which condition guarantees that the factors acquired by the products of σ-function all cancel out when their arguments are augmented by points in Λ.

There are many interesting relations between the $\wp$-functions and the σ-functions. Two notable examples are the addition law,

$$\wp(u) - \wp(v) = -\frac{\sigma(u+v)\sigma(u-v)}{\sigma^2(u)\sigma^2(v)},$$

and its generalization, the *Frobenius–Stickelberger relations* [57]: if $\wp_i^{(j)}$ denotes the value of $\frac{d^j}{du^j}\wp(u)|_{u=u_i}$, then

$$\begin{vmatrix} 1 & \wp_1 & \wp_1^{(1)} & \cdots & \wp_1^{(n-2)} \\ 1 & \wp_2 & \wp_2^{(1)} & \cdots & \wp_2^{(n-2)} \\ \vdots & & & & \vdots \\ 1 & \wp_n & \wp_n^{(1)} & \cdots & \wp_n^{(n-2)} \end{vmatrix} =$$

$$(-1)^{1/2n(n-1)} \prod_{i=1}^{n}(i!) \times \frac{\sigma(\sum_{i=1}^{n} u_i)\prod_{i<j}\sigma(u_i - u_j)}{\prod_{i=1}^{n}\sigma^n(u_i)}.$$

Now imagine we are considering the whole family of nonsingular elliptic curves,

$$V = \{(x, y, g_2, g_3)| \\ y^2 = 4x^2 - g_2x - g_3, g_2^3 - 27g_3^2 \neq 0\}.$$

The field, F, of abelian functions is generated by the parameters g_2 and g_3 and the $\wp$ and $\wp'$ functions (themselves functions of the parameters). Weierstraß showed that the σ function satisfies differential identities in the parameters and in u:

$$Q_0\sigma = 0, \quad Q_2\sigma = 0,$$

where

$$Q_0 = 4g_2\partial_{g_2} + 6g_3\partial_{g_3} - u\partial_u + 1,$$
$$Q_2 = 6g_3\partial_{g_2} + 1/3g_2^2\partial_3 - \frac{1}{2}\partial_u^2 - \frac{1}{21}g_2u^2.$$

The derivations of F, $Der(F)$, are generated by the three operators

$$L_0 = -u\partial_u + 4g_2\partial_{g_2} + 6g_3\partial_{g_3},$$
$$L_1 = \partial_u,$$
$$L_2 = -\zeta\partial_u + 6g_3\partial_{g_2} + 1/3g_2^2\partial_{g_3}.$$

These operators further satisfy the commutation relations of a Lie algebra over F,

$$[L_0, L_1] = L_1,\ [L_0, L_2] = 2L_2,$$
$$[L_1, L_2] = \wp L_1,$$

exhibiting in this way a fundamental symmetry structure underlying these functions [58].

The $\wp$-function appears in a number of physical applications. One is as the solution to the *nonlinear* pendulum equations:

$$\ddot{\theta} + \sin\theta = 0$$

and similar contexts. Another is as a symmetry reduction of some classes of PDE, notably the Korteweg–de Vries (KdV) equation:

$$u_t + u_{xxx} + 6uu_x = 0.$$

The symmetry is translational. One seeks solutions depending on the invariant $\eta = x - ct$ and finds exactly, after two integrations, solutions

$$u(x,t) = f(\eta)$$

where

$$f'^2 = -2f^3 + cf^2 + C_1 f + C_2,$$

C_1 and C_2 being arbitrary constants of integration. These can be written easily in terms of $\wp$-functions and represent quasiperiodic nonlinear wave solutions on a finite space interval (typically waves in a wave tank.) In the limit that both $C_1 = C_2 = 0$, to accommodate the boundary conditions in an infinite tank

$$f, f' \to 0, \quad |\eta| \to 0,$$

we obtain the single soliton solution

$$u(x,t) = \frac{1}{2}\operatorname{sech}^2\left(\frac{\sqrt{c}}{2}(x - ct - \eta_0)\right)$$

for constant η_0.

Analogues of the Weierstraß $\wp$-function can be defined for plane curves of genus g greater than one (the Weierstraß case). They satisfy addition formulae and are periodic with $2g$ complex periods in g variables.

For example, the quintic curve

$$y^2 = 4x^5 + a_4x^4 + a_3x^3 + a_2x^2 + a_1x + a_0$$

is a genus two plane curve and there are three $\wp$-functions of two variables u_1 and u_2, written

$$\wp_{11}(u_1,u_2),\ \wp_{12}(u_1,u_2),\ \wp_{22}(u_1,u_2).$$

Integrability conditions on the $\wp$-functions imply the existence of an entire function $\sigma(u_1,u_2)$ such that

$$\wp_{ij}(u_1,u_2) = -2\partial_{u_i}\partial_{u_j}\log\sigma(u_1,u_2).$$

Further indices represent higher derivatives: $\wp_{111} = \partial/\partial u_1 \wp_{11}$, and so on.

The analogue of the second-order equation satisfied by the Weierstrass function is a set of five equations for the $\wp_{ij}$:

$$\begin{aligned}
&\wp_{1111} - 6\wp_{11}^2 = -\frac{1}{2}a_0a_4 + \frac{1}{8}a_1a_3 \\
&-3a_0\wp_{22} + a_1\wp_{12} + a_2\wp_{11} \\
&\wp_{1112} - 6\wp_{11}\wp_{12} = -a_0 \\
&-\frac{1}{2}a_1\wp_{22} + a_2\wp_{12} \\
&\wp_{1122} - 4\wp_{12}^2 - 2\wp_{11}\wp_{22} = \frac{1}{2}a_3\wp_{12} \\
&\wp_{1222} - 6\wp_{12}\wp_{22} = a_4\wp_{12} - 2\wp_{11} \\
&\wp_{2222} - 6\wp_{22}^2 = \frac{1}{2}a_3 + a_4\wp_{22} + 4\wp_{12}
\end{aligned}$$

Such equations have been studied for a long time [59] and are still an active area of research [60].

The Weierstraß $\wp$-function arises in an important class of linear, Fuchsian equations with coefficients doubly periodic on $\mathbb{C}$:

7.6.1.1 Lamé Equations

The second-order family of *Lamé equations,*

$$w'' - (h + n(n+1)\wp(z))w = 0,$$

for integer n, has regular singular points at the lattice points of Λ and exponents $n+1$ and $-n$. Although the exponents differ by an integer $(2n+1)$, no logarithmic terms in fact occur and, depending on the value of h, solutions algebraic in the $\wp$-function may exist [9, 22].

For the case $n = 1$,

$$w'' - (h + 2\wp(z))w = 0,$$

one can show that two linearly independent solutions

$$w_1 = e^{-z\zeta(a)} \frac{\sigma(z+a)}{\sigma(z)}$$

and

$$w_2 = e^{z\zeta(a)} \frac{\sigma(z-a)}{\sigma(z)}$$

exist provided $h = \wp(a)$ is not a root of the Weierstraß cubic, that is, provided $\wp'(a) \neq 0$. If $h = e_1$ *is* a root of the cubic, then the above solutions are equal and a second linearly independent solution must be chosen:

$$w_2 = e^{-z\eta_1} \frac{\sigma(z+\omega_1)}{\sigma(z)} \left(\zeta(z+\omega_1) + e_1 z\right).$$

7.6.2 Jacobian Elliptic Functions

A genus one plane curve may also be written in quartic form

$$y^2 = a_4x^4 + 4a_3x^3 + 6a_2x^2 + 4a_1x + a_0$$

for which all branch points lie in the finite part of the complex plane and an analogous treatment leads to parameterization by alternative doubly periodic functions due to Jacobi [20, 21].

While $\wp(u)$ has a double pole in each period parallelogram, the Jacobi functions have a pair of distinct single poles. Owing to the fact that the degree of a divisor of a meromorphic function on a compact Riemann surface vanishes, there must be, up to multiplicity, two corresponding zeros for each function.

Defining the Jacobi sn function by

$$sn(u) = t,$$
$$u = \int_0^t \frac{dx}{\sqrt{(1-x^2)(1-k^2x^2)}},$$

so that

$$(x')^2 = (1-x^2)(1-k^2x^2),$$

we define cn and dn by

$$cn^2(u) + sn^2(u) = 1,$$
$$dn^2(u) + k^2 sn^2(u) = 1,$$

choosing the signs in such a way that

$$cn(u) \to 0, \quad dn(u) \to 0$$

as $u \to 0$.

From these definitions one shows that

$$sn'(u) = cn(u)dn(u);$$
$$cn'(u) = -sn(u)dn(u);$$
$$dn'(u) = -k^2 sn(u)cn(u).$$

In the limit that $k \to 0$, sn and cn become circular functions.

Further, if K is the value of the complete elliptic integral

$$K = \int_0^{\pi/2} \frac{d\phi}{\sqrt{1-k^2\sin^2\phi}}$$

then

$$sn(K) = 1,$$
$$cn(K) = 0,$$
$$dn(K) = \sqrt{1-k^2}.$$

K determines the periods of the Jacobi elliptic functions:

$$sn(u+4K) = sn(u);$$
$$cn(u+4K) = cn(u);$$
$$dn(u+2K) = dn(u).$$

As with the $\wp$-functions, and for the same underlying algebro-geometric reasons, the Jacobi functions obey addition laws:

$$sn(u+v) = \frac{sn(u)cn(v)dn(v) + sn(v)cn(u)dn(u)}{1-k^2sn^2(u)sn^2(v)};$$
$$cn(u+v) = \frac{cn(u)cn(v) - sn(uv)sn(v)dn(u)dn(v)}{1-k^2sn^2(u)sn^2(v)};$$
$$dn(u+v) = \frac{dn(u)dn(v) - k^2sn(u)sn(v)cn(u)cn(v)}{1-k^2sn^2(u)sn^2(v)}.$$

7.6.3 Theta Functions

In many ways, the most natural approach to doubly (and multi-) periodic functions in general is via the ϑ-function [12, 22, 60–62]. We shall present the simplest, genus one case and indicate roughly how higher genus ϑ-functions are defined.

Let τ be a complex number with a strictly positive imaginary part and let

$$q = e^{i\pi\tau}.$$

The (fourth) ϑ-function with parameter τ and variable u is defined by an infinite sum:

$$\vartheta_4(u|\tau) = \sum_{n=-\infty}^{\infty} (-1)^n q^{n^2} e^{2niu}.$$

Similar to the σ-function, which is a modified ϑ-function, it is entire and satisfies a *modularity* condition (periodicity up to multiplicative factors):

$$\vartheta_4(u+\pi|\tau) = \vartheta_4(u|\tau);$$
$$\vartheta_4(u+\tau\pi|\tau) = -\frac{1}{q}e^{-2iu}\vartheta_4(u|\tau).$$

The third and first ϑ-functions are defined as

$$\vartheta_3(u|\tau) = \vartheta_4(u+\frac{1}{2}\pi|\tau)$$
$$\vartheta_1(u|\tau) = -ie^{iu+1/4\pi i\tau}\vartheta_4(u+\frac{1}{2}\tau\pi|\tau).$$

and finally the second,

$$\vartheta_2(u|\tau) = \vartheta_1(u+\frac{1}{2}\pi|\tau).$$

One should think of these apparently redundant variants of ϑ as analogues of circular function relations,

$$\cos(u) = \sin(u+\frac{1}{2}\pi),$$

which serve to simplify writing compact, polynomial identities.

Consider a fundamental parallelogram in the complex plane with corners at 0, π, $\tau\pi$ and $\pi+\tau\pi$, closed on two nonparallel sides, open on the other two. Any translation in $\mathbb{C}$ of this parallelogram will contain a single, simple zero of any of the $\vartheta_i(u|\tau)$ functions and they may be used to construct abelian functions in the same way as was $\sigma(u)$.

The ϑ's satisfy many polynomial (algebraic) addition laws, one example of which, the τ dependence being understood, is

$$\vartheta_3(u+v)\vartheta_3(u-v)\vartheta_3(0)^2 = \vartheta_3(u)^2\vartheta_3(v)^2 + \vartheta_1(u)^2\vartheta_1(v)^2.$$

They also satisfy a simple, linear PDE of the heat equation type, in which the parameter τ plays the rôle of "imaginary time":

$$\frac{\partial \vartheta_i}{\partial \tau} + \frac{i\pi}{4}\frac{\partial^2 \vartheta_i}{\partial u^2} = 0.$$

The higher genus, $g \geq 1$, generalization of the ϑ function to g independent, complex variables, $\mathbf{u} \in \mathbb{C}^g$, involves a $g \times g$ symmetric, complex-valued matrix τ whose entries a have positive definite imaginary part. Then the (Riemann) ϑ-function is,

$$\vartheta(\mathbf{u}|\tau) = \sum_{\mathbf{N}\in\mathbb{Z}^g} \exp 2\pi i \left(\frac{1}{2}\mathbf{N}^T\tau\mathbf{N} + \mathbf{N}^T\mathbf{u}\right),$$

where $\mathbf{N}$ and $\mathbf{u}$ are column vectors and $(\cdot)^T$ denotes the row vector transpose.

Such ϑ functions are associated with algebraic curves (on complex dimensional manifolds) by taking a canonical homology basis of closed loops (Chapter 6)

$$\{a_1, \ldots, a_g, b_1, \ldots, b_g\}$$

and a basis of holomorphic one-forms

$$\{\omega_1, \ldots, \omega_g\},$$

satisfying

$$\int_{a_i} \omega_j = \delta_{ij},$$

and

$$\tau_{ij} = \int_{b_i} \omega_j.$$

The modular properties of θ are summarized in the statement that, for μ and μ' column vectors in $\mathbb{Z}^g$,

$$\vartheta(\mathbf{u} + \mu' + \tau\mu, \tau) = \exp 2\pi i \left(-\mu'^T\mathbf{u} - \frac{1}{2}\mu'^T\tau\mu\right)\vartheta(\mathbf{u}, \tau).$$

The analogues of the ϑ_i are the so-called ϑ-functions with *characteristics*

$$\vartheta\begin{bmatrix}\epsilon\\ \epsilon'\end{bmatrix}(\mathbf{u}, \tau) = \sum_{\mathbf{N}\in\mathbb{Z}^g} \exp 2\pi i \left(\frac{1}{2}\mathbf{N}_\epsilon^T\tau\mathbf{N}_\epsilon + \mathbf{N}_\epsilon^T\mathbf{u}_{\epsilon'}\right).$$

Here the notation $(\cdot)_\epsilon$ stands for augmentation by the additive constant $\epsilon/2$, for example, $\mathbf{N}_\epsilon = \mathbf{N} + \frac{1}{2}\epsilon$.

When the entries in ϵ and ϵ' are integers, the function is called the *first-order ϑ-function with integer characteristics.*

The variables $\mathbf{u}$ are properly said to live on a g complex dimensional algebraic (torus) variety, the Jacobi variety, *Jac*(X), of the algebraic curve. This is $\mathbb{C}^g$ factored by the lattice $\Lambda = \{\mu' + \tau\mu | \mu, \mu' \in \mathbb{Z}^g\}$. There is a natural map, the *Abel map*, from any n-fold symmetric product of the algebraic curve with itself to *Jac*(X) :

$$\phi_n : X^n = X_1 \times \cdots \times X_n \to \mathbb{C}^g/\Lambda,$$

given, for some choice of $P_0 \in X$, by

$$\phi_n(P_1, \ldots, P_n) = \sum_{i=1}^{n} \int_{P_0}^{P_i} \omega, \quad mod\,\Lambda.$$

ω is the column vector of holomorphic differentials, $d\mathbf{u}$.

In the case that $n = g$, the Abel map is a local homomorphism and the Jacobi inversion problem, still only implicitly solved in general, is to recover the *divisor* $P_1 + P_2 + \cdots + P_g \in X^g$ from a given point in *Jac*(X).

Of particular importance is Riemann's theorem describing the locus of zeros (the ϑ-divisor) of the ϑ-function:

$$\vartheta(e) = 0 \Leftrightarrow e \in c + \phi_{g-1}(X^{g-1})$$

where c is a very specific vector – the *Riemann vector* of constants [12].

7.6.4 Painlevé Transcendents

Modifying the Weierstraß equation a little further, one may consider the simple nonautonomous case:

$$y'' = 6y^2 + x,$$

x being the independent variable. This is the simplest of the Painlevé equations. Similar to the Weierstraß equation, such equations arise as symmetry reductions of integrable PDEs such as the KdV, modified KdV, Boussinesq, and sine-Gordon equations, under scaling [63–67].

For example, the KdV equation,

$$u_t + u_{xxx} + 6uu_x = 0$$

has a symmetry

$$x \mapsto \lambda x$$
$$t \mapsto \lambda^3 t$$
$$u \mapsto \lambda^{-2} u$$

so that in terms of symmetry invariants we seek solutions, w satisfying

$$u(x,t) = (3t)^{-2/3} w\left(\frac{x}{(3t)^{1/3}}\right)$$

which are invariant under the symmetry (*scaling* solutions). Put $\eta = \frac{x}{(3t)^{1/3}}$. Then the KdV reduces to

$$w''' - 2w - \eta w' + 6ww' = 0,$$

prime denoting $d/d\eta$. Using a Miura transformation

$$w = v' - v^2.$$

This equation can be solved subject to knowledge of solutions to

$$v'' = 2v^3 + \eta v + \alpha,$$

for constant values of α. This last equation is a Painlevé equation of type II.

The solutions share properties of the $\wp$-function in being analytic except at movable poles in the finite part of the complex plane, but they are not periodic.

Equations characterized by the property that their only movable singularites (that is, singularities whose locations depend on initial values or constants of integration) are poles are said to have the *Painlevé* property. A necessary but insufficient criterion for a nonlinear ODE to possess the Painlevé property is that the exponents ν of a series expansion about an arbitrary point x_0,

$$(x - x_0)^\nu \sum_0^\infty a_n (x - x_0)^n,$$

analogous to the Frobenius expansions described earlier, be integers.

First-order ODEs of this type, polynomial in the dependent variable and its derivative, are either linear, Riccati

$$y' = a(x)y^2 + b(x)y + c(x)$$

or elliptic [68]. Second-order equations were roughly classified by Painlevé [66] (more thoroughly by Gambier and coworkers [69]) and hence go by his name. A list of the six second-order Painlevé equations is given below:

PI

$$y'' = 6y^2 + x$$

PII

$$y'' = 2y^3 + xy + a$$

PIII

$$y'' = \frac{y'^2}{y} - \frac{y'}{x} + \frac{ay^2 + b}{x} + cy^3 + \frac{d}{y}$$

PIV

$$y'' = \frac{y'^2}{2y} + 3/2y^3 + 4xy^2 + 2\left(x^2 - a\right) y + \frac{b}{y}$$

PV

$$y'' = \left(\frac{1}{2y} + \frac{1}{y-1}\right) y'^2 - \frac{y'}{x} + \frac{(y-1)^2}{x^2}\left(ay + \frac{b}{y}\right) + c\frac{y}{x} + d\frac{y(y+1)}{y-1}$$

PVI

$$y'' = \frac{1}{2}\left(\frac{1}{y} + \frac{1}{y-1} + \frac{1}{y-x}\right) y'^2 - \left(\frac{1}{x} + \frac{1}{x-1} + \frac{1}{y-x}\right) y' + \frac{y(y-1)(y-x)}{x^2(x-1)^2} \times \left(a + b\frac{x}{y^2} + c\frac{x-1}{(y-1)^2} + d\frac{x(x-1)}{(y-x)^2}\right)$$

The parameters a, b, c, and d are arbitrary, but specific relations between them allow specialization. PVI contains all the other cases as reductions. Generically, the solutions are transcendents not to be found among the (linear) elementary and special functions, except for certain special values of the parameters.

Painlevé equations are associated with *linear* ODEs with regular singular points in the following way.

Consider a family of linear, second-order ODEs with regular singular points in $\mathbb{C}$ depending on a complex parameter, λ. We can write this as a first-order equation in system form,

$$\Psi_x(\lambda, x) = B(\lambda, x, y)\Psi(\lambda, x),$$

where Ψ is a two-component column vector of functions and B a 2×2 matrix with at worst poles of order one in x and λ, polynomial in y. The monodromy group of this system will be generated, as before, by linear maps

$$\Psi \mapsto M_i \Psi,$$

associated with each pole in B, $i = 1, 2, \ldots$

We will require that the monodromy properties, for example, the group exponents, be invariant under variation in the parameter λ. This is the *isomonodromy condition* and it means that the above system is compatible with a system of the form

$$\Psi_\lambda(\lambda, x) = A(\lambda, x, y)\Psi(\lambda, x),$$

A having, in general, first-order poles in λ. The statement that these two systems are compatible is that they share a common general solution, so that

$$\partial_x \left(\partial_\lambda \Psi\right) = \partial_\lambda \left(\partial_x \Psi\right),$$

which in turn implies

$$\partial_x A - \partial_\lambda B + [A, B] = 0.$$

This system constitutes a (generally nonlinear) ODE in y with independent variable x.

All Painlevé equations are realized as isomonodromy conditions in this manner [70, 71]. For example, in the case of PI, we may (but this is not unique) choose,

$$A = (4\lambda^2 + 2y^2 + x)\begin{pmatrix} 1 & 0 \\ 0 & -1 \end{pmatrix}$$

$$+(4\lambda^2 y + 2y^2 + x)\begin{pmatrix} 0 & -1 \\ 1 & 0 \end{pmatrix}$$

$$-(2\lambda y_x + \frac{1}{2\lambda})\begin{pmatrix} 0 & 1 \\ 1 & 0 \end{pmatrix},$$

$$B = (\lambda + \frac{y}{\lambda})\begin{pmatrix} 1 & 0 \\ 0 & -1 \end{pmatrix}$$

$$+\frac{y}{\lambda}\begin{pmatrix} 0 & -1 \\ 1 & 0 \end{pmatrix}.$$

Painlevé equations also come equipped with Bäcklund transformations. Most generally, a *Bäcklund map* is a transformation mapping solutions of one differential equation to those of another or the same equation (*autoBäcklund map*) perhaps with different parameter values. We have seen examples for linear equations already: the Laplace, Moutard, and Darboux maps.

Since PI has no parameters, we consider the case of PII. An obvious Bäcklund map is:

$$y(x, a) \mapsto \tilde{y} = -y(x, -a).$$

A less obvious one, for $a \neq \pm 1/2$, is, writing $y(x, a) = y_a$:

$$y_a \mapsto y_{a\pm 1} = -y_a - \frac{2a \pm 1}{2y_a^2 \pm 2y_{a,x} + x}.$$

The sequence of $y(x, a)$ so generated satisfies the three-term recurrence relation:

$$\frac{a + \frac{1}{2}}{y_a + y_{a+1}} + \frac{a - \frac{1}{2}}{y_a + y_{a-1}} + 2y_a^2 + x = 0.$$

In fact, the Bäcklund maps detailed above comprise a representation of an extended affine Weyl group, the infinite dihedral group generated by a reflection and a translation.

One further aspect of the richness of the Painlevé theory is the existence of families of simple, nontranscendental, solution for certain parameter values. Let us consider PII once more. For $a \in \mathbb{Z}$ there exist solutions

$$y_n = \frac{\mathrm{d}}{\mathrm{d}x} \ln\left(\frac{Q_{n-1}(x)}{Q_n(x)}\right)$$

the Q_n satisfying a second-order differential-difference recurrence relation,

$$\frac{Q_n'^2 - 4Q_n Q_n''}{Q_{n+1}Q_{n-1} - zQ_n^2} = -\frac{1}{4}.$$

with initial terms $Q_0 = 1$, $Q_1 = x$, $Q_2 = x^3 + 4$ and so on. The corresponding rational solutions to PII are

$$y_1 = -\frac{1}{x},$$

$$y_2 = \frac{1}{x} - \frac{3x^2}{x^3 + 4},$$

etc.

Another class of rational solutions to PII are obtained from the generating function

$$\mathrm{e}^{\lambda x - (4/3)\lambda^3} = \sum_{m=0}^{\infty} p_m(x)\lambda^m.$$

Put

$$\tau_n = \det\left(\frac{d^j p_{2i-1}(x)}{\mathrm{d}x^j}\right)_{i,j=1,\ldots,n},$$

and the solutions are

$$y(x,n) = \frac{\mathrm{d}}{\mathrm{d}x} \ln\left(\frac{\tau_{n-1}}{\tau_n}\right).$$

There are also solutions expressible using classical special functions. Thus for $a = n + (1/2)$, $n \in \mathbb{Z}$, a family of solutions to PII is obtained using the Bäcklund transformation, $y_a \mapsto y_{a\pm1}$ above, starting from one of

$$y(x, \pm\frac{1}{2}) = \mp\frac{\phi'(x)}{\phi(x)},$$

ϕ being an Airy function combination

$$\phi(x) = C_1 Ai(-2^{-1/3}x) + C_2 Bi(-2^{-1/3}x).$$

The situations described above for PI and PII are replicated for other Painlevé equations.

7.7 Discrete Special Functions

7.7.1 ∂–δ Theory

An attempt at a universal theory of special functions was made in the middle of the last century by imposing a relationship between differential and difference operations [72]. In this approach, the majority of special function recurrence relations are shown to be reducible to a relation of the form,

$$\partial_z F(z, \alpha) = F(z, \alpha + 1).$$

This universal equation has many solutions for F. For instance,

- $F(z, \alpha) = \mathrm{e}^z$;
- $F(z, \alpha) = \sin(z - \frac{1}{2}\alpha z)$;
- $F(z, \alpha) = \mathrm{e}^{-z^2} H_\alpha(-z)$;
- $F(z, \alpha) = \mathrm{e}^{i\alpha\pi} z^{-\alpha/2} J_\alpha(2\sqrt{z})$.

The equation can be shown to admit unique solutions, to be expandable in the parameters z and in α, to give rise to other solely differential identities, and to provide generating functions of almost all the classical special functions.

It also naturally leads in to the ideas of difference calculus.

7.7.2 Quantum Groups

Properly speaking these are not groups, the name *quantum group* is nevertheless established, particularly in the mathematical physics literature, for certain *Hopf bialgebras* that play a rôle in the theory of a whole new class of discrete or q-difference special functions analogous to that played by Lie groups in the classical theory [73, 74]. Most importantly, it is their representations that are studied in applications.

For q a nonzero complex number with $q^2 \neq 1$, the quantum group $U_q(\mathfrak{sl}_2)$ is an algebra generated over $\mathbb{C}[q]$ by elements E, F, K, and K^{-1} defined by the relations

$$KK^{-1} = K^{-1}K = 1,$$
$$KEK^{-1} = q^2 E, \quad KFK^{-1} = q^{-2}F,$$

and

$$EF - FE = \frac{K - K^{-1}}{q - q^{-1}}.$$

It carries a *comultiplication*, a $\mathbb{C}$-linear map from the algebra to its two-fold tensor self-product,

$$\Delta(E) = E \otimes K + 1 \otimes E,$$
$$\Delta(F) = F \otimes 1 + K^{-1} \otimes F$$
$$\Delta(K) = K \otimes K;$$

a *counit*, ϵ,

$$\epsilon(E) = \epsilon(F) = 0,$$
$$\epsilon(K) = 1;$$

and an *antipode*, S,

$$S(K) = K^{-1},$$
$$S(E) = -EK^{-1},$$
$$S(F) = -KF.$$

The comultiplication is consistent with the defining relations in the sense that, for instance,

$$\Delta(E)\Delta(F) - \Delta(F)\Delta(E) = \frac{\Delta(K) - \Delta(K^{-1})}{q - q^{-1}}.$$

In the limit that $q \to 1$, one obtains (almost) the universal enveloping algebra $U(\mathfrak{sl}_2)$, so in that sense the quantum group is a *deformation* of the Lie algebra, which is morally its classical limit.

In the representation theory of quantum groups, the $U_q(\mathfrak{sl}_2)$ generators are realized as difference operators on functions.

7.7.3 Difference Operators

The classical special functions have discrete analogues in which differential operators are replaced by *difference* operators of which the most studied are the q-difference operator [30, 75, 76]

$$D_q(f(x)) = \frac{f(x) - f(qx)}{x - qx},$$

the h-derivative

$$d_h(f(x)) = \frac{f(x+h) - f(x)}{h},$$

and the Askey–Wilson difference operator

$$D_q(f(x)) = \frac{\breve{f}(q^{1/2}e^{i\theta}) - \breve{f}(q^{-1/2}e^{i\theta})}{\frac{1}{2}(q^{1/2} - q^{-1/2})(z - 1/z)},$$

where $\breve{f}(z)$ is defined by,

$$\breve{f}(z) = f\left(\frac{1}{2}\left(z + \frac{1}{z}\right)\right),$$

x and z are related as

$$z^2 - 2xz + 1 = 0,$$

and we take the branch $\sqrt{x+1} > 0$, $x >\geq -1$.

The corresponding processes, by analogy with the classical derivative, are sometimes called *quantum calculus*.

Although each difference operator tends to the classical limit (on appropriate function classes) as $q \to 1$ and $h \to 0$, such operators are not quite derivations. Instead, for example,

$$D_q(fg)(x) =$$
$$f(x)D_q(g(x)) + g(qx)D_q(f(x)).$$

That such definitions are natural is evidenced by simple observations of the following sort. The notation for a q-integer

$$[n]_q = \frac{1 - q^n}{1 - q}$$

allows us to write the q-derivative of x^n as

$$D_q(x^n) = [n]_q x^{n-1}$$

and suggests the q-factorial definition

$$[n]_q! = [n]_q[n-1]_q \cdots [1]_q$$

with $[0]_q! = 1$ as usual. We can then define q-binomial coefficients,

$$\left[\begin{matrix} n \\ j \end{matrix}\right]_q = \frac{[n]_q!}{[n-j]_q![j]_q!}$$

The q-Pascal rules for the q-binomial coefficients are

$$\left[\begin{matrix} n \\ j \end{matrix}\right]_q = \left[\begin{matrix} n-1 \\ j-1 \end{matrix}\right]_q + q^j \left[\begin{matrix} n-1 \\ j \end{matrix}\right]_q$$

and

$$\left[\begin{matrix} n \\ j \end{matrix}\right]_q = \left[\begin{matrix} n-1 \\ j \end{matrix}\right]_q + q^{n-j} \left[\begin{matrix} n-1 \\ j-1 \end{matrix}\right]_q$$

Assume x and y are noncommutative objects satisfying $yx = qxy$ (x and y are coordinates on the *quantum plane*). There is a q-binomial theorem,

$$(x+y)^n = \sum_{j=0}^{n} \left[\begin{matrix} n \\ j \end{matrix}\right]_q x^j y^{n-j},$$

taking due regard of the ordering of the x's and y's.

Such ideas go back to Gauss and Euler.

There are two q-analogues of the classical exponential function:

$$e_q^x = \sum_{j=0}^{\infty} \frac{x^j}{[j]_q!}$$

and

$$E_q^x = \sum_{j=0}^{\infty} q^{j(j-1)/2} \frac{x^j}{[j]_q!}.$$

Elementary calculations confirm q-difference identities:

$$D_q e_q^x = e_q^x$$
$$D_q E_q^x = E_q^{qx};$$

the reciprocal relation

$$e_q^x E_q^{-x} = 1$$

and, subject again to the commutation relation $yx = qxy$, the addition law

$$e_q^x e_q^y = e_q^{x+y}. \qquad (7.1)$$

q-Analogues of classical special functions can be constructed. As a simple example we consider the q-Hermite polynomials.

7.7.4 q-Hermite Polynomials

The nth (continuous) q-Hermite polynomial, denoted $H(x|q)$, satisfies a three-term recurrence relation:

$$2xH_n(x|q) = H_{n+1}(x|q) + (1-q^n)H_{n-1}(x|q), \qquad (7.2)$$

with initial terms

$$H_0(x|q) = 1, \quad H_1(x|q) = 2x.$$

Let us define the symbols,

$$(a;q)_0 = 1,$$
$$(a;q)_n = \prod_{k=1}^{n} (1 - aq^{k-1}).$$

Here n may take any value in $\mathbb{N}$ or ∞, to signify an infinite product. Note that if $a = q$, we recover the symbol $[n]_q!$ defined earlier. We also define

$$(a, b; q)_n = (a, q)_n (b, q)_n.$$

Using this symbol, the q-Hermite polynomial generating functions are

$$\frac{1}{(te^{i\theta}, te^{-i\theta}|q)} = \sum_{0}^{\infty} H_n(\cos\theta|q) \frac{t^n}{(q|q)_n}.$$

They may also be themselves considered as generating functions for the q-binomial coefficients:

$$z^{-n}H_n(\cos\theta|q) = \sum_{k=0}^{n} \left[\begin{array}{c} n \\ k \end{array} \right]_q z^{-2k},$$

$$z = e^{i\theta}.$$

The $H_n(x|q)$ have the classical polynomials as limits in the following sense:

$$\lim_{q\to 1^-} \left(\frac{2}{1-q}\right)^{n/2} H_n(x\sqrt{\frac{1-q}{2}}|q)) \to H_n(x).$$

They also satisfy ladder difference relations similar to the differential ladder relations of their classical counterparts:

$$D_q H_n(x|q) = \frac{2(1-q^n)}{1-q} q^{(1-n)/2} H_{n-1}(x|q);$$

$$\frac{1}{w(x|q)} D_q \left(w(x|q)H_n(x|q)\right) = -\frac{2q^{n/2}}{1-q} H_{n+1}(x|q).$$

7.7.5 Discrete Painlevé Equations

There is a list of *seven* discrete Painlevé equations (there are two variants of PV) of which the first three are

δ-PI

$$x_{n+1} + x_{n-1} = -x_n + \frac{\alpha n + \beta}{x_n} + 1;$$

δ-PII

$$x_{n+1} + x_{n-1} = \frac{(\alpha n + \beta)x_n + a}{1 - x_n^2};$$

q-PIII

$$x_{n+1}x_{n-1} = \frac{(x_n - aq\lambda^n)(x_n - bq\lambda^n)}{(1 - cx_n)(1 - \frac{x_n}{c})}.$$

Here, α, β, q, and λ are constant parameters and the notation δ- or q- refers to the additive or multiplicative structures of the difference scheme [77].

These equations are of interest in several respects.

Firstly, they reduce to the classical Painlevé equations in appropriate "continuum limits:" we allow that $x_n = f(\epsilon n)$ for differentiable f and let $\epsilon \to 0$ with some adjustment of the free parameters [78, 79].

Secondly, they exhibit a property called "singularity confinement" [77], which means that although certain initial conditions may give rise to indeterminate values of x_n for some n, this indeterminacy is not stable under small variations in those initial conditions.

Thirdly, the initial conditions can be assigned "degrees" in which the order of growth of the solutions is merely polynomial rather than exponential [80]. This *algebraic entropy* is an unusual property for difference schemes.

Finally, in the sense of the *Nevanlinna theory*, the solutions are well controlled [81].

Discrete Painlevé equations share with their classical limits the availability of Bäcklund maps and Lax pairs (isomonodromy).

Finally, they are classified by the extended affine Weyl groups [82] that, so far as this article is concerned, nicely returns us to our starting point in the representations of discrete symmetries.

References

1. Rose, J.S. (1978) *A course on Group Theory*, Cambridge University Press.

2. Benson, D.J. (1993) *Polynomial Invariants of Finite Groups, LMS Lecture Note Series 190*, CUP.
3. Borovik, A.V. and Borovik, A. (2010) *Mirrors and Reflections: The Geometry of Finite Reflection Groups*, Springer.
4. Humphreys, J.E. (1990) *Reflection Groups and Coxeter Groups, Studies in Advanced Mathematics 29*, CUP.
5. Fulton, W. (1997) *Young Tableaux: With Applications to Representation Theory and Geometry*, Cambridge University Press.
6. MacDonald, I.G. (1998) *Symmetric Functions and Hall Polynomials*, 2nd edn, OUP.
7. Forsyth, A.R. (1959) *Theory of Differential Equations*, Dover.
8. Gray, J. (1996) *Linear Differential Equations and Group Theory from Riemann to Poincaré*, 2nd edn, Birkhäuser.
9. Ince, E.L. (1956) *Ordinary Differential Equations*, Dover.
10. Klein, F.C. (1914) Vorlesungen über das Ikosaeder und die Auflösung der Gleichungen vom 5ten Grade, Teubner (1884), *Lectures on the Icosahedron, and the Solution of Equations of the Fifth Degree*, 2nd revised edn (ed. English translation by G.G. Morrice), Kegan Paul & Co., London.
11. Kuga, M. (1994) *Galois' Dream: Group Theory and Differential Equations*, Corr. 2nd printing, Birkhäuser.
12. Farkas, H.M. and Kra, I. (1992) *Riemann Surfaces*, Springer.
13. Lie, S. (1970) *Transformationsgruppen*, Chelsea, New York.
14. Humphreys, J.E. (1972) *Introduction to Lie Algebras and Representation Theory*, Springer.
15. Olver, P.J. (1986) *Applications of Lie Groups to Differential Equations*, Springer.
16. Ovsiannikov, L.V. (1982) *Group Analysis of Differential Equations*, Academic Press.
17. Warner, F.W. (2010) *Foundations of Differentiable Manifolds and Lie Groups*, Springer.
18. Fulton, W. and Harris, J. (1999) *Representation Theory: A First Course*, Springer.
19. Jeffreys, H. and Jeffreys, B. (2000) *Methods of Mathematical Physics*, 3rd edn, Cambridge University Press.
20. Wang, Z.X. and Guo, D.R. (1989) *Special Functions*, World Scientific.
21. Abramowitz, M. and Stegun, I.A. (1965) *Handbook of Mathematical Functions*, Dover Publications, Inc.
22. Whitaker, E.T. and Watson, G.N. (1996) *A Course of Modern Analysis*, 4th edn, CUP.
23. Hitchen, N. (1982) Monopoles and Geodesics. *Commun. Math. Phys.*, **83**, 579–602.
24. Miller, W. (1977) *Symmetry and Separation of Variables (Encyclopedia of Mathematics and its Applications)*, Addison-Wesley.
25. Gradshteyn, I.S. and Ryzhik, I.M. (2007) *Gradshteyn and Ryzhik's Table of Integrals, Series, and Products*, 7th edn (eds A. Jeffrey and D. Zwillinger), Academic Press.
26. Talman, J.D. (1968) *Special Functions: A Group Theoretic Approach (based on lectures by E.P. Wigner)*, W.A.Benjamin.
27. Vilenkin, N.J. (1968) *Special Functions and the Theory of Group Representations, Translations of Mathematical Monographs 22*, AMS.
28. Bultheel, A., Gonzalez-Vera, P., Hendriksen, E., and Njastad, O. (1999) *Orthogonal Rational Functions*, Cambridge University Press.
29. Khrushchev, S. (2008) *Orthogonal Polynomials and Continued Fractions*, CUP.
30. Ismail, M.E.H. (2005) *Classical and Quantum Orthogonal Polynomials in one Variable, Encyclopedia of Mathematics and its Applications 98*, CUP.
31. Infeld, L. and Hull, T.E. (1953) The factorization method. *Rev. Mod. Phys.*, **23**, 21–68.
32. Weyl, H. (1950) *The Theory of Groups and Quantum Mechanics*, Dover.
33. Cartan, H. (2003) *The Theory of Spinors*, Dover.
34. Darboux, G. (1890) *Leçons Sur La Theorie Generale Des Surfaces*, Gauthier Villars Et Fils.
35. (a) Moutard, Th.F. (1895) *C. R. Acad. Sci. Paris*, **80**, 129; (b) deLEcole, J. (1878) *Polytech. Cahier*, **45**, 1.
36. Novikov, S., Manakov, S.V., Pitaevskii, L.P., and Zakharov, V.E. (1984) *Theory of Solitons: The Inverse Scattering Method*, Consultants Bureau, New York.
37. Kamran, N. (2002) *Selected Topics in the Geometrical Study of Differential Equations, CBMS 96*, AMS.
38. Toda, M. (1989) *Theory of Nonlinear Lattices*, Springer.

39. Matveev, V.B. and Salle, M.A. (1991) *Darboux Transformations and Solitons*, Springer.
40. Dunkl, C.F. (2003) Special functions and generating functions associated with reflection groups. *J. Comput. Appl. Math.*, **153**, 181–190.
41. Opdam, E.M. (2000) *Lecture Notes on Dunkl Operators for Real and Complex Reflection Groups, Mathematical Society of Japan Memmoirs*, vol. **8**, Mathematical Society of Japan.
42. Rösler, M. (2003) Dunkl operators: theory and applications, in *Orthogonal Polynomials and Special Functions, Springer Lecture Notes in Mathematics*, vol. **1817**, Springer.
43. Bender, C.M. and Orsag, S.A. (1999) *Advanced Mathematical Methods for Scientists and Engineers*, Springer.
44. Hille, E. (1976) *Ordinary Differential Equations in the Complex Domain*, John Wiley & Sons, Inc.
45. Meyer, R.E. (1989) A simple explanation of the Stokes phenomenon. *SIAM Rev.*, **31**, 435–445.
46. Ritt, J.F. (1950) *Differential Algebra*, AMS.
47. Ritt, J.F. (1948) *Integration in Finite Terms: Liouville's Theory of Elementary Methods*, Columbia University Press.
48. Abel, N.H. (2010) *Oeuvres Complètes*, Nabu Press.
49. Kaplansky, I. (1976) *An Introduction to Differential Algebra*, 2nd edn, Hermann.
50. Kolchin, E.R. (1973) *Differential Algebra and Algebraic Groups, Pure and Applied Mathematics 54*, Academic Press, Boston, MA.
51. Magid, A.R. (1994) *Lectures on Differential Galois Theory, University Lecture Series*, vol. 7, AMS.
52. van der Put, M. and Singer, M.F. (1997) *Galois Theory of Difference Equations, Lecture Notes in Mathematics 1666*, Springer.
53. Artin, E. (1971) *Galois Theory*, 6th printing, University of Notre Dame.
54. Armitage, J.V. and Eberlien, W.F. (2006) *Elliptic Functions, LMS student tects 67*, Cambridge University Press.
55. Moll, V. (1999) *Elliptic curves: Function theory, Geometry, Arithmetic*, CUP.
56. Weierstraß, K. (1856) Theorie der Abel'schen Funktionen, *Jour. reine & angewandte Math.*, **52**, 285–379.
57. Frobenius, G. and Stickelberger, L. (1877) Zur Theorie der elliptischen Funktionen. *Journal für Math.*, **33**, 179.
58. Buchstaber, V.M. and Leykin, D.V. (2008) Solution of problem of differentiation of abelian functions over parameters for families of (n, s)-curves. *Funct. Anal. Appl*, **42**, 268–278.
59. Baker, H.F. (1907) *Abelian Functions: Abel's Theorem and the Allied Theory of Theta Functions*, Cambridge University Press, Cambridge 1897 (reprinted in 1995); *An Introduction to the Theory of Multiply Periodic Functions*, Cambridge University Press.
60. Buchstaber, V.M., Enolski, V., and Leykin, D. (1997) Kleinian functions, hyperelliptic Jacobians and applications. *Rev. Math. Math. Phys.*, **10**, 1–125.
61. Mumford, D. (1983/4) *Tata Lectures on Theta I & II*, Birkhäuser.
62. Weierstraß, K. (1894) *Zur Theorie der elliptischen Funktionen, Mathematische Werke*, Bd 2, Teubner, Berlin, pp. 245–255.
63. Bobenko, A.I. and Eitner, U. (2000) *Painlevé Equations in the Differential Geometry of Surfaces, LNM 1753*, Springer.
64. Conte, R. and Musette, M. (2008) *The Painlevé Handbook*, Springer.
65. Clarkson, P. *Painlevé Transcendents*, Chapter 32 in Digital Library of Mathematical Functions, *http://dlmf.nist.gov/32* (accessed on 24 February 2014).
66. Painlevé, P. (1897) *Lecons, sur la theorie analytique des equations differentielles, professees a Stockholm*, Libraire Scientifique à Hermann, Paris.
67. Noumi, M. (2004) *Painlevé Equations through Symmetry*, American Mathematical Society.
68. Goursat, E. (1917) in *A Course in Mathematical Analysis* (eds translation by E.R. Hedrick and O. Dunkel), Ginn & Co.
69. Gambier, B. (1910) Sur les équations différentielles du second ordre et du premier degré dont l'intégrale générale est à points critique fixés. *Acta Math.*, **33**, 1–55.
70. Flaschka, H. and Newell, A.C. (1980) Monodromy- and spectrum-preserving deformations. I. *Commun. Math. Phys.*, **76**, 65–116.
71. Jimbo, M. and Miwa, T. (1981) Monodromy preserving deformation of linear ordinary

differential equations with rational coefficients. *Phys. D* **2**, 407–448.

72. Truesdell, C. (1948) *An Essay Towards a Unified Theory of Special Functions*, Princeton University Press.
73. Brown, K.A. and Goodearl, K.R. (2002) *Lectures on Algebraic Quantum Groups*, Birkhäuser.
74. Klimik, A. and Smüdgen, K. (1997) *Quantum Groups and their representations*, Springer.
75. Kac, V. and Cheung, P. (2002) *Quantum Calculus*, Springer.
76. van der Put, M. and Singer, M.F. (2003) *Galois Theory of Linear Differential Equations*, Springer.
77. Grammaticos, B. and Ramani, A. (2004) *Discrete Painlevé Equations: A Review, Lecture Notes in Physics*, vol. **644**, Springer, pp. 245–321.
78. Conte, R. and Musette, M. (1996) A new method to test discrete Painlevé equations. *Phys. Lett. A*, **223**, 439–448.
79. Fokas, A.S., Grammaticos, B., and Ramani, A. (1993) From continuous to discrete Painlevé equations. *J. Math. Anal. Appl.*, **180**, 342–3360.
80. Veselov, A.P. (1992) Growth and integrability in the dynamics of mappings. *Commun. Math. Phys.*, **145**, 181–193.
81. Ablowitz, M.J. and Halburd, R. (2000) *Nevanlinna theory and difference equations of Painlevé type. Proceedings of the Workshop on Nonlinearity, Integrability and All That: Twenty Years after NEEDS '79 (Gallipoli, 1999)*, 3–11, 2000, World Science Publishing, River Edge, NJ.
82. Sakai, H. (2001) Rational surfaces associated with affine root systems and geometry of the Painlevé equations. *Commun. Math. Phys.*, **220**, 165.

8
Computer Algebra

James H. Davenport

8.1
Introduction

Computer algebra, sometimes also called symbolic computation to distinguish it from numeric computation, is exactly what the name suggests: it is the use of computers to do algebra. Though sometimes referred to as "computer mathematics," it is not necessarily that, as mathematically incorrect results can be algebraically "correct," as in the school howler

$$1 = \sqrt{1} = \sqrt{(-1)^2} = -1. \tag{8.1}$$

This problem is further discussed in Section 8.7.2.

In fact, it is not even algebra for which we need software packages, computers by themselves cannot actually do arithmetic: only a limited subset of it. It is the case that $e^{\pi\sqrt{163}} \approx 2.625\,374\,126\,407\,687\,43.9\,999\,999\,999\,992\,500\,725$. If we ask Excel (or any similar software package) to compute $e^{\pi\sqrt{163}} - 262\,537\,412\,640\,768\,744$, we will be told that the answer is 256 (or −10944, depending on the version of Excel). More mysteriously, if we go back and look at the formula in the cell, we see that it is now $e^{\pi\sqrt{163}} - 262\,537\,412\,640\,768\,800$, or even $e^{\pi\sqrt{163}} - 262\,537\,412\,640\,768\,000$. In fact, 262 537 412 640 768 744 is too large a whole number (or integer, as mathematicians say) for Excel to handle, and Excel has converted it into floating-point (what Excel terms *scientific*) notation. Excel, or any other software using the IEEE standard [1] representation for floating-point numbers, can only store them to a given accuracy, about (we say "about" because the internal representation is binary, rather than decimal) 16 decimal places.[1] In fact,

1) In fact, Excel is more complicated even than this, as the calculations in this table show.

i		1	2	3	4	…	10	11	12	…	15	16
a	10^i	10	100	1000	1…0	…	1…0	10^{11}	10^{12}	…	10^{15}	10^{16}
b	a-1	9	99	999	9999	9…9	…	9…9	10^{12}	…	10^{15}	10^{16}
c	a-b	1	1	1	1	1	…	1	1	…	1	0

We can see that the printing changes at 12 decimal digits, but that actual accuracy is not lost until we subtract 1 from 10^{16}.

Mathematical Tools for Physicists, Second Edition. Edited by Michael Grinfeld.

it requires twice this precision to show that $e^{\pi\sqrt{163}} \neq 262\,537\,412\,640\,768\,744$. As $e^{\pi\sqrt{163}} = (-1)^{\sqrt{-163}}$, it follows from deep results of transcendental number theory [2], that not only is $e^{\pi\sqrt{163}}$ not an integer, it is not a fraction (or rational number), nor even the solution of a polynomial equation with integer coefficients: essentially it is a "new" number.

Although computers had been used to support mathematics earlier (in 1951, the 79-digit number $180\left(2^{127}-1\right)^2-1$ was shown to be prime and in 1952, the great mathematician Emil Artin had the equally great computer scientist John von Neumann performed an extensive calculation relating to elliptic curves on the MANIAC computer [3, p. 119]), actual computer algebra is 60 years old, since in 1953, two theses [4, 5] described programs to differentiate expressions, and Haselgrove [6] showed that algorithms in group theory could be implemented on computers.

8.2 Computer Algebra Systems

While these early programs could only do one thing, they soon evolved into more general-purpose *systems* that could do a variety of algebraic calculations. The descendants of [4, 5] are the "polynomial" systems that manipulate (fractions of) polynomials in "indeterminates" such as x, y or "pseudo-indeterminates" such as $\sin x$, e^y, and so on, early examples of which are the Polynomial Manipulator PM [7], the Symbolic Mathematical Laboratory [8] and its successor Macsyma [9][2)], the Cambridge Algebra Manipulation Language CAMAL [10], and Reduce [11]. The group theory of Haselgrove [6] gave rise to CAYLEY [12], then Magma [13] as well as a range of more specialist systems.

This split survives to this day, with major "polynomial" systems, now generally described as "calculus" systems, being Maple, Mathematica, and SAGE (the latter incorporates much of MAXIMA) and major group theory systems being GAP and MAGMA. Why the split? The difference seems one of mathematical attitude, if one can call it that. The designer of a calculus system envisages it being used to compute an integral, factor a polynomial, multiply two matrices, or otherwise operate on a mathematical *datum*. The designer of a group theory system, while he will permit the user to multiply, say, permutations or matrices, does not regard this as the object of the system: rather the object is to manipulate whole groups (etc.) of permutations (or matrices, or other mathematical data), that is, mathematical *structures*, and take, say, the quotient of two groups, rather than of two polynomials. More recently, we have seen the rise of systems to manipulate *polynomial structures* (generally *ideals*–see Definition 8.3) such as CoCoA [14, 15], SINGULAR [16] and Macaulay [17, 18], and "calculus" systems incorporate more of this, so the distinction is blurring.

2) And various offshoots MAXIMA, Vaxima, and so on.

8.3 "Elementary" Algorithms

8.3.1 Representation of Polynomials

The "obvious" way to represent a polynomial, say $3x^3 + 2x - 5$, in a computer would be to store a vector of coefficients, that is, `[3,0,2,-5]`. Addition of polynomials is then just addition of vectors, and multiplication is a convolution. Polynomials in several variables are then stored either as multidimensional arrays (a method known as *distributed*) or as polynomials in x whose coefficients are polynomials in y and so on (a method known as *recursive*).

The problem with this approach can already be seen in the previous paragraph: we had to write $3x^3 + 2x - 5$ as $3x^3 + 0x^2 + 2x - 5$ in order to get a complete set of coefficients. This might seem harmless, but a dense representation of $abcde - 1$ would be

$$a^1b^1c^1d^1e^1 + \underbrace{0 \cdot a^1b^1c^1d^1e^0 + \cdots}_{5 \text{ terms}}$$
$$+ \underbrace{0 \cdot a^1b^1c^1d^0e^0 + \cdots}_{10 \text{ terms}}$$
$$+ \cdots + a^0b^0c^0d^0e^0$$

for a total of 32 terms. $a^2b^2c^2d^2e^2 - 1$ has 243 terms, and, in general, a polynomial of degree d in n variables has $(d+1)^n$ terms. Adding two such polynomials requires $(d+1)^n$ operations, and multiplying two takes $(d+1)^{2n}$ operations.[3] Storing, or at least counting, all potential terms is known as the *dense model*.

Hence most systems adopt a *sparse* approach (analogous to the storage of sparse matrices in MatLab, etc.) and only store the nonzero coefficients (and the corresponding exponents). Hence, $3x^3 + 2x - 5$ becomes `[(3,3),(2,1),(-5,0)]` and $a^2b^2c^2d^2e^2 - 1$ becomes `[(1,2,2,2,2,2),(-1,0,0,0,0,0)]` (sparse distributed) or `[a,(z,2),(-1,0)]` where `z` is the representation of $b^2c^2d^2e^2$ (sparse recursive). If there are t_f and t_g nonzero terms in f and g, addition takes $\leq \min(t_f, t_g)$ *arithmetic operations* and multiplication takes $\leq 2t_ft_g$.

3) "Fast" algorithms, such as Karatsuba multiplication [19] or fast Fourier transform ones [20] can improve on these running times, but not on the storage requirements.

We have emphasized *arithmetic operations* because the real challenge now becomes that of arranging the output in a useful order (nice), and combining/canceling duplicates (necessary to preserve sparseness). This can be viewed as a sorting problem, and techniques such as HeapSort [21], BucketSort [22], or hashing (as in Maple [23]) are used. In the sparse-distributed representation, order is not quite so obvious: do we put x^2y before (greater x power) or after (lower total degree) xy^3? See Section 8.5.4 for more details.

Once we move beyond addition and multiplication, just counting nonzero terms *in the input* no longer bounds the complexity. The division of two two-term polynomials can be arbitrarily large, because $(x^n - 1)/(x - 1) = x^{n-1} + x^{n-2} + \cdots + x + 1$. However, it is possible to compute $h = f/g$ with the number of operations depending only on the number of terms t_f, t_g, and t_h, and not on the degrees [24]. Counting only the nonzero terms in the input and output is known as the *sparse model*.

It is worth noting that neither the dense model nor the sparse one quite captures our intuitive notion of "small" – most people would think of $1 + (x - 1)^{100}$ as a "small expression," but neither the dense nor sparse models would think so. This is the domain of the "straight-line program" model, but this is beyond the scope of this article. An example showing the superior performance, both in time and more especially in space, of the straight-line program over conventional data structures is given in [25].

8.3.2 Greatest Common Divisors

The computation of polynomial common divisors has many uses, not least in the simplification of rational functions, as in $(x^2 + 3x + 2)/(x + 1) \to x + 2$. The problem might seem to be solved by Euclid's Algorithm (8.1), and, indeed, conceptually it is. In practice, though, there are two serious problems with this.

Algorithm 8.1 (Euclid)

Input: $a_0, a_1 \in K[x]$.
Output: $h \in K[x]$ *a greatest common divisor of* a_0 *and* a_1

$i := 1$;
while $a_i \neq 0$ **do**
  $a_{i+1} = \mathrm{rem}(a_{i-1}, a_i)$; # Let q_i be the corresponding quotient
  $i := i + 1$;
return a_{i-1};

8.3.2.1 Intermediate Expression Swell

As we go round the loop in Algorithm 8.1, the degree (in x) of the polynomials does keep decreasing. But nothing is said about the size of the coefficients. If we apply this algorithm to two fairly innocuous polynomials:

$$\left.\begin{aligned} A(x) &= x^8 + x^6 - 3x^4 - 3x^3 + 8x^2 + 2x - 5; \\ B(x) &= 3x^6 + 5x^4 - 4x^2 - 9x - 21. \end{aligned}\right\} \tag{8.2}$$

we get the sequence

$$\frac{-5}{9}x^4 + \frac{127}{9}x^2 - \frac{29}{3},$$
$$\frac{50157}{25}x^2 - 9x - \frac{35847}{25},$$
$$\frac{93\,060\,801\,700}{1\,557\,792\,607\,653}x + \frac{23\,315\,940\,650}{173\,088\,067\,517}$$

and finally

$$\frac{761\,030\,000\,733\,847\,895\,048\,691}{86\,603\,128\,130\,467\,228\,900}.$$

As this is a number, it follows that no *polynomial* can divide both A and B, that is, that $\gcd(A, B) = 1$. The reader might think that "clearing fractions" would help: it does not, and we end up with

$$7\,436\,622\,422\,540\,486\,538\,114\,177\,255$$
$$855\,890\,572\,956\,445\,312\,500.$$

Had the coefficients not been integers, but rather polynomials in $y, z, \ldots$, the growth would have been even more spectacular: see the worked example at *http://staff.bath.ac.uk/masjhd/JHD-CA/GCD3var.html*. The good news is that advanced algorithms (Section 8.4) have pretty much solved this problem: we can compute greatest common divisors (gcds) efficiently, working all the time with data whose polynomial degrees in $y, z, \ldots$ are no larger than those in the input, and with numerical coefficients that are not much larger than in the input. We say "not much larger" because a gcd *can* have larger coefficients than the input, for example

$$\begin{aligned} A &= x^5 + 3x^4 + 2x^3 - 2x^2 - 3x - 1 \\ &= (x+1)^4(x-1); \\ B &= x^6 + 3x^5 + 3x^4 + 2x^3 + 3x^2 + 3x + 1 \\ &= (x+1)^4(x^2 - x + 1); \\ \gcd(A, B) &= x^4 + 4x^3 + 6x^2 + 4x + 1 \\ &= (x+1)^4. \end{aligned} \tag{8.3}$$

These algorithms tend to be probabilistic, and produce an answer g, which, if it is a common divisor of A and B at all (and this needs to be verified, either by checking that g divides A and B, or by computing A/g and B/g as well as g), is guaranteed to be a greatest common divisor.

8.3.2.2 Sparsity

The "good news" above is indeed good as far as it goes, and shows that, *in the dense model*, we can do about as well as possible, but it does not speak to the sparse model. In particular, it still leaves open the possibilities that

1. the gcd g might be much denser than the inputs A and B
2. even if the gcd g is not much denser than the inputs A and B, the cofactors A/g and B/g computed as part of the verification might be much denser
3. even if neither of these happens, various intermediate results might be much denser than A and B.

Problem 2 is easy to demonstrate: consider the cofactors in $\gcd(x^p - 1, x^q - 1)$, where p and q are distinct primes. Problem 1 is demonstrated by the following elegant example of [26]

$$\gcd(x^{pq} - 1, x^{p+q} - x^p - x^q + 1) = x^{p+q-1} - x^{p+q-2} \pm \cdots - 1, \tag{8.4}$$

and indeed just knowing *whether* two polynomials have a nontrivial gcd is hard, by the following result.

Theorem 8.1 [27] *It is NP-hard to determine whether two sparse polynomials (in the standard encoding) have a nontrivial common divisor.*

This theorem, like the examples above, relies on the factorization of $x^p - 1$, and it is an open question [24, Challenge 3] whether this is the only obstacle.

More precisely, we have the following equivalent of the problem solved for division.

Challenge 8.1 [24, Challenge 5] *Find an algorithm for computing $h = \gcd(f, g)$ which is polynomial-time in t_f, t_g, and t_h.*

A weaker problem, but still unsolved, is

Challenge 8.2 *Find an algorithm for computing $h = \gcd(f, g)$ which is polynomial-time in t_f, t_g, t_h, and $t_{f/h}$, $t_{g/h}$.*

Problem 3 is particularly acute in the case of multivariate polynomials. Further developments [28, Section 4.4] of the ideas in Section 8.4, using methods from [29], have produced algorithms whose time seems empirically to vary as d^2t rather than d^n for polynomials of degree d and t terms, in n variables. Hence this problem is largely solved in practice, but theoretical analysis is still lacking, as we cannot solve Challenges 8.1 and 8.2.

8.3.3 Square-free Decomposition

One particular use of gcds is given by the following result.

Proposition 8.1 *If α is a repeated root of the polynomial f, then it is a root of f', and hence of $\gcd(f, f')$. In fact, the roots of $\gcd(f, f')$ are* precisely *the repeated roots of f.*

Definition 8.1 *A polynomial f is said to be* square-free *if it has no repeated roots. The* square-free decomposition *of f is $f = \prod f_i^i$, where each f_i is square-free and $\gcd(f_i, f_j) = 1$ if $i \neq j$.*

Theorem 8.2 [30] *Over* **Z** *and in the standard sparse encoding, the two problems*

1. *deciding if a polynomial is square-free and*
2. *deciding if two polynomials have a nontrivial gcd.*

are equivalent under randomized polynomial-time reduction.

Hence, in the light of Theorem 8.1, determining square-freeness is hard, at least when polynomials with factors of the form $x^p - 1$ are involved. Again, it is an open question whether these (and scaled variants) are the only obstacles.

8.3.4 Extended Euclidean Algorithm

If we look at Algorithm 8.1, we see that a_0 and a_1 are, trivially, each a linear combination of a_0 and a_1. $a_{i+1} = a_{i-1} - q_i a_i$, so is a linear combination of a_0 and a_1 if each of a_{i-1} and a_i are. Hence by induction every a_i is a linear combination of a_0 and a_1, and, in particular, the gcd is.

Proposition 8.2 (Bezout's identity)
There exist polynomials λ *and* μ *such that* $\gcd(f, g) = \lambda f + \mu g$.

Note that Algorithm 8.1, as we ran it, generated fractions, so λ and μ may well have fractions as coefficients.

A useful application of Bezout's identity is the theory of partial fractions: if f and g are relatively prime, then

$$\frac{h}{fg} = \frac{h(\lambda f + \mu g)}{fg} = \frac{\lambda h}{g} + \frac{\mu h}{f}. \tag{8.5}$$

Even if h/fg is reduced ($\deg(h) < \deg(fg)$), the summands may not be, but we can adjust them:

$$\frac{h}{fg} = \frac{\operatorname{rem}(\lambda h, g)}{g} + \frac{\operatorname{rem}(\mu h, f)}{f}, \tag{8.6}$$

and the summands are now reduced.

8.4 Advanced Algorithms

There are two main classes of "advanced" algorithms in computer algebra for solving, not only the gcd problem and its variants, but many others.

8.4.1 Modular Algorithms – Integer

We described the polynomials in (8.2) as "innocuous." They *are* innocuous if we consider them, and apply Algorithm 8.1, modulo 5, writing P_5 to signify the polynomial P considered as a polynomial with coefficients modulo 5.

$$\begin{aligned}
A_5(x) &= x^8 + x^6 + 2x^4 + 2x^3 + 3x^2 + 2x;\\
B_5(x) &= 3x^6 + x^2 + x + 1;\\
C_5(x) &= \text{remainder}(A_5(x), B_5(x))\\
&= A_5(x) + 3(x^2+1)\,B_5(x) = 4x^2 + 3;\\
D_5(x) &= \text{remainder}(B_5(x), C_5(x))\\
&= B_5(x) + (x^4 + 4x^2 + 3)\,C_5(x) = x;\\
E_5(x) &= \text{remainder}(C_5(x), D_5(x))\\
&= C_5(x) + x\,D_5(x) = 3.
\end{aligned}$$

But if P divides A, then $A = PQ$ for some polynomial Q, and $A_5 = P_5 Q_5$. Similarly, if P divides B, then $B = PR$ for some polynomial R, and $B_5 = P_5 R_5$. Hence if $P = \gcd(A, B)$, P_5 has to be a common factor of A_5 and B_5, but there no a nontrivial one, so P_5 has to be 1, and, because the leading coefficient of P has to be ± 1, $P = 1$. Of course, there is nothing special about 5: we could have picked almost any prime – 2 does not work as $\gcd(A_2, B_2) \neq 1$ and 3 is dubious as B_3 has lower degree than B. However, we have to answer several questions before we can make this into a complete gcd algorithm replacing Algorithm 8.1.

1. How do we calculate a nontrivial gcd?
2. what do we do if the gcd modulo p is not the modular image of the gcd (as with A_2 and B_2)? In this case, we say that we have *bad reduction*, or that p is *bad*. In particular, what do we do if this keeps on happening?
3. How much does this method cost?

The first question already makes us think: if the gcd is $x + 42$, we are never going to decide this by working modulo 5 (which might suggest $x + 2$), or indeed any prime smaller than $85 = 2 \times 42 + 1$, because we have to allow for the sign, as working modulo 43 would suggest $x - 1$ as the answer, rather than $x + 42$. Furthermore, we know from (8.3) that the largest coefficient in a gcd is not, as we might hope, at most the largest coefficient in the inputs. Fortunately, we can nevertheless bound the largest coefficient in a gcd.

Proposition 8.3 (**Landau–Mignotte bound [31, 32]**) *Every coefficient of the gcd of* $A = \sum_{i=0}^{\alpha} a_i x^i$ *and* $B = \sum_{i=0}^{\beta} b_i x^i$ *(with* a_i *and* b_i *integers) is bounded by*

$$LM(A, B) := 2^{\min(\alpha,\beta)} \gcd(a_\alpha, b_\beta) \min\left(\frac{1}{|a_\alpha|}\sqrt{\sum_{i=0}^{\alpha} a_i^2}, \frac{1}{|b_\beta|}\sqrt{\sum_{i=0}^{\beta} b_i^2}\right). \quad (8.7)$$

Furthermore, the number 2 in $2^{\min(\alpha,\beta)}$ *cannot be replaced by any smaller number* [33].

Once we know this, we have two choices.

- Work modulo a prime $p > 2LM(A, B)$, so that, unless problem 2 occurs, P_p will, interpreting its coefficients as integers in $-p/2, \ldots, p/2$, be P itself. This is perfectly correct, but often inefficient as, although (8.7) cannot be improved on as a *worst case* bound, it is normally far too pessimistic.
- Work modulo several different small primes p_i, use the Chinese Remainder Theorem [28, Section A.3] to compute $P_{\prod p_i}$ from the various P_{p_i}. As before, if $\prod p_i > 2LM(A, B)$, $P_{\prod p_i}$ will, interpreting its coefficients as integers in $-(\prod p_i)/2, \ldots, (\prod p_i)/2$, be P itself. This method lends itself to various shortcuts if (8.7) is pessimistic in a given case, as it nearly always is.

Problem 2 is nevertheless a genuine problem. Again, we can say something.

Proposition 8.4 (**Good Reduction Theorem (Z) [28]**) *The bad* p *are those that divide either* $\gcd(a_\alpha, b_\beta)$ *or a certain nonzero determinant* D *(in fact, the resultant* $\mathrm{Res}_x(A/P, B/P)$*). Furthermore, if* p *divides* $\mathrm{Res}_x(A/P, B/P)$ *but not* $\gcd(a_\alpha, b_\beta)$, *then the gcd computed modulo* p *has a* larger *degree than the true result.*

In particular, there are only a finite number of bad primes.

We can test the first half of this (p dividing $\gcd(a_\alpha, b_\beta)$) and not even try such primes. After this, if we have two primes that give us different degrees for the gcd, we know that the prime with the higher degree has bad reduction, from the "Furthermore ..." statement, so we can discard it. The corresponding algorithm is given as [Algorithm Modular GCD (small prime version)] [28].

As far as the running time is concerned, if l bounds the length of the coefficients in the input *and* output, and d the degree of the polynomials, then the running time is $C(d + 1)(l^2 + (d + 1))$ for a suitable constant C, depending on the computer, software, and so on, provided we ignore the possibility of unlucky primes [34, (95)]. We note that this depends on the degree, not the number of terms, as even modulo p we cannot solve Challenges 8.1 and 8.2.

8.4.2 Modular Algorithms–Polynomial

Despite our failure to address the sparsity issue, that algorithm is still substantially better than Euclid's, because the coefficient growth is bounded, by Proposition 8.3. Can we do the same for multivariate polynomials, and avoid the spectacular growth seen in *http://staff.bath.ac.uk/masjhd/JHD-CA/GCD3var.html*? We consider the case of polynomials in x and y, regarding x as the main variable: the generalization to more variables is easy. There is one preliminary remark: the gcd might actually not depend on x, but be a polynomial in y alone. In that case, it would have to divide every coefficient of the input polynomials, regarded as polynomials in x whose coefficients are polynomials in y. Hence, it can be determined by gcds of polynomials in y alone, and we will not concern ourselves further with this (though the software implementer clearly has to), and consider only the part of the gcd that depends on x.

The answer is that we can find this efficiently, and the idea is similar to the previous one: we replace "working modulo p" with "evaluating y at v." Just as we wrote P_5 to signify the polynomial P considered as a polynomial with coefficients modulo 5, we write $P_{y=v}$ to signify the polynomial P with y evaluated at the value v. But if P divides A, then $A = PQ$ for some polynomial Q, and $A_{y=v} = P_{y=v}Q_{y=v}$. Similarly, if P divides B, then $B = PR$ for some polynomial R, and $B_{y=v} = P_{y=v}R_{y=v}$. Hence if $P = \gcd(A, B)$, $P_{y=v}$ has to be a common factor of $A_{y=v}$ and $B_{y=v}$. In particular, if $A_{y=v}$ and $B_{y=v}$ have no common factor, the gcd of A and B cannot depend on x. Analogous questions to those posed above appear.

1. How do we calculate a nontrivial gcd?
2. What do we do if the gcd of the evaluated polynomials is not the evaluation of the true gcd? In this case, we say that we have *bad reduction*, or that v is a *bad* value. In particular, what do we do if this keeps on happening?
3. How much does this method cost?

The answers are similar, or indeed easier. For question 1, we use the Chinese Remainder Theorem for polynomials [28, Section A.4] to compute the true result from several evaluations. How many? In the integer coefficient case, we had to rely on Proposition 8.3, but the equivalent is trivial.

Proposition 8.5 (Polynomial version of Proposition 8.3) $\deg_y(\gcd(f,g)) \leq \min(\deg_y(f), \deg_y(g))$.

For question 2, we have a precise analogue of Proposition 8.4.

Proposition 8.6 (Good Reduction Theorem (polynomial) [28]) *If* $y - v$ *does not divide* $\gcd(a_\alpha, b_\beta)$ *(which can be checked for in advance) or* $\mathrm{Res}_x(A/C, B/C)$, *then* v *is of good reduction. Furthermore, if* $y - v$ *divides* $\mathrm{Res}_x(A/C, B/C)$ *but not* $\gcd(a_\alpha, b_\beta)$, *then the gcd computed modulo* $y - v$ *has a* larger *degree than the true result.*

This gives us algorithms precisely equivalent to those in the previous section, and then do indeed generalize to n variables. Question 3 has the following answer. If l bounds the length of the coefficients in the input *and* output, and d the degree of the polynomials, then the running time is $C(d+1)^n(l^2 + l(d+1))$ for a suitable constant C, depending on the computer, software, and so on, provided we ignore the possibility of unlucky primes and values [34, (95)].

8.4.3 The Challenge of Factorization

If we want to factor polynomials, we might as well first do a square-free factorization (Section 8.3.3): as well as efficient, this will turn out to be necessary. This means that $\gcd(f, f') = 1$. It follows from Proposition 8.4 that there are only finitely many primes for which $\gcd(f_p, f_p') \neq 1$, and we check for, and discard, these primes.

Then the obvious solution is to follow this approach, and start with factoring modulo p. For small primes p and degrees d, we can just enumerate the $p^{(d+1)/2}$ possibilities,[4] but it turns out we can do much better. There are algorithms due to Berlekamp [35, 36] and Cantor–Zassenhaus [37], which take time proportional to d^3.

Modular methods, as described in the previous two sections, can indeed solve many other problems besides the gcd one. Indeed, no computer algebra conference goes past without a new modular algorithm appearing somewhere. However, there is one problem they cannot solve efficiently, and that is polynomial factorization.

This might seem strange, for we have the same modular/integer relationship as we had in the gcd case: if $f = f^{(1)} \cdot f^{(2)} \cdot \cdots \cdot f^{(k)}$ over the integers, then $f_p = f_p^{(1)} \cdot f_p^{(2)} \cdot \cdots \cdot f_p^{(k)}$. In particular, if f and f_p have the same degree, and f_p is irreducible, then f is also irreducible. We may as well remark that the converse is not true: it is possible for an irreducible f to factor modulo every prime.

Example 8.1 The polynomial $x^4 + 1$, which is irreducible over the integers, factors into two quadratics (and possibly further) modulo *every* prime.

$p = 2$ Then $x^4 + 1 = (x + 1)^4$.

$p = 4k + 1$ In this case, -1 is always a square, say $-1 = q^2$. This gives us the factorization $x^4 + 1 = (x^2 - q)(x^2 + q)$.

$p = 8k \pm 1$ In this case, 2 is always a square, say $2 = q^2$. This gives us the factorization $x^4 + 1 = (x^2 - (2/q)x + 1)(x^2 + (2/q)x + 1)$. In the case $p = 8k + 1$, we have this factorization and the factorization given in the previous case. As these two factorizations are not equal, we can calculate the gcds of the factors, in order to find a factorization as the product of four linear factors.

$p = 8k + 3$ In this case, -2 is always a square, say $-2 = q^2$. This is a result of the fact that -1 and 2 are not squares, and so their product must be a square. This property of -2 gives us the factorization $x^4 + 1 = (x^2 - (2/q)x - 1)(x^2 + (2/q)x - 1)$

4) The exponent $(d + 1)/2$ comes from the fact that we need only look for factors of degree $\leq d/2$.

This polynomial is not an isolated oddity: Swinnerton-Dyer [38] and Kaltofen *et al.* [39] proved that there are whole families of polynomials with this property of being irreducible, but of factorizing modulo every prime. Furthermore, these polynomials have the annoying habit of being generated disproportionately often [40]. However, this is not the main problem.

Suppose f_p factors as $g_1 g_2 g_3$, and f_q factors as $h_1 h_2 h_3$. We may be in luck, and these factorizations may be so incompatible that we can deduce that f has to be irreducible. However, they may be compatible, say all have degree k. Then it would be natural, following Section 8.4.2, to apply the Chinese Remainder Theorem, and deduce that

$$f_{pq} = \mathrm{CRT}(g_1, h_1)\mathrm{CRT}(g_2, h_2)\mathrm{CRT}(g_3, h_3). \tag{8.8}$$

This would indeed be correct, but it is not the only possibility. Our labels 1–3 are purely arbitrary, and it would be just as consistent to deduce

$$f_{pq} = \mathrm{CRT}(g_1, h_2)\mathrm{CRT}(g_2, h_3)\mathrm{CRT}(g_3, h_1), \tag{8.9}$$

or any of the other four possibilities. If we take $f = x^3 + 41x^2 + 551x + 2431$, $f_5 = (x + 2)(x + 1)(x + 3)$ and $f_7 = (x + 4)\ (x + 3)(x + 6)$. Combining $(x + 2)_5$ and $(x + 4)_7$

gives $(x+32)_{35}$, and indeed this divides f_{35}. If we combine $(x+2)_5$ and $(x+3)_7$, we get $(x+17)_{35}$, and again this divides f_{35}. In fact, we deduce that f_{35}, though of degree 3, has nine linear factors. An algebraist would tell us that polynomials modulo a composite, such as 35, do not possess unique factorization.

However, not all the possibilities (8.8), (8.9), and so on correspond to the true factorization. Although we have no a priori way of knowing it, a factor modulo p corresponds (apart from the bad reduction problem) to one, and only one, factor over the integers, and this corresponds to one, and only one, factor modulo q. Hence, if there are m rather than just three factors, there may be $m!$ possible matchings of the results modulo p and q, only one of which is right.

Worse, if pq is not bigger than twice the Landau–Mignotte bound, we will need to use a third prime r, and we will now have $(m!)^2$ possibilities, only one of which is right, and so on. This is clearly unsustainable, and we need a different approach.

8.4.4 *p*-adic Algorithms–Integer

Instead of using several primes, and deducing the answer modulo p, then pq and pqr, and so on, we will use one prime, and deduce the answer modulo p, then p^2, p^3, and so on. For simplicity, we will assume that $f_p = g_p h_p$ modulo p–more factors do not pose more challenges, but make the notation difficult. We will also suppose that f is monic: this is a genuine simplification, but solving the nonmonic case as well would distract us–details are in [28, Chapter 5].

Let us suppose that $f_{p^2} = g_{p^2} h_{p^2}$ where, and this is the difference, we insist that g_{p^2} corresponds to g_p, that is, that they are equal modulo p. We can therefore write $g_{p^2} = g_p + p\hat{g}_{p^2}$, and similarly for the h's. Then

$$f_{p^2} = \left(g_p + p\hat{g}_{p^2}\right)\left(h_p + p\hat{h}_{p^2}\right),$$

so

$$\frac{f_{p^2} - g_p h_p}{p} \equiv g_p \hat{h}_{p^2} + h_p \hat{g}_{p^2} \pmod{p}, \quad (8.10)$$

where the cross-product $g_{p^2} h_{p^2}$ disappears as it is multiplied by an extra p, and the fraction on the left is not really a fraction because $g_p h_p \equiv f_p \equiv f_{p^2} \pmod{p}$.

As we have made f_p square-free (remark at the start of Section 8.4.3), $\gcd(g_p, h_p) = 1$, Bezout's identity (Proposition 8.2) tells us that there are λ, μ such that $\lambda g_p + \mu h_p = 1$. Hence there are $\hat{\lambda}$, $\hat{\mu}$ such that

$$\hat{\lambda} g_p + \hat{\mu} h_p \equiv \frac{f_{p^2} - g_p h_p}{p} \pmod{p}. \quad (8.11)$$

The obvious choice is $\hat{\lambda} = \lambda f_{p^2} - g_p h_p / p$, but this has too large a degree. However, if we take the remainder (this is the same idea as (8.6)) of this with respect to h_p, and similarly with μ, we obtain $\hat{\lambda}$, $\hat{\mu}$ satisfying (8.11) and with $\deg \hat{\lambda} < \deg h_p$, $\deg \hat{\mu} < \deg g_p$. Hence we can take $g_{p^2} = g_p + p\hat{\mu}$, $h_{p_2} = h_p + p\hat{\lambda}$.

We now suppose that $f_{p^3} = g_{p^3} h_{p^3}$ where we insist that g_{p^3} corresponds to g_{p^2}, that is, that they are equal modulo p^2. We can therefore write $g_{p^3} = g_{p^2} + p^2 \hat{g}_{p^3}$, and similarly for the h's. Then

$$f_{p^3} = \left(g_{p^2} + p^2 \hat{g}_{p^3}\right)\left(h_{p^2} + p^2 \hat{h}_{p^3}\right),$$

so

$$\frac{f_{p^3} - g_{p^2} h_{p^2}}{p^2} \equiv g_p \hat{h}_{p^3} + h_p \hat{g}_{p^3} \pmod{p} \quad (8.12)$$

as before. Using the same Bezout's identity as before, we get that there are $\hat{\hat{\lambda}}$, $\hat{\hat{\mu}}$ such that

$$\hat{\hat{\lambda}} g_p + \hat{\hat{\mu}} h_p \equiv \frac{f_{p^3} - g_{p^2} h_{p^2}}{p^2} \pmod{p} \quad (8.13)$$

with $\deg \hat{\hat{\lambda}} < \deg h_p$, $\deg \hat{\hat{\mu}} < \deg g_p$. Hence we can take $g_{p^3} = g_{p^2} + p^2 \hat{\hat{\mu}}$, $h_{p_3} = h_{p^2} + p^2 \hat{\hat{\lambda}}$, and so on until we have $f_{p^n} = g_{p^n} h_{p^n}$ with p^n greater than twice the Landau–Mignotte bound on factors of f. Then, regarded as polynomials with integer coefficients in the range $(-p^n/2, p^n/2)$, g_{p^n} and h_{p^n} should be the factors of f.

However, as we saw in Example 8.1, f might factor more modulo p than it does over the integers. Nevertheless, all factorizations of f over the integers must correspond to factorizations, not necessarily into irreducibles, modulo p. Hence we need merely take all subsets of the factors modulo p, and see if their product corresponds to a factor over the integers. In principle, this process is exponential in the degree of f if it factors into low-degree polynomials modulo p but is in fact irreducible over the integers. In practice, it is possible to make the constant in front of the exponential very small, and to do better than looking at every subset [41].

8.4.5
p-adic Algorithms–Polynomial

It is also possible to use these methods in a multivariate setting, going from a solution modulo $(y - v)$ to one modulo $(y - v)^2$, then modulo $(y - v)^3$, and so on, but the details are beyond the scope of this chapter: see [28, Section 5.8]

8.5
Solving Polynomial Systems

8.5.1
Solving One Polynomial

Consider the quadratic equation

$$ax^2 + bx + c = 0. \quad (8.14)$$

The solutions are well known to most schoolchildren: there are two of them, of the form

$$x = \frac{-b \pm \sqrt{b^2 - 4ac}}{2a}. \quad (8.15)$$

However, if $b^2 - 4ac = 0$, that is, $c = b^2/4a$ then there is only one solution: $x = -b/2a$. In this case, the equation becomes $ax^2 + bx + b^2/4a = 0$, which can be rewritten as $a(x + b/2a)^2 = 0$, making it more obvious that there is a repeated root, and that the polynomial is not square-free (Definition 8.1).

Mathematicians dislike the sort of anomaly in "this equations has two solutions except when $c = b^2/4a$," especially as there are two roots as c tends to the value $b^2/4a$. We therefore say that, in this special case, $x = -b/2a$ is *a double root* of the equation. When we say we are counting the roots of f *with multiplicity,* we mean that $x = \alpha$ should be counted i times if $(x - \alpha)^i$ divides f.

Proposition 8.7 (**Fundamental Theorem of Algebra**) *The number of roots of a polynomial equation over the complex numbers, counted with multiplicity, is equal to the degree of the polynomial.*

There is a formula for the solutions of the cubic equation

$$x^3 + ax^2 + bx + c = 0, \quad (8.16)$$

$$
\begin{array}{lll}
S & := & \sqrt{12\,b^3 + 81\,c^2}\,; \\
T & := & \sqrt[3]{-108\,c + 12\,S}; \\
\textbf{return} & & \frac{1}{6}T - \frac{2b}{T};
\end{array}
$$

Figure 8.1 Program for computing solutions to a cubic.

albeit less well known to schoolchildren:

$$
\frac{1}{6}\sqrt[3]{36\,ba - 108\,c - 8\,a^3 + 12\,\sqrt{12\,b^3 - 3\,b^2a^2 - 54\,bac + 81\,c^2 + 12\,ca^3}}
- \frac{2b - \frac{2}{3}a^2}{\sqrt[3]{36\,ba - 108\,c - 8\,a^3 + 12\,\sqrt{12\,b^3 - 3\,b^2a^2 - 54\,bac + 81\,c^2 + 12\,ca^3}}} - \frac{1}{3}a. \tag{8.17}
$$

We can simplify this by making a transformation[5] to (8.16): replacing x by $x - a/3$. This transforms it into an equation

$$x^3 + bx + c = 0 \tag{8.18}$$

(where b and c have changed). This has solutions of the form

$$
\frac{1}{6}\sqrt[3]{-108\,c + 12\,\sqrt{12\,b^3 + 81\,c^2}} - \frac{2b}{\sqrt[3]{-108\,c + 12\,\sqrt{12\,b^3 + 81\,c^2}}}. \tag{8.19}
$$

Many texts (among those who discuss the cubic at all) stop here, but in fact the computational analysis of (8.19) is nontrivial. A cubic has (Proposition 8.7) three roots, but a naïve look at (8.19) shows two cube roots, each with three values, and two square roots, each with two values, apparently giving a total of $3 \times 3 \times 2 \times 2 = 36$ values. Even if we decide that the two occurrences of the square root should have the same sign, and similarly the cube root should have the same value, that is, we effectively execute the program in Figure 8.1, we would still seem to have six possibilities. In fact, however, the choice in the first line is only apparent, because

$$
\frac{1}{6}\sqrt[3]{-108\,c - 12\,\sqrt{12\,b^3 + 81\,c^2}} = \frac{2b}{\sqrt[3]{-108\,c + 12\,\sqrt{12\,b^3 + 81\,c^2}}}. \tag{8.20}
$$

In the case of the quadratic with real coefficients, there were two real solutions if $b^2 - 4ac > 0$, and complex solutions otherwise. However, the case of the cubic is more challenging. If we consider $x^3 - 1 = 0$, we compute (in Figure 8.1)

$$S := 9; \qquad T := 6; \qquad \textbf{return } 1;$$

(or either of the complex cube roots of unity if we choose different values of T). If we consider $x^3 + 1 = 0$, we get

$$S := 9; \qquad T := 0; \qquad \textbf{return } \text{“}\tfrac{0}{0}\text{”};$$

but we can (and must!) take advantage of (8.20) and compute

$$S := -9; \qquad T := -6; \qquad \textbf{return} -1;$$

(or either of the complex variants).

5) This is the simplest case of the Tschirnhaus transformation [42], which can always eliminate the x^{n-1} term in a polynomial of degree n.

For $x^3 + x$, we compute

$$S := \sqrt{12}; \qquad T := \sqrt{12}; \qquad \textbf{return } 0;$$

and the two complex roots come from choosing the complex roots in the computation of T, which is really $\sqrt[3]{12\sqrt{12}}$. $x^3 - x$ is more challenging: we compute

$$S := \sqrt{-12}; \quad T := \sqrt{-12}; \quad \textbf{return } \{-1, 0, 1\}; \tag{8.21}$$

that is, three real roots which can only be computed (at least via this formula) by means of complex numbers. In fact, it is clear that any other formula must have the same problem, because the only choices of ambiguity lie in the square and cube roots, and with the cube root, the ambiguity involves complex cube roots of unity.

Hence even solving the humble cubic makes three points.

1. The formulae (8.17) and (8.19) are ambiguous, and a formulation such as Figure 8.1 is to be preferred.
2. Even so we need to take care of $0/0$ issues.
3. Expressing real roots in terms of radicals may need complex numbers.

These points are much clearer with the quartic [28, Section 3.1.3]. Higher degrees produce an even more fundamental issue.

Theorem 8.3 (Abel, Galois [43]) *The general polynomial equation of degree 5 or more is not solvable in radicals (i.e. in terms of kth roots).*

Hence, in general, the *only* description is "a root of the polynomial $x^{\cdots} + \cdots$." As (8.19), and even more its quartic equivalent, is nontrivial, it turns out to be preferable to use this description for cubics and quartics internally as well, only converting to radicals on output (and possibly only if explicitly requested). See the example at (8.33), and contrast it with (8.27).

8.5.2 Real Roots

While a polynomial of degree n always has n roots (counted with multiplicity), it usually[6] has fewer *real* roots. While the roots α_i of an (irreducible) polynomial f are algebraically indistinguishable, because they all share the property that $f(\alpha_i) = 0$, this ceases to be true once we start asking questions about *real* roots. $f(x) = x^2 - r$ has real roots if, and only if, r is nonnegative, so asking questions about the reality of roots means that we can ask inequalities as well. Conversely, inequalities only make sense if we are talking about real, rather than complex numbers.

Just as we have seen that we can, in general, do no better than say "a root of the polynomial $f(x) = x^n + \cdots$," we cannot do much better than say "a real root of the polynomial $f(x) = x^n + \cdots$," but we can at least make it clear *which* of the roots we are talking about. There are two ways of doing this, which in practice we always apply to square-free f.

1. "The (unique) real root of $f(x)$ between a and b." (a, b) is then known as an *isolating interval*, and can be computed by a variety of means: see [28, Section 3.1.9]. Once we have an isolating interval, standard techniques

6) Quite what meaning should be attached to "usually" is surprisingly difficult. The "obvious" definitions would be "normally distributed" or "uniformly distributed in some range," and for these Kac [44] shows that the average number is $(2/\pi)\log n$. A definition with better geometric invariance properties gives $\sqrt{n(n+2)/3}$: very different [45].

of numerical analysis can refine it to be as small as we wish.

2. "The (unique) x such that $f(x) = 0$ and the derivatives of f satisfy the following sign conditions at x." It is a result known as *Thom's lemma* [46, Proposition 1.2] that specifying the sign conditions defines a root uniquely, if at all: see [28, Section 3.1.10].

8.5.3 Linear Systems

The same algorithms we are familiar with from numeric linear algebra are, in principle, applicable here. Hence, given a system of linear equations, written in matrix form as $\mathbf{M}{\cdot}\mathbf{x} = \mathbf{b}$, we can (possibly with pivoting), apply Gaussian elimination (row operations – adding a multiple of one row to another) to get $\mathbf{U}{\cdot}\mathbf{x} = \mathbf{b}'$, where $\mathbf{u}$ is upper triangular, and do back substitution. Similarly, we can compute the inverse of a matrix.

The real problem with symbolic, as opposed to numeric, linear algebra is not that the algorithms produce the output inefficiently so much as the fact that the output is huge: the inverse of an $n \times n$ matrix of numbers is an $n \times n$ matrix of numbers, but the inverse of an $n \times n$ matrix of small formulae may well be an $n \times n$ matrix of very large formulae. While we are all happy with

$$\begin{pmatrix} a & b \\ c & d \end{pmatrix}^{-1} = \begin{pmatrix} \frac{d}{ad-bc} & -\frac{b}{ad-bc} \\ -\frac{c}{ad-bc} & \frac{a}{ad-bc} \end{pmatrix}, \tag{8.22}$$

the reader is invited to experiment with the equivalent for larger matrices. In fact, the determinant of a generic $n \times n$ symbolic matrix is the sum of $n!$ terms, and this is the denominator of a generic inverse, while the numerators, of which there are n^2, are the sum of $(n-1)!$ terms, giving a total of $(n+1)!$ summands. Hence our aim here should be to *avoid* solving the generic problem, and solve systems with as few variables as possible. A further snag is that the inverse of a sparse matrix tends to be denser, often completely dense. For further details, see [28, Section 3.2].

8.5.4 Multivariate Systems

If nonlinear polynomial equations in one variable are more complicated than we might think, and generic linear systems in several variables are horrendous, we might despair of nonlinear systems in several variables. While these might be difficult, they are not without algorithmic approaches. One major one,[7] Gröbner bases, is based fundamentally on the *sparse-distributed* representation.

In this representation, a nonzero polynomial is represented as a sum of terms $\sum_i c_i m_i$ where the c_i are nonzero coefficients and the m_i are monomials, that is, products of powers of the variables. As mentioned earlier, when we asked whether x^2y came before or after xy^3, a key question is how we order the monomials. While there are many options [28, Section 3.3.3], three key ways of comparing $A = x_1^{a_1} \cdots x_n^{a_n}$ and $B = x_1^{b_1} \cdots x_n^{b_n}$ are the following.

Purely lexicographic – `plex` . We first compare a_1 and b_1. If they differ, this tells us whether $A > B$ ($a_1 > b_1$) or $A < B$ ($a_1 < b_1$). If they are the same, we go on to look at a_2 versus b_2 and so on. The order is similar to looking up words in a dictionary/lexicon – we look at the first letter, and after finding this, look at the second letter, and so on. In this order, x^2 is more important than xy^{10}.

7) The other principal one, *regular chains* is introduced in Section 8.5.6 and described in [28, Section 3.4].

Total degree, then lexicographic
−`grlex`. We first look at the total degrees $a = \sum a_i$ and $b = \sum b_i$: if $a > b$, then $A > B$, and $a < b$ means $A < B$. If $a = b$, then we look at lexicographic comparison. In this order xy^{10} is more important than x^2, and x^2y more important than xy^2.

Total degree, then reverse lexicographic
−`tdeg`. This order is the same as the previous, except that, if the total degrees are equal, we look lexicographically, then *take the opposite*. Many systems, in particular Maple and Mathematica, reverse the order of the variables first. The reader may ask "if the order of the variables is reversed, and we then reverse the sense of the answer, what's the difference?" Indeed, for two variables, there is no difference. However, with more variables it does indeed make a difference. For three variables, the monomials of degree three are ordered as

$$x^3 > x^2y > x^2z > xy^2 > xyz > xz^2 \\ > y^3 > y^2z > yz^2 > z^3$$

under `grlex`, but as

$$x^3 > x^2y > xy^2 > y^3 > x^2z > xyz > \\ y^2z > xz^2 > yz^2 > z^3$$

under `tdeg`. One way of seeing the difference is to say that `grlex` with $x > y > z$ discriminates *in favor of* x, whereas `tdeg` with $z > y > x$ discriminates *against* z. This metaphor reinforces the fact that there is no difference with two variables.

It seems that `tdeg` is, in general, the most efficient order.

Once we have fixed such an order, we have such concepts as the *leading monomial* or *leading term* of a polynomial, denoted $\text{lm}(p)$ and $\text{lt}(p)$. The coefficient of the leading term is the *leading coefficient*, denoted $\text{lc}(p)$.

8.5.5 Gröbner Bases

Even though the equations are nonlinear, Gaussian elimination is still available to us. So, given the three equations

$$x^2 - y = 0 \qquad x^2 - z = 0 \qquad y + z = 0,$$

we can subtract the first from the second to get $y - z = 0$, hence (*linear* algebra on this and the third) $y = 0$ and $z = 0$, and we are left with $x^2 = 0$, so $x = 0$, albeit with multiplicity 2.

However, we can do more than this. Given the two equations

$$x^2 - 1 = 0 \quad \text{and} \quad xy - 1 = 0, \qquad (8.23)$$

there might seem to be no row operation available. But in fact we can subtract x times the second equation from y times the first, to get $x - y = 0$. Hence the solutions are $x = \pm 1$, $y = x$.

This might seem to imply that, given two equations $f = 0$ and $g = 0$, we need to consider replacing $f = 0$ by $Ff + Gg = 0$ for arbitrary polynomials F and G. This would be erroneous on two counts.

(a) We do not need to consider arbitrary polynomials F and G, just terms (monomials with leading coefficients). These terms might as well be relatively prime, otherwise we are introducing a spurious common factor.

(b) However, we must not *replace* f, because $Ff + Gg$ might have zeros that are not zeros of f and g: we need to *add* $Ff + Gg$ to the set of equations under consideration.

Definition 8.2 *This generalization of row operations leads us to the following concepts.*

1. *If* $\mathrm{lm}(g)$ *divides* $\mathrm{lm}(f)$*, then we say that g reduces f to* $h = \mathrm{lc}(g)f - (\mathrm{lt}(f)/\mathrm{lm}(g))g$*, written* $f \rightarrow^g h$*. Otherwise, we say that f is* reduced *with respect to g. In this construction of h, the leading terms of both* $\mathrm{lc}(g)f$ *and* $(\mathrm{lt}(f)/\mathrm{lm}(g))g$ *are* $\mathrm{lc}(f)\mathrm{lc}(g)\mathrm{lm}(f)$*, and so cancel. Hence* $\mathrm{lm}(f)$ *comes before* $\mathrm{lm}(h)$ *in our order.*
2. *Any chain of reductions is finite (Dickson's Lemma), so any chain* $f_1 \rightarrow^g f_2 \rightarrow^g f_3 \cdots$ *terminates in a polynomial h reduced with respect to g. We write* $f_1 \overset{*\,g}{\rightarrow} h$*.*
3. *This extends to reduction by a set G of polynomials, where* $f \rightarrow^G h$ *means* $\exists g \in G : f \rightarrow^g h$*. We must note that a polynomial can have several reductions with respect to G (one for each element of G whose leading monomial divides the leading monomial of f).*
4. *Let* $f, g \in R[x_1, \ldots, x_n]$*. The S-polynomial of f and g, written S(f, g) is defined as*

$$S(f,g) = \frac{\mathrm{lt}(g)}{\gcd(\mathrm{lm}(f), \mathrm{lm}(g))}f - \frac{\mathrm{lt}(f)}{\gcd(\mathrm{lm}(f), \mathrm{lm}(g))}g. \qquad (8.24)$$

The computation after (8.23), $y(x^2 - 1) - x(xy - 1)$ is in fact $S(x^2 - 1, xy - 1)$. Hence we might want to think about computing S-polynomials, but then what about S-polynomials of S-polynomials, and so on? In fact, this is the right idea. We first note that the precise polynomials are not particularly important: the zeros of $G = \{f_1, f_2, \ldots\}$ are not changed if we add pf_1 to G for any polynomial p, or $f_1 + f_2$, and so on. Hence this definition.

Definition 8.3 *Let S be a set of polynomials in the variables* $x_1, \ldots, x_n$*, with coefficients from R. The* ideal generated by *S, denoted (S), is the set of all finite sums* $\sum f_i s_i$*:* $s_i \in S$*,* $f_i \in R[x_1, \ldots, x_n]$*. If S generates I, we say that S is a* basis *for I.*

What really matters is the (infinite) ideal, and not the particular finite basis that generates it. Nevertheless, some bases are nicer than others.

Theorem 8.4 [47, Proposition 5.38, Theorem 5.48] *The following conditions on a set* $G \in R[x_1, \ldots, x_n]$*, with a fixed ordering > on monomials, are equivalent.*

1. $\forall f, g \in G, S(f,g) \overset{*\,G}{\rightarrow} 0$.
2. *If* $f \overset{*\,G}{\rightarrow} g_1$ *and* $f \overset{*\,G}{\rightarrow} g_2$*, then* g_1 *and* g_2 *differ at most by a multiple in R, that is,* $\overset{*\,G}{\rightarrow}$ *is essentially well defined.*
3. $\forall f \in (G), f \overset{*\,G}{\rightarrow} 0$.
4. $(\mathrm{lm}(G)) = (\mathrm{lm}((G)))$*, that is, the leading monomials of G generate the same ideal as the leading monomials of the whole of (G).*

If G satisfies these conditions, G is called a Gröbner base (or standard basis).

These are very different kinds of conditions, and the strength of Gröbner theory lies in their interplay. Condition 2 underpins the others: $\overset{*\,G}{\rightarrow}$ is well defined. Condition 1 looks technical, but has the great advantage that, for finite G, it is finitely checkable: if G has k elements, we take the $k(k-1)/2$ unordered pairs from G, compute the S-polynomials, and check that they reduce to zero. This gives us either a proof or an explicit counterexample (which is the key to the following algorithm). As $f \stackrel{*\,G}{\rightarrow} 0$ means that $f \in (G)$, Condition 3 means that ideal membership is testable if we have a Gröbner base for the ideal.

Algorithm 8.2 (Buchberger)

Input: *finite* $G_0 \subset R[x_1, \ldots, x_n]$; *monomial ordering* $>$.
Output: G *a Gröbner base for* (G_0) *with respect to* $>$.

$G := G_0$; $n := |G|$;
we consider G as $\{g_1, \ldots, g_n\}$
$P := \{(i,j) : 1 \le i < j \le n\}$
while $P \neq \emptyset$ **do**
  Pick $(i,j) \in P$;
  $P := P \setminus \{(i,j)\}$;
  Let $S(g_i, g_j) \stackrel{*\,G}{\rightarrow} h$
  If $h \neq 0$ **then**
    # $\mathrm{lm}(h) \notin (\mathrm{lm}(G))$
    $g_{n+1} := h$; $G := G \cup \{h\}$;
    $P := P \cup \{(i, n+1) : 1 \le i \le n\}$;
    $n := n + 1$;

Given that, every time G grows, we add more possible S-polynomials to P, it is not obvious that this process terminates, but it does – essentially $(\mathrm{lm}(G))$ cannot grow forever. Estimating the running time is a much harder problem. There are doubly exponential (in the number of variables) *lower* bounds on the degree of the polynomials in a Gröbner base [48]. In practice, it is very hard to estimate the running time for a Gröbner base calculation, and the author has frequently been wrong by an order of magnitude – in both directions!

Proposition 8.8 *Given a Gröbner base G with respect to any ordering, it has a finite number of solutions if, and only if, each variable occurs alone (to some power) as the leading monomial of one of the elements of G. In this case, the number of solutions, counted with multiplicity, is the number of monomials irreducible under G*

While a Gröbner base can be computed with respect to any order, a Gröbner base with respect to a purely lexicographical order is particularly simple: essentially the nonlinear equivalent of a triangular matrix. If the system has only finitely many solutions,[8)] the Gianni – Kalkbrener algorithm [[28], Section 3.3.7] provides an equivalent of back substitution: solve the (unique) polynomial in x_n; for each root α_n solve the lowest-degree polynomial in x_{n-1}, x_n whose leading coefficient does not vanish at $x_n = \alpha_n$; for each root α_{n-1}, solve the lowest-degree polynomial in x_{n-2}, x_{n-1}, x_n whose leading coefficient does not vanish at $x_{n-1} = \alpha_{n-1}, x_n = \alpha_n$, and so on. The reader may be surprised that there can be a choice of polynomials in x_{n-1}, x_n, but consider

$$G = \{x_1^2 - 1, x_2^2 - 1, (x_1 - 1)(x_2 - 1)\}. \tag{8.25}$$

We have two choices for α_2: 1 and -1. When $x_2 = -1$, we use the polynomial $(x_1 - 1)(x_2 - 1)$ and deduce $x_1 = 1$. But when $x_2 = 1$, the leading coefficient (and in fact all) of $(x_1 - 1)(x_2 - 1)$ vanishes, and we are forced to use $x_1^2 - 1$, which tells us that $x_1 = \pm 1$.

Unfortunately, `plex` Gröbner bases are among the most expensive to compute. Hence practitioners normally use the following process.

1. Compute a `tdeg` Gröbner basis G_1.
2. Check (Proposition 8.8) that it has only finitely many solutions.
3. Maybe give up if there are too many.

8) This condition is unfortunately necessary: see [49] for an example.

4. Use the Faugère–Gianni–Lazard–Mora algorithm [28, Section 3.3.8] to convert G_1 to a `plex` Gröbner basis G_2.
5. Apply the Gianni–Kalkbrener Algorithm to G_2.

8.5.6 Regular Chains

Although not (yet?) quite so well known as Gröbner bases, the theory of *regular chains* [50], also known, slightly incorrectly, as triangular sets, provides an alternative, based instead on a (sparse) recursive view of polynomials. The fundamental algorithm takes as input a finite set S of polynomials, and rather than returning one Gröbner basis, returns a finite set of finite sets $\{S_1, \dots, S_k\}$ such that

1. $(\alpha_1, \dots, \alpha_n)$ is a solution of S if, and only if, it is a solution of some S_i.
2. Each S_i is *triangular* and a *regular chain*, which in particular means it can be solved by straightforward back substitution.

Applied to (8.25), this would produce two regular chains: $(x_1^2 - 1, x_2 - 1)$ and $(x_1 - 1, x_2 + 1)$, from which the solutions can indeed be read off.

Among widespread systems, this theory is currently only implemented in Maple. It has two advantages over the Gröbner basis approach: it produces triangular, that is, easy to understand systems even when there are infinitely many solutions, and it can be adapted to looking just for real solutions [51].

It does have its drawbacks, the most significant of which is probably that there is no guarantee that the roots of S_i and S_j are disjoint.

8.6 Integration

We have previously (Section 8.3.3) made use of the concept of f', the derivative of f. The reader may object that this is a construct of calculus, not of algebra. However, it is possible to define differentiation (which we will regard as being with respect to x) purely algebraically.

Definition 8.4 *A map* $'\colon K \to K$ *is a differentiation (with respect to x) if it satisfies*

1. $(f+g)' = f' + g'$;
2. $(fg)' = fg' + f'g$;
3. $x' = 1$.

It follows from these that

(a) $0' = 0$ by expanding $(x+0)' = x'$;
(b) $1' = 0$ by expanding $(1 \cdot x)' = x'$
(c) $\left(\frac{p}{q}\right)' = \frac{p'q - pq'}{q^2}$ by expanding $\left(q \cdot \left(\frac{p}{q}\right)\right)' = p'$ according to 2.

From this point of view, indefinite[9] integration is simply anti-differentiation:

$$\text{solve} \int f \mathrm{d}x = F \Leftrightarrow \text{find F such that } F' = f. \tag{8.26}$$

It is clear that F is only determined up to adding a constant c, that is, $c' = 0$.

8.6.1 Rational Functions

When faced with $\int (8x^7/(x^8+1))\mathrm{d}x$, most of us would spot that this is $\int (f'/f)\mathrm{d}x$, with $\log f$ as the answer. But we would struggle with $\int ((8x^7+1)/(x^8+1))\mathrm{d}x$, whose answer is

9) For definite integration, see Section 8.7.1.

$$
\begin{aligned}
&\left(\tfrac{1}{16}\sqrt{2+\sqrt{2}}+\tfrac{i}{16}\sqrt{2-\sqrt{2}}+1\right)\log\left(x+\tfrac{1}{2}\sqrt{2+\sqrt{2}}+\tfrac{i}{2}\sqrt{2-\sqrt{2}}\right)+\\
&\left(\tfrac{1}{16}\sqrt{2-\sqrt{2}}+\tfrac{i}{16}\sqrt{2+\sqrt{2}}+1\right)\log\left(x+\tfrac{1}{2}\sqrt{2-\sqrt{2}}+\tfrac{i}{2}\sqrt{2+\sqrt{2}}\right)+\\
&\left(-\tfrac{1}{16}\sqrt{2-\sqrt{2}}+\tfrac{i}{16}\sqrt{2+\sqrt{2}}+1\right)\log\left(x-\tfrac{1}{2}\sqrt{2-\sqrt{2}}+\tfrac{i}{2}\sqrt{2+\sqrt{2}}\right)+\\
&\left(-\tfrac{1}{16}\sqrt{2+\sqrt{2}}+\tfrac{i}{16}\sqrt{2-\sqrt{2}}+1\right)\log\left(x-\tfrac{1}{2}\sqrt{2+\sqrt{2}}+\tfrac{i}{2}\sqrt{2-\sqrt{2}}\right)+\\
&\left(-\tfrac{1}{16}\sqrt{2+\sqrt{2}}-\tfrac{i}{16}\sqrt{2-\sqrt{2}}+1\right)\log\left(x-\tfrac{1}{2}\sqrt{2+\sqrt{2}}-\tfrac{i}{2}\sqrt{2-\sqrt{2}}\right)+\\
&\left(-\tfrac{1}{16}\sqrt{2-\sqrt{2}}-\tfrac{i}{16}\sqrt{2+\sqrt{2}}+1\right)\log\left(x-\tfrac{1}{2}\sqrt{2-\sqrt{2}}-\tfrac{i}{2}\sqrt{2+\sqrt{2}}\right)+\\
&\left(\tfrac{1}{16}\sqrt{2-\sqrt{2}}-\tfrac{i}{16}\sqrt{2+\sqrt{2}}+1\right)\log\left(x+\tfrac{1}{2}\sqrt{2-\sqrt{2}}-\tfrac{i}{2}\sqrt{2+\sqrt{2}}\right)+\\
&\left(\tfrac{1}{16}\sqrt{2+\sqrt{2}}-\tfrac{i}{16}\sqrt{2-\sqrt{2}}+1\right)\log\left(x+\tfrac{1}{2}\sqrt{2+\sqrt{2}}-\tfrac{i}{2}\sqrt{2-\sqrt{2}}\right).
\end{aligned}
\tag{8.27}
$$

It seems clear that the sort of pattern-matching we have seen, and used effectively in school, will not scale to this level.[10)]

If we look at the integration of rational functions $\int(q(x)/r(x))\mathrm{d}x$ with $\deg q < \deg r$ (else we have a polynomial part which is trivial to integrate), we can conceptually

1. perform a square-free decomposition (Definition 8.1) of $r = \prod_{i=1}^{n} r_i^i$;
2. factorize each r_i completely, as $r_i(x) = \prod_{j=1}^{n_i}(x-\alpha_{i,j})$, where in general the $\alpha_{i,j}$ will be RootOf constructs;
3. perform a partial fraction decomposition (8.6) of q/r as
$$
\begin{aligned}
\frac{q}{r} &= \frac{q}{\prod_{i=1}^{n} r_i^i} = \sum_{i=1}^{n}\frac{q_i}{r_i^i}\\
&= \sum_{i=1}^{n}\sum_{j=i}^{n_i}\sum_{k=1}^{j}\frac{\beta_{i,j,k}}{(x-\alpha_{i,j})^k};
\end{aligned}
\tag{8.28}
$$
4. integrate this term by term, obtaining

$$
\begin{aligned}
\int\frac{q}{r} &= \sum_{i=1}^{n}\sum_{j=i}^{n_i}\sum_{k=2}^{j}\frac{-\beta_{i,j,k}}{(k-1)(x-\alpha_{i,j})^{k-1}}\\
&\quad+\sum_{i=1}^{n}\sum_{j=i}^{n_i}\beta_{i,j,1}\log(x-\alpha_{i,j}).
\end{aligned}
\tag{8.29}
$$

This is not a good idea computationally, as it would solve the trivial $\int((8x^7)/(x^8+1))\mathrm{d}x$ the same way as (8.27), only to see all the logarithms combine at the end, but shows us what the general form of the answer has to be – a rational function plus a sum of logarithms with constant coefficients We may need to introduce as algebraic numbers all the roots of the denominator, but no more algebraic numbers than that. The denominator of the rational function is going to have[11)] the same factors as the original r, but with multiplicity reduced by one, i.e. $\prod r_i^{n_i-1}$. This is, in fact, $R := \gcd(r, r')$ and

10) However, intelligent transformations do have a rôle to play in producing good human-readable answers: see [52].

11) Strictly speaking, this argument only proves that the denominator is at most this big. But it must be at least this big, else its derivative, the integrand, would not have the denominator it has.

so could be computed without the explicit factorization of step 2 above.

It can be shown, either by Galois theory or algorithmically, that, in fact, the numerator of the rational part of the integral can be written without any of the algebraic numbers we conceptually added in step 2 either. Hence

$$\int \frac{q}{r} = \frac{Q}{R} + \sum_{i=1}^{n} \sum_{j=i}^{n_i} \beta_{i,j,1} \log(x - \alpha_{i,j}). \quad (8.30)$$

Differentiating this shows that

$$\left(\sum_{i=1}^{n} \sum_{j=i}^{n_i} \beta_{i,j,1} \log(x - \alpha_{i,j}) \right)'$$

is also a rational function not needing any of the algebraic numbers we conceptually added in step 2: call it $S/(r/R)$, but we should note that this time we have *not* proved that the denominator in lowest terms is exactly r/R, and indeed if there are no logarithmic terms (all $\beta_{i,j,1} = 0$), we may have denominator just 1.

Differentiating (8.30) gives us

$$\frac{q}{r} = \left(\frac{Q}{R}\right)' + \frac{S}{r/R} = \frac{Q'(r/R) - Q\frac{R'r}{R^2} + SR}{r}, \quad (8.31)$$

and if we write Q and S as polynomials in x with unknown coefficients, (8.31) gives us a set of linear equations for these unknowns – a method known as the Horowitz–Ostrogradski algorithm [53–55].

Having found Q/R comparatively easily, we are left with $\int (S/(r/R))\mathrm{d}x$, whose integral may or may not involve individual roots of r/R – see the discussion around (8.27). Hence the real challenge is to integrate $(S/(r/R))$ without adding any unnecessary algebraic numbers. This was solved, and in a very satisfactory manner, independently in [56, 57]. It produces a set of polynomials Q_i, all of whose roots are *necessary* for the expression of $\int (S/(r/R))\mathrm{d}x$ in terms of logarithms, and such that the integral is

$$\sum_i \sum_{\rho=\mathrm{RootOf}(Q_i)} \gcd\left(S - \rho\left(\frac{r}{R}\right)', \frac{r}{R}\right). \quad (8.32)$$

Applying this algorithm to (8.27), we actually get

$$\frac{1}{8} \sum \rho \log(x + \rho + 8), \quad (8.33)$$
$$\rho = \mathrm{RootOf}(\alpha^8 - 64\,\alpha^7 + 1792\,\alpha^6 - 28672\,\alpha^5 + 286720\,\alpha^4 - 1835008\,\alpha^3 + 7340032\,\alpha^2 - 16777216\,\alpha + 16777217)$$

and (8.27) is the result of forcing a conversion of the RootOf into radicals: the polynomial is in fact $(\alpha - 8)^8 + 1$, another example of a small expression that is not small in either the dense or the sparse model.

8.6.2 More Complicated Functions

Equation (8.29) shows that every rational function *has* an integral that can be expressed as a rational function plus a sum of logarithms with constant coefficients. Conversely, we are used in calculus to statements like

$$\mathrm{e}^{-x^2} \text{ has no integral}, \quad (8.34)$$

which is nonsense as a statement of analysis, let alone numerical computation, and is really the *algebraic* statement

$$\text{there is no formula } f(x) \text{ such that } f'(x) = \mathrm{e}^{-x^2}. \quad (8.35)$$

Again, this is not correct, as Maple will tell us

$$\int \mathrm{e}^{-x^2} \mathrm{d}x = \frac{1}{2}\sqrt{\pi}\,\mathrm{erf}(x). \quad (8.36)$$

The key point really is that this statement is an oxymoron, because the definition of erf is

$$(\mathrm{erf}(x))' = \sqrt{\frac{4}{\pi}}\mathrm{e}^{-x^2}; \quad \mathrm{erf}(0) = 0. \quad (8.37)$$

Equation (8.36) is qualitatively different from

$$\int x^3 \mathrm{e}^{-x^2}\mathrm{d}x = \frac{-1}{2}\left(x^2+1\right)\mathrm{e}^{-x^2}, \quad (8.38)$$

which expresses an integral in terms of things "we already know about." Hence the trick is formalizing that last phrase. Just as Definition 8.4 provided a purely algebraic definition of differentiation, we can do the same with functions normally thought of as defined by calculus.

Definition 8.5 *We say that the abstract symbol θ is*

- *a* logarithm *of u if $\theta' = u'/u$;*
- *an* exponential *of u if $\theta' = u'\theta$.*

Definition 8.6 *We say that $\mathbf{C}(x, \theta_1, \dots, \theta_n)$ is a field of* elementary functions *if each θ_i is*

1. *algebraic over $\mathbf{C}(x, \theta_1, \dots, \theta_{i-1})$;*
2. *a* logarithm *of $u \in \mathbf{C}(x, \theta_1, \dots, \theta_{i-1})$;*
3. *an* exponential *of $u \in \mathbf{C}(x, \theta_1, \dots, \theta_{i-1})$.*

As $\sin(x) = (1/2i)(\exp(ix) - \exp(-ix))$ and $\arcsin(x) = -i\log(\sqrt{1-x^2}+ix)$, and so on, the trigonometric and inverse trigonometric functions are also included in this class. The textbook "$\int (1/(1+x^2))\mathrm{d}x = \arctan x$" is really

$$\text{“}\int \frac{1}{1+x^2}\mathrm{d}x = \frac{i}{2}\log\left(\frac{1-ix}{1+ix}\right).\text{”} \quad (8.39)$$

We can now make (8.35) precise.

There is no elementary field containing a u with $u' = \exp(-x^2)$. (8.35′)

How does one, or indeed a computer, prove such a result? And yet computers can and do: when Maple, say, responds to $\int \exp(\log^4(x))\mathrm{d}x$ by just echoing back the formula, it is not merely saying "I couldn't find an integral," it is also asserting that it has a proof that the integral is not elementary.

The key result is a major generalization of (8.29).

Theorem 8.5 (Liouville's Principle) *If $u \in \mathbf{C}(x, \theta_1, \dots, \theta_n)$ has an elementary integral v in some larger field $\mathbf{C}(x, \theta_1, \dots, \theta_m)$, it is possible to write*

$$\int u = v = w + \sum_{i=1}^{k} c_i \log w_i, \quad (8.40)$$

where w and the $w_i \in \mathbf{C}(x, \theta_1, \dots, \theta_n)$.

The proof, while somewhat technical, is based on the facts that differentiation cannot eliminate new exponentials, or new algebraics, and can only eliminate a new logarithm if it has a constant coefficient. If u itself is elementary, it is possible to go further and produce an algorithm that will find v, or *prove* that no such elementary v exists [58–60].

8.6.3 Linear Ordinary Differential Equations

If we actually apply the algorithm we have just stated exists, it will state that, if elementary, $\int \exp(-x^2)\mathrm{d}x = w(x)\exp(-x^2)$, and hence, equating coefficients of $\exp(-x^2)$ in $\exp(-x^2) = \left(w(x)\exp(-x^2)\right)'$, that $1 = w' - 2xw$. The algorithm thus also has to solve this linear differential equation, and prove that no rational function can solve it.

In fact, the method of integrating factors shows that there is a complete equivalence between integration and solving first-order linear differential equations. We can extend Definition 8.6 to allow arbitrary integrals, and therefore solutions of linear first-order differential equations, in the class of allowable functions.

Definition 8.7 *We say that* $\mathbf{C}(x, \theta_1, \dots, \theta_n)$ *is a field of* Liouvillian functions *if each* θ_i *is*

1. algebraic *over* $\mathbf{C}(x, \theta_1, \dots, \theta_{i-1})$*;*
2. *an* integral *of* $u \in \mathbf{C}(x, \theta_1, \dots, \theta_{i-1})$*, that is,* $\theta_i' = u$*;*
3. *an* exponential *of* $u \in \mathbf{C}(x, \theta_1, \dots, \theta_{i-1})$*.*

We can then ask whether second- or higher-order differential equations have solutions in terms of Liouvillian functions: a question solved in [61, 62].

8.7 Interpreting Formulae as Functions

The previous section treated $\exp(x)$ and $\log(x)$ as abstract symbolic expressions, and attached no meaning to exp and log as functions $\mathbf{R} \to \mathbf{R}$ or $\mathbf{R}^+ \to \mathbf{R}$. Similarly, we have regarded $\sqrt{2}$ as a number α with $\alpha^2 = 2$, have ignored the inconvenient fact that there are two such numbers, and attached no meaning to $\sqrt{\ }$ as a function $\mathbf{R}^+ \to \mathbf{R}$. These interpretation questions are looked at in the next two subsections.

8.7.1 Fundamental Theorem of Calculus Revisited

Definitions 8.4 and 8.5 defined differentiation *of formulae* in a purely algebraic way, and therefore (8.26) seemed to reduce the problem of integration to that of reversing differentiation. From the algebraic point of view, this is correct. From the analytic point of view, though, there is something to prove, and that proof imposes side-conditions.

Theorem 8.6 (Fundamental Theorem of Calculus, e.g., [Section 5.3, 63]) *Let f and F be functions defined on a closed interval* $[a, b]$ *such that* $F' = f$*. If f is Riemann-integrable on* $[a, b]$*, then*

$$\int_a^b f(x)\mathrm{d}x = F(b) - F(a) \text{ written } [F]_a^b .$$

Though this is the classical statement, we emphasize that $F' = f$ *must hold* throughout $[a, b]$*, and therefore F is differentiable, hence continuous, throughout this interval.*

The condition about f being Riemann-integrable is necessary to prevent calculations such as

$$\int_{-1}^{1} \frac{1}{x^3}\mathrm{d}x \stackrel{?}{=} \left[\frac{-1}{2x^2}\right]_{-1}^{1} = \frac{-1}{2} - \frac{-1}{2} = 0, \tag{8.41}$$

whereas, in fact, both $\int_{-1}^{0}(1/(x^3))\mathrm{d}x$ and $\int_0^1(1/(x^3))\mathrm{d}x$ are undefined.

The warning about continuity would also cover this case, as $-1/2x^2$ is not continuous, and might otherwise seem unnecessary: after all, are not exp and log continuous?

For a first example, consider

$$\int \frac{1}{2x^2 - 6x + 5}\mathrm{d}x = \begin{cases} F_1 := -\arctan\left(\frac{x-1}{x-2}\right) \\ F_2 := \arctan(2x - 3) \end{cases}, \tag{8.42}$$

where $\arctan'(x) = 1/(1 + x^2)$. From the point of view of Section 8.6, both are equally valid; F_1 and F_2 both differentiate to $1/(2x^2 - 6x + 5)$. However, there is a big difference from the point of view of Theorem 8.6: F_1 is discontinuous at $x = 2$, while F_2 is not, as seen in Figure 8.2, which

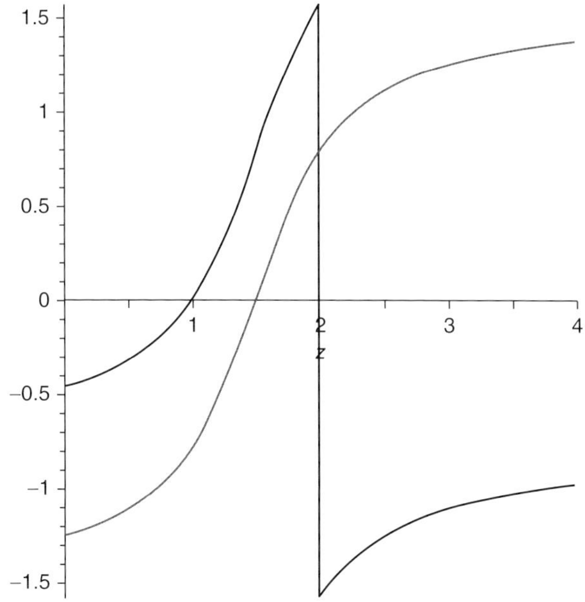

Figure 8.2 F_1 and F_2 from (8.42).

also shows a "constant" difference as well. In fact $\left[F_2\right]_1^3 = \arctan(3) + \pi/4$, whereas $\left[F_1\right]_1^3 = -\arctan(2)$, a silly result for the integral of a positive function. In fact, the two differ by π, which is the magnitude of the discontinuity in arctan "at infinity,"[12] that is, when $x = 2$ in F_2. While F_1 and F_2 both differentiate correctly at most values of x, F_1 is differentiable at $x = 2$, and therefore satisfies the differentiability hypothesis of Theorem 8.6, while F_2 is not continuous, and therefore not differentiable, at $x = 2$, and does not satisfy the hypothesis.

As arctan is built from log, we must question our bland assertion that "log is continuous." In fact, while it is continuous as a function $\mathbf{R}^+ \to \mathbf{R}$, it is not continuous $\mathbf{C} \to \mathbf{C}$, having a branch cut on the negative real axis.[13] As x goes from $-\infty$ to ∞, the argument of the logarithm in (8.39) goes from -1 clockwise round the unit circle, passing through 1 when $x = 0$ and arriving back at -1, but from the other side of the branch cut from that from which it departed.

Hence, while the methods of Section 8.6 can produce *formulae* that differentiate (in the sense of Definitions 8.4 and 8.5) correctly to the *formulae* being integrated, interpreting these formulae as functions, and doing definite integration, requires us to check that these formulae actually define *continuous* functions over the relevant range.

8.7.2 Simplification of Functions

As it is possible to define $\sqrt{x} = \exp\left(1/2\log(x)\right)$, it follows that this function has a branch cut in the same place as log, traditionally along the negative real axis. As with log, this function is continuous $\mathbf{R}^+ \to \mathbf{R}^+$, but is not continuous $\mathbf{C} \to \mathbf{C}$. We write $\mathbf{R}^+ \to \mathbf{R}^+$ because, as $\mathbf{R}^+ \to \mathbf{R}$,

12) $\lim_{x\to\infty} \arctan(x) = \pi/2$, whereas $\lim_{x\to-\infty} \arctan(x) = -\pi/2$.

13) This is the conventional location of the branch cut these days: see [64].

$\sqrt{}$ is ambiguous: is $\sqrt{4}$ equal to 2 or -2? In fact, it is also ambiguous $\mathbf{C} \to \mathbf{C}$, and we normally consider $\sqrt{} : \mathbf{C} \to \mathbf{C}^+$, where $\mathbf{C}^+ = \{x + iy : x > 0 \vee (x = 0 \wedge y \geq 0)\}$, the "positive half-plane."

This resolves ambiguity, but does not solve all our problems. The putative equation

$$\sqrt{x}\sqrt{y} \stackrel{?}{=} \sqrt{xy}, \tag{8.43}$$

which is true for $\sqrt{} : \mathbf{R}^+ \to \mathbf{R}^+$, fails for $\sqrt{} : \mathbf{C} \to \mathbf{C}^+$: consider $x = y = -1$. Manipulating such putative identities is dangerous, and most computer algebra systems will, by default, refuse to do this: for example Maple's `simplify` does not, but `simplify(...,symbolic)` (sometimes) will. However, just because (8.43) is not universally true does not mean that some relatives are not true. Consider the apparently similar

$$\sqrt{z-1}\sqrt{z+1} \stackrel{?}{=} \sqrt{z^2-1} \tag{8.44}$$

$$\sqrt{1-z}\sqrt{1+z} \stackrel{?}{=} \sqrt{1-z^2}. \tag{8.45}$$

Equation (8.45) is in fact true throughout $\mathbf{C}$, whereas (8.44) is only true on $\mathbf{C}^+ \cup [-1, 0]$. The difference is that the branch cuts of (8.44) include the imaginary axis, and divide $\mathbf{C}$ into distinct regions, whereas the branch cuts of (8.45) do not separate the complex plane. The apparently bizarre fact that (8.44) is also valid on $[-1, 0]$ is due to the fact that multiple branch cuts include this, and their effects cancel. Equally, though the components of (8.45) have branch cuts, they lie on top of each other, and the effects cancel, meaning that (8.45) itself has only a spurious cut [65]. See [66, 67] for the general theory, though it has to be admitted that implementations in computer algebra systems are few.

These problems are not confined to $\sqrt{}$. log is also not uniquely defined $\mathbf{C} \to \mathbf{C}$, because $1 = \exp(0) = \exp(2\pi i) = \exp(4\pi i) = \ldots$ Many mathematicians are content to live with this ambiguity, which is of course anathema to computer programmers – see [64].

$$\log z_1 + \log z_2 \stackrel{?}{=} \log z_1 z_2. \tag{8.46}$$

The equation merely states that the sum of one of the (infinitely many) logarithms of z_1 and one of the (infinitely many) logarithms of z_2 can be found among the (infinitely many) logarithms of $z_1 z_2$, and conversely every logarithm of $z_1 z_2$ can be represented as a sum of this kind (with a suitable choice of $\log z_1$ and $\log z_2$).

[68, pp. 259–260] (our notation)

It is more normal to use capital letters to denote the multivalued functions, so that $\mathrm{Log}(z) = \{\log(z) + 2k\pi i : k \in \mathbf{Z}\}$, and instead of (8.46) to write

$$\mathrm{Log} z_1 + \mathrm{Log} z_2 = \mathrm{Log} z_1 z_2, \tag{8.47}$$

interpreting "+" as elementwise addition of sets. With this notation, $\mathrm{Log}(1/z) = -\mathrm{Log}(z)$, but the single-valued equivalent $\log(1/z) \stackrel{?}{=} -\log(z)$ is true except on the negative real axis, where the branch cuts do not cancel. Again, it is fair to say that implementations in computer algebra systems lag significantly behind the theory.

8.7.3 Real Problems

It would be tempting to dismiss these issues as only affecting complex numbers. This is not true: consider the often-quoted identity

$$\arctan(x) + \arctan(y) = \arctan\left(\frac{x+y}{1-yx}\right). \tag{8.48}$$

It is certainly true at $x = y = 0$, and the partial derivatives of the two sides are equal, so we might be tempted to conclude that it is true everywhere "by analytic continuation." However, when $x = y = 2$, the left-hand side is positive and the right-hand side negative, so something is clearly wrong. The problem is the same "discontinuity at infinity" of arctan, that is, when $xy = 1$, as we saw in Figure 8.2.

In fact, the correct version of (8.48) is (8.49):

$$\arctan(x) + \arctan(y) = \arctan\left(\frac{x+y}{1-yx}\right) + \begin{cases} -\pi & xy > 1; x < 0 \\ 0 & < 1 \\ \pi & xy > 1; x > 0 \end{cases}. \tag{8.49}$$

8.8 Conclusion

We have seen various examples of where the *ad hoc* methods we have generally learnt at school, and which process smallish examples well, can be converted into algorithms for doing algebraic manipulation, be it gcd computations (Section 8.4), factoring polynomials (Section 8.4.4), solving nonlinear polynomial systems (Section 8.5.5), or indefinite integration (Section 8.6). Equally, we have seen that this is not always possible (Theorem 8.3).

Section 8.7 also reminds us that there is more to mathematics than just algebra, and that we need to be careful when interpreting algebraic objects as actual functions $\mathbf{C} \to \mathbf{C}$, or even $\mathbf{R} \to \mathbf{R}$, as in (8.49).

References

1. IEEE (1985) *IEEE Standard 754 for Binary Floating-Point Arithmetic*, IEEE.
2. Baker, A. (1975) *Transcendental Number Theory*, Cambridge University Press.
3. Silverman, J.H. and Tate, J. (1992) *Rational Points on Elliptic Curves*, Springer-Verlag.
4. Kahrimanian, H.G. (1953) Analytic differentiation by a digital computer. Master's thesis, Temple University Philadelphia.
5. Nolan, J. (1953) Analytic differentiation on a digital computer. Master's thesis, Mathematics Department M.I.T.
6. Haselgrove, C.B. (1953) Implementations of the Todd-Coxeter algorithm on EDSAC-1, Unpublished.
7. Collins, G.E. (1966) PM, a system for polynomial multiplication. *Commun. ACM*, **9**, 578–589.
8. Martin, W.A. (1967) Symbolic mathematical laboratory. PhD thesis, M.I.T. & Project MAC TR-36.
9. Moses, J. (2010) Macsyma: a personal history. *J. Symb. Comput.*, **47**, 123–130.
10. Fitch, J.P. (1974) *CAMAL Users Manual*, University of Cambridge Computer Laboratory.
11. Hearn, A.C. (1973) REDUCE-2 User's Manual. Technical Report UCP-19, Computational Physics Group University of Utah.
12. Cannon, J.J. (1974) A general purpose group theory program. Proceedings 2nd International Conference on Theory of Groups, pp. 204–217.
13. Bosma, W., Cannon, J., and Playoust, C. (1997) The Magma algebra system. I: the user language. *J. Symb. Comput.*, **24**, 235–265.
14. Giovini, A. and Niesi, G. (1990) CoCoA: a user-friendly system for commutative algebra. Proceedings DISCO '90, Springer, pp. 20–29.
15. Abbott, J.A. (2004) CoCoA: a laboratory for computations in commutative algebra. *ACM SIGSAM Bull. 1*, **38**, 18–19.
16. Greuel, G.-M., Pfister, G., and Schönemann, H. (2001) SINGULAR – a computer algebra system for polynomial computations, in *Proceedings Calculemus 2000* (eds M. Kerber and M. Kohlhase), A.K. Peters, Boston Mass, pp. 227–234.

17. Bayer, D. and Stillman, M. (1986) The design of Macaulay: a system for computing in algebraic geometry and commutative algebra. *Proceedings SYMSAC 86*, pp. 157–162.
18. Grayson, D. and Stillman, M. (2009) Macaulay2, a software system for research in algebraic geometry, *http://www.math.uiuc.edu/Macaulay2/* (accessed on 25 February 2014).
19. Karatsuba, A. and Ofman, J. (1963) Multiplication of multidigit numbers on automata. *Sov. Phys. Dokl.*, **7**, 595–596.
20. Schönhage, A. and Strassen, V. (1971) Schnelle Multiplikation großer Zahlen. *Computing*, **7**, 282–292.
21. Johnson, S.C. (1974) Sparse polynomial arithmetic. *Proceedings EUROSAM 74*, pp. 63–71.
22. Yan, T. (1998) The geobucket data structure for polynomials. *J. Symb. Comput.*, **25**, 285–294.
23. Char, B.W., Geddes, K.O., Gentleman, M.W., and Gonnet, G.H. (1983) The design of MAPLE: a compact, portable and powerful computer algebra system. *Proceedings EUROCAL 83*, pp. 101–115.
24. Davenport, J.H. and Carette, J. (2010) The sparsity challenges, in *Proceedings SYNASC 2009* (eds S. Watt *et al.*), IEEE Press, pp. 3–7.
25. Castaño, B., Heintz, J., Llovet, J., and Martínez, R. (2000) On the data structure straight-line program and its implementation in symbolic computation. *Math. Comput. Simulat.*, **51**, 497–528.
26. Schinzel, A. (2003) On the greatest common divisor of two univariate polynomials, I. *A Panorama of Number Theory or the View from Baker's Garden*, pp. 337–352.
27. Plaisted, D.A. (1977) Sparse complex polynomials and irreducibility. *J. Comput. Syst. Sci.*, **14**, 210–221.
28. Davenport, J.H. (2015) Computer algebra, To be published by C.U.P. in 2014.
29. Zippel, R.E. (1993) *Effective Polynomial Computation*, Kluwer Academic Publishers.
30. Karpinski, M. and Shparlinski, I. (1999) On the computational hardness of testing square-freeness of sparse polynomials, in *Proceedings AAECC-13* (eds M. Fossorier, H. Imai, S. Lin, and A. Poli), Springer, pp. 492–497.
31. Landau, E. (1905) Sur Quelques Théorèmes de M. Petrovic Relatif aux Zéros des Fonctions Analytiques. *Bull. Soc. Math. France*, **33**, 251–261.
32. Mignotte, M. (1974) An inequality about factors of polynomials. *Math. Comput.*, **28**, 1153–1157.
33. Mignotte, M. (1981) Some inequalities about univariate polynomials. *Proceedings SYMSAC 81*, pp. 195–199.
34. Brown, W.S. (1971) On Euclid's algorithm and the computation of polynomial greatest common divisors. *J. ACM*, **18**, 478–504.
35. Berlekamp, E.R. (1967) Factoring polynomials over finite fields. *Bell Syst. Tech. J.*, **46**, 1853–1859.
36. Berlekamp, E.R. (1970) Factoring polynomials over large finite fields. *Math. Comput.*, **24**, 713–735.
37. Cantor, D.G. and Zassenhaus, H. (1981) A new algorithm for factoring polynomials over finite fields. *Math. Comput.*, **36**, 587–592.
38. Swinnerton-Dyer, H.P.F. (1970) Letter to E.H. Berlekamp. Mentioned in [36].
39. Kaltofen, E., Musser, D.R., and Saunders, B.D. (1983) A generalized class of polynomials that are hard to factor. *SIAM J. Comput.*, **12**, 473–483.
40. Abbott, J.A., Bradford, R.J., and Davenport, J.H. (1985) A remark on factorisation. *SIGSAM Bull. 2*, **19**, 31–33.
41. Abbott, J.A., Shoup, V., and Zimmermann, P. (2000) Factorization in $\mathbb{Z}[x]$: the searching phase, in *Proceedings ISSAC 2000* (ed. C. Traverso), ACM, New York, pp. 1–7.
42. von Tschirnhaus, E.W. (1683) Methodus auferendi omnes terminos intermedios ex data aeqvatione. *Acta Eruditorium*, **2**, 204–207.
43. Galois, É. (1879) Œuvres mathématiques. Gauthier-Villars (sous l'auspices de la SMF).
44. Kac, M. (1943) On the average number of real roots of a random algebraic equation. *Bull. A.M.S.*, **49**, 314–320.
45. Lerario, A. and Lundberg, E. (2012) Statistics on Hilbert's Sixteenth Problem, *http://arxiv.org/abs/1212.3823*.
46. Coste, M. and Roy, M.F. (1988) Thom's Lemma, the coding of real algebraic numbers and the computation of the topology of semi-algebraic sets. *J. Symb. Comput.*, **5**, 121–129.
47. Becker, T., Weispfenning, V. (with Kredel, H.) *Groebner Bases: A Computational*

Approach to Commutative Algebra, Springer-Verlag, 1993.

48. Mayr, E.W. and Ritscher, S. (2010) Degree bounds for Gröbner bases of low-dimensional polynomial ideals, in *Proceedings ISSAC 2010* (ed. S.M. Watt), ACM, New York, pp. 21–28.
49. Fortuna, E., Gianni, P., and Trager, B. (2001) Degree reduction under specialization. *J. Pure Appl. Algebra*, **164**, 153–163.
50. Aubry, P., Lazard, D., and Moreno Maza, M. (1999) On the theories of triangular sets. *J. Symb. Comput.*, **28**, 105–124.
51. Chen, C., Davenport, J.H., May, J.P., Moreno Maza, M., Xia, B., and Xiao, R. (2013) Triangular decomposition of semi-algebraic systems. *J. Symb. Comput.*, **49**, 3–26.
52. Rich, A.D. and Jeffrey, D.J. (2009) A knowledge repository for indefinite integration based on transformation rules, in *Proceedings Intelligent Computer Mathematics* (eds J. Carette *et al.*), Springer, pp. 480–485.
53. Horowitz, E. (1969) Algorithm for symbolic integration of rational functions. PhD thesis, University of Wisconsin.
54. Horowitz, E. (1971) Algorithms for partial fraction decomposition and rational function integration. *Proceedings Second Symposium on Symbolic and Algebraic Manipulation*, pp. 441–457.
55. Ostrogradski, M.W. (1845) De l'intégration des fractions rationelles. *Bull. Acad. Imp. Sci. St. Petersburg (Class Phys.-Math.)*, **4**, 145–167.
56. Rothstein, M. (1976) *Aspects of symbolic integration and simplification of exponential and primitive functions*. PhD thesis, University of Wisconsin.
57. Trager, B.M. (1976) Algebraic factoring and rational function integration, in *Proceedings SYMSAC 76* (ed. R.D. Jenks), ACM, New York, pp. 219–226.
58. Risch, R.H. (1969) The problem of integration in finite terms. *Trans. A.M.S.*, **139**, 167–189.
59. Davenport, J.H. (1981) *On the Integration of Algebraic Functions, Springer Lecture Notes in Computer Science 102*, Springer.
60. Bronstein, M. (2005) *Symbolic Integration I*, 2nd edn, Springer-Verlag.
61. Kovacic, J.J. (1986) An algorithm for solving second order linear homogeneous differential equations. *J. Symb. Comput.*, **2**, 3–43.
62. Singer, M.F. (1981) Liouvillian solutions of n-th order homogeneous linear differential equations. *Am. J. Math.*, **103**, 661–682.
63. Apostol, T.M. (1967) *Calculus*, Vol. I, 2nd edn, Blaisdell.
64. Davenport, J.H. (2010) The challenges of multivalued "Functions", in *Proceedings AISC/Calculemus/MKM 2010* (eds S. Autexier *et al.*), Springer, pp. 1–12.
65. England, M., Bradford, R., Davenport, J.H., and Wilson, D.J. (2013) Understanding branch cuts of expressions, in *Proceedings CICM 2013* (eds J. Carette *et al.*), Springer, pp. 136–151.
66. Bradford, R.J. and Davenport, J.H. (2002) Towards better simplification of elementary functions, in *Proceedings ISSAC 2002* (ed. T. Mora), ACM, New York, pp. 15–22.
67. Davenport, J.H. (2003) The geometry of $\mathbf{C}^n$ is important for the algebra of elementary functions. Algebra Geometry and software systems, Springer, pp. 207–224.
68. Carathéodory, C. (1958) *Theory of Functions of a Complex Variable*, Chelsea Publishing.

9
Differentiable Manifolds

Marcelo Epstein

9.1
Introduction

In his authoritative *Physics*,[1] Aristotle (384–322 BC) establishes that space "has three dimensions, length, breadth, depth, the dimensions by which all body also is bounded."[2] Time is regarded intuitively as one dimensional. Moreover, both space and motion are considered as being continuous in the sense that they are "divisible into divisibles that are infinitely divisible."[3] The continuity of motion and space implies, therefore, the continuity of time.[4] From these modest beginnings, it would take 23 centuries to arrive at a rigorous mathematical definition of the most general entity that combines the intuitive notions of continuity and constancy of dimension. Accordingly, we introduce first the notion of topological space, which is the most general entity that can sustain continuous functions, and, subsequently, the notion of topological manifold, which is a topological space that locally resembles $\mathbb{R}^n$.

9.2
Topological Spaces

9.2.1
Definition

A *topological space* is a set A in which a topology has been introduced. A *topology* $\mathcal{T}(A)$ on the set A is a collection of subsets of A, called the *open sets* of A, with the following properties:

1. The empty set $\emptyset$ and the *total space* A are open sets.
2. The union of any arbitrary collection of open sets is an open set.
3. The intersection of any *finite* collection of open sets is an open set.

Given a point $a \in A$, a *neighborhood* of a is an open set $N(a) \in \mathcal{T}(A)$ containing a. By

1) Aristotle, *Physics*, Hardie R P and Gaye R K (translators), The Internet Classics Archive.
2) Ibid., Book IV, Part 1.
3) The Greek term used is $\sigma\upsilon\nu\epsilon\chi\acute{\eta}\varsigma$, which literally means "holding together."
4) Ibid., Book IV, Part 11.

Mathematical Tools for Physicists, Second Edition. Edited by Michael Grinfeld.

Property 1 above, every point of A has at least one neighborhood.

9.2.2 Continuity

A function $f : A \to B$ between the topological spaces A and B is *continuous at* $a \in A$ if for any neighborhood V of $f(a) \in B$ there exists a corresponding neighborhood U of $a \in A$ such that $f(U) \subset V$. A function that is continuous at every point of its domain is said to be *continuous*. A function is continuous if, and only if, the inverse image of every open set (in B) is open (in A).

A bijective (i.e., one-to-one and onto) function $f : A \to B$ between topological spaces is a *homeomorphism* if it is continuous and its inverse $f^{-1} : B \to A$ is also continuous. Topology can be understood as the study of those properties that are preserved under homeomorphisms.

9.2.3 Further Topological Notions

A subset $B \subset A$ of a topological space A with a topology $\mathcal{T}(A)$ *inherits* a topology $\mathcal{T}(B)$ as follows: The open sets of B are defined as the intersections of the open sets of A with B. The topology $\mathcal{T}(B)$ obtained in this way is called the *relative* or *subset* topology.

A topological space A is a *Hausdorff space* if, given any two different points $a, b \in A$, there exist respective *disjoint* neighborhoods $N(a)$ and $N(b)$, that is, $N(a) \cap N(b) = \emptyset$.

A *base* $\mathcal{B}$ of a topology $\mathcal{T}(A)$ is a collection of open sets such that every open set of A is a union of elements of $\mathcal{B}$. The topological space is said to be *generated* by a base. Thus, the open intervals of the real line $\mathbb{R}$ constitute a base of the ordinary topology of $\mathbb{R}$.

Recall that a set is *countable* if it can be put in a one-to-one correspondence with a subset of the natural numbers $\mathbb{N}$. A topology is said to be *second countable* (or to satisfy the second axiom of countability) if it has a countable basis. Second-countable topologies enjoy many special properties.

A subset of a topological space A is *closed* if its complement is open. Notice that an arbitrary subset of a topological space need not be either open or closed, or it may be both open and closed. A topological space is said to be *connected* if the only subsets that are both open and closed are the empty set and the total space.

An *open cover* of a topological space A is a collection C of open sets whose union is the total space. An *open subcover* of C is a subcollection of C that is itself an open cover. A topological space is *compact* if every open cover has a finite open subcover.

The *product topology* of two topological spaces, A and B, with respective topologies $\mathcal{S}(A)$ and $\mathcal{T}(B)$, is the topology on the Cartesian product $A \times B$ generated by the base of all the Cartesian products of the form $S \times T$, where $S \in \mathcal{S}(A)$ and $T \in \mathcal{T}(B)$.

9.3 Topological Manifolds

9.3.1 Motivation

As, as we have shown, a topological space is all one needs to define continuous functions, it might appear that this is the proper arena for the formulation of physical theories involving the notion of a *continuum*. Nevertheless, even from the Aristotelian viewpoint, it seems that we also need to convey the idea of a fixed number of underlying dimensions, a concept that is not automatically embodied in the definition of a topological space. If we think of the surface of the Earth and its cartographic representations, we understand intuitively that, although the surface of a sphere cannot be continuously mapped once and for all onto the plane $\mathbb{R}^2$, it can be so represented in a piecewise manner. In generalizing this example, we look for the notion of a topological space that is piecewise homeomorphic to $\mathbb{R}^n$.

9.3.2 Definition

An *n-dimensional topological manifold* $\mathcal{M}$ is a Hausdorff second-countable topological space such that each of its points has a neighborhood homeomorphic to an open set of $\mathbb{R}^n$.

The *standard topology* assumed in $\mathbb{R}^n$ is the one generated by all open balls in $\mathbb{R}^n$. Recall that the *open ball* of center $(c^1, \dots, c^n) \in \mathbb{R}^n$ and radius $r \in \mathbb{R}$ is the subset of $\mathbb{R}^n$ defined as $\{(x^1, \dots, x^n) \in \mathbb{R}^n : (x^1 - c^1)^2 + \dots + (x^n - c^n)^2 < r^2\}$. The empty set is obtained setting $r \leq 0$.

Although we have defined only finite-dimensional topological manifolds, which are modeled locally on $\mathbb{R}^n$, it is also possible to define topological manifolds modeled on infinite-dimensional Banach spaces.

9.3.3 Coordinate Charts

It follows from the definition of topological manifold that there exists an open cover each of whose elements is homeomorphic to an open set of $\mathbb{R}^n$. If we denote by $\mathcal{U}_\alpha$ the generic constituent of the open cover and by ϕ_α the corresponding homeomorphism, where α denotes a running index, we can identify the pair $(\mathcal{U}_\alpha, \phi_\alpha)$ with a *coordinate chart*. The collection of all these pairs is called an *atlas* of the topological manifold.

The terminology of coordinate charts arises from the fact that a chart introduces a *local coordinate system*. More specifically, the homeomorphism $\phi_\alpha : \mathcal{U}_\alpha \to \phi_\alpha(\mathcal{U}_\alpha) \subset \mathbb{R}^n$ assigns to each point $p \in \mathcal{U}_\alpha$ an ordered n-tuple $(x^1(p), \dots, x^n(p))$, called the *local coordinates* of p.

Whenever two charts, $(\mathcal{U}_\alpha, \phi_\alpha)$ and $(\mathcal{U}_\beta, \phi_\beta)$, have a nonempty intersection, we define the *transition function* $\phi_{\alpha\beta}$ as

$$\phi_{\alpha\beta} = \phi_\beta \circ \phi_\alpha^{-1} : \phi_\alpha(\mathcal{U}_\alpha \cap \mathcal{U}_\beta) \to \phi_\beta(\mathcal{U}_\alpha \cap \mathcal{U}_\beta). \tag{9.1}$$

Each transition function is a homeomorphism between open sets of $\mathbb{R}^n$. The inverse of $\phi_{\alpha\beta}$ is the transition function $\phi_{\beta\alpha} = \phi_{\alpha\beta}^{-1} = \phi_\alpha \circ \phi_\beta^{-1}$. Denoting by x^i and y^i $(i = 1, \dots, n)$, respectively, the local coordinates of $\mathcal{U}_\alpha$ and $\mathcal{U}_\beta$, a transition function boils down to the specification of n continuous and continuously invertible real functions of the form

$$y^i = y^i(x^1, \dots, x^n), \qquad i = 1, \dots, n. \tag{9.2}$$

9.3.4 Maps and Their Representations

If $\mathcal{M}$ and $\mathcal{N}$ are topological manifolds of dimensions m and n, respectively, a map $f : \mathcal{M} \to \mathcal{N}$ is *continuous* if it is a continuous map between the underlying topological spaces. A nice feature of topological manifolds, as opposed to general topological spaces, is the possibility of representing continuous maps locally as real functions of real variables. Let $p \in \mathcal{M}$ and denote $q = f(p) \in \mathcal{N}$. By continuity, we can always choose a chart $(\mathcal{U}, \phi)$ containing p such that its image $f(\mathcal{U})$ is contained in a chart $(\mathcal{V}, \psi)$ containing q. The map

$$\hat{f} = \psi \circ f \circ \phi^{-1} : \phi(\mathcal{U}) \to \psi(\mathcal{V}) \quad (9.3)$$

maps an open set in $\mathbb{R}^m$ to an open set in $\mathbb{R}^n$. This continuous map $\hat{f}$ is the *local coordinate representation* of f in the coordinate charts chosen.

9.3.5
A Physical Application

Lagrange's (1736–1813) conception of mechanics was purportedly purely analytical. In the Preface to the first edition of his Mécanique Analitique,[5] he explicitly states that "On ne trouvera point de Figures dans cet Ouvrage. Les méthodes que j'y expose ne demandent ni constructions, ni raisonnements géométriques ou méchaniques, mais seulemnet des opérations algébriques, assujeties à une marche régulière et uniforme. Ceux qui aiment l'Analyse verront avec plaisir la Méchanique en devenir une nouvelle branche, et me sauront gré d'en avoir étendu ainsi le domain." Nevertheless, it is not an exaggeration to say that in laying down the foundations of Analytical Mechanics, Lagrange was actually inaugurating the differential geometric approach to Physics. In Lagrange's view, a mechanical system was characterized by a finite number n of *degrees of freedom* to each of which a *generalized coordinate* is assigned. A *configuration* of the system is thus identified with an ordered n-tuple of real numbers. But, is this assignment unique? And, anyway, what are these numbers coordinates of?

Consider the classical example of a (rigid) double pendulum oscillating in a vertical plane under gravity. Clearly, this system can be characterized by two independent degrees of freedom. If we were to adopt as generalized coordinates the horizontal displacements, x_1 and x_2, of the two masses from, say, the vertical line through the point of suspension, we would find that to an arbitrary combination of these two numbers, there may correspond as many as 4 different configurations. If, to avoid this problem, we were to adopt as generalized coordinates the angular deviations θ_1 and θ_2, we would find that a given configuration can be characterized by an infinite number of combinations of values of these coordinates, owing to the additive freedom of 2π. If we attempt to solve this problem by limiting the range of these coordinates to the interval $[0, 2\pi]$, we lose continuity of the representation (because two neighboring configurations would correspond to very distant values of the coordinates).

Let us, therefore, go against Lagrange's own advice and attempt to draw a mental picture of the geometry of the situation. As the first mass (attached to the main point of suspension) is constrained to move along a circle, thus constituting a simple pendulum, we conclude that its configurations can be homeomorphically mapped onto a circumference (or any other closed curve in the plane). We say that this circumference is the *configuration space* of a simple pendulum. Now, the second mass can describe a circumference around any position of the first. It is not difficult to conclude that the configuration space of the double pendulum is given by the surface of a torus. Now that this basic geometric (topological) question has been settled, we realize that an atlas of this torus must consist of several charts. But the central conceptual gain of the geometrical approach is that the configuration space of a mechanical system, whose configurations are defined with continuity in mind, can be faithfully represented by a unique topological manifold, up to a homeomorphism.

5) Lagrange (1788), *Mécanique Analitique* [sic], chez la Veuve Desaint, Libraire, Paris.

9.3.6 Topological Manifolds with Boundary

Consider the *upper half space* defined as the subset $\mathbb{H}^n$ of $\mathbb{R}^n$ consisting of all points satisfying the inequality $x^n \geq 0$. Moreover, endow this subset with the subset topology induced by the standard topology of $\mathbb{R}^n$. An *n-dimensional topological manifold with boundary* is a Hausdorff second-countable topological space such that each of its points has a neighborhood homeomorphic to an open set of $\mathbb{H}^n$.

The *manifold boundary* ∂M of a topological manifold M with boundary is defined as the set of all points of M whose last coordinate x^n vanishes. Notice that this definition is independent of the atlas used. Thus, a topological manifold is a particular case of a topological manifold with boundary, namely, when the manifold boundary is empty.

As a physical example, consider the case of a plane pendulum suspended by means of a wrinkable, but inextensible, thread. Its configuration space consists of a solid disk, including the circumference and the interior points. Equivalently, any other solid simply connected plane figure can be used as configuration space of this mechanical system.

9.4 Differentiable Manifolds

9.4.1 Motivation

As we have learned, topological manifolds provide the most general arena for the definition of continuous functions. Continuity alone, however, may not be enough to formulate physical problems in a quantitative manner. Indeed, experience with actual dynamical and field theories of Mechanics and Electromagnetism, to name only the classical theories, has taught us to expect the various phenomena to be governed by ordinary or partial differential equations. These theories, therefore, must be formulated on a substratum that has more structure than a topological manifold, namely, an entity that allows for the definition of differentiable functions. Differentiable manifolds are the natural generalization of topological manifolds to handle differentiability. A differentiable manifold corresponds roughly to what physicists call a *continuum*.

9.4.2 Definition

As a topological space does not possess in itself enough structure to sustain the notion of differentiability, the key to the generalization of a topological manifold is to be found in restrictions imposed upon the transition functions, which are clearly defined in $\mathbb{R}^n$. Two charts, $(\mathcal{U}_\alpha, \phi_\alpha)$ and $(\mathcal{U}_\beta, \phi_\beta)$, of a topological manifold $\mathcal{M}$ are said to be *C^k-compatible*, if the transition functions $\phi_{\alpha\beta}$ and $\phi_{\beta\alpha}$, as defined in (9.1), are of class C^k. In terms of the representation (9.2), this means that all the partial derivatives up to and including the order k exist and are continuous. By convention, a continuous function is said to be of class C^0 and a *smooth* function is of class C^∞.

In a topological manifold, all charts of all possible atlases are automatically C^0-compatible. An *atlas of class C^k of a topological manifold $\mathcal{M}$* is an atlas whose charts are C^k-compatible. Two atlases of class C^k are *compatible* if each chart of one is compatible with each chart of the other. The union of compatible C^k-atlases is a C^k atlas. Given a C^k atlas, one can define the corresponding *maximal compatible atlas of class* C^k as the union of all atlases that are C^k-compatible with the given one. A maximal atlas, thus, contains all its compatible atlases.

An *n-dimensional differentiable manifold of class* C^k is an n-dimensional topological manifold $\mathcal{M}$ together with a maximal atlas of class C^k. For $k = 0$ one recovers the topological manifold. The C^∞ case delivers a *smooth manifold*, or simply a *manifold*.

A maximal C^k-atlas is also called a C^k*-differentiable structure*. Thus, a C^k-manifold is a topological manifold with a C^k-differentiable structure. For the particular case of $\mathbb{R}^n$, we can choose the *canonical atlas* consisting of a single chart (the space itself) and the identity map. The induced C^∞-differentiable structure is the *standard differentiable structure of* $\mathbb{R}^n$.

A differentiable manifold is *oriented* if it admits an atlas, called an *oriented atlas*, such that all the transition functions have a positive Jacobian determinant. Two oriented atlases are either compatible or every transition function between charts of the two atlases has a negative determinant. An *oriented manifold* is an orientable manifold with and oriented maximal atlas. In other words, only those coordinate transformations that preserve the orientation are permitted.

Given two differentiable manifolds, $\mathcal{M}$ and $\mathcal{N}$, of dimensions m and n, respectively, we define the $(m + n)$-dimensional *product manifold* by endowing the Cartesian product $\mathcal{M} \times \mathcal{N}$ with an atlas made of all the Cartesian products of charts of an atlas of $\mathcal{M}$ and an atlas of $\mathcal{N}$. The underlying topological space of a product manifold inherits the product topology.

9.4.3 Differentiable Maps

Let $\mathcal{M}$ and $\mathcal{N}$ be (smooth) manifolds of dimensions m and n, respectively. A continuous map $f : \mathcal{M} \to \mathcal{N}$ is *differentiable of class* C^k at a point $p \in \mathcal{M}$ if, using charts $(\mathcal{U}, \phi)$ and $(\mathcal{V}, \psi)$ belonging to the respective maximal atlases of $\mathcal{M}$ and $\mathcal{N}$, the local coordinate representation $\hat{f}$ of f, as defined in (9.3), is of class C^k at $\phi(p) \in \mathbb{R}^m$. This definition is independent of chart, because the composition of differentiable maps in $\mathbb{R}^m$ is differentiable. Notice how the notion of differentiability within the manifolds has been cleverly deflected to the charts.

Maps of class C^∞ are said to be *smooth maps*, to which we will confine our analysis from now on. In the special case $\mathcal{N} = \mathbb{R}$, the map $f : \mathcal{M} \to \mathbb{R}$ is called a (real) *function*. When, on the other hand, $\mathcal{M}$ is an open interval $H = (a, b)$ of the real line, the map $\gamma : H \to \mathcal{N}$ is called a *(parameterized) curve in* $\mathcal{N}$. The name *diffeomorphism* is reserved for the case in which $\mathcal{M}$ and $\mathcal{N}$ are of the same dimension and both f and its (assumed to exist) inverse f^{-1} are smooth. Two manifolds of the same dimension are said to be *diffeomorphic* if there exists a diffeomorphism between them.

9.4.4 Tangent Vectors

Let H be an open interval of the real line and, without loss of generality, assume that $0 \in H$. Consider the collection of all (smooth) curves $\gamma : H \to \mathcal{M}$ such that $\gamma(0) = p$. Our aim is to define the notion of tangency of two such curves at p, an aim that we achieve by using the technique of deflecting to charts. Indeed, if $(\mathcal{U}, \phi)$ is a chart containing p, the composition $\hat{\gamma} = \phi \circ \gamma : H \to \mathbb{R}^m$ is a curve in $\mathbb{R}^m$, where m is the dimension of $\mathcal{M}$. The coordinate expression of $\hat{\gamma}$ is given by m smooth real functions $x^i = \gamma^i(t)$, where t is the natural coordinate of $\mathbb{R}$ and $i = 1, \ldots, m$. We say that two curves, γ_1 and γ_2, in our collection are *tangent at p* if

$$\left.\frac{\mathrm{d}\gamma_1^i}{\mathrm{d}t}\right|_{t=0} = \left.\frac{\mathrm{d}\gamma_2^i}{\mathrm{d}t}\right|_{t=0}, \qquad i = 1, \ldots, m. \tag{9.4}$$

It is a simple matter to verify that this definition is independent of chart.

Noting that tangency at p is an equivalence relation, we define a *tangent vector at p* as an equivalence class of (smooth, parameterized) curves tangent at p. A tangent vector is thus visualized as what the members of a collection of tangent (parameterized) curves have in common. More intuitively, one may say that what these curves have in common is a small piece of a curve.

Let $f : \mathcal{M} \to \mathbb{R}$ be a (differentiable) function and let $\mathbf{v}$ be a tangent vector at $p \in \mathcal{M}$. Choosing any representative γ in the equivalence class $\mathbf{v}$, the composition $f \circ \gamma$ is a real-valued function defined on H. The *derivative of f at p along* $\mathbf{v}$ is defined as

$$\mathbf{v}(f) = \left.\frac{\mathrm{d}(f \circ \gamma)}{\mathrm{d}t}\right|_{t=0}. \tag{9.5}$$

This notation suggests that a vector can be regarded as a linear operator on the collection of differentiable functions defined on a neighborhood of a point. The linearity is a direct consequence of the linearity of the derivative. Not every linear operator, however, is a tangent vector because, by virtue of (9.5), tangent vectors must also satisfy the Leibniz rule, namely, for any two functions f and g

$$\mathbf{v}(fg) = f\mathbf{v}(g) + \mathbf{v}(f)g, \tag{9.6}$$

where, on the right-hand side, f and g are evaluated at p.

9.4.5 Brief Review of Vector Spaces

9.4.5.1 Definition

Recall that a *(real) vector space V* is characterized by two operations: *vector addition* and *multiplication by a scalar*. Vector addition, denoted by +, is associative and commutative. Moreover, there exists a *zero vector*, denoted by $\mathbf{0}$, such that it leaves all vectors unchanged upon addition. For each vector $\mathbf{v}$ there exists a vector $-\mathbf{v}$ such that $\mathbf{v} + (-\mathbf{v}) = \mathbf{0}$. Multiplication by a scalar, indicated by simple apposition, satisfies the consistency condition $1\mathbf{v} = \mathbf{v}$ for all vectors $\mathbf{v}$. It is, moreover, associative, namely, $(\alpha\beta)\mathbf{v} = \alpha(\beta\mathbf{v})$, and distributive, namely, $\alpha(\mathbf{u} + \mathbf{v}) = \alpha\mathbf{u} + \alpha\mathbf{v}$ and $(\alpha + \beta)\mathbf{v} = \alpha\mathbf{v} + \beta\mathbf{v}$, for all scalars α and β and for all vectors $\mathbf{u}$ and $\mathbf{v}$.

9.4.5.2 Linear Independence and Dimension

A *linear combination* is an expression of the form $\alpha_1\mathbf{v}_1 + \cdots + \alpha_k\mathbf{v}_k$, where, for each $i = 1, \ldots, k$, α_i is a scalar and $\mathbf{v}_i$ is a vector. The vectors $\mathbf{v}_1, \ldots, \mathbf{v}_k$ are *linearly independent* if the only vanishing linear combination is the trivial one, namely, if $\alpha_1\mathbf{v}_1 + \cdots + \alpha_k\mathbf{v}_k = \mathbf{0}$ implies necessarily $\alpha_1 = \cdots = \alpha_k = 0$. A vector space is m-dimensional if there exists a maximal linearly independent set $\mathbf{e}_1, \ldots, \mathbf{e}_m$. Such a set, if it exists, is called a *basis* of the vector space. All bases have the same number of elements, namely, m. If no maximal linearly independent set exists, the vector space is *infinite dimensional*. Given a basis $\mathbf{e}_1, \ldots, \mathbf{e}_m$ of a finite-dimensional vector space, every vector $\mathbf{v}$ can be represented uniquely in terms of *components* v^i as

$$\mathbf{v} = v^1\mathbf{e}_1 + \cdots + v^m\mathbf{e}_m = \sum_{i=1}^{m} v^i\mathbf{e}_i. \tag{9.7}$$

In many contexts, it is convenient to use *Einstein's summation convention*, whereby the summation symbol is understood whenever a monomial contains a once-diagonally repeated index. Thus, we write

$$\mathbf{v} = v^i \mathbf{e}_i, \qquad (9.8)$$

where the summation in the range 1 to m is understood to take place.

9.4.5.3 The Dual Space

A *linear function* on a vector space V is a map $\omega : V \to \mathbb{R}$ such that

$$\omega(\alpha \mathbf{u} + \beta \mathbf{v}) = \alpha\omega(\mathbf{u}) + \beta\omega(\mathbf{v}), \qquad (9.9)$$

for arbitrary scalars $(\alpha,\ \beta)$ and vectors ($\mathbf{u}$ and $\mathbf{v}$). From the physical viewpoint, a linear function expresses the *principle of superposition.* A trivial example of a linear function is the zero function, assigning to all vectors in V the number 0.

If we consider the collection V^* of all linear functions on a given vector space V, it is possible to introduce in it operations of addition and of multiplication by a scalar, as follows. The addition of two linear functions ω and σ is defined as the linear function $\omega + \sigma$ that assigns to each vector $\mathbf{v}$ the sum $\omega(\mathbf{v}) + \sigma(\mathbf{v})$. Similarly, for a scalar α and a linear function ω, we define the linear function $\alpha\omega$ by $(\alpha\omega)(\mathbf{v}) = \alpha(\omega(\mathbf{v}))$. With these two operations, it is not difficult to show that V^* acquires the structure of a vector space, called the *dual space* of V. Its elements are called *covectors*. The action of a covector ω on a vector $\mathbf{v}$ is indicated as $\omega(\mathbf{v})$ or, equivalently, as $\langle \omega,\ \mathbf{v} \rangle$.

Given a basis $\mathbf{e}_1, \ldots, \mathbf{e}_m$ of V, we define the m linear functions $\mathbf{e}^i$ by the formula

$$\mathbf{e}^i(\mathbf{v}) = v^i, \qquad i = 1, \ldots, m. \qquad (9.10)$$

In other words, $\mathbf{e}^i$ is the linear function (i.e., the covector) that assigns to a vector $\mathbf{v} \in V$ its ith component in the given basis. It is not difficult to show that these covectors form a basis of the dual space V^*, which is, therefore, of the same dimension as the original space V. By construction, a basis and its dual satisfy the identity

$$\mathbf{e}^i(\mathbf{e}_j) = \delta^i_j, \qquad (9.11)$$

where δ^i_j is the Kronecker symbol (equal to 1 if $i = j$ and to zero otherwise). Any covector ω can be expressed uniquely in terms of components in a dual basis, namely, $\omega = \omega_i \mathbf{e}^i$, where the summation convention is used.

9.4.6 Tangent and Cotangent Spaces

The collection $T_p\mathcal{M}$ of all the tangent vectors at $p \in \mathcal{M}$ is called the *tangent space to the manifold at p*. It is not difficult to show that tangent vectors at a point p satisfy all the conditions of a vector space if we define their addition and multiplication by a scalar in the obvious way (for example, by using chart components). In other words, $T_p\mathcal{M}$ is a vector space. To find its dimension, we choose a local chart $(\mathcal{U}, \phi)$ with coordinates $x^1, \ldots, x^m$, such that the point p is mapped to the origin of $\mathbb{R}^m$. The inverse map ϕ^{-1}, when restricted to the natural coordinate lines of $\mathbb{R}^m$, delivers m curves at p. Each of these curves, called a *coordinate line in $\mathcal{U}$*, defines a tangent vector, which we suggestively denote by $(\partial/\partial x^i)_p$. It can be shown that these vectors constitute a basis of $T_p\mathcal{M}$, called the *natural basis associated with the given coordinate system.* The dimension of the tangent space at each point of a manifold is, therefore, equal to the dimension of the manifold itself. The *cotangent space at p*, denoted by $T^*_p\mathcal{M}$, is defined as the dual space of $T_p\mathcal{M}$.

9.4.7 The Tangent and Cotangent Bundles

If we attach to each point p of an m-dimensional manifold $\mathcal{M}$ its tangent space $T_p\mathcal{M}$, we obtain, intuitively speaking, a $2m$-dimensional entity, which we denote by $T\mathcal{M}$, called the *tangent bundle* of $\mathcal{M}$. A crude visualization of this entity can be gathered when $\mathcal{M}$ is a 2-sphere, such as a globe, at each point of which we have stuck a postal stamp or a paper sticker. The tangent bundle is not the globe itself but rather the collection of the stickers. This collection of tangent spaces, however, has the property that it *projects* on the original manifold. In our example, each sticker indicates the point at which it has been attached. In other words, the set $T\mathcal{M}$ is endowed, by construction, with a *projection map* τ onto the *base manifold* $\mathcal{M}$. More explicitly, a typical point of $T\mathcal{M}$ consists of a pair $(p, \mathbf{v}_p)$, where $p \in \mathcal{M}$ and $\mathbf{v}_p \in T_p\mathcal{M}$. The projection map

$$\tau : T\mathcal{M} \to \mathcal{M} \tag{9.12}$$

is given by the assignation

$$\tau(p, \mathbf{v}_p) = p. \tag{9.13}$$

To see that the set $T\mathcal{M}$ can be regarded as a manifold, we construct explicitly an atlas out of any given atlas of the base manifold. Let $(\mathcal{U}, \phi)$ be a chart in $\mathcal{M}$ with coordinates $x^i, \ldots, x^m$. Adopting, as we may, the natural basis $(\partial/\partial x^i)_p$ of $T_p\mathcal{M}$ at each point $p \in \mathcal{U}$, we can identify each vector $\mathbf{v}_p$ with its components v^i_p. Put differently, we assign to each point $(p, \mathbf{v}_p) \in \tau^{-1}(\mathcal{U}) \subset T\mathcal{M}$ the $2m$ numbers $(x^1, \ldots, x^m, v^1, \ldots, v^m)$, namely, a point in $\mathbb{R}^{2m}$. We have thus obtained a coordinate chart on $\tau^{-1}(\mathcal{U})$. It is now a formality to extend this construction to a whole atlas of $T\mathcal{M}$ and to show that $T\mathcal{M}$ is a differentiable manifold of dimension $2m$. In the terminology of general *fiber bundles*, the set $T_p\mathcal{M} = \tau^{-1}(p)$ is called the *fiber* at $p \in \mathcal{M}$. As each fiber is an m-dimensional vector space, we say that the *typical fiber* of $T\mathcal{M}$ is $\mathbb{R}^m$.

Upon a coordinate transformation represented by (9.2), the components $\hat{v}^i$ of a vector $\mathbf{v}$ at p in the new natural basis $(\partial/\partial y^i)_p$ are related to the old components v^i in the basis $(\partial/\partial x^i)_p$ by the formula

$$\hat{v}^i = \left(\frac{\partial y^i}{\partial x^j}\right)_p v^j, \tag{9.14}$$

while the base vectors themselves are related by the formula

$$\left(\frac{\partial}{\partial y^i}\right)_p = \left(\frac{\partial x^j}{\partial y^i}\right)_p \left(\frac{\partial}{\partial x^j}\right)_p. \tag{9.15}$$

Comparing these two formulas, we conclude that the components of vectors behave *contravariantly*. In traditional treatments, it was customary to *define* tangent vectors as indexed quantities that transform contravariantly under coordinate changes.

A similar construction can be carried out by attaching to each point of a manifold $\mathcal{M}$ its cotangent space $T^*_p\mathcal{M}$ to obtain the set $T^*\mathcal{M}$, called the *cotangent bundle* of $\mathcal{M}$. A typical point of $T^*\mathcal{M}$ is a pair (p, ω_p), where $p \in \mathcal{M}$ and $\omega_p \in T^*_p\mathcal{M}$. The *projection map* $\pi : T^*\mathcal{M} \to \mathcal{M}$ is given by

$$\pi(p, \omega_p) = p. \tag{9.16}$$

Given a chart, the local dual basis to the natural basis $(\partial/\partial x^i)_p$ is denoted by $(\mathrm{d}x^i)_p$, with $i = 1, \ldots, m$. The covector $\omega_p \in T^*_p\mathcal{M}$ can be uniquely expressed as $\omega_p = \omega_i \mathrm{d}x^i$, where the subscript p has been eliminated for clarity. Given a point $(p, \omega_p) \in \pi^{-1}(\mathcal{U}) \subset T^*\mathcal{M}$, we assign to it the $2m$ numbers $(x^1, \ldots, x^m, \omega_1, \ldots, \omega_m)$. In this way, it can be rigorously shown that $T^*\mathcal{M}$ is a manifold of dimension $2m$.

Upon a coordinate transformation, the components $\hat{\omega}_i$ of a covector ω at p transform according to

$$\hat{\omega}_i = \left(\frac{\partial x^j}{\partial y^i}\right)_p \omega_j, \tag{9.17}$$

while the dual base vectors themselves are related by the formula

$$\mathrm{d}y^i = \left(\frac{\partial y^i}{\partial x^j}\right)_p \mathrm{d}x^j. \tag{9.18}$$

The components of covectors behave *covariantly*.

9.4.8
A Physical Interpretation

In the context of the application presented in Section 9.3.5, what do the tangent and cotangent bundles represent? If the manifold Q is the configuration space of a system with m degrees of freedom, then a curve $\gamma : (a, b) \to Q$ represents a possible *trajectory* of the system as it evolves in the time interval $a < t < b$. At a point $q \in Q$, therefore, a "small" piece of a curve conveys the notion of a *virtual displacement* δq, that is, a small displacement compatible with the degrees of freedom of the system. In the limit, we obtain a tangent vector at q, representing a *velocity*. We conclude that the tangent space $T_q Q$ is the repository of all possible velocities (or virtual displacements) of the system at the configuration q. The tangent bundle TQ, accordingly, is the collection of all possible velocities of the system at all possible configurations. An element of TQ consists of an ordered pair made up of a configuration and a velocity at this configuration. The projection map τ assigns to this pair the configuration itself. In Lagrangian Mechanics, the fundamental geometric arena is precisely the tangent bundle TQ. Indeed, the *Lagrangian density* $\mathcal{L}$ of a mechanical system is given by a function $\mathcal{L} : TQ \to \mathbb{R}$, assigning to each configuration and each velocity (at this configuration) a real number.

A covector Q at q is a linear function that assigns to each tangent vector (virtual displacement δq) at q a real number $\delta W = \langle Q, \delta q \rangle$, whose meaning is the *virtual work* of the *generalized force* Q on the virtual displacement δq (or the *power* of the generalized force on the corresponding velocity). The terminology and the notation are due to Lagrange. The interesting feature of the geometric approach is that, once the basic geometric entity has been physically identified as a manifold, its tangent and cotangent bundles are automatically the carriers of physical meaning. In Hamiltonian Mechanics, covectors at $q \in Q$ can be regarded as *generalized momenta* of the system. Thus, the cotangent bundle T^*Q is identified with the *phase space* of the system, namely, the repository of all configurations and momenta. The *Hamiltonian function* of a mechanical system is a function $\mathcal{H} : T^*Q \to \mathbb{R}$.

9.4.9
The Differential of a Map

Given a differentiable map

$$g : \mathcal{M} \to \mathcal{N} \tag{9.19}$$

between two manifolds, $\mathcal{M}$ and $\mathcal{N}$, of dimensions m and n, respectively, we focus attention on a particular point $p \in \mathcal{M}$ and its image $q = g(p) \in \mathcal{N}$. Let $\mathbf{v}_p \in T_p\mathcal{M}$ be a tangent vector at p and let $\gamma : H \to \mathcal{M}$ be one of its representative curves. The composite map

$$g \circ \gamma : H \longrightarrow \mathcal{N} \tag{9.20}$$

is then a smooth curve in $\mathcal{N}$ passing through q. This curve (the image of γ by g) is, therefore, the representative of a tangent vector at q, which we will denote $(g_*)_p(\mathbf{v}_p)$. The vector $(g_*)_p(\mathbf{v}_p)$ is independent of the representative curve γ chosen for $\mathbf{v}_p$. Moreover, $(g_*)_p$ is a linear map on vectors at p.

The map $(g_*)_p$ just defined is called the *differential of g at p*. It is a linear map between the tangent spaces $T_p\mathcal{M}$ and $T_{g(p)}\mathcal{N}$. As this construction can be carried out at each and every point of $\mathcal{M}$, we obtain a map g_* between the tangent bundles, namely,

$$g_* : T\mathcal{M} \to T\mathcal{N}, \tag{9.21}$$

called the *differential of g*. Alternative notations for this map are Dg and Tg, and it is also known as the *tangent map*. One should note that the map g_* includes the map g between the base manifolds, because it maps vectors at a point p linearly into vectors at the image point $q = g(p)$, and not just to any vector in $T\mathcal{N}$. It is, therefore, a *fiber-preserving map*. This fact is best illustrated in the following commutative diagram:

$$\begin{array}{ccc} T\mathcal{M} & \xrightarrow{\;g_*\;} & T\mathcal{N} \\ {\scriptstyle \tau_{\mathcal{M}}}\downarrow & & \downarrow{\scriptstyle \tau_{\mathcal{N}}} \\ \mathcal{M} & \xrightarrow{\;g\;} & \mathcal{N} \end{array} \tag{9.22}$$

where $\tau_{\mathcal{M}}$ and $\tau_{\mathcal{N}}$ are the projection maps of $T\mathcal{M}$ and $T\mathcal{N}$, respectively. The differential is said to *push forward* tangent vectors at p to tangent vectors at the image point $g(p)$.

In the particular case of a function $f : \mathcal{M} \to \mathbb{R}$, the differential f_* can be interpreted somewhat differently. Indeed, the tangent space $T_r\mathbb{R}$ can be trivially identified with $\mathbb{R}$ itself, so that f_* can be seen as a real-valued function on $T\mathcal{M}$. This function is denoted by $\mathrm{d}f : T\mathcal{M} \to \mathbb{R}$. The differential of a function satisfies the identity

$$\mathrm{d}f(\mathbf{v}) = \mathbf{v}(f). \tag{9.23}$$

In local systems of coordinates x^i ($i = 1, \ldots, m$) and y^α ($\alpha = 1, \ldots, n$) around p and $g(p)$, respectively, the differential of g at p maps the vector with components v^i into the vector with components

$$[(g_*)_p(\mathbf{v}_p)]^\alpha = \left(\frac{\partial g^\alpha}{\partial x^i}\right)_p v^i, \tag{9.24}$$

where $g^\alpha = g^\alpha(x^1, \ldots, x^n)$ is the coordinate representation of g in the given charts. The $(m \times n)$-matrix with entries $\{\left(\partial g^\alpha/\partial x^i\right)_p\}$ is the *Jacobian matrix* at p of the map g in the chosen coordinate systems. The *rank* of the Jacobian matrix is independent of the coordinates used. It is called the *rank of g at p*.

Let $f : \mathcal{N} \to \mathbb{R}$ be a differentiable function and let $g : \mathcal{M} \to \mathcal{N}$ be a differentiable map between manifolds. Then,

$$((g_*)_p\mathbf{v}_p)(f) = \mathbf{v}_p(f \circ g), \quad p \in \mathcal{M}. \tag{9.25}$$

The differential of a composition of maps is equal to the composition of the differentials. More precisely, if $g : \mathcal{M} \to \mathcal{N}$ and $h : \mathcal{N} \to \mathcal{P}$ are differentiable maps, then

$$((h \circ g)_*)_p(\mathbf{v}_p) = (h_*)_{g(p)}((g_*)_p(\mathbf{v}_p)). \tag{9.26}$$

In coordinates, this formula amounts to the multiplication of the Jacobian matrices.

9.5 Vector Fields and the Lie Bracket

9.5.1 Vector Fields

A *vector field* $\mathbf{V}$ on a manifold $\mathcal{M}$ is an assignment of a tangent vector $\mathbf{V}_p = \mathbf{V}(p) \in T_p\mathcal{M}$ to each point $p \in \mathcal{M}$. We restrict our attention to *smooth vector fields,* whose components are smooth functions in any given chart. A vector field is, therefore, a smooth map

$$\mathbf{V} : \mathcal{M} \to T\mathcal{M}, \tag{9.27}$$

satisfying the condition

$$\tau \circ \mathbf{V} = id_{\mathcal{M}}, \tag{9.28}$$

where $id_{\mathcal{M}}$ is the identity map of $\mathcal{M}$. The meaning of this last condition is that the vector assigned to the point p is a tangent vector at p, rather than at any other point.

A geometrically convenient way to look at a vector field is to regard it as a *cross section* of the tangent bundle. This terminology arises from the pictorial representation depicted in Figure 9.1, where the base manifold is represented by a shallow arc and the fibers (namely, the tangent spaces) by straight lines hovering above it. Then, a cross section looks like a curve cutting through the fibers.

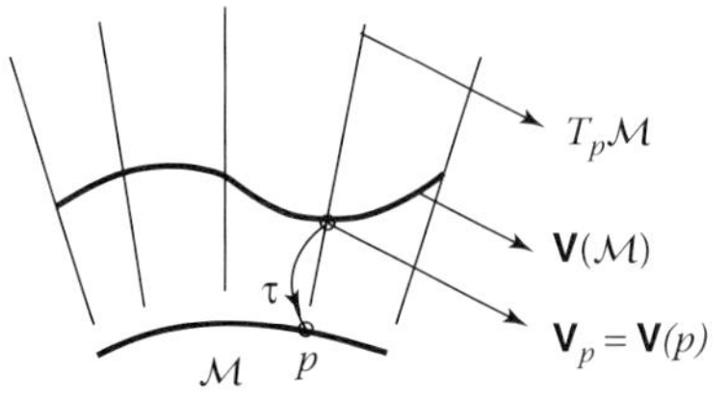

Figure 9.1 A vector field as a cross section of the tangent bundle.

9.5.2 The Lie Bracket

If $\mathbf{V}$ is a (smooth) vector field on a manifold $\mathcal{M}$ and $f : \mathcal{M} \to \mathbb{R}$ is a smooth function, then the map

$$\mathbf{V}f : \mathcal{M} \to \mathbb{R}, \tag{9.29}$$

defined as

$$p \mapsto \mathbf{V}_p(f) \tag{9.30}$$

is again a smooth map. It assigns to each point $p \in \mathcal{M}$ the directional derivative of the function f in the direction of the vector field at p. In other words, a vector field assigns to each smooth function another smooth function. Given, then, two vector fields $\mathbf{V}$ and $\mathbf{W}$ over $\mathcal{M}$, the iterated evaluation

$$h = \mathbf{W}(\mathbf{V}f) : \mathcal{M} \to \mathbb{R}, \tag{9.31}$$

gives rise to a legitimate smooth function h on $\mathcal{M}$.

On the basis of the above considerations, one may be tempted to define a composition of vector fields by declaring that the composition $\mathbf{W} \circ \mathbf{U}$ is the vector field that assigns to each function f the function h defined by (9.31). This wishful thinking, however, does not work. To see why, it is convenient to work in components in some chart with coordinates x^i. Let

$$\mathbf{V} = V^i \frac{\partial}{\partial x^i} \qquad \mathbf{W} = W^i \frac{\partial}{\partial x^i}, \tag{9.32}$$

where the components V^i and W^i ($i = 1, \ldots, m$) are smooth real-valued functions defined over the m-dimensional domain of the chart. Given a smooth function $f : \mathcal{M} \to \mathbb{R}$, the function $g = \mathbf{V}f$ is evaluated at a point $p \in \mathcal{M}$ with coordinates x^i ($i = 1, \ldots, m$) as

$$g(p) = V^i \frac{\partial f}{\partial x^i}. \qquad (9.33)$$

Notice the slight abuse of notation we incur into by identifying the function f with its representation in the coordinate system.

We now apply the same prescription to calculate the function $h = \mathbf{W}g$ and obtain

$$\begin{aligned} h(p) = W^i \frac{\partial g}{\partial x^i} &= W^i \frac{\partial \left(V^j \frac{\partial f}{\partial x^j} \right)}{\partial x^i} \\ &= \left(W^i \frac{\partial V^j}{\partial x^i} \right) \frac{\partial f}{\partial x^j} + W^i V^j \frac{\partial^2 f}{\partial x^i \partial x^j}. \end{aligned} \qquad (9.34)$$

The last term of this expression, by involving second derivatives, will certainly not transform as the components of a vector should under a change of coordinates. Neither will the first. This negative result, on the other hand, suggests that the offending terms could perhaps be eliminated by subtracting from the composition $\mathbf{WV}$ the opposite composition $\mathbf{VW}$, namely,

$$\begin{aligned} &(\mathbf{WV} - \mathbf{VW})(f) \\ &= \left(W^i \frac{\partial V^j}{\partial x^i} - V^i \frac{\partial W^j}{\partial x^i} \right) \frac{\partial f}{\partial x^j}. \end{aligned} \qquad (9.35)$$

The vector field thus obtained is called the *Lie bracket* of $\mathbf{W}$ and $\mathbf{V}$ (in that order) and is denoted by $[\mathbf{W}, \mathbf{V}]$. More explicitly, its components in the coordinate system x^i are given by

$$[\mathbf{W}, \mathbf{V}]^j = W^i \frac{\partial V^j}{\partial x^i} - V^i \frac{\partial W^j}{\partial x^i}. \qquad (9.36)$$

Upon a coordinate transformation, these components transform according to the rules of transformation of a vector.

The following properties of the Lie bracket are worthy of notice:

1. Skew symmetry:

$$[\mathbf{W}, \mathbf{V}] = -[\mathbf{V}, \mathbf{W}]. \qquad (9.37)$$

2. Jacobi identity:

$$\begin{aligned} &[[\mathbf{W}, \mathbf{V}], \mathbf{U}] + [[\mathbf{V}, \mathbf{U}], \mathbf{W}] \\ &\qquad + [[\mathbf{U}, \mathbf{W}], \mathbf{V}] = 0. \end{aligned} \qquad (9.38)$$

The collection of all vector fields over a manifold has the natural structure of an infinite-dimensional vector space, where addition and multiplication by a scalar are defined in the obvious way. In this vector space, the Lie bracket operation is bilinear. A vector space endowed with a bilinear operation satisfying conditions (1) and (2) is called a *Lie algebra*.

Vector fields can be multiplied by functions to produce new vector fields. Indeed, for a given function f and a given vector field $\mathbf{V}$, we can define the vector field $f\mathbf{V}$ by

$$(f\mathbf{V})_p = f(p)\mathbf{V}_p. \qquad (9.39)$$

It can be shown that

$$\begin{aligned} [g\mathbf{W}, f\mathbf{V}] &= gf[\mathbf{W}, \mathbf{V}] \\ &\quad + g\left(\mathbf{W}f\right)\, \mathbf{V} - f\left(\mathbf{V}g\right)\, \mathbf{W}, \end{aligned} \qquad (9.40)$$

where g, f are smooth functions and $\mathbf{W}, \mathbf{V}$ are vector fields over a manifold $\mathcal{M}$.

9.5.3
A Physical Interpretation: Continuous Dislocations

Let an atomic lattice be given by, say, all points with integer coordinates in $\mathbb{R}^2$. To

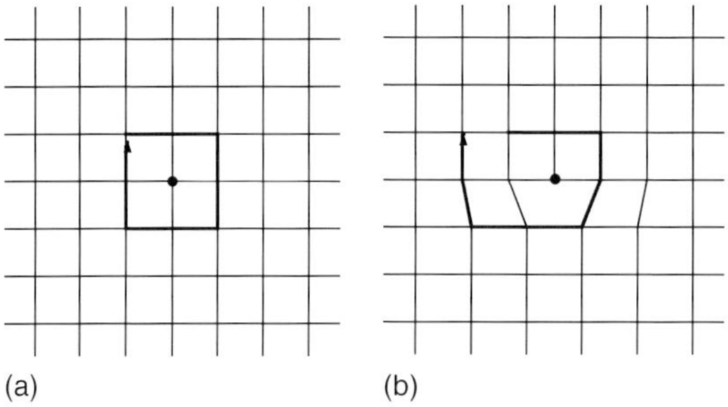

Figure 9.2 Dislocation in a crystal lattice. (a) Perfect lattice and (b) dislocated lattice.

each atom we can associate two vectors (in this instance, unit and orthogonal) determined by joining it to its immediate neighbors to the right and above, respectively. If the lattice is deformed regularly, these vectors will deform accordingly, changing in length and angle, but always remaining linearly independent at each atom. In the (not precisely defined) continuous limit, we can imagine that each point of $\mathbb{R}^2$ has been endowed with a basis or frame, the collection of which is called a *moving frame* (or *repère mobile*).[6)]

Returning to the discrete picture, if there is a dislocation (for example, a half-line of atoms is missing, as shown on the right-hand side of Figure 9.2), the local bases will be altered differently from the case of a mere deformation. The engineering way to recognize this is the so-called *Burgers' circuit*, which consists of a four-sided path made of the same number of atomic spacings in each direction. The failure of such a path to close is interpreted as the presence of a local dislocation in the lattice. We want to show that in the putative continuous limit, this failure is represented by the non-vanishing of a Lie bracket. What we have in the continuous case as the only remnant of the discrete picture is a smoothly distributed collection of bases, which we have called a moving frame, and which can be seen as two vector fields $\mathbf{E}_\alpha$ $(\alpha = 1, 2)$ over $\mathbb{R}^2$.

From the theory of ordinary differential equations, we know that each vector field gives rise, at least locally, to a well-defined family of parameterized integral curves, where the parameter is determined up to an additive constant. More specifically, these curves are obtained as the solutions $\mathbf{r} = \mathbf{r}(s^\alpha)$ of the systems of equations

$$\frac{d\mathbf{r}(s^\alpha)}{ds^\alpha} = \mathbf{E}_\alpha[\mathbf{r}(s^\alpha)], \qquad (\alpha = 1, 2; \quad \text{no sum on } \alpha), \tag{9.41}$$

where $\mathbf{r}$ represents the natural position vector in $\mathbb{R}^2$. The parameter s^α (one for each of the two families of curves) can be pinned down in the following way. Select a point p_0 as origin and draw the (unique) integral curve γ_1 of the first family passing through this origin. Adopting the value $s^1 = 0$ for the parameter at the origin, the value of s^1 becomes uniquely defined for all the remaining points of the curve. Each of the curves of the second family must intersect this curve of the first family. We adopt, therefore, for each of the curves of the second family the value $s^2 = 0$ at the corresponding point of intersection with that reference curve (of the first family). In this way, we obtain (at least locally) a new coordinate system s^1, s^2 in $\mathbb{R}^2$. By construction, the second natural base vector of this coordinate system is $\mathbf{E}_2$. But there is no

6) This idea was introduced mathematically by Cartan and, in a physical context, by the brothers Cosserat.

guarantee that the first natural base vector will coincide with $\mathbf{E}_1$, except at the curve γ_1 through the adopted origin. In fact, if we repeat the previous construction in reverse, that is, with the same origin but adopting the curve γ_2 of the *second* family as a reference, we obtain, in general, a different system of coordinates, which is well adapted to the basis vectors $\mathbf{E}_1$, but not necessarily to $\mathbf{E}_2$ (Figure 9.3).

Assume now that, starting at the adopted origin, we move an amount of Δs^1 along γ_1 to arrive at a point p' and thereafter we climb an amount of Δs^2 along the encountered curve of the second family through p'. We arrive at some point p_1. Incidentally, this is the point with coordinates $(\Delta s^1, \Delta s^2)$ in the coordinate system obtained by the first construction. If, however, starting at the same origin we move by Δs^2 along the curve γ_2 to a point $\hat{p}$ and then move by Δs^1 along the encountered curve of the first family, we will arrive at a point p_2 (whose coordinates are $(\Delta s^1, \Delta s^2)$ in the *second* construction) which is, in general, different from p_1. Thus, we have detected the failure of a four-sided circuit to close! The discrete picture has, therefore, its continuous counterpart in the noncommutativity of the flows along the two families of curves.

Let us calculate a first-order approximation to the difference between p_2 and p_1. For this purpose, let us evaluate, to the first order, the base vector $\mathbf{E}_2$ at the auxiliary point p'. The result is

$$\mathbf{E}_2' = \mathbf{E}_2(p_0) + \frac{\partial \mathbf{E}_2}{\partial x^i}\frac{dx^i}{ds^1}\,\Delta s^1, \tag{9.42}$$

where derivatives are calculated at p_0. The position vector of p_1, always to first-order approximation, is obtained, therefore, as

$$\mathbf{r}_1 = \Delta s^1 \mathbf{E}_1(p_0) + \Delta s^2\left(\mathbf{E}_2(p_0) + \frac{\partial \mathbf{E}_2}{\partial x^i}\frac{dx^i}{ds^1}\,\Delta s^1,\right). \tag{9.43}$$

In a completely analogous manner, we calculate the position vector of p_2 as

$$\mathbf{r}_2 = \Delta s^2 \mathbf{E}_2(p_0) + \Delta s^1\left(\mathbf{E}_1(p_0) + \frac{\partial \mathbf{E}_1}{\partial x^i}\frac{dx^i}{ds^2}\,\Delta s^2,\right). \tag{9.44}$$

By virtue of (9.41), however, we have

$$\frac{dx^i}{ds^\alpha} = E^i_\alpha, \tag{9.45}$$

where E^i_α is the ith component in the natural basis of $\mathbb{R}^2$ of the base vector $\mathbf{E}_\alpha$. From the previous three equations, we obtain

$$\mathbf{r}_2 - \mathbf{r}_1 = \left(\frac{\partial \mathbf{E}_1}{\partial x^i}E^i_2 - \frac{\partial \mathbf{E}_2}{\partial x^i}E^i_1\right)\Delta s^1 \Delta s^2 = [\mathbf{E}_1, \mathbf{E}_2]\,\Delta s^1 \Delta s^2. \tag{9.46}$$

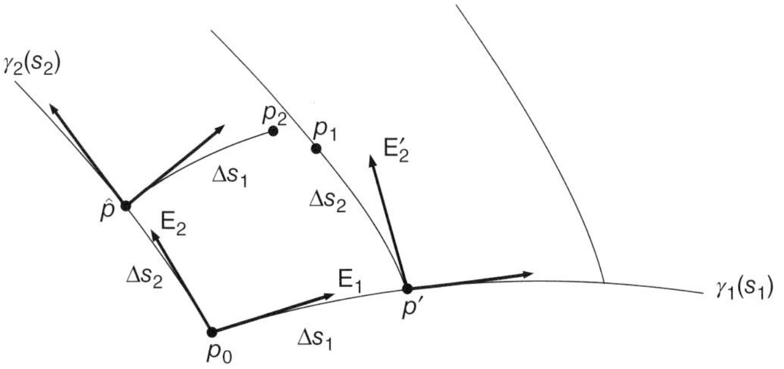

Figure 9.3 The continuous case.

We thus confirm that the closure of the infinitesimal circuits generated by two vectors fields is tantamount to the vanishing of their Lie bracket. This vanishing, in turn, is equivalent to the commutativity of the flows generated by these vector fields. For this reason, the Lie bracket is also called the *commutator* of the two vector fields. In physical terms, we may say that the vanishing of the Lie brackets between the vector fields representing the limit of a lattice is an indication of the absence of dislocations.

As in this example we have introduced the notion of a moving frame, that is, a smooth field of bases $\mathbf{E}_i$ $(i = 1, \ldots, n)$ over an n-dimensional manifold, it makes sense to compute all the possible Lie brackets between the base vectors and to express them in terms of components in the local basis. As a Lie bracket of two vector fields is itself a vector field, there must exist unique scalar fields c^k_{ij} such that

$$[\mathbf{E}_i, \mathbf{E}_j] = c^k_{ij}\mathbf{E}_k \qquad (i, j, k = 1, \ldots, n). \quad (9.47)$$

These scalars are known as the *structure constants* of the moving frame. The structure constants vanish identically if, and only if, the frames can be seen locally as the natural base vectors of a coordinate system.

9.5.4 Pushforwards

We have seen that the differential of a map between manifolds carries tangent vectors to tangent vectors. This operation is sometimes called a *pushforward*. Does a map also pushforward vector fields to vector fields? Let $\mathbf{V} : \mathcal{M} \to T\mathcal{M}$ be a vector field on $\mathcal{M}$ and let $g : \mathcal{M} \to \mathcal{N}$ be a smooth map. As the differential of g is a map of the form $g_* : T\mathcal{M} \to T\mathcal{N}$, the composition $g_* \circ \mathbf{V}$ makes perfect sense, but it delivers a (well-defined) map $g_*\mathbf{V}$ from $\mathcal{M}$ (and *not* from $\mathcal{N}$) into $T\mathcal{N}$. This is not a vector field, nor can it in general be turned into one. If the dimension of $\mathcal{M}$ is larger than that of $\mathcal{N}$, points in $\mathcal{N}$ will end up being assigned more than one vector. If the dimension of the source manifold is less than that of the target, on the other hand, even if the function is one-to-one, there will necessarily exist points in $\mathcal{N}$ to which no vector is assigned. The only case in which the pushforward of a vector field can be regarded as a vector field on the target manifold is the case in which both manifolds are of the same dimension and the map is a diffeomorphism.

Notwithstanding the above remark, let $g : \mathcal{M} \to \mathcal{N}$ be a smooth map. We say that the vector fields $\mathbf{V} : \mathcal{M} \to T\mathcal{M}$ and $\mathbf{W} : \mathcal{N} \to T\mathcal{N}$ are *g-related* if

$$g_*\mathbf{V}(p) = \mathbf{W}(g(p)) \qquad \forall p \in \mathcal{M}. \quad (9.48)$$

According to this definition, if g happens to be a diffeomorphism, then $\mathbf{V}$ and $g_*\mathbf{V}$ are automatically g-related. The pushed-forward vector field is then given by $g_* \circ \mathbf{V} \circ g^{-1}$.

Theorem 9.1 *Let $\mathbf{V}_1$ be g-related to $\mathbf{W}_1$ and let $\mathbf{V}_2$ be g-related to $\mathbf{W}_2$. Then the Lie bracket $[\mathbf{V}_1, \mathbf{V}_2]$ is g-related to the Lie bracket $[\mathbf{W}_1, \mathbf{W}_2]$, that is,*

$$[g_*\mathbf{V}_1, g_*\mathbf{V}_2] = g_*[\mathbf{V}_1, \mathbf{V}_2]. \quad (9.49)$$

9.6 Review of Tensor Algebra

9.6.1 Linear Operators and the Tensor Product

A *linear operator* T between two vector spaces U and V is a linear map $T : U \to V$ that respects the vector-space structure.

More precisely,

$$T(\alpha \mathbf{u}_1 + \beta \mathbf{u}_2) = \alpha T(\mathbf{u}_1) + \beta T(\mathbf{u}_2),$$
$$\forall\, \alpha, \beta \in \mathbb{R},\ \mathbf{u}_1, \mathbf{u}_2 \in U, \tag{9.50}$$

where the operations are understood in the corresponding vector spaces. When the source and target vector spaces coincide, the linear operator is called a *tensor*. Occasionally, the terminology of *two-point tensor* (or just tensor) is also used for the general case, particularly when the dimension of both spaces is the same. We will use these terms (linear operator, linear map, tensor, and so on) liberally.

Consider the collection $L(U, V)$ of *all* linear operators between two given vector spaces, and endow it with the natural structure of a vector space. To do so, we define the sum of two linear operators S and T as the linear operator $S + T$ whose action on an arbitrary vector $\mathbf{u} \in U$ is given by

$$(S + T)(\mathbf{u}) = S(\mathbf{u}) + T(\mathbf{u}). \tag{9.51}$$

Similarly, we define the product of a scalar α by a linear operator T as the linear operator αT given by

$$(\alpha T)(\mathbf{u}) = \alpha T(\mathbf{u}). \tag{9.52}$$

It is a straightforward matter to verify that the set $L(U, V)$, with these two operations, is a vector space. In the case of the dual space V^* (which can be identified with $L(V, \mathbb{R})$), we were immediately able to ascertain that it was never empty, because the zero map is linear. The same is true for $L(U, V)$, whose zero element is the linear map $\mathbf{0} : U \to V$ assigning to each vector of U the zero vector of V. Inspired by the example of the dual space, we will attempt now to construct a basis of $L(U, V)$ starting from given bases at U and V. This point takes some more work, but the result is, both conceptually and notationally, extremely creative.

Let $\omega \in U^*$ and $\mathbf{v} \in V$ be, respectively a covector of the source space and a vector of the target space of a linear operator $T : U \to V$. We define the *tensor product* of $\mathbf{v}$ with ω as the linear operator $\mathbf{v} \otimes \omega \in L(U, V)$ obtained as follows:

$$(\mathbf{v} \otimes \omega)\,(\mathbf{u}) = \langle \omega, \mathbf{u} \rangle\, \mathbf{v}, \qquad \forall\, \mathbf{u} \in U. \tag{9.53}$$

We emphasize that the tensor product is fundamentally noncommutative. We note, on the other hand, that the tensor product is a bilinear operation, namely, it is linear in each of the factors, namely,

$$(\alpha \mathbf{u}_1 + \beta \mathbf{u}_2) \otimes \omega = \alpha(\mathbf{u}_1 \otimes \omega) + \beta(\mathbf{u}_2 \otimes \omega) \tag{9.54}$$

and

$$\mathbf{u} \otimes (\alpha\omega_1 + \beta\omega_2) = \alpha(\mathbf{u} \otimes \omega_1) + \beta(\mathbf{u} \otimes \omega_2) \tag{9.55}$$

for all $\alpha, \beta \in \mathbb{R}$.

One of the reasons for the conceptual novelty of the tensor product is that it does not seem to have an immediately intuitive interpretation. In fact, it is a very singular linear operator, because, fixing the first factor, it squeezes the whole vector space U^* into an image consisting of a single line of V (the line of action of $\mathbf{v}$).

Let the dimensions of U and V be, respectively, m and n, and let $\{\mathbf{e}_\alpha\}$ $(\alpha = 1, \ldots, m)$ and $\{\mathbf{f}_i\}$ $(i = 1, \ldots, n)$ be respective bases. It makes sense to consider the $m \times n$ tensor products $\mathbf{f}_i \otimes \mathbf{e}^\alpha \in L(U, V)$. We want to show that these linear operators (considered as vectors belonging to the vector space $L(U, V)$) are in fact linearly independent. Assume that a vanishing linear combination has been found, namely, $\rho^i_{\ \alpha}\, \mathbf{f}_i \otimes \mathbf{e}^\alpha = \mathbf{0}$, where $\rho^i_{\ \alpha} \in \mathbb{R}$ and where the summation convention is appropriately used (Greek indices ranging from 1 to m, and Latin indices ranging from 1 to n). Applying this linear combination to the base vector $\mathbf{e}_\beta \in V$, we obtain $\rho^i_{\ \beta}\,\mathbf{f}_i = \mathbf{0}$, whence $\rho^i_{\ \beta} = 0$, proving that the only vanishing linear combination is the trivial one.

Let $T \in L(U, V)$ be an arbitrary linear operator. By linearity, we may write

$$T(\mathbf{u}) = T(u^\alpha \mathbf{e}_\alpha) = u^\alpha\, T(\mathbf{e}_\alpha). \tag{9.56}$$

Each $T(\mathbf{e}_\alpha)$, being an element of V, can be written as a unique linear combination of the basis, namely,

$$T(\mathbf{e}_\alpha) = T^i_{\ \alpha}\,\mathbf{f}_i, \quad T^i_{\ \alpha} \in \mathbb{R}. \tag{9.57}$$

We form now the linear operator $T^i_{\ \alpha}\,\mathbf{f}_i \otimes \mathbf{e}^\alpha$ and apply it to the vector $\mathbf{u}$, which yields

$$T^i_{\ \alpha}\,\mathbf{f}_i \otimes \mathbf{e}^\alpha\,(\mathbf{u}) = u^\alpha\, T^i_{\ \alpha}\,\mathbf{f}_i. \tag{9.58}$$

Comparing this result with (9.56, 9.57) we conclude that the original operator T and the operator $T^i_{\ \alpha}\,\mathbf{f}_i \otimes \mathbf{e}^\alpha$ produce the same result when operating on an arbitrary vector $\mathbf{u} \in U$. They are, therefore, identical and we can write

$$T = T^i_{\ \alpha}\,\mathbf{f}_i \otimes \mathbf{e}^\alpha. \tag{9.59}$$

In other words, every linear operator in $L(U, V)$ can be written as a linear combination of the $m \times n$ linearly independent operators $\mathbf{f}_i \otimes \mathbf{e}^\alpha$, showing that these operators form a basis of $L(U, V)$, whose dimension is, therefore, the product of the dimensions of U and V. For these reasons, the vector space $L(U, V)$ is also called the *tensor-product space of V and U^**, and is denoted as $V \otimes U^*$. The unique coefficients $T^i_{\ \alpha}$ are called the *components* of the tensor T in the corresponding basis.

The *composition* of linear operators is a particular case of the composition of functions. Let $T : U \to V$ and $S : V \to W$ be linear operators between the vector spaces U, V, and W. The composition $S \circ T : U \to W$ is usually denoted as ST and is called the *product* of the operators. Choosing bases in the vector spaces U, V, and W and expressing the operators $T : U \to V$ and $S : V \to W$ in components, the composition $S \circ T$ is represented in components by the product $[S][T]$ of the matrices of components of S and T. This is the best justification of the, at first sight odd, rule for multiplying matrices.

We have spoken about the conceptual novelty of the tensor product. No less important is its notational convenience. In the case of a vector, say $\mathbf{v} \in V$, it is obviously advantageous to be able to express the master concept of a vector in terms of the subsidiary notion of its components in a particular basis by simply writing: $\mathbf{v} = v^i \mathbf{f}_i$. If we change the basis, for instance, the fact that what we have just written is an invariant expression, with a meaning beyond the particular basis chosen, can be exploited, as we have already done. In the case of linear operators, we have now obtained, according to (9.59), a similar way to express the "real thing" invariantly in terms of its components on a basis arising from having chosen arbitrary bases in both the source and the target spaces. We can now *show* a tensor itself, much in the same way as we are able to show a vector itself. So powerful is this idea that, historically, the notation was invented before the concept of tensor product had been rigorously defined. In old Physics texts it was called the *dyadic* notation, and it consisted of simply apposing the elements of the bases involved (usually in the same (Cartesian) vector space: **ii**, **ij**, etc.). It is also interesting to recall that in Quantum Mechanics the prevailing notation for covectors and tensors is the ingenious device introduced by Dirac in terms of "bras" and "kets."

9.6.2
Symmetry and Skew Symmetry

Let $T : U \to V$ be a linear map. We define the *transpose* of T as the map $T^T : V^* \to U^*$ obtained by the prescription

$$\langle T^T(\omega), \mathbf{u} \rangle = \langle \omega, T(\mathbf{u}) \rangle. \quad (9.60)$$

If $\{\mathbf{e}_\alpha\}$ $(\alpha = 1, \ldots, m)$ and $\{\mathbf{f}_i\}$ $(i = 1, \ldots, n)$ are, respectively, bases of U and V, whereby T is expressed as

$$T = T^i_{\ \alpha}\, \mathbf{f}_i \otimes \mathbf{e}^\alpha, \quad (9.61)$$

then the transpose of T is expressed as

$$T^T = T^i_{\ \alpha}\, \mathbf{e}^\alpha \otimes \mathbf{f}_i. \quad (9.62)$$

In other words, the transpose of a tensor is obtained by leaving the components unchanged and switching around the base vectors. On the other hand, we may want to express the transpose in its own right by the standard formula

$$T^T = (T^T)_\alpha^{\ i}\, \mathbf{e}^\alpha \otimes \mathbf{f}_i, \quad (9.63)$$

applicable to the components of a tensor in terms of a basis. Comparing the last two equations, we conclude that

$$(T^T)_\alpha^{\ i} = T^i_{\ \alpha}. \quad (9.64)$$

Notice the precise order of the indices in each case.

A linear operator T is said to be *symmetric* if $T = T^T$ and *skew-* (or *anti-*) *symmetric* if $T = -T^T$. Recall, however, that in general T and T^T operate between different spaces. This means that the notion of *symmetry* should be reserved to the very special case in which the target space is precisely the dual of the source space, namely, when the linear operator belongs to some $L(U, U^*)$ or, equivalently to $U^* \otimes U^*$. We conclude that, as expected from the very tensor-product notation, a linear operator and its transpose are of the same nature (and may, therefore, be checked for symmetry) if, and only if, they belong to a tensor product of the form $V \otimes V$. Having said this, it is clear that if some artificial (noncanonical) isomorphism is introduced between a space and its dual (by means of an inner product, for example, as we shall eventually do), then the concept of symmetry can be extended.

9.6.3
The Algebra of Tensors on a Vector Space

Although a more general situation may be envisioned, we now consider the collection of all possible tensor products involving any finite number of factors, each factor being equal to a given vector space V or its dual V^*. The order of the factors, of course, matters, but it is customary to say that a tensor product is *of type (r,s)* if it is obtained by multiplying r copies of V and s copies of V^*, regardless of the order in which these copies appear in the product. An element of such tensor product is also called a tensor of type (r, s). Another common terminology is to refer to r and s, respectively, as the contravariant and covariant degrees of the tensor. Thus, a vector is a tensor of type $(1, 0)$, while a covector is of type $(0, 1)$. By convention, a tensor of type $(0, 0)$ is identified with a scalar. As the field of scalars $\mathbb{R}$ has the natural structure of a vector space (whose elements are tensors of type $(0, 0)$), it makes sense to take its tensor product with a vector space. Note that $\mathbb{R} \otimes V = V$.

The tensor product of a tensor of type (r_1, s_1) with a tensor of type (r_2, s_2) is a tensor of type $(r_1 + r_2, s_1 + s_2)$. A map from a Cartesian product of vector spaces into a vector space is said to be *multilinear* if it is linear in each of the arguments. A tensor T of type (r, s) can be considered as a multilinear map such that $T(\omega_1, \dots, \omega_r, \mathbf{v}_1, \dots, \mathbf{v}_s) \in \mathbb{R}$, where $\mathbf{v}_i$ and ω_j belong, respectively, to V and V^*, for each $i = 1, \dots, r$ and each $j = 1, \dots, s$.

The collection of all tensors of all orders defined on a vector space V can be given the formal structure of an *algebra* (with the operations of *direct sum* and tensor product) known as the *algebra of tensors* on V. Considering only tensors of covariant degree zero, namely, tensors of type $(r, 0)$, we obtain the *contravariant tensor algebra* of V. When written in components, all indices of tensors in this algebra are superscripts.

In a similar way, one can define the *covariant tensor algebra* by considering tensors of type $(0, s)$. On the other hand, considering V^* as a vector space in its own right, we could form its contravariant tensor algebra, and these two objects turn out to be the same. The contravariant and covariant algebras can be considered dual to each other in the sense that there exists a canonical way to evaluate an element of one over an element of the other to produce a real number linearly. Considering a tensor T of type $(k, 0)$ and a tensor S of type $(0, k)$ and using a basis in V, this evaluation reads

$$\langle S, T\rangle = S_{i_1 \cdots i_k} T^{i_1 \cdots i_k}. \tag{9.65}$$

If the tensors are of different orders (that is, a tensor of type $(r, 0)$ and a tensor of type $(0, s)$ with $s \neq r$), we define the evaluation as zero.

A tensor T of type $(r, 0)$ can be seen as a multilinear map

$$\begin{aligned} T : V^*, \dots, V^* &\longrightarrow \mathbb{R} \\ (\omega_1, \dots, \omega_r) &\mapsto T(\omega_1, \dots, \omega_r), \\ &\qquad \omega_1, \dots, \omega_r \in V^*. \end{aligned} \tag{9.66}$$

For tensors in the contravariant or covariant algebras it makes sense to speak about *symmetry* and *skew symmetry*.

A tensor of type $(r, 0)$ is said to be *(completely) symmetric* if the result of the operation (9.66) is independent of the order of the arguments. Put in other words, exchanging any two arguments with each other produces no effect in the result of the multilinear operator T. A similar criterion applies for completely symmetric tensors of order $(0, s)$, except that the arguments are vectors rather than covectors. Choosing a basis in V, symmetry boils down to indifference to index swapping.

Analogously, a tensor of type $(r, 0)$ is *(completely) skew symmetric* if every mutual exchange of two arguments alters the sign of the result, leaving the absolute value unchanged. By convention, all tensors of type $(0, 0)$ (scalars), $(1, 0)$ (vectors), and $(0, 1)$ (covectors) are considered to be both symmetric and skew symmetric. Notice that a completely skew-symmetric tensor of type $(r, 0)$ with r larger than the dimension of the vector space of departure must necessarily vanish.

The collections of all symmetric or skew-symmetric tensors (whether contravariant or covariant) do not constitute a subalgebra of the tensor algebra, for the simple reason that the tensor multiplication of two symmetric (or skew-symmetric) tensors is not symmetric (skew symmetric) in general. Nevertheless, it is possible, and convenient, to define algebras of symmetric and skew-symmetric tensors by modifying the multiplicative operation so that the results stay within the algebra. The case of skew-symmetric tensors is the most fruitful. It gives rise to the so-called *exterior algebra* of a vector space, which we will now explore. It will permit us to answer many intriguing questions such as: is there anything analogous to the cross-product of vectors in dimensions other than 3? What is an area and what is the meaning of flux?

9.6.4 Exterior Algebra

The space of skew-symmetric contravariant tensors of type $(r, 0)$ will be denoted by $\Lambda^r(V)$. The elements of $\Lambda^r(V)$ will be also called *r-vectors* and, more generally, *multivectors*. The number r is the *order* of the multivector. As before, the space $\Lambda^0(V)$ coincides with the scalar field $\mathbb{R}$, while $\Lambda^1(V)$ coincides with the vector space V.

Consider the ordered r-tuple of covectors $(\omega_1, \omega_2, \dots, \omega_r)$ and let π denote a *permutation* of this set. Such a permutation is even (odd) if it is obtained by an even (odd) number of exchanges between pairs of elements in the original set. An even (odd) permutation π has a *signature*, denoted by $\text{sign}(\pi)$, equal to 1 (-1).

Given an arbitrary tensor T of type $(r, 0)$, we define its *skew-symmetric part* $\mathcal{A}^r(T)$ as the multilinear map defined by the formula

$$\mathcal{A}^r(T)(\omega_1, \omega_2, \dots, \omega_r) = \frac{1}{r!} \sum_\pi \text{sign}(\pi)\ T(\pi). \tag{9.67}$$

As an example, for the case of a contravariant tensor of degree 3, namely, $T = T^{ijk}\mathbf{e}_i \otimes \mathbf{e}_j \otimes \mathbf{e}_k$, where $\mathbf{e}_h$ $(h = 1, \dots, n \geq r)$ is a basis of V, the skew-symmetric part is obtained as

$$\mathcal{A}^3(T) = \frac{1}{6}\big(T^{ijk} + T^{jki} + T^{kij} - T^{ikj} - T^{jik} - T^{kji}\big)\mathbf{e}_i \otimes \mathbf{e}_j \otimes \mathbf{e}_k. \tag{9.68}$$

Given two multivectors a and b, of orders r and s, respectively, we define their *exterior product* or *wedge product* as the multivector $a \wedge b$ of order $(r + s)$ obtained as

$$a \wedge b = \mathcal{A}^{r+s}(a \otimes b). \tag{9.69}$$

What this definition in effect is saying is that in order to multiply two skew-symmetric tensors and obtain a skew-symmetric result, all we have to do is take their tensor product and then project back into the algebra (that is, skew-symmetrize the result).[7)] As $\mathcal{A}$ is, by definition, a linear operator, the wedge product is linear in each of the factors.

We have seen that the tensor product is not commutative. But, in the case of the exterior product, exchanging the order of the factors can at most affect the sign. The general result is

$$b \wedge a = (-1)^{rs} a \wedge b, \quad a \in \Lambda^r V, b \in \Lambda^s(V). \tag{9.70}$$

Thus, for example, the wedge product with itself of a multivector of odd order must necessarily vanish. With some work, it is possible to show that the wedge product is associative, namely, $(a \wedge b) \wedge c = a \wedge (b \wedge c)$.

To calculate the dimension of $\Lambda^r(V)$, we note that, being a tensor, every element in $\Lambda^k(V)$ is expressible as a linear combination of the n^r tensor products $\mathbf{e}_{i_1} \otimes \cdots \otimes \mathbf{e}_{i_r}$, where $\mathbf{e}_i$, $i = 1, \dots, n$, is a basis of V. Because of the skew symmetry, however, we need to consider only products of the form $\mathbf{e}_{i_1} \wedge \cdots \wedge \mathbf{e}_{i_r}$. Two such products involving the same factors in any order are either equal or differ in sign, and a product with a repeated factor vanishes. This means that we need only count all possible combinations of n symbols taken r at a time without

7) In spite of the natural character of this definition of the wedge product, many authors adopt a definition that includes a combinatorial factor. Thus, the two definitions lead to proportional results. Each definition has some advantages, but both are essentially equivalent. Our presentation of exterior algebra follows closely that of Sternberg S (1983), *Lectures on Differential Geometry*, 2nd ed., Chelsea.

repetition. The number of such combinations is $n!/(n-r)!r!$. One way to keep track of all these combinations is to place the indices $i_1, \ldots, i_k$ in strictly increasing order. These combinations are linearly independent, thus constituting a basis. Therefore, the dimension of $\Lambda^r(V)$ is $n!/(n-r)!r!$.

We note that the spaces of r-vectors and $(n-r)$-vectors have the same dimension. There is a kind of fusiform dimensional symmetry around the middle, the dimension starting at 1 for $r = 0$, increasing to a maximum toward $r = n/2$ (say, if n is even) and then going back down to 1 for $r = n$. This observation plays an important role in the identification (and sometimes confusion) of physical quantities. For example, an n-vector functions very much like a scalar, but with a subtle difference.

Let a skew-symmetric contravariant tensor $a \in \Lambda^r(V)$ be given by means of its components on the basis of $\mathbf{e}_{i_1} \otimes \cdots \otimes \mathbf{e}_{i_r}$ inherited from a basis $\mathbf{e}_1, \ldots, \mathbf{e}_n$ of V as

$$a = a^{i_1,\ldots,i_r}\, \mathbf{e}_{i_1} \otimes \cdots \otimes \mathbf{e}_{i_r}. \tag{9.71}$$

Recalling that the skew-symmetry operator $\mathcal{A}^r$ is linear, we obtain

$$\begin{aligned} a = \mathcal{A}^r(a) &= a^{i_1,\ldots,i_r}\, \mathcal{A}^r\left(\mathbf{e}_{i_1} \otimes \cdots \otimes \mathbf{e}_{i_r}\right) \\ &= a^{i_1,\ldots,i_r}\, \mathbf{e}_{i_1} \wedge \cdots \wedge \mathbf{e}_{i_r}. \end{aligned} \tag{9.72}$$

In these expressions, the summation convention is implied. We have obtained the result that, given a skew-symmetric tensor in components, we can substitute the wedge products for the tensor products of the base vectors. On the other hand, if we would like to express the r-vector a in terms of its components on the basis of $\Lambda^r(V)$ given by the wedge products of the base vectors of V taken in strictly increasing order of the indices, a coefficient of $r!$ will have to be included, namely,

$$\begin{aligned} &a^{i_1,\ldots,i_r}\, \mathbf{e}_{i_1} \wedge \cdots \wedge \mathbf{e}_{i_r} \\ &\quad = r! \sum_{i_1<\cdots<i_r} a^{i_1,\ldots,i_r}\, \mathbf{e}_{i_1} \wedge \cdots \wedge \mathbf{e}_{i_r}. \end{aligned} \tag{9.73}$$

This means that the components on the basis (with strictly increasing indices) of the skew-symmetric part of a contravariant tensor of type $(k, 0)$ are obtained without dividing by the factorial $k!$ in the projection algorithm. This, of course, is a small advantage to be gained at the expense of the summation convention.

Consider the n-fold wedge product $a = \mathbf{v}_1 \wedge \mathbf{v}_2 \wedge \cdots \wedge \mathbf{v}_n$, where the $\mathbf{v}$'s are elements of an n-dimensional vector space V. Let $\{\mathbf{e}_1, \mathbf{e}_2, \ldots, \mathbf{e}_n\}$ be a basis of V. As each of the $\mathbf{v}$'s is expressible uniquely in this basis, we may write

$$\begin{aligned} a &= (v_1^{i_1}\mathbf{e}_{i_1}) \wedge (v_2^{i_2}\mathbf{e}_{i_2}) \wedge \cdots \wedge (v_n^{i_n}\mathbf{e}_{i_n}) \\ &= v_1^{i_1} v_2^{i_2} \cdots v_n^{i_n}\, \mathbf{e}_{i_1} \wedge \mathbf{e}_{i_2} \wedge \cdots \wedge \mathbf{e}_{i_n}, \end{aligned} \tag{9.74}$$

where the summation convention is in full swing. Out of the possible n^n terms in this sum, there are exactly $n!$ that can survive, because each of the indices can attain n values, but repeated indices in a term kill it. However, because each of the surviving terms consists of a scalar coefficient times the exterior product of all the n elements of the basis, we can collect them all into a single scalar coefficient A multiplied by the exterior product of the base vectors arranged in a strictly increasing ordering of the indices, namely, we must have that $a = A\mathbf{e}_1 \wedge \mathbf{e}_2 \wedge \cdots \wedge \mathbf{e}_n$. This scalar coefficient consists of the sum of all the products $v_1^{i_1} v_2^{i_2} \cdots v_n^{i_n}$ with no repeated indices and with a minus sign if the superscripts form an odd permutation of $1, 2, \ldots, n$. This is precisely the definition of the determinant of the matrix whose entries are v_i^j. We conclude that, using in $\Lambda^n(V)$ the basis induced by a basis in V, the component of the exterior product of n vectors in an n-dimensional space is equal to the determinant of the matrix of the components of the individual vectors. Apart from providing a neat justification for the notion of determinant, this formula correctly suggests that the geometrical meaning of an n-vector is some measure of the ability of the (n-dimensional) parallelepiped subtended by the vectors to contain a volume. As we have not yet introduced any metric notion, we cannot associate a number to this volume. Notice on the other hand that, although we cannot say how large a volume is, we can certainly tell that a given n-parallelepiped is, say, twice as large as another. Notice, finally, that changing the order of two factors, or reversing the sense of one factor, changes the sign of the multivector. So, n-vectors represent *oriented* n-dimensional parallelepipeds.

The collection of all multivectors of all orders (up to the dimension of V), with the exterior product replacing the tensor product, constitutes the *exterior algebra* of V. In a similar way, starting from the dual space V^*, we can construct the algebra of multicovectors.

9.7 Forms and General Tensor Fields

9.7.1 1-Forms

Let $f : \mathcal{U} \to \mathbb{R}$ be a smooth function defined in a neighborhood $\mathcal{U} \subset \mathcal{M}$ of the point p, and let $df_p : T\mathcal{U} \to \mathbb{R}$ denote its differential at p. We can regard this differential as an element of $T_p^*\mathcal{M}$ by defining its value $\langle df_p, \mathbf{v}_p \rangle$ on any vector $\mathbf{v}_p \in T_p\mathcal{M}$ as

$$\langle df_p, \mathbf{v}_p \rangle = df_p(\mathbf{v}_p) = \mathbf{v}_p(f). \tag{9.75}$$

In other words, the action of the evaluation of the differential of the function on a tangent vector is equal to the directional derivative of the function in the direction of the vector.

A smooth assignment of a covector ω_p to each point $p \in \mathcal{M}$ is called a *differential 1-form on the manifold*. It can be regarded as a cross section of the cotangent bundle, namely, a map

$$\mathbf{\Omega} : \mathcal{M} \to T^*\mathcal{M}, \tag{9.76}$$

such that $\pi \circ \mathbf{\Omega} = id_{\mathcal{M}}$.

As we have seen, the differential of a function at a point defines a covector. It follows that a smooth scalar function $f : \mathcal{M} \to \mathbb{R}$ determines, by pointwise differentiation, a differential 1-form $\mathbf{\Omega} = df$. It is important to remark that *not all differential 1-forms can be obtained as differentials of functions.* The ones that can are called *exact.*

A differential 1-form $\mathbf{\Omega}$ (that is, a cross section of $T^*\mathcal{M}$) can be regarded as acting on vector fields $\mathbf{V}$ (cross sections of $T\mathcal{M}$) to deliver functions $\langle \mathbf{\Omega}, \mathbf{V} \rangle : \mathcal{M} \to \mathbb{R}$, by pointwise evaluation of a covector on a vector.

9.7.2 Pullbacks

Let $f : \mathcal{N} \to \mathbb{R}$ be a smooth function. We define its pullback by g as the map $g^*f : \mathcal{M} \to \mathbb{R}$ given by the composition

$$g^*f = f \circ g. \tag{9.77}$$

For a differential 1-form $\mathbf{\Omega}$ on $\mathcal{N}$, we define the pullback $g^*\mathbf{\Omega} : \mathcal{M} \to T^*\mathcal{M}$ by showing how it acts, point by point, on tangent vectors, namely,

$$\langle [g^*\mathbf{\Omega}](p), \mathbf{v}_p \rangle = \langle \mathbf{\Omega}(g(p)), (g_*)_p \mathbf{v}_p \rangle, \tag{9.78}$$

which can be more neatly written in terms of vector fields as

$$\langle g^*\mathbf{\Omega}, \mathbf{V}\rangle = \langle \mathbf{\Omega} \circ g,\ g_*\mathbf{V}\rangle. \tag{9.79}$$

Expressed in words, this means that the pullback by g of a 1-form in $\mathcal{N}$ is the 1-form in $\mathcal{M}$ that assigns to each vector the value that the original 1-form assigns to the image of that vector by g_*.

It is important to notice that the pullbacks of functions and differential 1-forms are always well defined, regardless of the dimensions of the spaces involved. This should be contrasted with the push-forwards of vector fields, which fail in general to be vector fields on the target manifold.

9.7.3 Tensor Bundles

Given a point p of a manifold $\mathcal{M}$, we may identify the vector space V with the tangent space $T_p\mathcal{M}$ and construct the corresponding spaces of tensors of any fixed type. Following the same procedure as for the tangent and cotangent bundles, which will thus become particular cases, one can define *tensor bundles* of any type by adjoining to each point of a manifold the tensor space of the corresponding type. A convenient notational scheme is $C^k(\mathcal{M}), C_k(\mathcal{M})$, respectively, for the bundles of contravariant and covariant tensors of order k. Similarly, the bundles of k-vectors and of k-forms can be denoted, respectively, by $\Lambda^k(\mathcal{M}), \Lambda_k(\mathcal{M})$. Each of these bundles can be shown (by a procedure identical to that used in the case of the tangent and cotangent bundles) to have a natural structure of a differentiable manifold of the appropriate dimension. A (smooth) section of a tensor bundle is called a *tensor field over* $\mathcal{M}$, of the corresponding type. A (smooth) section of the bundle $\Lambda_k(\mathcal{M})$ of k-forms is also called a *differential k-form*. A scalar function on a manifold is also called a *differential* 0*-form*.

In a chart of the m-dimensional manifold $\mathcal{M}$ with coordinates x^i, a contravariant tensor field $\mathbf{T}$ of order r is given as

$$\mathbf{T} = T^{i_1,\dots,i_r} \frac{\partial}{\partial x^{i_1}} \otimes \cdots \otimes \frac{\partial}{\partial x^{i_r}}, \tag{9.80}$$

where $T^{i_1,\dots,i_r} = T^{i_1,\dots,i_r}(x^1, \dots, x^m)$ are r^m smooth functions of the coordinates. Similarly, a covariant tensor field $\mathbf{U}$ of order r is given by

$$\mathbf{U} = U_{i_1,\dots,i_r}\, \mathrm{d}x^{i_1} \otimes \cdots \otimes \mathrm{d}x^{i_r}, \tag{9.81}$$

and a differential r-form ω by

$$\omega = \omega_{i_1,\dots,i_r}\, \mathrm{d}x^{i_1} \wedge \cdots \wedge \mathrm{d}x^{i_r}. \tag{9.82}$$

Notice that, in principle, the indexed quantity $\omega_{i_1,\dots,i_r}$ need not be specified as skew symmetric with respect to the exchange of any pair of indices, because the exterior product of the base forms will do the appropriate skew symmetrization job. As an alternative, we may suspend the standard summation convention in (9.82) and consider only indices in ascending order. As a result, if $\omega_{i_1,\dots,i_r}$ *is* skew symmetric *ab initio*, the corresponding components are to be multiplied by $r!$.

Of particular interest for the theory of integration on manifolds are differential m-forms, where m is the dimension of the manifold. From our treatment of the algebra of r-forms, we know that the dimension of the space of m-covectors is exactly 1. In a coordinate chart, a basis for differential m-forms is, therefore, given by: $\mathrm{d}x^1 \wedge \cdots \wedge \mathrm{d}x^m$. In other words, the representation of a differential m-form ω in a chart is

$$\omega = f(x^1, \dots, x^m)\, \mathrm{d}x^1 \wedge \cdots \wedge \mathrm{d}x^m, \tag{9.83}$$

where $f(x^1, \dots, x^m)$ is a smooth scalar function of the coordinates in the patch. Consider now another coordinate patch with coordinates $y^1, \dots, y^m$, whose domain has a nonempty intersection with the domain of the previous chart. In this chart, we have

$$\omega = \hat{f}(y^1, \dots, y^m)\, \mathrm{d}y^1 \wedge \cdots \wedge \mathrm{d}y^m. \tag{9.84}$$

We want to find the relation between the functions f and $\hat{f}$. As the transition functions $y^i = y^i(x^1, \dots, x^m)$ are smooth, we can write

$$\begin{aligned}\omega &= \hat{f}(y^1, \dots, y^m)\, \mathrm{d}y^1 \wedge \cdots \wedge \mathrm{d}y^m \\ &= \hat{f}\, \frac{\partial y^1}{\partial x_{j_1}} \cdots \frac{\partial y^m}{\partial x_{j_m}}\, \mathrm{d}x^{j_1} \wedge \cdots \wedge \mathrm{d}x^{j_m}\end{aligned} \tag{9.85}$$

or, by definition of determinant

$$\begin{aligned}\omega &= \det \left\{ \frac{\partial y^1, \dots, y^m}{\partial x^1, \dots, x^m} \right\} \hat{f}\, \mathrm{d}x^1 \wedge \cdots \wedge \mathrm{d}x^m \\ &= J_{y,x} \hat{f}\, \mathrm{d}x^1 \wedge \cdots \wedge \mathrm{d}x^m,\end{aligned} \tag{9.86}$$

where the Jacobian determinant $J_{y,x}$ does not vanish at any point of the intersection of the two coordinate patches. Comparing with (9.83), we conclude that

$$f = J_{y,x} \hat{f}. \tag{9.87}$$

A nowhere vanishing differentiable m-form on a manifold $\mathcal{M}$ of dimension m is called a *volume form* on $\mathcal{M}$. It can be shown that a manifold is orientable if, and only if, it admits a volume form.

The notion of pullback can be naturally generalized for covariant tensors of any order. For a contravariant tensor field $\mathbf{U}$ of order r on $\mathcal{N}$ (and, in particular, for differential r-forms on $\mathcal{N}$), the pullback by a smooth function $g : \mathcal{M} \to \mathcal{N}$ is a corresponding field on $\mathcal{M}$ obtained by an extension of the case $r = 1$, as follows:

$$g^*\mathbf{U}\,(\mathbf{V}_1, \dots, \mathbf{V}_r) = (\mathbf{U} \circ g)\,(g_*\mathbf{V}_1, \dots, g_*\mathbf{V}_r), \tag{9.88}$$

where $\mathbf{U}$ is regarded as a multilinear function of r vector fields $\mathbf{V}_i$.

9.7.4 The Exterior Derivative

The *exterior derivative* of differential forms is an operation that generalizes the gradient, curl, and divergence operators of classical vector calculus. The exterior derivative of a differential r-form on a manifold $\mathcal{M}$ is a differential $(r+1)$-form defined over the same manifold. Instead of introducing, as one certainly could, the definition of exterior differentiation in an intrinsic axiomatic manner, we will proceed to define it in a coordinate system and show that the definition is, in fact, coordinate independent. Let, therefore, x^i $(i = 1, \dots, m)$ be a coordinate chart and let ω be an r-form given as

$$\omega = \omega_{i_1,\dots,i_r}\, \mathrm{d}x^{i_1} \wedge \cdots \wedge \mathrm{d}x^{i_r}, \tag{9.89}$$

where $\omega_{i_1,\dots,i_r} = \omega_{i_1,\dots,i_r}(x^1, \dots, x^m)$ are smooth functions of the coordinates. We define the exterior derivative of ω, denoted by $\mathrm{d}\omega$, as the differential $(r+1)$-form obtained as

$$\mathrm{d}\omega = \mathrm{d}\omega_{i_1,\dots,i_r} \wedge \mathrm{d}x^{i_1} \wedge \cdots \wedge \mathrm{d}x^{i_r}, \tag{9.90}$$

where the d on the right-hand side denotes the ordinary differential of functions. More explicitly,

$$\mathrm{d}\omega = \frac{\partial \omega_{i_1,\dots,i_r}}{\partial x^k}\, \mathrm{d}x^k \wedge \mathrm{d}x^{i_1} \wedge \cdots \wedge \mathrm{d}x^{i_r}. \tag{9.91}$$

Note that for each specific combination of (distinct) indices $i_1, \dots, i_r$, the index k ranges only on the remaining possibilities, because the exterior product is skew symmetric. Thus, in particular, if ω is a differential m-form defined over an m-dimensional manifold, its exterior derivative vanishes identically (as it should, being an $(m+1)$-form).

Let y^i $(i = 1, \dots, m)$ be another coordinate chart with a nonempty intersection with the previous chart. We have

$$\omega = \hat{\omega}_{i_1,\dots,i_r}\, \mathrm{d}y^{i_1} \wedge \cdots \wedge \mathrm{d}y^{i_r}, \tag{9.92}$$

for some smooth functions $\hat{\omega}_{i_1,\dots,i_r}$ of the y^i-coordinates. The two sets of components are related by

$$\omega_{i_1,\dots,i_r} = \hat{\omega}_{j_1,\dots,j_r} \frac{\partial y^{j_1}}{\partial x^{i_1}} \cdots \frac{\partial y^{j_r}}{\partial x^{i_r}}. \tag{9.93}$$

Notice that we have not troubled to collect terms by, for example, prescribing a strictly increasing order. The summation convention is in effect. We now apply the prescription (9.90) and obtain

$$\mathrm{d}\omega = \mathrm{d}\left(\hat{\omega}_{j_1,\dots,j_r} \frac{\partial y^{j_1}}{\partial x^{i_1}} \cdots \frac{\partial y^{j_r}}{\partial x^{i_r}}\right) \wedge \mathrm{d}x^{i_1} \wedge \cdots \wedge \mathrm{d}x^{i_r}. \tag{9.94}$$

The crucial point now is that the terms containing the second derivatives of the coordinate transformation will evaporate as a result of their intrinsic symmetry, because they are contracted with an intrinsically skew-symmetric wedge product of two 1-forms. We have, therefore,

$$\mathrm{d}\omega = \frac{\partial \hat{\omega}_{j_1,\dots,j_r}}{\partial y^m} \frac{\partial y^m}{\partial x^k} \frac{\partial y^{j_1}}{\partial x^{i_1}} \cdots \frac{\partial y^{j_r}}{\partial x^{i_r}} \mathrm{d}x^k \wedge \mathrm{d}x^{i_1} \wedge \cdots \wedge \mathrm{d}x^{i_r}, \tag{9.95}$$

or, finally,

$$\mathrm{d}\omega = \frac{\partial \hat{\omega}_{j_1,\dots,j_r}}{\partial y^m} \mathrm{d}y^m \wedge \mathrm{d}y^{j_1} \wedge \cdots \wedge \mathrm{d}y^{j_r}, \tag{9.96}$$

which is exactly the same prescription in the coordinate system y^i as (9.90) is in the coordinate system x^i. This completes the proof of independence from the coordinate system.

From this definition, we can deduce a number of important properties of the exterior derivative, namely,

1. *Linearity*

$$d(a\,\alpha + b\,\beta) = a\,d\alpha + b\,d\beta \quad \forall a, b \in \mathbb{R} \quad \alpha, \beta \in \Lambda_r(\mathcal{M}). \tag{9.97}$$

2. *Quasi-Leibniz rule*

$$d(\alpha \wedge \beta) = d\alpha \wedge \beta + (-1)^r \alpha \wedge d\beta \quad \forall \alpha \in \Lambda_r(\mathcal{M}),\ \beta \in \Lambda_s(\mathcal{M}). \tag{9.98}$$

3. *Nilpotence*

$$d^2(.) = d(d(.)) = 0. \tag{9.99}$$

Moreover, it can be shown that the exterior derivative commutes with pullbacks. Finally, the exterior derivative of a 1-form has the following interesting interaction with the Lie bracket. If α is a differential 1-form and $\mathbf{u}$ and $\mathbf{v}$ are smooth vector fields on a manifold $\mathcal{M}$, then

$$\langle d\alpha \mid \mathbf{u} \wedge \mathbf{v}\rangle = \mathbf{u}\left(\langle \alpha \mid \mathbf{v}\rangle\right) - \mathbf{v}\left(\langle \alpha \mid \mathbf{u}\rangle\right) - \langle \alpha \mid [\mathbf{u}, \mathbf{v}]\rangle. \tag{9.100}$$

A differential form ω is *closed* if $\mathrm{d}\omega = 0$. Thus, all m-forms in an m-dimensional manifold are automatically closed. An r-form (with $r > 1$) is *exact* if there exists an $(r-1)$-form σ such that $\omega = \mathrm{d}\sigma$. By Property 3 above, all exact forms are closed. The converse is true *locally*. In other words, for every point in a manifold, there exists an open neighborhood on which the restriction of a closed form is exact. But this property may fail globally. An example is the 1-form given by $\omega = (x\,dy - y\,dx)/(x^2 + y^2)$ defined on an annular region of $\mathbb{R}^2$ with center at the origin $x = y = 0$. This form is closed but not exact. The existence of forms of this type reflects the presence of topological invariants (such as holes) in the manifold.

9.8 Symplectic Geometry

9.8.1 Symplectic Vector Spaces

A tensor T of type $(0, r)$ on V is a multilinear function acting on r vector arguments, $(\mathbf{v}_1, \dots, \mathbf{v}_r)$. Fixing one argument, say $\mathbf{v}_1$, we obtain a tensor $T_{\mathbf{v}_1}$ of type $(0, r-1)$. In particular, a tensor T of type $(0, 2)$ assigns to each vector $\mathbf{u} \in V$ the covector $T_{\mathbf{u}}$ defined by

$$T_{\mathbf{u}}(\mathbf{v}) = T(\mathbf{u}, \mathbf{v}) \quad \forall \mathbf{v} \in V. \tag{9.101}$$

The tensor T of type $(0, 2)$ is *nondegenerate* if $T_{\mathbf{u}} = \mathbf{0}$ implies that $\mathbf{u} = \mathbf{0}$. Since, in a given basis, the components of the covector $T_{\mathbf{u}}$ are $T_{ij}u^i$, we conclude that a necessary and sufficient condition for T to be nondegenerate is that the matrix with entries $[T_{ij}]$ must have a non-vanishing determinant, a condition that is independent of the basis chosen.

A *symplectic vector space* is a vector space in which a nondegenerate 2-covector ω has been singled out. The standard example is provided by a vector space of even dimension $2m$. Choosing a basis $\{\mathbf{e}_1, \dots, \mathbf{e}_m, \mathbf{f}_1, \dots, \mathbf{f}_m\}$, the 2-covector

$$\omega_{\mathbf{ef}} = \sum_{i=1}^{m} \mathbf{e}^i \wedge \mathbf{f}^i \tag{9.102}$$

is nondegenerate. It can be shown that every symplectic vector space is necessarily even-dimensional and that there exists a basis for which ω has the form (9.102).

An important property of a symplectic vector space is that, owing to the nondegeneracy of the 2-covector ω, there exists a natural correspondence between vectors and covectors.

9.8.2 Symplectic Manifolds

Recall that an r-form ω on a manifold $\mathcal{M}$ is a smooth r-covector field, namely, a smooth assignment of an r-covector ω_p at each point $p \in \mathcal{M}$. Equivalently, ω is a (smooth) section of the bundle $\Lambda_r(\mathcal{M})$. A *symplectic form* on $\mathcal{M}$ is a nondegenerate closed 2-form ω. A *symplectic manifold* $(\mathcal{M}, \omega)$ is a manifold in which a symplectic form ω has been singled out. According to our discussion above, a symplectic manifold is necessarily even-dimensional.

Given an m-dimensional manifold Q (for example, the configuration space of a mechanical system), the tangent and cotangent bundles are manifolds of even dimension $2m$. It is a remarkable fact that the cotangent bundle T^*Q of any manifold is automatically endowed with a *canonical symplectic form*. By "canonical," we mean that this form is defined intrinsically (i.e., independently of any coordinate chart). It is not surprising, therefore, that this canonical structure results in a corresponding physical interpretation. For a mechanical system, the cotangent bundle represents the phase space (of positions and momenta) and the canonical form plays a fundamental role in Hamiltonian mechanics.

A generic point $s \in T^*Q$ has the form $s = (q, p)$, where $q = \pi(s) \in Q$ and $p \in T_q^*Q$. Put differently, a point in the cotangent bundle consists of a point q in the base manifold and a 1-covector p at q. Let $\mathbf{V}$ be a tangent vector to T^*Q at the point $s = (q, p) \in T^*Q$, namely, $\mathbf{V} \in T(T^*Q)$. As the projection $\pi : T^*Q \to Q$ is a differentiable map, its differential $\pi_* : T(T^*Q) \to TQ$ is well defined. In particular, $\pi_*(\mathbf{V}_s) \in T_qQ$. But the tangent bundle $T(T^*Q)$, as a tangent bundle, has its own projection $\hat{\tau} : T(T^*Q) \to T^*Q$. In particular, $\hat{\tau}(\mathbf{V}_s) = s = (q, p)$. As this is a covector at $q \in Q$, it makes sense to evaluate it on the tangent vector $\pi_*(\mathbf{V}_s) \in T_qQ$.

Recall that a 1-form on T^*Q is a smooth assignment of a covector θ_s at each point $s = (q, p) \in T^*Q$. We define the *canonical 1-form* θ on T^*Q by the formula

$$\theta(\mathbf{V}_s) = \langle \hat{\tau}(\mathbf{V}_s), \pi_*(\mathbf{V}_s) \rangle. \tag{9.103}$$

The *canonical symplectic form* ω on T^*Q is defined as

$$\omega = -\mathrm{d}\theta. \tag{9.104}$$

Thus, ω is exact and, therefore, closed. Moreover, it is nondegenerate. It is, in fact, not difficult to obtain a coordinate expression of the canonical symplectic form. We have seen that a chart $(q^1, \ldots, q^m)$ in Q induces a chart in T^*Q. Indeed, any 1-form p on Q has the coordinate expression $p = p_i \mathrm{d}q^i$, where the summation convention is in force. The induced chart in T^*Q uses as coordinates the $2m$ numbers $(q^1, \ldots, q^m, p_1, \ldots, p_m)$. The canonical 1-form θ is given by $\theta = p_i \mathrm{d}q^i$. It follows that the canonical symplectic form is expressed as $\omega = -\mathrm{d}p_i \wedge \mathrm{d}q^i = \mathrm{d}q^i \wedge \mathrm{d}p_i$.

9.8.3 Hamiltonian Systems

A *Hamiltonian system* consists of a symplectic manifold $(\mathcal{M}, \omega)$ and a smooth real-valued function $\mathcal{H} : \mathcal{M} \to \mathbb{R}$ called the system *Hamiltonian*. In Classical Mechanics, the symplectic manifold is identified with the phase space $\mathcal{M} = T^*Q$ of the underlying configuration manifold Q.

A key concept in Hamiltonian systems is that of *Hamiltonian vector field*. As the Hamiltonian $\mathcal{H}$ is differentiable, its differential $\mathrm{d}\mathcal{H}$ is a well-defined 1-form on $\mathcal{M}$. In a symplectic manifold, on the other hand, to each 1-form, we can assign uniquely a vector field, by exploiting the pointwise nondegeneracy of the symplectic form. We thus obtain the associated Hamiltonian vector field $\mathbf{V}_H$. More explicitly, at each point $s \in \mathcal{M}$ we have

$$\langle \mathrm{d}\mathcal{H}, \mathbf{U} \rangle = \omega(\mathbf{V}_H, \mathbf{U}) \qquad \forall \mathbf{U} \in T_s\mathcal{M}. \tag{9.105}$$

A curve γ in $\mathcal{M}$ is a *trajectory* of the Hamiltonian system if it satisfies *Hamilton's equations*, namely, if it is an integral curve of the Hamiltonian vector field, namely,

$$\frac{\mathrm{d}\gamma}{\mathrm{d}t} = \mathbf{V}_H(\gamma(t)). \tag{9.106}$$

In the natural coordinates of a cotangent bundle, the curve γ consists of the $2m$ functions $q^i = q^i(t)$ and $p_i = p_i(t)$, with $i = 1, \ldots, m$. The Hamiltonian vector field has the components $\partial H/\partial p_i$ and $-\partial H/\partial q^i$. We thus recover the standard form of Hamilton's equations, that is,

$$\frac{\mathrm{d}q^i}{\mathrm{d}t} = \frac{\partial H}{\partial p_i}, \tag{9.107}$$

and

$$\frac{\mathrm{d}p_i}{\mathrm{d}t} = -\frac{\partial H}{\partial q^i}, \tag{9.108}$$

Notice that the construction (9.105) applies to any smooth real-valued function defined on $\mathcal{M}$, not just the Hamiltonian. Namely, to any such function $\mathcal{G}$ we can uniquely assign a vector field $\mathbf{V}_G$. We can thus define an operation between any two scalar fields $\mathcal{G}$ and $\mathcal{K}$, called the *Poisson bracket* $\{\mathcal{G}, \mathcal{K}\}$, by any of the equivalent prescriptions:

$$\{\mathcal{G}, \mathcal{K}\} = \mathbf{V}_K(\mathcal{G}) = \langle \mathrm{d}\mathcal{G}, \mathbf{V}_K\rangle = \omega(\mathbf{V}_G, \mathbf{V}_K). \tag{9.109}$$

The derivative of a scalar function $\mathcal{G}$ along a trajectory γ of the Hamiltonian system $(\mathcal{M}, \mathcal{H})$ is obtained as

$$\frac{\mathrm{d}\mathcal{G}}{\mathrm{d}t} = \frac{\mathrm{d}\gamma}{\mathrm{d}t}(\mathcal{G}) = \langle \mathrm{d}\mathcal{G}, \frac{\mathrm{d}\gamma}{\mathrm{d}t}\rangle = \langle \mathrm{d}\mathcal{G}, \mathbf{V}_H\rangle = \{\mathcal{G}, \mathcal{H}\}. \tag{9.110}$$

Thus, the Poisson bracket of a function $\mathcal{G}$ (representing some physical property of the system) with the Hamiltonian function describes the time evolution of $\mathcal{G}$. The vanishing of this Poisson bracket indicates, therefore, a conserved quantity.

9.9 The Lie Derivative

9.9.1 The Flow of a Vector Field

Let $\mathbf{V} : \mathcal{M} \to T\mathcal{M}$ be a (smooth) vector field. A (parameterized) curve $\gamma : H \to \mathcal{M}$ is called an *integral curve* of the vector field if its tangent at each point coincides with the vector field at that point. In other words, denoting by s the curve parameter, the following condition holds:

$$\frac{\mathrm{d}\gamma(s)}{\mathrm{d}s} = \mathbf{V}(\gamma(s)) \qquad \forall s \in H \subset \mathbb{R}. \tag{9.111}$$

As a consequence of the fundamental theorem of existence and uniqueness of local solutions of systems of ordinary differential equations, it is possible to prove the following fundamental theorem for vector fields on manifolds.

Theorem 9.2 *If $\mathbf{V}$ is a vector field on a manifold $\mathcal{M}$, then for every $p \in \mathcal{M}$, there exists an integral curve $\gamma(s,p) : I_p \to \mathcal{M}$ such that (i) I_p is an open interval of $\mathbb{R}$ containing the origin $s = 0$; (ii) $\gamma(0,p) = p$; and (iii) I_p is maximal in the sense that there exists no integral curve starting at p and defined on an open interval of which I_p is a proper subset. Moreover,*

$$\gamma(s, \gamma(s', x)) = \gamma(s + s', x) \qquad \forall s, s', s + s' \in I_p. \tag{9.112}$$

The map given by

$$p, s \mapsto \gamma(s, p) \tag{9.113}$$

is called the *flow* of the vector field $\mathbf{V}$ whose integral curves are $\gamma(s,p)$. In this definition, the map is expressed in terms of its action on pairs of points belonging to two different manifolds, $\mathcal{M}$ and $\mathbb{R}$, respectively. Not all pairs, however, are included in the domain, because I_p is not necessarily equal to $\mathbb{R}$. Moreover, because the intervals I_p are point dependent, the domain of the flow is not even a product manifold. One would be tempted to take the intersection of all such intervals so as to work with a product manifold given by $\mathcal{M}$ times the smallest interval I_p. Unfortunately, as we know from elementary calculus, this (infinite) intersection may consist of a single point. All that can be said about the domain of the flow is that it is an open subset of the Cartesian product $\mathcal{M} \times \mathbb{R}$. When the domain is equal to this product manifold, the vector field is said to be *complete* and the corresponding flow is called a *global flow*. It can be shown that if $\mathcal{M}$ is compact, or if the vector field is smooth and vanishes outside a compact subset of $\mathcal{M}$, the flow is necessarily global.

9.9.2 One-parameter Groups of Transformations Generated by Flows

Given a point $p_0 \in \mathcal{M}$, it is always possible to find a small enough neighborhood $U(p_0) \subset \mathcal{M}$ such that the intersection of all the intervals I_p with $p \in U(p_0)$ is an open interval J containing the origin. For each value $s \in J$, the flow $\gamma(s, p)$ can be regarded as a map

$$\gamma_s : U(p_0) \longrightarrow \mathcal{M}, \tag{9.114}$$

defined as

$$\gamma_s(p) = \gamma(s, p), \qquad p \in U(p_0). \tag{9.115}$$

This map is clearly one-to-one, because otherwise we would have two integral curves intersecting each other, against the statement of the fundamental theorem. Moreover, again according to the fundamental theorem, this is a smooth map with a smooth inverse over its image. The inverse is, in fact, given by

$$\gamma_s^{-1} = \gamma_{-s}, \tag{9.116}$$

where γ_{-s} is defined over the image $\gamma_s(U(p_0))$. Notice that γ_0 is the identity map of $U(p_0)$. Finally, for the appropriate range of values of s and r, we have the composition law

$$\gamma_r \circ \gamma_s = \gamma_{r+s}. \tag{9.117}$$

The set of maps γ_s is said to constitute the *one-parameter local pseudo-group* generated by the vector field (or by its flow). If the neighborhood $U(p_0)$ can be extended to the whole manifold for some open interval J (no matter how small), each map γ_s is called a *transformation* of $\mathcal{M}$. In that case, we speak of a *one-parameter pseudo-group of transformations* of $\mathcal{M}$. Finally, in the best of all possible worlds, if $J = \mathbb{R}$ the one-parameter subgroup of transformations becomes elevated to a *one-parameter group of transformations*. This is an Abelian (i.e., commutative) group, as is clearly shown by the composition law (9.117). We may say that every complete vector field generates a one-parameter group of transformations of the manifold.

The converse construction, namely, the generation of a vector field out of a given one-parameter pseudo-group of transformations, is also of interest. It can be shown that every one-parameter pseudo-group of transformations γ_s is generated by the vector field

$$\mathbf{V}(p) = \frac{d\gamma_s(p)}{ds}\Big|_{s=0}. \tag{9.118}$$

9.9.3 The Lie Derivative

We have learned that a vector field determines at least a one-parameter pseudo-group in a neighborhood of each point of the underlying manifold. For each value of the parameter s within a certain interval containing the origin, this neighborhood is mapped diffeomorphically onto another neighborhood. Having at our disposal a diffeomorphism, we can consider the pushed-forward or pulled-back versions of tensors of every type, including multivectors and differential forms. Physically, these actions represent how the various quantities are *convected* (or dragged) by the flow. To elicit a mental picture, we show in Figure 9.4 a vector $\mathbf{w}_p$ in a manifold as a small segment $\overrightarrow{pq}$ (a small piece of a curve, say), and we draw the integral curves of a vector field $\mathbf{V}$ emerging from each of its end points, p and q. These curves are everywhere tangent to the underlying vector field $\mathbf{V}$, which

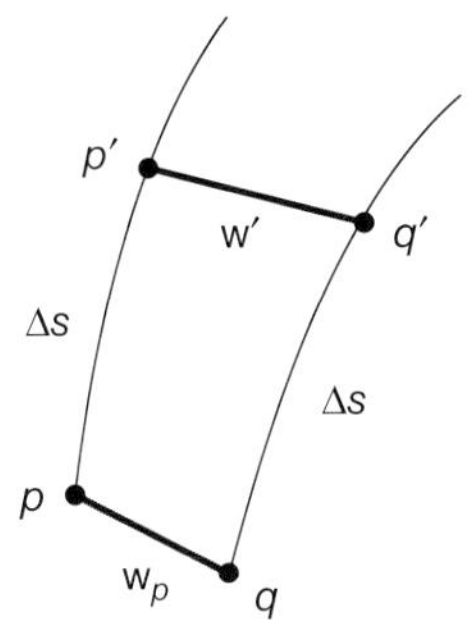

Figure 9.4 Dragging of a vector by a flow.

we do not show in the figure. If s denotes the (natural) parameter along these integral curves, an increment of Δs applied from each of these points along the corresponding integral curve, will result in two new points p' and q', respectively. The (small) segment $\overrightarrow{p'q'}$ can be seen as a vector $\mathbf{w}'$, which we regard as the convected counterpart of $\mathbf{w}_p$ as it is dragged by the flow of $\mathbf{V}$ by an amount Δs. If $\mathbf{w}_p$ happens to be part of a vector field $\mathbf{W}$ defined in a neighborhood of p', so that $\mathbf{w}_p = \mathbf{W}(p)$, we have that at the point p' there is, in addition to the dragged vector $\mathbf{w}'$, a vector $\mathbf{W}(p')$. There is no reason why these two vectors should be equal. The difference $\mathbf{W}(p') - \mathbf{w}'$ (divided by Δs) gives us an idea of the meaning of the Lie derivative of $\mathbf{W}$ with respect to $\mathbf{V}$ at p'.

The idea behind the definition of the Lie derivative of a tensor field with respect to a given vector field at a point p is the following. We consider a small value s of the parameter and convect the tensor field back to $s = 0$ by using the appropriate pullback or pushforward. This operation will, in particular, provide a value of the convected tensor field at the point p. We then subtract from this value the original value of the field at that point (a legitimate operation, because both tensors operate on the same tangent and/or cotangent space), divide by s and compute the limit as $s \to 0$. To understand how to calculate a Lie derivative, it is sufficient to make the definition explicit for the case of functions, vector fields, and 1-forms. The general case is then inferred from these three basic cases, as we shall demonstrate. We will also prove that the term "derivative" is justified. Notice that a Lie derivative is defined with respect to a given vector field. It is not an intrinsic property of the tensor field being differentiated. The Lie derivative of a tensor field at a point is a tensor of the same type.

9.9.3.1 The Lie Derivative of a Scalar

Let $g : \mathcal{M} \to \mathcal{N}$ be a mapping between two manifolds and let $f : \mathcal{N} \to \mathbb{R}$ be a function. Recall that, according to (9.77), the pullback of f by g is the map $g^*f : \mathcal{M} \to \mathbb{R}$ defined as the composition

$$g^*f = f \circ g. \tag{9.119}$$

Let a (time-independent, for now) vector field $\mathbf{V}$ be defined on $\mathcal{M}$ and let $\gamma_s : U \to \mathcal{M}$ denote the action of its flow on a neighborhood of a point $p \in \mathcal{M}$. If a function $f : \mathcal{M} \to \mathbb{R}$ is defined, we can calculate

$$\gamma_s^* f := f \circ \gamma_s. \tag{9.120}$$

The Lie derivative at the point p is given by

$$L_V f(p) = \lim_{s\to 0} \frac{(\gamma_s^* f)(p) - f(p)}{s} = \lim_{s\to 0} \frac{f(\gamma_s(p)) - f(p)}{s} \tag{9.121}$$

Thus, we obtain

$$L_V f(p) = \mathbf{v}_p(f). \tag{9.122}$$

In simple words, the Lie derivative of a scalar field with respect to a given vector field coincides, at each point, with the directional derivative of the function in the direction of the field at that point.

9.9.3.2 The Lie Derivative of a Vector Field

Vectors are pulled forward by mappings. Thus, given the map $g : \mathcal{M} \to \mathcal{N}$, to bring a tangent vector from $\mathcal{N}$ back to $\mathcal{M}$, we must use the fact that g is invertible and that the inverse is differentiable, such as when g is a diffeomorphism. Let $\mathbf{W} : \mathcal{N} \to T\mathcal{N}$ be a vector field on $\mathcal{N}$. The corresponding vector field on $\mathcal{M}$ is then given by $g_*^{-1} \circ \mathbf{W} \circ g : \mathcal{M} \to T\mathcal{M}$. Accordingly, the Lie derivative of the vector field $\mathbf{W}$ with respect to the vector field $\mathbf{V}$, with flow γ_s, at a point $p \in \mathcal{M}$ is defined as

$$L_V \mathbf{W}(p) = \lim_{s\to 0} \frac{\gamma_{s*}^{-1} \circ \mathbf{W} \circ \gamma_s(p) - \mathbf{W}(p)}{s}. \tag{9.123}$$

It can be shown that the Lie derivative of a vector field coincides with the Lie bracket

$$L_V \mathbf{W} = [\mathbf{V}, \mathbf{W}]. \tag{9.124}$$

9.9.3.3 The Lie Derivative of a 1-form

As 1-forms are pulled back by a map, we define the Lie derivative of the 1-form $\omega : \mathcal{M} \to T^*\mathcal{M}$ at the point p as

$$L_V \omega(p) = \lim_{s\to 0} \frac{\gamma_s^* \circ \omega \circ \gamma_s(p) - \omega(p)}{s}. \tag{9.125}$$

9.9.3.4 The Lie Derivative of Arbitrary Tensor Fields

It is clear that, by virtue of their definition by means of limits, the Lie derivatives defined so far are linear operators. To extend the definition of the Lie derivative to tensor fields of arbitrary order, we need to make sure that the Leibniz rule with respect to the tensor product is satisfied. Otherwise, we would not have the right to use the term "derivative" to describe it. It is enough to consider the case of a monomial such as

$$\mathbf{T} = \omega_1 \otimes \cdots \otimes \omega_m \otimes \mathbf{W}_1 \otimes \cdots \otimes \mathbf{W}_n, \tag{9.126}$$

where ω_i are m 1-forms and $\mathbf{W}_j$ are n vector fields. We define

$$L_V \mathbf{T}(p) = \lim_{s\to 0} \frac{\gamma_s^* \circ \omega_1 \circ \gamma_s(p) \otimes \cdots \otimes \gamma_{s*}^{-1} \circ \mathbf{W}_1 \circ \gamma_s(p) \otimes \cdots - \mathbf{T}(p)}{s}. \tag{9.127}$$

Let us verify the satisfaction of the Leibniz rule for the case of the tensor product of a 1-form by a vector.

$$L_V(\omega \otimes \mathbf{W})(p) = \lim_{s\to 0} \frac{\gamma_s^* \circ \omega \circ \gamma_s(p) \otimes \gamma_{s*}^{-1} \circ \mathbf{W} \circ \gamma_s(p) \otimes - \omega(p) \otimes \mathbf{W}(p)}{s}. \tag{9.128}$$

Subtracting and adding to the denominator the expression $\omega(p) \otimes \gamma_{s*}^{-1} \circ \mathbf{W} \circ \gamma_s(p)$ the Leibniz rule follows suit.

An important property of the Lie derivative is the following: The Lie derivative of a differential form (of any order) commutes with the exterior derivative, that is,

$$L_V(\mathrm{d}\omega) = \mathrm{d}(L_V \omega), \tag{9.129}$$

for all vector fields $\mathbf{V}$ and for all differential forms ω.

9.9.3.5 The Lie Derivative in Components

Taking advantage of the Leibniz rule, it is not difficult to calculate the components of the Lie derivative of a tensor in a given coordinate system x^i, provided the components of the Lie derivative of the base vectors $\partial/\partial x^i$ and the base 1-forms dx^i are known. A direct application of the formula (9.36) for the components of the Lie bracket, yields

$$L_V\left(\frac{\partial}{\partial x^i}\right) = -\frac{\partial V^k}{\partial x^i}\frac{\partial}{\partial x^k}. \qquad (9.130)$$

To obtain the Lie derivative of dx^i, we recall that the action of dx^i (as a covector) on $\partial/\partial x^j$ is simply δ^i_j, whose Lie derivative vanishes. This action can be seen as the contraction of their tensor product. We obtain, therefore,

$$0 = L_V\left\langle dx^i, \frac{\partial}{\partial x^j}\right\rangle = \left\langle dx^i, -\frac{\partial V^k}{\partial x^j}\frac{\partial}{\partial x^k}\right\rangle + \left\langle L_V dx^i, \frac{\partial}{\partial x^j}\right\rangle, \qquad (9.131)$$

whence

$$L_V dx^i = \frac{\partial V^i}{\partial x^k} dx^k. \qquad (9.132)$$

Further Reading

Some general treatises on Differential Geometry:

Chern, S.S., Chern, W.H., and Lam, K.S. (1999) *Lectures on Differential Geometry*, World Scientific.

Kobayashi, S. and Nomizu, K. (1996) *Foundations of Differential Geometry*, Wiley Classics Library Edition.

Lee, J.M. (2003) *Introduction to Smooth Manifolds*, Springer.

Sternberg, S. (1983) *Lectures on Differential Geometry*, 2nd edn, Chelsea Publishing Company.

Warner, F.W. (1983) *Foundations of Differentiable Manifolds and Lie Groups*, Springer.

Some books that emphasize physical applications or deal with particular physical theories in a geometric way are

Abraham, R. and Marsden, J.E. (2008) *Foundations of Mechanics*, 2nd edn, AMS Chelsea Publishing.

Arnold, V.I. (1978) *Mathematical Methods of Classical Mechanics*, Springer.

Choquet-Bruhat, Y., de Witt-Morette, C., and Dillard-Beck, M. (1977) *Analysis, Manifolds and Physics*, North-Holland.

Frankel, T. (2004) *The Geometry of Physics: An Introduction*, 2nd edn, Cambridge University Press.

Misner, W., Thorne, K.S., and Wheeler, J.A. (1973) *Gravitation*, W H Freeman and Company.

Much of the material in this article is reproduced, with permission, from

Epstein, M. (2010) *The Geometrical Language of Continuum Mechanics*, Cambridge University Press.

10
Topics in Differential Geometry

Marcelo Epstein

10.1
Integration

10.1.1
Integration of n-Forms in $\mathbb{R}^n$

The simplest n-dimensional manifold is $\mathbb{R}^n$ itself with the standard topology and the standard notion of differentiability. Accordingly, we present the standard notion of integration over a domain of $\mathbb{R}^n$ in terms of differential forms so as to be able to extend this notion to arbitrary manifolds.

Let $x^1, \dots, x^n$ be the standard global chart of $\mathbb{R}^n$, and let ω be a smooth n-form defined over some open set $D \subset \mathbb{R}^n$. There exists, then, a smooth function $f : D \to \mathbb{R}$ such that

$$\omega = f\, \mathrm{d}x^1 \wedge \cdots \wedge \mathrm{d}x^n. \tag{10.1}$$

For any regular domain of integration $\mathcal{A} \subset D$, we define

$$\int_{\mathcal{A}} \omega = \underbrace{\int \int \cdots \int}_{\mathcal{A}} f \mathrm{d}x^1 \mathrm{d}x^2 \cdots \mathrm{d}x^n, \tag{10.2}$$

where the right-hand side is the ordinary n-fold Riemann integral in $\mathbb{R}^n$.

It is important to check that this definition is independent of the coordinate system adopted in D. For this purpose, let

$$\phi : D \longrightarrow \mathbb{R}^n \tag{10.3}$$

be a coordinate transformation expressed in components as the n smooth functions

$$x^1, \dots, x^n \mapsto y^1(x^1, \dots, x^n), \dots, y^n(x^1, \dots, x^n). \tag{10.4}$$

Recall that for (10.4) to qualify as a coordinate transformation, the Jacobian determinant

$$J = \det \left[\frac{\partial(y^1, \dots, y^n)}{\partial(x^1, \dots, x^n)} \right], \tag{10.5}$$

must be nonzero throughout D. For definiteness, we will assume that it is strictly positive (so that the change of coordinates is orientation preserving). According to the formulas of transformation of variables under a multiple Riemann integral, we

Mathematical Tools for Physicists, Second Edition. Edited by Michael Grinfeld.

must have

$$\underbrace{\int\int\cdots\int}_{\mathcal{A}} f(x^i)\,\mathrm{d}x^1\cdots\mathrm{d}x^n = \underbrace{\int\int\cdots\int}_{\mathcal{A}} f(x^i(y^j))\,J^{-1}\mathrm{d}y^1\cdots\mathrm{d}y^n. \tag{10.6}$$

But, because ω is an n-form in an n-dimensional manifold, its representation in the new coordinate system is precisely

$$\omega = f(x^i(y^j))\,J^{-1}\mathrm{d}y^1\wedge\cdots\wedge\mathrm{d}y^n, \tag{10.7}$$

which shows that the definition (10.2) is indeed independent of the coordinate system adopted in D.

A more fruitful way to exploit the coordinate independence property is to regard $\phi : D \to \mathbb{R}^n$ not as a mere coordinate transformation but as an actual change of the domain of integration. In this case, the transformation formula is interpreted readily in terms of the pullback of ω as

$$\int_{\phi(\mathcal{A})} \omega = \int_{\mathcal{A}} \phi^*(\omega), \tag{10.8}$$

for every n-form ω defined over an open set containing $\phi(\mathcal{A})$.

10.1.2 Integration of Forms on Oriented Manifolds

Let $\mathcal{M}$ be an oriented manifold of dimension m and let $(\mathcal{U}, \psi)$ be a (consistently) oriented chart. The integral of an m-form ω over $\mathcal{U}$ is defined as

$$\int_{\mathcal{U}} \omega = \int_{\psi(\mathcal{U})} \left(\psi^{-1}\right)^* \omega. \tag{10.9}$$

Notice that the right-hand side is a standard Riemann (or Lebesgue) integral of a function in $\mathbb{R}^m$, according to (10.2). In other words, given an m-form defined over the domain of a chart in an m-dimensional manifold, we simply pull back this form to the codomain of the chart (an open subset of $\mathbb{R}^m$) and integrate. The result is independent of the chart used.

To integrate over a domain covered by more than one chart, we need to use the concept of *partition of unity*, whose detailed presentation we omit. Briefly and imprecisely stated, a partition of unity is a collection of real-valued nonnegative smooth functions, one for each of the members of an open cover and vanishing outside it. Moreover, we assume that the cover is *locally finite*, in the sense that each point of the manifold belongs to only a finite number of members of the cover. Finally, the (finite) sum of all these functions at each point is equal to 1 (hence the name). It can be shown that partitions of unity exist provided the manifold is *paracompact*, namely, it admits a locally finite cover. Denoting by ϕ_i the functions making up the partition of unity, we define the integral as

$$\int_{\mathcal{M}} \omega = \sum_i \int_{\mathcal{M}} \phi_i \omega, \tag{10.10}$$

where the integrand on the right-hand side is just the product of a function times a differential form. Each integral on the right-hand side is well defined by (10.9).

For the definition implied by (10.10) to make sense, we must prove that the result is independent of the choice of charts and of the choice of partition of unity. This can be done quite straightforwardly by expressing each integral of the right-hand side of one choice in terms of the quantities of the second choice, and vice versa.

10.1.3 Stokes' Theorem

The boundary of an m-dimensional $\mathcal{M}$ with boundary will be denoted by $\partial\mathcal{M}$, which we consider as a manifold of dimension $m-1$. For example, $\mathcal{M}$ may be a closed ball in $\mathbb{R}^m$ and $\partial\mathcal{M}$ the bounding spherical surface. The boundary of an oriented manifold can be consistently oriented. Given an $(m-1)$-form ω on $\mathcal{M}$, it makes sense to calculate its integral over the (oriented) boundary ∂D. Stokes' theorem asserts that

$$\int_{\partial D} \omega = \int_D \mathrm{d}\omega. \tag{10.11}$$

We omit the proof and limit ourselves to remark that this elegant formula encompasses all the integral theorems of ordinary vector calculus.

10.2 Fluxes in Continuum Physics

One of the basic notions of Continuum Physics is that of an *extensive property*, a term that describes a property that may be assigned to *subsets* of a given universe, such as the mass of various parts of a material body, the electrical charge enclosed in a certain region of space, and so on. Mathematically speaking, therefore, an extensive property is expressed as a real-valued *set function* p, whose argument ranges over subsets $\mathcal{R}$ of a universe $\mathcal{U}$. It is usually assumed, on physical grounds, that the function p is additive, namely,

$$p(\mathcal{R}_1 \cup \mathcal{R}_2) = p(\mathcal{R}_1) + p(\mathcal{R}_2), \quad \text{whenever} \quad \mathcal{R}_1 \cap \mathcal{R}_2 = \emptyset. \tag{10.12}$$

With proper regularity assumptions, additivity means that, from the mathematical standpoint, p is a *measure* in $\mathcal{U}$.

In the appropriate space–time context, the *balance* of an extensive property expresses a relation between the rate of change of the property in a given region and the causes responsible for that change. Of particular importance is the idea of *flux* of the property through the boundary of a region, which is an expression of the rate of change of the property as a result of interaction with other regions. It is a common assumption that the flux between regions takes place through, and only through, common boundaries. In principle, the flux is a set function on the boundaries of regions. In most physical theories, however, this complicated dependence can be greatly simplified by means of the so-called *Cauchy postulates* and Cauchy's theorem.

10.2.1 Extensive-Property Densities

We will identify the universe $\mathcal{U}$ as an m-dimensional differentiable manifold. Under appropriate continuity assumptions, a set function such as the extensive property p is characterized by a *density*. Physically, this means that the property at hand cannot be concentrated on subsets of dimension lower than m. More specifically, we assume that the density ρ of the extensive property p is a smooth m-form on $\mathcal{U}$ such that

$$p(\mathcal{R}) = \int_{\mathcal{R}} \rho, \tag{10.13}$$

for any subset $\mathcal{R} \subset \mathcal{U}$ for which the integral is defined. Clearly, the additivity condition (10.12) is satisfied automatically.

We introduce the time variable t as if space–time were just a product manifold $\mathbb{R} \times \mathcal{U}$. In fact, this trivialization is observer dependent, but it will serve for our present purposes. The density ρ of the extensive property p should, accordingly, be conceived as a function $\rho = \rho(t, x)$, where $x \in \mathcal{U}$. Notice that, because for fixed x and variable t, ρ belongs to the same vector space $\Lambda^m\left(T_x^*\mathcal{U}\right)$, it makes sense to take the partial derivative with respect to t to obtain the new m-form

$$\beta = \frac{\partial \rho}{\partial t}, \tag{10.14}$$

defined on $\mathcal{U}$. For a fixed (i.e., time-independent) region $\mathcal{R}$, we may write

$$\frac{\mathrm{d}p(\mathcal{R})}{\mathrm{d}t} = \int_{\mathcal{R}} \beta. \tag{10.15}$$

In other words, the integral of the m-form β over a fixed region measures the rate of change of the content of the property p inside that region.

10.2.2 Balance Laws, Flux Densities, and Sources

In the classical setting of continuum physics, it is assumed that the change of the content of a smooth extensive property p within a fixed region $\mathcal{R}$ can be attributed to just two causes: (i) the rate at which the property is produced (or destroyed) within $\mathcal{R}$ by the presence of sources and sinks and (ii) the rate at which the property enters or leaves $\mathcal{R}$ through its boundaries, namely, the *flux* of p. For the sake of definiteness, in this section we adopt the convention that the production rate is positive for sources (rather than sinks) and that the flux is positive when there is a an outflow (rather than an inflow) of the property. The *balance equation* for the extensive property p states that *the rate of change of p in a fixed region $\mathcal{R}$ equals the difference between the production rate and the flux.* A good physical example is the balance of internal energy in a rigid body due to volumetric heat sources and heat flux through the boundaries.

As we have assumed continuity for p as a set function, we will do the same for both the production and the flux. As a result, we postulate the existence of an m-form s, called the *source density* such that the production rate in a region $\mathcal{R}$ is given by the integral

$$\int_{\mathcal{R}} s. \tag{10.16}$$

Just as ρ itself, the m-form s is defined over all of $\mathcal{U}$ and is independent of $\mathcal{R}$. Thus, from the physical point of view, we are assuming that the phenomena at hand can be described locally. This assumption excludes interesting phenomena, such as internal actions at a distance or surface-tension effects.

As far as the flux term is concerned, we also assume that it is a continuous function of subsets of the boundary $\partial\mathcal{R}$. We postulate the existence, for each region $\mathcal{R}$, of a smooth $(m-1)$-form $\tau_{\mathcal{R}}$, called the *flux density,* such that the flux of p is given by

$$\int_{\partial\mathcal{R}} \tau_{\mathcal{R}}. \tag{10.17}$$

Thus, the classical balance law of the property p assumes the form

$$\int_{\mathcal{R}} \beta = \int_{\mathcal{R}} s - \int_{\partial\mathcal{R}} \tau_{\mathcal{R}}. \tag{10.18}$$

An equation of balance is said to be a *conservation law* if both s and $\tau_{\mathcal{R}}$ vanish identically.

10.2.3 Flux Forms and Cauchy's Formula

We note that (beyond the obvious fact that β and s are m-forms, whereas $\tau_{\mathcal{R}}$ is an $(m-1)$-form), there is an essential complication peculiar to the flux densities $\tau_{\mathcal{R}}$. Indeed, in order to specify the flux for the various regions of interest, it seems that one has to specify the form $\tau_{\mathcal{R}}$ for each and every region $\mathcal{R}$. In other words, while the rate of change of the property and the production term are specified by forms whose domain (for each time t) is the entire space $\mathcal{U}$, the flux term must be specified by means of a set function, whose domain is the collection of all regions. We refer to the set function $\mathcal{R} \mapsto \tau_{\mathcal{R}}$ as a *system of flux densities*. Consider, for example, a point $x \in \mathcal{U}$ belonging simultaneously to the boundaries of two different regions. Clearly, we do not expect that the flux density will be the same for both. The example of suntanning should be sufficiently convincing in this regard. Consider, however, the following particular case. Let the natural *inclusion* map

$$\iota : \partial\mathcal{R} \longrightarrow \mathcal{U} \tag{10.19}$$

be defined by

$$\iota(x) = x \quad \forall x \in \partial\mathcal{R}. \tag{10.20}$$

Notice that this formula makes sense, because $\partial\mathcal{R} \subset \mathcal{U}$. Moreover, the map ι is smooth. It can, therefore, be used to pullback forms of any order on $\mathcal{U}$ to forms of the same order on $\partial\mathcal{R}$. In particular, we can define

$$\int_{\partial\mathcal{R}} \phi = \int_{\partial\mathcal{R}} \iota^*(\phi), \tag{10.21}$$

for any form ϕ on $\mathcal{U}$. Let us now assume the existence of a globally defined $(m-1)$-*flux form* Φ on $\mathcal{U}$ and let us define the associated system of flux densities by means of the formula

$$\tau_{\mathcal{R}} = \iota^*_{\partial\mathcal{R}}(\Phi), \tag{10.22}$$

where we use the subscript $\partial\mathcal{R}$ to emphasize the fact that each region requires its own inclusion map. Equation (10.22) is known as *Cauchy's formula*. Clearly, this is a very special system of flux densities (just as a conservative force field is a special vector field derivable from a single scalar field). Nevertheless, it is one of the fundamental results of classical Continuum Mechanics that, under rather general assumptions (known as *Cauchy's postulates*), every system of flux densities can be shown to derive from a unique flux form using Cauchy's formula (10.22). We will omit the general proof of this fact, known as *Cauchy's theorem*.

In less technical terms, Cauchy's formula is the direct result of assuming that the flux is given by a *single* 2-form defined over the three-dimensional domain of the body. The fact that one and the same form is to be used for a given location, and integrated over the given boundary, is trivially seen to imply (and generalize) the linear dependence of the flux on the normal to the boundary, as described in the standard treatments.

10.2.4 Differential Expression of the Balance Law

Assuming the existence of a flux form Φ, the general balance law (10.18) can be written as

$$\int_{\mathcal{R}} \beta = \int_{\mathcal{R}} s - \int_{\partial\mathcal{R}} \iota^*_{\partial\mathcal{R}}(\Phi). \tag{10.23}$$

Using Stokes' theorem (10.11), we can rewrite the last term as

$$\int_{\partial\mathcal{R}} \iota^*_{\partial\mathcal{R}}(\Phi) = \int_{\mathcal{R}} \mathrm{d}\Phi, \tag{10.24}$$

where the dependence on $\partial\mathcal{R}$ has evaporated. Using this result, we write (10.23) as

$$\int_{\mathcal{R}} \beta = \int_{\mathcal{R}} s - \int_{\mathcal{R}} \mathrm{d}\Phi. \tag{10.25}$$

As this balance law should be valid for arbitrary $\mathcal{R}$, and because the forms β, s, and Φ

are defined globally and independently of the region of integration, we obtain

$$\beta = s - \mathrm{d}\Phi. \tag{10.26}$$

This equation is known as the *differential expression of the general balance law.*

10.3 Lie Groups

10.3.1 Definition

Recall that a *group* is a set $\mathcal{G}$ endowed with a binary associative internal operation, called *group multiplication* or *group product*, which is usually indicated by simple apposition, namely, if $g, h \in \mathcal{G}$ then the product is $gh \in \mathcal{G}$. Associativity means that $(gh)k = g(hk)$, for all $g, h, k \in \mathcal{G}$. Moreover, there exists an *identity element* $e \in \mathcal{G}$ such that $eg = ge = g$ for all $g \in \mathcal{G}$. Finally, for each $g \in \mathcal{G}$ there exists an *inverse* $g^{-1} \in \mathcal{G}$ such that $gg^{-1} = g^{-1}g = e$. The identity can be shown to be unique, and so is also the inverse of each element of the group. If the group operation is also *commutative*, namely, if $gh = hg$ for all $g, h \in \mathcal{G}$, the group is said to be *commutative* or *Abelian*. In this case, it is customary to call the operation *group addition* and to indicate it as: $g + h$. The identity is then called the *zero element* and is often denoted as 0. Finally, the inverse of g is denoted as $-g$. This notation is easy to manipulate as it is reminiscent of the addition of numbers, which is indeed a particular case.

A *subgroup* of a group $\mathcal{G}$ is a subset $\mathcal{H} \subset \mathcal{G}$ closed under the group operations of multiplication and inverse. Thus, a subgroup is itself a group.

Given two groups, $\mathcal{G}_1$ and $\mathcal{G}_2$, a *group homomorphism* is a map $\phi : \mathcal{G}_1 \to \mathcal{G}_2$ that preserves the group multiplication, namely,

$$\phi(gh) = \phi(g)\,\phi(h) \quad \forall\, g, h \in \mathcal{G}_1, \tag{10.27}$$

where the multiplications on the left- and right-hand sides are, respectively, the group multiplications of $\mathcal{G}_1$ and $\mathcal{G}_2$.

The group structure is a purely algebraic concept, whereby nothing is assumed as far as the nature of the underlying set is concerned. The concept of a *Lie group* arises from making such an assumption. More specifically, a *Lie group* is a (smooth) manifold $\mathcal{G}$ with a group structure that is compatible with the differential structure, namely, such that the multiplication $\mathcal{G} \times \mathcal{G} \to \mathcal{G}$ and the inversion $\mathcal{G} \to \mathcal{G}$ are smooth maps.

A homomorphism ϕ between two Lie groups is called a *Lie-group homomorphism* if ϕ is C^∞. If ϕ happens to be a diffeomorphism, we speak of a *Lie-group isomorphism*. An isomorphism of a Lie group with itself is called a *Lie-group automorphism*.

Let V be an n-dimensional vector space. The collection $L(V, V)$ of all linear operators from V to V can be considered as a differentiable manifold. Indeed, fixing a basis in V, we obtain a global chart in $\mathbb{R}^{n^2}$. The collection $GL(V)$ of all *invertible* linear operators, with the operation of multiplication as the composition of linear operators, can be shown to be a Lie group, known as the *general linear group of* V. The general linear group of $\mathbb{R}^n$ is usually denoted by $GL(n; \mathbb{R})$. Its elements are the nonsingular square matrices of order n, with the unit matrix I acting as the group unit. Its various Lie subgroups are known as the *matrix groups*. The group operation is the usual matrix multiplication.

10.3.2 Group Actions

Let $\mathcal{G}$ be a group (not necessarily a Lie group) and let X be a set (not necessarily a differentiable manifold). We say that the group $\mathcal{G}$ *acts on the right* on the set X if for each $g \in \mathcal{G}$ there is a map $R_g : X \to X$ such that: (i) $R_e(x) = x$ for all $x \in X$, where e is the group identity; (ii) $R_g \circ R_h = R_{hg}$ for all $g, h \in \mathcal{G}$. When there is no room for confusion, we also use the notation xg for $R_g(x)$. Each of the maps R_g is a bijection of X. Moreover, $R_{g^{-1}} = (R_g)^{-1}$. The *orbit* through $x \in X$ is the subset $x\mathcal{G}$ of X consisting of all the elements of X of the form xg, where $g \in \mathcal{G}$.

The action of $\mathcal{G}$ on X is said to be *effective* if the condition $R_g(x) = x$ for every $x \in X$ implies $g = e$. The action is *free* if $R_g(x) = x$ for *some* $x \in X$ implies $g = e$. Finally, the action is *transitive* if for every $x, y \in X$, there exists $g \in \mathcal{G}$ such that $R_g(x) = y$.

In a completely analogous manner, we can say that $\mathcal{G}$ *acts on the left* on X if for each $g \in \mathcal{G}$ there is a map $L_g : X \to X$ such that (i) $L_e(x) = x$ for all $x \in X$, where e is the group identity; (ii) $L_g \circ L_h = L_{gh}$ for all $g, h \in \mathcal{G}$. The order of the composition is the essential difference between a right and a left action. We may also use the notation gx for $L_g(x)$.

The notion of group action can naturally be applied when $\mathcal{G}$ is a Lie group. In this instance, a case of particular interest is that for which the set on which $\mathcal{G}$ acts is a differentiable manifold and the induced bijections are transformations of this manifold. A *transformation* of a manifold $\mathcal{M}$ is a diffeomorphism $\phi : \mathcal{M} \to \mathcal{M}$. The definition of the action is then supplemented with a smoothness condition. More explicitly, A Lie group $\mathcal{G}$ is said to *act on the right* on a manifold $\mathcal{M}$ if

1. Every element $g \in \mathcal{G}$ induces a transformation $R_g : \mathcal{M} \to \mathcal{M}$.
2. $R_g \circ R_h = R_{hg}$, namely, $(ph)g = p(hg)$ for all $g, h \in \mathcal{G}$ and $p \in \mathcal{M}$.
3. The *right action* $R : \mathcal{G} \times \mathcal{M} \to \mathcal{M}$ is a smooth map. In other words, $R_g(p)$ is differentiable in *both* variables (g and p).

With these conditions, the Lie group $\mathcal{G}$ is also called a *Lie group of transformations of* $\mathcal{M}$. Just as in the general case, we have used the alternative notation pg for $R_g(p)$, with $p \in \mathcal{M}$, wherever convenient. A similar definition can be given for the *left action* of a Lie group on a manifold.

If e is the group identity, then R_e and L_e are the identity transformations of $\mathcal{M}$. Indeed, because a transformation is an invertible map, every point $p \in \mathcal{M}$ can be expressed as qg for some $q \in \mathcal{M}$ and some $g \in \mathcal{G}$. Using Property (2) of the right action, we have $R_e(p) = pe = (qg)e = q(ge) = qg = p$, with a similar proof for the left action.

It is convenient to introduce the following (useful, though potentially confusing) notation. We denote the right action *as a map from* $\mathcal{G} \times \mathcal{M}$ *to* $\mathcal{M}$ by the symbol R. Thus, $R = R(g, p)$ has two arguments, one in the group and the other in the manifold. Fixing, therefore, a particular element g in the group, we obtain a function of the single variable x, which we have already denoted by $R_g : \mathcal{M} \to \mathcal{M}$. But we can also fix a particular element p in the manifold and thus obtain another function of the single variable g. We will denote this function by $R_p : \mathcal{G} \to \mathcal{M}$. A similar scheme of notation can be adopted for a left action L. Notice that the image of R_p (respectively L_p) is nothing but the orbit $p\mathcal{G}$ (respectively $\mathcal{G}p$). The potential for confusion arises when the manifold $\mathcal{M}$ happens to coincide with the group $\mathcal{G}$, as described below. Whenever an ambiguous situation arises, we will resort to the full action function of two variables.

Recall that a Lie group is both a group and a manifold. Thus, it is not surprising that every Lie group $\mathcal{G}$ induces two canonical groups of transformations on itself, one by right action and one by left action, called, respectively, *right translations* and *left translations* of the group. They are defined, respectively, by $R_g(h) = hg$ and $L_g(h) = gh$, with $g, h \in \mathcal{G}$, where the right-hand sides are given by the group multiplication itself. For this reason, it should be clear that these actions are both free (and, hence, effective) and transitive.

10.3.3 One-Parameter Subgroups

A one-parameter subgroup of a Lie group $\mathcal{G}$ is a differentiable curve

$$\begin{aligned}\gamma : \mathbb{R} &\longrightarrow \mathcal{G}\\ t &\mapsto g(t),\end{aligned} \qquad (10.28)$$

satisfying

$$g(0) = e, \qquad (10.29)$$

and

$$g(t_1)\, g(t_2) = g(t_1 + t_2), \qquad \forall t_1, t_2 \in \mathbb{R}. \qquad (10.30)$$

If the group $\mathcal{G}$ acts (on the left, say) on a manifold $\mathcal{M}$, the composition of this action with a one-parameter subgroup determines a one-parameter group of transformations of $\mathcal{M}$, namely,

$$\gamma_t(p) = L_{g(t)}(p) \qquad p \in \mathcal{M}. \qquad (10.31)$$

We know that associated with this flow, there exists a unique vector field $\mathbf{v}^\gamma$. More precisely, we have

$$\mathbf{v}^\gamma(p) = \left.\frac{\mathrm{d}\gamma_t(p)}{\mathrm{d}t}\right|_{t=0}. \qquad (10.32)$$

Fixing the point p, we obtain the map L_p from the group to the manifold. The image of the curve γ under this map is obtained by composition as

$$t \mapsto L_p(g(t)) = L(g(t), p) = L_{g(t)}(p) = \gamma_t(p), \qquad (10.33)$$

where we have used (10.31). In other words, the image of the curve γ (defining the one-parameter subgroup) by the map L_p is nothing but the integral curve of the flow passing through p. By definition of derivative of a map between manifolds, we conclude that the tangent $\mathbf{g}$ to the one-parameter subgroup γ at the group identity e is mapped by L_{p*} to the vector $\mathbf{v}^\gamma(p)$

$$\mathbf{v}^\gamma(p) = \left(L_{p*}\right)_e \mathbf{g}. \qquad (10.34)$$

This means that a one-parameter subgroup $g(t)$ appears to be completely characterized by its tangent vector $\mathbf{g}$ at the group identity. We will shortly confirm this fact more fully. The vector field induced on $\mathcal{M}$ by a one-parameter subgroup is called the *fundamental vector field* associated with the corresponding vector $\mathbf{g}$ at the group identity.

Let us now identify the manifold $\mathcal{M}$ with the group $\mathcal{G}$ itself. In this case, we have, as already discussed, two canonical actions giving rise to the left and right translations of the group. We want to reinterpret (10.34) in this particular case. For this purpose, and to avoid the notational ambiguity alluded to above, we restore the fully fledged notation for the action as a function of two variables. We thus obtain

$$\mathbf{v}^\gamma(h) = \left(\frac{\partial L(g, h)}{\partial g}\right)_{g=e} \mathbf{g}. \qquad (10.35)$$

Notice that, somewhat puzzlingly, but consistently, this can also be written as

$$\mathbf{v}^\gamma(h) = \left(R_{h*}\right)_e \mathbf{g}. \qquad (10.36)$$

Thus, when defining the action of a one-parameter subgroup from the left, it is the right action whose derivative delivers the corresponding vector field, and vice versa.

10.3.4 Left- and Right-Invariant Vector Fields on a Lie Group

A vector field $\mathbf{v} : \mathcal{G} \to T\mathcal{G}$ is said to be *left invariant* if

$$\mathbf{v}(L_g h) = L_{g*}\mathbf{v}(h), \qquad \forall g, h \in \mathcal{G}. \quad (10.37)$$

In other words, vectors at one point are dragged to vectors at any other point by the derivative of the appropriate left translation. A similar definition, but replacing L with R, applies to *right-invariant vector fields*.

A vector field is left invariant if, and only if,

$$\mathbf{v}(g) = \left(L_{g*}\right)_e \mathbf{v}(e), \quad \forall g \in \mathcal{G}. \quad (10.38)$$

Another way of expressing this result is by saying that there exists a one-to-one correspondence between the set of left- (or right-) invariant vector fields on $\mathcal{G}$ and the tangent space $T_e\mathcal{G}$ at the group identity. This correspondence is linear. Moreover, one can show that the Lie bracket of two left- (right-) invariant vector fields is itself left (right) invariant. The set $\mathfrak{g}$ of left-invariant vector fields (or, equivalently, the tangent space $T_e\mathcal{G}$) with the Lie bracket operation is called the *Lie algebra of the group* $\mathcal{G}$. From an intuitive point of view, the elements of the Lie algebra of a Lie group represent infinitesimal approximations, which Sophus Lie himself called *infinitesimal generators* of the elements of the group. Although the infinitesimal generators are in principle commutative (sum of vectors), the degree of noncommutativity of the actual group elements is captured, to first order, by the Lie bracket.

10.4 Fiber Bundles

10.4.1 Introduction

Fiber bundles are differentiable manifolds with extra structure. Its points have, as it were, a double allegiance – not only to the manifold itself but also to a smaller entity called a *fiber*. We have already encountered two important instances of fiber bundles that clearly exhibit this feature: the tangent and cotangent bundles of a manifold. The property of belonging to a specific fiber is, in those two examples, represented by the existence of a projection map. Given a tangent vector or a covector, these maps tell us to which point of the base manifold they are attached. Moreover, we have found in those two cases that, once a chart $(\mathcal{U}, \phi)$ is chosen in the original manifold, a chart for the bundle can be constructed. Effectively, therefore, the part of the bundle sitting above $\mathcal{U}$ becomes assimilated to the Cartesian product of $\mathcal{U}$ and the typical fiber. In the case of the tangent bundle, for example, the chunk sitting above $\mathcal{U}$ is $\tau^{-1}(\mathcal{U})$, while the typical fiber is $\mathbb{R}^m$. These two properties (existence of a projection and local equivalence to a Cartesian product) are the two essential ingredients of the definition of a general fiber bundle. A third ingredient consists of the permitted transition functions along the fibers, a degree of freedom that is controlled by a given group of transformations of the typical fiber.

10.4.2 Definition

A fiber bundle with base manifold $\mathcal{B}$, typical fiber manifold $\mathcal{F}$ and structure group $\mathcal{G}$, is a manifold $\mathcal{C}$ and a smooth surjective bundle-projection map $\pi : \mathcal{C} \to \mathcal{B}$ such that there exists an open covering $\mathcal{U}_\alpha$ of $\mathcal{B}$ and respective *local trivializations*

$$\psi_\alpha : \pi^{-1}(\mathcal{U}_\alpha) \longrightarrow \mathcal{U}_\alpha \times \mathcal{F}, \tag{10.39}$$

with the property $\pi = pr_1 \circ \psi_\alpha$, as illustrated in the following commutative diagram:

$$\begin{array}{ccc} \pi^{-1}(\mathcal{U}_\alpha) & \xrightarrow{\psi_\alpha} & \mathcal{U}_\alpha \times F \\ & \searrow^{\pi} & \downarrow pr_1 \\ & & \mathcal{U}_\alpha \end{array} \tag{10.40}$$

Moreover, as illustrated in Figure 10.1, whenever $b \in \mathcal{U}_\alpha \bigcap \mathcal{U}_\beta \neq \emptyset$, the transition maps $\tilde{\psi}_{\beta,\alpha}(b) = \tilde{\psi}_{\alpha,b} \circ \tilde{\psi}_{\beta,b}^{-1}$ belong to the structure group $\mathcal{G}$ and depend smoothly on position throughout the intersection.

Consider now, for the same $\mathcal{C}$, $\mathcal{B}$, $\mathcal{F}$, π, and $\mathcal{G}$, a different open covering $\mathcal{V}_\beta$ with local trivializations ϕ_β. We say that it defines the same fiber bundle as before if, on nonvanishing intersections, the transition maps $\tilde{\psi}_{\alpha,b} \circ \tilde{\phi}_{\beta,b}^{-1}$ belong to the structure group $\mathcal{G}$ and depend smoothly on position b throughout the intersection. The two trivializations are said to be compatible. In this sense, we can say that the union of the two trivialization coverings becomes itself a new trivialization covering of the fiber bundle. When there is no room for confusion, a fiber bundle is denoted as a pair $(\mathcal{C}, \pi)$ indicating just the total space and the projection. An alternative notation is $\pi : \mathcal{C} \to \mathcal{B}$. A more complete notation would be $(\mathcal{C}, \pi, \mathcal{B}, \mathcal{F}, \mathcal{G})$.

The *fundamental existence theorem* of fiber bundles states that, given the manifolds $\mathcal{B}$ and $\mathcal{F}$ and a Lie group $\mathcal{G}$ acting effectively to the left on $\mathcal{F}$, and given, moreover, an open covering $\mathcal{U}_\alpha$ of $\mathcal{B}$ and a smooth assignment of an element of $\mathcal{G}$ to each point in every nonvanishing intersection of the covering, there exists a fiber bundle $(\mathcal{C}, \pi)$ with local trivializations based on that covering and with the assigned elements of $\mathcal{G}$ as transition maps. Furthermore, any two bundles with this property are equivalent.

An important application of the fundamental existence theorem is that, given a bundle $(\mathcal{C}, \pi, \mathcal{B}, \mathcal{F}, \mathcal{G})$, we can associate to it other bundles with the same base manifold and the same structure group, but with different typical fiber $\mathcal{F}'$, in a precise way. Indeed, we can choose a trivialization covering of the given bundle, calculate the transition maps, and then define the *associated bundle* $(\mathcal{C}', \pi', \mathcal{B}, \mathcal{F}', \mathcal{G})$, modulo an equivalence, by means of the assertion of the fundamental theorem. A case of particular interest is that in which the new fiber is identified with the structure group. This gives rise to the so-called *associated principal bundle*.

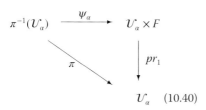

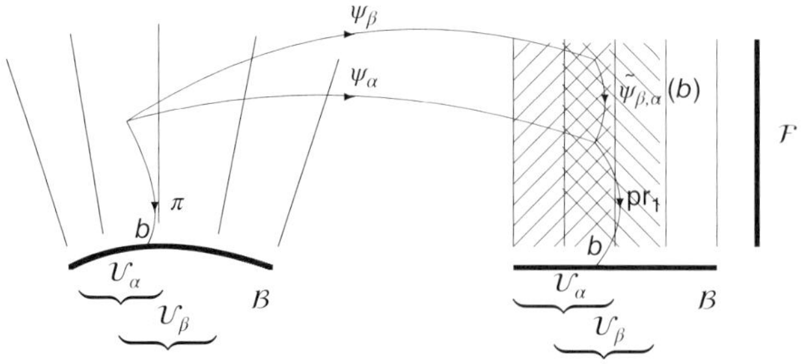

Figure 10.1 A general fiber bundle.

10.4.3 Simultaneity in Classical Mechanics

An important example of the use of fiber bundles arises in the notion of classical space–time. Starting from a four-dimensional manifold of events, Classical Physics assumes that all observers agree on whether or not two events happened simultaneously, regardless of their locations. As a consequence, a time-projection operator arises naturally in this context. As a result, a Physics that abides by the principle of absolute simultaneity must of necessity be formulated upon a space–time manifold that has the structure of a fiber bundle, the base manifold being a one-dimensional manifold. The typical fiber, representing space, is a three-dimensional manifold. In the Galilean formulation, this typical fiber is the Euclidean space $\mathbb{E}^3$. The structure group is the group of Galilean transformations of $\mathbb{E}^3$ (those preserving Euclidean length). An observer is a bundle trivialization. From this point of view, it can be claimed that the theory of Relativity is simpler than its Classical counterpart from at least the following point of view: the structure of relativistic space–time is less involved than that of Galilean space–time. The extra structure in the latter is provided by the notion of *absolute simultaneity*. In contrast, in Relativity, space–time is just a four-dimensional manifold endowed with a special metric structure.

10.4.4 Adapted Coordinate Systems

The total space C of a smooth fiber bundle $(C, \pi, \mathcal{B}, \mathcal{F}, \mathcal{G})$ is a differentiable manifold in its own right and, as such, can sustain a whole class of equivalent atlases. Among these atlases, however, there are some that enjoy the property of being in some sense adapted to the fibered structure of C seen as a fiber bundle. When working in coordinates, these *adapted atlases* or, more particularly, the corresponding *adapted coordinate charts* are used almost exclusively. The construction of these special charts mimics the construction of charts in a product manifold, by taking advantage of the local triviality of fiber bundles. Thus, let p be a point in C and let $\{\mathcal{U}, \psi\}$ be a local trivialization, namely, a diffeomorphism $\psi : \pi^{-1}(\mathcal{U}) \to \mathcal{U} \times \mathcal{F}$, such that $\pi(p) \in \mathcal{U}$. Without loss of generality, we may assume that the trivialization is subordinate to a chart in the base manifold $\mathcal{B}$ in the sense that $\mathcal{U}$ is the domain of a chart in $\mathcal{B}$ with coordinates x^i $(i = 1, \ldots, m)$. Moreover, because the typical fiber $\mathcal{F}$ is itself a differentiable manifold of dimension n, the point $f = pr_2(\psi(p))$ belongs to some chart of $\mathcal{F}$ with domain $\mathcal{V} \subset \mathcal{F}$ and coordinates u^α $(\alpha = 1, \ldots, n)$. The induced adapted coordinate chart in C is the map $u_\psi : \psi^{-1}(\mathcal{U} \times \mathcal{V}) \to \mathbb{R}^{m+n}$ whose value at p is $(x^1, \ldots, x^m, u^1, \ldots, u^n)$. The proof that in this way an atlas can be constructed in C is straightforward. It is implicitly assumed that the differential structure of C is compatible with the one induced by the adapted atlases.

10.4.5 The Bundle of Linear Frames

The *bundle of linear frames*, $F\mathcal{B}$, of a base n-dimensional manifold $\mathcal{B}$ can be defined constructively in the following way. At each point $b \in \mathcal{B}$ we form the set $F_b\mathcal{B}$ of all ordered n-tuples $\{e\}_b = (e_1, \ldots, e_n)$ of linearly independent vectors e_i in $T_b\mathcal{B}$, namely, the set of all bases of $T_b\mathcal{B}$. Our total space will consist of all ordered pairs of the form $(b, \{e\}_b)$ with the obvious projection onto $\mathcal{B}$. The pair $(b, \{e\}_b)$ is called *a linear frame at b*. Following a procedure identical to the one used for the tangent bundle, we obtain that each basis $\{e\}_b$ is expressible uniquely as

$$e_j = p^i_{\ j} \frac{\partial}{\partial x^i} \tag{10.41}$$

in a coordinate system x^i, where $\{p^i_{\ j}\}$ is a nonsingular matrix. We conclude that the typical fiber in this case is $GL(n; \mathbb{R})$. But so is the structure group. Indeed, in another coordinate system, y^i, we have

$$e_j = q^i_{\ j} \frac{\partial}{\partial y^i}, \tag{10.42}$$

where

$$q^i_{\ j} = \frac{\partial y^i}{\partial x^m} p^m_j = a^i_{\ m} p^m_j. \tag{10.43}$$

This is an instance of a *principal fiber bundle*, namely, a fiber bundle whose typical fiber and structure group coincide. The action of the group on the typical fiber is the natural left action of the group on itself. One of the interesting features of a principal bundle is that the structure group has also a natural *right* action *on the bundle itself*, and this property can be used to provide an alternative definition of principal bundles, which we shall pursue later. In the case of FB, for example, the right action is defined, in a given coordinate system x^i, by

$$R_a\{e\} = p^k_{\ i} a^i_{\ j} \frac{\partial}{\partial x^k}, \quad j = 1, \ldots, n, \tag{10.44}$$

which sends the basis (10.41) at b to another basis at b, that is, the action is fiber preserving. One can verify that this definition of the action is independent of the system of coordinates adopted. The principal bundle of linear frames of a manifold is associated to all the tensor bundles, including the tangent and the cotangent bundles, of the same manifold. By a direct application of the fundamental existence theorem, we know that the associated principal bundle is defined uniquely up to an equivalence. Many properties of bundles can be better understood by working first on the associated principal bundle.

10.4.6 Bodies with Microstructure

The modeling of complex materials, such as liquid crystals and granular media, requires the introduction of extra kinematic degrees of freedom. The *matrix* or *macromedium* is, in this case, an ordinary manifold that becomes the base manifold of a fiber bundle whose typical fiber represents the *micromedium*. In the case of granular media (such as concrete), each grain is assumed to undergo a homogeneous deformation. It is natural, therefore, to regard each of the smoothly distributed grains as the collection of all possible local bases of the tangent space to the base manifold. In other words, the granular medium is represented by the linear frame bundle of the macromedium.

10.4.7 Principal Bundles

The existence of a free right action on a manifold is strong enough to provide an independent definition of a principal fiber bundle which, although equivalent to the one already given, has the merit of being independent of the notion of transition maps. Moreover, once this more elegant and constructive definition has been secured, a subsidiary definition of the associated (nonprincipal) bundles becomes available, again without an explicit mention of the transition maps. Finally, this more abstract definition brings out intrinsically the nature and meaning of the associated bundles.

Let $\mathcal{P}$ be a differentiable manifold (the *total space*) and $\mathcal{G}$ a Lie group (the *structure group*), and let $\mathcal{G}$ act freely to the right on $\mathcal{P}$. This means that there exists a smooth map

$$\begin{aligned} R_g &: \mathcal{P}\times\mathcal{G} \longrightarrow \mathcal{P} \\ &(p,g) \mapsto R_g p = pg, \end{aligned} \tag{10.45}$$

such that, for all $p \in \mathcal{P}$ and all $g, h \in \mathcal{G}$, we have

$$\begin{aligned} R_{gh}p &= R_h R_g p = pgh, \\ R_e p &= p, \end{aligned} \tag{10.46}$$

where e is the group identity. The fact that the action is free means that if, for some $p \in \mathcal{P}$ and some $g \in \mathcal{G}$, $R_g p = p$, then necessarily $g = e$. Define now the quotient $\mathcal{B} = \mathcal{P}/\mathcal{G}$ and check that $\mathcal{B}$ is a differentiable manifold and that the canonical projection $\pi_P : \mathcal{P} \to \mathcal{P}/\mathcal{G}$ is differentiable. The set $\pi_P^{-1}(b)$ is called the fiber over $b \in \mathcal{B}$.

Recall that an element of the quotient $\mathcal{B} = \mathcal{P}/\mathcal{G}$ is, by definition of quotient, an equivalence class in $\mathcal{P}$ by the action of the group $\mathcal{G}$. In other words, each element b of the quotient (namely, of the base manifold $\mathcal{B}$) can be regarded as representing an orbit. The projection map assigns to each element of $\mathcal{P}$ the orbit to which it belongs. The fiber over b consists of all the elements of $\mathcal{P}$ that belong to the specific orbit represented by b.

To complete the definition of a principal bundle, we need only to add the condition that $\mathcal{P}$ be locally trivial, namely, that for each $b \in \mathcal{B}$, there exists a neighborhood $\mathcal{U} \subset \mathcal{P}$ such that $\pi_P^{-1}(\mathcal{U})$ is isomorphic to the product $\mathcal{U} \times \mathcal{G}$. More precisely, there exists a fiber-preserving diffeomorphism

$$\begin{aligned} \psi &: \pi_P^{-1}(\mathcal{U}) \longrightarrow \mathcal{U}\times\mathcal{G} \\ &p \mapsto (b, \tilde{\psi}_b), \end{aligned} \tag{10.47}$$

where $b = \pi_P(p)$, with the additional property that it must be *consistent with the group action*, namely (see Figure 10.2),

$$\tilde{\psi}_b(pg) = \tilde{\psi}_b(p)g, \qquad \forall p \in \pi_P^{-1}(\mathcal{U}), g \in \mathcal{G}. \tag{10.48}$$

This completes the definition of the principal bundle. The right action is fiber preserving and every fiber is diffeomorphic to $\mathcal{G}$. Moreover, every fiber coincides with an orbit of the right action of $\mathcal{G}$.

10.4.8 Associated Bundles

The concept of associated bundle has already been defined and used to introduce the notion of the principal bundle associated with any given fiber bundle. On the other hand, in the preceding section, we have introduced an independent definition of principal bundles by means of the idea of a right action of a group on a given total manifold. We want now to show that this line of thought can be pursued to obtain

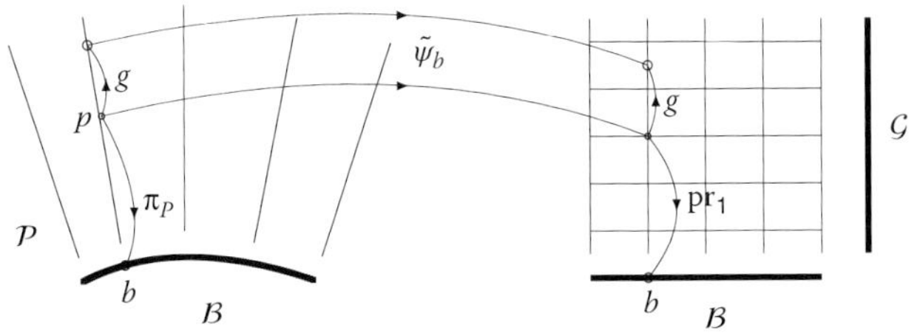

Figure 10.2 The group consistency condition.

another view of the collection of all (non-principal) fiber bundles associated with a given principal bundle.

As a more or less intuitive motivation for this procedure, it is convenient to think of the example of the principal bundle of linear frames $F\mathcal{B}$ of a manifold $\mathcal{B}$. We already know that this bundle is associated to the tangent bundle $T\mathcal{B}$. Consider now a pair (f, v), where $f \in F\mathcal{B}$ and $v \in T\mathcal{B}$, such that $\pi_P(f) = \pi(v) = b$. In other words, f and v represent, respectively, a basis and a vector of the tangent space at some point $b \in \mathcal{B}$. We can, therefore, identify v with its components on the linear frame f, namely, with an element of the typical fiber ($\mathbb{R}^n$) of $T\mathcal{B}$. If we consider now a pair $(\hat{f}, v)$, where v is the same as before but $\hat{f}$ is a new linear frame at b, the corresponding element of the typical fiber representing the *same* vector v changes. More explicitly, with an obvious notation, if $\hat{f}_j = a^i{}_j f_i$, then $v^i = a^i{}_j \hat{v}^j$ or $\hat{v}^i = (a^{-1})^i{}_j v^j$. We conclude that to represent the *same object* under a change of frame, there needs to be some kind of compensatory action in the change of the components. The object itself (in this case, the tangent vector) can be identified with the collection (or equivalence class) of all pairs made up of a frame and a matrix related in this compensatory way. In terms of the group actions on the typical fibers, if $\hat{f} = R_a f$, then the representative r of the vector v in $\mathbb{R}^n$ changes according to $\hat{r} = L_{a^{-1}} r$. We may, therefore, think of a vector as an equivalence class of elements of the Cartesian product $\mathcal{G} \times R^n$, corresponding to the following equivalence relation: $(g, r) \sim (\hat{g}, \hat{r})$ if, and only if, there exists $a \in \mathcal{G}$ such that $\hat{g} = ga$ and $\hat{r} = L_{a^{-1}} r$.

With the above motivation in mind, the following construction of a fiber bundle associated to a given principal bundle will seem less artificial than it otherwise would. We start from the principal bundle $(\mathcal{P}, \pi_P, \mathcal{B}, \mathcal{G}, \mathcal{G})$ and a manifold $\mathcal{F}$, which we want to construe as the typical fiber of a new fiber bundle $(\mathcal{C}, \pi, \mathcal{B}, \mathcal{F}, \mathcal{G})$ associated with $\mathcal{P}$. For this to be possible, we need to have an effective left action of $\mathcal{G}$ on $\mathcal{F}$, which we assume to have been given. To start off, we form the Cartesian product $\mathcal{P} \times \mathcal{F}$ and notice that the structure group $\mathcal{G}$ acts on it with a right action induced by its right action on $\mathcal{P}$ and its left action on $\mathcal{F}$. To describe this new right action, we will keep abusing the notation in the sense that we will use the same symbols for all the actions in sight, because the context should make clear which action is being used in each particular expression. Let (p, f) be an element of the product $\mathcal{P} \times \mathcal{F}$, and let $a \in \mathcal{G}$. We define the effective right action

$$R_a(p, f) = (R_a p, L_{a^{-1}} f). \tag{10.49}$$

The next step toward the construction of the associated bundle with typical fiber $\mathcal{F}$ consists of taking the quotient space $\mathcal{C}$ generated by this action. In other words, we want to deal with a set whose elements are equivalence classes in $\mathcal{C} \times \mathcal{F}$ by the equivalence relation "$(p_1, f_1) \sim (p_2, f_2)$ if, and only if, there exists $a \in \mathcal{G}$ such that $(p_2, f_2) = R_a(p_1, f_1)$." The motivation for this line of attack should be clear from the introductory remarks to this section. Recalling that the right action of $\mathcal{G}$ on $\mathcal{P}$ is fiber preserving, it becomes obvious that all the pairs (p, f) in a given equivalence class have first components p with the same projection $\pi_P(p)$ on $\mathcal{B}$. This means that we have a perfectly well-defined projection π in the quotient space $\mathcal{C}$, namely, $\pi : \mathcal{C} \to \mathcal{B}$ is a map that assigns to each equivalence class the common value of the projection of the first component of all its constituent pairs.

Having a projection, we can now define the fiber of $\mathcal{C}$ over $b \in \mathcal{B}$ naturally as

$\pi^{-1}(b)$. We need to show now that each such fiber is diffeomorphic to the putative typical fiber $\mathcal{F}$. More precisely, we want to show that for each local trivialization $(\mathcal{U}, \psi)$ of the original principal bundle $\mathcal{P}$, we can also construct a local trivialization of $\pi^{-1}(\mathcal{U})$, namely, a diffeomorphism $\rho : \pi^{-1}(\mathcal{U}) \to \mathcal{U} \times \mathcal{F}$. To understand how this works, let us fix a point $b \in \mathcal{U}$ and recall that, given the local trivialization $(\mathcal{U}, \psi)$, the map $\tilde{\psi}_b$ provides us with a diffeomorphism of the fiber $\pi_{\mathcal{P}}^{-1}(b)$ with $\mathcal{G}$. We now form the product map of $\tilde{\psi}_b$ with the identity map of $\mathcal{F}$, namely, $(\tilde{\psi}_b, \mathrm{id}_{\mathcal{F}}) : \pi_{\mathcal{P}}^{-1}(b) \times \mathcal{F} \to \mathcal{G} \times \mathcal{F}$. Each equivalence class by the right action (10.49) is mapped by the product map $(\tilde{\psi}_b, \mathrm{id}_{\mathcal{F}})$ into an orbit, as shown in Figure 10.3.

These orbits do not intersect with each other. Moreover, they can be seen as graphs of single-valued $\mathcal{F}$-valued functions of $\mathcal{G}$. Therefore, choosing any particular value $g \in \mathcal{G}$, we see that these orbits can be parameterized by $\mathcal{F}$. This provides the desired one-to-one and onto relation between the fiber $\pi^{-1}(b)$ and the manifold $\mathcal{F}$, which can now legitimately be called the typical fiber of $\mathcal{C}$. To complete the construction of the desired fiber bundle, we need to guarantee that the fiberwise isomorphism that we have just constructed depends differentiably on b, a requirement that we assume fulfilled.

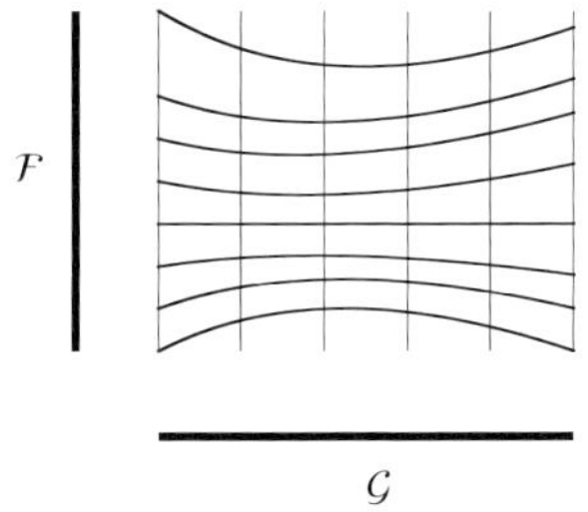

Figure 10.3 Images of equivalence classes.

10.5 Connections

10.5.1 Introduction

All the fibers of a fiber bundle are, by definition, diffeomorphic to each other. In the absence of additional structure, however, there is no canonical way to single out a particular diffeomorphism between fibers. In the case of a product bundle, for example, such a special choice is indeed available because of the existence of the second projection map onto the typical fiber. In this extreme case, we may say that we are in the presence of a *canonical distant parallelism* in the fiber bundle. An equivalent way to describe this situation is by saying that we have a canonical family of nonintersecting smooth cross sections such that each point in the fiber bundle belongs to one, and only one, of them. In a general fiber bundle we can only afford this luxury noncanonically and locally. A *connection* on a fiber bundle is, roughly speaking, an additional structure defined on the bundle that permits to establish intrinsic fiber diffeomorphisms for fibers lying along curves in the base manifold. In other words, a connection can be described as a curve-dependent parallelism. Given a connection, it may so happen that the induced fiber parallelisms turn out to be curve independent. A quantitative measure of this property or the lack thereof is provided by the vanishing, or otherwise, of the *curvature* of the connection.

10.5.2 Ehresmann Connection

Consider the tangent bundle TC of the total space C of an arbitrary fiber bundle $(C, \pi, \mathcal{B}, \mathcal{F}, \mathcal{G})$, and denote by $\tau_C : TC \to C$ its natural projection. If the dimensions of the base manifold $\mathcal{B}$ and the typical fiber $\mathcal{F}$ are, respectively, m and n, the dimension of C is $m+n$, and the typical fiber of (TC, τ_C) is $\mathbb{R}^{m+n}$, with structure group $GL(m+n; \mathbb{R})$. At each point $c \in C$, the tangent space T_cC has a canonically defined *vertical subspace* V_c, which can be identified with the tangent space $T_cC_{\pi(c)}$ to the fiber of C at c. The dimension of V_c is n. A vector in T_cC belongs to the vertical subspace V_c (or is *vertical*) if, and only if, its projection by π_* is the zero vector of $T_{\pi(c)}\mathcal{B}$. If a vector in T_cC is not vertical, there is no canonical way to assign to it a vertical component. It is this deficiency, and only this deficiency, that the Ehresmann connection remedies. Formally, an Ehresmann connection consists of a smooth *horizontal distribution* in C. This is a smooth assignment to each point $c \in C$ of an (m-dimensional) subspace $H_c \subset T_cC$ (called the *horizontal subspace at* c), such that

$$T_cC = H_c \oplus V_c. \tag{10.50}$$

In this equation, $\oplus$ denotes the direct sum of vector spaces. Each tangent vector $\mathbf{u} \in T_cC$ is, accordingly, uniquely decomposable as the sum of a horizontal part $h(\mathbf{u})$ and a vertical part $v(\mathbf{u})$. A vector is *horizontal,* if its vertical part vanishes. The only vector that is simultaneously horizontal and vertical is the zero vector. As H_c and $T_{\pi(c)}\mathcal{B}$ have the same dimension (m), the restriction $\pi_*|_{H_c} : H_c \to T_{\pi(c)}\mathcal{B}$, is a vector-space isomorphism. We denote its inverse by Γ_c. Thus, given a vector $\mathbf{v}$ tangent to the base manifold at a point $b \in \mathcal{B}$, there is a unique horizontal vector $\Gamma_c\mathbf{v}$ at $c \in \pi^{-1}(\{b\})$ such that $\pi_*(\Gamma_c\mathbf{v}) = \mathbf{v}$. This unique vector is called the *horizontal lift* of $\mathbf{v}$ to c. In particular, $\Gamma_c\left(\pi_*(\mathbf{u})\right) = \Gamma_c\left(\pi_*(h(\mathbf{u}))\right) = h(\mathbf{u})$. These ideas are schematically illustrated in Figure 10.4.

10.5.3 Parallel Transport along a Curve

Let

$$\gamma : (-\epsilon, \epsilon) \longrightarrow \mathcal{B} \tag{10.51}$$

be a smooth curve in the base manifold $\mathcal{B}$ of the fiber bundle (C, π), and let $c \in C_{\gamma(0)}$ be a point in the fiber at $\gamma(0)$. A *horizontal lift* of

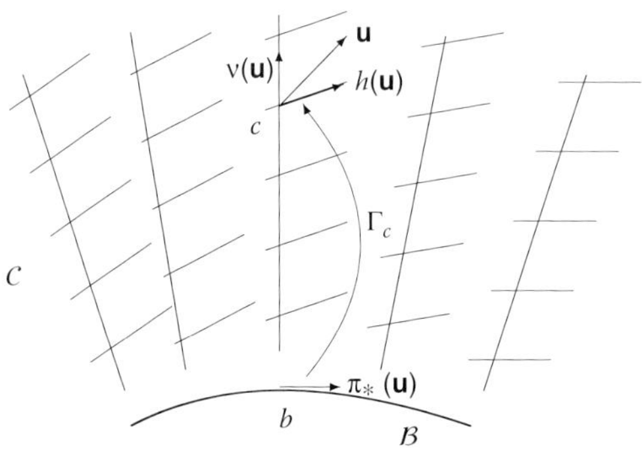

Figure 10.4 Ehresmann connection.

γ through c is defined as a curve

$$\hat{\gamma} : (-\epsilon, \epsilon) \longrightarrow \mathcal{C}, \tag{10.52}$$

such that

$$\hat{\gamma}(0) = c, \tag{10.53}$$

$$\pi\left(\hat{\gamma}(t)\right) = \gamma(t), \qquad \forall t \in (-\epsilon, \epsilon), \tag{10.54}$$

and

$$\hat{\gamma}'(t) \in H_{\hat{\gamma}(t)}, \qquad \forall t \in (-\epsilon, \epsilon), \tag{10.55}$$

where a prime denotes the derivative with respect to the curve parameter t. A horizontal lift is thus a curve that projects onto the original curve and, moreover, has a horizontal tangent throughout.

Consider the "cylindrical" subbundle $\gamma^*\mathcal{C}$ obtained by pulling back the bundle $\mathcal{C}$ to the curve γ or, less technically, by restricting the base manifold to the curve γ. The tangent vector field of γ has a unique horizontal lift at each point of this bundle. In other words, the curve generates a (horizontal) vector field throughout this restricted bundle. By the fundamental theorem of the theory of ordinary differential equations (ODEs), it follows that, at least for small enough ϵ, there is a unique horizontal lift of γ through any given point in the fiber at $\gamma(0)$, namely, the corresponding integral curve of the horizontal vector field. We conclude, therefore, that the horizontal lift of a curve through a point in a fiber bundle exists and is locally unique. As the horizontal curve issuing from c cuts the various fibers lying on γ, the point c is said to undergo a *parallel transport* relative to the given connection and the given curve. Thus, given a point $c \in \mathcal{C}$ and a curve γ through $\pi(c) \in \mathcal{B}$, we obtain a unique parallel transport of c along γ by solving a system of ODEs (so as to travel always horizontally). These concepts are illustrated schematically in Figure 10.5

10.5.4 Connections in Principal Bundles

A connection in a principal bundle $(\mathcal{P}, \pi, \mathcal{B}, \mathcal{G}, \mathcal{G})$ is an Ehresmann connection that is compatible with the right action R_g of $\mathcal{G}$ on $\mathcal{P}$, namely,

$$(R_g)_*(H_p) = H_{R_g p}, \qquad \forall g \in \mathcal{G}, \quad p \in \mathcal{P}. \tag{10.56}$$

This condition can be stated verbally as follows: the horizontal distribution is invariant under the group action.

Recall that the group $\mathcal{G}$ acts freely (to the right) on $\mathcal{P}$. Consequently, the fundamental vector field $\mathbf{v_g}$ associated with any nonzero vector $\mathbf{g}$ in the Lie algebra $\mathfrak{g}$ of $\mathcal{G}$ does not vanish anywhere on $\mathcal{P}$. Moreover, because the action of $\mathcal{G}$ maps fibers into themselves,

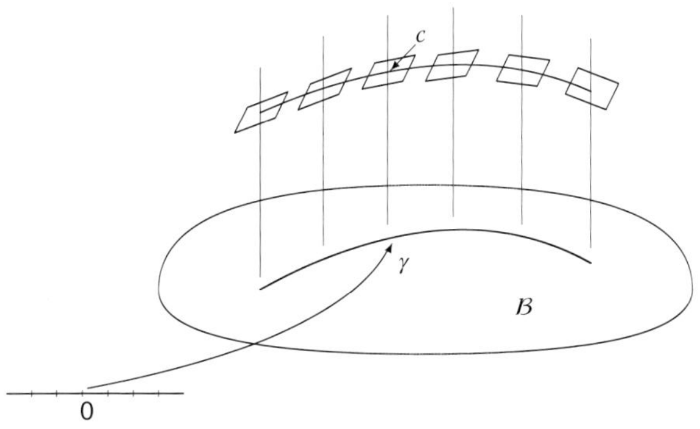

Figure 10.5 Parallel transport along a curve.

the fundamental vector fields are all vertical. The correspondence between vectors in the Lie algebra and tangent vectors to the fiber at any point is clearly linear and one-to-one. As the dimension of $\mathcal{G}$ is equal to the dimension of each fiber, we conclude that the map $\mathfrak{g} \to V_p$ given by $\mathbf{g} \mapsto \mathbf{v}_{\mathbf{g}}(p)$ is a linear isomorphism between the Lie algebra and each of the vertical subspaces of the principal bundle.

Let $\mathbf{v} \in T\mathcal{P}$ be any tangent vector to the fiber bundle. A connection Γ assigns to it a unique vertical part and, as we have just seen, the action of the group assigns to this vertical part an element of the Lie algebra $\mathfrak{g}$. This means that we have a well-defined linear map

$$\omega : T\mathcal{P} \longrightarrow \mathfrak{g}, \qquad (10.57)$$

associated with a given connection in a principal bundle. This map can be regarded as a *Lie-algebra-valued 1-form.* It is called the *connection form* associated with Γ.

10.5.5 Distributions and the Theorem of Frobenius

We have mentioned in Section 10.5.1 the notion of curvature of a connection as an indication of how the parallel transport of an entity along a curve depends on the curve itself. Before giving a precise definition of this concept, however, it may prove useful to introduce the more general concept of *involutivity* of a distribution. The reason for this digression is that a connection can always be regarded as a (horizontal) distribution. A k-dimensional distribution of an m-dimensional manifold $\mathcal{M}$ (with $m \geq k$) is defined as a smooth assignment of a k-dimensional subspace $\mathcal{D}_x$ of the tangent space $T_x\mathcal{M}$ to each point $x \in \mathcal{M}$. A fundamental question in the theory of distributions is whether or not there exist *integral embedded submanifolds,* namely, embedded submanifolds of dimension k whose tangent space at each point x coincides with $\mathcal{D}_x$.

An *embedded submanifold* of dimension k is defined as a subset $\mathcal{S} \subset \mathcal{M}$ such that for each point $s \in \mathcal{S}$ one can find a chart of $\mathcal{M}$ with coordinates x^i $(i = 1, \dots, m)$ such that s belongs to this chart and such that the intersection of the set $\mathcal{S}$ with the chart coincides with the set obtained by keeping the last $m - k$ coordinates constant. This idea becomes clear if one thinks of the particular case of $\mathbb{R}^2$ as embedded in $\mathbb{R}^3$ with coordinates x, y, z. The equation of the embedded submanifold $\mathbb{R}^2$ can be given as $z = 0$.

In some sense, the question of existence of integral submanifolds can be regarded as a generalization to many dimensions of the question of integrability of systems of ODE, which would correspond to the case $k = 1$, namely, to the case in which the subspaces of the distribution are mere lines. While in the particular case $k = 1$, we are assured, by the fundamental theorem of ODEs, of the (local) existence of integral curves, the answer in the general case $k > 1$ is usually negative. A k-dimensional distribution is said to be *completely integrable* if for each point of the manifold $\mathcal{M}$ there exists a chart x^i $(i = 1, \dots, m)$ such that each set obtained by keeping the last $n - k$ coordinates thereat constant is an integral submanifold (of dimension k). Assume that a completely integrable distribution has been given. Then, according to our definition, the first k natural vectors of the local coordinate system just described belong to the distribution and constitute a basis of $\mathcal{D}_x$ at each point x in the chart. Any vector fields $\mathbf{v}_\alpha$ $(\alpha = 1, \dots, k)$ with this property (of constituting a basis of the distribution) are said to *span* the distribution. Within the chart, any vector fields $\mathbf{v}_\alpha$ $(\alpha = 1, \dots, k)$ that span the distribution must be expressible, therefore, as

$$\mathbf{v}_\alpha = v_\alpha^{\ \beta} \frac{\partial}{\partial x^\beta}, \tag{10.58}$$

where the summation convention applies for Greek indices within the range $1, \ldots, k$. We now calculate the Lie bracket of any pair of the spanning vectors as

$$[\mathbf{v}_\alpha, \mathbf{v}_\beta] = v_\alpha^{\ \rho} \frac{\partial v_\beta^{\ \sigma}}{\partial x^\rho} \frac{\partial}{\partial x^\sigma} - v_\beta^{\ \sigma} \frac{\partial v_\alpha^{\ \rho}}{\partial x^\sigma} \frac{\partial}{\partial x^\rho}. \tag{10.59}$$

Notice that, in calculating the Lie brackets, we have used the fact that the components of the vectors $\mathbf{v}_\alpha$ vanish on the natural base vectors $\partial/\partial x^i$ with $i > k$. Moreover, because the given vectors are linearly independent, the matrix with entries $v_\alpha^{\ \beta}$ is nonsingular. Inverting, therefore, (10.58), we can express the natural base vectors $\partial/\partial x^\alpha$ ($\alpha = 1, \ldots, k$) in terms of the vectors $\mathbf{v}_\beta$, with the result that the Lie brackets are themselves linear combinations of these vectors, namely, there exist scalars $C^\gamma_{\alpha\beta}$ such that

$$[\mathbf{v}_\alpha, \mathbf{v}_\beta] = C^\gamma_{\alpha\beta}\, \mathbf{v}\gamma. \tag{10.60}$$

A distribution with this property (namely, that the Lie bracket of any two vector fields in the distribution is also in the distribution) is said to be *involutive*. We have proven, therefore, that every completely integrable distribution is involutive. The converse of this result (that is, that every involutive distribution is completely integrable) is also true, and is the content of the theorem of Frobenius, whose proof we omit.

10.5.6 Curvature

Suppose that we draw through a point b of the base manifold $\mathcal{B}$ a small closed curve γ. If we now choose a point p in the fiber on b, we have learned that there exists a unique horizontal lift $\tilde{\gamma}$, namely, a horizontal curve containing p and projecting on γ. Is this curve closed? To clarify the meaning of this question and its possible answer, recall that a connection on a principal fiber bundle is a special case of a distribution, which we have called horizontal (the dimension of the horizontal distribution equals the dimension of the base manifold and is thus strictly smaller than the dimension of the fiber bundle, assuming that the typical fiber is of dimension greater than zero). Clearly, if the horizontal distribution is involutive, any horizontal lift of a small curve in the base manifold will lie entirely on an integral surface and, therefore, will be closed. This observation suggests that a measure of the lack of closure of the horizontal lift of closed curves is the fact that the Lie bracket between horizontal vector fields has a vertical component. We want to see now how to extract this information from the connection itself. More particularly, because a connection is specified by its connection form ω, we want to extract this information from ω alone.

Consider two horizontal vector fields $\mathbf{u}$ and $\mathbf{v}$. Let us evaluate the 2-form[1)] $\mathrm{d}\omega$ on this pair as

$$\langle \mathrm{d}\omega \mid \mathbf{u} \wedge \mathbf{v} \rangle = \mathbf{u}\left(\langle \omega \mid \mathbf{v} \rangle\right) - \mathbf{v}\left(\langle \omega \mid \mathbf{u} \rangle\right) - \langle \omega \mid [\mathbf{u}, \mathbf{v}] \rangle, \tag{10.61}$$

which, in view of the fact that $\mathbf{u}$ and $\mathbf{v}$ are assumed to be horizontal, yields

$$\langle \mathrm{d}\omega \mid \mathbf{u} \wedge \mathbf{v} \rangle = -\langle \omega \mid [\mathbf{u}, \mathbf{v}] \rangle. \tag{10.62}$$

The right-hand side of this equation will vanish if, and only if, the Lie bracket is horizontal. This means that we have found a

1) Notice that this is a Lie-algebra-valued differential form.

way to extract the right information from ω by just taking its exterior derivative and applying it to two horizontal vector fields. Notice, however, that $\mathrm{d}\omega$ can be applied to arbitrary pairs of vector fields, not necessarily horizontal. To formalize this point, given a connection, we define the *exterior covariant derivative* $D\alpha$ of an r-form α as the $(r+1)$-form given by

$$\langle D\alpha \mid \mathbf{U}_1 \wedge \cdots \wedge \mathbf{U}_{r+1}\rangle = \langle \mathrm{d}\alpha \mid h(\mathbf{U}_1) \wedge \cdots \wedge h(\mathbf{U}_{r+1})\rangle, \quad (10.63)$$

where $h(.)$ denotes the horizontal component of a vector. Accordingly, we define the *curvature 2-form* Ω of a connection ω on a principal fiber bundle as

$$\Omega = D\omega. \quad (10.64)$$

10.5.7 Cartan's Structural Equation

Our definition of curvature, by using both the connection 1-form and the horizontal projection map, is a hybrid that mixes both (equivalent) definitions of a connection. It is possible, on the other hand, to obtain an elegant formula that involves just the connection 1-form. This formula, known as *Cartan's structural equation*, reads

$$\Omega = \mathrm{d}\omega + \frac{1}{2}\,[\omega, \omega], \quad (10.65)$$

or, more precisely, for any two vectors $\mathbf{u}$ and $\mathbf{v}$ at a point[2)] of the frame bundle,

$$\langle \Omega \mid \mathbf{U} \wedge \mathbf{V}\rangle = \langle \mathrm{d}\omega \mid \mathbf{U} \wedge \mathbf{V}\rangle + \frac{1}{2}\,[\omega(\mathbf{U}), \omega(\mathbf{V})]. \quad (10.66)$$

2) Notice that this formula is valid pointwise, because the Lie bracket on the right-hand side is evaluated in the Lie algebra, not in the manifold.

The proof of this formula, whose details we omit, is based on a careful examination of three cases: (i) $\mathbf{u}$ and $\mathbf{v}$ are horizontal, whereby the formula is obvious; (ii) $\mathbf{u}$ is horizontal and $\mathbf{v}$ is vertical, in which case one can extend them, respectively, to a horizontal and a fundamental (vertical) vector field; (iii) $\mathbf{u}$ and $\mathbf{v}$ are both vertical, in which case they can both be extended to fundamental fields.

10.5.8 Bianchi Identities

Unlike the ordinary exterior derivative d, the operator D (of exterior covariant differentiation) is not necessarily nilpotent, namely, in general, $D^2 \neq 0$. Therefore, there is no reason to expect that $D\Omega$, which is equal to $D(D\omega)$, will vanish identically. But, in fact, it does. To see that this is the case, notice that, by definition of D, we need only verify the vanishing of $D\Omega$ on an arbitrary triple of *horizontal* vectors. It can be shown that

$$D\Omega = 0. \quad (10.67)$$

In terms of components, we obtain differential identities to be satisfied by any curvature form. They are known as the *Bianchi identities*.

10.5.9 Linear Connections

A connection on the bundle of linear frames $F\mathcal{B}$ is called a *linear connection* on $\mathcal{B}$. Among principal bundles, the bundle of linear frames occupies a special position for various reasons. In the first place, the bundle of linear frames is canonically defined for any given base manifold $\mathcal{B}$. Moreover, the associated bundles include all the tensor bundles, thus allowing for a unified treatment of all such entities.

Another way to express this peculiar feature of the bundle of linear frames is that, whereas the quantities parallel-transported along curves in a general principal bundle are of a nature not necessarily related to the base manifold, in the case of the bundle of linear frames, the quantities transported are precisely the very frames used to express the components of vectors and forms defined on the base manifold. An elegant manifestation of this property is the existence of a canonical 1-form that ties everything together. A direct consequence of the existence of this 1-form is the emergence of the idea of the *torsion* of a connection. We start the treatment of linear connections by lingering for a while on the definition of the canonical 1-form.

10.5.10 The Canonical 1-Form

Given a tangent vector $\mathbf{v} \in T_x\mathcal{B}$ at a point x in the base manifold, and a point $p \in F_x\mathcal{B}$ in the fiber over x, and recalling that p consists of a frame (or basis) $\{\mathbf{e}_1, \dots, \mathbf{e}_m\}$ of $T_x\mathcal{B}$, we can determine uniquely the m components of $\mathbf{v}$ in this frame, namely,

$$\mathbf{v} = v^a \mathbf{e}_a. \tag{10.68}$$

In other words, at each point $p \in F\mathcal{B}$, we have a well-defined nonsingular linear map[3)]

$$u(p) : T_{\pi(p)}\mathcal{B} \longrightarrow \mathbb{R}^m. \tag{10.69}$$

The *canonical 1-form* θ on $F\mathcal{B}$ is defined as

$$\theta(\mathbf{V}) = u(p)\circ\pi_*(\mathbf{V}), \qquad \forall\, \mathbf{V} \in T_p(F\mathcal{B}). \tag{10.70}$$

Note that this is an $\mathbb{R}^m$-valued form. The canonical 1-form of the frame bundle is a particular case of a more general construct known as a *soldering form*.

It may prove instructive to exhibit the canonical form in components. Let $x^1, \dots, x^m$ be a local coordinate system on $\mathcal{U} \subset \mathcal{B}$. Every frame $\{\mathbf{e}_1, \dots, \mathbf{e}_m\}$ at $x \in \mathcal{U}$ can be expressed uniquely by means of a nonsingular matrix with entries x^i_j as

$$\mathbf{e}_a = x^i_a \frac{\partial}{\partial x^i}. \tag{10.71}$$

This means that the $m + m^2$ functions $\{x^i, x^i_a\}$ constitute a coordinate system for the linear frame bundle $\pi^{-1}(\mathcal{U})$. We call it the coordinate system *induced* by x^i. The projection map $\pi : F\mathcal{B} \to \mathcal{B}$ has the coordinate representation

$$(x^i, x^i_a) \mapsto \pi(x^i, x^i_a) = (x^i), \tag{10.72}$$

with some notational abuse.

Consider now the tangent bundle $TF(\mathcal{B})$ with projection $\tau : TF(\mathcal{B}) \to F(\mathcal{B})$. The coordinate system $\{x^i, x^i_a\}$ induces naturally a coordinate system in $TF(\mathcal{B})$. A vector $\mathbf{X} \in TF(\mathcal{B})$ is expressed in these coordinates as follows:

$$\mathbf{X} \mapsto \left(x^i, x^i_a, X^i \frac{\partial}{\partial x^i} + X^i_a \frac{\partial}{\partial x^i_a}\right) = \left(x^i, x^i_a, X^i, X^i_a\right). \tag{10.73}$$

The derivative of the projection π is a map $\pi_* TF(\mathcal{B}) \to T\mathcal{B}$. Its coordinate representation is

$$\left(x^i, x^i_a, X^i, X^i_a\right) \mapsto \left(x^i, X^i\right). \tag{10.74}$$

The map u defined in (10.69) is given in coordinates by

$$[u(x^i, x^i_a)](x^j, w^j) = x^{-a}_i w^i \quad (a = 1, \dots, m), \tag{10.75}$$

where we have denoted by x^{-a}_i the entries of the inverse of the matrix with entries x^i_a.

3) This map is, in fact, an alternative definition of a linear frame at a point of a manifold $\mathcal{B}$.

Combining (10.74) and (10.75), we obtain from (10.70) the following coordinate representation of the ($\mathbb{R}^m$)-valued canonical form θ:

$$\theta^a = x_i^{-a}\,\mathrm{d}x^i \qquad (a = 1, \ldots, m). \tag{10.76}$$

10.5.11
The Christoffel Symbols

The canonical form θ exists independently of any connection. Let us now introduce a connection on $F(\mathcal{B})$, that is, a linear connection on $\mathcal{B}$. If we regard a connection as a horizontal distribution, there must exist nonsingular linear maps $\Gamma(x, p)$ from each $T_x\mathcal{B}$ to each of the tangent spaces $T_pF(\mathcal{B})$ (with $\pi(p) = x$) defining the distribution. Noticing that the same distribution may correspond to an infinite number of such maps, we pin down a particular one by imposing the extra condition that they must be also horizontal lifts. In other words, we demand that

$$\pi_* \circ \Gamma(x, p) = \mathrm{id}_{T_x\mathcal{B}}. \tag{10.77}$$

The implication of this condition is that, when written in components, we must have

$$\Gamma(x, p)\left(v^i \frac{\partial}{\partial x^i}\right) = v^i \frac{\partial}{\partial x^i} - \hat{\Gamma}^j_{ia}(x, p)\, v^i \frac{\partial}{\partial x^j_a}, \tag{10.78}$$

where $\hat{\Gamma}^j_{ia}(x, p)$ are smooth functions of x and p. The minus sign is introduced for convenience.

These functions, however, cannot be arbitrary, because they must also satisfy the compatibility condition (10.56). It is not difficult to verify that this is the case if, and only if, the functions Γ^j_{ik} defined by

$$\Gamma^j_{ik} = \hat{\Gamma}^j_{ia}(x, p)\, x_k^{-a}(p) \tag{10.79}$$

are independent of p along each fiber.

We conclude that a linear connection is completely defined (on a given coordinate patch) by means of m^3 smooth functions. These functions are known as the *Christoffel symbols* of the connection.

10.5.12
Parallel Transport and the Covariant Derivative

Now that we are in possession of explicit coordinate expressions for the horizontal distribution, we can write explicitly the system of ODEs that effects the horizontal lift of a curve in $\mathcal{B}$. A solution of this system is a one-parameter family of frames being parallel-transported along the curve. Let the curve γ in the base manifold be given by

$$x^i = x^i(t). \tag{10.80}$$

On this curve, the connection symbols are available as functions of t, by composition. The nontrivial part of the system of equations is given by

$$\frac{\mathrm{d}x^i_a(t)}{\mathrm{d}t} = -\Gamma^i_{jk}(t)\, x^k_a(t)\, \frac{\mathrm{d}x^j(t)}{\mathrm{d}t}. \tag{10.81}$$

The local solution of this system with given initial condition (say, $x^i_a(0) = \bar{x}^i_a$) is the desired curve in $F(\mathcal{B})$, representing the parallel transport of the initial frame along the given curve.

Let now $\bar{\mathbf{v}}$ be a vector in $T_{x^i(0)}\mathcal{B}$, that is, a vector at the initial "time" $t = 0$. We say that the curve $\mathbf{v} = \mathbf{v}(t)$ in $T\mathcal{B}$ is the parallel transport of $\bar{\mathbf{v}}$ if it projects on γ, with $\mathbf{v}(0) = \bar{\mathbf{v}}$, and if the components of $\mathbf{v}(t)$ in a parallel-transported frame along γ are constant.[4] For this definition to make sense, we must make sure that the constancy of the

4) The same criterion for parallel transport that we are using for the tangent bundle can be used for any associated bundle.

components is independent of the particular initial frame chosen. This, however, is a direct consequence of the fact that our linear connection is, by definition, consistent with the right action of the group.

To obtain the system of ODEs corresponding to the parallel transport of $\bar{\mathbf{v}}$ along γ, we enforce the constancy conditions

$$v^i(t)\, x_i^{-a}(t) = C^a, \tag{10.82}$$

where each C^a $(a = 1, \ldots, m)$ is a constant and v^i denotes components in the coordinate basis. Differentiating this equation with respect to t and invoking (10.81), we obtain

$$\frac{dv^i}{dt} + \Gamma^i_{jk}\,\frac{dx^j}{dt}\,v^k = 0. \tag{10.83}$$

A vector field along γ satisfying this equation is said to be *covariantly constant.* For a given vector field $\mathbf{w}$ on $\mathcal{B}$, the expression on the left-hand side makes sense in a pointwise manner whenever a vector $\mathbf{u}$ is defined at a point (whereby the curve γ can be seen as a representative at $t = 0$). The expression

$$\nabla_{\mathbf{u}}\mathbf{w} = \left(\frac{dw^i}{dt} + \Gamma^i_{jk}\,u^j\,w^k\right)\frac{\partial}{\partial x^i}, \tag{10.84}$$

is called the *covariant derivative* of $\mathbf{v}$ in the direction of $\mathbf{u}$. From the treatment above, it can be seen that the covariant derivative is precisely the limit

$$\nabla_{\mathbf{u}}\mathbf{w} = \lim_{t\to 0}\frac{\rho_{t,0}\mathbf{w} - \mathbf{w}(0)}{t}, \tag{10.85}$$

where $\rho(a, b)$ denotes the parallel transport along γ from $t = b$ to $t = a$.

10.5.13 Curvature and Torsion

To obtain an explicit equation for the curvature form Ω, we should start by elucidating the connection form ω on the basis of the connection symbols Γ. Given a vector $\mathbf{X} \in T_pF\mathcal{B}$, we know that its horizontal component $h(\mathbf{X})$ is given by

$$h(\mathbf{X}) = \Gamma(\pi(p), p)\circ\pi_*\,(\mathbf{X}). \tag{10.86}$$

Its vertical component must, therefore, be given by

$$v(\mathbf{X}) = \mathbf{X} - h(\mathbf{X}) = \mathbf{X} - \Gamma(\pi(p), p)\circ\pi_*\,(\mathbf{X}). \tag{10.87}$$

Recall that the connection form ω assigns to $\mathbf{X}$ the vector in $\mathfrak{g}$ such that $v(\mathbf{X})$ belongs to its fundamental vector field. Let the coordinates of p be (x^i, x^i_a). The right action of $GL(m;\mathbb{R})$ is given by

$$\left(R_g(p)\right)^i_a = x^i_b\, g^b_a, \tag{10.88}$$

where we have shown only the action on the fiber component and g^b_a is the matrix corresponding to $g \in GL(m;\mathbb{R})$. Consequently, if $g(t)$ is a one-parameter subgroup represented by the vector

$$\hat{g}^a_b = \left.\frac{dg^a_b(t)}{dt}\right|_{t=0}, \tag{10.89}$$

the value of the corresponding fundamental vector field at p is

$$\tilde{g}^i_a = x^i_b\,\hat{g}^b_a. \tag{10.90}$$

The coordinate expression of (10.87) is

$$(v(\mathbf{X}))^i_a = X^i_a - h(\mathbf{X}) = \mathbf{X} - \Gamma(\pi(p), p)\circ\pi_*\,(\mathbf{X}). \tag{10.91}$$

Let the main part of the vector **X** be given by

$$\mathbf{X} = v^i \frac{\partial}{\partial x^i} + X^i_a \frac{\partial}{\partial x^i_a}. \quad (10.92)$$

Then, (10.91) delivers

$$(v(\mathbf{X}))^i_a = X^i_a + \Gamma^j_{ik}\, v^i\, x^k_a. \quad (10.93)$$

According to (10.90), the corresponding element of the Lie algebra is

$$\hat{g}^b_a = \left(X^j_a + \Gamma^j_{ik}\, v^i\, x^k_a\right) x^{-b}_j. \quad (10.94)$$

Accordingly, the Lie-algebra-valued connection form ω is given by

$$\omega^b_a = \Gamma^j_{ik}\, x^k_a\, x^{-b}_j\, \mathrm{d}x^i + x^{-b}_j\, \mathrm{d}x^j_a. \quad (10.95)$$

The exterior derivative is given by

$$\begin{aligned}\mathrm{d}\omega^b_a &= \frac{\partial \Gamma^j_{ik}}{\partial x^m}\, x^k_a\, x^{-b}_j\, \mathrm{d}x^m \wedge \mathrm{d}x^i \\ &+\Gamma^j_{ik} x^{-b}_j\, \mathrm{d}x^k_a \wedge \mathrm{d}x^i - \Gamma^j_{ik} x^k_a x^{-b}_s x^{-c}_j \mathrm{d}x^s_c \wedge \mathrm{d}x^i \\ &-x^{-c}_j x^{-b}_s\, \mathrm{d}x^s_c \wedge \mathrm{d}x^j_a. \quad (10.96)\end{aligned}$$

A vector such as (10.92) has the following horizontal component:

$$h(\mathbf{X}) = v^i \frac{\partial}{\partial x^i} - \Gamma^j_{ik} x^k_a v^i \frac{\partial}{\partial x^j_a}. \quad (10.97)$$

With a similar notation, the horizontal component of another vector **Y** is given by

$$h(\mathbf{Y}) = w^i \frac{\partial}{\partial x^i} - \Gamma^j_{ik} x^k_a w^i \frac{\partial}{\partial x^j_a}. \quad (10.98)$$

Consider now the following evaluations:

$$\langle \mathrm{d}x^j \wedge \mathrm{d}x^i \mid h(\mathbf{X}) \wedge h(\mathbf{Y})\rangle = v^j w^i - v^i w^j, \quad (10.99)$$

$$\begin{aligned}&\langle \mathrm{d}x^k_a \wedge \mathrm{d}x^i \mid h(\mathbf{X}) \wedge h(\mathbf{Y})\rangle \\ &\quad = -\Gamma^k_{rs} x^s_a\, (v^r w^i - v^i w^r), \quad (10.100)\end{aligned}$$

and

$$\begin{aligned}&\langle \mathrm{d}x^j_c \wedge \mathrm{d}x^s_a \mid h(\mathbf{X}) \wedge h(\mathbf{Y})\rangle \\ &\quad = -\Gamma^j_{rn} x^n_c\, \Gamma^s_{ik} x^k_a\, (v^r w^i - v^i w^r). \quad (10.101)\end{aligned}$$

Putting all these results together, we obtain

$$\begin{aligned}\langle \Omega \mid \mathbf{X} \wedge \mathbf{Y}\rangle &= \langle \omega \mid h(\mathbf{X}) \wedge h(\mathbf{Y})\rangle \\ &= x^k_a x^{-b}_j R^j_{kri} v^r w^i, \quad (10.102)\end{aligned}$$

where

$$R^j_{kri} = \frac{\Gamma^j_{ik}}{\partial x^r} - \frac{\Gamma^j_{rk}}{\partial x^i} + \Gamma^j_{rh}\Gamma^h_{ik} - \Gamma^j_{ih}\Gamma^h_{rk} \quad (10.103)$$

is called the *curvature tensor* of the linear connection.

In analogy with the concept of curvature form, we define the *torsion form* of a connection as

$$\Theta = D\theta. \quad (10.104)$$

Notice that the coupling with the connection is in the fact that the operator D is the exterior *covariant* derivative, which involves the horizontal projection. To understand the meaning of the torsion, consider a case in which the curvature vanishes. This means that there exists a *distant* (or curve-independent) parallelism in the manifold $\mathcal{B}$. Thus, fixing a basis of the tangent space at any one point $x_0 \in \mathcal{B}$, a field of bases is uniquely determined at all other points. The question that the torsion tensor addresses is the following: does there exist a coordinate system such that these bases coincide at each point with its natural basis? An interesting example can be constructed in $\mathbb{R}^3$ as follows. Starting from the standard coordinate system, move up the x^3 axis and, while so doing, apply a linearly increasing rotation to the horizontal planes, imitating the action of a corkscrew. Thus, we obtain a system of (orthonormal) bases that are perfectly

Cartesian plane by horizontal plane, but twisted with respect to each other as we ascend. These frames can be used to define a distant parallelism (two vectors are parallel if they have the same components in the local frame). It is not difficult to show (or to see intuitively) that there is no coordinate system that has these as natural bases (use, for example, Frobenius' theorem). This example explains the terminology of "torsion."

To obtain the coordinate expression of the torsion form, we start by calculating the exterior derivative of (10.76) as

$$d\theta^a = dx_i^{-a} \wedge dx^i = -x_j^{-a} x_i^{-b} dx_b^j \wedge dx^i. \tag{10.105}$$

Using (10.100), we obtain

$$\begin{aligned}\langle D\theta \mid \mathbf{X} \wedge \mathbf{Y} \rangle &= \langle d\theta \mid h(\mathbf{X}) \wedge h(\mathbf{Y}) \rangle \\ &= x_j^{-a}\, T_{ri}^j\, v^r w^i,\end{aligned} \tag{10.106}$$

where

$$T_{ri}^j = \Gamma_{ri}^j - \Gamma_{ir}^j \tag{10.107}$$

are the components of the *torsion tensor* of the connection.

Suppose that a linear connection with vanishing curvature has been specified on the manifold $\mathcal{B}$, and let $\mathbf{e}_1, \ldots, \mathbf{e}_n$ be a field of parallel frames on the manifold. Then the Christoffel symbols of the connection are given by the formula

$$\Gamma_{kj}^i = -e_k^{-a} \frac{\partial e_a^i}{\partial x^j}, \tag{10.108}$$

where e_a^i are the components of the frame in the natural basis of a coordinate system $x^1, \ldots, x^m$. The components of the torsion tensor are proportional to the components of the Lie brackets of corresponding pairs of vectors of the frames.

10.6 Riemannian Manifolds

10.6.1 Inner-Product Spaces

We have come a long way without the need to speak about metric concepts, such as the length of a vector or the angle between two vectors. That even the concept of power of a force can be introduced without any metric background may have seemed somewhat surprising, particularly to those accustomed to hear about "the magnitude of the force multiplied by the magnitude of the velocity and by the cosine of the angle they form." It is very often the case in applications to particular fields (Mechanics, Theoretical Physics, Chemistry, Engineering, etc.) that there is much more structure to go around than really needed to formulate the basic concepts. For the particular application at hand, there is nothing wrong in taking advantage of this extra structure. Quite to the contrary – the extra structure may be the carrier of implicit assumptions that permit, consciously or not, the formulation of the physical laws. The most dramatic example is perhaps the adherence to Euclidean Geometry as the backbone of Newtonian Physics. On the other hand, the elucidation of the minimal (or nearly so) structure necessary for the formulation of a fundamental notion, has proven time and again to be the beginning of an enlightenment that can lead to further developments and, no less importantly, to a better insight into the old results.

We have seen how the concept of the space dual to a given vector space arises naturally from the consideration of linear functions on the original vector space. On the other hand, we have learned that, intimately related as they are, there is no natural isomorphism between these two spaces. In other words, there is no natural way to associate a covector to a given vector, and vice versa. In Newtonian Mechanics, however, the assumption of a Euclidean metric, whereby the theorem of Pythagoras holds globally, provides such identification. In Lagrangian Mechanics, it is the kinetic energy of the system that can be shown to provide such extra structure, at least locally. In General Relativity, this extra structure (but of a somewhat different nature) becomes the main physical quantity to be found by solving Einstein's equations. In all these cases, the identification of vectors with covectors is achieved by means of the introduction of a new operation called an *inner product* (or a *dot product* or, less felicitously, a *scalar product*).

A vector space V is said to be an *inner-product space* if it is endowed with an operation (called an inner product)

$$\begin{aligned} \cdot : V \times V &\longrightarrow \mathbb{R} \\ (\mathbf{u}, \mathbf{v}) &\mapsto \mathbf{u} \cdot \mathbf{v}, \end{aligned} \tag{10.109}$$

satisfying the following properties:[5)]

1. Commutativity

$$\mathbf{u} \cdot \mathbf{v} = \mathbf{v} \cdot \mathbf{u}, \quad \forall\, \mathbf{u}, \mathbf{v} \in V; \tag{10.110}$$

2. Bilinearity[6)]

$$(\alpha \mathbf{u}_1 + \beta \mathbf{u}_2) \cdot \mathbf{v} = \alpha(\mathbf{u}_1 \cdot \mathbf{v}) + \beta(\mathbf{u}_2 \cdot \mathbf{v}), \quad \forall\, \alpha, \beta \in \mathbb{R},\ \mathbf{u}_1, \mathbf{u}_2, \mathbf{v} \in V; \tag{10.111}$$

3. Positive definiteness[7)]

$$\mathbf{v} \neq \mathbf{0} \implies \mathbf{v} \cdot \mathbf{v} > 0. \tag{10.112}$$

One can show that $\mathbf{0} \cdot \mathbf{v} = 0$, for all $\mathbf{v}$. The *magnitude* or *length* of a vector $\mathbf{v}$ is defined as the nonnegative number $\sqrt{\mathbf{v} \cdot \mathbf{v}}$. Two vectors $\mathbf{u}, \mathbf{v} \in V$ are called *orthogonal* (or *perpendicular*) to each other if $\mathbf{u} \cdot \mathbf{v} = 0$.

We want now to show how the existence of an inner product induces an isomorphism between a space and its dual (always in the finite-dimensional case). Let $\mathbf{v} \in V$ be a fixed element of V. By the linearity of the inner product, the product $\mathbf{v} \cdot \mathbf{u}$ is linear in the second argument. Accordingly, we define the covector $\omega_v \in V^*$ corresponding to the vector $\mathbf{v} \in V$, by

$$\langle \omega_v, \mathbf{u} \rangle = \mathbf{v} \cdot \mathbf{u}, \quad \forall\, \mathbf{u} \in V. \tag{10.113}$$

It is not difficult to prove that this linear map from V to V^* is one-to-one and that, therefore, it constitutes an isomorphism between V and V^*. We conclude that in an inner-product space there is no need to distinguish notationwise between vectors and covectors.

We call *reciprocal basis* the basis of V that corresponds to the dual basis in the isomorphism induced by the inner product. We already know that the dual basis operates on vectors in the following way:

$$\langle \mathbf{e}^i, \mathbf{v} \rangle = v^i, \quad \forall \mathbf{v} \in V, \tag{10.114}$$

where v^i is the ith component of $\mathbf{v} \in V$ in the basis $\{\mathbf{e}_j\}$ $(j = 1, \ldots, n)$. The reciprocal basis, therefore, consists of *vectors* $\{\mathbf{e}^j\}$ $(j = 1, \ldots, n)$ such that

$$\mathbf{e}^i \cdot \mathbf{v} = v^i, \quad \forall \mathbf{v} \in V. \tag{10.115}$$

5) It is to be noted that in the case of a complex vector space, such as in Quantum Mechanics applications, these properties need to be altered somewhat.

6) The term bilinearity refers to the fact that the inner product is linear in each of its two arguments. Nevertheless, given that we have already assumed commutativity, we need only to show linearity with respect to one of the arguments.

7) In Relativity, this property is removed.

Let the components of the reciprocal base vectors be expressed as

$$\mathbf{e}^i = g^{ij}\mathbf{e}_j. \tag{10.116}$$

In other words, we denote by g^{ij} the jth component of the ith member of the reciprocal basis we are seeking. It follows from (10.115) that

$$\begin{aligned}\mathbf{e}^i \cdot \mathbf{v} &= (g^{ij}\mathbf{e}_j) \cdot (v^k\mathbf{e}_k) \\ &= g^{ij}\,(\mathbf{e}_j \cdot \mathbf{e}_k)\, v^k = v^i, \quad \forall v^k \in \mathbb{R}.\end{aligned} \tag{10.117}$$

Looking at the very last equality, it follows that

$$g^{ij}\,(\mathbf{e}_j \cdot \mathbf{e}_k) = \delta^i_k. \tag{10.118}$$

Indeed, regarded as a matrix equation, (10.117) establishes that the matrix with entries $[g^{ij}\,(\mathbf{e}_j \cdot \mathbf{e}_k)]$ (summation convention understood), when multiplied by an arbitrary column vector, leaves it unchanged. It follows that this matrix must be the identity. This is only possible if the matrix with entries

$$g_{ij} = \mathbf{e}_i \cdot \mathbf{e}_j, \tag{10.119}$$

is the inverse of the matrix with entries g^{ij}. So, the procedure to find the reciprocal basis is the following: (i) Construct the (symmetric) square matrix with entries $g_{ij} = \mathbf{e}_i \cdot \mathbf{e}_j$. (ii) Invert this matrix to obtain the matrix with entries g^{ij}. (iii) Define $\mathbf{e}^i = g^{ij}\mathbf{e}_j$. Note that the *metric matrix* $\{g_{ij}\}$ is always invertible, as it follows from the linear independence of the basis.

A basis of an inner-product space is called *orthonormal* if all its members are of unit length and mutually orthogonal. The reciprocal of an orthonormal basis coincides with the original basis.

Having identified an inner-product space with its dual, and having brought back the dual basis to the original space under the guise of the reciprocal basis, we have at our disposal contravariant and covariant components of vectors. Recall that before the introduction of an inner product, the choice of a basis in V condemned vectors to have contravariant components only, while the components of covectors were covariant.

Starting from $\mathbf{v} = v^i\mathbf{e}_i = v_i\mathbf{e}^i$ and using (10.118) and (10.119), the following formulas can be derived:

$$v^i = g^{ij}v_j, \tag{10.120}$$

$$v_i = g_{ij}v^j, \tag{10.121}$$

$$\mathbf{e}_i = g_{ij}\mathbf{e}^j, \tag{10.122}$$

$$v^i = \mathbf{v} \cdot \mathbf{e}^i, \tag{10.123}$$

$$v_i = \mathbf{v} \cdot \mathbf{e}_i, \tag{10.124}$$

$$\mathbf{e}^i \cdot \mathbf{e}^j = g^{ij}, \tag{10.125}$$

$$\mathbf{e}^i \cdot \mathbf{e}_j = \delta^i_j. \tag{10.126}$$

A linear map $Q : U \to V$ between inner-product spaces is called *orthogonal* if $QQ^T = \mathrm{id}_V$ and $Q^TQ = \mathrm{id}_U$, where id stands for the identity map in the subscript space. The components of an orthogonal linear map in orthonormal bases of both spaces comprise an *orthogonal matrix*. A linear map T between inner-product spaces preserves the inner product if, and only if, it is an orthogonal map. By preservation of inner product, we mean that $T(\mathbf{u}) \cdot \mathbf{T}(\mathbf{v}) = \mathbf{u} \cdot \mathbf{v}, \ \forall \mathbf{u}, \mathbf{v} \in \mathrm{U}$.

10.6.2 Riemannian Manifolds

If each tangent space $T_x\mathcal{M}$ of the manifold $\mathcal{M}$ is endowed with an inner product, and if this inner product depends smoothly on $x \in \mathcal{M}$, we say that $\mathcal{M}$ is a *Riemannian manifold*. To clarify the concept of smoothness, let $\{\mathcal{U}, \phi\}$ be a chart in $\mathcal{M}$ with coordinates $x^1, \ldots, x^n$. This chart induces the (smooth) basis field $\partial/\partial x^1, \ldots, \partial/\partial x^n$. We define the *contravariant components of the metric tensor* $\mathbf{g}$ associated with the given inner product (indicated by $\cdot$) as

$$g_{ij} = \left(\frac{\partial}{\partial x^i}\right) \cdot \left(\frac{\partial}{\partial x^j}\right). \tag{10.127}$$

Smoothness means that these components are smooth functions of the coordinates within the patch. The *metric tensor* itself is given by

$$\mathbf{g} = g_{ij}\, \mathrm{d}x^i \otimes \mathrm{d}x^j. \tag{10.128}$$

We have learned how an inner product defines an isomorphism between a vector space and its dual. When translated to Riemannian manifolds, this result means that the tangent and cotangent bundles are naturally isomorphic (via the pointwise isomorphisms of the tangent and cotangent spaces induced by the inner product).

A nontrivial physical example is found in Lagrangian Mechanics, where the kinetic energy (assumed to be a positive-definite quadratic form in the generalized velocities) is used to view the configuration space Q as a Riemannian manifold.

10.6.3 Riemannian Connections

The theory of Riemannian manifolds is very rich in results. Classical differential geometry was almost exclusively devoted to their study and, more particularly, to the study of two-dimensional surfaces embedded in $\mathbb{R}^3$, where the Riemannian structure is derived from the Euclidean structure of the surrounding space.

A *Riemannian connection* is a linear connection on a Riemannian manifold. The most important basic result for Riemannian connections is contained in the following theorem:

Theorem 10.1 *On a Riemannian manifold there exists a unique linear connection with vanishing torsion and such that the covariant derivative of the metric vanishes identically.*

We omit the proof. The Christoffel symbols of this connection are given in terms of the metric tensor by the formula

$$\Gamma^k_{ij} = \frac{1}{2}\, g^{kh} \left(\frac{\partial g_{ih}}{\partial x^j} + \frac{\partial g_{jh}}{\partial x^i} - \frac{\partial g_{ij}}{\partial x^h} \right). \tag{10.129}$$

The curvature tensor associated with this special connection is called the *Riemann–Christoffel* curvature tensor. A Riemannian manifold is said to be *locally flat* if, for each point, a coordinate chart can be found such that the metric tensor components everywhere in the chart reduce to the identity matrix. It can be shown that local flatness is equivalent to the identical vanishing of the Riemann–Christoffel curvature tensor.

Further Reading

Some general treatises on Differential Geometry:

Chern, S.S., Chern, W.H., and Lam, K.S. (1999) *Lectures on Differential Geometry*, World Scientific.

Kobayashi, S. and Nomizu, K. (1996) *Foundations of Differential Geometry*, Wiley Classics Library Edition.

Lee, J.M. (2003) *Introduction to Smooth Manifolds*, Springer.

Sternberg, S. (1983) *Lectures on Differential Geometry*, 2nd edn, Chelsea Publishing Company.

Warner, F.W. (1983) *Foundations of Differentiable Manifolds and Lie Groups*, Springer.

Some books that emphasize physical applications or deal with particular physical theories in a geometric way are

Abraham, R. and Marsden, J.E. (2008) *Foundations of Mechanics*, 2nd edn, AMS Chelsea Publishing.
Arnold, V.I. (1978) *Mathematical Methods of Classical Mechanics*, Springer.
Choquet-Bruhat, Y., de Witt-Morette, C., and Dillard-Beck, M. (1977) *Analysis, Manifolds and Physics*, North-Holland.
Frankel, T. (2004) *The Geometry of Physics: An Introduction*, 2nd edn, Cambridge University Press.
Misner, W., Thorne, K.S., and Wheeler, J.A. (1973) *Gravitation*, W H Freeman and Company.

Much of the material in this article is reproduced, with permission, from

Epstein, M. (2010) *The Geometrical Language of Continuum Mechanics*, Cambridge University Press.

Part III
Analysis

Mathematical Tools for Physicists, Second Edition. Edited by Michael Grinfeld.

11
Dynamical Systems

David A.W. Barton

11.1
Introduction

11.1.1
Definition of a Dynamical System

Dynamical systems are defined using three key ingredients: a *state space*, a *time set*, and an *evolution operator*.

The *state space* X of a system is the set of all possible states of a system. For many purposes $X = \mathbb{R}^n$ is a suitable state space, though for some dynamical systems (notably partial differential equations or delay differential equations) the state space may be a more general Banach space.

The *time set* T is the set of times at which a system is defined. The two standard cases are $T = \mathbb{R}$ for continuous-time dynamical systems or $T = \mathbb{Z}$ for discrete-time systems. Only continuous-time systems are considered here, though many of the results carry through directly to discrete time systems.

The *evolution operator* forms the core of a dynamical system. The evolution operator $\phi^t : X \to X$ maps an initial state forward by $t \in T$ time units, that is, $x(t) = \phi^t x_0$ where x_0 is the initial state at time $t = 0$. In certain contexts (e.g., discrete maps), the evolution operator is given explicitly. In most contexts, however, it is usual for the evolution operator to be defined implicitly, for example, through the solution of a differential equation.

Evolution operators have two defining characteristics, namely,

$$\phi^0 x_0 = x_0 \qquad \text{("no time, no evolution")} \tag{11.1}$$

and

$$\phi^{t+s} x_0 = \phi^t(\phi^s x_0) \qquad \text{("determinism").} \tag{11.2}$$

Note that nonautonomous or stochastic systems are excluded by this definition; however, similar formalisms can be derived for those types of systems [1].

Finally, a *dynamical system* is formally defined as the triple $\{X, T, \phi^t\}$, where X is the state space, T is the time set, and ϕ^t is the evolution operator of the dynamical system.

Two further concepts, which are important in the study of dynamical systems, are orbits and ω-limit sets. For a given point in

Mathematical Tools for Physicists, Second Edition. Edited by Michael Grinfeld.

state space x_0, the *orbit* that passes through this point is given by the set $\{\phi^t(x_0) | \forall t \in T\}$. Conversely, the ω-limit set of a point x_0 is the set of points $\{\lim_{t\to\infty} \phi^t(x_0)\}$.

For a more detailed introduction to the theory of dynamical systems than this chapter permits, see [2–4].

11.1.2 Invariant Sets

The main objects of interest when studying dynamical systems are the *invariant sets* of a system, that is, the subsets of state space that are invariant under the action of the evolution operator (e.g., equilibria). The study of these invariant sets, and the associated long-time dynamics, provides extensive information about the behavior of the system without the need for determining transient behavior, which can be a difficult task.

The two basic types of invariant set considered in this chapter are as follows.

1. *Equilibria.* The simplest type of invariant set is an equilibrium $x^* \in X$ such that $x^* = \phi^t(x^*)$ for all t. A trivial example of an equilibrium can be found in the (linear) dynamical system defined by $\mathrm{d}x/\mathrm{d}t = \mu x$ for $\mu \neq 0$ (i.e., $x^* = 0$).
2. *Periodic orbits.* These are defined by a function $x(t) = f(\omega t)$, where f is a continuous and periodic (with period 2π) function that satisfies the underlying dynamical system, and ω is the frequency, of the orbit. Periodic orbits that are isolated (i.e., no points neighboring the periodic orbit are part of another periodic orbit) are also known as *limit cycles.*

This is by no means an exhaustive list. Several obvious omissions are given below.

- *Homoclinic* and *heteroclinic orbits.* These are defined by the points x such that $\phi^t(x) \to x_\omega$ as $t \to \infty$ and $\phi^t(x) \to x_\alpha$ as $t \to -\infty$. For homoclinic orbits, $x_\alpha = x_\omega$, and for heteroclinic orbits, $x_\alpha \neq x_\omega$.
- *Quasi-periodic orbits.* These are defined by a multivariate function $x(t) = f(\omega_1 t, \dots, \omega_n t)$ that satisfies the underlying dynamical system.
- *Strange attractors.* These are associated with chaotic dynamics.

However, these and other invariant sets are beyond the scope of this chapter.

This chapter focuses on the dynamics local to these invariant sets and the conditions under which the dynamics change (so-called bifurcations).

11.2 Equilibria

11.2.1 Definition and Calculation

Consider the ordinary differential equation (ODE)

$$\frac{\mathrm{d}x}{\mathrm{d}t} = f(x, \mu), \qquad x \in \mathbb{R}^n, \ \mu \in \mathbb{R}, \quad (11.3)$$

where $f : \mathbb{R}^n \times \mathbb{R} \to \mathbb{R}^n$, x is the state of the system, and μ is a system parameter. The equilibria of (11.3) are defined as the points where $\mathrm{d}x/\mathrm{d}t = 0$. As such, for a particular choice of the parameter μ, there may exist any number of equilibria. The calculation of equilibria is a root-finding problem $f(x^*, \mu) = 0$ and, for any nontrivial f, numerical root-finding methods are often needed.

11.2.2 Stability

In the context of dynamical systems, there are several different notions of stability for equilibria. The two key notions are asymptotic stability and Lyapunov stability.

An equilibrium x^* is said to be *asymptotically stable* if the ω-limit set of all the neighboring points is the single point x^*. Thus, if the system is perturbed slightly away from the equilibrium position, the system will always return to the equilibrium.

Lyapunov stability is a weaker concept. An equilibrium x^* is said to be *Lyapunov stable* if for every neighborhood U of x^* there exists a smaller neighborhood U_1 of x^* contained within U such that all solutions starting in U_1 remain within U for all time. Thus if the system is perturbed slightly away from the equilibrium position, the system will always stay in the vicinity of the equilibrium. Clearly, asymptotic stability implies Lyapunov stability.

11.2.3 Linearization

In the vicinity of an equilibrium x^*, the ODE (11.3) can be linearized to give

$$\frac{\mathrm{d}\tilde{x}}{\mathrm{d}t} = J\tilde{x}, \tag{11.4}$$

where J is the Jacobian matrix defined by

$$J = \left.\left[J_{i,j}\right]\right|_{x=x^*} = \left.\left[\frac{\partial f_i}{\partial x_j}\right]\right|_{x=x^*}. \tag{11.5}$$

The Jacobian matrix contains a great deal of information about the equilibrium, in particular, about its stability. An exponential solution of (11.4) reveals that the growth or decay of solutions is determined by the eigenvalues of J. As such, the asymptotic stability of the equilibria is determined by the sign of the real part of the eigenvalues; for an equilibrium to be stable all the eigenvalues must have negative real parts. An equilibrium with eigenvalues that have both positive and negative real parts is called a *saddle* (see, e.g., Figure 11.1b). Should any of the eigenvalues be complex, the resulting solutions will have spiraling behavior (see, e.g., Figure 11.1d and e).

If there are no eigenvalues on the imaginary axis (i.e., none with zero real part) the equilibrium is said to be *hyperbolic*.

In the case of a hyperbolic equilibrium, it is possible to formalize the link between the original nonlinear dynamical system and the linearized system through the idea of topological equivalence, as described in Section 11.2.5.

Associated with stable equilibria (and other stable invariant sets) is the notion of a *basin of attraction*, that is, the set of points in state space that, when evolved forward in time, approach the equilibrium. Formally, the basin of attraction for an equilibrium x^* is defined as

$$B(x^*) = \left\{ x : \lim_{t\to\infty} \phi^t(x) = x^* \right\}. \tag{11.6}$$

The basins of attraction for different equilibria become particularly important when considering the initial value problem of a dynamical system. The equilibrium state reached in the long-time limit (assuming the system equilibrates) is determined by the basin of attraction that the initial conditions lie in. Various methods exist for finding basins of attraction including manifold computations (see Section 11.2.6), cell mapping [5], and subdivision [6]; typically, these methods are only applied to low-dimensional systems as they do not scale well to larger systems.

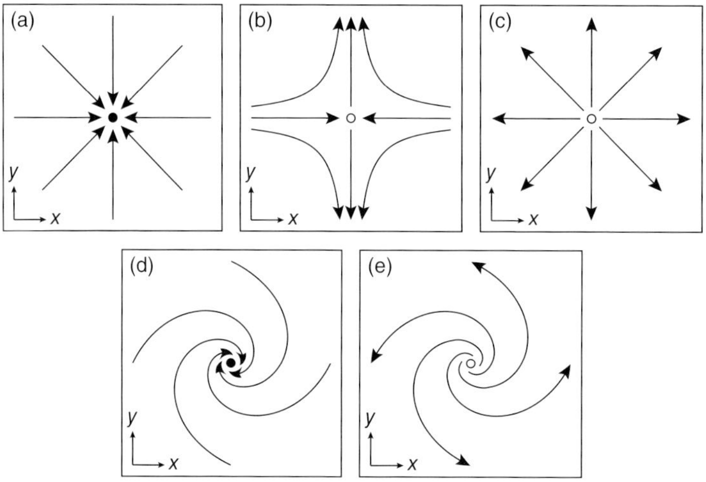

Figure 11.1 Possible phase portraits of a 2D linear dynamical system, which show the evolution in state space of several orbits. The equilibria are classified as (a) a stable node; (b) a saddle; (c) an unstable node; (d) a stable spiral; (e) an unstable spiral. Only three of the phase portraits are topologically distinct as (a) and (d) are topologically equivalent, as are (c) and (e).

11.2.4 Lyapunov Functions

While the Jacobian matrix provides information about the local stability of an equilibrium, for nontrivial systems it is often difficult, if not impossible, to calculate the eigenvalues analytically and so determine the stability. Furthermore, the Jacobian only provides a means to determine the local dynamics and not the global dynamics as is sometimes desired.

A common alternative approach to studying the stability of an equilibrium is to construct a so-called *Lyapunov function*. A Lyapunov function V is a continuous scalar function of the state variables that has the following properties (assuming the equilibrium has been shifted to the origin).

- V is positive definite, that is, $V(x) > 0 \quad \forall x \in X \backslash \{0\}$.
- $\mathrm{d}V/\mathrm{d}t$ is negative definite, that is, $(\mathrm{d}/\mathrm{d}t)V(x) < 0 \quad \forall x \in X \backslash \{0\}$.

If these conditions are satisfied, then the equilibrium is globally asymptotically stable. If the conditions are weakened such that V (respectively $\mathrm{d}V/\mathrm{d}t$) is locally positive definite (respectively locally negative definite), then the equilibrium will be locally asymptotically stable. It is possible to weaken the second condition further and require only that $\mathrm{d}V/\mathrm{d}t$ be negative semi-definite (i.e., $(\mathrm{d}/\mathrm{d}t)V(x) \leq 0 \quad \forall x \in X \backslash \{0\}$); in this case the equilibrium is said to be Lyapunov stable.

The main difficulty of using a Lyapunov function approach is that there are no universally applicable rules for constructing them; trial and error, in general, is the only way. However, for some systems, it is possible to appeal to physical conservation laws to find a Lyapunov function. One example is the Duffing equation

$$\frac{\mathrm{d}^2 y}{\mathrm{d}t^2} + \xi \frac{\mathrm{d}y}{\mathrm{d}t} + y + \alpha y^3 = 0, \qquad (11.7)$$

where the state space is $[y, \dot{y}]$. When $\xi = 0$, (11.7) can be multiplied by $\dot{y}$ and integrated with respect to time to arrive at

$$E(t) = \frac{1}{2}\left(\dot{y}^2 + y^2 + \frac{1}{2}\alpha y^4\right) = \text{const.} \quad (11.8)$$

When $\xi > 0$, $\alpha > 0$, this function is a Lyapunov function. Clearly, with the given constraints, $E(t) > 0$ for all nonzero $[y, \dot{y}]$ and is only zero at the equilibrium position $[0, 0]$. Differentiating this expression with respect to time gives

$$\frac{\mathrm{d}E}{\mathrm{d}t} = \frac{\mathrm{d}y}{\mathrm{d}t}\left(\frac{\mathrm{d}^2y}{\mathrm{d}t^2} + y + \alpha y^3\right) \quad (11.9)$$

$$= \frac{\mathrm{d}y}{\mathrm{d}t}\left(-\left(\xi\frac{\mathrm{d}y}{\mathrm{d}t} + y + \alpha y^3\right) + y + \alpha y^3\right) \quad (11.10)$$

$$= -\xi\left(\frac{\mathrm{d}y}{\mathrm{d}t}\right)^2. \quad (11.11)$$

Thus the time derivative is negative for all values of $\dot{y} > 0$. However, $\mathrm{d}E/\mathrm{d}t = 0$ regardless of the value of y when $\dot{y} = 0$. As such, the conditions for asymptotic stability are not satisfied but the equilibrium can be said to be Lyapunov stable. For $\xi > 0$, the origin is actually asymptotically stable (as can be verified by the linearization); the choice of Lyapunov function is deficient in this case.

One further benefit of using Lyapunov functions to prove stability is that Lyapunov functions can be used even when the equilibrium in question is nonhyperbolic (unlike linearization). Take for example the following system

$$\frac{\mathrm{d}x}{\mathrm{d}t} = y - x^3 \quad (11.12)$$

$$\frac{\mathrm{d}y}{\mathrm{d}t} = -x - y^3. \quad (11.13)$$

The linearization of this system yields purely imaginary eigenvalues. However, the Lyapunov function $V = (1/2)\left(x^2 + y^2\right)$ can be used to show that the equilibrium at $x, y = 0$ is asymptotically stable.

11.2.5 Topological Equivalence

Another key concept in dynamical systems is the idea of *topological equivalence*; it is a relationship between two dynamical systems that ensures that their dynamics are equivalent in a certain sense. Thus, provided certain criteria are met, the dynamics of a seemingly complicated system can be replaced by an equivalent but simpler system. Put formally, two dynamical systems are topologically equivalent if there exists a homeomorphism that maps orbits of one onto the orbits of the other, preserving the direction of time. Typically, this homeomorphism is not known explicitly but its existence can be inferred using various theorems. Furthermore, topological equivalence is not necessarily a global property, applying to the whole of state space, but instead may only be a local property, applying to the neighborhood of an equilibrium (or other invariant set).

Two important theorems relating to topological equivalence are as follows.

Equivalence of linear flows. Two linear dynamical systems are topologically equivalent if, and only if, they have the same number of eigenvalues with positive real part, the same number of eigenvalues with negative real part, and the same number of eigenvalues with zero real part.

Hartman–Grobman theorem. A (nonlinear) dynamical system is topologically equivalent to its linearization in the neighborhood of an equilibrium provided the equilibrium is hyperbolic.

Thus, the dynamics near a hyperbolic equilibrium are essentially linear.

The Hartman–Grobman theorem combined with the equivalence of linear flows is a very powerful tool when analyzing nonlinear systems. For example, the dynamics near a hyperbolic equilibrium of the 2D dynamical system

$$\begin{aligned}\dot{x} &= f(x, y)\\ \dot{y} &= g(x, y)\end{aligned} \tag{11.14}$$

will be topologically equivalent to one of the phase portraits shown in Figure 11.1 as determined by the eigenvalues of its linearization. Similar eigenvalue-based characterizations are possible in higher dimensions, but graphical representations become more difficult.

Furthermore, these two theorems ensure that, in the neighborhood of a hyperbolic equilibrium, small changes in the system parameters will not (topologically) change the dynamics. For the dynamics to change, the equilibrium must first become nonhyperbolic (i.e., one of the eigenvalues lies on the imaginary axis); in this case, the Hartman–Grobman theorem fails. When this occurs, the system is said to be *structurally unstable*, that is, small changes in the equation can cause topological changes in the local dynamics. If these small changes can be realized by varying system parameters, then the system is said to be at a *bifurcation point*.

11.2.6 Manifolds

A hyperbolic saddle-type equilibrium x_s has eigenvalues that have both negative and positive real parts and so it has directions in which it is attracting and directions in which it is repelling. As such, it possesses a *stable manifold* $W_\text{s}(x_\text{s})$, corresponding to the set of points in state space that approach the equilibrium in forward time, and an *unstable manifold* $W_\text{u}(x_\text{s})$, corresponding to the set of points in state space that approach the equilibrium in backward time. The stable and unstable manifolds are thus defined as follows

$$W_\text{s}(x_\text{s}) = \left\{ x : \lim_{t\to\infty} \phi_t(x) = x_\text{s} \right\} \quad \text{(stable manifold)} \tag{11.15}$$

and

$$W_\text{u}(x_\text{s}) = \left\{ x : \lim_{t\to-\infty} \phi_t(x) = x_\text{s} \right\} \quad \text{(unstable manifold).} \tag{11.16}$$

The dimension of the stable and unstable manifolds is equal to the number of eigenvalues with negative or positive real parts, respectively.

These manifolds are said to be *invariant manifolds*; an orbit of the dynamical system that starts in the (un)stable manifold will remain contained within the manifold for all time.

Should an equilibrium be nonhyperbolic (i.e., it has eigenvalues with zero real part), it will also possess a *center manifold*; these play a vital part when looking at *bifurcations* and will be discussed later in Section 11.2.11.

Stable manifolds act as *separatrices* in state space; that is, they form boundaries between the basins of attraction of different attractors in the system. This is most easily seen and exploited in 2D dynamical systems, see, for example, Figure 11.2, but also holds true for higher-dimensional systems. (Note that while a stable manifold can act as a separatrix, not all separatrices are stable manifolds.)

Both stable and unstable manifolds are global objects and in general can only be

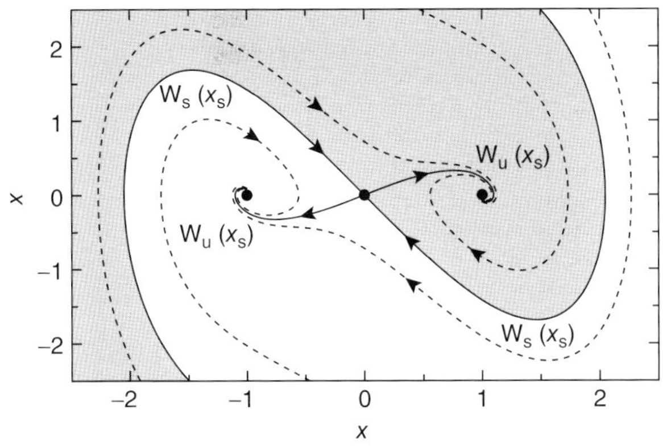

Figure 11.2 A phase portrait of the bistable Duffing equation $\ddot{x} + \dot{x} - x + x^3 = 0$ that shows how the stable manifold W_s of the saddle equilibrium acts as a separatrix, dividing the basins of attraction of the two stable equilibria. The stable and unstable manifolds of the saddle equilibrium are marked as solid curves and representative orbits are shown as dashed curves.

found numerically. However, a local (linear) approximation to the stable and unstable manifolds of a saddle equilibrium x_s is given by the eigenvectors corresponding to the stable and unstable eigenvalues, respectively.

This local approximation suggests a straightforward method for globalizing a one-dimensional unstable manifold; simply choose a starting point close to the equilibrium along the unstable eigendirection and evolve the dynamical system forward in time, that is

$$W_u(x_s) \approx \left\{\phi^t(x_s + \epsilon e_u) \quad \forall t > 0\right\}, \quad (11.17)$$

where e_u is the normalized unstable eigenvector. The accuracy of the approximation is determined by the size of ϵ and the accuracy of the numerical integrator used. A one-dimensional stable manifold can be found in a similar manner by reversing the direction of time to give

$$W_s(x_s) \approx \left\{\phi^t(x_s + \epsilon e_s) \quad \forall t < 0\right\}, \quad (11.18)$$

where e_s is the normalized stable eigenvector.

This approach of globalizing a one-dimensional manifold can naively be extended to two-dimensional manifolds (and higher) by evolving forward (or backward) in time a set of points chosen from a circle around the equilibrium in the plane spanned by the corresponding eigenvectors. This may work in simple cases, but when the manifold has nontrivial geometry or the dynamics on the manifold are not uniform (e.g., when the corresponding eigenvalues are significantly different in size), this method will not produce accurate results. More sophisticated methods are required. For a comprehensive overview of methods for calculating manifolds, see [7].

11.2.7 Local Bifurcations

Often the main interest when studying a particular dynamical system is to understand what happens to the dynamics as the system parameters change. As stated in Section 11.2.3, the dynamics in the neighborhood of an equilibrium can only change topologically if the equilibrium becomes nonhyperbolic as a parameter is varied, causing the Hartman–Grobman theorem to fail. Thus the presence of a nonhyperbolic equilibrium is a sign that the dynamical system is *structurally unstable,* that is, arbitrarily small parameter perturbations will lead to topological changes in the dynamics.

In the case of an equilibrium becoming nonhyperbolic, the loss of structural stability indicates the presence of a *local bifurcation*. The bifurcation is said to be *local* because away from the neighborhood of the equilibrium, the system dynamics will remain topologically unchanged. To bring about a global change in the dynamics, a *global bifurcation* is required (often involving the stable and unstable manifolds of a saddle equilibrium). Global bifurcations are beyond the scope of this chapter but are covered in detail in [4].

There are two generic local bifurcations of equilibria: the saddle-node bifurcation (otherwise known as a fold owing to its relationship to singularity theory) and the Hopf bifurcation. They are generic in the sense that they can be expected to occur in an arbitrary dynamical system as a single parameter is varied. They are *codimension one* bifurcations; if the system has two parameters, then there will be a one-dimensional curve of parameter values at which the bifurcation occurs within the two-dimensional parameter plane.

All other bifurcations either require special properties, such as symmetry, or are nongeneric (of codimension higher than one), that is, they can only be expected to occur in an arbitrary dynamical system when two or more parameters are varied simultaneously. Again, if the system has two parameters, then for a codimension-two bifurcation there will be a zero-dimensional point at which the bifurcation occurs within the two-dimensional parameter plane.

As many physical systems contain symmetries (at least approximately), the symmetric pitchfork bifurcation, although nongeneric, is of particular interest and so is also covered below, along with the saddle-node bifurcation and Hopf bifurcation.

11.2.8 Saddle-Node Bifurcation

At a *saddle-node bifurcation* (*fold*) two equilibria collide at a point in state space and are destroyed. This is most easily seen in the simple dynamical system defined by

$$\frac{\mathrm{d}x}{\mathrm{d}t} = \mu - x^2, \qquad \mu, x \in \mathbb{R}. \tag{11.19}$$

Equation (11.19) has two equilibria for $\mu > 0$, one stable ($x = \sqrt{\mu}$) and one unstable ($x = -\sqrt{\mu}$). For $\mu < 0$, it does not possess any equilibria; instead $x \to -\infty$ as $t \to \infty$ for all initial conditions. This is shown graphically in Figure 11.3a and b. When $\mu = 0$, the equilibria coincide at $x = 0$ and the linearization of (11.19) indicates that they are nonhyperbolic. Thus ($\mu = 0, x = 0$) is the bifurcation point.

In terms of the eigenvalues of the associated Jacobian of an equilibrium, a saddle-node bifurcation corresponds to an eigenvalue passing through zero. It should be noted that this characterization in terms of eigenvalues is not unique; in the presence of other properties (e.g., symmetry), an eigenvalue passing through zero can correspond to a transcritical or pitchfork bifurcation. As such, to be sure that a bifurcation is a saddle-node bifurcation, certain *genericity* conditions must be met.

The saddle-node bifurcation theorem is as follows.

Theorem 11.1 *Consider the one-dimensional dynamical system*

$$\frac{\mathrm{d}x}{\mathrm{d}t} = f(x, \mu), \quad x \in \mathbb{R}, \mu \in \mathbb{R}, \tag{11.20}$$

where f is smooth and has an equilibrium $x = 0$ at $\mu = 0$. Furthermore, let $\lambda = f_x(0,0) = 0$. If the genericity conditions (1) $f_{xx}(0,0) \neq 0$ and (2) $f_\mu(0,0) \neq 0$ are satisfied, then (11.20) *is topologically*

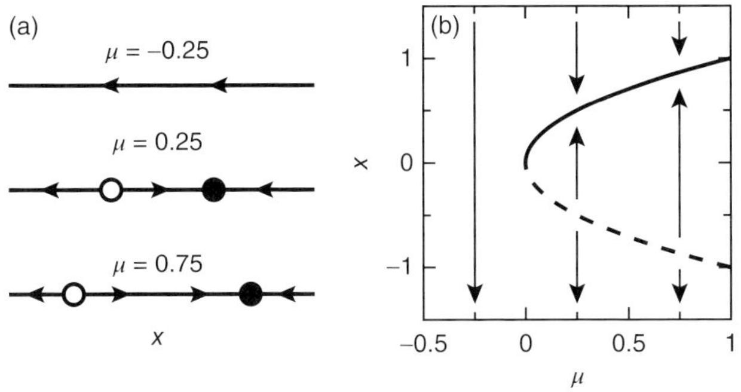

Figure 11.3 (a) A one-dimensional phase portrait for (11.19) as μ is varied; the equilibria are marked as circles and the arrows show the direction of flow. (b) The corresponding one-parameter bifurcation diagram with the saddle-node bifurcation occurring at $\mu = 0$, $x = 0$. Solid curves mark the stable equilibria and dashed lines mark the unstable equilibria.

equivalent near the origin to the normal form

$$\frac{\mathrm{d}y}{\mathrm{d}t} = \alpha \pm y^2. \qquad (11.21)$$

The power of this theorem is that the dynamics of *any* system satisfying the bifurcation conditions will be topologically equivalent to (11.21).

This theorem only applies to one-dimensional systems; however, multi-dimensional systems that have a single eigenvalue passing through zero can be reduced to a one-dimensional system using a center manifold reduction as described in Section 11.2.11.

A common tool to visualize the effect of parameter changes on the dynamics is the use of a *bifurcation diagram* whereby the equilibria (and other invariant sets) of the system are plotted using a suitable norm against the parameter value. The bifurcation diagram associated with (11.19) is shown in Figure 11.3b where it can clearly be seen that the two equilibria collide and disappear at $\mu = 0$.

Hysteresis loops in a system are often the result of a pair of saddle-node bifurcations. A common example is that of a periodically forced mass—spring–damper system with hardening spring characteristics $(\mathrm{d}^2x/\mathrm{d}t^2) + 2\xi\mathrm{d}x/\mathrm{d}t + x + \beta x^3 = \Gamma\sin(\omega t)$. A simpler, analytically treatable equation (with exact solutions) containing this type of behavior is

$$\frac{\mathrm{d}x}{\mathrm{d}t} = \mu + \sigma x - x^3. \qquad (11.22)$$

A one-parameter bifurcation diagram for (11.22) showing x plotted against μ for $\sigma = 1$ is shown in Figure 11.4a. The second parameter, σ, controls the relative positions (in terms of μ) of the saddle-node bifurcations. As $\sigma \to 0$ from above, the saddle-node bifurcations meet at a point before disappearing in a codimension-two *cusp bifurcation* as shown in the two-parameter bifurcation diagram Figure 11.4b, where μ is plotted against σ.

11.2.9 Hopf Bifurcation

At a *Hopf bifurcation* (also known as an *Andronov–Hopf bifurcation*), an equilibrium becomes unstable as a pair of complex

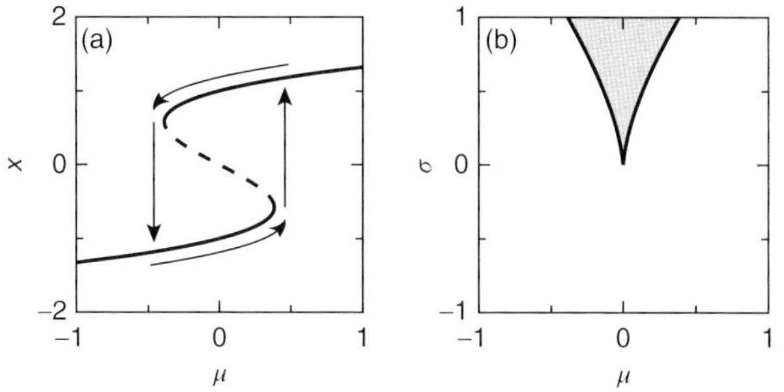

Figure 11.4 (a) A one-parameter bifurcation diagram for (11.22) as μ varies. Solid curves mark the stable equilibria and dashed lines mark the unstable equilibria. A potential hysteresis loop that occurs as μ is varied is marked by arrows. (b) A two-parameter bifurcation diagram for the same equation with the solid curve being a saddle-node bifurcation curve. A cusp bifurcation occurs at $\mu = 0$, $\sigma = 0$. Inside the gray shaded region there exist three equilibria (two stable and one unstable); everywhere else there exists only one (stable) equilibrium.

conjugate eigenvalues pass through the imaginary axis. At the bifurcation point, a limit cycle is created, which then grows in amplitude as the system parameters are changed further. Thus Hopf bifurcations are commonly associated with vibration problems, particularly the onset of vibrations. A prototype example of a Hopf bifurcation is given by

$$\frac{dx}{dt} = \mu x - \omega y + \alpha x(x^2 + y^2),$$
$$\frac{dy}{dt} = \omega x + \mu y + \alpha y(x^2 + y^2). \tag{11.23}$$

Alternatively, (11.23) can be written in complex form

$$\frac{dz}{dt} = (\mu + i\omega)z + \ell_1 |z|^2 z. \tag{11.24}$$

In both forms, μ is the bifurcation parameter, ω is the frequency of the bifurcating limit cycle (which is equal to the imaginary part of the eigenvalues passing through the imaginary axis) and ℓ_1 is a parameter known as the *first Lyapunov coefficient*, which determines the *criticality* of the bifurcation. The Hopf bifurcation can be either subcritical ($\ell_1 > 0$) or supercritical ($\ell_1 < 0$).

Figure 11.5 shows a one-parameter bifurcation diagram for (11.23) in the supercritical case; to obtain the subcritical case, simply reverse all the directions of flow indicated in the figure. Thus, in the supercritical case, a stable limit cycle is created and, in the subcritical case, an unstable limit cycle is created.

If (11.23) is rewritten in polar form, $x = r\cos(\theta)$ and $y = r\sin(\theta)$, it is clear that a Hopf bifurcation is really a pitchfork bifurcation of the equation governing the growth of r. As such, a one-parameter bifurcation diagram showing the growth in amplitude of the limit cycles that emerge from the Hopf bifurcation is shown by Figure 11.7a for the supercritical case and Figure 11.7b for the subcritical case.

In the neighborhood of a Hopf bifurcation (either sub- or supercritical), the bifurcating limit cycle is well described by a single sine function; away from the Hopf bifurcation, however, its shape can change. Furthermore, close to the bifurcation point, the amplitude of the limit cycle grows proportional to $\sqrt{\mu}$.

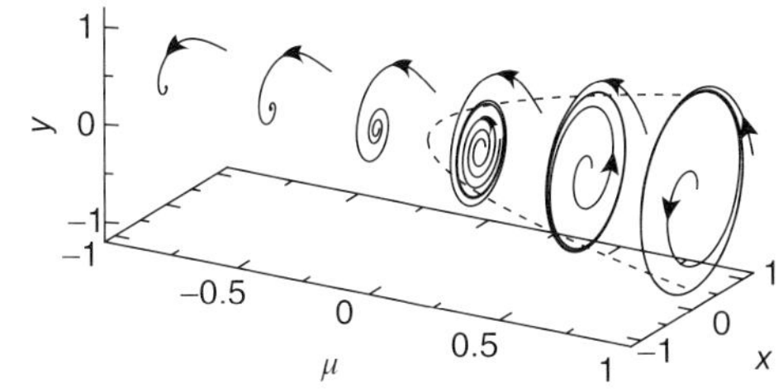

Figure 11.5 A one-parameter bifurcation diagram of (11.23) showing the onset of limit cycle oscillations as μ increases. The (supercritical) Hopf bifurcation occurs at $\mu = x = y = 0$ and the amplitude of the resulting limit cycle is shown as a dashed line.

The Hopf bifurcation theorem itself is stated below.

Theorem 11.2 *Consider the two-dimensional dynamical system*

$$\dot{x} = f(x, \mu), \quad x \in \mathbb{R}^2, \eta \in \mathbb{R}, \tag{11.25}$$

where f is smooth and has an equilibrium $x = 0$ at $\mu = 0$. Furthermore, let $\lambda_{1,2} = \sigma(\mu) \pm i\omega(\mu)$ be the eigenvalues of the Jacobian matrix such that $\sigma = 0$ when $\mu = 0$. If the genericity conditions (i) $\ell_1 \neq 0$, where ℓ_1 is the first Lyapunov coefficient [4], and (ii) $\sigma_\mu(0) \neq 0$ are satisfied, then (11.25) *has the topological normal form $z \in \mathbb{C}$*

$$\frac{dz}{dt} = (\sigma + i\omega)z + \ell_1 |z|^2 z.$$

As with the saddle-node bifurcation, higher-dimensional systems can be reduced to this two-dimensional form using a center manifold reduction.

Take for example the *van der Pol equation*

$$\frac{d^2y}{dt^2} + y = (\mu - y^2)\frac{dy}{dt} \tag{11.26}$$

where μ is the bifurcation parameter. Equation (11.26) has an equilibrium $y = 0$, which undergoes a Hopf bifurcation when $\mu = 0$ as seen by a pair of complex conjugate eigenvalues of the corresponding Jacobian passing through the imaginary axis. (Note that this is not the typical form of the van der Pol equation – if μ multiplies the y^2 term as well, a *singular Hopf bifurcation* occurs where the transition from an infinitesimal limit cycle to a large limit cycle happens over an exponentially small range of parameter values.) To determine the behavior of the limit cycle near the bifurcation point, perturbation methods can be used.

Close to the bifurcation point, the limit cycle will be small, so rescale y such that $\epsilon x = y$, which gives

$$\frac{d^2x}{dt^2} + x = (\mu - \epsilon^2 x^2)\frac{dy}{dt}. \tag{11.27}$$

Then consider small perturbations to the bifurcation parameter μ about the bifurcation point. It turns out that these small perturbations must be of order ϵ^2 for a Hopf bifurcation (doing the following analysis with μ of $O(\epsilon)$ will only find the equilibrium solution). Thus write $\mu = \epsilon^2 \tilde{\mu}$; abusing notation slightly, the tilde is immediately dropped to give

$$\frac{d^2x}{dt^2} + x = \epsilon^2(\mu - x^2)\frac{dx}{dt}. \tag{11.28}$$

As with most Hopf bifurcation problems, (11.28) is directly amenable to perturbation methods such as the Poincaré–Lindstedt method or the method of multiple scales [8, 8]. For simplicity, the Poincaré–Lindstedt approach is shown here. First, rescale time such that $\tau = \omega t$ where $\omega = 1 + \omega_1\epsilon + \omega_2\epsilon^2 + O(\epsilon^3)$; ω_1 and

ω_2 are constants to be determined. Next, assume a perturbation solution of the form $x = x_0 + x_1\epsilon + x_2\epsilon^2 + O(\epsilon^3)$. Expanding out (11.28) and collecting terms of the same order in ϵ gives the equations

$$O(1):\ \frac{d^2x_0}{d\tau^2} + x_0 = 0, \tag{11.29}$$

$$O(\epsilon):\ \frac{d^2x_1}{d\tau^2} + x_1 = -2\omega_1\frac{d^2x_0}{d\tau^2}, \tag{11.30}$$

$$O(\epsilon^2):\ \frac{d^2x_2}{d\tau^2} + x_2 = -2\omega_1\frac{d^2x_1}{d\tau^2} - \left(\omega_1^2 + 2\omega_2\right)\frac{d^2x_0}{d\tau^2} + \left(\mu - x_0^2\right)\frac{dx_0}{d\tau}. \tag{11.31}$$

With the initial conditions $x_0(0) = A$ and $x_0'(0) = 0$ (because the system is autonomous, orbits can be arbitrarily shifted to meet these conditions), (11.29) admits a solution $x_0 = A\cos(t)$. The secular terms in (11.30) (the $\cos(\tau)$ terms that arise from x_0 and lead to a resonance effect) can then be eliminated by setting $\omega_1 A = 0$; because $A = 0$ will produce a trivial solution, $\omega_1 = 0$ is used instead.

Finally, the elimination of the secular terms in (11.31) (both $\cos(\tau)$ and $\sin(\tau)$ terms) gives two constraints

$$\mu A - \frac{1}{4}A^3 = 0 \quad \text{and} \quad \omega_2 A = 0. \tag{11.32}$$

Again, seeking a nontrivial solution for A gives $A = 2\sqrt{\mu}$ and $\omega_2 = 0$. Thus, as expected, the amplitude of the limit cycle grows proportionally to $\sqrt{\mu}$. A comparison of the perturbation solution and numerical simulations give excellent agreement as shown in Figure 11.6.

11.2.10 Pitchfork Bifurcation

The *pitchfork bifurcation* is a nongeneric bifurcation (it is of codimension 3 [10]); however, for systems with $\mathbb{Z}^2$ symmetry, it becomes generic. Many physical systems possess this symmetry, at least approximately, and so it is often of interest. A common example is the buckling of a simple Euler strut.

The pitchfork bifurcation is also known as a *symmetry breaking* bifurcation because before the bifurcation the equilibrium solution is invariant under the action of the symmetry group, whereas the bifurcating solutions are not. A simple example is

$$\frac{dx}{dt} = \mu x - \alpha x^3. \tag{11.33}$$

As with a saddle-node bifurcation, a pitchfork bifurcation is characterized by an eigenvalue passing through zero. For $\alpha = +1$ and $\mu < 0$, this equation has a

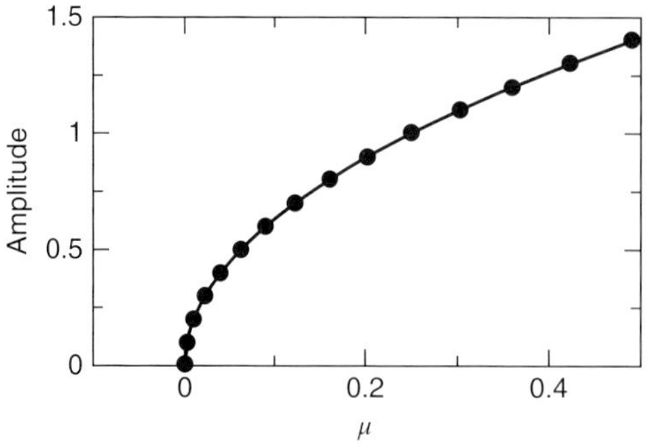

Figure 11.6 The growth of a limit cycle from the Hopf bifurcation of the van der Pol equation (11.26) as μ varies; the solid curve is the perturbation solution, whereas the dots are from numerical simulations.

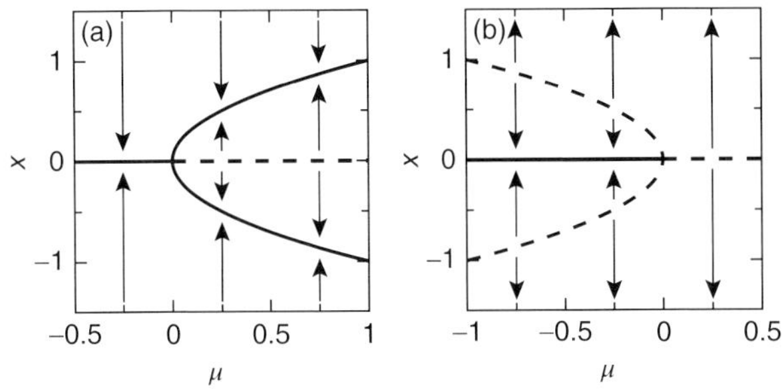

Figure 11.7 One-parameter bifurcation diagrams for (11.33) showing the existence of a pitchfork bifurcation at $\mu = 0$ and $x = 0$. Stable equilibria are denoted by solid curves and unstable equilibria are denoted by dashed curves. (a) The supercritical case ($\alpha = 1$) and panel (b) the subcritical case ($\alpha = -1$).

single (stable) equilibrium at $x = 0$ and it undergoes a pitchfork bifurcation at $\mu = 0$, which results in three equilibria, two stable ($x = \pm\sqrt{\mu}$) and one unstable ($x = 0$); this is the supercritical case. Alternatively, for $\alpha = -1$ and $\mu < 0$ this equation has three equilibria, two unstable ($x = \pm\sqrt{-\mu}$) and one stable ($x = 0$), and it undergoes a pitchfork bifurcation at $\mu = 0$, which results in a single unstable equilibrium at $x = 0$; this is the subcritical case.

The supercritical case is shown in Figure 11.7a and the subcritical case in Figure 11.7b.

The pitchfork bifurcation theorem is as follows.

Theorem 11.3 *Consider the one-dimensional dynamical system*

$$\frac{dx}{dt} = f(x, \mu), \quad x \in \mathbb{R}, \mu \in \mathbb{R}, \qquad (11.34)$$

where f is smooth, has $\mathbb{Z}^2$ symmetry and has an equilibrium $x = 0$ at $\mu = 0$. Furthermore, let $\lambda = f_x(0, 0) = 0$. If the genericity conditions (i) $f_{xxx}(0, 0) \neq 0$ and (ii) $f_{x\mu} \neq 0$ are satisfied, then (11.34) *is topologically equivalent near the origin to the normal form*

$$\frac{dy}{dt} = \alpha y \pm y^3, \qquad (11.35)$$

where the sign of the y^3 term determines the criticality of the bifurcation.

Again, higher-dimensional systems can be reduced to this one-dimensional form using a center manifold reduction.

As many physical systems only possess the $\mathbb{Z}^2$ symmetry approximately, the *unfoldings* of the pitchfork bifurcation become important. Arbitrary asymmetric perturbations to (11.33) give rise to

$$\frac{dx}{dt} = \mu x - x^3 + \alpha_1 + \alpha_2 x^2, \qquad (11.36)$$

where α_1 and α_2 are unfolding parameters. (In general, a codimension-n bifurcation requires n parameters to unfold the behavior.) The unfolding of the pitchfork bifurcation reveals the bifurcation scenarios that can occur when a system is close to having a pitchfork bifurcation. These scenarios are shown in Figure 11.8. For small values of α_1 and α_2, the pitchfork bifurcation typically unfolds into a single saddle-node bifurcation and a disconnected branch of solutions as shown in Figure 11.8a and d. However, when α_1 lies between 0 and $\alpha_2^3/27$, the pitchfork bifurcation unfolds into three saddle-node bifurcations as shown in Figure 11.8b and c.

This type of unfolding procedure is important for higher codimension bifurcations because they often act as *organizing centers* for the dynamics, that is, while the higher codimension bifurcations may

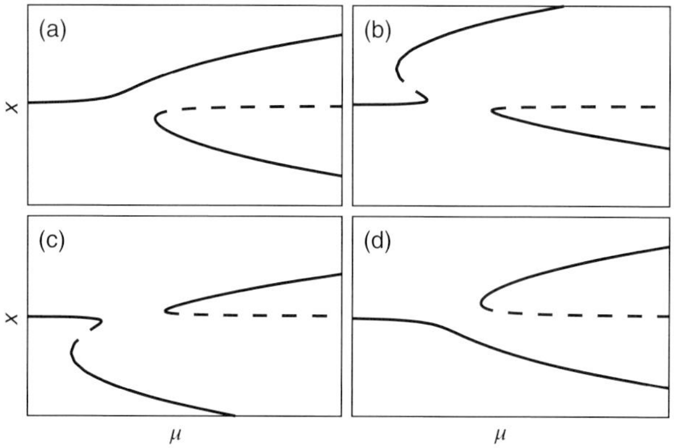

Figure 11.8 The four different unfoldings of the pitchfork bifurcation. In (a) and (d) the pitchfork bifurcation unfolds into a single saddle-node bifurcation. In contrast, in (b) and (c) it unfolds into three saddle-node bifurcations. The parameter ranges in (11.36) corresponding to the different unfoldings are (a) $\alpha_1 > 0$ and $\alpha_1 > \alpha_2^3/27$, (b) $\alpha_1 > 0$ and $\alpha_1 < \alpha_2^3/27$, (c) $\alpha_1 < 0$ and $\alpha_1 > \alpha_2^3/27$, (d) $\alpha_1 < 0$ and $\alpha_1 < \alpha_2^3/27$.

never be seen in the system of interest, their proximity may be obvious.

11.2.11 Center Manifolds

Consider the two systems of differential equations

$$\frac{dx}{dt} = xy, \qquad \frac{dy}{dt} = -y - x^2, \tag{11.37}$$

and

$$\frac{dx}{dt} = x^2y - x^5, \qquad \frac{dy}{dt} = -y + x^2. \tag{11.38}$$

Both of these systems have an equilibrium at $x = y = 0$ and the corresponding Jacobian matrices are identical:

$$J = \begin{bmatrix} 0 & 0 \\ 0 & -1 \end{bmatrix}. \tag{11.39}$$

As J has an eigenvalue at 0, the Hartman–Grobman theorem fails. Thus the stability of the equilibria cannot be deduced from the Jacobian. In this case, the equilibrium of (11.37) is asymptotically stable, whereas the equilibrium of (11.38) is asymptotically unstable.

Center manifold theory is the means by which the stability of (11.37) and (11.38) can be determined.

Consider the system

$$\frac{dx}{dt} = Ax + f(x, y) \qquad \frac{dy}{dt} = By + g(x, y), \tag{11.40}$$

where $x \in \mathbb{R}^n$, $y \in \mathbb{R}^m$ and A and B are matrices such that the eigenvalues of A have zero real parts and the eigenvalues of B have negative real parts.

To put a given system into the form of (11.40) is simply a case of applying a sequence of coordinate transformations that first shift the equilibrium of interest to the origin and then second put the system into Jordan normal form.

If $f \equiv g \equiv 0$, then (11.40) contains two trivial invariant manifolds: $x = 0$ is the stable manifold of the equilibrium, and $y = 0$ is the center manifold of the equilibrium. Furthermore, the dynamics of the system will collapse exponentially quickly onto the center manifold and thus the dynamics on the center manifold will determine the stability of the equilibrium.

In general, with nonzero f and g, there exists a center manifold of (11.40) that can be written as $y = h(x)$ where $h(0) = 0$ and $h'(0) = 0$ [11]. While this does not permit a direct solution for the dynamics on the center manifold, it does permit a polynomial approximation to be calculated to any desired order.

Take (11.37) and consider the polynomial expansion of h to fifth order, that is, $y = h(x) = ax^2 + bx^3 + cx^4 + dx^5 + O(x^6)$. Substituting this into $\mathrm{d}y/\mathrm{d}t$ gives

$$\begin{aligned} \frac{\mathrm{d}y}{\mathrm{d}t} &= h'(x)\frac{\mathrm{d}x}{\mathrm{d}t} = xh'(x)h(x), \\ &= 2a^2x^4 + 5abx^5 + O(x^6). \end{aligned} \qquad (11.41)$$

Similarly, substituting the same expansion into the right-hand side of the differential equation governing y in (11.37) gives

$$\begin{aligned} \frac{\mathrm{d}y}{\mathrm{d}t} &= -h(x) - x^2, \\ &= -(a+1)x^2 - bx^3 - cx^4 - dx^5 + O(x^6). \end{aligned} \qquad (11.42)$$

Equating the coefficients of both expressions leads to $a = -1$, $b = 0$, $c = -2$, and $d = 0$. Thus the center manifold approximation is $y = h(x) = -x^2 - 2x^4 + O(x^6)$ and the resulting dynamics on the center manifold are

$$\frac{\mathrm{d}x}{\mathrm{d}t} = -x^3 - 2x^5 + O(x^7). \qquad (11.43)$$

Hence the equilibrium at $x = 0$, $y = 0$ is asymptotically stable.

A similar calculation can be performed for (11.38) to show that the equilibrium is asymptotically unstable.

11.3 Limit Cycles

11.3.1 Definition and Calculation

Once again consider the ODE

$$\frac{\mathrm{d}x}{\mathrm{d}t} = f(x, \mu), \qquad x \in \mathbb{R}^n,\ \mu \in \mathbb{R}, \quad (11.44)$$

where $f : \mathbb{R}^n \times \mathbb{R} \to \mathbb{R}^n$. Many of the concepts developed for equilibria of (11.44) apply directly to limit cycles of (11.44) as well, although the underlying proofs of the corresponding theorems may be quite different.

A limit cycle is an isolated closed orbit of (11.44) that contains no rest points (points where $\mathrm{d}x/\mathrm{d}t = 0$). Any solution that passes through a point on the limit cycle is a periodic solution and has the property that $x(t + T) = x(t)$ for all t, where the period is T.

As limit cycles are a function of time, they are typically more difficult to calculate than equilibria. However, in the case of weak nonlinearities, perturbation methods may give a reasonable approximation to a limit cycle, for example the method of multiple scales is often used to good effect. (In particular, when analyzing the limit cycles emerging from a Hopf bifurcation; see for example, Section 11.2.9.)

11.3.1.1 Harmonic Balance Method

Another analytical approach is the *harmonic balance method*, whereby a harmonic solution to the equations of motion is postulated, substituted into the ODE of interest, and the resulting expression expanded in terms of $\sin(n\omega t)$ and $\cos(n\omega t)$. The orthogonality of the sin and cos terms over one period gives a number of algebraic equations that can then be solved. For example, consider the periodically forced Duffing equation

$$y'' + \mu y' + \alpha y + \beta y^3 = \Gamma \cos(\omega t). \quad (11.45)$$

Assume a solution in the form of a truncated Fourier series with coefficients that slowly vary with time $y = p(t) + a(t)\cos(\omega t) + b(t)\sin(\omega t)$. The first and second derivatives become

$$\begin{aligned} y' &= p' + \left(a' + b\omega\right)\cos(\omega t) \\ &\quad + \left(b' - a\omega\right)\sin(\omega t), \end{aligned} \quad (11.46a)$$

$$\begin{aligned} y'' &= p'' + \left(2b' - a\omega\right)\omega\cos(\omega t) \\ &\quad + \left(b\omega - 2a'\right)\omega\sin(\omega t), \end{aligned} \quad (11.46b)$$

where it is presumed that $p'' = a'' = b'' = 0$ owing to the slowly varying approximation. The expressions for y, y', and y'' are then substituted into (11.45) and the constant coefficients, along with the coefficients of $\cos(\omega t)$ and $\sin(\omega t)$, are balanced on each side. Balancing the constant coefficients gives

$$\mu p' = -p\left(\beta p^2 + \alpha + \frac{3}{2}\beta r^2\right), \quad (11.47)$$

and balancing the coefficients of $\cos(\omega t)$ and $\sin(\omega t)$ gives

$$\begin{pmatrix} \mu & 2\omega \\ -2\omega & \mu \end{pmatrix}\begin{pmatrix} a' \\ b' \end{pmatrix} = \begin{pmatrix} a\left(\omega^2 - \alpha - 3\beta p^2 - \frac{3}{4}\beta r^2\right) - \mu\omega b + \Gamma \\ b\left(\omega^2 - \alpha - 3\beta p^2 - \frac{3}{4}\beta r^2\right) + \mu\omega a \end{pmatrix}, \quad (11.48)$$

where $r^2 = a^2 + b^2$. The steady-state response, $\tilde{p}, \tilde{r}$, is obtained from the equilibria of (11.47) and (11.48), which requires $a' = b' = p' = 0$; squaring and summing the result from (11.48) gives the frequency response equation

$$\left[\left(\omega^2 + \alpha - 3\beta\tilde{p}^2 - \frac{3}{4}\beta\tilde{r}^2\right)^2 + (\mu\omega)^2\right]\tilde{r}^2 = \Gamma^2, \quad (11.49)$$

where $\tilde{p}$ may take on multiple values. More specifically, after setting $p' = 0$ in (11.47), the steady-state solution $\tilde{p}$ reveals that both $\tilde{p} = 0$ and $\tilde{p}^2 = -\alpha/\beta - 3/2\tilde{r}^2$ are equilibria. When the Duffing equation is bistable, where $\alpha = -1$ and $\beta = 1$, the latter solution is interesting because it restricts the values of $\tilde{r}$ that provide a physical solution for $\tilde{p}$ to $\tilde{r}^2 \leq 2/3$.

The harmonic balance method is particularly useful in the case of the bistable Duffing equation because this is a situation in which perturbation methods are not applicable owing to the strength of the nonlinearity. However, higher-order harmonic approximations often result in algebraic equations that can only be solved numerically and so the harmonic balance method is often limited to the first-order equations.

In general, when dealing with strong nonlinearities or limit cycles that are far from harmonic, numerical methods are the way forward. There are many different methods for calculating limit cycles numerically; two common methods discussed here are numerical shooting and collocation. Both methods treat the problem of calculating a limit cycle as a periodic boundary value problem (BVP).

11.3.1.2 **Numerical Shooting**

Numerical shooting is a straightforward method for solving BVPs where a numerical integrator is available to calculate solutions of (11.44) for given initial conditions. For a limit cycle, the BVP of interest is

$$\frac{\mathrm{d}x}{\mathrm{d}t} = f(x, \mu), \qquad \text{with} \qquad x(0) - x(T) = 0, \tag{11.50}$$

where the period of the limit cycle is T. However, (11.50) is not directly amenable to numerical methods. To calculate a particular limit cycle, time in (11.50) must first be rescaled so that the period T appears as an explicit parameter to be determined. This results in

$$\frac{\mathrm{d}x}{\mathrm{d}t} = Tf(x, \mu), \qquad \text{with} \qquad x(0) - x(1) = 0. \tag{11.51}$$

Furthermore, owing to the autonomous nature of (11.51), there exists a one-parameter family of solutions that are invariant under time (phase) shifts. As such a *phase condition* is needed to restrict the problem to finding an isolated limit cycle and so ensure the correct behavior of the numerical method.

There are a wide range of possible phase conditions. For simple implementations, it may be desirable to use known properties of the system of interest. For example, if all limit cycles of interest pass through the point $x = C$, then the phase condition $x(0) = C$ may be used. A more robust, but harder to implement, condition is $x(0) = x(1/2)$ which holds for all limit cycles; however, this condition is still not completely robust and may become singular.

The most common and robust phase condition is the integral condition

$$\int_0^1 \frac{\mathrm{d}\hat{x}}{\mathrm{d}t} \cdot [x(t) - \hat{x}(t)]\mathrm{d}t = 0, \tag{11.52}$$

which minimizes the $\mathcal{L}_2$ distance between the limit cycle and a reference limit cycle $\hat{x}$ (typically taken as the last successfully computed limit cycle). The condition (11.52) is implemented in many bifurcation analysis packages.

The final nonlinear BVP to be solved is the system formed by (11.51) and (11.52) where the time series of x is calculated by numerical integration from the initial condition $x(0)$. The BVP is then solved using a nonlinear root-finding method such as a Newton iteration starting with an initial guess $\{x^{(0)}(0), T_0\}$:

$$\begin{bmatrix} x^{(i+1)}(0) \\ T^{(i+1)} \end{bmatrix} = \begin{bmatrix} x^{(i)}(0) \\ T^{(i)} \end{bmatrix} - J^{-1} \begin{bmatrix} x^{(i)}(0) - x^{(i)}(1) \\ \int_0^1 \frac{\mathrm{d}\hat{x}}{\mathrm{d}t} \cdot [x^{(i)}(t) - \hat{x}(t)]\mathrm{d}t \end{bmatrix}, \tag{11.53}$$

where J is the corresponding Jacobian matrix, which can be approximated by a finite-difference scheme or calculated directly from the variational equations (see Section 11.3.2). The integral phase condition can be discretized using a simple scheme such as the trapezoid method without a loss of accuracy.

Numerical shooting is suitable for both stable and unstable limit cycles, though increasingly better initial guesses are required as the limit cycle becomes more unstable. To overcome stability problems, *multiple shooting* is often used.

11.3.1.3 **Collocation**

Collocation is another method for solving BVPs of the form (11.51). Similar to numerical shooting, a phase condition (similar to (11.52)) is required. Collocation differs from numerical shooting in how the time series is computed/represented.

In collocation, the time series is approximated as a series expansion in terms of certain basis functions that can be differentiated exactly, for example, $x(t) = \sum_{i=0}^{n} x_i P_i(t)$ in the case of Lagrange polynomials. Common examples include harmonic functions (Fourier series), piecewise-defined Lagrange polynomials, and globally defined Chebyshev polynomials. This series expansion is then substituted into the system of interest and evaluated at discrete-time points (*collocation points*). The idea is to determine the coefficients of the basis functions (using a nonlinear root finder) such that, at these collocation points, the series expansion of the limit cycle exactly satisfies (11.51) and (11.52); in general, these are the only points where they are satisfied exactly and so globally the series expansion is an approximate solution.

(In contrast, a Galerkin-based method would use a similar series expansion of the solution but, instead of pointwise evaluations of (11.51) and (11.52), integrals over the whole period would be calculated. When the integrals can be calculated analytically, this has benefits over collocation. However, when the integrals must be calculated numerically using quadrature, these benefits disappear.)

For more details see [12] or [13].

11.3.2 Linearization

Following the treatment of equilibria in Section 11.2.3, the dynamical system defined by (11.44) can similarly be linearized in the vicinity of a limit cycle $x^*(t)$. The linearization is inherently time varying and is given by

$$\frac{\mathrm{d}\tilde{x}}{\mathrm{d}t} = J(t)\tilde{x}(t), \tag{11.54}$$

where J is the (time-varying) Jacobian given by

$$J(t) = \left[J_{i,j}(t)\right]\Big|_{x(t)=x^*(t)} = \left[\frac{\partial f_i}{\partial x_j}\right]\Bigg|_{x(t)=x^*(t)}. \tag{11.55}$$

Equation (11.54) is also known as the *first variational equation* and can be derived formally by considering the growth/decay of small perturbations to the limit cycle.

To determine the stability of a limit cycle with period T, a suitable discrete time map is constructed, the fixed points of which correspond to the limit cycles of the original system. There are two methods for constructing the discrete time map (both equivalent); these are illustrated in Figure 11.9. Either consider the time-T map that evolves the point $x(t)$ to $x(t+T)$, or alternatively consider the dynamics of a map from an arbitrary section that intersects with the limit cycle, back to itself (a so-called *Poincaré map* and *Poincaré section*). When either of the maps are linearized, the end result is a map of the form $\tilde{x}^{(i+1)} = M\tilde{x}^{(i)}$ where M is the Jacobian matrix of the map; M is also known as the *monodromy matrix*. The stability of the fixed point (and correspondingly the limit cycle) is determined by the eigenvalues of M, which are known as *Floquet multipliers*.

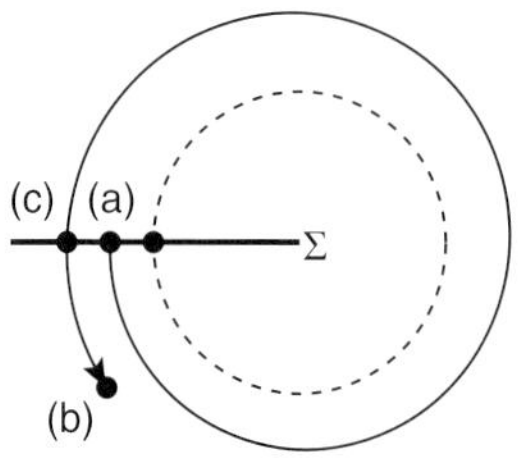

Figure 11.9 A schematic of the discrete-time maps associated with the linearization of a limit cycle. Point (a) is a starting point near a given limit cycle (marked by the dashed closed curve). It is then either evolved forward in time by T time units to point (b) as with the time-T map, or it is evolved forward in time until it hits the Poincaré section Σ once more at point (c) as with the Poincaré map.

A limit cycle is linearly stable if all the Floquet multipliers lie inside the unit circle in the complex plane. Furthermore, a limit cycle is *hyperbolic* if no Floquet multipliers lie on the unit circle.

(Note that the time-T map is a mapping from $\mathbb{R}^n$ to $\mathbb{R}^n$, whereas the Poincaré map is a mapping from $\mathbb{R}^{n-1}$ to $\mathbb{R}^{n-1}$; the extra Floquet multiplier associated with the time-T map is a so-called trivial multiplier and will always be $+1$ because the limit cycle can be phase-shifted arbitrarily.)

In the case of the time-T map, the monodromy matrix M can be calculated by integrating the first variational equation (11.54) with the initial condition $\tilde{x}(0) = I$ where I is the $n \times n$ identity matrix. The matrix M is then the matrix $\tilde{x}(T)$.

11.3.3 Topological Equivalence

In the vicinity of a hyperbolic equilibrium, the notion of topological equivalence combined with the Hartman–Grobman theorem is a powerful one. It states that the dynamics near an equilibrium are equivalent to those of the corresponding linearization. The straightforward calculation of stability is one of the possibilities that emerges from this result.

For limit cycles, a similar proposition holds; the dynamics in the neighborhood of a hyperbolic limit cycle (none of its Floquet multipliers lie on the unit circle) are topologically equivalent to the linearized dynamics given by the first variational equation. This follows from the discrete-time version of the Hartman–Grobman theorem for the fixed points of a map, because for any limit cycle, a time-T map or a Poincaré map can be defined as described in Section 11.3.2.

11.3.4 Manifolds

By analogy with equilibria (see (11.15) and (11.16)), stable and unstable manifolds for a limit cycle can be defined as the set of points that approach the limit cycle in forward time and backward time, respectively.

An important difference between the manifolds of an equilibrium and the manifolds of a limit cycle is their respective dimensions. For an equilibrium, the dimension of the stable manifold (unstable manifold) is given by the number of eigenvalues with negative (positive) real parts. For a limit cycle in an autonomous system, the dimensions of the stable and unstable manifolds are increased by one due to the presence of the trivial multiplier at $+1$.

Figure 11.10 shows the two possible configurations for the stable and unstable manifolds of a saddle-type limit cycle in three-dimensional space. The first case

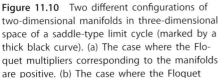

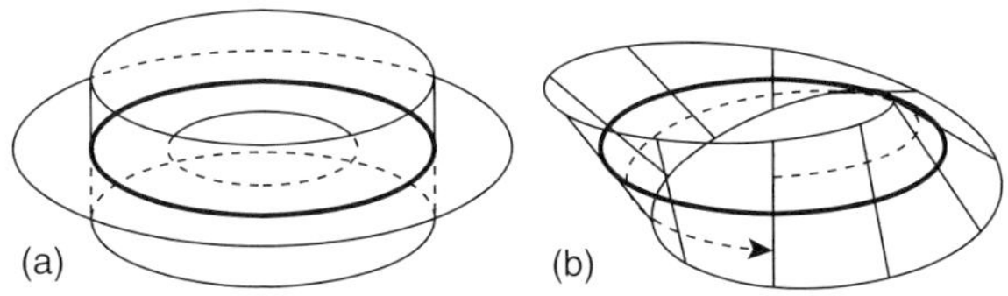

Figure 11.10 Two different configurations of two-dimensional manifolds in three-dimensional space of a saddle-type limit cycle (marked by a thick black curve). (a) The case where the Floquet multipliers corresponding to the manifolds are positive. (b) The case where the Floquet multipliers are negative and the resulting manifold is twisted, having the form of a Möbius strip. (Only one manifold is shown in (b) for clarity.) A representative orbit is marked in (b) by a dashed curve.

is when the Floquet multipliers of both manifolds are positive; the manifolds form simple bands around the limit cycle. The second case is when the Floquet multipliers of both manifolds are negative; in this case, both the manifolds form twisted bands around the limit cycle.

It should be noted that it is not possible (in three dimensions) to have a twisted stable (unstable) manifold and an untwisted unstable (stable) manifold. This can proven directly from Liouville's formula, which ensures that the product of all the Floquet multipliers is positive.

A local (linear) approximations to the stable and unstable manifolds at a single point on the limit cycle are given by the eigenvectors of the monodromy matrix. These local approximations can then be extended to the complete limit cycle by integrating forward the first variational equation (11.54) for time T, using the (normalized) eigenvector as the initial condition. Globalization of the manifolds can then be achieved using the same numerical methods as mentioned in Section 11.2.6.

11.3.5 Local Bifurcations

It follows from the discrete-time Hartman–Grobman theorem that the dynamics in the neighborhood of a limit cycle can only change qualitatively (i.e., topologically) if the limit cycle becomes nonhyperbolic; that is, a Floquet multiplier lies on the unit circle. Thus, as with equilibria, a nonhyperbolic limit cycle signifies a local bifurcation.

Limit cycles can undergo the same basic bifurcations as equilibria can. In particular, the saddle-node and pitchfork bifurcations are essentially the same with limit cycles taking the place of the equilibria. These occur when a Floquet multiplier passes through the unit circle in the complex plane at +1. (Again, the pitchfork bifurcation requires $\mathbb{Z}^2$ symmetry for it to be a generic bifurcation.)

Hopf bifurcations (called *secondary Hopf bifurcations* or *Neimark-Sacker bifurcations*) of limit cycles also occur. They are associated with the onset of quasi-periodic motion (the existence of an invariant torus) and they occur when a complex conjugate pair of eigenvalues passes through the unit circle. The details of the bifurcation are significantly complicated if strong resonances occur between the frequency of the underlying limit cycle and the frequency of the bifurcation. The details of this bifurcation are beyond the scope of this chapter and the interested reader is referred to [4].

Finally, limit cycles can undergo one final (generic) local bifurcation known as a *period-doubling bifurcation*, which occurs when a single Floquet multiplier passes through the unit circle at −1. Period-doubling bifurcations are often associated with the onset of chaos. In particular, a period-doubling cascade (a sequence of period-doubling bifurcations one after the other) is a well known route to chaos.

11.3.6 Period-Doubling Bifurcation

At a period-doubling bifurcation, a limit cycle becomes nonhyperbolic as a Floquet multiplier passes through the unit circle in the complex plane at −1. At the bifurcation point, a new limit cycle is created that has a period of twice that of the original limit cycle and, furthermore, the original limit cycle changes stability.

As with Hopf and pitchfork bifurcations, a period-doubling bifurcation can be sub- or supercritical. In the subcritical case, an unstable limit cycle becomes stable at the bifurcation point and coexists with an unstable period-doubled limit cycle. In the supercritical case, a stable limit cycle becomes unstable at the bifurcation point and coexists with a stable period-doubled limit cycle.

Period-doubling bifurcations are often a precursor to chaos. For example, the Rössler system,

$$\begin{aligned}\frac{dx}{dt} &= -y - z,\\ \frac{dy}{dt} &= x + ay,\\ \frac{dz}{dt} &= b + z(x - c),\end{aligned} \tag{11.56}$$

becomes chaotic through a sequence of supercritical period-doubling bifurcations as shown in Figure 11.11. It should be noted that an infinite number of period-doubling bifurcations exist in a finite parameter interval. The parameter values at which the period-doubling bifurcations occur tend toward a geometric progression as the period increases; the ratio of the distances between successive period-doubling bifurcations is known as *Feigenbaum's constant.*

In the midst of the chaotic attractor shown in Figure 11.11, a period-3 orbit becomes stable. It then also undergoes a period-doubling sequence before the system becomes fully chaotic again. Windows of periodic behavior like this are typical in chaotic attractors of smooth dynamical systems.

11.4 Numerical Continuation

11.4.1 Natural Parameter Continuation

The preceding sections have expounded the different dynamical effects that can occur as parameter values are changed in a system. If the system of interest is sufficiently simple, it is enough to find the equilibria analytically and compute quantities such as the Jacobian to determine when bifurcations occur.

Many nontrivial systems are not amenable to such analysis and require the application of numerical methods from the start. In such cases, numerical simulation is often the first recourse, swiftly followed by brute-force bifurcation diagrams such as Figure 11.11 in Section 11.3.6. A great deal of information can be gathered using these simple techniques.

However, such use of numerical simulation does not reveal much about the nature of the bifurcations and transitions that occur as the parameters are varied because unstable invariant sets are missed. It is branches of these unstable invariant

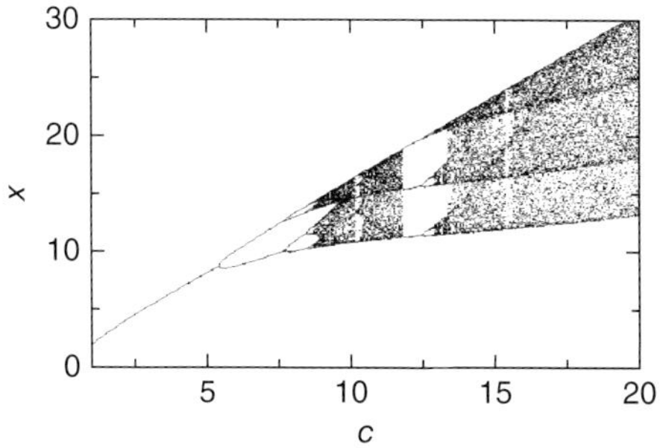

Figure 11.11 A one-parameter brute-force bifurcation diagram of the Rössler system (11.56) for $a = b = 0.1$ showing a period-doubling route to chaos. At each parameter value, the system is simulated until it reaches steady state and then the values of x at which the orbit intersects the Poincaré map defined by $dx/dt = 0$ are recorded.

sets that connect the different branches of stable invariant sets and so reveal what bifurcations are occurring.

Numerical continuation is a means to overcome these difficulties. At heart, it is a path-following method that tracks the solutions of

$$f(x, \mu) = 0, \quad x \in \mathbb{R}^n, \mu \in \mathbb{R}^p, \qquad (11.57)$$

as the system parameters μ vary. This type of problem arises naturally from equilibrium problems but many other types of problem can also be put into this form, for example, the calculation of limit cycles using numerical shooting or collocation.

Numerical continuation relies on the implicit function theorem, which states that the solutions of (11.57) can be written explicitly as $x = g(\mu)$ provided the matrix of partial derivatives of f with respect to x is not singular (i.e., the system is not at a bifurcation point). Thus, when the implicit function theorem holds, a continuous branch of solutions can be tracked with respect to μ.

A simple version of numerical continuation is an iterative, predictor-corrector algorithm called *natural parameter continuation*. The algorithm is as follows. To start the algorithm an initial point $\{x_0, \mu_0\}$ is required, which can be generated using numerical simulation (or some other method).

1. Take a step in the system parameter of interest to get $\mu_{i+1} = \mu_i + \Delta$.
2. Generate a predicted solution $\tilde{x}_{i+1}$ of (11.57) for $\mu = \mu_{i+1}$ from the previous solution x_i.
3. Generate a corrected solution x_{i+1} by using a nonlinear root finder (e.g., a Newton iteration) on $f(x, \mu_{i+1}) = 0$ using $\tilde{x}_{i+1}$ as an initial guess for the root finder.

Repeat this process until the desired parameter range is covered; Δ should be sufficiently small to ensure convergence of the nonlinear root finder and to reveal the desired level of detail in the solution branch.

As the calculation of solutions uses a nonlinear root finder rather than numerical simulation, it does not matter whether the solutions are stable or unstable. In fact, to determine the stability of equilibria, the Jacobian matrix must be computed separately. Similarly, for limit cycles the calculation of the monodromy matrix is also a separate step. (In practice, it is often possible to use the Jacobian previously calculated by the Newton iteration for stability information.)

There are a number of different ways to generate predicted solutions. Two common ways are *tangent prediction* and *secant prediction*. Tangent prediction calculates the local derivative $v = f_\mu(x_i, \mu_i)$ and applies the prediction $\tilde{x}_{i+1} = x_i + \Delta v$. Secant prediction, on the other hand, estimates the local derivative using a secant estimation, that is, $\tilde{v} = (x_i - x_{i-1})/(\mu_i - \mu_{i-1})$, and then calculates the predicted solution as $\tilde{x}_{i+1} = x_i + \Delta\tilde{v}$. Secant prediction is simpler in that it does not require any derivative information; however, it requires two initial points to start the algorithm.

Figure 11.12 shows an example of natural parameter continuation using the Rössler system (compare with Figure 11.11). The limit cycle is discretized using collocation.

Natural parameter continuation is inherently flawed. As it relies directly on the implicit function theorem, it fails at saddle-node bifurcations because past

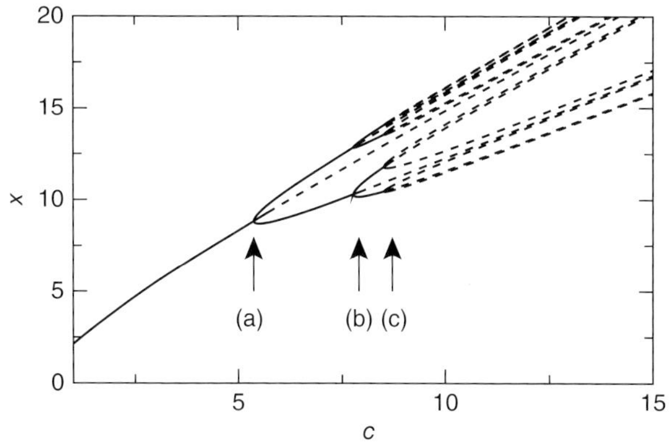

Figure 11.12 Numerical continuation of limit cycles in the Rössler system. A series of period-doubling bifurcations occur and the bifurcating branches are continued up to the period 8 limit cycle. The first three period-doubling bifurcations are marked by (a), (b) and (c). (Compare with Figure 11.11.)

the bifurcation point, there is no solution for x. Instead, a different approach is required.

11.4.2 Pseudo-Arc-Length Continuation

Pseudo-arc-length continuation overcomes the difficulties posed by saddle-node bifurcations by reparameterizing the zero problem (11.57) in terms of the arc length along the solution curve. By doing so, the implicit function theorem no longer fails at saddle-node bifurcations. Thus the zero problem becomes

$$f(x(s), \mu(s)) = 0, \quad x \in \mathbb{R}^n, \mu \in \mathbb{R}^p, \quad (11.58)$$

and both x and μ are solved for simultaneously in terms of the arc length. To do this, an additional equation is required (because μ is now an additional unknown). A true arc-length parameterization would require the addition of the nonlinear algebraic equation $(\mathrm{d}s)^2 = (\mathrm{d}x)^2 + (\mathrm{d}\mu)^2$; in practice it is better to use a linear approximation to this nonlinear equation, namely, the *pseudo-arc-length* equation

$$(\hat{x}')^T(x - \hat{x}) + (\hat{\mu}')^T(\mu - \hat{\mu}) = \Delta, \quad (11.59)$$

where $\hat{x}$ and $\hat{\mu}$ are previously computed solutions and the primes $()'$ denote differentiation with respect to arc length. (Secant approximations to the derivatives with respect to arc-length work as well as the actual derivatives.)

The pseudo-arc-length continuation algorithm is thus as follows. (Use $\hat{x} = x_i$ and $\hat{\mu} = \mu_i$ in (11.59).)

1. Generate a predicted solution $\{\tilde{x}_{i+1}, \tilde{\mu}_{i+1}\}$ of (11.58) and (11.59) from the previous solution $\{x_i, \mu_i\}$.
2. Generate a corrected solution $\{x_{i+1}, \mu_{i+1}\}$ by using a nonlinear root finder (e.g., a Newton iteration) on the combined system of (11.58) and (11.59) using $\{\tilde{x}_{i+1}, \tilde{\mu}_{i+1}\}$ as an initial guess for the root finder.

This is illustrated in contrast to natural parameter continuation in Figure 11.13. The prediction step is the same as before except that derivatives are now with respect to the arc length and predictions are needed for μ_{i+1} as well as x_{i+1}.

11.4.3 Continuation of Bifurcations

As well as tracking invariant sets of a dynamical system in general, it is also possible to track bifurcations in terms of

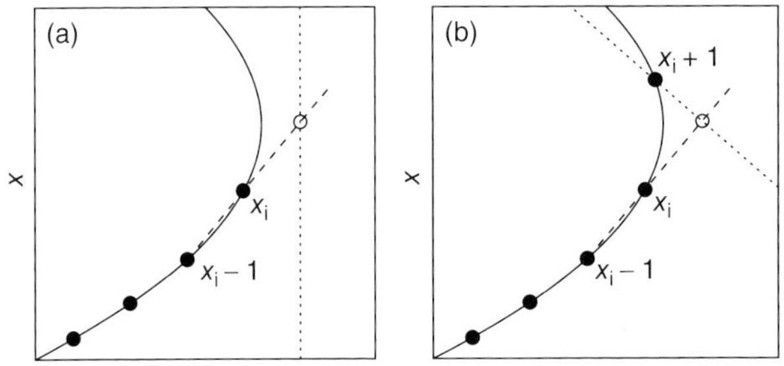

Figure 11.13 (a), (b) Natural parameter continuation and pseudo-arc-length continuation, respectively, close to a saddle-node bifurcation. Predicted solutions are denoted by open circles and corrected solutions are denoted by dots. The dashed lines denote the tangent line x_i' (using a secant approximation) and the dotted lines indicate the search direction of the nonlinear root finder. The addition of the pseudo-arc-length equation (11.59) in (b) skews the search direction in order to pass around the saddle-node bifurcation.

the system parameters. For example, it is possible to track a saddle-node bifurcation as two parameters vary to trace out bifurcation diagrams as in Figure 11.4. All that is required is to augment the zero problem of (11.58) and (11.59) by a third equation that encompasses the bifurcation condition. This third equation is known as a bifurcation *test function*.

A saddle-node bifurcation of equilibria occurs when an eigenvalue of the Jacobian matrix passes through zero. As such, a suitable test function is

$$\det(J) = 0, \tag{11.60}$$

where J is the Jacobian matrix associated with the equilibrium. While this is conceptually a simple test function, for numerical purposes, it is less useful because derivatives of this function are not straightforward to calculate. (A finite-difference approximation can be used but this can be computationally expensive to calculate for large systems.) Instead, a better approach is to track the eigenvector corresponding to the zero eigenvalue. As eigenvectors can be scaled arbitrarily, a further condition is required to normalize the eigenvector. Thus the system of equations needed to continue saddle-node bifurcations of equilibria is formed from (11.58), (11.59) and

$$\begin{aligned} Jv &= 0, \\ \|v\| &= 1, \end{aligned} \tag{11.61}$$

where v is the eigenvector corresponding to the zero eigenvalue. Thus the variables to solve for are x, μ_1, μ_2, and v. (Two parameters must be allowed to vary to continue a curve of saddle-node bifurcations.)

For test functions of other bifurcations, and the bifurcations of limit cycles, see [3, 4].

References

1. Arnold, L. (1998) *Random Dynamical Systems*, Springer-Verlag, Berlin.
2. Strogatz, S.H. (1994) *Nonlinear Dynamics and Chaos*, Perseus, New York.
3. Seydel, R. (2010) *Practical Bifurcation and Stability Analysis, Interdisciplinary Applied Mathematics*, vol. 5, 3rd edn, Springer-Verlag, New York.
4. Kuznetsov, Y.A. (1998) *Elements of Applied Bifurcation Theory, Applied Mathematical*

Sciences, 2nd edn, Springer-Verlag, New York.

5. Hsu, C.S. (1987) *Cell-to-Cell Mapping: A Method of Global Analysis for Nonlinear Systems*, Springer-Verlag.
6. Dellnitz, M. and Hohmann, A. (1997) A subdivision algorithm for the computation of unstable manifolds and global attractors. *Numer. Math.*, **75** (3), 293–317.
7. Krauskopf, B., Osinga, H.M., Doedel, E.J., Henderson, M.E., Guckenheimer, J., Vladimirsky, A., Dellnitz, M., and Junge, O. (2005) A survey of methods for computing (un)stable manifolds of vector fields. *Int. J. Bifurcation Chaos*, **15** (3), 763–791.
8. Bender, C.M. and Orszag, S.A. (1978) *Advanced Mathematical Methods for Scientists and Engineers*, Springer-Verlag.
9. Murdock, J. (2013) *Perturbation Methods*, John Wiley & Sons, Ltd.
10. Golubitsky, M. and Schaeffer, D.G. (1985) *Singularities and Groups in Bifurcation Theory, Applied Mathematical Sciences* (eds F. John, J.E. Marsden, and L. Sirovich), Springer-Verlag, Berlin.
11. Carr, J. (1981) *Applications of Centre Manifold Theory*, Springer-Verlag.
12. Trefethen, L.N. (2000) *Spectral Methods in Matlab*, SIAM.
13. Boyd, J.P. (2001) *Chebyshev and Fourier Spectral Methods*, 2nd edn, Dover, New York.

12 Perturbation Methods

James Murdock

12.1 Introduction

Perturbation theory arises when a situation is given that admits of a mathematical description, and one asks how this description changes when the situation is varied slightly or "perturbed." This could result either in a continuation of the original situation with only small quantitative changes or in an abrupt qualitative change in the nature of the situation. Among the possible "abrupt" changes are the formation of a transition layer and the creation of various types of bifurcations; although bifurcation theory is usually treated as a separate subject from perturbation theory, the two areas are closely related. The specific subject matter of this chapter will be the following two topics.

1. A system of ordinary or partial differential equations is given, together with initial or boundary conditions. The system contains a small parameter, and is explicitly solvable when the parameter is zero. One asks how to construct approximate solutions (in explicit analytic form) when the parameter is small but nonzero; one asks for error estimates for these approximate solutions, and whether the approximate solutions exhibit the same qualitative behavior as the unknown exact solutions.
2. A matrix or linear transformation is given, depending on a small parameter. The eigenvalues and eigenvectors (or the Jordan normal form) are known when the parameter is zero, and one asks for approximate calculations of the eigenvalues or normal form when the parameter is small but nonzero.

The origins of perturbation theory lie in three classical problems, planetary motion, viscous fluid flow past a wall, and changes in the spectrum as a matrix or linear operator is varied. The present chapter is structured in the same threefold way: after an initial section presenting basic concepts common to the three areas, we take up in turn dynamical systems (Section 12.3), transition layer problems (Section 12.4), and spectra (Section 12.1.5); the intervening Section 12.5 deals with a recent

Mathematical Tools for Physicists, Second Edition. Edited by Michael Grinfeld.

method applicable to the problems of both Section 12.3 and 12.4. To conclude this introduction, we briefly describe the three classical problems.

Isaac Newton showed that the inverse square law of gravitational force implies that a single planet will move around the sun in an ellipse, satisfying Kepler's laws of planetary motion. The same law of gravity, however, implies that the several planets will exert attractive forces on each other, which will "perturb" their orbits. Laplace computed the perturbations, to a certain degree of approximation, and found that they were periodic, so that the solar system was "stable" (in Laplace's sense) and would not destroy itself. His techniques were laborious, and can be much simplified today by Hamiltonian mechanics; furthermore, he did not entirely prove that the solar system is stable, since his method, if carried to higher orders, does not always converge. However, a great many of the ideas of modern perturbation theory originated here: variation of parameters, averaging, multiple scales, and the problems that in the twentieth century led to the Kolmogorov–Arnol'd–Moser theorem, the Nekhoroshev theorem, and other topics in dynamical systems theory.

In theory, a fluid that is viscous (even to the smallest degree) will adhere to the walls of any container (such as a pipe) in which it is flowing. However, at any reasonable distance from the walls, the fluid flows almost as if the wall were not there. In order to resolve this apparent paradox, L. Prandtl introduced the idea of a "boundary layer," a thin layer of fluid against the wall in which the fluid passes from rest to rapid motion. Here the "unperturbed" problem is the inviscid flow (which does not adhere to the wall), the "perturbation" is the small viscosity, and the effect of the perturbation is not a small correction of the motion but a quite drastic correction confined to a small region (the boundary layer). This example leads to the ideas of stretched or scaled coordinates, inner and outer solutions, and matching.

According to quantum mechanics, all observable quantities are the eigenvalues of operators on a Hilbert space. In simple problems, the Hilbert space will be finite dimensional and the operators are representable as matrices; in other cases, they are partial differential operators. In modeling an atom, for instance, the eigenvalues will be related to the frequencies of light (the spectrum) emitted by the atom when electrons jump from one shell to another. These frequencies can be perturbed, for instance, if the atom is placed in a weak magnetic field. Mathematically, the resulting problem is to determine the changes in the "spectrum" (the set of eigenvalues) of a matrix or other operator when the operator is slightly perturbed. One striking difference between this problem and the first two is that quantum mechanics is entirely a linear theory, whereas in both the dynamical systems problems and the boundary layer problems nonlinearities can play a crucial role.

12.2 Basic Concepts

12.2.1 Perturbation Methods versus Numerical Methods

A system of differential equations that is not explicitly solvable calls for an approximate solution method of some type. The most useful approximate methods are numerical methods (implemented on a digital computer) and perturbation methods. Perturbation methods are usable only if the system is "close" to an explicitly solvable system, in the sense that the system would become solvable if certain small changes were made, such as deleting small terms or averaging some term over a rapidly rotating angle. In such a case, perturbation theory takes the solution of the simplified problem as a "zeroth-order approximation" that can be successively improved, giving higher-order approximations having explicit formulas. Numerical methods operate without the restriction that the problem be nearly solvable, but give a solution in the form of numerical tables or graphs. The advantage of a formula is that one can see by inspection the manner in which each variable and parameter affects the solution. As both numerical and perturbation solution are approximate, it is often helpful to verify a perturbation solution by comparing it with numerical ones (or directly with experimental data), especially when a mathematically valid error estimate for the perturbation solution is missing.

12.2.2 Perturbation Parameters

Mathematical models of physical phenomena typically contain several variables, which are divided into "coordinates" (of space or time) and "parameters." A spring/mass system with a cubic nonlinearity, for instance, will contain position and time coordinates as well as parameters for the mass and for the coefficients of the linear and cubic terms in the restoring force. The first step in preparing such a system for perturbation analysis is to nondimensionalize these variables. The second step is to look for solvable special cases that can serve as the starting point for approximating the solution of nearby cases. Most often, these solvable cases will be obtained by setting some of the parameters equal to zero. For instance, the forced and damped nonlinear oscillator given by

$$\ddot{y} + C\dot{y} + k^2 y + Ay^3 = B\cos\omega t \tag{12.1}$$

becomes easily solvable if $A = B = C = 0$; it is still solvable if only $A = 0$, but not quite as simply, and the case $C = B = 0$ is solvable with elliptic functions. It is therefore plausible to look for approximate solutions by perturbation theory if A, B, and C are small, and it may also be possible if only A is small, or if only B and C are small.

Suppose that we choose to investigate the case when A, B, and C are small. Ideally, we could treat these as three small independent parameters, and considerable work is now being devoted to the investigation of such "multiparameter" perturbation methods, especially in the context of bifurcation theory (see [1]). However, most classical perturbation methods are developed only for single-parameter problems. Therefore, it is necessary to *make a choice* as to how to reduce (12.1) to a single-parameter problem. The simplest way is to write $A = a\varepsilon$, $B = b\varepsilon$, and $C = c\varepsilon$, obtaining

$$\ddot{y} + \varepsilon c\dot{y} + k^2 y + \varepsilon ay^3 = \varepsilon b\cos\omega t. \tag{12.2}$$

We appear to have added a parameter instead of reducing the number, but now a, b, and c are regarded as constants, whereas ε, the *perturbation parameter*, is taken as a small, but variable, quantity; typically, we expect the perturbation solution to have the form of a power series in ε. We have, in effect, chosen to investigate a particular *path* leading from the origin in the space of variables A, B, C. It is at once clear that other paths are possible, for instance

$$\ddot{y} + \varepsilon^2 c\dot{y} + k^2 y + \varepsilon ay^3 = \varepsilon b\cos\omega t. \tag{12.3}$$

One might choose (12.3) over (12.2) if the goal is to investigate systems in which the damping is *extremely* small, small even compared to the cubic term and the forcing. But it is not clear, without experience, what the best formulation will be for a given problem. As an example of the role of experience, one might expect (knowing something about resonance in the linear case) that the results of studying (12.2) will be different if ω is close to k than if it is far away. But how do you express mathematically the idea that "ω is close to k"? Recalling that the only parameter available to express "smallness" is ε, the best answer turns out to be

$$\omega^2 = k^2 + \varepsilon d, \tag{12.4}$$

where d is another constant. Substituting (12.4) into (12.2) leads to the "correct" formulation of the near-resonance problem. One can see that the setting up of perturbation problems is sometimes an art rather than a science. In this chapter, we will for the most part assume that a perturbation problem has been chosen. Mathematical analysis of the problem may then suggest the use of *stretched* or otherwise *rescaled* variables, which, in fact, amount to a modification of the initial perturbation problem.

In recent years, a further consideration has come to the fore regarding the choice of parameters in a mathematical model. Physical considerations may have led to a model such as (12.1) above, and yet we know that this model is not exactly correct; there may, for instance, be very small nonlinearities other than the cubic term in the restoring force, or nonlinearities in the damping, or additional harmonics in the forcing. How many such effects should be included in the model? In the past, it was simply a matter of trial and error. But in certain cases, there now exists a mathematical theory that is able to determine just which additional small terms might make a *qualitative* (rather than just a tiny quantitative) difference in the solution. In these cases, it is sometimes best to *add all such significant small terms to the equation before attempting the solution, even if there is no evident physical reason for doing so.* The process of adding these additional terms is called finding the *universal unfolding* of the system. The advantage is that the universal unfolding will account for all possible qualitative behaviors that may be observed as a result of unknown perturbations. For instance, in the past, many bifurcations were not observed to occur exactly as predicted. They were called *imperfect bifurcations,* and each situation required that the specific perturbation responsible for the imperfection be discovered and incorporated into the model. Now it is often possible to determine all possible imperfections in advance by examining the universal unfolding of the original model. See [1].

In addition to the types of problems already discussed, there exist problems that do not contain a perturbation parameter but nonetheless allow treatment by perturbation methods. For instance, a system of differential equations may have a particular solution (often an equilibrium solution or a periodic solution) that can be computed exactly, and one may wish to study the solutions lying in a neighborhood of this one. One way to treat such problems is called *coordinate perturbations*; the coordinates themselves (or more precisely, the differences between the coordinates of the known and unknown solutions) are treated as small quantities, in place of a perturbation parameter. Another approach is to introduce a parameter ε as a scale factor multiplying these coordinate differences. Both ideas will be illustrated in the discussion of *normal forms* in Section 12.3.6.

12.2.3 Perturbation Series

Let us suppose that a perturbation problem has been posed, and let the exact solution (which we wish to approximate) be denoted by $u(\mathbf{x}, \varepsilon)$. Here ε is the (scalar) perturbation parameter, $\mathbf{x}$ is a vector consisting of all other variables in the problem including coordinates and other parameters, and u is the quantity being solved for. (In the case of (12.2) above, $u = y$ and $\mathbf{x} = (t, a, b, c, \omega)$, or if (12.4) is used, then $\mathbf{x} = (t, a, b, c, d)$.) The simplest form in which to seek an approximation is that of a (truncated) *power series* in ε:

$$u(\mathbf{x}, \varepsilon) \cong u_0(\mathbf{x}) + \varepsilon u_1(\mathbf{x}) + \cdots + \varepsilon^k u_k(\mathbf{x}). \qquad (12.5)$$

There are times when this is insufficient and we require a *Poincaré series*

$$u(\mathbf{x}, \varepsilon) \cong \delta_0(\varepsilon)u_0(\mathbf{x}) + \delta_1(\varepsilon)u_1(\mathbf{x}) + \cdots + \delta_k(\varepsilon)u_k(\mathbf{x}), \qquad (12.6)$$

where each δ_i is a monotone function of ε defined for $\varepsilon > 0$ satisfying

$$\lim_{\varepsilon \to 0} \frac{\delta_{i+1}(\varepsilon)}{\delta_i(\varepsilon)} = 0; \qquad (12.7)$$

such functions δ_i are called *gauges*. (Of course, a Poincaré series reduces to a power series if $\delta_i(\varepsilon) = \varepsilon^i$.) Finally, there are times when not even a Poincaré series is sufficient and a *generalized series* is needed:

$$u(\mathbf{x}, \varepsilon) \cong \delta_0(\varepsilon)u_0(\mathbf{x}, \varepsilon) + \delta_1(\varepsilon)u_1(\mathbf{x}, \varepsilon) + \cdots + \delta_k(\varepsilon)u_k(\mathbf{x}, \varepsilon). \qquad (12.8)$$

With such a series, it might appear that we could delete the gauges, or rather assimilate them into the u_i because these are now allowed to depend upon ε; but the intention is that the dependence of u_i on ε should not affect its order of magnitude. An example is the following two-term generalized series in which the vector $\mathbf{x}$ consists only of the time t:

$$u(t, \varepsilon) \cong \sin(1 + \varepsilon)t + \varepsilon \cos(1 + \varepsilon)t. \qquad (12.9)$$

Here $u_0(t, \varepsilon) = \sin(1 + \varepsilon)t$ and $u_1(t, \varepsilon) = \cos(1 + \varepsilon)t$; the dependence of these coefficients upon ε modifies their period but not their amplitude, and the second term still has the order of magnitude of its gauge ε.

Notice that we have written only truncated series in the previous paragraph. While most perturbation methods allow, in principle, for the computation of infinite series, these series seldom converge, and in practice it is impossible to calculate more than a few terms. The type of accuracy that we hope for in a perturbation solution is not convergence (improvement in accuracy as the number of terms increases), but rather *asymptotic validity*, which means improvement in accuracy as ε approaches zero. To explain this concept, let us consider a generalized series

$$u(\varepsilon) \cong \delta_0(\varepsilon)u_0(\varepsilon) + \delta_1(\varepsilon)u_1(\varepsilon) + \cdots + \delta_k(\varepsilon)u_k(\varepsilon) \qquad (12.10)$$

that contains no variables other than ε. We will say that this series is an *asymptotic approximation* to $u(\varepsilon)$ if

$$u(\varepsilon) = \delta_0(\varepsilon)u_0(\varepsilon) + \delta_1(\varepsilon)u_1(\varepsilon) + \cdots + \delta_k(\varepsilon)u_k(\varepsilon) + R(\varepsilon), \qquad (12.11)$$

where the remainder or error $R(\varepsilon)$ satisfies a bound of the form

$$|R(\varepsilon)| \le c\delta_{k+1}(\varepsilon) \qquad (12.12)$$

for some constant $c > 0$ and some gauge δ_{k+1} that approaches zero more rapidly than

δ_k as $\varepsilon \to 0$. (If u, and hence R, are vector-valued, then $|R(\varepsilon)|$ denotes a vector norm.) Equation (12.12) is abbreviated with the "big-oh" notation

$$R(\varepsilon) = \mathcal{O}(\delta_{k+1}(\varepsilon)). \tag{12.13}$$

The series (12.10) is called *asymptotically valid*, or *an asymptotic series*, if it is an asymptotic approximation (in the above sense) and in addition every truncation of (12.10) is also an asymptotic approximation, with the error being "big-oh" of the first omitted gauge. The case in which $u(\mathbf{x}, \varepsilon)$ depends upon variables $\mathbf{x}$ in addition to ε will be discussed in the following section on uniformity.

As a technical matter, any bound such as (12.12) is not intended to hold for all ε, but only for ε in some interval $0 \le \varepsilon \le \varepsilon_0$. A perturbation solution is never expected to be valid for large values of the perturbation parameter, and the meaning of "large" is relative. In this chapter, we will not continue to mention ε_0, but it is always lurking in the background.

Although Fourier series may arise in perturbation theory when dealing with oscillatory problems, a Fourier series is never a perturbation series. Rather, if a perturbation series such as $u(t, \varepsilon) \cong u_0(t) + \varepsilon u_1(t)$ depends periodically on time t, then the coefficients $u_k(t)$ may be expressed as Fourier series in t.

12.2.4 Uniformity

In the previous section we have defined asymptotic validity of a perturbation series for a function $u(\varepsilon)$ depending only on ε. This is adequate for a problem such as finding a root of a polynomial (supposing that the polynomial contains a perturbation parameter ε), because the root is a single number. But for most perturbation problems (such as differential equations), the solution is a function of space and/or time coordinates, and possibly various parameters, in addition to ε. For such problems, the previous definition of asymptotic validity is insufficient.

Let us return to the generalized series (12.8), and denote the error by $R(\mathbf{x}, \varepsilon)$. Now we may require that *for each fixed* $\mathbf{x}$ this error is of order $\delta_{k+1}(\varepsilon)$:

$$|R(\mathbf{x}, \varepsilon)| \le c(\mathbf{x})\delta_{k+1}(\varepsilon). \tag{12.14}$$

Notice that the "constant" c here may change when we move to a new point $\mathbf{x}$. In this case, we say that the error is *pointwise* of order δ_{k+1}.

Alternatively, we can choose a domain D of $\mathbf{x}$ and require that the error be *uniformly* of order δ_{k+1} for all $\mathbf{x}$ in D:

$$|R(\mathbf{x}, \varepsilon)| \le c\delta_{k+1}(\varepsilon). \tag{12.15}$$

Here the constant c is truly constant. In this case, we say $R(\mathbf{x}, \varepsilon) = \mathcal{O}(\delta_{k+1}(\varepsilon))$ *uniformly in* $\mathbf{x}$ *for* $\mathbf{x}$ *in* D. If every truncation of a perturbation series is uniformly of the order of the first omitted gauge, we say that the series is *uniformly valid*. (In the last sentence, we neglected to say "in D," but it is important to remember that such an expression has no meaning unless the domain D is understood.) Obviously, uniform asymptotic validity is stronger than pointwise validity, and it is safe to say that every method used in perturbation theory has been introduced in order to gain uniform validity for a problem for which previous methods only gave pointwise validity.

Now the definition of uniform validity calls for an estimate of the error of an approximation, and such an estimate is a difficult thing to come by. It would be convenient if there were an easier test for uniformity. In actuality, there is not. However, it is possible to obtain a simple *necessary* (but not sufficient) condition for a series to be uniformly valid; namely, it is not difficult to show that if a series (12.8) is uniformly valid in a domain D, then each coefficient $u_k(\mathbf{x})$ with $k \geq 1$ is bounded on D, that is, there exist constants c_k such that

$$|u_k(\mathbf{x}, \varepsilon)| \leq c_k \tag{12.16}$$

for $\mathbf{x}$ in D. If this is true, we say that the series (12.8) is *uniformly ordered*. We have already encountered the concept of a uniformly ordered series when discussing (12.9) above. A uniformly ordered series is one in which each term after the first is of no greater order than is indicated by its gauge. It is easy to inspect a perturbation series, once it has been constructed, and determine whether it is uniformly ordered. If not, the series is called *disordered*, and it cannot be uniformly valid. On the other hand, if it is uniformly ordered, it does not follow that it is uniformly valid, because one has done nothing toward estimating the error. Almost all textbooks are misleading on this point, because they almost invariably claim to be showing the uniform validity of a series when, in fact, they are only showing that it is uniformly ordered. However if a perturbation series is constructed on the basis of good intuitive insight into a problem, and if it is uniformly ordered, then it frequently turns out to be uniformly valid as well. (An elementary example in which this is *not* the case will be given in Section 12.3.2.)

With regard to uniform ordering, Poincaré series occupy a special place. Recall that a series is a Poincaré series if its coefficients do not depend on ε; see (12.6). In this case, if the domain D is compact and the coefficients $u_k(\mathbf{x})$ are continuous, then the coefficients are automatically bounded and the series is uniformly ordered (but still not automatically uniformly valid). However, even a Poincaré series may fail to be uniformly ordered if D is not compact or if D is allowed to depend on ε. We have not considered this latter possibility until now, but in fact, in many problems one must consider domains $D(\varepsilon)$ that depend on ε. An important example is a boundary layer, a thin domain near the boundary of some larger region, whose thickness depends on ε. For such domains, the definitions (12.15) and (12.16) of uniform validity and uniform ordering are the same, except that they are to hold for all $\mathbf{x}$ in $D(\varepsilon)$; that is, for each value of ε, (12.15) or (12.16) hold for a different range of $\mathbf{x}$.

It is now possible to explain one of the principal divisions in the subject of perturbation theory, the division of perturbation problems into *regular* and *singular*. A perturbation problem is regular if there exists a Poincaré series that is uniformly valid on the intended domain; it is singular if it is necessary to use a generalized series in order to obtain a uniformly valid solution on the intended domain. This is the only correct definition. Many textbooks give partial definitions such as "a perturbation problem for a differential equation is singular if ε multiplies the highest derivative." Such a definition, which refers only to the differential equation without stating an intended domain, cannot be correct. The presence of an ε multiplying the highest derivative does, of course, affect the domain on which a problem can be regular; we will see below that a problem such as $\ddot{u} + u + \varepsilon u^3 = 0$ is regular on any fixed interval $[0, T]$ but singular on an expanding interval $[0, 1/\varepsilon]$, while a problem such as $\varepsilon\ddot{u} + (t^2 + 1)\dot{u} + u = 0$ is regular only on a shrinking interval $[0, \varepsilon]$ and singular on a fixed interval.

12.3 Nonlinear Oscillations and Dynamical Systems

In this section, we discuss the major perturbation methods used in the study of nonlinear ordinary differential equations (dynamical systems). Typical problems include the location and bifurcation of rest (or equilibrium) points and periodic or quasiperiodic solutions; the approximation of solutions close to these; the solution of initial value problems for systems that become solvable (usually either linear or integrable Hamiltonian) when a small parameter vanishes; and the splitting of a homoclinic orbit under perturbation. In all of these problems, there is an interplay between qualitative and quantitative behavior. Advance knowledge of the qualitative behavior may assist the choice of a successful perturbation method. Alternatively, if the qualitative behavior is unknown, perturbation methods may be used in an exploratory fashion, as long as it is kept in mind that very frequently, especially in nonlinear problems, a perturbation solution may appear correct but may predict qualitative features (such as periodicity, stability, or presence of chaos) erroneously. As a general rule, qualitative results obtained from perturbation solutions (or any other approximate solutions) should be taken as conjectural, until supported by some combination of numerical or experimental evidence and rigorous mathematical proof.

12.3.1 Rest Points and Regular Perturbations

Given a system of differential equations of the form

$$\dot{\mathbf{x}} = \mathbf{f}(\mathbf{x}, \varepsilon) = \mathbf{f}_0(\mathbf{x}) + \varepsilon\mathbf{f}_1(\mathbf{x}) + \mathcal{O}(\varepsilon^2), \qquad (12.17)$$

with $\mathbf{x} = (x_1, \ldots, x_n) \in \mathbb{R}^n$ and $\varepsilon \in \mathbb{R}$, the *rest points* are the solutions of the system of equations $\mathbf{f}(\mathbf{x}, \varepsilon) = 0$. If a rest point $\mathbf{a}_0$ is known when $\varepsilon = 0$, it may continue to exist as a function $\mathbf{a}(\varepsilon)$ with

$$\mathbf{f}(\mathbf{a}(\varepsilon), \varepsilon) = 0 \qquad (12.18)$$

for ε near zero, or it may bifurcate into two or more rest points, or disappear altogether; the results may differ for $\varepsilon > 0$ and $\varepsilon < 0$. A crucial role is played by the matrix $A_0 = \mathbf{f}_\mathbf{x}(\mathbf{x}_0, 0)$ of partial derivatives of $\mathbf{f}$ at the unperturbed rest point, and especially its eigenvalues. If A_0 is invertible (zero is not an eigenvalue), then, by the implicit function theorem, a unique continuation $\mathbf{a}(\varepsilon)$ exists and is computable as a series

$$\mathbf{a}(\varepsilon) = \mathbf{a}_0 + \varepsilon\mathbf{a}_1 + \mathcal{O}(\varepsilon^2). \qquad (12.19)$$

This is the simplest example of a perturbation series. Putting (12.19) into (12.18) and expanding gives $\mathbf{f}_0(\mathbf{a}_0) + \varepsilon(A_0\mathbf{a}_1 + \mathbf{f}_1(\mathbf{a}_0)) + \cdots = 0$, or (because $\mathbf{f}_0(\mathbf{a}_0) = 0$)

$$\mathbf{a}_1 = -A_0^{-1}\mathbf{f}_1(\mathbf{a}_0). \qquad (12.20)$$

The solution may be continued to higher order, and the matrix function

$$A(\varepsilon) = \mathbf{f}_\mathbf{x}(\mathbf{a}(\varepsilon), \varepsilon) \qquad (12.21)$$

may also be studied, because (in most cases) it determines the stability of the rest point. If all of the eigenvalues of A_0 are off the imaginary axis (A_0 is hyperbolic), the same will be true for $A(\varepsilon)$ for small ε, and the stability type (dimensions of the stable and unstable manifolds) of the rest point will not change. When this is not the case, the methods of Section 12.6 determine the spectrum of $A(\varepsilon)$, and thus usually suffice to decide how the stability changes.

When A_0 is not invertible, bifurcation (change in the number of rest points) usually occurs. Even when A_0 has only one zero eigenvalue, various possibilities (such as saddle-node and pitchfork bifurcations) exist, and it is not possible to give details here. A reference treating the subject from the standpoint of perturbation theory is [2]; a quite different modern treatment is [1].

Now suppose that a solution

$$\mathbf{x}(t,\varepsilon) = \mathbf{x}_0(t) + \varepsilon\mathbf{x}_1(t) + \mathcal{O}(\varepsilon^2) \qquad (12.22)$$

of the system (12.17) is to be found, with initial condition $\mathbf{x}(0,\varepsilon) = \mathbf{a}(\varepsilon)$ (no longer a rest point) given by (12.19). Substituting (12.22) into (12.17), expanding, and equating terms of the same degree in ε yields

$$\begin{aligned} \dot{\mathbf{x}}_0 &= \mathbf{f}_0(\mathbf{x}_0) \\ \dot{\mathbf{x}}_1 &= \mathbf{f}_{0\mathbf{x}}(\mathbf{x}_0(t))\mathbf{x}_1 + \mathbf{f}_1(\mathbf{x}_0(t)). \end{aligned} \qquad (12.23)$$

If the first equation of (12.23), which is the same as (12.17) with $\varepsilon = 0$, can be solved with initial condition $\mathbf{x}_0(0) = \mathbf{a}_0$, its solution $\mathbf{x}_0(t)$ can be placed in the second equation of (12.23), which then becomes an inhomogeneous *linear* equation for $\mathbf{x}_1$; it is to be solved with initial condition $\mathbf{x}_1(0) = \mathbf{a}_1$. Equations of this type are not necessarily easy to solve, but are certainly easier than the nonlinear equation (12.17). If this is successful, the procedure may be continued to higher order.

This is usually called the *regular perturbation method* or the method of *straightforward expansion*. According to our earlier definition, a perturbation method is regular if it provides a Poincaré series that is uniformly valid on the intended domain. Here (12.22) is a Poincaré series, and it can be shown to be uniformly valid on any *finite* interval $[0, T]$; that is, the error bound is of the order of the first omitted term, and once T is chosen, the coefficient in the error bound is fixed. So the term *"regular"* is justified if this is the intended domain. In many problems (below), one seeks a solution valid on an "expanding" interval such as $[0, 1/\varepsilon]$; the straightforward expansion is usually not valid for this purpose, and so is often called a *naive expansion*.

There are situations in which straightforward expansion is valid for much longer than finite intervals. For instance, if a solution is approaching a sink (a rest point with all eigenvalues in the left half-plane), the straightforward expansion is valid for all $t \geq 0$; see [3, Chapter 5]. More generally, if the first equation of (12.23) has a solution that connects two hyperbolic rest points, then a straightforward expansion (to any order) beginning with that solution will be *shadowed* by an exact solution of (12.17) connecting two hyperbolic rest points of that system; that is, the approximate and exact solutions will remain close (to the order of the first omitted term) for all time, both past and future, although the two solutions may not have any point in common. (In particular, the approximate and shadowing solutions will not satisfy the same initial conditions.) See [3, Chapter 6].

12.3.2 Simple Nonlinear Oscillators and Lindstedt's Method

The "hard" nonlinear spring or unforced Duffing equation is given by

$$\ddot{u} + u + \varepsilon u^3 = 0 \qquad (12.24)$$

for $\varepsilon > 0$. It can be expressed as a first-order system in the $(u, \dot{u})$ phase plane in the form $\dot{u} = v, \dot{v} = -u - \varepsilon u^3$. In the phase plane, the orbits are closed curves surrounding the rest point at the origin; this may be seen from the conservation of energy. (The "soft" spring with $\varepsilon < 0$ behaves differently.) As the solutions of (12.24) with any initial conditions $u(0) = a, \dot{u}(0) = b$ are smooth functions of ε, they may be expanded as Taylor series having the form (if we retain two terms)

$$u(t, \varepsilon) \cong u_0(t) + \varepsilon u_1(t). \qquad (12.25)$$

The coefficients may be determined by substituting (12.25) into (12.24), expanding the u^3 term, dropping powers of ε higher than the first, and setting each order in ε separately equal to zero. This gives two *linear* equations that can be solved (recursively) for u_0 and u_1. The result, for $b = 0$ (and it is enough to consider this case because every solution is at rest momentarily when its amplitude is at its maximum), is

$$u(t, \varepsilon) \cong a \cos t - \varepsilon \frac{a^3}{32}(\cos t + 12t \sin t - \cos 3t). \qquad (12.26)$$

Upon examining (12.26) for uniform ordering, we discover that all functions of t appearing there are bounded for all t except for $12t \sin t$, which becomes unbounded as t approaches infinity. This is an example of a so-called "secular" term, one which grows over the "ages" (*saeculae* in Latin). We conclude from this that (12.26) is uniformly ordered for t in any finite interval $D = [0, T]$ but not for $D = [0, \infty)$. According to the general principles discussed in Section 12.2.4, this shows that (12.26) is not uniformly valid on $[0, \infty)$, and it *leads us to suspect, but does not prove,* that (12.16) is uniformly valid on $[0, T]$. In the present case, this conjecture is correct. If the intended domain D for the solution of (12.24) is a finite interval, then we have obtained a uniform approximation in the form of a Poincaré series, and the problem is a regular one. If a solution valid for a longer time is desired, the problem will prove to be singular.

In an effort to extend the validity of the solution, we recall that the actual solutions of (12.24) are periodic, whereas (12.26) is not. The problem is that the period of the exact solution depends upon ε, and there is no way that a Poincaré series can have such a period because the coefficients are not allowed to depend on ε. To remedy this, we seek to approximate the (unknown) frequency of the solution in the form

$$\nu(\varepsilon) \cong \tilde{\nu}(\varepsilon) = \nu_0 + \varepsilon \nu_1 + \varepsilon^2 \nu_2 \qquad (12.27)$$

with the solution itself being represented as

$$u(t, \varepsilon) \cong u_0(\tilde{\nu}(\varepsilon)t) + \varepsilon u_1(\tilde{\nu}(\varepsilon)t), \qquad (12.28)$$

which is now a generalized series. Notice that we have carried (12.27) to one more order than (12.28). Now we substitute (12.27) and (12.28) into (12.24) and attempt to determine $\nu_0, u_0, \nu_1, u_1, \nu_2$ recursively, in that order, in such a way that each u_i is periodic. The mechanics of doing this will be explained in the next paragraph; the result is

$$u(t, \varepsilon) \cong a \cos t^+ + \varepsilon \left(-\frac{1}{32} a^3 \cos t^+ + \frac{1}{32} a^3 \cos 3t^+ \right), \qquad (12.29)$$

where

$$t^+ = \tilde{\nu}(\varepsilon)t = \left(1 + \varepsilon \frac{3}{8} a^2 - \varepsilon^2 \frac{21}{256} a^4 \right) t. \qquad (12.30)$$

Examining (12.29), we see that it is uniformly ordered for all time, because the coefficients are bounded (there are no secular terms). One might therefore conjecture that the solution is uniformly valid for all time, *but this would be incorrect!* (This example is an excellent warning as to the need for proofs of validity in perturbation theory.) The difficulty is that t^+ uses the approximate frequency $\tilde{\nu}(\varepsilon)$ in place of the exact frequency $\nu(\varepsilon)$; there is no escape from this, as the exact frequency remains unknown. Therefore, (12.29) gradually gets out of phase with the exact solution. The reason for taking (12.27) to one higher order than (12.28) is to minimize this effect. It can be shown that (12.29) is uniformly valid on the *expanding interval* $D(\varepsilon) = [0, 1/\varepsilon]$. (This is our first example of a domain that depends on ε, as discussed in Section 12.2.4) If the intended domain is such an expanding interval, then (12.29) provides a uniformly valid generalized series, and the problem is seen to be singular. (If the intended domain is all t, the problem is simply impossible to approximate asymptotically.)

In order to complete the example in the last paragraph, we must indicate how to obtain (12.29) and (12.30). The easiest way is to substitute $\tau = \nu(\varepsilon)t$ into (12.24) to obtain

$$\nu(\varepsilon)^2 \frac{d^2 u}{d\tau^2} + u + \varepsilon u^3 = 0. \qquad (12.31)$$

Then substitute (12.27) and (12.28) into (12.31), expand, and set the coefficient of each power of ε equal to zero as usual. It is easy to find that $\nu_0 = 1$ and $u_0 = a\cos\tau$ (which in the end becomes $a\cos t^+$ because we do not know the exact frequency ν). The crucial step arises when examining the equation for u_1, which is (writing $' = d/d\tau$)

$$\begin{aligned} u_1'' + u_1 &= -a^3\cos^3\tau + 2\nu_1 a\cos\tau \qquad (12.32)\\ &= \left(-\frac{3}{4}a^3 + 2\nu_1 a\right)\cos\tau \\ &\quad -\frac{1}{4}a^3\cos 3\tau. \end{aligned}$$

From the Fourier series expansion (the second line of (12.32)), we see that the term in $\cos\tau$ will be resonant with the free frequency, and hence produce unbounded (secular) terms in u_1, unless the coefficient of $\cos\tau$ vanishes. In this way, we conclude that

$$\nu_1 = \frac{3}{8}a^2 \qquad (12.33)$$

and, after deleting the $\cos\tau$ term from (12.32), we solve it for u_1. This procedure is repeated at each subsequent stage.

The previous example is typical of *unforced conservative* oscillators, in which every solution (at least in a certain region) is periodic. There are two additional classes of oscillators that must be mentioned, although we cannot give them as much space as they deserve: self-excited oscillators and forced oscillators.

The standard example of a self-excited oscillator is the *van der Pol equation*

$$\ddot{u} + \varepsilon(u^2 - 1)\dot{u} + u = 0. \qquad (12.34)$$

Instead of a region in the phase plane filled with periodic solutions, this equation has a single periodic solution (limit cycle) for $\varepsilon > 0$. The Lindstedt method, described above, can be used to approximate the periodic solution, but must be modified slightly: the initial condition can no longer be assigned arbitrarily, because to do so will, in general, yield a nonperiodic solution for which the Lindstedt method fails. (These solutions can be found by averaging or multiple scales; see below.) Suppose that the limit cycle crosses the positive x axis (in the phase plane) at $(a(\varepsilon), 0)$ and has frequency $\nu(\varepsilon)$. Then the solution is sought in the form of (12.27) and (12.28) together with an additional expansion $a(\varepsilon) = \alpha_0 + \varepsilon a_1 + \cdots$; the coefficients u_i, ν_i, and a_i are determined recursively, choosing ν_i and a_i so that no secular terms arise in u_{i+1}. This example shows the effect of the dynamics of a system on the correct formulation of a perturbation problem.

The general nearly linear, periodically forced oscillator can be written

$$\ddot{u} + u = \varepsilon f(u, \dot{u}, \omega t), \qquad (12.35)$$

where $f(u, v, \theta)$ is 2π-periodic in θ; thus the period of the forcing is $2\pi/\omega$. The dynamics of (12.35) can be complicated, and we will limit ourselves to one type of periodic solution, the *harmonic* oscillations, which are *entrained* by the forcing so that they have the same frequency ω; these harmonic solutions occur for ε small and ω close to 1 (the frequency of the solutions when $\varepsilon = 0$). As the problem contains two parameters (ε and ω) and we are limited to one-parameter methods, it is necessary to express the statement "ω is close to 1" in terms of the perturbation parameter ε. (A study using two independent parameters would uncover the phenomenon of "resonance horns" or "resonance tongues." See [4, Section 5.5]) It turns out that an efficient way to do so is to write

$$\omega^2 = 1 + \varepsilon\beta, \qquad (12.36)$$

where β is a new parameter that is considered fixed (not small). With (12.36), (12.35) can have one or more isolated periodic solutions with unknown initial conditions $(a(\varepsilon), b(\varepsilon))$. (We can no longer assume $b = 0$.) On the other hand, the frequency of the solution this time is not unknown but equals ω. So the Lindstedt method undergoes another modification dictated by the dynamics: u, a, and b are expanded in ε and solved recursively, choosing the coefficients of a and b to eliminate secular terms from u. In contrast with the previous cases, there is no accumulating phase error because the frequency is known, and the perturbation approximations are uniformly valid for all time.

12.3.3 Averaging Method for Single-Frequency Systems

All of the systems discussed in the previous section, and a great many others, can be expressed in *periodic standard form*,

$$\dot{\mathbf{x}} = \varepsilon \mathbf{f}(\mathbf{x}, t, \varepsilon), \qquad (12.37)$$

where $\mathbf{x} = (x_1, \ldots, x_n) \in \mathbb{R}^n$ and where $\mathbf{f}$ is 2π-periodic in t. The solutions of (12.37) may be approximated by the *method of averaging*, which not only locates the periodic solutions (and proves their existence), but also determines their stability or instability and approximates the transient (nonperiodic) solutions. The method of averaging has been rediscovered many times and exists (with slight variations) under a variety of names, including: method of van der Pol; method of Krylov–Bogoliubov–Mitropolski (KBM method); method of slowly varying amplitude and phase; stroboscopic method; Struble's method; von Zeipel's method (in the Hamiltonian case); method of Lie series or Lie transforms. Some of the differences in these "methods" pertain to how the original system is put into periodic standard form, and others to details about how the near-identity transformations (described below) are handled.

To illustrate how a system may be put into periodic standard form, consider the van der Pol equation (12.34), or, written as a system,

$$\dot{u} = v \qquad (12.38)$$
$$\dot{v} = -u + \varepsilon(1 - u^2)v.$$

Rotating polar coordinates (r, φ) may be introduced by $u = r\cos(\varphi - t)$, $v = r\sin(\varphi - t)$, giving

$$\dot{r} = \varepsilon\left(1 - r^2\cos^2(\varphi - t)\right) \times r\sin^2(\varphi - t) \qquad (12.39)$$
$$\dot{\varphi} = \varepsilon\left(1 - r^2\cos^2(\varphi - t)\right)\sin(\varphi - t) \times \cos(\varphi - t),$$

which is in periodic standard form with $x = (r, \varphi)$. The same result may be achieved by seeking a solution of (12.34) by variation of parameters, in the form $u = r\cos(\varphi - t)$ where r and φ are variables, and imposing the requirement that $\dot{u} = r\sin(\varphi - t)$; the motivation for these choices is that with r and φ constant, these solve (12.34) for $\varepsilon = 0$. The transformation to periodic standard form is merely a change of coordinates, not an assumption about the nature of the solutions.

In its crudest form, the method of averaging simply consists in replacing (12.37) by

$$\dot{\mathbf{y}} = \varepsilon\bar{\mathbf{f}}(\mathbf{y}), \qquad (12.40)$$

where

$$\bar{\mathbf{f}}(\mathbf{y}) = \frac{1}{2\pi}\int_0^{2\pi} \mathbf{f}(\mathbf{y}, t, 0)\,\mathrm{d}t. \qquad (12.41)$$

System (12.40) is easier to solve than (12.37), because it is autonomous. The form of (12.40) can be motivated by the fact that for small ε, $\mathbf{x}$ in (12.37) is slowly varying and therefore nearly constant over one period; therefore, to a first approximation, we might hold $\mathbf{x}$ constant while integrating over one period in (12.37) to find the "average" influence due to $\mathbf{f}$. But this sort of motivation gives no idea how to estimate the error or to extend the method to higher-order approximations. A much better procedure is to return to (12.37) and perform a near-identity change of variables of the form

$$\mathbf{x} = \mathbf{y} + \varepsilon\mathbf{u}_1(\mathbf{y}, t, \varepsilon), \qquad (12.42)$$

where $\mathbf{u}_1$ is a periodic function of t, which is to be determined so that the transformed equation has the form

$$\dot{\mathbf{y}} = \varepsilon\mathbf{g}_1(\mathbf{y}) + \varepsilon^2\hat{\mathbf{g}}(\mathbf{y}, t, \varepsilon), \qquad (12.43)$$

where $\mathbf{g}_1$ is independent of t. It turns out that such a transformation is possible only if we take $\mathbf{g}_1 = \bar{\mathbf{f}}$; by doing so, (12.40) can be obtained from (12.43) simply by deleting the ε^2 term. The solution $\mathbf{y}(t)$ of (12.40) is called the *first approximation* to the solution of (12.37); if this is substituted into (12.42), the result $\mathbf{x}(t)$ is called the *improved first approximation*. This "improvement" consists in adding (an approximation to) small periodic fluctuations around $\mathbf{y}(t)$ that are actually present, but are of the same order as the error of the first approximation, so while the numerical results are better, the *asymptotic* accuracy of the improved first approximation is not better than that of the first approximation.

When it is formulated in this way, the first-order method of averaging is seen to consist of nothing but coordinate changes (first into periodic standard form, then into form (12.43)), followed by truncation (dropping $\hat{\mathbf{g}}$); it is only the truncation that introduces error, and this error can be estimated using Gronwall's inequality. It is also clear how to proceed to higher orders:

simply replace (12.42) by

$$\mathbf{x} = \mathbf{y} + \varepsilon \mathbf{u}_1(\mathbf{y}, t, \varepsilon) + \cdots + \varepsilon^k \mathbf{u}_k(\mathbf{y}, t, \varepsilon) \tag{12.44}$$

and (12.43) by

$$\dot{\mathbf{y}} = \varepsilon \mathbf{g}_1(\mathbf{y}) + \cdots + \varepsilon^k \mathbf{g}_k(\mathbf{y}) + \varepsilon^{k+1} \hat{\mathbf{g}}(\mathbf{y}, t, \varepsilon); \tag{12.45}$$

the averaged equations are obtained by deleting $\hat{\mathbf{g}}$. It is, of course, necessary to determine the $\mathbf{u}_i$ and $\mathbf{g}_i$ recursively in such a way that the $\mathbf{u}_i$ are periodic and the $\mathbf{g}_i$ are independent of t; this is where the technical details of various versions of averaging come into play. (Warning: $\mathbf{g}_i$ for $i > 1$ are not simply averages of higher-order terms of $\mathbf{f}$, but of expressions involving these and $\mathbf{u}_j$ for $j < i$.) Once (12.45), without $\hat{\mathbf{g}}$, has been determined, it must be solved. Setting $\tau = \varepsilon t$, it becomes

$$\frac{\mathrm{d}\mathbf{y}}{\mathrm{d}\tau} = \mathbf{g}_1(\mathbf{y}) + \varepsilon \mathbf{g}_2(\mathbf{y}) + \cdots + \varepsilon^{k-1} \mathbf{g}_k(\mathbf{y}), \tag{12.46}$$

which can often be solved by regular perturbation theory on the interval $0 \le \tau \le 1$, that is, $0 \le t \le 1/\varepsilon$. Then the solution must be put back into the transformation (12.44); the resulting approximate solutions of (12.37) will differ from the exact solutions (with the same initial condition) by $\mathcal{O}(\varepsilon^k)$ during a time interval of length $\mathcal{O}(1/\varepsilon)$. The *kth approximation* and *improved kth approximation* are obtained by using (12.44) without, or with, its kth term; again, these are asymptotically equivalent, although the improved approximation is numerically more accurate.

For additional information about the method of averaging, see [3]. As in regular perturbation theory, it is possible under special conditions to obtain results on half-infinite or infinite intervals of time; see Chapter 5 for systems with attractors and Chapter 6 for shadowing. See Appendix E for averaging applied to partial differential equations.

12.3.4 Multifrequency Systems and Hamiltonian Systems

Oscillatory problems that cannot be put into periodic standard form can often be put into the following *angular standard form*:

$$\dot{\mathbf{r}} = \varepsilon \mathbf{f}(\mathbf{r}, \boldsymbol{\theta}, \varepsilon) \tag{12.47}$$
$$\dot{\boldsymbol{\theta}} = \boldsymbol{\Omega}(\mathbf{r}) + \varepsilon \mathbf{g}(\mathbf{r}, \boldsymbol{\theta}, \varepsilon),$$

where $\mathbf{r} = (r_1, \ldots, r_n)$ is a vector of amplitudes and $\boldsymbol{\theta} = (\theta_1, \ldots, \theta_m)$ a vector of angles (so that $\mathbf{f}$ and $\mathbf{g}$ are periodic in each θ_i with period 2π). This form includes the periodic standard form, by taking $m = 1$ and $\dot{\theta} = 1$. The "naive" method of averaging for (12.47) would be to replace $\mathbf{f}$ and $\mathbf{g}$ by their averages over $\boldsymbol{\theta}$, for instance

$$\bar{\mathbf{f}}(\mathbf{r}) = \frac{1}{(2\pi)^m} \int_0^{2\pi} \cdots \times \int_0^{2\pi} \mathbf{f}(\mathbf{r}, \boldsymbol{\theta}, 0)\, \mathrm{d}\theta_1 \cdots \mathrm{d}\theta_m. \tag{12.48}$$

To justify this process, and to extend the method to higher order, one tries to make a near-identity change of variables $(\mathbf{r}, \boldsymbol{\theta}) \to (\boldsymbol{\rho}, \boldsymbol{\varphi})$ that will render the system independent of $\boldsymbol{\varphi}$ up through a given order k in ε. However, one encounters at once the famous difficulty of "small divisors" which make the existence of such a transformation doubtful. If $\mathbf{f}$ is expanded in a convergent multiple Fourier series

$$\mathbf{f}(\mathbf{r}, \boldsymbol{\theta}, 0) = \sum_{\nu} \mathbf{a}_{\nu}(\mathbf{r}) \mathrm{e}^{\mathrm{i}(\nu_1\theta_1 + \cdots + \nu_m\theta_m)}, \tag{12.49}$$

then the transformation to averaged form necessarily involves the series

$$\sum_{\nu\neq 0} \frac{a_\nu(\mathbf{r})}{i(\nu_1\Omega_1(\mathbf{r}) + \cdots + \nu_m\Omega_m(\mathbf{r}))} e^{i(\nu_1\theta_1+\cdots+\nu_m\theta_m)}, \tag{12.50}$$

which may not converge because the denominators $i(\nu_1\theta_1 + \cdots + \nu_m\theta_m)$ may be small (or even zero), causing the coefficients to become large. It is of no use at this point to say that "perturbation theory is not concerned with the convergence of series," because the series in question is not being used for approximation, but to prove the existence of a transformation needed in order to justify the method.

Some preliminary progress can be made by considering the case in which the series (12.49), and hence (12.50), are finite. In this case, convergence difficulties cannot arise, but there is still the difficulty that for some $\mathbf{r}$, one or more of the denominators of (12.50) may become zero. As $\Omega_i(r)$ are the frequencies of the free oscillations ($\varepsilon = 0$) of (12.47), the vanishing of a denominator indicates a *resonance relationship* among these frequencies. In general, for each $\boldsymbol{\nu}$ there will be a *resonance manifold*, or *resonance hypersurface*, consisting of those r for which the denominator involving $\boldsymbol{\nu}$ vanishes. On (or near) any such surface it is not permissible to average over all angles $\boldsymbol{\theta}$, although it may be possible to average over a subset of these angles or over certain integral linear combinations of them.

Results beyond these have been obtained in the important special case of Hamiltonian systems; the *Kolmogorov–Arnol'd—Moser (* or *KAM) theorem* and the *Nekhoroshev theorem* are a high point of modern perturbation theory and together give the definitive answer to the problem of the stability (in the sense of Laplace) of the (idealized Newtonian) solar system, with which this chapter began. Consider a system defined by a Hamiltonian function of the form

$$H(\mathbf{r},\boldsymbol{\theta},\varepsilon) = H_0(\mathbf{r}) + \varepsilon H_1(\mathbf{r},\boldsymbol{\theta}) + \varepsilon^2 H_2(\mathbf{r},\boldsymbol{\theta}) + \cdots, \tag{12.51}$$

where $\mathbf{r}$ and $\boldsymbol{\theta}$ are as before except that $m = n$. Written in the form (12.47), this system is

$$\dot{\mathbf{r}} = \varepsilon \frac{\partial H_1}{\partial \boldsymbol{\theta}} + \cdots \tag{12.52}$$
$$\dot{\boldsymbol{\theta}} = -\frac{\partial H_0}{\partial \mathbf{r}} - \varepsilon \frac{\partial H_1}{\partial \mathbf{r}} + \cdots .$$

As H_1 is assumed to be periodic in the components of $\boldsymbol{\theta}$, it may be expanded in a multiple Fourier series like (12.49); differentiating with respect to any component of $\boldsymbol{\theta}$ then eliminates the constant term ($\mathbf{a}_0(r)$, which is the average). It follows that $\partial H_1/\partial\boldsymbol{\theta}$ has zero average, so that the (naive) first-order averaged equation for $\mathbf{r}$ becomes

$$\dot{\mathbf{r}} \cong 0. \tag{12.53}$$

This suggests that the motion is oscillatory with nearly constant amplitudes. The KAM theorem states that (if a certain determinant does not vanish) the great majority of initial conditions will lead to quasiperiodic motion on an invariant torus close to a torus $\mathbf{r}$ = constant. The Nekhoroshev theorem states that even for those initial conditions that do not lie on invariant tori, the drift in $\mathbf{r}$ (called *Arnol'd diffusion*) takes place exponentially slowly (as $\varepsilon \to 0$). (Notice that an n-dimensional torus in $2n$-dimensional space does not form the boundary of an open set if $n > 1$, so the presence of many such invariant tori does not prevent other solutions from slowly drifting off to infinity. For $n = 2$, the invariant 2-tori do not bound in 4-space, but they do in the three-dimensional submanifolds of constant energy; because solutions have constant energy, Arnol'd discussion cannot take place.) For details see [5]. The proofs involve very deep mathematics, such as the Moser–Nash implicit function theorem, with implications for partial differential equations, Riemannian geometry, and elsewhere.

In all applications of averaging and related methods to Hamiltonian systems, it is necessary to have a means of handling near-identity transformations that preserve the Hamiltonian form of the equations; that is, one needs to construct near-identity transformations that are *canonical* (or *symplectic*). Classically, such transformations can be constructed from their *generating functions* (in the sense of Hamilton–Jacobi theory); averaging procedures carried out in this way are called *von Zeipel's method.* At the present time, this approach can be regarded as obsolete. It has been replaced by the method of *Lie transforms,* in which near-identity canonical transformations are generated as the flows of Hamiltonian systems in which ε takes the place of time. (The Lie method is not limited to Hamiltonian systems, but is particularly useful in this context.) Algorithmic procedures for handling near-identity transformations in this way have been developed, and they are considerably simpler than using generating functions. See [6, Section 5.7] and [3, Chapter 3].

12.3.5 Multiple-Scale Method

The earliest perturbation problem, that of planetary motion, illustrates the appeal of the idea of multiple scales. A single planet under the influence of Newtonian gravitation would travel around the sun in an elliptic orbit characterized by certain quantities called the "Keplerian elements" (the eccentricity, major axis, and certain angles giving the position of the ellipse in space). As the actual (perturbed) motion of the planets fits this same pattern for long periods of time, it is natural to describe the perturbed motion as "elliptical motion with slowly varying Keplerian elements." A simpler example would be a decaying oscillation of the form $e^{-\varepsilon t}\sin t$, which could be described as a periodic motion with slowly decaying amplitude. Solutions of nonlinear oscillations obtained by the method of averaging frequently have this form, in which time appears both as t and as εt, the latter representing slow variation; sometimes other combinations such as $\varepsilon^2 t$ appear.

This leads to the question whether it is possible to arrive at such solutions more directly, by postulating the necessary time scales in advance. The "method of multiple scales" is the result of such an approach, and is sometimes regarded as the most flexible general method in perturbation theory, because it is applicable both to oscillatory problems (such as those covered by averaging) and to boundary layer problems (discussed below). However, this very flexibility is also its drawback, because the "method" exists in an immense variety of ad hoc formulations adapted to particular problems. (See [6] for examples of many of these variations.) There are two-scale methods using fast time t and slow time εt; two-scale methods using strained fast time $(\nu_0 + \varepsilon\nu_1 + \varepsilon^2\nu_2 + \cdots)t$ (similar to the strained time in the Lindstedt method) and slow time εt; multiple-scale methods using $t, \varepsilon t, \varepsilon^2 t, \ldots, \varepsilon^n t$; and methods using scales that are nonlinear functions of t. The scales to be used must be selected in advance by intuition or experience, while in other methods (averaging and matching), the required scales are generated automatically. Sometimes, the length of validity of a solution can be increased by increasing the number of scales, but (contrary to a common impression) this is by no means always the case; see [3, Section 3.5]. Some problems come with more than one time scale from the beginning, for instance, problems that contain a "slowly varying parameter" depending on εt. It may seem natural to treat such a system by the method of multiple scales, but another possibility is to introduce $\tau = \varepsilon t$ as an additional independent variable subject to $\dot{\tau} = \varepsilon$. In summary, the popularity of multiple scales results from its shorter calculations, but this aside, other methods have greater power.

The general outlines of the method are as follows. Suppose the chosen time scales are t, τ, σ with $\tau = \varepsilon t, \sigma = \varepsilon^2 t$. An approximate solution is sought as a series taken to a certain number of terms, such as

$$x_0(t, \tau, \sigma) + \varepsilon x_1(t, \tau, \sigma) + \varepsilon^2 x_2(t, \tau, \sigma). \tag{12.54}$$

In substituting (12.54) into the differential equations to be solved, the definitions of the scales (such as $\tau = \varepsilon t$) are used, so that ordinary derivatives with respect to t are replaced by combinations of partial derivatives with respect to the different scales; thus

$$\frac{\mathrm{d}}{\mathrm{d}t} = \frac{\partial}{\partial t} + \varepsilon \frac{\partial}{\partial \tau} + \varepsilon^2 \frac{\partial}{\partial \sigma}. \tag{12.55}$$

From this point on (until the very end, when τ and σ are again replaced by their definitions), the separate time scales are treated as independent variables. This has the effect of changing ordinary differential equations into partial differential equations that are highly underdetermined, so that various free choices are possible in expressing the solution. The point of the method is now to make these choices skillfully so that the final series (12.54) is uniformly ordered (and, it is hoped, uniformly valid) on the desired domain.

As an illustration, we return to the van der Pol equation (12.34) with initial conditions $u(0) = a$, $\dot{u}(0) = 0$. Choosing time scales t and $\tau = \varepsilon t$, and writing the solution as $u \cong u_0(t, \tau) + \varepsilon u_1(t, \tau)$, one finds recursively that (with subscripts denoting partial derivatives) $u_{0tt} + u_0 = 0$ and $u_{1tt} + u_1 = -2u_{0t\tau} - u_0^2 u_{0t}$. The first equation gives $u_0(t, \tau) = A(\tau)\cos t + B(\tau)\sin t$, a modulated oscillation with slowly varying coefficients. The solution remains underdetermined, because there is nothing here to fix $A(\tau)$ and $B(\tau)$. The solution for u_0 is now substituted into the right-hand side of the differential equation for u_1, and $A(\tau)$ and $B(\tau)$ are chosen to eliminate resonant terms so that the solution for u_1 will remain bounded. (This is similar to the way the undetermined quantities are fixed in the Lindstedt method.) The result is

$$u(t) \cong u_0(t, \varepsilon t) = \frac{2a}{\sqrt{a^2 + (4 - a^2)\mathrm{e}^{-\varepsilon t}}} \cos t. \tag{12.56}$$

This is the same result (to first order) as would be found by applying averaging to (12.39), but it has been found without any preliminary coordinate transformations. On the other hand, the possibility of constructing the solution depended on the correct initial guess as to the time scales to be used; the method of averaging generates the needed time scales automatically. The solution (12.56) exhibits oscillations tending toward a limit cycle that is a simple harmonic motion of amplitude 2. This is qualitatively correct, but the motion is not simple harmonic; carrying the solution to higher orders will introduce corrections.

12.3.6 Normal Forms

Suppose that the origin is a rest point for a system $\dot{\mathbf{x}} = \mathbf{f}(\mathbf{x})$, $\mathbf{x} \in \mathbb{R}^n$, and it is desired to study solutions of the system near this point. (Any rest point can be moved to the origin by a shift of coordinates.) The system can be expanded in a (not necessarily convergent) series

$$\dot{\mathbf{x}} = A\mathbf{x} + \mathbf{f}_2(\mathbf{x}) + \mathbf{f}_3(\mathbf{x}) + \cdots, \qquad (12.57)$$

where A is a matrix, $\mathbf{f}_2$ consists of homogeneous quadratic terms, and so forth. The matrix A can be brought into real canonical form by a change of coordinates (or into Jordan canonical form, if one is willing to allow complex variables and keep track of the conditions guaranteeing reality in the original variables). The object of normal form theory is to continue this simplification process into the higher-order terms. This is usually done recursively, one degree at a time, by applying changes of coordinates that differ from the identity by terms having the same degree as the term to be simplified. This is an example of a *coordinate perturbation* (Section 12.2.2), because it is $\|\mathbf{x}\|$ that is small, not a perturbation parameter. However, writing $\mathbf{x} = \varepsilon\xi$ turns (12.57) into

$$\dot{\xi} = A\xi + \varepsilon\mathbf{f}_2(\xi) + \varepsilon^2\mathbf{f}_3(\xi) + \cdots, \qquad (12.58)$$

which is an ordinary perturbation of a linear problem.

When A is *semisimple* (diagonalizable using complex numbers), it is possible to bring all of the terms $\mathbf{f}_2, \mathbf{f}_3, \ldots$ (up to any desired order) into a form that exhibits symmetries determined by A. For instance,

$$\begin{aligned}\begin{bmatrix} \dot{x} \\ \dot{y} \end{bmatrix} = \begin{bmatrix} 0 & -1 \\ 1 & 0 \end{bmatrix} \begin{bmatrix} x \\ y \end{bmatrix} &+\alpha_1 \left(x^2+y^2\right) \begin{bmatrix} x \\ y \end{bmatrix} \\ &+\beta_1 \left(x^2+y^2\right) \begin{bmatrix} -y \\ x \end{bmatrix} \\ &+\alpha_2 \left(x^2+y^2\right)^2 \begin{bmatrix} x \\ y \end{bmatrix} \\ &+\beta_2 \left(x^2+y^2\right)^2 \begin{bmatrix} -y \\ x \end{bmatrix} + \cdots \end{aligned} \qquad (12.59)$$

is the normal form for any system having this 2×2 matrix for its linear part; all terms of even degree have been removed, and all remaining terms of odd degree are symmetrical under the group of rotations about the origin, which is just the group generated by the linear part of (12.59). Because of this symmetry, the system is quite simple in polar coordinates:

$$\begin{aligned} \dot{r} &= \alpha_1 r^3 + \alpha_2 r^5 + \cdots + \alpha_k r^{2k+1}, \\ \dot{\theta} &= 1 + \beta_1 r^2 + \beta_2 r^4 + \cdots + \beta_k r^{2k}. \end{aligned} \qquad (12.60)$$

This is solvable by quadrature and (even without integration) the first nonzero α_j determines the stability of the origin. In general, when A is semisimple, the system in normal form always gains enough symmetry to reveal certain geometrical structures called *preserved foliations*, and frequently is solvable by quadrature. These solutions have error estimates (due to truncation) similar to those of the method of averaging, to which the method of normal forms is closely related.

When A is not semisimple (its Jordan form has off-diagonal ones), the results of normalization are not so easy to explain or to use, because the nonlinear terms acquire a symmetry different from that of the linear term. Nevertheless, the normal form in such cases has proven essential to the study of such problems as the Takens–Bogdanov and Hamiltonian Hopf bifurcations. A full exposition of normal form theory is given in

[7]. Popular expositions covering only the semisimple case are [8] and [9].

12.3.7 Perturbation of Invariant Manifolds; Melnikov Functions

With the steadily increasing importance of nonlinear phenomena such as chaos and strange attractors, finding solutions of specific initial value problems often becomes less important than finding families (manifolds) of solutions characterized by their qualitative behavior. Many of these problems are accessible by means of perturbation theory. We will briefly describe one example. If a dynamical system has a rest point of saddle type, there will exist a *stable manifold* and an *unstable manifold* of the saddle point; the former consists of all points that approach the saddle point as $t \to \infty$, the latter of points approaching the saddle as $t \to -\infty$. In some cases, the stable and unstable manifold will coincide; that is, points that approach the saddle in the distant future also emerged from it in the distant past. (The simplest case occurs in the plane when the stable and unstable manifolds form a figure-eight pattern with the saddle at the crossing point.) If such a system is perturbed, it is important to decide whether the stable and unstable manifolds separate, or continue to intersect; and if they intersect, whether they are transverse. (The latter case leads to chaotic motion.) The criterion that in many cases decides between these alternatives is based on the *Melnikov function*; if this function has simple zeros, the manifolds will intersect transversely and there will be a chaotic region. The Melnikov function is an integral over the homoclinic orbit of the normal component of the perturbation; the form of the integral is determined by applying regular perturbation methods to the solutions in the stable and unstable manifolds and measuring the distance between the approximate solutions. For details see [10].

12.4 Initial and Boundary Layers

The problems considered in Sections 12.3.2–12.3.5 are regular perturbation problems when considered on a fixed interval of time, but become singular when considered on an expanding interval such as $0 \le t \le 1/\varepsilon$. We now turn to problems that are singular even on a fixed interval. It is not easy to solve these problems even numerically, because for sufficiently small ε they are what numerical analysts call "stiff." Each of these problems has (in some coordinate system) a small parameter multiplying a (highest-order) derivative.

12.4.1 Multiple-Scale Method for Initial Layer Problems

As a first example, we consider initial value problems of the form

$$\begin{aligned} &\varepsilon \ddot{u} + b(t)\dot{u} + c(t)u = 0 \\ &u(0) = \alpha \\ &\dot{u}(0) = \frac{\beta}{\varepsilon} + \gamma. \end{aligned} \tag{12.61}$$

One may think, for instance, of an object of small mass ε subjected to a time-dependent restoring force and friction; at time zero, the position and velocity have just reached α and γ when the object is subjected to an impulse-imparting momentum β, increasing the velocity by an amount β/ε, which is large because ε is small. We will use this example to explain two methods that are applicable to many problems in which a small parameter multiplies the highest derivative.

In approaching any perturbation problem, one first tries to understand the case $\varepsilon = 0$, but here it does not make sense to set $\varepsilon = 0$. On one hand, the differential equation drops from second order to first, and can no longer accept two initial conditions; on the other hand, the second initial condition becomes infinite. Progress can be made, however, by introducing the "stretched" time variable

$$\tau = \frac{t}{\varepsilon}. \tag{12.62}$$

Upon substituting (12.62) into (12.61) and writing $' = \mathrm{d}/\mathrm{d}\tau$, we obtain

$$\begin{aligned} &u'' + b(\varepsilon\tau)u' + \varepsilon c(\varepsilon\tau)u = 0 \qquad (12.63)\\ &u(0) = \alpha\\ &u'(0) = \beta + \varepsilon\gamma. \end{aligned}$$

This problem is regular (for a fixed interval of τ) and can be solved readily. For a first approximation, it suffices to set $\varepsilon = 0$ in (12.63) to obtain $u'' + b_0u' = 0$ with $b_0 = b(0)$; the solution is

$$u_0^{\mathrm{i}} = -\frac{\beta}{b_0}\mathrm{e}^{-b_0\tau} + \alpha + \frac{\beta}{b_0}, \tag{12.64}$$

called the first-order *inner solution.* (Higher-order approximations can be found by substituting a perturbation series $u_0^{\mathrm{i}} + \varepsilon u_1^{\mathrm{i}} + \cdots$ into (12.63).) The name "inner solution" comes from the fact that (12.64) is only uniformly valid on an interval such as $0 \le \tau \le 1$, which translates into $0 \le t \le \varepsilon$ in the original time variable; this is a narrow "inner region" close to the initial conditions. It is necessary somehow to extend this to a solution valid for a fixed interval of t. This is, of course, equivalent to an expanding interval of τ, and one might attempt to solve (12.63) on such an expanding interval by previously discussed methods. The equation cannot be put in a form suitable for averaging. However, the method of multiple scales is flexible enough to be adapted to this situation. One takes as time scales τ and t, and seeks a solution in the form

$$\begin{aligned} u \cong &\{u_0^{\mathrm{i}}(\tau) + u_0^{\mathrm{cor}}(t)\}\\ &+\varepsilon\{u_1^{\mathrm{i}}(\tau) + u_1^{\mathrm{cor}}(t)\} + \cdots. \end{aligned} \tag{12.65}$$

(We could have taken $u_0(\tau, t) + \varepsilon u_1(\tau, t) + \cdots$, but the solution turns out to be the sum of the previously calculated u^{i} and a "correction" u^{cor}, so it is convenient to postulate this form initially.) One can differentiate (12.65) with respect to τ using (12.62) and substitute it into (12.63), or equivalently differentiate with respect to t and substitute into (12.61). Assuming that $u^{\mathrm{cor}}(0) = 0$ (because the inner part u^{i} should suffice initially), one finds that u^{i} must satisfy (12.63) as expected, and that u^{cor} satisfies a first-order differential equation together with the assumed initial condition $u^{\mathrm{cor}}(0) = 0$; thus u^{cor} is fully determined. At the first order, u_0^{cor} in fact satisfies the differential equation obtained from (12.61) by setting $\varepsilon = 0$; this is the very equation that we initially discarded as unlikely to be meaningful. Upon solving this equation (with zero initial conditions) and adding the result to u_0^{i}, we obtain the *composite solution*

$$\begin{aligned} u_0^{\mathrm{c}} = u_0^{\mathrm{i}} + u_0^{\mathrm{cor}} = &-\frac{\beta}{b_0}\mathrm{e}^{-b_0t/\varepsilon}\\ &+\left(\alpha + \frac{\beta}{b_0}\right)\exp\left[-\int_0^t \frac{c(s)}{b(s)}\,\mathrm{d}s\right]. \end{aligned} \tag{12.66}$$

This solution is uniformly valid on any fixed interval of t.

12.4.2 Matching for Initial Layer Problems

Although the method of multiple scales is successful for problems of this type, it is not used as widely as the *method of matched asymptotic expansions*, probably because multiple scales requires that the choices of gauges and scales be made in advance, whereas matching allows for the discovery of the correct gauges and scales as one proceeds. (Recall that gauges are the functions $\delta_i(\varepsilon)$, usually just powers ε^i, that multiply successive terms of a perturbation series; scales are the stretched time or space variables used.) To apply the matching method to (12.61), begin with the first-order inner solution (12.64) that is valid near the origin. Assume that at some distance from the origin, a good first approximation should be given by setting $\varepsilon = 0$ in (12.61) and discarding the initial conditions (which we have already seen do not make sense with $\varepsilon = 0$); the result is

$$u_0^{\text{o}} = A \exp\left[-\int_0^t \frac{c(s)}{b(s)}\,\mathrm{d}s\right], \qquad (12.67)$$

called the first-order *outer solution*. As we have discarded the initial conditions, the quantity A in (12.67) remains undetermined at this point. Now one compares the inner solution (12.64) with the outer solution (12.67) in an effort to determine the correct value of A so that these solutions "match." In the present instance, the inner solution decays rapidly (assuming $b_0 > 0$) to $\alpha + \beta/b_0$, while the outer solution has A as its initial value (at $t = 0$). One might try to determine where the "initial layer" ends, and choose A so that u_0^{i} and u_0^{o} agree at that point; but in fact it is sufficient to set $A = \alpha + \beta/b_0$ on the assumption that the inner solution reaches this value at a point close enough to $t = 0$ to allow taking it as the initial condition for the outer solution. Finally, we note that adding the inner and outer solutions would duplicate the quantity $\alpha + \beta/b_0$ with which one ends and the other begins, so we subtract this "common part" u^{io} of the inner and outer solutions to obtain the composite solution

$$u^{\text{c}} = u^{\text{i}} + u^{\text{o}} - u^{\text{io}}, \qquad (12.68)$$

which is equal to the result (12.66) obtained by multiple scales.

In the last paragraph, we have cobbled together the inner and outer solution in a very ad hoc manner. In fact, several systematic procedures exist for carrying out the matching of u^{i} and u^{o} to any order and extracting the common part u^{io}. The most common procedure consists of what are sometimes called the *van Dyke matching rules*, details of which will be given below. Although this procedure is simple to use, it does not always lead to correct results, in particular, when it is necessary to use logarithmic gauges. The other methods, *matching in an intermediate variable* and *matching in an overlap domain*, are too lengthy to explain here (see [11]), but give better results in difficult cases. None of these methods has a rigorous justification *as a method*, although it is often possible to justify the results for a particular problem or class of problems. Occasionally, one encounters problems in which the inner and outer solutions cannot be matched. These cases sometimes require a "triple deck," that is, a third (or even fourth) layer time scale. In other cases, there does not exist a computable asymptotic approximation to the exact solution.

To explain the van Dyke matching rules, we will first assume that the inner and outer solutions $u^{\text{i}}(\tau, \varepsilon)$ and $u^{\text{o}}(t, \varepsilon)$ have been computed to some order ε^k. In the problem we have been studying, the outer solution contains undetermined constants whose value must be determined, and the inner solution contains none, but in more general problems to be considered below there may be undetermined constants in both. It is important to understand that the inner and outer solutions are naturally computed in such a way that u^{i} is "expanded in powers of ε with τ fixed" while u^{o} is "expanded in powers of ε with t fixed." We are about to re-expand each solution with the opposite variable fixed. *The first step* is to express u^{i} in the outer variable t by setting $\tau = \varepsilon t$. The resulting function of t and ε is then expanded in powers of ε to order ε^k, holding t fixed. This new expansion is called u^{io}, the *outer expansion of the inner solution.* Notice that in computing u^{io} we retain only the terms of degree $\leq k$, so that in effect part of u^{i} is discarded because it moves up to order higher than k; the meaning of this is that the discarded terms of u^{i} are insignificant, at the desired order of approximation, in the outer region. *The second step* is to express u^{o} in the inner variable τ by setting $t = \tau/\varepsilon$ and expand the resulting function of τ and ε in powers of ε to order k holding τ constant. The result, called u^{oi} or the *inner expansion of the outer solution*, contains those parts of u^{o} that are significant in the inner region (to order k), arranged according to their significance in the inner region. *The third step* is to set $u^{\mathrm{io}} = u^{\mathrm{oi}}$ and use this equation to determine the unknown constants. The rationale for this is that if the domains of validity of the inner and outer regions overlap, then, because the inner solution is valid in the overlap, but the overlap belongs to the outer region, u^{io}, which is the inner solution stripped of the part that is insignificant in the outer region, should be valid there; similarly, because the outer solution is valid in the overlap, but the overlap belongs to the inner region, u^{oi} should be valid there. Now in setting $u^{\mathrm{io}} = u^{\mathrm{oi}}$, it is not possible to carry out the necessary computations unless both are expressed in the same variable, so it is necessary to choose either t or τ and express both sides in that variable before attempting to determine the unknown constants. *The fourth step* is to compute the composite solution $u^{\mathrm{c}} = u^{\mathrm{i}} + u^{\mathrm{o}} - u^{\mathrm{io}}$. At this stage, u^{io} (which is equal to u^{oi}) is known as the *common part* of u^{i} and u^{o}; it is subtracted because otherwise it would be represented twice in the solution.

12.4.3 Slow–Fast Systems

The systems considered above, and many others, can be put into the form

$$\begin{aligned}\dot{\mathbf{x}} &= \mathbf{f}(\mathbf{x}, \mathbf{y}, \varepsilon) \\ \varepsilon\dot{\mathbf{y}} &= \mathbf{g}(\mathbf{x}, \mathbf{y}, \varepsilon)\end{aligned} \tag{12.69}$$

with $\mathbf{x} \in \mathbb{R}^n$ and $\mathbf{y} \in \mathbb{R}^m$, which is called a *slow–fast system.* When $\varepsilon = 0$, the second equation changes drastically, ceasing to be differential; the motion is confined to the set $\mathbf{g}(\mathbf{x}, \mathbf{y}) = 0$, called the *slow manifold.*

We now assume for simplicity that $n = m = 1$, so that the vectors $\mathbf{x}$ and $\mathbf{y}$ become scalars x and y, and the slow manifold becomes the *slow curve.* For $\varepsilon \neq 0$, the entire (x, y) plane is available, but (assuming $\partial g/\partial y < 0$, in which case we say the slow curve is *stable*) any point moves rapidly toward the slow curve and then slowly "along" it (close to, but not actually on it). These two stages of the motion can be approximated separately as inner and outer solutions and then matched. To obtain the inner solution (the rapid part) one rescales time by setting $t = \varepsilon\tau$ and obtains (with $' = \mathrm{d}/\mathrm{d}\tau$)

$$\begin{aligned}x' &= \varepsilon f(x, y, \varepsilon) \\ y' &= g(x, y, \varepsilon),\end{aligned} \tag{12.70}$$

in which ε no longer multiplies a derivative. This problem is regular (Section 12.3.1) on finite intervals of τ, which are short intervals of t. For details of the matching see [4, Section 7.7] and [12, Chapters 6 and 7].

An interesting case arises when the slow curve is S-shaped, with the upper and lower branches stable and the middle section (the doubled-over part) unstable. A point can move along a stable branch until it reaches a vertical tangent point, then "fall off" and move rapidly to the other stable branch, then move along that branch to the other vertical tangent point and "fall off" the other way, leading to a cyclic motion called a *relaxation oscillation*. In a further, very unusual scenario, the point may actually turn the corner at the vertical tangent point and follow the unstable branch for some distance before "falling." This rather recently discovered phenomenon is called a *canard*. The explanation of canards is that several time scales become involved; the solution is actually "falling" away from the unstable branch all the time, but doing so at a rate that is slow even compared to the already slow motion along the branch. For relaxation oscillations and canards, see [13].

Recently, an approach to slow–fast systems (in any number of dimensions) called *geometric singular perturbation theory* has come into prominence. Initiated by Fenichel, the idea is to prove that (12.69) for ε near zero has an actual invariant manifold close to the slow manifold defined above, and that solutions near this manifold are (in the stable case) asymptotic to solutions in the manifold, with asymptotic phase. The emphasis is on a clear geometric description of the motion rather than on computation, but computational aspects are included. A good introduction is [14].

12.4.4 Boundary Layer Problems

Problems in which a small parameter multiplies the highest derivative are encountered among boundary value problems at least as frequently as among initial value problems. As the basic ideas have been covered in the previous sections, it is only necessary to point out the differences that arise in the boundary value case. Either the multiple-scale or matching methods may be used; we will use matching. The method will be illustrated here with an ordinary differential equation; a partial differential equation will be treated in Section 12.4.5.

Consider the problem

$$\varepsilon y'' + b(x)y' + c(x)y = 0 \qquad (12.71)$$
$$y(0) = \alpha$$
$$y(1) = \beta$$

on the interval $0 \le x \le 1$. The differential equation here is the same as (12.61), only the independent variable is a space variable rather than time in view of the usual applications. If $b(x)$ is positive throughout $0 < x < 1$, there will be a boundary layer at the left end point $x = 0$; if negative, the boundary layer will be at the right end point; and if $b(x)$ changes sign, there may be internal transition layers as well. We will consider the first case. To the first order, the outer solution y^{o} will satisfy the first-order equation $b(x)y' + c(x)y = 0$ obtained by setting $\varepsilon = 0$ in (12.71); it will also satisfy the right-hand boundary condition $y(1) = \beta$. Therefore, the outer solution is completely determined. The first-order inner solution y^{i} will satisfy the equation $\mathrm{d}^2y/\mathrm{d}\xi^2 + b_0y = 0$ with $b_0 = b(0)$, obtained by substituting the stretched variable $\xi = x/\varepsilon$ into (12.71) and setting $\varepsilon = 0$; it will also satisfy the left-hand boundary condition $y(0) = \alpha$. As this is a second-order equation with only one boundary condition, it will contain an undetermined constant that must be identified by matching the inner and outer solutions. The differential equations satisfied by the inner and outer solutions are the same as in the case of (12.61), the only difference being that this time the constant that must be fixed by matching belongs to the inner solution rather than the outer.

12.4.5 WKB Method

There are a great variety of problems that are more degenerate than the one we have just discussed, which can exhibit a wide range of exotic behaviors. These include internal layers, in which a stretched variable such as $\xi = (x - x_0)/\varepsilon$ must be introduced around a point x_0 in the interior of the domain; triple decks, in which two stretched variables such as x/ε and x/ε^2 must be introduced at one end; and problems in which the order of the differential equation drops by more than one. The simplest example of the latter type is

$$\varepsilon^2 y'' + a(x)y = 0. \qquad (12.72)$$

This problem is usually addressed by a technique called the *WKB* or *WKBJ method.* This method is rather different in spirit from the others we have discussed, because it depends heavily on the linearity of the perturbed problem. Rather than pose initial or boundary value problems, one finds approximations for two linearly independent solutions of the linear equation (12.72) on the whole real line. The general solution then consists of the linear combinations of these two. If $a(x) = k^2(x) > 0$, these approximate solutions are

$$y^{(1)} \cong \frac{1}{\sqrt{k(x)}} \cos \frac{1}{\varepsilon} \int k(x)\, dx \qquad (12.73)$$

and

$$y^{(2)} \cong \frac{1}{\sqrt{k(x)}} \sin \frac{1}{\varepsilon} \int k(x)\, dx. \qquad (12.74)$$

If $a(x) = -k^2(x) < 0$, the two solutions are

$$y^{(1),(2)} \cong \frac{1}{\sqrt{k(x)}} \exp \frac{1}{\varepsilon} \int \pm k(x)\, dx. \qquad (12.75)$$

A recent derivation of (12.75) is given in Section 12.5; for classical approaches, see [15] and [4].

If $a(x)$ changes sign, one has a difficult situation called a *turning point problem.* This can be addressed in various ways by matching solutions of these two types or by using Airy functions. The latter are solutions of the differential equation $y'' + xy = 0$, which is the simplest problem with a turning point at the origin. These Airy functions can be considered as known (they can be expressed using Bessel functions of order 1/3) and solutions to more general turning point problems can be expressed in terms of them. For an introduction to turning point problems, see [16, Chapter 2], and for theory see [17, Chapter 8].

12.4.6 Fluid Flow

We will conclude this section with a brief discussion of the problem of fluid flow past a flat plate, because of its historical importance (see Section 12.1) and because it illustrates two aspects of perturbation theory that we have avoided so far: the use of perturbation theory for partial differential equations and the need to combine undetermined scales with undetermined gauges. The classic reference for this material is [18]. Consider a plane fluid flow in the upper half-plane, with a "flat plate" occupying the interval $0 \le x \le 1$ on the x-axis; that is, the fluid will adhere to this interval, but not to the rest of the x-axis. The stream function $\psi(x, y)$ of such a fluid will satisfy

$$\varepsilon(\psi_{xxxx} + 2\psi_{xxyy} + \psi_{yyyy}) - \psi_y(\psi_{xxx} + \psi_{xyy}) \\ +\psi_x(\psi_{xxy} + \psi_{yyy}) = 0 \tag{12.76}$$

with $\psi(x, 0) = 0$ for $-\infty < x < \infty$, $\psi_y(x, 0) = 0$ for $0 \leq x \leq 1$, and $\psi(x, y) \to y$ as $x^2 + y^2 \to \infty$. The latter condition describes the flow away from the plate, and this in fact gives the leading order outer solution as

$$\psi^{\mathrm{o}}(x, y) = y. \tag{12.77}$$

To find an inner solution, we stretch y by an undetermined scale factor,

$$\eta = \frac{y}{\mu(\varepsilon)}, \tag{12.78}$$

and expand the inner solution using undetermined gauges, giving (to first order)

$$\psi^{\mathrm{i}} = \delta(\varepsilon)\Psi(x, \eta). \tag{12.79}$$

Substituting this into (12.76) and discarding terms that are clearly of higher order yields

$$\frac{\varepsilon}{\mu}\Psi_{\eta\eta\eta\eta} + \delta\left[\Psi_x\Psi_{\eta\eta\eta} - \Psi_\eta\Psi_{x\eta\eta}\right] = 0. \tag{12.80}$$

The relative significance of ε/μ and δ has not yet been determined, but if either of them were dominant, the other term would drop out of (12.80) to first order, and the resulting solution would be too simple to capture the behavior of the problem. So we must set

$$\frac{\varepsilon}{\mu} = \delta \tag{12.81}$$

and conclude that the first-order inner solution satisfies

$$\Psi_{\eta\eta\eta\eta} + \Psi_x\Psi_{\eta\eta\eta} - \Psi_\eta\Psi_{x\eta\eta} = 0. \tag{12.82}$$

It is not possible to solve (12.82) in closed form, but it is possible to express the solution as

$$\Psi(x, \eta) = \sqrt{2x}f\left(\frac{\eta}{\sqrt{2x}}\right), \tag{12.83}$$

where f is the solution of the ordinary differential equation $f''' + ff'' = 0$ with $f(0) = 0, f'(0) = 0$, and $f'(\infty) = 1$. In attempting to match the inner and outer solutions, it is discovered that this is only possible if $\delta = \mu$. Together with (12.81), this finally fixes the undetermined scales and gauges as

$$\delta = \mu = \sqrt{\varepsilon}. \tag{12.84}$$

Upon attempting to continue the solution to higher orders, obstacles are encountered that can only be overcome by introducing triple decks and other innovations. See [19] and [20].

12.5
The "Renormalization Group" Method

For many years, applied mathematicians have wished for a unified treatment of the types of perturbation problems studied in Sections 12.3 and 12.4. Over the past 20 years, some progress has been made in this direction under the name *renormalization group* or *RG* method. This *mathematical* RG method has its roots in an earlier *physical* RG method (in quantum field theory and statistical mechanics) that we do not discuss here; see [21] for the relations between these. The RG method provides a single heuristic that works for most types of perturbation problems in ordinary and partial differential equations. While the common perturbation methods abandon the naive (or straightforward) expansion when it only has pointwise validity, the

RG method is built on the idea that the naive solution contains all the information necessary to construct a uniformly ordered formal solution; this information is merely arranged incorrectly. The RG method does not (at least so far) automatically prove that the uniformly ordered solution is uniformly valid (on an intended domain). Therefore proofs of validity have not been unified, but are still distinct for different types of problems. The name "RG method" is somewhat unfortunate, because no group is involved, but it is firmly established.

The RG method introduces two new operations into perturbation theory, the *absorption process* and the *envelope process*. There are three forms of the RG method, which we classify as follows.

1. The *mixed method*, which uses both the absorption and envelope processes, was introduced by Chen *et al.* [22] on the basis of the RG method in physics, and later simplified by Ziane [23] and Lee DeVille *et al.* [24].
2. The *pure envelope method*, which uses only the envelope process, was introduced by Woodruff [25] under the name *invariance method*, before the (mathematical) RG method (and with no reference to the physical one). It was rediscovered by Paquette [26] as an improvement on the mixed method, because it can handle problems in which absorption does not work. Nevertheless, it has not become well known.
3. The *pure absorption method*, which avoids the envelope process, was introduced by Kirkinis [27] and popularized in [28].

The envelope process was referred to by other names (RG equation, invariance condition) until it was identified by Kunihiro [29, 30] as equivalent to the classical notion of an envelope of a family of curves.

Our presentation is limited to ordinary differential equations, and (mostly) to leading order approximations. For important applications to partial differential equations in fluid mechanics, see [31].

12.5.1 Initial and Boundary Layer Problems

We begin with the simplest initial layer problem

$$\begin{aligned} &\varepsilon \ddot{u} + \dot{u} + u = 0 \qquad (12.85)\\ &u(0) = \alpha \\ &\dot{u}(0) = \frac{\beta}{\varepsilon} + \gamma, \end{aligned}$$

of the form (12.61). We immediately convert this to

$$\begin{aligned} &u'' + u' + \varepsilon u = 0 \qquad (12.86)\\ &u(0) = \alpha \\ &u'(0) = \beta + \varepsilon\gamma, \end{aligned}$$

which is of the form (12.63). Recall that $\dot{} = \mathrm{d}/\mathrm{d}t$, $' = \mathrm{d}/\mathrm{d}\tau$, and $\tau = t/\varepsilon$ is the fast variable; it is a general rule in the RG method to work in the fastest natural independent variable. Another feature of all RG methods is to set the initial (or boundary) conditions aside temporarily and instead seek a general solution with integration constants. For methods that use absorption (methods 1 and 3), this is done as follows. Putting $u = u_0 + \varepsilon u_1 + \cdots$, the naive perturbation method gives

$$\begin{aligned} &u_0'' + u_0' = 0 \qquad (12.87)\\ &u_1'' + u_1 = -u_0 \end{aligned}$$

so that

$$u_0 = a + be^{-\tau} \qquad (12.88)$$
$$u_1 = -a\tau + b\tau e^{-\tau} + c + de^{-\tau},$$

where a, b, c, and d are arbitrary. Here $-a\tau + b\tau e^{-t}$ is a particular solution for u_1, while $c + de^{-\tau}$ is the general solution of the associated homogeneous problem $u_1'' + u_1 = 0$. It is convenient to delete this "homogeneous part" from u_1 (and all higher u_i) by allowing a and b to be functions of ε, with

$$a = a(\varepsilon) = a_0 + \varepsilon a_1 + \cdots,$$
$$b = b(\varepsilon) = b_0 + \varepsilon b_1 + \cdots; \qquad (12.89)$$

thus a_1 and b_1 replace c and d. Then a straightforward (or naive) approximation to the general solution is

$$\tilde{u}(\tau, \varepsilon) = a + be^{-\tau} + \varepsilon(-a\tau + b\tau e^{-\tau}). \qquad (12.90)$$

By regular perturbation theory, this approximation is uniformly valid with error $\mathcal{O}(\varepsilon^2)$ for $0 \leq \tau \leq 1$, that is, in the initial layer $0 \leq t \leq \varepsilon$. But it is not uniformly ordered on $0 \leq \tau \leq 1/\varepsilon$ $(0 \leq t \leq 1)$, so it cannot be uniformly valid there. (The term $-\varepsilon a\tau$ is secular, becoming unbounded as $\tau \to \infty$; for $\tau = 1/\varepsilon$, this term is formally $\mathcal{O}(\varepsilon)$ but actually $\mathcal{O}(1)$. The term $\varepsilon b\tau e^{-\tau}$ is bounded, achieving its maximum at $\tau = 1$, so it is not truly secular, but is often classified as secular anyway because it is "secular relative to $be^{-\tau}$").

Continuing by the mixed method (method 1), let $\tau_0 > 0$ be arbitrary, and split the secular terms of (12.90) as follows:

$$\tilde{u}(\tau, \tau_0, \varepsilon) = a + be^{-\tau} + \varepsilon[-a\tau_0 + b\tau_0 e^{-\tau}$$
$$-a(\tau - \tau_0) + b(\tau - \tau_0)e^{-\tau}]. \qquad (12.91)$$

Now the terms in $\tau - \tau_0$ are secular, while those in τ_0 alone are not (because τ_0 is considered constant). Next we perform an *absorption operation* by setting

$$a_r(\tau_0, \varepsilon) = a - \varepsilon a\tau_0 + \mathcal{O}(\varepsilon^2),$$
$$b_r(\tau_0, \varepsilon) = b + \varepsilon b\tau_0 + \mathcal{O}(\varepsilon^2), \qquad (12.92)$$

remembering that (12.89) is still in effect, and rewriting (12.91) as

$$\tilde{u}(\tau, \tau_0, \varepsilon) = a_r + b_r e^{-\tau} + \varepsilon[-a_r(\tau - \tau_0)\tau_0$$
$$+b_r(\tau - \tau_0)e^{-\tau}]. \qquad (12.93)$$

This is called "absorbing the secular terms into the integration constants a and b to produce renormalized constants a_r and b_r." ("Constant" means independent of τ. Later these constants will become slowly varying functions of τ.) This produces a *family* of naive solutions of (12.86), one naive solution for each τ_0 (ignoring initial conditions). Each of the solutions in this family is pointwise asymptotically valid with error $\mathcal{O}(\varepsilon^2)$ for all t, but is uniformly valid only on a bounded interval around τ_0; we say that $\tilde{u}(t, \tau_0, \varepsilon)$ is a *family of locally valid approximate solutions.*

At this point, we stop to observe that (12.93) could have been obtained at the start, in place of (12.90). Instead of omitting the "homogeneous part" of u_1, we could have included any "homogeneous part" that we like in u_1 (satisfying $u_1'' + u_1' = 0$), for instance, $a\tau_0 - b\tau_0 e^{-\tau}$, giving $u_1 = -a(\tau - \tau_0) + b(\tau - \tau_0)e^{-\tau}$, which immediately produces

$$\tilde{u}(\tau, \tau_0, \varepsilon) = a + be^{-\tau} + \varepsilon(-a(\tau - \tau_0)\tau_0$$
$$+b(\tau - \tau_0)e^{-\tau}). \qquad (12.94)$$

This is the same as (12.93) except that there are no "renormalized variables" and no (12.92). (This equation would never be used again anyway.) The idea of this "pure envelope method" (method 2) is to select the homogeneous part of each u_i so that the secular terms vanish at τ_0. Note that a and b in (12.90) are functions of ε by (12.89) and of τ_0 because they can be chosen independently for each choice of τ_0.

From this point on, methods 1 and 2 coincide; we use the notation of method 2. The *goal* is to perform a kind of asymptotic matching of the (continuumly many) local solutions (12.94) to produce a single, smoothed-out solution that is as good as each local solution where that local solution is good. The *procedure* to achieve this goal is to differentiate $\tilde{u}$ with respect to τ_0, set $\tau_0 = \tau$, and set the result equal to zero. The *justification* for this procedure varies from author to author; the best are due to Kunihiro and Paquette. Kunihiro shows that this procedure amounts to taking the envelope of the local solutions in such a way that the envelope is tangent to each local solution at the point where $\tau = \tau_0$, that is, the point where that local solution is best. Paquette shows that the procedure amounts to choosing the τ_0-dependent coefficients in (12.94) in such a way that two local solutions with nearby values of τ_0 have an overlap domain in which both are equally valid asymptotically. (Woodruff's approach is equivalent, but more complicated. Chen et. al. claimed vaguely that a_r and b_r in (12.93) should be chosen so that $\tilde{u}$ does not depend on τ_0, because τ_0 was introduced artificially into the problem and was not part of the original data. Thus the derivative with respect to τ_0 should vanish, and, again because $\tilde{u}$ should not depend on τ_0, we should be free to set $\tau_0 = \tau$. This argument was quickly recognized as inadequate, because the derivative does not vanish in general, but only when $\tau_0 = \tau$.)

Now we carry out the *envelope procedure* on the example at hand. Differentiating (12.94) gives

$$\frac{\partial \tilde{u}}{\partial \tau_0} = \frac{\partial a}{\partial \tau_0} + \frac{\partial b}{\partial \tau_0} e^{-\tau} + \varepsilon \left(-\frac{\partial a}{\partial \tau_0}(\tau - \tau_0) + a + \frac{\partial b}{\partial \tau_0}(\tau - \tau_0) e^{-\tau} - b e^{-\tau} \right). \quad (12.95)$$

Setting $\tau = \tau_0$ and equating to zero gives

$$\left(\frac{\partial a}{\partial \tau} + \varepsilon a \right) + \left(\frac{\partial b}{\partial \tau} - \varepsilon b \right) e^{-\tau} = 0. \quad (12.96)$$

(Notice how setting $\tau_0 = \tau$ automatically makes a and b into functions of τ when previously they were "constants" depending only on τ_0 and ε). The only plausible way to solve (12.96) is to solve

$$\frac{\partial a}{\partial \tau} + \varepsilon a = 0, \qquad \frac{\partial b}{\partial \tau} - \varepsilon b = 0 \quad (12.97)$$

separately, giving

$$a(\tau, \varepsilon) = \overline{a}(\varepsilon) e^{-\varepsilon \tau}, \qquad b(\tau, \varepsilon) = \overline{b}(\varepsilon) e^{\varepsilon \tau}. \quad (12.98)$$

(Separating (12.96) into (12.97) can be avoided by creating two families of solutions, $\tilde{u}^{(1)}$ with $b = 0$ and $\tilde{u}^{(2)}$ with $a = 0$, and applying the envelope process to each family separately, but this does not seem to have been noticed in the literature.) Note that a and b turn out to depend on τ only through the combination $\varepsilon\tau$, and are therefore slowly varying; this shows how the RG method is able to generate the required time scales without postulating them in advance, as must be done in the multiple-scale method. Now we insert (12.98) into (12.94) with $\tau_0 = \tau$ to obtain

$$\tilde{u}(\tau, \varepsilon) = \overline{a}(\varepsilon) e^{-\varepsilon \tau} + \overline{b}(\varepsilon) e^{\varepsilon \tau} e^{-\tau}. \quad (12.99)$$

Writing $\overline{a}(\varepsilon) = \overline{a}_0 + \varepsilon \overline{a}_1 + \cdots$, $\overline{b}(\varepsilon) = \overline{b}_0 + \varepsilon \overline{b}_1 + \cdots$, with undetermined coefficients, we can now seek to satisfy the initial conditions in (12.86), finding for instance that $a_0 = \alpha + \beta$ and $b_0 = \beta$, so that the leading order solution is

$$\tilde{u} = (\alpha + \beta)e^{-\tau} - \beta e^{(-1+\varepsilon)\tau}. \tag{12.100}$$

It is easy to check (rigorously) that $e^{(-1+\varepsilon)\tau} = e^{-\tau} + \mathcal{O}(\varepsilon)$ uniformly for $0 \le \tau < 0$; if this replacement is made, (12.100) coincides with the approximation given by (12.66). As expected, γ does not appear in the leading order. Higher-order approximations can be carried out as well, and in this simple example, they coincide with truncations of the exact solution (obtained by elementary means and expanded in ε).

To solve the same problem by Kirkinis's method 3 (pure absorption), we begin again from (12.90) as in method 1. This time, however, we absorb the entire secular (and relative secular) terms into a and b without introducing τ_0, by setting

$$a_r(\tau, \varepsilon) = a(\varepsilon) - \varepsilon a(\varepsilon)\tau + \cdots,$$
$$b_r(\tau, \varepsilon) = b(\varepsilon) + \varepsilon b(\varepsilon)\tau + \cdots \tag{12.101}$$

and obtaining

$$\tilde{u}(\tau, \varepsilon) = a_r + b_r e^{-\tau}. \tag{12.102}$$

(In this simple problem, all terms of order ε are secular, so the renormalized form of $\tilde{u}$ looks like the leading term of the original form. In the problems treated below, there are usually nonsecular terms remaining at order ε.) Then we invert the equations in (12.101) to obtain

$$a(\varepsilon) = a_r(\varepsilon, \tau) + \varepsilon a_r(\varepsilon, \tau),$$
$$b(\varepsilon) = b_r(\varepsilon, \tau) - \varepsilon b_r(\varepsilon, \tau), \tag{12.103}$$

omitting terms that are formally of order ε^2. (In a more complicated problem, equations (12.101) might be coupled and need to be inverted as a system. Kirkinis writes the absorption in the inverted form from the beginning, with undetermined coefficients, which he finds recursively.) Next we differentiate (12.103) with respect to τ and truncate, which gives

$$0 = \frac{da_r}{d\tau} + \varepsilon a_r, \qquad 0 = \frac{db_r}{d\tau} - \varepsilon b_r, \tag{12.104}$$

and immediately brings us to (12.97); this differentiation step replaces the envelope process as a way of finding the "RG equations." The solution proceeds from here as before.

More general initial and boundary layer problems, including nonlinear ones, may be solved the same methods illustrated above in the simplest case.

12.5.2 Nonlinear Oscillations

Now we consider a broad class of oscillatory problems that are usually solved by the method of averaging, namely, the class of systems of the form

$$\dot{\mathbf{x}} = A\mathbf{x} + \varepsilon \mathbf{f}(\mathbf{x}), \quad \mathbf{x}(0) = \mathbf{c}, \tag{12.105}$$

where $\mathbf{x} \in \mathbb{C}^n$, A is a diagonal matrix with eigenvalues λ_i on the imaginary axis, and $\mathbf{f}(\mathbf{x})$ is a vector polynomial, expressible as a finite sum

$$\mathbf{f}(\mathbf{x}) = \sum_{\alpha,i} C_{\alpha i}\mathbf{x}^{\boldsymbol{\alpha}}\mathbf{e}_i, \tag{12.106}$$

with $\boldsymbol{\alpha} = (\alpha_1, \dots, \alpha_n)$ a nonnegative integer vector, $\mathbf{x}^{\boldsymbol{\alpha}} = x_1^{\alpha_1} \cdots x_n^{\alpha_n}$, and $\mathbf{e}_i$ the standard unit vectors in $\mathbb{C}^n$. This class of problems includes all single and multiple oscillators with weak polynomial nonlinearities and weak coupling, when expressed as first-order systems in complex variables with reality conditions. We will solve this to leading order by methods 2 and 3, although our calculations are based on those of Ziane, who uses method 1. All methods begin by writing $\mathbf{x} = \mathbf{x}_0 + \varepsilon \mathbf{x}_1 + \cdots$ and obtaining

$$\dot{\mathbf{x}}_0 = A\mathbf{x}_0, \qquad \dot{\mathbf{x}}_1 = A\mathbf{x}_1 + \varepsilon \mathbf{f}(\mathbf{x}_0). \quad (12.107)$$

For method 2, we take $\mathbf{x}_0$ and $\mathbf{x}_1$ to be

$$\begin{aligned} \mathbf{x}_0(t, \mathbf{a}) &= \mathrm{e}^{At}\mathbf{a}, \\ \mathbf{x}_1(t, t_0, \mathbf{a}) &= \mathrm{e}^{A(t-t_0)} \int_{t_0}^{t} \mathrm{e}^{-As}\mathbf{f}(\mathrm{e}^{As}\mathbf{a})\mathrm{d}s, \end{aligned} \quad (12.108)$$

where $\mathbf{a}$ is an arbitrary vector of integration constants and $\mathbf{x}_1$ is the particular solution with zero initial conditions at an arbitrary point t_0. (Recall that by letting $\mathbf{a}$ depend on ε, we may choose only particular solutions at higher orders, and because $\mathbf{x}_1$ vanishes at t_0, so do any secular terms it may contain.) The integrand in (12.108) has the form

$$\begin{aligned} \mathrm{e}^{-As}\mathbf{f}(\mathrm{e}^{As}\mathbf{a}) &= \sum_{\alpha,i} C_{\alpha i}\mathrm{e}^{(\langle \lambda,\alpha\rangle - \lambda_i)s}\mathbf{a}^{\alpha}\mathbf{e}_i \quad (12.109) \\ &= \mathbf{R}(\mathbf{a}) + \mathbf{Q}(s, \mathbf{a}), \quad (12.110) \end{aligned}$$

where $\langle \lambda, \boldsymbol{\alpha} \rangle = \lambda_1\alpha_1 + \cdots \lambda_n\alpha_n$, $\mathbf{a}^{\alpha} = a_1^{\alpha_1} \cdots a_n^{\alpha_n}$, $\mathbf{R}$ contains the terms with $\langle \lambda, \boldsymbol{\alpha} \rangle - \lambda_i = 0$, and $\mathbf{Q}$ the remaining terms. It follows by examining the integral in (12.108) that

$$\mathbf{x}_1(t, t_0, \mathbf{a}) = \mathrm{e}^{A(t-t_0)}[\mathbf{R}(\mathbf{a})(t - t_0) + \mathbf{S}(t, t_0, \mathbf{a})], \quad (12.111)$$

where $\mathbf{S}$ is bounded (nonsecular). Thus

$$\begin{aligned} \tilde{\mathbf{x}}(t, t_0, \mathbf{a}, \varepsilon) &= \mathbf{x}_0 + \varepsilon\mathbf{x}_1 = \mathrm{e}^{At}\mathbf{a} + \varepsilon \mathrm{e}^{A(t-t_0)} \\ &\quad \times [\mathbf{R}(\mathbf{a})(t - t_0) + \mathbf{S}(t, t_0, \mathbf{a})] \end{aligned} \quad (12.112)$$

is a family of local approximate solutions, and we proceed to apply the envelope process:

$$\left.\frac{\partial \tilde{\mathbf{x}}}{\partial t_0}\right|_{t_0=t} = \mathrm{e}^{At}\left[\frac{\partial \mathbf{a}}{\partial t} - \varepsilon\mathbf{R}(\mathbf{a})\right] = 0. \quad (12.113)$$

It follows that

$$\frac{\partial \mathbf{a}}{\partial t} = \varepsilon \mathbf{R}(\mathbf{a}). \quad (12.114)$$

If this equation can be solved for $\mathbf{a} = \mathbf{a}(t, \varepsilon)$, the solution is inserted into

$$\tilde{\mathbf{x}}(t, \mathbf{a}, \varepsilon) = \mathrm{e}^{At}\mathbf{a} + \varepsilon[\mathbf{S}(t, t, \mathbf{a})], \quad (12.115)$$

which is obtained from (12.112) by setting $t_0 = t$.

To compare this with the method of averaging, note that (12.105) is converted into periodic standard form by setting $\mathbf{x} = \mathrm{e}^{At}\xi$ and obtaining

$$\dot{\xi} = \mathrm{e}^{-At}\mathbf{f}(\mathrm{e}^{At}\xi). \quad (12.116)$$

This corresponds to (12.37), with ξ for $\mathbf{x}$. The averaged equation (12.40) then coincides with (12.114), with $\mathbf{y}$ for $\mathbf{a}$. Finally, substituting $\mathbf{a}$ into (12.115) corresponds to first substituting $\mathbf{y}$ into (12.42) to obtain (in our present notation) ξ, and then multiplying that by e^{At} to get $\mathbf{x}$. Thus the first-order RG solution for this class of systems is exactly the same as the *improved* first-order solution by averaging (rather than the usual first-order averaging solution). This illustrates the remark made by several authors that the solution given by the RG method is sometimes *better* than the ones given by more familiar perturbation methods, even though these are asymptotically equivalent.

To solve the same problem by method 3, write the naive solution of (12.107) to

$\mathcal{O}(\varepsilon)$ as

$$\tilde{\mathbf{x}}(t, \mathbf{a}, \varepsilon) = \mathrm{e}^{At}[\mathbf{a} + \varepsilon(\mathbf{R}(\mathbf{a})t + \mathbf{S}(t, \mathbf{a}))], \tag{12.117}$$

where $\mathbf{a}$ depends on ε. Absorb the entire secular term into $\mathbf{a}$ by writing $\mathbf{a}_\mathrm{r}(t, \varepsilon) = \mathbf{a}(\varepsilon) + \varepsilon t\mathbf{R}(\mathbf{a}(\varepsilon))$ and then invert this to get

$$\mathbf{a}(\varepsilon) = \mathbf{a}_\mathrm{r}(t, \varepsilon) - \varepsilon t\mathbf{R}(\mathbf{a}_\mathrm{r}(t, \varepsilon)), \tag{12.118}$$

with

$$\tilde{\mathbf{x}}(t, \mathbf{a}_\mathrm{r}, \varepsilon) = \mathrm{e}^{At}[\mathbf{a}_\mathrm{r} + \varepsilon\mathbf{S}(t, \mathbf{a})]. \tag{12.119}$$

Differentiating (12.118) gives

$$\frac{\partial \mathbf{a}_\mathrm{r}}{\partial t} = \varepsilon\left(I - \varepsilon t\frac{\partial \mathbf{R}}{\partial \mathbf{a}_\mathrm{r}}\right)^{-1} \times\mathbf{R}(\mathbf{a}_\mathrm{r}) = \varepsilon\mathbf{R}(\mathbf{a}_\mathrm{r}) + \mathcal{O}(\varepsilon^2), \tag{12.120}$$

in agreement with (12.114) and with the first-order averaged equation. As before, the solution of (12.120) is inserted into (12.119).

The class of problems we have considered does not include all problems that can be put into periodic standard form, so the RG method does not yet completely encompass the averaging method, even for the periodic case.

12.5.3 WKB Problems

Finally, we look briefly at the WKB problem

$$\ddot{y} + k^2(\varepsilon t)y = 0, \tag{12.121}$$

which is equivalent to (12.72) with $a(x) = k^2(x)$ and $x = \varepsilon t$. This problem cannot be solved directly by methods 1 or 3 that use absorption, but can be solved by method 2, illustrating Paquette's claim that the pure envelope method is the strongest of the three forms of RG. (Chen et. al. have given a very interesting treatment of WKB problems, including those with turning points, by method 1, but they begin by changing variables to achieve a form for which absorption is possible.) The central idea is to construct local solutions by forming locally valid differential equations and solving these by "naive" expansions that are nevertheless already *generalized* asymptotic expansions (that is, ε appears in the coefficients, not just the gauges), unlike the local solutions used in previous examples. These local solutions are then "matched" by the envelope process. This technique is used by both Woodward and Paquette. We follow Paquette's (simpler) method, but where he takes only the leading term of $\tilde{y}$ (and so achieves a less accurate solution than the standard WKB approximation), we follow Woodward in taking the first two terms.

First we expand $k^2(\varepsilon t)$ in a Taylor series in $t - t_0$ for arbitrary t_0, keeping the first two terms. This results in a *local equation* approximating (12.121) near t_0:

$$\ddot{y} + [k^2(\varepsilon t_0) + 2\varepsilon k(\varepsilon t_0)k'(\varepsilon t_0)(t - t_0)]y = 0. \tag{12.122}$$

An approximate solution of the local equation is sought in the form $y = y_0 + \varepsilon y_1$; we find

$$\begin{aligned} &\ddot{y}_0 + k^2(\varepsilon t_0)y_0 = 0 \qquad (12.123)\\ &\ddot{y}_1 + k^2(\varepsilon t_0)y_1 = -2k(\varepsilon t_0)k'(\varepsilon t_0)(t - t_0)y_0. \end{aligned}$$

(Here t_0 is a constant.) Two linearly independent solutions for y_0 are

$$y_0^{(1),(2)}(t, t_0, \varepsilon) = \mathrm{e}^{\pm \mathrm{i}k(\varepsilon t_0)(t-t_0)}. \tag{12.124}$$

Choosing $y_0^{(1)}$, the inhomogeneous linear equation for $y_1^{(1)}$ is solvable by undetermined coefficients, with solution

$$y_1(t, t_0, \varepsilon) = \left[\frac{\mathrm{i}k'(\varepsilon t_0)}{2}(t - t_0)^2 - \frac{k'(\varepsilon t_0)}{2k(\varepsilon t_0)}(t - t_0) \right] \mathrm{e}^{\mathrm{i}k(\varepsilon t_0)(t - t_0)}, \tag{12.125}$$

and $y_1^{(2)}$ is similar (but will not be needed). We create from these the family of local solutions

$$\tilde{y} = a(t_0, \varepsilon) \left[1 + \varepsilon \left(\frac{\mathrm{i}k'(\varepsilon t_0)}{2}(t - t_0)^2 - \frac{k'(\varepsilon t_0)}{2k(\varepsilon t_0)}(t - t_0) \right) \right] \mathrm{e}^{\mathrm{i}k(\varepsilon t_0)(t - t_0)}, \tag{12.126}$$

where $a(t_0, \varepsilon)$ is to be found by the envelope process: compute $\partial \tilde{y}/\partial t_0$, put $t_0 = t$, and set the result equal to zero, obtaining

$$\frac{\partial a(t, \varepsilon)}{\partial t} = a(t, \varepsilon) \left[\mathrm{i}k(\varepsilon t) - \varepsilon \frac{k'(\varepsilon t)}{2k(\varepsilon t)} \right]. \tag{12.127}$$

A nonzero solution of this equation (by separation of variables) is

$$a(t, \varepsilon) = \frac{1}{\sqrt{k(\varepsilon t)}} \exp \mathrm{i} \int_0^t k(\varepsilon s) \mathrm{d}s. \tag{12.128}$$

Inserting this into (12.125) with $t_0 = t$ leads (remarkably) to exactly the same expression for $\tilde{y}$:

$$\tilde{y}(t, \varepsilon) = \frac{1}{\sqrt{k(\varepsilon t)}} \exp \mathrm{i} \int k(\varepsilon t) \mathrm{d}t. \tag{12.129}$$

The real and imaginary parts of this solution are linearly independent, real, approximate solutions of (12.121) that coincide with (12.73) and (12.74) when expressed in terms of x.

12.6 Perturbations of Matrices and Spectra

In this section, we address the question, if

$$A(\varepsilon) = A_0 + \varepsilon A_1 + \cdots \tag{12.130}$$

is a matrix or linear operator depending on a small parameter, and the spectrum of A_0 is known, can we determine the spectrum of $A(\varepsilon)$ for small ε? For the case of a matrix, the spectrum is simply the set of eigenvalues (values of λ for which $A\mathbf{v} = \lambda\mathbf{v}$ for some nonzero column vector $\mathbf{v}$ called an *eigenvector*). More generally, the spectrum is defined as the set of λ for which $A - \lambda I$ is not invertible; for linear transformations on infinite-dimensional spaces (such as Hilbert or Banach spaces), this need not imply the existence of an eigenvector. Our attention here will be focused on the matrix case, but many of the procedures (excluding those that involve the determinant or the Jordan normal form) are applicable as well to any operators whose spectrum consists of eigenvalues. The classical reference for the general infinite dimensional case is [32]. For matrices that are not diagonalizable, one can (and should) ask not only for the eigenvalues and eigenvectors, but also generalized eigenvectors $\mathbf{v}$ for which $(A - \lambda I)^k \mathbf{v} = 0$ for some integer $k > 1$.

The most direct approach (which we do not recommend) to finding the eigenvalues of (12.130) in the matrix case would be to examine the *characteristic equation*

$$P(\lambda, \varepsilon) = \det(A(\varepsilon) - \lambda I) = 0, \tag{12.131}$$

having the eigenvalues as roots. There are standard perturbation methods for finding the roots of polynomials (see [4, Chapter 1]), the simplest of which is to substitute

$$\lambda(\varepsilon) = \lambda_0 + \varepsilon \lambda_1 + \cdots \tag{12.132}$$

into (12.131) and solve recursively for λ_i. This method works if λ_0 is a simple root of $P(\lambda, 0) = 0$; that is, it will work if A_0 has distinct eigenvalues. If there are repeated eigenvalues, then in general it is necessary to replace (12.132) with a *fractional power series* involving gauges $\delta_i(\varepsilon) = \varepsilon^{i/q}$ for some integer q that is most readily determined by using Newton's diagram. Although these are useful perturbation methods for finding roots of general polynomials, they have two drawbacks in the case of eigenvalues: if the matrices are large, it is difficult to compute the characteristic polynomial; more importantly, these methods do not take account of the special features of eigenvalue problems. For instance, if $A(\varepsilon)$ is a symmetric matrix, then its eigenvalues will be real, and fractional powers will not be required (even if A_0 has repeated eigenvalues). But the fact that A is symmetric is lost in passing to the characteristic polynomial, and one cannot take advantage of these facts.

For these reasons, it is best to seek not only the eigenvalues but also at the same time the eigenvectors that go with them. The general procedure (which must be refined in particular situations) is to seek solutions of

$$A(\varepsilon)\mathbf{v}(\varepsilon) = \lambda(\varepsilon)\mathbf{v}(\varepsilon) \qquad (12.133)$$

in the form

$$\begin{aligned} \lambda(\varepsilon) &= \lambda_0 + \varepsilon\lambda_1 + \varepsilon^2\lambda_2 + \cdots \\ \mathbf{v}(\varepsilon) &= \mathbf{v}_0 + \varepsilon\mathbf{v}_1 + \varepsilon^2\mathbf{v}_2 + \cdots . \end{aligned} \qquad (12.134)$$

In the first two orders, the resulting equations are

$$\begin{aligned} (A_0 - \lambda_0 I)\mathbf{v}_0 &= 0 \\ (A_0 - \lambda_0 I)\mathbf{v}_1 &= (\lambda_1 I - A_1)\mathbf{v}_0 . \end{aligned} \qquad (12.135)$$

We will now discuss how to solve (12.135) under various circumstances.

The simplest case occurs if A_0 is real and symmetric (or complex and Hermitian), and also has distinct eigenvalues. In this case, the first equation of (12.135) can be solved simply by choosing an eigenvector $\mathbf{v}_0$ for each eigenvalue λ_0 of A_0. It is convenient to normalize $\mathbf{v}_0$ to have length one, that is, $(\mathbf{v}_0, \mathbf{v}_0) = 1$ where $(\cdot, \cdot)$ is the inner (or "dot") product. Now we fix a choice of λ_0 and $\mathbf{v}_0$ and insert these into the second equation of (12.135). The next step is to choose λ_1 so that the right-hand side lies in the image of $A_0 - \lambda_0 I$; once this is accomplished, it is possible to solve for $\mathbf{v}_1$. To determine λ_1, we rely upon special properties of the eigenvectors of a symmetric matrix; namely, they are orthogonal (with respect to the inner product). Thus there exists an orthogonal basis of eigenvectors in which A_0 is diagonal; examining $A_0 - \lambda_0 I$ in this basis, we see that its kernel (or null space) is spanned by $\mathbf{v}_0$ and its image (or range) is spanned by the rest of the eigenvectors. Therefore, the image is perpendicular to the kernel. It follows that $(\lambda_1 I - A_1)\mathbf{v}_0$ will lie in the image of $A_0 - \lambda_0 I$ if and only if its orthogonal projection onto $\mathbf{v}_0$ is zero, that is, if and only if $(\lambda_1\mathbf{v}_0 - A_1\mathbf{v}_0, \mathbf{v}_0) = 0$, or, using $(\mathbf{v}_0, \mathbf{v}_0) = 1$,

$$\lambda_1 = (A_1\mathbf{v}_0, \mathbf{v}_0). \qquad (12.136)$$

It is not necessary to find $\mathbf{v}_1$ unless it is desired to go on to the next stage and find λ_2. (There is a close similarity between these steps and those of the Lindstedt method in Section 3.1, in which each term in the frequency expansion is determined to make the next equation solvable.)

If A_0 has distinct eigenvalues but is not symmetric, most of the last paragraph still applies, but the eigenvectors of A_0 need not be orthogonal. The vector space still decomposes as a direct sum of the image and kernel of $A_0 - \lambda_0 I$, but the inner product can no longer be used to effect the decomposition; λ_1 can still be determined but cannot be written in the form (12.136).

If A_0 does not have distinct eigenvalues, the situation can become quite complicated. First, suppose $A(\varepsilon)$ is symmetric for all ε, so that all A_i are symmetric. In this case, every eigenvalue has a "full set" of eigenvectors (as many linearly independent eigenvectors as its algebraic multiplicity). However, suppose that A_0 has an eigenvalue λ_0 of multiplicity two, with eigenvectors $\mathbf{z}$ and $\mathbf{w}$. It is likely that for $\varepsilon \neq 0$, the eigenvalue λ_0 splits into two distinct eigenvalues having separate eigenvectors. In this case, it is not possible to choose an arbitrary eigenvector $\mathbf{v}_0$ from the plane of $\mathbf{z}$ and $\mathbf{w}$ to use in the second equation of (12.135); only the limiting positions (as $\varepsilon \to 0$) of the two eigenvectors for $\varepsilon \neq 0$ are suitable candidates for $\mathbf{v}_0$. As these are unknown in advance, one must put $\mathbf{v}_0 = a\mathbf{z} + b\mathbf{w}$ (for unknown real a and b) into (12.135), then find two choices of a, b, and λ_1 that make the second equation solvable. It also may happen that the degeneracy cannot be resolved at this stage but must be carried forward to higher stages before the eigenvalues split; or, of course, they may never split.

If $A(\varepsilon)$ is not symmetric, and hence not necessarily diagonalizable, the possibilities become even worse. The example

$$A(\varepsilon) = \begin{bmatrix} 1 & \varepsilon \\ 0 & 1 \end{bmatrix} \tag{12.137}$$

shows that a full set of eigenvectors may exist when $\varepsilon = 0$ but not for $\varepsilon \neq 0$; the contrary case (diagonalizable for $\varepsilon \neq 0$ but not for $\varepsilon = 0$) is exhibited by

$$A(\varepsilon) = \begin{bmatrix} 1 & 1 \\ 0 & 1+\varepsilon \end{bmatrix}. \tag{12.138}$$

These examples show that the Jordan normal form of $A(\varepsilon)$ is not, in general, a continuous function of ε.

There is a normal form method, closely related to that of Section 3.6, that is successful in all cases. It consists in simplifying the terms of (12.130) by applying successive coordinate transformations of the form $I + \varepsilon^k S_k$ for $k = 1, 2, \ldots$ or a single coordinate transformation of the form $T(\varepsilon) = I + \varepsilon T_1 + \varepsilon^2 T_2 + \cdots$; the matrices S_k or T_k are determined recursively. It is usually assumed that A_0 is in Jordan canonical form, hence is diagonal if possible. If A_0 is diagonal and $A(\varepsilon)$ is diagonalizable, the normalized A_k will be diagonal for $k \geq 1$, so that (12.130) will give the asymptotic expansion of the eigenvalues and $T(\varepsilon)$ the asymptotic expansion of all the eigenvalues (as its columns). In more complicated cases, the normalized series (12.130) will belong to a class of matrices called the *Arnol'd unfolding* of A_0, and although it will not always be in Jordan form, it will be in the simplest form compatible with smooth dependence on ε. Still further simplifications (the *metanormal form*) can be obtained using fractional powers of ε. This theory is described in [7, Chapter 2].

Glossary

Asymptotic approximation: An approximate solution to a perturbation problem that increases in accuracy at a known rate as the perturbation parameter approaches zero.

Asymptotic series: A series, the partial sums of which are asymptotic approximations of some function to successively higher order.

Averaging: A method of constructing asymptotic approximations to oscillatory problems. In the simplest case, it involves replacing periodic functions by their averages to simplify the equations to be solved.

Bifurcation: Any change in the number or qualitative character (such as stability) of the solutions to an equation as a parameter is varied.

Boundary layer: A transition layer located near the boundary of a region where boundary values are imposed.

Composite solution: A solution uniformly valid on a certain domain, created by matching an inner solution and an outer solution each valid on part of the domain.

Gauge: A monotonic function of a perturbation parameter used to express the order of a term in an asymptotic series.

Generalized series: An asymptotic series of the form $\sum \delta_i(\varepsilon)u_i(\mathbf{x}, \varepsilon)$ in which the perturbation parameter ε appears both in the gauges and in the coefficients. See Poincaré Series.

Initial layer: A transition layer located near the point at which an initial value is prescribed.

Inner solution: An approximate solution uniformly valid within a transition layer.

Lie series: A means of representing a near-identity transformation by a function called a *generator*. There are several forms; in Deprit's form, if $W(\mathbf{x}, \varepsilon)$ is the generator, then the solution of $\mathrm{d}\mathbf{x}/\mathrm{d}\varepsilon = W(\mathbf{x}, \varepsilon)$ with $\mathbf{x}(0) = \mathbf{y}$ for small ε is a near-identity transformation of the form $\mathbf{x} = \mathbf{y} + \varepsilon \mathbf{u}_1(\mathbf{y}) + \cdots$.

Lindstedt method: A method of approximating periodic solutions whose frequency varies with the perturbation parameter by using a scaled time variable.

Matching: Any of several methods for choosing the arbitrary constants in an inner and an outer solution so that they both approximate the same exact solution.

Multiple scales: The simultaneous use of two or more variables having the same physical significance (for instance, time or distance) but proceeding at different "rates" (in terms of the small parameter), for instance "normal time" t and "slow time" $\tau = \varepsilon t$. The variables are treated as if they were independent during part of the discussion, but at the end are reduced to a single variable again.

Naive expansion: A Poincaré series obtained by the regular perturbation method (straightforward expansion), proposed as a solution to a singular problem, where a generalized series is required. See RG method.

Outer solution: An approximate solution uniformly valid in a region away from a transition layer.

Overlap domain: A region in which both an inner and an outer approximation are valid, and where they can be compared for purposes of matching.

Perturbation parameter: A parameter, usually denoted ε, occurring in a mathematical problem, such that the problem has a known solution when $\varepsilon = 0$ and an approximate solution is sought when ε is small but nonzero.

Perturbation series: A finite or infinite series obtained as a formal approximate solution to a perturbation problem, in the hope that it will be uniformly asymptotically valid on some domain.

Poincaré series: An asymptotic series of the form $\sum \delta_i(\varepsilon)u_i(x)$ in which the perturbation parameter ε appears only in the gauges. See Generalized Series.

Regular perturbation problem: A perturbation problem having an approximate solution in the form of a Poincaré series that is uniformly valid on the entire intended domain.

Relaxation oscillation: A self-sustained oscillation characterized by a slow buildup of tension (in a spring, for instance) followed by a rapid release or relaxation. The rapid phase is an example of a transition layer.

Rescaled coordinate: A coordinate that has been obtained from an original variable by a transformation depending on the perturbation parameter, usually by multiplying by a scaling factor. For instance time t may be rescaled to give a "slow time" εt (see multiple scales) or a "strained time" $(\omega_0 + \varepsilon\omega_1 + \dots)t$ (see Lindstedt method).

Resonance: In linear problems, an equality of two frequencies. In nonlinear problems, any integer relationship holding between two or more frequencies, of the form $\nu_1\omega_1 + \cdots + \nu_k\omega_k = 0$, especially one involving small integers or one that produces zero denominators in a Fourier series.

RG (renormalization group) method: A method (not involving group theory) that converts a naive expansion for a singular problem into a generalized expansion that may be uniformly valid on the intended domain.

Self-excited oscillation: An oscillation about an unstable equilibrium that occurs because of instability and has its own natural frequency, rather than an oscillation in response to an external periodic forcing.

Singular perturbation problem: A perturbation problem that cannot be uniformly approximated by a Poincaré series on the entire intended domain, although this may be possible over part of the domain. For singular problems, one seeks a solution in the form of a generalized series.

Transition layer: A small region in which the solution of a differential equation changes rapidly and in which some approximate solution (outer solution) that is valid elsewhere fails.

Triple deck: A problem that exhibits a transition layer within a transition layer and that therefore requires the matching of three approximate solutions rather than only two.

Unfolding: A family of perturbations of a given system obtained by adding several small parameters. An unfolding is universal if (roughly) it exhibits all possible qualitative behaviors for perturbations of the given system using the least possible number of parameters.

Uniform approximation: An approximate solution whose error is bounded by a constant times a gauge function everywhere on an intended domain.

References

1. Golubitsky, M. and Schaeffer, D.G. (1985) *Singularities and Groups in Bifurcation Theory*, vol. 1, Springer-Verlag, New York.
2. Iooss, G. and Joseph, D.D. (1980) *Elementary Stability and Bifurcation Theory*, Springer-Verlag, New York.
3. Sanders, J.A., Verhulst, F., and Murdock, J. (2007) *Averaging Methods in Nonlinear Dynamical Systems*, Springer-Verlag, New York.
4. Murdock, J.A. (1999) *Perturbations: Theory and Methods*, Society for Industrial and Applied Mathematics, Philadelphia, PA.
5. Lochak, P. and Meunier, C. (1988) *Multiphase Averaging for Classical Systems*, Springer-Verlag, New York.
6. Nayfeh, A. (1973) *Perturbation Methods*, John Wiley & Sons, Inc., New York.
7. Murdock, J.A. (2003) *Normal Forms and Unfoldings for Local Dynamical Systems*, Springer-Verlag, New York.
8. Nayfeh, A. (1993) *Method of Normal Forms*, John Wiley & Sons, Inc., New York.
9. Kahn, P.B. and Zarmi, Y. (1998) *Nonlinear Dynamics: Exploration through Normal Forms*, John Wiley & Sons, Inc., New York.
10. Wiggins, S. (2003) *Introduction to Applied Nonlinear Dynamical Systems and Chaos*, Springer-Verlag, New York.
11. Lagerstrom, P.A. (1988) *Matched Asymptotic Expansions*, Springer-Verlag, New York.
12. Smith, D.R. (1985) *Singular-Perturbation Theory*, Cambridge University Press, Cambridge.

13. Grasman, J. (1987) *Asymptotic Methods for Relaxation Oscillations and Applications*, Springer-Verlag, New York.
14. Jones, C.K.R.T. (1994) Geometric singular perturbation theory, in *Dynamical Systems (Montecatini Terme, 1994), Lecture Notes in Mathematics 1609*, Springer-Verlag, New York, pp. 44–118.
15. Bender, C.M. and Orszag, S.A. (1999) *Advanced Mathematical Methods for Scientists and Engineers*, Springer-Verlag, New York.
16. Lakin, W.D. and Sanchez, D.A. (1970) *Topics in Ordinary Differential Equations*, Dover, New York.
17. Wasow, W. and Robert, E. (1976) *Asymptotic Expansions for Ordinary Differential Equations*, Krieger Publishing Co., Huntington, NY.
18. van Dyke, M. (1975) *Perturbation Methods in Fluid Mechanics*, Annotated Edition, Parabolic Press, Stanford, CA.
19. Sychev, V.V., Ruban, A.I., Sychev, V.V., and Korolev, G.L. (1998) *Asymptotic Theory of Separated Flows*, Cambridge University Press, Cambridge.
20. Rothmayer, A.P. and Smith, F.T. (1998) Incompressible triple- deck theory, in *The Handbook of Fluid Dynamics* (ed. R.W. Johnson), CRC Press, Boca Raton, FL.
21. Oono, Y. (2000) Renormalization and asymptotics. *Int. J. Mod. Phys. B*, **14**, 1327–1361.
22. Chen, L.-Y., Goldenfeld, N., and Oono, Y. (1996) Renormalization group and singular perturbations. *Phys. Rev. E*, **54**, 376–394.
23. Ziane, M. (2000) On a certain renormalization group method. *J. Math. Phys.*, **41**, 3290–3299.
24. Lee DeVille, R.E., Harkin, A., Holzer, M., Josić, K., and Kaper, T.J. (2008) Analysis of a renormalization group method for solving perturbed ordinary differential equations, *Physica D*, **237**, 1029–1052.
25. Woodruff, S.L. (1993) The use of an invariance condition in the solution of multiple-scale singular perturbation problems: ordinary differential equations. *Stud. Appl. Math.*, **90**, 225–248.
26. Paquette, G.C. (2000) Renormalization group analysis of differential equations subject to slowly modulated perturbations. *Physica A*, **276**, 122–163.
27. Kirkinis, E. (2008) The renormalization group and the implicit function theorem for amplitude equations. *J. Math. Phys.*, **49**, 1–16. article 073518.
28. Kirkinis, E. (2012) The renormalization group: a perturbation method for the graduate curriculum. *SIAM Rev.*, **54**, 374–388.
29. Kunihiro, T. (1995) A geometrical formulation of the renormalization group method for global analysis. *Progr. Theor. Phys.*, **94**, 503–514; errata, same volume (**94**), 835.
30. Kunihiro, T. (1997) The renormalization-group method applied to asymptotic analysis of vector fields. *Progr. Theor. Phys.*, **97**, 179–200.
31. Veysey, J. and Goldenfeld, N. (2007) Simple viscous flows: from boundary layers to the renormalization group. *Rev. Mod. Phys.*, **79**, 883–927.
32. Kato, T. (1966) *Perturbation Theory for Linear Operators*, Springer-Verlag, New York.

Further Reading

Andrianov, I.V. and Manevitch, L.I. (2002) *Asymptotology*, Kluwer, Dordrecht.

Bender, C.M. and Orszag, S.A. (1999) *Advanced Mathematical Methods for Scientists and Engineers*, Springer-Verlag, New York.

Bush, A.W. (1992) *Perturbation Methods for Engineers and Scientists*, CRC Press, Boca Raton, FL.

Hinch, E.J. (1991) *Perturbation Methods*, Cambridge University Press, Cambridge.

Kevorkian, J. and Cole, J.D. (1981; corrected second printing 1985) *Perturbation Methods in Applied Mathematics*, Springer-Verlag, New York.

Nayfeh, A. (1981) *Introduction to Perturbation Techniques*, John Wiley & Sons, Inc., New York.

O'Malley, R.E. (1991) *Singular Perturbation Methods for Ordinary Differential Equations*, Springer-Verlag, New York.

13
Functional Analysis

Pavel Exner

The area covered by the chapter title is huge. We describe the part of functional analysis dealing with linear operators, in particular, those on Hilbert spaces and their spectral decompositions, which are important for applications in quantum physics. The claims are presented without proofs, often highly nontrivial, for which we refer to the literature mentioned at the end of the chapter.

13.1
Banach Space and Operators on Them

The first section summarizes notions from the general functional analysis that we will need in the following.

13.1.1
Vector and Normed Spaces

The starting point is the notion of a *vector space* V over a field $\mathbb{F}$ as a family of *vectors* equipped with the operations of *summation* and *multiplication by a scalar* $\alpha \in \mathbb{F}$. The former is *commutative* and *associative*, and the two operations are mutually *distributive*. We consider here the case when $\mathbb{F} = \mathbb{C}$, the set of complex numbers, but other fields are also used, typically the sets of reals or quaternions.

Standard examples are $\mathbb{C}^n$, the space of n-tuples of complex numbers, or ℓ^p, the space of complex sequences $\{x_j\}_{j=1}^{\infty}$ satisfying $\sum_{j=1}^{\infty} |x_j|^p < \infty$ for a fixed $p \geq 1$. Other frequently occurring examples are *function spaces* such as $C(\mathcal{I})$ consisting of continuous functions on the interval $\mathcal{I}$, or $L^p(\mathcal{I})$ the elements of which are classes of functions differing on a zero-measure set and satisfying $\int_{\mathcal{I}} |f(x)|^p \, \mathrm{d}x < \infty$. The *dimension* of a vector space is given by the maximum number of linearly independent vectors one can find in it. In the examples, the space $\mathbb{C}^n$ is n-dimensional, the other indicated ones are (countably) infinite-dimensional.

A map $f : V \to \mathbb{C}$ is called a *functional*. We say that it is *linear* if $f(\alpha x) = \alpha f(x)$ and *antilinear* if $f(\alpha x) = \overline{\alpha} f(x)$ for all $\alpha \in \mathbb{C}$ and $x \in V$. Similarly, a *seminorm* is a real functional p satisfying $p(x + y) \leq p(x) + p(y)$ and $p(\alpha x) = |\alpha| p(x)$. A map $F : V \times V \to \mathbb{C}$ is called a *form*. Its (anti)linearity is defined as above; it

Mathematical Tools for Physicists, Second Edition. Edited by Michael Grinfeld.

is called *sesquilinear* if it is linear in one argument and antilinear in the other.

A *norm* on a vector space is a seminorm, usually denoted as $\|\cdot\|$, with the property that $\|x\| = 0$ implies $x = 0$. A space equipped with it is called a *normed space*. The same vector space can be equipped with different norms, for example both $\|x\|_\infty := \sup_{1\le j\le n} |x_j|$ and $\|x\|_p := \left(\sum_{j=1}^{\infty} |x_j|^p\right)^{1/p}$ are norms on $\mathbb{C}^n$. Analogous norms can be introduced on the other vector spaces mentioned above, for instance, $\|f\|_\infty := \sup_{x\in\mathcal{I}} |f(x)|$ on $C(\mathcal{I})$ or $\|f\|_p := \left(\int_{\mathcal{I}} |f(x)|^p \,\mathrm{d}x\right)^{1/p}$ on $L^p(\mathcal{I})$.

The space can be equipped with an *inner* (or *scalar*) *product*, which is a positive sesquilinear form $(\cdot,\cdot)$ with the property that (x, x) implies $x = 0$. In such a case, one way to define a norm is through the formula $\|x\| := \sqrt{(x, x)}$. This norm is said to be *induced* by the scalar product; it satisfies the *Schwarz inequality*

$$|(x, y)| \le \|x\|\|y\| . \tag{13.1}$$

On the other hand, a norm $\|\cdot\|$ can be induced by a scalar product if and only if it satisfies the *parallelogram identity*, $\|x + y\|^2 + \|x - y\|^2 = 2\|x\|^2 + 2\|y\|^2$.

In this way, a vector space is equipped with a metric structure with the distance of points given by $d(x, y) := \|x - y\|$. Then one can ask about *completeness* of such a space, that is, whether any *Cauchy sequence*, $\{x_n\} \subset V$ satisfying $d(x_n, x_m) \to 0$ as $n, m \to \infty$, has a limit in V. A (metrically) complete normed space is called a *Banach space*; similarly, a complete inner-product space is called a *Hilbert space*. The spaces ℓ^p and $L^p(\mathcal{I})$ are examples of Banach spaces, they are Hilbert if $p = 2$.; The metric induces, *mutatis mutandis*, a *topology* on a vector space making it an example of *topological vector space* in which the vector and topological properties are combined in such a way that the operations of summation and scalar multiplication are continuous. A vector space of infinite dimension can be equipped with different, mutually inequivalent topologies. Most important are *locally convex topologies* generated by a family of seminorms with the property that for each nonzero $x \in V$ there is a seminorm p such that $p(x) \ne 0$. The norm topology is a particular case.

The presence of topology allows us to investigate various properties of subsets and subspaces of V such as compactness, openness and closeness [14, Chap. II; 20, Chap. 1], or convergence of sequences [1, Chap. IV; 2, Chap. 1]. If the space is equipped with different topologies, the same sequence may converge with respect to some of them and have no limit with respect to others.

13.1.2
Operators on Banach Spaces

A map between two Banach spaces, $B : \mathcal{X} \to \mathcal{Y}$, is called an *operator*. It is *linear* if $B(\alpha x + y) = \alpha Bx + By$ holds; we will consider here linear operators only and therefore drop the adjective. Such an operator is *continuous* iff it is *bounded*, which means there exists a number c such that $\|Bx\|_{\mathcal{Y}} \le c\|x\|_{\mathcal{X}}$ holds for all $x \in \mathcal{X}$. The infimum of such numbers is called the *norm* $\|B\|$ of B. One has $\|B\| = \sup_{S_1} \|Bx\|_{\mathcal{Y}}$, where $S_1 := \{x \in \mathcal{X} : \|x\|_{\mathcal{X}} = 1\}$ is the unit sphere in $\mathcal{X}$.

The functional $\|\cdot\|$ is indeed a norm; in particular, the space $\mathcal{B}(\mathcal{X}, \mathcal{Y})$ of all bounded operators from $\mathcal{X}$ to $\mathcal{Y}$ equipped with it becomes a Banach space. In addition, for the composition of operators $B \in \mathcal{B}(\mathcal{Y}, \mathcal{Z})$ and $C \in \mathcal{B}(\mathcal{X}, \mathcal{Y})$, the operator norm satisfies the inequality $\|BC\| \le \|B\|\|C\|$.

In general, an operator can be defined on a subspace of $\mathcal{X}$ only, which is then called the *domain* of B, denoted by $D(B)$. If $D(B)$ is *dense* in $\mathcal{X}$ and B is *bounded*, then there is a unique extension $\tilde{B} \in \mathcal{B}(\mathcal{X}, \mathcal{X}) =: \mathcal{B}(\mathcal{X})$ of the operator B to the space $\mathcal{X}$, and moreover $\|\tilde{B}\| = \|B\|$ holds, hence bounded operators may be without loss of generality supposed to be defined on the whole $\mathcal{X}$.

On the other hand, a domain can be dense in different Banach spaces having thus different extensions. As an example, consider the *Fourier transformation* defined by

$$\hat{f}(y) = (2\pi)^{-1/2} \int_{\mathbb{R}} e^{-ixy} f(x)\, dx \qquad (13.2)$$

on the set $\mathcal{S}(\mathbb{R})$ of infinitely differentiable functions $f : \mathbb{R} \to \mathbb{C}$, which have, together with all their derivatives, a faster-than-powerlike decay. $\mathcal{S}(\mathbb{R})$ is dense in $L^1(\mathbb{R})$, hence the map $f \mapsto \hat{f}$ extends to a unique operator from $L^1(\mathbb{R})$ to the space $C_\infty(\mathbb{R})$ of continuous functions satisfying $\lim_{|y|\to\infty} \hat{f}(y) = 0$ (this property is usually referred to as the *Riemann–Lebesgue lemma*), which is bounded, $\|\hat{f}\|_\infty \le (2\pi)^{-1/2}\|f\|_1$.

However, $\mathcal{S}(\mathbb{R})$ is dense also in $L^2(\mathbb{R})$ and $f \mapsto \hat{f}$ can be thus uniquely extended to a map $F : L^2(\mathbb{R}) \to L^2(\mathbb{R})$ called *Fourier–Plancherel operator* with the norm $\|F\| = 1$. The right-hand side of (13.2) may not be defined for a general $f \in L^2(\mathbb{R})$, hence we have to write the action of F as

$$(Ff)(y) = \underset{n\to\infty}{\text{l.i.m.}}\ (2\pi)^{-1/2} \int_{|x|\le n} e^{-ixy} f(x)\, dx\,, \qquad (13.3)$$

where the symbol l.i.m., *limes in medio*, means convergence with respect to the norm of $L^2(\mathbb{R})$. In a similar way, one defines the Fourier transformation and Fourier–Plancherel operator on functions $f : \mathbb{R}^n \to \mathbb{C}$, $n > 1$.

Functionals represent a particular case of operators with $\mathcal{Y} = \mathbb{C}$. The space $\mathcal{B}(\mathcal{X}, \mathbb{C})$ of bounded functionals is called the *dual* to the Banach space $\mathcal{X}$ and denoted as $\mathcal{X}^*$. It is a Banach space with the norm $\|f\| = \sup_{\|x\|=1} |f(x)|$. In some cases, the dual space can be determined explicitly, for instance, $(\ell^p)^*$ is isomorphic to ℓ^q, where $q = p/(p-1)$ for $1 < p < \infty$ and $q = \infty$ for $p = 1$.

The metric completeness on Banach spaces implies the *uniform boundedness principle* for an operator family $\mathcal{F} \subset \mathcal{B}(\mathcal{X}, V)$ where V is a normed space with the norm $\|\cdot\|$: if $\sup_{B\in\mathcal{F}} \|Bx\| < \infty$, then there is a $c > 0$ such that $\sup_{B\in\mathcal{F}} \|B\| < c$. This result has several important consequences, in particular,

Theorem 13.1 (Open-map theorem) *If an operator $\mathcal{B}(\mathcal{X}, \mathcal{Y})$ maps to the whole Banach space $\mathcal{Y}$ and $G \subset \mathcal{X}$ is an open set, then the image set BG is open in $\mathcal{Y}$.*

Theorem 13.2 (Inverse-mapping theorem) *If $\mathcal{B}(\mathcal{X}, \mathcal{Y})$ is a one-to-one map, then B^{-1} is a continuous linear map from $\mathcal{Y}$ to $\mathcal{X}$.*

One more consequence concerns *closed operators*, which are those whose *graph* $\Gamma(T) = \{[x, Tx] : x \in D(T)\}$ is a closed set in $\mathcal{X} \oplus \mathcal{Y}$. Alternatively, they can be characterized by sequences: an operator T is closed *iff* for any sequence $\{x_n\} \subset D(T)$ such that $x_n \to x$ and $Tx_n \to y$ as $n \to \infty$, we have $x \in D(T)$ and $y = Tx$.

Theorem 13.3 (Closed-graph theorem) *A closed linear operator $T : \mathcal{X} \to \mathcal{Y}$ defined on the whole of $\mathcal{X}$ is continuous.*

13.1.3 Spectra of Closed Operators

Consider now the set $C(\mathcal{X})$ of closed linear operators mapping the Banach space $\mathcal{X}$ to itself. It follows from the closed-graph theorem that if $T \in C(\mathcal{X})$ is unbounded, its domain $D(T) \neq \mathcal{X}$. A number $\lambda \in \mathbb{C}$ is called an *eigenvalue* of T if there is a nonzero $x \in D(T)$, called an *eigenvector*, such that $Tx = \lambda x$. The span of such vectors is a subspace in $\mathcal{X}$ called *eigenspace* (related to the eigenvalue λ) and its dimension is the (geometric) *multiplicity* of λ. Any eigenspace of a closed operator $T \in C(\mathcal{X})$ is closed.

This is a familiar way to describe the spectrum known from linear algebra. An alternative way is to notice that the inverse $(T-\lambda)^{-1}$ is not defined if λ is an eigenvalue. John von Neumann was the first to notice that this offers a finer tool in case of unbounded operators when $(T-\lambda)^{-1}$ may exist as an unbounded operator, which means that $T-\lambda$ does not map $\mathcal{X}$ onto itself but to its proper subspace only.

Consequently, we define the *spectrum* $\sigma(T)$ of T as the set consisting of three parts: the *point spectrum* $\sigma_{\mathrm{p}}(T)$ being the family of all eigenvalues, the *continuous spectrum* $\sigma_{\mathrm{c}}(T)$ for which $\mathrm{Ran}(T-\lambda) \neq \mathcal{X}$ but it is dense in $\mathcal{X}$, and finally the *residual spectrum* $\sigma_{\mathrm{r}}(T)$ for which $\overline{\mathrm{Ran}(T-\lambda)} \neq \mathcal{X}$, where the symbol $\overline{\mathcal{Y}}$ conventionally denotes the closure of the set $\mathcal{Y}$. The spectrum $\sigma(T)$ is a closed set, its complement $\rho(T) = \mathbb{C} \setminus \sigma(T)$ is called the *resolvent set* of the operator T and the map $R_T : \rho(T) \to \mathcal{B}(\mathcal{X})$ defined by $R_T(\lambda) := (T-\lambda)^{-1}$ is the *resolvent* of T.

In general, the spectrum may be empty. For a bounded operator B on a Banach space $\mathcal{X}$, however, the spectrum is always a nonempty compact set and its radius $r(B) = \sup\{|\lambda| : \lambda \in \sigma(T)\} = \lim_{n\to\infty} \|B^n\|^{1/n} \leq \|B\|$. The resolvent can be in this case expressed through the *Neumann series*,

$$R_B(\lambda) = -\lambda^{-1} I - \sum_{k=1}^{\infty} \lambda^{-(k+1)} B^k, \qquad (13.4)$$

which converges outside the circle of spectral radius, that is, for $|\lambda| > \|B\|$.

13.2 Hilbert Spaces

As a particular case of Banach spaces, Hilbert spaces share the general properties listed above but the existence of inner product gives them specific features. Hilbert spaces are important especially from the viewpoint of quantum theory where they are used as *state spaces*, the *pure states* of such systems being identified with one-dimensional subspaces of an appropriate Hilbert space.

13.2.1 Hilbert-Space Geometry

Hilbert spaces are distinguished by the existence of *orthogonal projection*: given a closed subspace $\mathcal{G} \subset \mathcal{H}$ and a vector $x \in \mathcal{H}$, there is a unique vector $y_x \in \mathcal{H}$ such that the distance $d(x, \mathcal{G}) := \inf_{y\in\mathcal{G}} \|x-y\| = \|x-y_x\|$. It implies that each vector can be uniquely decomposed into a sum, $x = y + z$ with $y \in \mathcal{G}$ and z belonging to $\mathcal{G}^{\perp}$, the orthogonal complement of $\mathcal{G}$ in $\mathcal{H}$. It also provides a criterion for a set $M \subset \mathcal{H}$ to be *total*, that is, such that its linear envelope is dense in $\mathcal{H}$: it happens iff $M^{\perp} = \{0\}$.

Another characteristic feature of a Hilbert space is the relation to its dual: by the *Riesz lemma* to any $f \in \mathcal{H}^*$, there is a unique $y_f \in \mathcal{H}$ such that $f(x) = (y_f, x)$; the map $f \mapsto y_f$ is an antilinear isometry of $\mathcal{H}$

and $\mathcal{H}^*$. Functionals allow us to introduce a convergence in a Hilbert space different from the one determined by the norm of $\mathcal{H}$: a sequence $\{x_n\} \subset \mathcal{H}$ converges *weakly* if $(y, x_n) \to (y, x)$ holds for all $y \in \mathcal{H}$. If one has in addition $\|x_n\| \to \|x\|$, then $\{x_n\}$ converges also in the norm.

A family $\{e_\alpha\} \subset \mathcal{H}$ of pairwise orthogonal vectors of unit length is called an *orthonormal basis* of $\mathcal{H}$ if it is total. Such a basis always exists and any two orthonormal bases in a given $\mathcal{H}$ have the same cardinality, which is called the *dimension* of $\mathcal{H}$. Hilbert spaces $\mathcal{H}, \mathcal{H}'$ are isomorphic iff $\dim \mathcal{H} = \dim \mathcal{H}'$. A Hilbert space is *separable* if its dimension is at most countable.

Given an orthonormal basis $\{e_\alpha\}$, we define *Fourier coefficients* of a vector $x \in \mathcal{H}$ as (e_α, x). Using them, we can write the Fourier expansion

$$x = \sum_\alpha (e_\alpha, x) e_\alpha \tag{13.5}$$

of x; it always makes sense as a convergent series because the set of nonzero Fourier coefficients is at most countable even if the basis $\{e_\alpha\}$ is uncountable, and because the *Parseval identity* is valid,

$$\|x\|^2 = \sum_\alpha |(e_\alpha, x)|^2. \tag{13.6}$$

The orthonormal basis most often used in the space ℓ^2 consists of the vectors $\phi_n := \{0, \dots, 0, 1, 0, \dots\}$ with the nonzero entry on the nth place. In case of spaces $L^2(\mathcal{I})$, the choice depends on the interval. For the finite interval $\mathcal{I} = (0, 2\pi)$, the *trigonometric basis*, $\{e_k : k = 0, \pm 1, \pm 2, \dots\}$ with $e_k(x) = (2\pi)^{-1/2} \mathrm{e}^{\mathrm{i}kx}$, is used. On the whole line, $\mathcal{I} = \mathbb{R}$, the functions

$$h_n(x) = (2^n n!)^{-1/2} \pi^{-1/4} \mathrm{e}^{-x^2/2} H_n(x), \quad n = 0, 1, 2, \dots, \tag{13.7}$$

where $H_n(x) = (-1)^N \mathrm{e}^{x^2} \mathrm{d}^n/\mathrm{d}x^n \mathrm{e}^{-x^2}$ are *Hermite polynomials*, form an orthonormal basis. Similarly, one can construct a basis for the semi-infinite interval $\mathcal{I} = (0, \infty)$ using Laguerre polynomials.

Let us mention two other examples of Hilbert spaces. Given a set M equipped with measure μ and a Hilbert space $\mathcal{G}$, we consider measurable *vector-valued functions* $f : M \to \mathcal{G}$ such that $\int_M \|f(x)\|_\mathcal{G}^2 \,\mathrm{d}\mu(x) < \infty$. We regard classes of such functions differing on a set of zero measure as elements of a new space. One can check that it is a Hilbert space with respect to the inner product

$$(f, g) := \int_M (f(x), g(x))_\mathcal{G}^2 \,\mathrm{d}\mu(x); \tag{13.8}$$

we denote this space as $L^2(X, \mathrm{d}\mu; \mathcal{G})$. Such spaces appear in various applications, for instance, in quantum mechanics, $\mathcal{G}$ is often a finite-dimensional space associated with *spin states* of the system.

Our second example is the *Hilbert space of analytic functions* $f : \mathbb{C} \to \mathbb{C}$ equipped this time with the "weighted" inner product,

$$(f, g) = \frac{1}{\pi} \int_\mathbb{C} \overline{f(z)} g(z) \,\mathrm{e}^{-|z|^2} \,\mathrm{d}z, \tag{13.9}$$

where $\mathrm{d}z$ is a shorthand for the "two-dimensional" measure $\mathrm{d}(\mathrm{Re} z)\,\mathrm{d}(\mathrm{Im} z)$ in the complex plane. An orthonormal basis in it is formed by the monomials, $u_n(z) = (z!)^{-1/2} z^n$, $n = 0, 1, \dots$. Denoting $e_z(w) = \mathrm{e}^{\bar{z}w}$, we have the identity

$$f(w) = \frac{1}{\pi} \int_\mathbb{C} (e_z, f)\, e_z(w) \,\mathrm{e}^{-|z|^2} \,\mathrm{d}z, \tag{13.10}$$

which can be regarded as a continuous analogue of the Fourier expansion with respect to an "overcomplete basis;" this relation plays the central role in description of quantum-mechanical *coherent states*.

13.2.2
Direct Sums and Tensor Products

Applications require many different spaces, hence it is useful to know how to construct new Hilbert spaces from given ones. One possibility are the L^2 spaces of vector-valued functions mentioned above; here we add two more methods. The first is based on *direct sums*. Let $\{\mathcal{H}_k\}$ be a family of Hilbert spaces, for simplicity supposed to be at most countable, with norm of $\mathcal{H}_k$ being denoted by $\|\cdot\|_k$. We consider the set of sequences $X = \{x_k : x_k \in \mathcal{H}_k\}$ such that $\sum_k \|x_k\|_k^2 < \infty$, equip it with vector operations defined componentwise and define the inner product on it by

$$(X, Y) = \sum_k (x_k, y_k)_k, \tag{13.11}$$

where $(\cdot,\cdot)_k$ is the inner product in $\mathcal{H}_k$, obtaining a new Hilbert space, which we denote as $\bigoplus_k \mathcal{H}_k$ or $\sum_k^{\oplus} \mathcal{H}_k$. The dimension of the direct product is clearly the sum of the dimension of its components. In a similar way, one can define, under appropriate measurability hypotheses, a *direct integral* $\int_M^{\oplus} \mathcal{H}(x)\,\mathrm{d}\mu(x)$ with respect to a measure μ of a family of Hilbert spaces dependent on the integration variable x.

The definition of a *tensor product* of Hilbert spaces $\mathcal{H}_1, \mathcal{H}_2$ is more involved. We introduce first a tensor product *realization* as a bilinear map $\otimes : \mathcal{H}_1 \times \mathcal{H}_2 \to \mathcal{H}$, which associates with $x \in \mathcal{H}_1$ and $y \in \mathcal{H}_2$ an element $x \otimes y \in \mathcal{H}$ in such a way that $(x \otimes y, x' \otimes y') = (x, x')_1 (y, y')_2$ holds for all $x, x' \in \mathcal{H}_1$ and $y, y' \in \mathcal{H}_2$ and the set $\mathcal{H}_1 \otimes \mathcal{H}_2$ is total in $\mathcal{H}$. For each pair $\mathcal{H}_1, \mathcal{H}_2$, a realization of their tensor product exists, and furthermore, all such realizations are isomorphic to each other.

This allows us to investigate the tensor product through a fixed realization of it, in particular, to write $\mathcal{H}_1 \otimes \mathcal{H}_2 = \mathcal{H}$ having in mind a concrete map $\otimes$. Frequently appearing examples are the relations

$$L^2(M, \mathrm{d}\mu) \otimes L^2(N, \mathrm{d}\nu) = L^2(M \times N, \mathrm{d}(\mu \otimes \nu)), \tag{13.12}$$
$$L^2(X, \mathrm{d}\mu) \otimes \mathcal{G} = L^2(X, \mathrm{d}\mu; \mathcal{G}),$$

where $\mu \otimes \nu$ denotes the product measure of μ and ν, defined through the maps $(f \otimes g)(x, y) = f(x)g(y)$ and $(f \otimes \phi)(x) = f(x)\phi$, respectively. If $\{e_\alpha^{(1)}\}$ and $\{e_\beta^{(2)}\}$ are orthonormal bases in the two Hilbert spaces, then the vectors $e_\alpha^{(1)} \otimes e_\beta^{(2)}$ constitute an orthonormal basis in $\mathcal{H}_1 \otimes \mathcal{H}_2$, in particular, $\dim(\mathcal{H}_1 \otimes \mathcal{H}_2) = \dim \mathcal{H}_1 \cdot \dim \mathcal{H}_2$.

The construction of tensor product extends naturally to any finite family of Hilbert spaces. Tensor products of infinite families can also be constructed, however, the procedure is more involved. In quantum mechanics, tensor products serve to describe *composite systems* the state space of which is of the form $\bigotimes_j \mathcal{H}_j$ where $\mathcal{H}_j$ is the state spaces of the jth component; the latter can be either a real physical system or a formal "subsystem" coming from a separation of variables.

13.3
Bounded Operators on Hilbert Spaces

As we have mentioned, without loss of generality, we may regard bounded operators as defined on the whole Hilbert space. We can specify various classes of them.

13.3.1
Hermitean Operators

Given an operator $B \in \mathcal{B}(\mathcal{H})$, we define its *adjoint* as the unique operator B^* satisfying $(y, Bx) = (B^*y, x)$ for all $x, y \in \mathcal{H}$. The map $B \mapsto B^*$ is an antilinear isometry, $\|B^*\| = \|B\|$. The adjoint satisfies

$B^{**} = B$ and $(BC)^* = C^*B^*$, and moreover, $(B^*)^{-1} = (B^{-1})^*$ holds provided B^{-1} exists. Another useful relation is the expression for the subspace on which B^* vanishes, $\mathrm{Ker}\, B^* = (\mathrm{Ran}\, B)^\perp$.

An operator $A \in \mathcal{B}(\mathcal{H})$ is called *Hermitean* if it coincides with its adjoint, $A = A^*$. The spectrum of A is a subset of real axis situated between $m_A = \inf(x, Ax)$ and $m_A = \sup(x, Ax)$; we have $\|A\| = \max(|m_A|, |M_A|)$. The operator is called *positive* if $m_A \geq 0$. For any $B \in \mathcal{B}(\mathcal{H})$, the product B^*B is a Hermitean operator which, in addition satisfies $\|B^*B\| = \|B\|^2$. The last property means, in particular, that $\mathcal{B}(\mathcal{H})$ has also the structure of a C^**-algebra* [3].

To any positive A, there is a unique positive operator $\sqrt{A}$ having the meaning of its *square root*. In particular, one can associate with any $B \in \mathcal{B}(\mathcal{H})$ the operator $|B| := \sqrt{B^*B}$. While this offers an analogy with the modulus of a complex number, caution is needed, for instance, *none of the relations* $|BC| = |B||C|$, $|B^*| = |B|$, $|B + C| \leq |B| + |C|$ *are valid* in general.

An important class of positive operators are *projections* assigning to any vector $x \in \mathcal{H}$ its orthogonal projection to a given subspace $\mathcal{G} \subset \mathcal{H}$. An operator $E \in \mathcal{B}(\mathcal{H})$ is a projection iff $E^2 = E = E^*$; its spectrum contains only the eigenvalues 0 and 1 except for the trivial cases, $E = 0, I$, when only one of them is present.

Projections E, F are *orthogonal* if $\mathrm{Ran}\, E \perp \mathrm{Ran}\, F$, which is equivalent to the condition $EF = FE = 0$. The sum $E + F$ is a projection iff E and F are orthogonal. The product of the projections is a projection iff $EF = FE$ in which case $\mathrm{Ran}\, EF = \mathrm{Ran}\, E \cap \mathrm{Ran}\, F$. Furthermore, the difference $E - F$ is a projection iff $E \geq F$, that is, $E - F$ is positive, which is further equivalent to $\mathrm{Ran}\, E \supset \mathrm{Ran}\, F$, or to the relations $EF = FE = F$.

13.3.2 Unitary Operators

An operator $U \in \mathcal{B}(\mathcal{H})$ is called an *isometry* if its domain $D(U)$ and range $\mathrm{Ran}\, U$ are closed subspaces and the norm is preserved, $\|Ux\| = \|x\|$. If $D(U) = \mathcal{H}$ the operator is called *unitary*, the same name is used for linear isometries between different Hilbert spaces. If $D(U) \neq \mathcal{H}$ we speak about a *partial isometry*. A unitary operator preserves also the inner product, $(Ux, Uy) = (x, y)$, and satisfies the relation $U^{-1} = U^*$; its spectrum is a subset of the unit circle, $\sigma(U) \subset \{z \in \mathbb{C} : |z| = 1\}$.

If V is a partial isometry, the products V^*V and VV^* are the projections to its initial and final subspace, $D(V)$ and the range $\mathrm{Ran}\, V$, respectively. Using partial isometries, one can find a *polar decomposition* of a bounded operator: to any $B \in \mathcal{B}(\mathcal{H})$ there is a unique partial isometry W_B such that $B = W_B|B|$ and the relations $\mathrm{Ker}\, W_B = \mathrm{Ker}\, B$ and $\mathrm{Ran}\, W_B = \overline{\mathrm{Ran}\, B}$ hold. As another word of caution, the "opposite" decomposition, $B = |B|W$, may not exist if B is not Hermitean.

Both Hermitean and unitary operators are *normal*, which means they commute with their adjoints, $BB^* = B^*B$, and similarly the *real* and *imaginary part* of such an operator, $\mathrm{Re}\, B := \frac{1}{2}(B + B^*)$ and $\mathrm{Im}\, B := (1/2\mathrm{i})(B - B^*)$, commute. Normal operators have *empty residual spectrum* and their eigenspaces corresponding to different eigenvalues are orthogonal. An operator $B \in \mathcal{B}(\mathcal{H})$ is said to have a *pure point spectrum* if its eigenvectors form an orthonormal basis in $\mathcal{H}$. Every such operator is normal and $\sigma(B) = \overline{\sigma_{\mathrm{p}}(B)}$.

13.3.3 Compact Operators

An operator $C \in \mathcal{B}(\mathcal{H})$ is called *compact* if it maps any bounded subset of $\mathcal{H}$ into a precompact one, that is, a set the closure of which is compact. Equivalently, a compact operator maps any weakly convergent sequence $\{x_n\} \subset \mathcal{H}$ into a sequence convergent with respect to the norm of $\mathcal{H}$. The set $\mathcal{K}(\mathcal{H})$ of compact operators is a subspace in $\mathcal{B}(\mathcal{H})$, which is closed in the operator norm and, in addition, it has the $*$-*ideal property*, namely, that for any $C \in \mathcal{K}(\mathcal{H})$ and $B \in \mathcal{B}(\mathcal{H})$ the operators C^*, BC, CB are also compact. Compactness has several implications for the operator spectrum:

Theorem 13.4 (Riesz–Schauder theorem) *For a $C \in \mathcal{K}(\mathcal{H})$ any nonzero point of the spectrum is an eigenvalue of finite multiplicity and the only possible accumulation point of the spectrum is zero. The point spectrum is at most countable and the eigenvalue moduli can be ordered, namely,* $|\lambda_1| \geq \cdots \geq |\lambda_j| \geq |\lambda_{j+1}| \geq \cdots$, *with* $\lim_{j\to\infty} \lambda_j = 0$ *if* $\dim \mathcal{H} = \infty$.

Theorem 13.5 (Fredholm alternative) *Given an equation $x - \lambda Cx = y$ with $C \in \mathcal{K}(\mathcal{H})$, $\lambda \in \mathbb{C}$, and $y \in \mathcal{H}$, one and only of the following situations can occur: (i) the equation has a unique solution x_y for any $y \in \mathcal{H}$, in particular, $x_0 = 0$, or (ii) the equation without the right-hand side has a nontrivial solution.*

Theorem 13.6 (Hilbert–Schmidt theorem) *A normal compact operator has a pure point spectrum.*

A *finite-dimensional* operator, that is, an operator C such that $\dim \operatorname{Ran} C < \infty$, is compact, and a norm-convergent sequence of such operators has a compact limit. In fact, there is a *canonical form*, which makes it possible to express every compact operator as such a limit. The operator $|C|$ corresponding to $C \in \mathcal{K}(\mathcal{H})$ is compact and its eigenvalues, $|C|e_j = \mu_j e_j$, form a nonincreasing sequence in $[0, \infty)$; we call them *singular values* of C. Using Fourier expansion of $|C|x$ in combination with the polar decomposition, $C = W|C|$, and introducing the vectors $f_j := We_j$, we arrive at the formula

$$C = \sum_{j=1}^{J_C} \mu_j (e_j, \cdot) f_j, \tag{13.13}$$

where $J_C := \dim \operatorname{Ran} |C|$ and the series converges in the operator norm if $J_C = \infty$.

13.3.4 Schatten Classes

The set $\mathcal{K}(\mathcal{H})$ has some distinguished subsets. As mentioned above, we can associate with a $C \in \mathcal{B}(\mathcal{H})$ its singular values $\{\mu_j\}$. For a fixed $p \geq 1$, we denote by $\mathcal{J}_p$, or $\mathcal{J}_p(\mathcal{H})$, the set of compact operators C such that $\|C\|_p := \left(\sum_j \mu_j^p\right)^{1/p} < \infty$. Each $\mathcal{J}_p$ is a Banach space, which is a closure of the set of finite-dimensional operators with respect to the norm $\|\cdot\|_p$. The inclusion $\mathcal{J}_p \subset \mathcal{J}_{p'}$ holds for $p < p'$ and the set $\mathcal{K}(\mathcal{H})$ is alternatively denoted as $\mathcal{J}_\infty(\mathcal{H})$.

Two of these so-called *Schatten classes* are of particular importance. $\mathcal{J}_2(\mathcal{H})$ is a Hilbert space with respect to the scalar product defined by $(B, C)_2 = \sum_j (Be_j, Ce_j)$ for any fixed orthonormal basis $\{e_j\} \subset \mathcal{H}$, its elements are called *Hilbert–Schmidt operators*. They form a $*$-*ideal* in $\mathcal{B}(\mathcal{H})$: for any $C \in \mathcal{J}_2(\mathcal{H})$ and $B \in \mathcal{B}(\mathcal{H})$, the operators C^*, BC, CB are also Hilbert–Schmidt. We have a useful criterion for integral operators, $B \in \mathcal{B}(L^2(M, \mathrm{d}\mu))$ acting as $(Bf)(x) = \int_M g_B(x, y) f(y)\, \mathrm{d}\mu(y)$: such an operator belongs to the Hilbert–Schmidt class iff its kernel is square integrable,

$$\|B\|_2^2 = \int_{M\times M} |g_B(x,y)|^2 \, \mathrm{d}(\mu\otimes\mu)(x,y) < \infty, \tag{13.14}$$

with respect to the product measure $\mu\otimes\mu$ [2, App. A].

Another important class, again an $*$-*ideal* in $\mathcal{B}(\mathcal{H})$, is $\mathcal{J}_1(\mathcal{H})$ the elements of which are called *trace-class operators*. An operator $B\in\mathcal{B}(\mathcal{H})$ belongs to $\mathcal{J}_1$ iff it is a product of two Hilbert–Schmidt operators. For any $C\in\mathcal{J}_1(\mathcal{H})$, one can define its *trace* by $\mathrm{Tr}\, C := \sum_j (e_j, Be_j)$ where $\{e_j\}$ is any fixed orthonormal basis in $\mathcal{H}$; the $\mathcal{J}_1$-norm is given by $\|C\|_1 = \mathrm{Tr}\,|C|$. The trace is a functional of the unit norm satisfying

$$\mathrm{Tr}\, C^* = \overline{\mathrm{Tr}\, C}\,, \quad \mathrm{Tr}\,(BC) = \mathrm{Tr}\,(CB) \quad \text{for} \quad C\in\mathcal{J}_1(\mathcal{H})\,,\ B\in\mathcal{B}(\mathcal{H})\,. \tag{13.15}$$

Important trace-class operators are *density matrices* used in quantum physics to describe *mixed states*. They are positive $W\in\mathcal{J}_1(\mathcal{H})$ satisfying the normalization condition $\mathrm{Tr}\, W = 1$. Note that one-dimensional projections describing pure states are a particular case of density matrices; a state is *not* pure iff $\mathrm{Tr}\, W^2 < 1$. The subset of pure states can be also characterized geometrically: the set of all density matrices on a given $\mathcal{H}$ is *convex* and one-dimensional projections are its *extremal points*.

13.4 Unbounded Operators

Unbounded operators are considerably more difficult to deal with because one has to pay attention to their definition domains; however, they appear in numerous applications.

13.4.1 Operator Adjoint and Closure

An operator T on $\mathcal{H}$ is said to be *densely defined* if $\overline{D(T)} = \mathcal{H}$; in contrast to the previous section, there is no standard way to extend such an operator to the whole $\mathcal{H}$. The set of densely defined operators on $\mathcal{H}$ will be denoted as $\mathcal{L}(\mathcal{H})$. To a given $T\in\mathcal{L}(\mathcal{H})$ and $y\in\mathcal{H}$, there is at most one vector $y^*\in\mathcal{H}$ such that the relation $(y,Tx) = (y^*,x)$ holds. We denote by $D(T^*)$ the subspace of y for which such a y^* exists and introduce the *adjoint* T^* of T acting as $T^*y = y^*$.

As in the bounded case, we have $\mathrm{Ker}\, T^* = (\mathrm{Ran}\, T)^\perp$, so $\mathrm{Ker}\, T^*$ is a closed subspace. If $T^{-1}\in\mathcal{L}(\mathcal{H})$, then T^* is also invertible and $(T^*)^{-1} = (T^{-1})^*$. On the other hand, if S is an extension of T, $S\supset T$, then $S^*\subset T^*$. Other relations valid for bounded operators hold generally in a weaker form only, for instance, $T^{**}\supset T$ provided $T^*\in\mathcal{L}(\mathcal{H})$; similarly, the relations $(S+T)^*\supset S^*+T^*$ and $(TS)^*\supset S^*T^*$ are valid; in both cases, they turn to identities if $T\in\mathcal{B}(\mathcal{H})$.

This allows to introduce two important notions. An operator $A\in\mathcal{L}(\mathcal{H})$ is *symmetric* if $A\subset A^*$, in other words, if $(y,Ax) = (Ay,x)$ holds for all $x,y\in D(A)$. If, in addition, $A = A^*$, the operator is called *self-adjoint*. If A is bounded, the two notions coincide and we speak about a Hermitean operator; in the unbounded case, it is better avoid this term because it may cause misunderstanding.

As an example, let us mention the multiplication operator Q on $\mathcal{H} = L^2(\mathcal{I})$ acting as $(Q\psi)(x) = \int_{\mathcal{I}} x\psi(x)\,\mathrm{d}x$, which one associates in quantum mechanics with the particle *position*. It is bounded iff the interval $\mathcal{I}$ is bounded; if it is not the case, we put $D(Q) = \{\psi\in\mathcal{H} : \int_{\mathcal{I}} x^2|\psi(x)|^2\,\mathrm{d}x < \infty\}$. It is easy to check that the operator Q defined in this way is not only symmetric but also self-adjoint.

The fact that the domain of an unbounded symmetric operator is not the whole $\mathcal{H}$ seen in this example is valid generally, because the closed-graph theorem has the following consequence:

Theorem 13.7 (Hellinger–Toeplitz theorem) *A symmetric operator A with $D(A) = \mathcal{H}$ is bounded.*

The notion of adjoint operator allows us to formulate new properties of closed operators, in addition to the general ones mentioned above. The adjoint to any $T \in \mathcal{L}(\mathcal{H})$ is a closed operator. The *closure* $\overline{T}$ of an operator T is its smallest closed extension, or equivalently, the operator the graph of which is the closure of $\Gamma(T)$ in $\mathcal{H} \oplus \mathcal{H}$. If the adjoint T^* is densely defined, the closure of T exists and the relations $T^{**} = \overline{T}$ and $(\overline{T})^* = T^*$ are valid. The set of closed densely defined operators on a given $\mathcal{H}$ will be denoted as $\mathcal{L}_c(\mathcal{H})$.

In particular, any self-adjoint operator is closed. As the closure represents a unique way of extending an operator, it is useful to define an *essentially self-adjoint (e.s.a.)* operator A as such that its closure is self-adjoint, $\overline{A} = A^*$. This concept is useful in applications because it is often easier to prove that an operator A is e.s.a., in which case we know it has a unique self-adjoint extension even if we may not know the domain $D(\overline{A})$ of the latter explicitly. More generally, a subspace D in the domain of an operator $T \in \mathcal{L}_c(\mathcal{H})$ is called a *core* if the restriction $T \upharpoonright D$ of T to D satisfies $\overline{T \upharpoonright D} = T$. It means that D is a core of a self-adjoint A iff $A \upharpoonright D$ is e.s.a.; one usually says that A *is e.s.a.* on D.

For any $T \in \mathcal{L}_c(\mathcal{H})$, the product T^*T is self-adjoint and positive, that is, $(x, T^*Tx) = \|Tx\|^2 \geq 0$ holds for any $x \in D(T^*T)$, and $D(T^*T)$ is a core for T; similarly, TT^* is self-adjoint and positive, and $D(TT^*)$ is a core for T^*.

For a closed Hilbert-space operator T, we introduce its *essential spectrum*, $\sigma_{\mathrm{ess}}(T)$, as the set of all $\lambda \in \mathbb{C}$ to which one can find a sequence $\{x_n\} \subset D(T)$ of unit vectors having no convergent subsequence and satisfying $(T - \lambda)x_n \to 0$ as $n \to \infty$. The spectrum of the operator T is then the union of $\sigma_{\mathrm{ess}}(T)$ with $\sigma_{\mathrm{p}}(T) \cup \sigma_{\mathrm{r}}(T)$ and one has $\sigma_{\mathrm{c}}(T) = \sigma_{\mathrm{ess}}(T) \setminus (\sigma_{\mathrm{p}}(T) \cup \sigma_{\mathrm{r}}(T))$. However, in distinction to the spectral decomposition mentioned in Section 13.1.3, the present one is not disjoint, for instance, an eigenvalue of infinite multiplicity belongs simultaneously to $\sigma_{\mathrm{p}}(T)$ and $\sigma_{\mathrm{ess}}(T)$.

13.4.2 Normal and Self-Adjoint Operators

As in the bounded case, an operator $T \in \mathcal{L}_c(\mathcal{H})$ is said to be *normal* if $T^*T = TT^*$. The set of all such operators on a given $\mathcal{H}$ is denoted as $\mathcal{L}_n(\mathcal{H})$. An operator T is normal iff $D(T) = D(T^*)$ and $\|Tx\| = \|T^*x\|$ holds for all $x \in D(T)$.

A typical example is that of *operators of multiplication by a function* generalizing the operator Q described above. Given a function $f : M \to \mathbb{C}$, measurable with respect to a measure μ on M, we define the operator Tf acting as $(T_f\psi)(x) = f(x)\psi(x)$ with the domain $D(T_f) = \{\psi \in L^2(M, \mathrm{d}\mu) : \int_M |f(x)\psi(x)|^2\, \mathrm{d}\mu(x) < \infty\}$. Such an operator is densely defined and normal. It is self-adjoint iff f is real-valued almost everywhere (a.e.) in M and bounded iff f is essentially bounded, $\|T_f\| = \|f\|_\infty$.

As in the bounded case, the residual spectrum of a normal operator is void and its resolvent set coincides with its *regularity domain*, that is, $\lambda \notin \sigma(T)$ holds iff there is a constant $c = c(\lambda) > 0$ such that $\|(T - \lambda)x\| \geq c\|x\|$, which is further equivalent to $\mathrm{Ran}\,(T - \lambda) = \mathcal{H}$. This implies, in particular, that the spectrum of a self-adjoint operator A is a subset of the real axis satisfying the relation $\inf \sigma(A) = \inf\{(x, Ax) : x \in D(A), \|x\| = 1\}$. In the above example, the spectrum $\sigma(T_f)$ coincides with the *essential range* of the function f, which is the set of all points such that the f-preimage of each of their neighborhoods has a nonzero μ measure.

Self-adjointness is of fundamental importance for quantum physics because such operators are used to describe *observables* of quantum systems.

Theorem 13.8 (Basic self-adjointness criterion) *For a symmetric operator A, the following claims are equivalent: (i) A is self-adjoint, (ii) A is closed and* $\mathrm{Ker}\,(A^* \pm \mathrm{i}) = \{0\}$, *and (iii)* $\mathrm{Ran}\,(A \pm \mathrm{i}) = \mathcal{H}$. *In a similar way, essential self-adjointness of A is equivalent to* $\mathrm{Ker}\,(A^* \pm \mathrm{i}) = \{0\}$, *or to* $\overline{\mathrm{Ran}\,(A \pm \mathrm{i})} = \mathcal{H}$.

In view of the above-mentioned role that self-adjointness plays in quantum mechanics, it is important to have other sufficient conditions. An often used one is based on a *perturbation argument*. Given linear operators A, S on $\mathcal{H}$, we say that S is *A-bounded* if $D(S) \supset D(A)$ and there are $a, b \geq 0$ such that

$$\|Sx\| \leq a\|Ax\| + b\|x\| \tag{13.16}$$

holds for any $x \in D(A)$. The infimum of the a's for which the inequality (13.16) is satisfied with some $b \geq 0$ is called the *A-bound* of S. The A-boundedness can be used to prove self-adjointness of operators if a is small enough.

Theorem 13.9 (Kato–Rellich theorem) *Let A be self-adjoint and S symmetric and A-bounded with the A-bound* less than one, *then the sum $A + S$ is self-adjoint. Moreover, if $D \subset D(A)$ is a core for A, then $A + S$ is e.s.a. on D.*

Another useful criterion relies on analytical vectors. A vector x that belongs to $D(A^j)$ for all $j \geq 1$, is called *analytic* (with respect to the operator A) if the power series $\sum_j \|A^j x\| z_j / j!$ has a nonzero convergence radius.

Theorem 13.10 (Nelson theorem) *A symmetric operator is e.s.a. if it has a total set of analytic vectors.*

Sometimes, one can analyze an operator by mapping it to another one the structure of which is more simple. Operators T on $\mathcal{H}$ and S on $\mathcal{G}$ are *unitarily equivalent* if there is a unitary $U : \mathcal{G} \to \mathcal{H}$ such that $T = USU^{-1}$. Unitary equivalence preserves numerous operator properties. If S is densely defined, the same is true for T and $T^* = US^*U^{-1}$; in particular, if S is symmetric or self-adjoint, then T is again symmetric or self-adjoint, respectively. In a similar way, U preserves operator invertibility, closedness, and other properties.

As an example, consider the operator P on $L^2(\mathbb{R})$ acting by $P\psi = -\mathrm{i}\psi'$ with the domain consisting of functions $\psi : \mathbb{R} \to \mathbb{C}$ with the derivatives $\psi' \in L^1 \cap L^2$. In quantum mechanics, this operator describes *momentum* of a one-dimensional particle. It is related to the operator Q on $L^2(\mathbb{R})$ introduced above by the relation

$$P = F^{-1} Q F, \tag{13.17}$$

where F is the Fourier–Plancherel operator (13.3). Both operators are self-adjoint and also e.s.a. on various subsets of their domain, for instance, on $\mathcal{S}(\mathbb{R})$. Moreover, because the unitary equivalence preserves the spectrum, we have also $\sigma(P) = \sigma(Q) = \mathbb{R}$.

13.4.3 Tensor Products of Operators

Next we recall how to construct operators on tensor products of Hilbert spaces. We will consider a product $\mathcal{H}_1 \otimes \mathcal{H}_2$; an extension to any finite number of Hilbert spaces is straightforward. Given operators $B_j \in \mathcal{B}(\mathcal{H}_j)$ we define the map $B_1 \otimes B_2$ on $\mathcal{H}_1 \times \mathcal{H}_2$ by $(B_1 \otimes B_2)(x_1 \otimes x_2) = B_1 x_1 \otimes B_2 x_2$ and extend it first linearly, then continuously. As both the B_j are bounded, the resulting operator is defined on the whole of $\mathcal{H}_1 \otimes \mathcal{H}_2$, satisfying $\|B_1 \otimes B_2\| = \|B_1\| \|B_2\|$.

Tensor products of bounded operators satisfy the usual rules known from matrix algebra, $B_1 C_1 \otimes B_2 C_2 = (B_1 \otimes B_2)(C_1 \otimes C_2)$ and $(B_1 \otimes B_2)^* = B_1^* \otimes B_2^*$, and furthermore, $(B_1 \otimes B_2)^{-1} = B_1^{-1} \otimes B_2^{-1}$ provided that both the B_j are invertible. If the component operators B_j are normal (unitary, Hermitean, projections) the same is respectively true for their tensor product.

If we have operators T_j on $\mathcal{H}_j$, in general unbounded, attention has to be paid to their domains. It is natural to define $T_1 \otimes T_2$ on its "minimal" domain, that is, the linear hull of $D(T_1) \times D(T_2)$. If the T_j's are densely defined, the same is true for their tensor product and $(T_1 \otimes T_2)^* \supset T_1^* \otimes T_2^*$; if the T_j's are closable, so is $T_1 \otimes T_2$ and $\overline{T_1 \otimes T_2} \supset \overline{T}_1 \otimes \overline{T}_2$. One has $(T_1 + S_1) \otimes T_2 = T_1 \otimes T_2 + S_1 \otimes T_2$; however, for the product, in general, the inclusion $(T_1 S_1) \otimes (T_2 S_2) \subset (T_1 \otimes T_2)(S_1 \otimes S_2)$ holds only.

We have mentioned the use of Hilbert-space tensor products to describe composite quantum systems. If a self-adjoint operator A describes an observable of a subsystem with state space $\mathcal{H}_1$ and the complement state space $\mathcal{H}_2$, then the same observable related to the composite system is described by the operator $\overline{A \otimes I}$ on $\mathcal{H}_1 \otimes \mathcal{H}_2$, where I is the unit operator. For example, the first momentum component of a three-dimensional particle is described by the operator $P_1 = \overline{P \otimes I \otimes I}$ on $L^2(\mathbb{R}^3)$. In view of (13.3), we can express it alternatively as $P_1 = F_3^{-1} Q_1 F_3$, where Q_1 is related in the same way to the operator Q on $L^2(\mathbb{R})$ and $F_3 = F \otimes F \otimes F$ is the three-dimensional Fourier–Plancherel operator.

In addition to tensor products themselves, some operators constructed from them are of importance. Given a pair of self-adjoint operators A_j on $\mathcal{H}_j$, we consider operators of the polynomial form,

$$P[A_1, A_2] = \sum_{k=0}^{n_1} \sum_{l=0}^{n_2} a_{kl} (A_1^k \otimes A_2^l) \qquad (13.18)$$

with real-valued coefficients a_{kl}. Without loss of generality, we may suppose that the senior coefficient $a_{n_1 n_2}$ is nonzero and take the linear hull of $D(A_1^{n_1}) \times D(A_2^{n_2})$ as the domain of $P[A_1, A_2]$; one can check that such an operator is e.s.a.

A frequently occurring example is that of self-adjoint operators of the form $\overline{\underline{A}_1 + \underline{A}_2}$, where $\underline{A}_1 := A_1 \otimes I_2$ and $\underline{A}_2 := I_1 \otimes A_2$, which describe sums of observables related to the respective subsystems. In the above example of a three-dimensional quantum particle, the operator $H_0 = \overline{P_1^2 + P_2^2 + P_3^2}$ on $L^2(\mathbb{R}^3)$ describes the kinetic energy (up to a multiplicative factor); we note that it is e.s.a. on $\mathcal{S}(\mathbb{R}^3)$ and acts on its elements as the negative Laplacian, $H_0 \psi = -\Delta \psi$.

13.4.4 Self-Adjoint Extensions

If A' is a symmetric extension of an operator $A \in \mathcal{L}(\mathcal{H})$, we have $A \subset A' \subset (A')^* \subset A^*$. One naturally asks under which conditions one can close the gap by choosing a self-adjoint A'. To show that there are operators for which self-adjoint extensions may not exist, consider again $\psi \mapsto -i\psi'$, this time on the halfline $\mathbb{R}^+ = (0, \infty)$. We denote by $\tilde{P}$ such an operator defined on all $\psi \in (L^2 \cap L^1)(\mathbb{R}^+)$ and by P_0 its restriction to functions satisfying $\psi(0+) = 0$. Using integration by parts, it is easy to see that $P_0^* = \tilde{P}$, and furthermore, that P_0 is symmetric, while $\tilde{P}$ is not; there is obviously no self-adjoint P such that $P_0 \subset P \subset \tilde{P}$ would hold.

To solve the problem in its generality, we introduce *deficiency subspaces* of an operator T to be $\mathrm{Ker}(T^* - \overline{\lambda})$, where λ is a fixed number from its regularity domain as $\mathrm{Ker}(T^* - \overline{\lambda})$. The map $\lambda \mapsto \dim \mathrm{Ker}(T^* - \overline{\lambda})$ is constant on any arcwise-connected component of the regularity domain. In particular, for a symmetric operator A any nonreal number belongs to its regularity domain which thus the latter has at most two connected components; this allows us to define its *deficiency indices* $n_\pm(A) = \dim \mathrm{Ker}(A^* \mp i)$.

The deficiency indices in turn determine whether self-adjoint extensions of a symmetric A exist. By the basic self-adjointness criterion, we know that A is e.s.a. iff $n_\pm(A) = 0$. If it is not the case, all *symmetric* extensions of A can be parameterized, according to the theory constructed by John von Neumann, by isometric maps from $\mathrm{Ker}(A^* - i)$ to $\mathrm{Ker}(A^* + i)$. Consequently, nontrivial self-adjoint extensions exist iff the deficiency indices are nonzero and equal to each other, $n_+(A) = n_-(A)$. If both of them are finite, any maximal symmetric extension is symmetric, otherwise there may exist maximal extensions that are not self-adjoint.

Returning to the example, it is now easy to see why there is no self-adjoint momentum operator on $L^2(\mathbb{R}^+)$; it follows from the fact that the deficiency indices of P_0 are $(1, 0)$. On the other hand, the operator $\psi \mapsto -i\psi'$ on $L^2(\mathcal{I})$, where $\mathcal{I} = (a, b)$ is a finite interval, defined on functions $\psi \in L^2 \cap L^1$ satisfying $\psi(a+) = \psi(b-) = 0$ has deficiency indices $(1, 1)$ and thus a family of self-adjoint extensions. We can denote them by P_θ as each of them can be characterized by the boundary condition $\psi(b-) = e^{i\theta}\psi(a+)$ for some $\theta \in [0, 2\pi)$.

Characterizing self-adjoint extensions by means of boundary conditions is common for differential operators. As another example, let us mention again a one-dimensional particle on a halfline. While its momentum operator does not exist, the operator of kinetic energy does, acting as $\psi \mapsto -\psi''$ modulo a multiplicative constant. If we choose for its domain all $\psi \in L^2(\mathbb{R}^+)$ such that $\psi(0+) = \psi'(0+) = 0$, we obtain a symmetric operator with deficiency indices $(1, 1)$. Its self-adjoint extensions T_λ are characterized by the condition $\psi'(0+) - \lambda\psi(0+) = 0$ for $\lambda \in \mathbb{R}$, or by the Dirichlet condition $\psi(0+) = 0$, which formally corresponds to $\lambda = \infty$.

This example is a particular case of *one-dimensional Schrödinger operator* on $L^2(\mathcal{I})$, where the interval $\mathcal{I}$ can be finite, semifinite, or $\mathcal{I} = \mathbb{R}$; such an operator acts as $\psi \mapsto -\psi'' + V\psi$, where the *potential* V is a locally integrable function. Asking whether this formal differential operator can be made self-adjoint, one has to inspect solutions to the equation

$$-\psi''(x) + V(x)\psi(x) = \lambda\psi(x)\,, \qquad \lambda \in \mathbb{C}\,. \tag{13.19}$$

Theorem 13.11 (Weyl alternative) *At each end point of $\mathcal{I}$, just one of the following possibilities is valid: (i)* limit-circle case: *for any $\lambda \in \mathbb{C}$* all *solutions are L^2 in the vicinity of the end point, or (ii)* limit-point case: *for any $\lambda \in \mathbb{C}$ there is at least one solution which is* not *L^2 in the vicinity of the end point.*

If one or both end points of $\mathcal{I}$ are of the limit-circle type, one has to impose boundary conditions there to make the operator self-adjoint.

A useful tool to find relations between self-adjoint extensions of a given symmetric operator is the *Krein formula*: if the maximal common part A of self-adjoint operators A_1, A_2 has deficiency indices (n, n), then the relation

$$(A_1 - z)^{-1} - (A_2 - z)^{-1} = \sum_{j,k=1}^{n} \lambda_{jk}(z)(y_k(\bar{z}), \cdot)y_j(z) \qquad (13.20)$$

holds for any $z \in \rho(A_1) \cap \rho(A_2)$, where the matrix (λ_{jk}) is nonsingular and $y_j(z)$, $j = 1, \dots, n$, are linearly independent vectors from $\mathrm{Ker}(A^* - z)$; the functions $\lambda_{jk}(\cdot)$ and $y_j(\cdot)$ can be chosen to be analytic in $\rho(A_1) \cap \rho(A_2)$. Kreins's formula has numerous applications, in particular, when constructing and analyzing solvable models of quantum systems [7].

13.5 Spectral Theory of Self-Adjoint Operators

It is well known from linear algebra that any symmetric matrix possesses a unique finite family of eigenvalues and that the corresponding eigenvectors can be chosen to form an orthonormal basis in the appropriate vector space. Now we are going to show how these properties extend to self-adjoint operators on an arbitrary Hilbert space.

13.5.1 Functional Calculus

Our goal here is to define spectral decomposition of an operator. To that end, we first have to introduce integration with respect to a particular type of operator measures. They are defined on a set X equipped with a family (a σ-field, to be exact) of measurable subsets; typically, we have in mind $\mathbb{R}$, the real line, and the family $\mathcal{B}$ of all its Borel subsets.

Given a Hilbert space $\mathcal{H}$, we introduce a *projection-valued* (or *spectral*) *measure*, a map that associates with any measurable set M a projection $E(M) \in \mathcal{B}(\mathcal{H})$ such that $E(X) = I$ and the map is *σ-additive*, that is, for any at most countable *disjoint* family $\{M_n\}$ of measurable sets, the relation $E\left(\bigcup_n M_n\right) = \sum_n E(M_n)$ is valid; as a consequence of additivity, we have $E(\emptyset) = 0$.

If $X = \mathbb{R}^d$, one can construct spectral measures starting from projections assigned to rectangular sets – in particular, intervals if $d = 1$ – in a way fully similar to that used in the conventional measure theory. With a spectral measure E on the real line, one can associate also a *spectral decomposition* as a right-continuous map $\lambda \mapsto E_\lambda$, which is nondecreasing and satisfies the relations

$$\operatorname*{s-lim}_{\lambda \to -\infty} E_\lambda = 0\,, \quad \operatorname*{s-lim}_{\lambda \to +\infty} E_\lambda = I, \qquad (13.21)$$

where the strong convergence means that $E_\lambda x$ converges to 0 or I, respectively, for any $x \in \mathcal{H}$; the relation between the two notions is given by $E_\lambda = E((-\infty, \lambda])$.

On a finite-dimensional $\mathcal{H}$, any spectral measure is supported by a finite set of points, $\{\lambda_j\} \subset \mathbb{R}$, so that $E(M) = E\big(M \cap \{\lambda_j\}\big)$ holds for any measurable M; the corresponding spectral decomposition is the steplike function $E_\lambda = E\big(\{\lambda_j : \lambda_j \le \lambda\}\big)$. Another simple example is the spectral measure E_Q of the multiplication operator Q on $L^2(\mathbb{R})$ introduced in Section 13.4.1. It acts as multiplication by the characteristic function, $E_Q(M)\psi = \chi_M \psi$, for any $M \in \mathcal{B}$.

Having defined the spectral measure, we are able to introduce integration with respect to it as the map which assigns to a measurable function f on X the operator

$$\mathcal{T}(f) = \int_X f(\lambda)\,\mathrm{d}E(\lambda) \qquad (13.22)$$

on the corresponding Hilbert space $\mathcal{H}$. To define the action of $\mathcal{T}(f)$, we use a construction analogous to the one employed in the usual calculus. First, we define the integral on *simple functions* having a finite number of values, $f = \sum_j \lambda_j \chi_{M_j}$ for a finite family $\{M_j\}$ of measurable sets; we set $\mathcal{T}(f) := \sum_j \lambda_j E(M_j)$. In the next step, we use the fact that any bounded measurable function f can be approximated by simple functions $\{f_n\}$ in the sense that $\|f - f_n\|_\infty \to 0$ as $n \to \infty$, which allows us in the next step to define the operator $\mathcal{T}(f) := \lim_{n\to\infty} \mathcal{T}(f_n)$ for such functions, where the convergence is understood in the sense of operator norm.

The map $f \mapsto \mathcal{T}(f)$ introduced in this way is linear and multiplicative, $\mathcal{T}(fg) = \mathcal{T}(f)\mathcal{T}(g) = \mathcal{T}(g)\mathcal{T}(f)$. Denoting by $L^\infty(X, \mathrm{d}E)$ the family of functions bounded a.e. with respect to the measure E, we have $\|\mathcal{T}(f)\| = \|f\|_\infty$ and $\mathcal{T}(f)^* = \mathcal{T}(\bar{f})$ for any such function. Moreover, if a sequence $\{f_n\} \subset L^\infty(X, \mathrm{d}E)$ converges pointwise to a function f and the set $\{\|f_n\|_\infty\}$ is bounded, one has $f \in L^\infty(X, \mathrm{d}E)$ and the limit can be interchanged with the integral.

Integrating unbounded functions is more involved. First, we notice that for any spectral measure E and $x \in \mathcal{H}$ the relation $\mu_x(\cdot) = (x, E(\cdot)x)$ defines a numerical measure on X. Using it, we can associate with a measurable function f the set

$$D_f := \left\{ x \in \mathcal{H} : \int_X |f(\lambda)|^2\,\mathrm{d}\mu_x(\lambda) < \infty \right\}, \qquad (13.23)$$

which is dense in $\mathcal{H}$. Furthermore, to the function f one can construct a sequence $\{f_n\} \subset L^\infty(X, \mathrm{d}E)$ such that $f_n(\lambda) \to f(\lambda)$ and $|f_n(\lambda)| \le |f(\lambda)|$ holds E-a.e. in X. We use it to define the operator $\mathcal{T}(f) = \int_X f(\lambda)\,\mathrm{d}E(\lambda)$ by

$$\mathcal{T}(f)x = \lim_{n\to\infty} \mathcal{T}(f_n)x\,, \quad x \in D_f\,; \qquad (13.24)$$

the chosen domain is natural because $\|\mathcal{T}(f)x\|^2 = \int_X |f(\lambda)|^2\,\mathrm{d}\mu_x(\lambda)$.

The map $f \mapsto \mathcal{T}(f)$ is homogeneous and $\mathcal{T}(f+g) \supset \mathcal{T}(f) + \mathcal{T}(g)$; similarly, for the multiplication of functions, we have in general $\mathcal{T}(fg) \supset \mathcal{T}(f)\mathcal{T}(g)$. It also holds that $\mathcal{T}(f) = \mathcal{T}(g)$ implies that $f(\lambda) = g(\lambda)$ holds almost everywhere with respect to the measure E. Furthermore, we have $\mathcal{T}(f)^* = \mathcal{T}(\bar{f})$ as in the bounded case and the relation $\mathcal{T}(f)^{-1} = \mathcal{T}(f^{-1})$ holds provided F is nonzero E-a.e.

The operator $\mathcal{T}(f)$ defined by (13.24) is normal for any measurable f, it is self-adjoint if $f(t)$ is real with a possible exception of an E-zero measure set. The function f also determines the spectrum $\sigma(\mathcal{T}(f))$ that coincides with the essential range of f with respect to the measure E, as defined in Section 13.4.2. In particular, $\lambda \in \mathbb{C}$ is an eigenvalue of $\mathcal{T}(f)$ iff $E(f^{(-1)}(\{\lambda\})) \neq 0$ and, in this case, $\operatorname{Ran} E(f^{(-1)}(\{\lambda\}))$ is the corresponding eigenspace.

13.5.2
Spectral Theorem

Armed with the notions introduced above, we are ready to state the fundamental structural result about self-adjoint operators.

Theorem 13.12 (The spectral theorem) *To any self-adjoint operator A on a Hilbert space $\mathcal{H}$, there is a unique spectral measure E_A on $\mathcal{H}$ such that*

$$A = \int_{\mathbb{R}} \lambda\,\mathrm{d}E_A(\lambda)\,. \qquad (13.25)$$

Furthermore, a bounded operator B commutes with A, $BA \subset AB$, iff it commutes with its spectral decomposition, that is, with $E_\lambda^{(A)} = E_A((-\infty, \lambda])$ for any $\lambda \in \mathbb{R}$.

The second claim allows us to specify the meaning of commutativity, which is, in general, not easy to introduce if both the operators involved are unbounded. If A_1, A_2 are self-adjoint we say that they *commute* if any element of the spectral decompositions of A_1 commutes with any element of the spectral decompositions of A_2.

The spectral theorem has various corollaries and modifications. For instance, a bounded normal operator, which we can write as $B = A_1 + iA_2$ with commuting Hermitean A_1, A_2, can be expressed as $B = \int_{\mathbb{C}} z\, dF(z)$, where F is the projection-valued measure on $\mathbb{C}$ obtained as the product measure of E_{A_1} and E_{A_2}. Furthermore, with a unitary operator U, one can associate a unique spectral measure E_U with the support in the interval $[0, 2\pi) \subset \mathbb{R}$ such that

$$U = \int_{\mathbb{R}} e^{i\lambda}\, dE_U(\lambda)\,. \tag{13.26}$$

These claims are used in one of the possible ways to prove the spectral theorem, in which one checks its validity subsequently for Hermitean, bounded normal, and unitary operators, passing finally to the general case through the appropriate "change of variables" in the integral.

As an example of the decomposition (13.26), let us mention the Fourier–Plancherel operator (13.3). It acts on the elements of the orthonormal basis (13.7) as $Fh_n = (-i)^n h_n$, $n = 0, 1, \ldots$, hence its spectrum is pure point, $\sigma(F) = \{1, -i, -1, i\}$. The spectral measure can be in this case expressed explicitly,

$$E_F(M) = \frac{1}{4} \sum_{j,k=0}^{3} \chi_M\left(\frac{\pi k}{2}\right)(-i)^{jk} F^j\,. \tag{13.27}$$

The spectral theorem provides a tool to describe and classify spectra of self-adjoint operators. To begin with, a real λ belongs to $\sigma(A)$ iff $E_A(\lambda - \epsilon, \lambda + \epsilon) \neq 0$ for any $\epsilon > 0$; it is an eigenvalue iff the point λ itself has a nonzero E_A measure. In particular, any isolated point of the spectrum is an eigenvalue. The results of the previous section imply that the spectrum of a self-adjoint A is always nonempty and that such an operator is bounded iff its spectrum is bounded.

The essential spectrum of a closed operator has been defined in Section 13.4.1. If A is self-adjoint, $\lambda \in \sigma_{ess}(A)$ iff $\dim \operatorname{Ran} E_A(\lambda - \epsilon, \lambda + \epsilon) = \infty$ holds for any $\epsilon > 0$ and the essential spectrum is a closed set. Points of $\sigma_{ess}(A)$ fall into three, mutually nonexclusive categories: such a λ can belong to $\sigma_c(A)$, be an eigenvalue of infinite multiplicity, or an accumulation point of eigenvalues. The complement of $\sigma_{ess}(A)$ consists of isolated eigenvalues of finite multiplicity; it is a subset of the point spectrum for which we use the term *discrete spectrum*.

We have seen in Section 13.3.3 that the spectrum of compact operators away from zero consists of isolated eigenvalues of finite multiplicity. This is the reason why the essential spectrum of a self-adjoint A is stable with respect to compact perturbations; this fact is usually referred to as the *Weyl theorem*. It can be substantially generalized. Given a self-adjoint A, an operator T is called *A-compact* if $D(T) \supset D(A)$ and $T(A - i)^{-1}$ is compact; this property again guarantees spectral stability.

Theorem 13.13 (Generalized Weyl theorem) $\sigma_{ess}(A + T) = \sigma_{ess}(A)$ *holds if the operator T is symmetric and A-compact.*

Using the spectral theorem, we can introduce also another classification of the spectrum. We call $\mathcal{H}_{ac}(A)$ the subspace of all $x \in \mathcal{H}$ such that $\mu_x(N) = 0$ holds any Borel set N of zero Lebesgue measure, in other words, such that the measure μ_x is absolutely continuous with respect to the Lebesgue measure. The orthogonal complement of $\mathcal{H}_{ac}(A)$ is denoted by $\mathcal{H}_s(A)$. The projections to these subspace commute with the operator A, so we can write it as $A = A_{ac} \oplus A_s$, its spectrum being the union of the component spectra, which we call $\sigma_{ac}(A)$, the *absolutely continuous spectrum* of A, and $\sigma_s(A)$, the *singular* one, respectively. Furthermore, the complement $\sigma_{sc}(A) := \sigma_s(A) \setminus \sigma_p(A)$ is called the *singularly continuous spectrum* of A; any of these components may be nonempty if $\dim \mathcal{H} = \infty$.

Before closing this section, let us recall the role that spectral analysis plays in quantum physics. As we have said, one associates a self-adjoint operator A with any observable of a quantum system. In contrast to classical physics, measurement has a probabilistic character here: its possible outcomes coincide with the spectrum of A and the probability of finding the measured value in a set $M \subset \mathbb{R}$ is given by $w(M, A; W) = \mathrm{Tr}(E_A(M)W)$ if the state of the system before the measurement is described by a density matrix W, in particular, by

$$w(M, A; \psi) = \int_M d(\psi, E_A(\lambda)) = \|E_A(M)\psi\|^2 \tag{13.28}$$

if the state is pure being described by a unit vector $\psi \in \mathcal{H}$. The postulate makes sense; it is easy to check that $\mathrm{Tr}\,(E_A(\cdot)W)$ is a probability measure on $\mathbb{R}$.

Measurement of this type can be regarded as the simplest, dichotomic observables with two possible outcomes, positive (the observed value is found in M) and negative (it is found outside M); one often uses the term *yes–no experiment* for them. Using this notion, we can also describe what happens with the system after the measurement. If the outcome is positive, the resulting state is described by the vector $E_A(M)\psi/\|E_A(M)\psi\|$, or more generally by the density matrix $E_A(M)WE_A(M)/\mathrm{Tr}(E_A(M)W)$; in the negative case, we replace M by $\mathbb{R} \setminus M$.

13.5.3 More about Spectral Properties

Functional calculus allows us to define *functions of a self-adjoint operator A* naturally using its spectral measure by the relation

$$f(A) := \mathcal{T}(f) = \int_{\mathbb{R}} f(\lambda)\, dE_A(\lambda)\,. \tag{13.29}$$

The term "function" has to be taken with a grain of salt here because it is f that is the "variable" in the above formula. A simple example is $f(Q)$ on $L^2(\mathbb{R})$, which acts as $(f(Q)\psi) = f(x)\psi(x)$ on L^2 functions satisfying $\int_{\mathbb{R}} |f(x)\psi(x)|^2\, dx < \infty$. The definition (13.29) is consistent in the sense that for elementary functions such as polynomials, it gives the same result as one would obtain without using the spectral theory.

A slightly more involved example concerns functions of the momentum operator P on $L^2(\mathbb{R})$. One can use the unitary equivalence (13.17) to express their action. For instance, for the exponential function, one obtains the unitary operators

$$\left(e^{iaP}\psi\right)(x) = \psi(x+a)\,, \quad a \in \mathbb{R}, \tag{13.30}$$

acting as the translation group in $L^2(\mathbb{R})$, and for $f \in L^2(\mathbb{R})$ we get the expression

$$\left(f(P)\psi\right)(x) = \frac{1}{\sqrt{2\pi}} \int_{\mathbb{R}} (Ff)(y-x)\psi(y)\, dy \tag{13.31}$$

with the integral kernel given by the Fourier–Plancherel image of f.

The limiting property of sequences $\{\mathcal{T}(f_n)\}$ mentioned above can be used to derive an expression of the spectral measure in terms of the resolvent $R_A(z) = (A-z)^{-1}$. Specifically, because the function $\arctan\left(b-\lambda/\epsilon\right) - \arctan\left(a-\lambda/\epsilon\right)$ for fixed $a<b$ tends to $\frac{1}{2}[\chi_{[a,b]} + \chi_{(a,b)}]$ as $\epsilon \to 0$, we obtain the relation

$$E_A([a,b]) + E_A((a,b)) = \frac{1}{\pi \mathrm{i}} \operatorname*{s-lim}_{\epsilon\to 0+} \int_a^b [R_A(\lambda+\mathrm{i}\epsilon) - R_A(\lambda - \mathrm{i}\epsilon)]\,\mathrm{d}\lambda\,, \tag{13.32}$$

which is known as the *Stone formula*. If $\sigma_{\mathrm{p}}(A) = \emptyset$, there is no difference between $E_A([a,b])$ and $E_A((a,b))$. The behavior of the integrand in the vicinity of the real axis determines also more subtle spectral properties, in particular, if

$$\sup_{0<\epsilon<1} \int_a^b \left|\mathrm{Im}\,(\psi, R_A(\lambda+\mathrm{i}\epsilon)\psi)\right|^p \mathrm{d}\lambda < \infty \tag{13.33}$$

holds for some $p>1$ and $\psi \in \mathcal{H}$, then $E_A((a,b))\psi$ belongs to $\mathcal{H}_{\mathrm{ac}}(A)$, the absolutely continuous spectral subspace of the operator A.

In Section 13.5.2, we have mentioned how a spectral measure on $\mathbb{C}$ can be associated with a bounded normal operator. In a similar way, one can treat a finite family of commuting self-adjoint operators $A_1,\dots,A_n$. One can construct a projection-valued measure E on $\mathbb{R}^n$ as the product measure of $E_{A_1},\dots,E_{A_n}$. This allows us to define functions of such a family of commuting operators by

$$f(A_1,\dots,A_n) := \int_{\mathbb{R}^n} f(\lambda_1,\dots,\lambda_n)\, \mathrm{d}E(\lambda_1,\dots,\lambda_n)\,; \tag{13.34}$$

it is easy to see that such an operator commutes with all the $A_1,\dots,A_n$.

In quantum physics, observables described by commuting self-adjoint operators are called *compatible*; they can be measured simultaneously, which means we can measure them in any order provided the measurements follow immediately after each other. The probability of finding the measured values in the set $M \subset \mathbb{R}^n$ is

$$w(M,\{A_1,\dots,A_n\};W) = \mathrm{Tr}(E_A(M)W) \tag{13.35}$$

if the system before the measurement was a mixed state described by a density matrix W. One can imagine that compatible observables are measured by a single apparatus. It can also be used to measure their functions (13.34) after a proper rescaling; the appropriate probability equals

$$w(M, f(A_1,\dots,A_n); W) = w(f^{(-1)}(M), \{A_1,\dots,A_n\}; W)\,, \tag{13.36}$$

assuming again that the initial state of the system is described by a density matrix W.

A particular case of commuting self-adjoint operators are those which can be expressed through tensor products as described in Section 13.4.3. The operators $\underline{A}_1 = \overline{A_1 \otimes I_2}$ and $\underline{A}_2$ commute with each other and $\sigma(\underline{A}_j) = \sigma(A_j)$ holds for $j=1,2$. The self-adjoint operator $\overline{P[A_1,A_2]}$ corresponding to (13.18) coincides with $P(\underline{A}_1,\underline{A}_2)$ defined according to (13.34) and its spectrum is given by $\sigma(\overline{P[A_1,A_2]}) = \overline{P(\sigma(A_1)\times\sigma(A_2))}$.

For example, the measured values of the total energy of a system consisting of two noninteracting subsystems are sums of measured values of the corresponding subsystem energies, i.e. $\sigma(\overline{\underline{H}_1 + \underline{H}_2}) = \{\lambda_1+\lambda_2 \,:\, \lambda_j \in \sigma(H_j)\}$.

13.5.4 Groups of Unitary Operators

Let us now consider families $\{U(s) : s \in \mathbb{R}\}$ of unitary operators on a given $\mathcal{H}$ such that the map $s \mapsto U(s)$ is *strongly continuous*, that is, $U(\cdot)x$ is continuous for any $x \in \mathcal{H}$, and the group property is satisfied, $U(t+s) = U(t)U(s)$ for any $t, s \in \mathbb{R}$. With such a group, one can associate its *generator* acting as

$$Tx = \lim_{s\to 0} \frac{U(s) - I}{is}\, x \tag{13.37}$$

on the domain consisting of all $x \in \mathcal{H}$ for which the limit exists. It is not difficult to check that $\{e^{isA} : s \in \mathbb{R}\}$ corresponding to a self-adjoint A is a strongly continuous unitary group and A is its generator; a deep result says that the converse is also true.

Theorem 13.14 (Stone theorem) *To any strongly continuous unitary group $\{U(s) : s \in \mathbb{R}\}$ there is a unique self-adjoint operator A such that $U(s) = e^{isA}$ holds for any $s \in \mathbb{R}$.*

The theorem has various consequences. Using functional calculus, for instance, it yields an alternative expression of commutativity: self-adjoint operators A_1, A_2 commute iff $[e^{isA_1}, e^{itA_2}] = 0$ holds for all $s, t \in \mathbb{R}$.

Elementary examples are operators by e^{isx} on $L^2(\mathbb{R})$ generated by the operator Q, or the group of translations of $L^2(\mathbb{R})$ which is in view of (13.30) generated by the momentum operator P. Another example is provided by the group of *dilations* associated with scaling of the real axis, $(U_d(s)\psi)(x) = e^{s/2}\psi(e^s x)$ on $L^2(\mathbb{R})$, which is generated by the symmetrized product $A_d = \frac{1}{2}\,\overline{PQ + QP}$.

Unitary groups arise in various contexts in quantum physics. The most important are the operators $U(t) = e^{-itH}$, where H is the operator of total energy, or *Hamiltonian*, which describe the time evolution of a conservative system (in the units where $\hbar = 1$). If such a system is undisturbed by measurements, the state vector ψ_0 at the initial time instant $t = 0$ is mapped to $\psi_t = U(t)\psi_0$ at the time t. The differential form of the evolution is known as the *Schrödinger equation*,

$$i\frac{d}{dt}\psi_t = H\psi_t \tag{13.38}$$

with the initial condition $\psi_0 \in D(H)$, or alternatively $i d/dt W_t\phi = [H, W_t]\phi$ for a mixed state described by density matrix W_t.

If Hermitean operators A, B commute, then the products $e^{isA}e^{isB}$ form a strongly continuous unitary group and $A + B$ is its generator. This need not be true if the operators are unbounded but the conclusion still makes sense in the functional-calculus sense. If A, B do not commute, the products may not form a group; however, we have a limiting relation called the *Trotter formula*: if $C = A + B$ is e.s.a., then

$$e^{it\overline{C}} = \operatorname*{s-lim}_{n\to\infty} \left(e^{itA/n}e^{itB/n}\right)^n. \tag{13.39}$$

This result is useful because often we have operators representing observables that are sums of self-adjoint parts, and we know explicit expressions of unitary groups of the latter. A prime example is that of quantum-mechanical Hamiltonians of the form $-\Delta + V(x)$ where Trotter's formula provides a way to express the corresponding evolution operator through the *Feynman path integral* [1, Chap. X; 4].

If $U_j(\cdot)$ is a strongly continuous unitary group on $\mathcal{H}_j$ with the generator A_j, $j = 1, 2$, then the operators $U_1(s) \otimes U_2(s)$ also form such a group and its generator is $\overline{A_1 + A_2}$. For instance, the evolution operator of a system consisting of two noninteraction subsystems with Hamiltonians H_j is $e^{-itH_1} \otimes e^{-itH_2}$. This conclusion extends easily to any finite number of subsystems; it naturally ceases to be valid if the subsystems interact so that the total Hamiltonian is no longer of the form $\overline{H_1 + H_2}$.

In a similar way, one can take tensor products of unitary group elements referring to different values of the parameters involved. For example, the operators $U(s) = e^{is_1 P_1} \otimes \cdots \otimes e^{is_n P_n}$ on $L^2(\mathbb{R}^n)$, where $s = (s_1, \dots, s_n)$, form the group of translations of the n-dimensional Euclidean space generalizing relation (13.30), $(U(s)\psi)(x) = \psi(x+s)$ for any $x = (x_1, \dots, x_n) \in \mathbb{R}^n$. By means of the n-dimensional Fourier–Plancherel operator, these operators are unitarily equivalent to $V(s)\psi)(x) = e^{is\cdot x}\psi(x)$, where $s \cdot x = s_1 x_1 + \cdots + s_n x_n$ is the inner product in $\mathbb{R}^n$.

The unitary groups mentioned in the last example are associated with important quantum-mechanical variables, coordinates of the position and momentum. It is easy to check that their products in a different order differ by an exponential factor,

$$U(t)V(s) = e^{is\cdot t} V(s)U(t), \quad s, t \in \mathbb{R}^n. \tag{13.40}$$

The relations (13.40) can be regarded a mathematically rigorous form of *canonical commutation relations*, often referred to as their *Weyl form*; the operators $U(t), V(s)$ on $L^2(\mathbb{R}^n)$ described in the previous paragraph define the so-called *Schrödinger representation* of (13.40). A fundamental question concerns the uniqueness of this representation; we ask, of course, about *irreducible* representations for which no nontrivial projection commutes with all the operators.

Theorem 13.15 (Stone–von Neumann theorem) *Any irreducible (unitary, strongly continuous) representation of the Weyl relations* (13.40) *is unitarily equivalent to the Schrödinger representation of the corresponding dimension.*

The reader should be warned that an analogue of this theorem in situations with infinite number of degrees of freedom, that is, when $\mathbb{R}^n$ is replaced with a real Hilbert space of infinite dimension, *is not valid*. Indeed, in quantum field theory one can find examples with infinite number of inequivalent representations of such generalized Weyl relations.

13.6 Some Applications in Quantum Mechanics

We have mentioned already some ways in which functional-analytic notions and results are used in quantum physics. In the last section of this chapter, we will briefly describe two specific applications. We do it mostly to whet the reader's appetite; there is a large number of related results for which we refer to the literature indicated at the end of the chapter.

13.6.1 Schrödinger Operators

In nonrelativistic quantum mechanics, the Hamiltonian of a spinless quantum particle, or a system of such particles, is often of the form

$$H = -\Delta + V(x) \tag{13.41}$$

on $L^2(\mathbb{R}^n)$. In general, the expression involves nontrivial coefficients, in particular, the e.s.a. operator describing the kinetic part is $\sum_{j=1}^n (1/2m_j)P_j^2$; however, one can always put $2m_j = 1$ by using suitable units.

The first question when analyzing operators (13.41) concerns their (essential) self-adjointness. The answer is easy if the potential V is bounded; unfortunately, most potentials we have to deal with in actual physical models are unbounded. One way to address the self-adjointness problem uses perturbation theory; the Kato–Rellich theorem can be used to make the following conclusion:

Theorem 13.16 (Sufficient self-adjointness condition) *Assume that the potential* $V \in L^p + L^\infty$, *that is,* $V = V_p + V_\infty$ *with* $V_\infty \in L^\infty(\mathbb{R}^n)$ *and* $V_p \in L^p(\mathbb{R}^n)$, *where* $p = 2$ *if* $n \le 3$ *and* $p > \frac{1}{2}n$ *for* $n \ge 4$, *then the operator* (13.41) *is e.s.a. on any core of* $H_0 = -\Delta$.

This result is not immediately applicable to operators (13.41) on $L^2(\mathbb{R}^{3N})$ describing systems on N particles, $N > 1$. In this case, the interaction is typically a sum of potentials with the property that there is a three-dimensional projection E in $\mathbb{R}^{3N}$ such that $V(x) = V(Ex)$. Such potentials describe either *one-particle forces* when E refers to coordinates of a single particle, or *two-particle* ones when E refers to relative coordinates of a pair of particles.

Theorem 13.17 (Kato theorem) *Let* $n = 3N$ *and* $V = \sum_{k=1}^{m} V_k$, *where each potential component* $V_k(E_k\,\cdot) \in (L^2 + L^\infty)(\mathbb{R}^3)$; *then the operator* (13.41) *is e.s.a. on any core of* H_0.

The main importance of this result is that it guarantees self-adjointness of atomic Hamiltonian with potentials of Coulomb type for electron charge e,

$$V(x) = -\sum_{j=1}^{Z} \frac{Ze^2}{|x_j - x_0|} + \sum_{1\le j<k\le Z} \frac{e^2}{|x_j - x_k|}, \qquad (13.42)$$

referring to a fixed nucleus of atomic number Z, and similar operators describing atoms with a finite nucleus mass as well as molecules, atomic and molecular ions, and so on; it means that the usual quantum-mechanical description of such objects can be justified from the first principles.

Analyzing the discrete spectrum of operators (13.41), one is often interested in relations between the number and position of the eigenvalues on the one hand and the potential on the other. There are many such results of which we will mention just two important ones. A frequently used estimate is expressed in the following way:

Theorem 13.18 (Birman–Schwinger bound) *Consider the operator* (13.41) *on* $L^2(\mathbb{R}^3)$ *with the potential such that the right-hand side of the relation* (13.43) *is finite; then the number of eigenvalues of* H *is finite and satisfies the inequality*

$$N(V) \le \frac{1}{16\pi^2} \int_{\mathbb{R}^6} \frac{|V(x)V(y)|}{|x-y|^2}\, dx\, dy\,. \qquad (13.43)$$

Note that no similar bound exists for $n = 1, 2$ because there an arbitrarily weak negative potential produces a *bound state*, that is, an isolated negative eigenvalue.

The bound (13.43) does not exhibit a correct *semiclassical behavior*, which means that it becomes poor when we replace V by gV and study the asymptotic behavior as $g \to \infty$. This is not case for the following more general result.

Theorem 13.19 (Lieb–Thirring inequality) *Suppose that* $\sigma_d(H) = \{\lambda_j\}$ *and fix* $\gamma \ge 0$ *for* $n \ge 3$, $\gamma > 0$ *for* $n = 2$, *and* $\gamma \ge \frac{1}{2}$ *for* $n = 1$; *then the inequality*

$$\mathrm{Tr}\,(H_-^\gamma) = \sum_j (-\lambda_j)^\gamma \le L_{\gamma,n} \int_{\mathbb{R}^n} V_-(x)^{\gamma+\frac{n}{2}}\, dx \qquad (13.44)$$

holds, where $L_{\gamma,n} \ge L^{\mathrm{cl}}_{\gamma,n} = \Gamma(\gamma+1)\left[2^n \quad \pi^{n/2}\Gamma(\gamma+\frac{n}{2}+1)\right]^{-1}$ *and* $V_-(x) = \max\{-V(x), 0\}$ *is the negative part of the potential.*

In fact, the constants are $L_{\gamma,n} = R(\gamma,n)L^{\text{cl}}_{\gamma,n}$ where $R(\gamma,n) = 1$ for $\gamma \geq 3/2$, $R(\gamma,n) \leq 2$ for $1 \leq \gamma < 3/2$ or $1/2 \leq \gamma < 1$ and $n = 1$, and $R(\gamma,n) \leq 4$ for $1/2 \leq \gamma < 1$ and $n \geq 2$. In dimensions $n \geq 3$, the inequality holds for $\gamma = 0$, which yields a bound to the number of bound states,

$$N(V) \leq L_{0,n} \int_{\mathbb{R}^n} V_-(x)^{n/2}\,\mathrm{d}x, \qquad (13.45)$$

known as the *Cwikel–Lieb–Rozeblium theorem.*

The number of bound states can be infinite if the potential has a slow enough decay at infinity. Suppose that $V = V_2 + V_\infty$ with $V_2 \in L^2(\mathbb{R}^3)$ and $V_\infty \in L^\infty(\mathbb{R}^3)$ such that $V_\infty(x) \to 0$ holds as $|x| \to \infty$. This ensures that $\sigma_{\text{ess}}(H) = [0,\infty)$. If there are $c \in \left[0, \frac{1}{4}\right)$ and $r_0 \geq 0$ such that $V(x) \geq -c|x|^{-2}$ holds $|x| \geq r_0$, then $\sigma_{\text{d}}(H)$ is finite. On the other hand, the discrete spectrum is infinite provided there are positive d, r_0, ϵ such that $V(x) \leq -d|x|^{-2+\epsilon}$ holds $|x| \geq r_0$.

Behavior of the potential at large distances is important in determining the essential spectrum also more generally. In some cases of physical importance, this task is not easy; in particular, for systems of N particles interacting through two-particle forces associated with the potential of the form

$$V(x_1,\ldots,x_N) = \sum_{1\leq j<k\leq Z} V_{jk}(x_j - x_k), \quad (13.46)$$

where $x_j = (x_{j1}, x_{j2}, x_{j3})$ are coordinates of the jth particle. We divide our N particles into $n(D)$ *clusters* C_i; if $N = 2$ there is only one such *partition*, for $N \geq 2$ there are different partitions $D = \{C_1,\ldots,C_{n(D)}\}$. For each partition we define ϵ^D_{jk} to be one if the indices j,k belong to the same cluster and zero otherwise. This allows us to define the Hamiltonian with the intercluster interaction switched off, $H_D = -\Delta + \sum_{j<k=1}^N \epsilon^D_{jk} V_{jk}(x_j - x_k)$, and the cluster Hamiltonians

$$H_{C_i} = -\Delta_{C_j} + \sum_{\{j,k\in C_j : j<k\}}^{N} V_{jk}(x_j - x_k)\,; \qquad (13.47)$$

in fact, we are interested in the Hamiltonians $H^{\text{rel}}_{C_i}$ with the center-of-mass motion separated acting in $L^2(\mathbb{R}^{3n_i-3})$, where n_i is the number of particles in the cluster C_i. Using them, we define for a given D the quantity $\lambda_D := \sum_{i=1}^{n(D)} \inf\sigma\big(H^{\text{rel}}_{C_i}\big)$.

Theorem 13.20 (Hunziker–van Winter–Zhislin theorem) *Suppose that the two-particle potentials are $V_{jk} = V_{jk,2} + V_{jk,\infty}$ with $V_{jk,2} \in L^2(\mathbb{R}^3)$ and $V_{jk,\infty} \in L^\infty(\mathbb{R}^3)$ such that $V_{jk,\infty}(x) \to 0$ holds as $|x| \to \infty$; then the essential spectrum of H is $[\Lambda,\infty)$ where*

$$\Lambda := \min_{n(D)\geq 2} \lambda_D = \min_{n(D)=2} \lambda_D\,. \qquad (13.48)$$

13.6.2 Scattering Theory

The second application we are going to discuss concerns one of the most frequently occurring situations in physics when we compare the dynamics of a quantum system with an asymptotic one. Suppose that we have a pair of Hamiltonians, the "full" one H and the "free" one H_0, together with the corresponding evolution operators, $U(t) = \mathrm{e}^{-\mathrm{i}tH}$ and $U_0(t) = \mathrm{e}^{-\mathrm{i}tH_0}$. The question is how to find to a given $\psi \in \mathcal{H}$ vectors $\psi_\pm$ such that the "trajectories" $U(t)\psi$ and $U_0(t)\psi_\pm$ coincide asymptotically as $t \to \pm\infty$.

Naturally, there may be vectors ψ for which no $\psi_\pm$ exist, for instance, *bound states* of the system described by eigenvectors of the operator H. The distinguishing property is whether a state remains localized or leaves a fixed space region. We consider an increasing family $\{M_r : r \geq 0\}$ of subspaces of the configuration space and denote by F_r the respective projections assuming that $\text{s-lim}_{r\to\infty} F_r = I$; for definiteness, think of concentric balls of radius r. Using these projections, we define the family

$$\mathcal{M}_s(H) = \Big\{ \psi \in \mathcal{H} : \lim_{|t|\to\infty} F_r U(t)\psi = 0 \text{ for any } r > 0 \Big\} \tag{13.49}$$

of *scattering states*, and the complementary family of *bound states*,

$$\mathcal{M}_b(H) = \Big\{ \psi \in \mathcal{H} : \lim_{r\to\infty} \sup_{t\in\mathbb{R}} \|(I - F_r) U(t)\psi\| = 0 \Big\}. \tag{13.50}$$

It is not difficult to check that $\mathcal{M}_b(H) \supset \mathcal{H}_p(H)$ and $\mathcal{M}_s(H) \supset \mathcal{H}_c(H)$ where we use the spectral subspaces notation introduced in Section 13.5.2. We identify conventionally the scattering states with the absolutely continuous part of H, setting $\mathcal{M}_s(H) \supset \mathcal{H}_{ac}(H)$. This means to exclude states associated with $\mathcal{H}_{sc}(H)$ that typically lie "between" the bound and scattering states: the probability of finding them in a fixed bounded region may not have zero limit as $|t| \to \infty$ but its mean value can be made arbitrarily small when averaged over a sufficiently long time interval. One usually aims to prove that such pathological states are absent, that is, $\sigma_{sc}(H) = \emptyset$.

In this setup, we can describe the asymptotic relation between the full and free dynamics by a pair of *wave operators* defined by

$$\Omega_\pm(H, H_0) := \text{s-lim}_{t\to\pm\infty} U(t)^* U_0(t) E_{ac}(H_0), \tag{13.51}$$

which map $\mathcal{H}_{ac}(H_0)$ into $\mathcal{H}_{ac}(H)$. In general, it may happen that $\text{Ran}\,\Omega_\pm(H, H_0) \neq \mathcal{M}_s(H)$, for instance, if the scattered particle is captured by the target and is unable to leave the interaction region. The wave operators are said to be *complete* if

$$\text{Ran}\,\Omega_+(H, H_0) = \text{Ran}\,\Omega_-(H, H_0) = \mathcal{H}_{ac}(H), \tag{13.52}$$

and *asymptotically complete* if, in addition, $\sigma_{sc}(H) = \emptyset$. Under the completeness assumption, one can define, in particular, the *scattering operator* (or *S-matrix*), $S(H, H_0) := \Omega_+(H, H_0)^* \Omega_-(H, H_0)$, which maps the "incoming" asymptotic states into the "outgoing" ones.

The wave operators are assigned to a pair of operators. They satisfy the *chain rule*, $\Omega_\pm(H, H_0) = \Omega_\pm(H, H_1)\Omega_\pm(H_1, H_0)$ for any self-adjoints H, H_1, H_0 for which the wave operators exist. Another important property is the *intertwining relation*, $\Omega_\pm(H, H_0)H_0 \subset H\Omega_\pm(H, H_0)$. It implies, in particular, that the scattering operator commutes with the free Hamiltonian, or equivalently, $U_0(t)S(H, H_0) = S(H, H_0)U_0(t)$ holds for all $t \in \mathbb{R}$. As a consequence, it can be expressed in the form of a direct integral,

$$S(H, H_0) = \int_{\sigma_{ac}(H_0)}^{\oplus} S(\lambda)\, d\lambda\,; \tag{13.53}$$

for the fiber operators $S(\lambda)$, we usually employ the name *on-shell scattering matrix*.

Many sufficient conditions have been derived for existence of wave operators. Some are abstract and rather simple such as the following one.

Theorem 13.21 (Kato–Rosenblum theorem) *Suppose that $H = H_0 + V$, where H_0 is self-adjoint and V is a Hermitean trace-class operator; then the wave operators $\Omega_\pm(H, H_0)H_)$ exist and are complete.*

The range of applications of this result is rather limited, because the interaction part of the Hamiltonian in physical models is rarely of a trace-class character. More often, one can use the following related result.

Theorem 13.22 (Birman–Kuroda theorem) *Let H, H_0 be self-adjoint and the resolvent difference $(H - z)^{-1} - (H_0 - z)^{-1} \in \mathcal{J}_1(\mathcal{H})$ for some $z \in \rho(H) \cap \rho(H_0)$; then the wave operators $\Omega_\pm(H, H_0)$ exist and are complete.*

As regards Schrödinger operators, one is usually interested in scattering caused by the potential; in other words, the situation where H is the operator (13.41) and $H_0 = -\Delta$ on $L^2(\mathbb{R}^n)$. Various existence conditions for the wave operators in terms of the potential V can be derived, for instance,

Theorem 13.23 (Hack–Cook theorem) *Let H be the operator (13.41) on $L^2(\mathbb{R}^3)$. If the potential $V \in (L^2 + L^s)(\mathbb{R}^3)$ with $s \in [2, 3)$, then the wave operators $\Omega_\pm(H, H_0)$ exist.*

Completeness of the wave operators requires additional hypotheses. As a sample result we mention the following sufficient conditions.

Theorem 13.24 *Let H be the operator (13.41) on $L^2(\mathbb{R}^3)$ with $V \in (L^1 \cap L^2)(\mathbb{R}^3)$, then the wave operators $\Omega_\pm(H, H_0)$ exist and are complete. If, in addition,*

$$\frac{1}{16\pi^2} \int_{\mathbb{R}^6} \frac{|V(x)V(y)|}{|x - y|^2} \, dx \, dy < 1 \qquad (13.54)$$

holds, then $\sigma_{sc}(H) = \emptyset$, so the wave operators are asymptotically complete.

The asymptotic completeness can be established for many other scattering systems described by Schrödinger operators. In particular, it has been demonstrated in the situation when (13.41) describes a system of N particles interacting through Coulomb potentials of the type (13.42), hence scattering processes in atomic and molecular physics can again be studied from the first principles.

Glossary

Banach space: a (metrically) complete normed space

Commutativity of self-adjoint operators: it means commutativity of their spectral decompositions

Compact operator: it maps each bounded set into a precompact one

Deficiency indices of a symmetric A**:** the numbers $\dim \operatorname{Ker}(A^* \mp i)$

Deficiency subspace of T**:** all solutions to the equation $T^* x = \overline{\lambda} x$

Density matrices: positive trace-class operators normalized by $\operatorname{Tr} W = 1$

Discrete spectrum: isolated eigenvalues of finite multiplicity

Essential spectrum: complement of the discrete spectrum

Essentially self-adjoint operator: its closure is self-adjoint

Fourier expansion: expression of a Hilbert-space vector with respect to an orthonormal basis

Fourier–Plancherel operator: the continuous extension of Fourier transformation to $L^2(\mathbb{R}^n)$.

Function spaces: families of functions, or equivalence classes of functions, with pointwise defined addition and multiplication

Hamiltonian: an operator describing energy of the system

Hermitean operator: a bounded self-adjoint operator

Hilbert space: a (metrically) complete inner-product space

Hilbert–Schmidt operator: an operator C such that $\sum_j \|Ce_j\|^2 < \infty$ holds for any orthonormal basis $\{e_j\}$
Inner product: a sesquilinear form on a vector space such that $(x, x) = 0$ implies $x = 0$
Limit point – limit circle: alternatives determining deficiency indices of a one-dimensional Schrödinger operator
Normal operator: it commutes with its adjoint
Operator: a linear map between (subspaces of) Banach (Hilbert) spaces
Operator with pure point spectrum: its eigenvectors form an orthonormal basis
Parseval identity: it expresses the vector norm through its Fourier coefficients
Point spectrum: eigenvalues of the operator
Projection: an operator assigning to any vector $x \in \mathcal{H}$ its orthogonal projection to a given subspace $\mathcal{G} \subset \mathcal{H}$
Pure state of a quantum system: a one-dimensional subspace in the state Hilbert space of the system
Riesz lemma: an expression of a bounded linear functional on a Hilbert space
Self-adjoint operator: it coincides with its adjoint
Spectral measure: a measure the values of which are projections on a given Hilbert space
Spectrum of an operator T**:** those λ for which the resolvent $R_T(\lambda) = (T - \lambda)^{-1}$ does not exist as a bounded operator
Stone formula: expression of the spectral measure in terms of the resolvent
Symmetric operator: it is a restriction of its adjoint
Trace-class operator: an operator C such that $\mathrm{Tr}\,|C| < \infty$
Unitarily equivalent operators: they can be mapped one to another with the help of a unitary operator
Unitary operator: an isometric map defined on the whole Hilbert space
Wave operators: asymptotic maps between the full and free dynamics in a scattering system
Weyl relations: expression of canonical commutation relations in terms of the associated unitary groups
Weyl theorem: stability of the essential spectrum with respect to compact perturbations

References

1. Reed, M. and Simon, B. (1972–1978) *Methods of Modern Mathematical Physics I–IV*, Academic Press, New York.
2. Blank, J., Exner, P., and Havlíček, M. (2008) *Hilbert Space Operators in Quantum Physics*, 2nd edn, Springer, Dordrecht.
3. Bratelli, O. and Robinson, D.W. (1979, 1981) *Operator Algebras and Quantum Statistical Mechanics I, II*, Springer-Verlag, New York.
4. Albeverio, S.A., Høegh-Krohn, R.J., and Mazzucchi, S. (2008) *Mathematical Theory of Feynman Path Integrals*, 2nd edn, Springer, Berlin.
5. Adams, R.A. and Fournier, J.J.F. (2003) *Sobolev Spaces*, 2nd edn, Elsevier, Oxford.
6. Akhiezer, N.I. and Glazman, I.M. (1993) *Theory of Linear Operators in Hilbert Space*, Dover, Mineola, NY.
7. Albeverio, S.A., Gesztesy, F., Høegh-Krohn, R., and Holden, H. (2005) *Solvable Models in Quantum Mechanics*, 2nd edn with appendix by P. Exner, AMS Chelsea, Providence, RI.
8. Amrein, W.O., Jauch, J.M., and Sinha, K.B. (1977) *Scattering Theory in Quantum Mechanics*, Benjamin, Reading, MA.
9. Birman, M.Š. and Solomyak, M.Z. (1987) *Spectral Theory of Self-Adjoint Operators in Hilbert Space*, Kluwer, Dordrecht.
10. Cycon, H.L., Froese, R.G., Kirsch, W., and Simon, B. (2007) *Schrödinger Operators, with Applications to Quantum Mechanics and Global Geometry*, 2nd printing, Springer, Berlin.
11. Davies, E.B. (2007) *Linear Operators and their Spectra*, Cambridge University Press, Cambridge.

12. Exner, P. (1985) *Open Quantum Systems and Feynman Integrals*, Reidel, Dordrecht.
13. Jauch, J.M. (1968) *Foundations of Quantum Mechanics*, Addison-Wesley, Reading, MA.
14. Kolmogorov, A.N. and Fomin, S.V. (1999) *Elements of the Theory of Functions and Functional Analysis*, Dover, Rochester, NY.
15. Lieb, E.H. and Loss, M. (2001) *Analysis*, 2nd edn, AMS, Providence, RI.
16. Lieb, E.H. and Seiringer, R. (2010) *The Stability of Matter in Quantum Mechanics*, Cambridge University Press, Cambridge.
17. von Neumann, J. (1996) *Mathematical Foundations of Quantum Mechanics*, 12th printing, Princeton University Press, Princeton, NJ.
18. Prugovečki, E. (1981) *Quantum Mechanics in Hilbert Space*, 2nd edn, Academic Press, New York.
19. Riesz, F. and Sz-Nagy, B. (1972) *Leçons d'analyse foncionelle*, 6me edn, Akademiai Kiadó, Budapest.
20. Rudin, W. (1991) *Functional Analysis*, 2nd edn, McGraw-Hill, New York.
21. Simon, B. (2005) *Trace Ideals and Their Applications*, 2nd edn, AMS Chelsea, Providence, RI.
22. Weidmann, J. (2000, 2003) *Lineare Operatoren in Hilberträumen I, II*, B.G. Teubner, Stuttgart.
23. Yafaev, D.R. (1992) *Mathematical Scattering Theory*, AMS, Providence, RI.

14 Numerical Analysis

Lyonell Boulton

14.1 Introduction

The foundations of the compendium of mathematical theories now known as numerical analysis can be traced back to the Plimton 322 clay tablet, which is believed to be over 3800 years old. However, arguably, it was only with the advent of machines capable of performing a prescribed set of mathematical tasks without direct human intervention, that numerical analysis consolidated as an independent subject of scientific interest. For some, numerical analysis is one of the success stories of the second half of the twentieth century.

In recent years, areas of numerical analysis such as numerical linear algebra, optimization, and numerical differential equations, have turned themselves into highly specialized research subjects. Language and methods are increasingly becoming only accessible, and often only of interest, to specialists. Yet some of these methods are of prime importance in theoretical and experimental contemporary science. As electronically assisted mathematics, ranging from the use of highly specialized scientific apparatus to large computer networks, permeated into most of our scientific experience nowadays, there is an increasing need for the development of mathematically rigorous frameworks capable of validating the data that is being produced by those means.

In this chapter, we only scratch the surface of some of the classical themes in numerical analysis. It is intended as a very rough introduction to the language and techniques in the theory for nonspecialists. It is by no means exhaustive; however, it may serve as a reading guide to more specialized literature on the subject. We miss many important aspects of the theory, including the crucial connection with concrete computational implementations. An excellent introduction to the latter is available in [1].

Section 14.2 is concerned with classical techniques for the solution of nonlinear equations and systems. This includes some qualitative aspects of the described methods and a brief introduction to numerical minimization. The particular case of linear systems of equations, which is far

Mathematical Tools for Physicists, Second Edition. Edited by Michael Grinfeld.

more specialized and developed, is considered separately in Section 14.3. There we describe the standard classification of methods for the solution of systems of linear equations into direct and iterative, both of which are important in their own right. We also discuss the classical approach to finite eigenvalue problems. Section 14.4 is devoted to approximation of continuous data. We only describe the classical theory of polynomial interpolation as a background to the highly developed theory of finite elements. In Sections 14.5 and 14.6, we focus on two concrete subjects. In the former, we describe in some detail the elements of the theory for the numerical solution of the initial-value (Cauchy) problems for ordinary differential equations (ODEs). In the latter, we describe the use of the Galerkin method for the numerical solution of spectral problems in the context of one-dimensional Schrödinger operators.

14.2 Algebraic Equations

The first type of mathematical problem that commonly requires a numerical approximation is an algebraic equation. By far the most successful general tools for the approximated solution of general scalar or vector algebraic equations are the *iteration methods*. The special case of linear systems of equations is of great importance and it will be dealt with separately in the next section.

14.2.1 Nonlinear Scalar Equations

Consider a general equation

$$f(x) = 0$$

for a given continuous scalar function $f : [a,b] \longrightarrow \mathbb{R}$. An analytic solution of this problem can only be found in close form for very particular functions f. Otherwise, approximation of possible solutions can be achieved by means of *fixed-point iteration* schemes: we pick an initial guess x_0 and iterate

$$x_{n+1} = g(x_n), \qquad n = 1, \dots,$$

to get the *fixed point*

$$x_* = \lim_{n\to\infty} x_n$$

such that $g(x_*) = x_*$. Here $x = g(x)$ is equivalent to $f(x) = 0$, so that $f(x_*) = 0$.

Example 14.1 [Newton's method] See [2, Section 2.1]. Kepler's model for the planetary motion seeks for solutions x_* (dependent on time) of the equation

$$x - \mathcal{E} \sin x = \tau.$$

Here $\mathcal{E}$ is the (fixed) eccentricity of the elliptical orbit and τ is proportional to time. Newton proposed iterating

$$x_{n+1} = x_n + \frac{\tau - x_n + \mathcal{E}\sin x_n}{1 - \mathcal{E}\cos x_n} \qquad n = 1, \dots,$$

in order to obtain x_*. This corresponds to choosing

$$g(x) = x - \frac{f(x)}{f'(x)} \quad \text{for } f(x) = \tau - x + \mathcal{E}\sin x$$

above, and it is the special case of what is now called Newton's method.

If the function g is *Lipschitz continuous*,

$$|g(x) - g(y)| \le \lambda |x - y|$$

for all x, y in a segment $[a, b]$ containing x_*, then the *absolute error* $e_n = x_* - x_n$ satisfies

$$|e_n| \leq \lambda |e_{n-1}| \leq \lambda^n |e_0| \qquad \text{for all} \quad n \geq 1.$$

If $\lambda < 1$ and $g(x) \in [a, b]$ for all $x \in [a, b]$, then there exists a unique fixed point $x_* \in [a, b]$ of $g(x)$ and $|e_n| \to 0$ for any choice of x_0. These conditions are satisfied in a neighborhood of x_* if, for example, $|g'(x_*)| < 1$. In this case, Taylor's theorem yields

$$\lim_{n\to\infty} \frac{|e_n|}{|e_{n-1}|} = |g'(x_*)| \quad \text{(linear convergence)}.$$

If further $|g'(x_*)| = 0$, then

$$\lim_{n\to\infty} \frac{|e_n|}{|e_{n-1}|^2} = \frac{1}{2}|g''(x_*)| \quad \text{(quadratic convergence)}.$$

An application of higher-order Taylor's approximation shows that the condition $|g^{(j)}(x_*)| = 0$ for $j = 1, \ldots, k-1$, implies

$$\lim_{n\to\infty} \frac{|e_n|}{|e_{n-1}|^k} = \frac{1}{k!}|g^{(k)}(x_*)| \quad \text{(}k\text{th order convergence)}.$$

Particular iteration methods for the solution of scalar algebraic equations are described in Table 14.1.

14.2.2 Nonlinear Systems

The idea of fixed-point iteration can be formally extended to multiple dimensions. Given $G : \Omega \longrightarrow \mathbb{R}^d$ where $\Omega \subset \mathbb{R}^d$ and an initial $\underline{x}_0 \in \Omega$, the iteration

$$\underline{x}_{n+1} = G(\underline{x}_n), \qquad n = 1, \ldots,$$

would converge to a fixed point $\underline{x}_* \in \Omega$ such that $G(\underline{x}_*) = \underline{x}_*$, if

$$\|G(\underline{x}) - G(\underline{y})\| \leq \lambda \|\underline{x} - \underline{y}\|, \qquad \underline{x}, \underline{y} \in \Omega,$$

Table 14.1 Most common iteration methods for the approximation of solutions of nonlinear scalar equations. See [3, Chapter 6].

Usual name	Formula for x_{n+1}	Order	Comments/pitfalls
Newton	$x_n - \frac{f(x_n)}{f'(x_n)}$	2	$f'(x_*) \neq 0$
Modified Newton	$x_n - \frac{kf(x_n)}{f'(x_n)}$	2	$\lvert f^{(j)}(x_*)\rvert = 0$ for $j < k$ and $\lvert f^{(k)}(x_*)\rvert \neq 0$
Steffensen	$x_n - \frac{f(x_n)^2}{f(x_n+f(x_n))-f(x_n)}$	2	No need to compute f'
Secant	$x_n - \frac{(x_n-x_{n-1})f(x_n)}{f(x_n)-f(x_{n-1})}$	$\frac{1+\sqrt{5}}{2}$	Iteration requires less work than Newton or Steffensen
(General) fixed point	$g(x_n)$	k	g is a general function such that $\lvert g^{(j)}(x_*)\rvert = 0$ for $j < k$ and $\lvert g^{(k)}(x_*)\rvert \neq 0$
(General) higher order	$g(g(x_n))$	$2k$	g is as in the general fixed-point method and k is its order of convergence (see previous row)

for $\lambda < 1$. Here $\|\cdot\|$ is a suitable norm of $\mathbb{R}^d$. The *error vector*, $\underline{e}_n = \underline{x}_n - \underline{x}_*$, satisfies the identity $\|\underline{e}_{n+1}\| \leq \lambda \|\underline{e}_n\|$. If the Jacobian

$$J_G(\underline{x}) = \left[\frac{\partial G_j}{\partial x_k}\right]_{jk=1}^d$$

exists for all $\underline{x} \in \Omega$, we can represent G in Taylor series,

$$G(\underline{y}) = G(\underline{x}) + J_G(\underline{x})(\underline{x} - \underline{y}) + R_G(\underline{x}, \underline{y}),$$
$$R_G(\underline{x}, \underline{y}) \leq c\|\underline{x} - \underline{y}\|^2.$$

This leads to the following general "local" convergence theorem around fixed points [4, pp. 299–301].

Theorem 14.1 *If $\|J_G(\underline{x}_*)\| < 1$, starting the iteration at any $\underline{x}_0$ sufficiently close to $\underline{x}_*$ ensures*

$$\lim_{n\to\infty} \|\underline{x}_n - \underline{x}_*\| = 0 \qquad \text{and}$$
$$\lim_{n\to\infty} \frac{\|\underline{e}_{n+1}\|}{\|\underline{e}_n\|} = \|J_G(\underline{x}_*)\|.$$

For a system of d algebraic equations in d unknowns,

$$F(\underline{x}) = 0,$$

a natural extension of the Newton method, sometimes called the *Newton–Simpson method*, can be formulated as follows. Suppose that $J_F : \Omega \longrightarrow \mathbb{R}^{d\times d}$ is not singular, for example, $\det(J_F(\underline{x})) \neq 0$ for all $\underline{x} \in \Omega$. Starting from $\underline{x}_0 \in \Omega$, find $\underline{x}_{n+1}$ the solution of

$$J_F(\underline{x}_n)\underline{x}_{n+1} = J_F(\underline{x}_n)\underline{x}_n - F(\underline{x}_n), \quad n = 1, \ldots \tag{14.1}$$

This iteration corresponds to picking

$$G(\underline{x}) = \underline{x} - J_F(\underline{x})^{-1}F(\underline{x}).$$

In practice, the former is preferable to the latter, as finding all entries of the inverse of the Jacobian matrix is in itself a numerically challenging problem for large dimensions and it is often unnecessary. See [3, Chapter 7] and [2, Chapter 7].

In concrete implementations, the iteration scheme (14.1) is usually written in coupled form

$$\begin{cases} J_F(\underline{x}_n)\underline{y}_n = -F(\underline{x}_n) \\ \underline{x}_{n+1} = \underline{x}_n - \underline{y}_n \end{cases} \qquad n = 1, \ldots$$

Arguments involving Theorem 14.1 show that, if $\|J_F(\underline{x})^{-1}F(\underline{x})\|$ and $\|J_F(\underline{x})^{-1}H_F(\underline{x})\|$ are uniformly bounded by a small enough constant on a region Ω (here $H_F(\underline{x})$ is the Hessian of F), then $\underline{x}_n$ converges quadratically to a solution to the corresponding homogeneous system for any initial $\underline{x}_0 \in \Omega$.

Example 14.2 [An iteration method for computing eigenvalues] See [2, Section 7.2]. Given a hermitian matrix $A \in \mathbb{R}^{d\times d}$, the eigenvalues of A can be estimated by solving the nonlinear equation in $d+1$ components,

$$F(\underline{x}, \lambda) = 0 \quad \text{for} \quad F(\underline{x}, \lambda) = \begin{bmatrix} A\underline{x} - \lambda\underline{x} \\ \frac{1}{2}(1 - \underline{x}^t \cdot \underline{x}) \end{bmatrix}.$$

The Newton–Simpson method for F reduces in this case to the coupled iteration

$$\begin{cases} (A - \lambda_n I)\underline{z}_n = \underline{x}_n \\ \lambda_{n+1} = \lambda_n + \frac{1 + \underline{x}_n^t \cdot \underline{x}_n}{\underline{x}_n^t \cdot \underline{z}_n} \\ \underline{x}_{n+1} = (\lambda_{n+1} - \lambda_n)\underline{z}_n \end{cases}, \qquad n = 1, \ldots,$$

which is equivalent to the *inverse power method*. Under fairly mild conditions on A, λ_n would converge to the eigenvalue of A that is smallest in modulus.

14.2.3 Numerical Minimization

Given a differentiable function $f : \mathbb{R}^d \longrightarrow \mathbb{R}$, we now consider the *unconstrained minimization problem* of finding $\underline{x}_* \in \mathbb{R}^d$ such that

$$f(\underline{x}_*) \leq f(\underline{x}) \qquad \forall \underline{x} \in \Omega.$$

See [5]. When $\Omega = \mathbb{R}^d$, $\underline{x}_*$ is called a *global minimizer*. When $\Omega \subsetneq \mathbb{R}^d$ is only a neighborhood of $\underline{x}_*$, it is called a *local minimizer*. As any minimizer is a *critical point*,

$$\operatorname{grad} f(\underline{x}_*) = 0,$$

the above problem is naturally viewed in the context of solution to nonlinear systems of equations.

Closely linked with the fixed-point iterative procedures described previously are the *descent methods* [3, Section 7.2.2]: pick an initial $\underline{x}_0 \in \Omega$ and iterate

$$\underline{x}_{n+1} = \underline{x}_n + \alpha_n \underline{\delta}_n, \qquad n = 1, \ldots$$

Here $\underline{\delta}_n$ are suitable *descent directions* satisfying

$$\underline{\delta}_n^{\mathrm{t}} \cdot \operatorname{grad} f(\underline{x}_n) < 0$$

and the scalar $\alpha_n > 0$ measures the *stepsize* along this direction. By Taylor's theorem

$$f(\underline{x} + \alpha\underline{\delta}) - f(\underline{x}) = \alpha \operatorname{grad} f(\underline{\xi})^{\mathrm{t}} \cdot \underline{\delta},$$

where $\underline{\xi}$ lies in the segment joining $\underline{x}$ with $\underline{x} + \alpha\underline{\delta}$. Hence we always find $\alpha_n > 0$ small enough such that

$$f(\underline{x}_n + \alpha_n \underline{\delta}_n) < f(\underline{x}_n), \qquad (14.2)$$

whenever $\operatorname{grad} f(\underline{x}_n) \neq 0$. If $\operatorname{grad} f(\underline{x}_n) = 0$, we have arrived at a critical point, so we may assign $\alpha_n = 0$ to stop the iteration.

Every choice of $\underline{\delta}_n$ and α_n gives rise to a different method, with different approximation properties. See Table 14.2.

Determining α_n such that (14.2) is satisfied without incurring into high computational expense, depends on the type of problem considered. One commonly used choice is to minimize

$$\phi(\alpha) = f(\underline{x}_n + \alpha \underline{\delta}_n),$$

which guarantees that

$$\operatorname{grad} f(\underline{x}_{n+1})^{\mathrm{t}} \cdot \underline{\delta}_n = 0.$$

Procedures leading toward finding minima of $\phi(\alpha)$ in this context are called *linear search techniques*. See [3, Section 7.2.3].

Example 14.3 [Descent methods for quadratic forms] Let $A \in \mathbb{R}^{d \times d}$ be a hermitian positive definite matrix and let $\underline{b} \in \mathbb{R}^d$

Table 14.2 Descent directions for some of the standard methods of unconstrained minimization. See [5].

Method	$\underline{\delta}_n$	Comments / pitfalls
Newton	$-H_f(\underline{x}_n)^{-1} \operatorname{grad} f(\underline{x}_n)$	Computing $H_f(\underline{x}_n)^{-1}$ is usually impractical
Inexact Newton	$-B_n^{-1} \operatorname{grad} f(\underline{x}_n)$	$B_n \approx H_f(\underline{x}_n)^{-1}$
Gradient	$-\operatorname{grad} f(\underline{x}_n)$	Inexact Newton with $B_n = I$ and standard for quadratic forms
Conjugate gradient	$-\operatorname{grad} f(\underline{x}_n) + \beta_n \underline{\delta}_{n-1}$	β_n chosen so that $\underline{\delta}_n$ are mutually orthogonal

be a fixed vector. Let

$$f(\underline{x}) = \frac{1}{2}\underline{x}^{\mathrm{t}} \cdot A\underline{x} - \underline{x}^{\mathrm{t}} \cdot \underline{b}.$$

As

$$\operatorname{grad} f(\underline{x}) = A\underline{x} - \underline{b},$$

the minimum of f will occur exactly at the solution of a linear system. Let

$$\underline{r}_n = \underline{b} - A\underline{x}_n.$$

The gradient method uses the direction of this residual as search direction and

$$\alpha_n = \frac{\underline{r}_n^{\mathrm{t}} \cdot \underline{r}_n}{\underline{r}_n^{\mathrm{t}} \cdot A\underline{r}_n}.$$

On the other hand, for the conjugate gradient method,

$$\alpha_n = \frac{\underline{r}_n^{\mathrm{t}} \cdot \underline{\delta}_n}{\underline{\delta}_n^{\mathrm{t}} \cdot A\underline{\delta}_n} \quad \text{and} \quad \beta_n = \frac{\underline{r}_n^{\mathrm{t}} \cdot \underline{\delta}_{n-1}}{\underline{\delta}_{n-1}^{\mathrm{t}} \cdot A\underline{\delta}_{n-1}}.$$

The latter uses a more clever choice of search direction, because it avoids reusing previously chosen directions.

14.3 Finite-Dimensional Linear Systems

Particular techniques for the approximation of solutions of the linear inhomogeneous equation

$$A\underline{x} = \underline{b}, \tag{14.3}$$

as well as the associated linear eigenvalue problem

$$A\underline{u} = \lambda\underline{u} \qquad \underline{u} \neq 0, \tag{14.4}$$

deserve special attention given their simple structure. Here $A \in \mathbb{C}^{d\times d}$ and $\underline{b} \in \mathbb{C}^d$ are data, and the unknowns are $\underline{x} \in \mathbb{C}^d$ in the former and $(\lambda, \underline{u}) \in \mathbb{C} \times \mathbb{C}^d$ in the latter problem.

If $\det A \neq 0$, the solution of (14.3) can be determined from *Cramer's rule,*

$$x_j = \frac{\det A_j}{\det A}, \qquad j = 1, \ldots, d,$$

where A_j is the matrix obtained by replacing the jth entry of A by $\underline{b}$. However, this approach is of practical interest only for matrices of small size, as the calculation of determinants requires a high number of *flops* (floating point arithmetic operations). For standard matrices this will be roughly $3(d+1)!$ operations, which is extremely inefficient for $n > 10$.

Two types of methods are available for large matrices, the *direct methods* and the *iterative methods.*

14.3.1 Direct Methods and Matrix Factorization

Direct methods determine the solution of a linear system in a finite number of steps.

Gaussian elimination reduces a general system of the form (14.3) to a triangular *row echelon form,*

$$\begin{aligned} A^1_{11}x_1 + A^1_{12}x_2 + \ldots + A^1_{1d}x_d &= b^1_1 \\ A^2_{22}x_2 + \ldots + A^2_{2d}x_d &= b^2_2 \\ \vdots \qquad & \quad \vdots \\ A^d_{dd}x_d &= b^d_d. \end{aligned}$$

We can find an explicit expression for the coefficients of this system via the *Gauss elimination algorithm* running for $r = 1, \ldots, d-1$:

$$\begin{aligned} A^1_{jk} &= A_{jk} & b^1_j &= b_j & j,k &= 1,\dots,d \\ A^{r+1}_{jk} &= A^r_{jk} - m_{jr}A^r_{rk} & b^{r+1}_j &= b^r_j - m_{jr}b^r_r & j,k &= r+1,\dots,d \\ m_{jr} &= \frac{A^r_{jr}}{A^r_{rr}} & & & j &= r+1,\dots,d. \end{aligned}$$

This assumes that the *pivot elements* A^r_{rr} are nonzero. The matrices

$$U = [A^j_{jk}]^d_{jk=1} \qquad \text{and} \qquad L = [m_{jk}]^d_{jk=1}$$
$$m_{jj} = 1,$$

render a decomposition of the form

$$A = LU,$$

which is usually called an *LU-factorization* of A. Here $L \in \mathbb{C}^{d\times d}$ is *lower triangular* ($L_{jk} = 0$ for $k < j$) and $U \in \mathbb{C}^{d\times d}$ is *upper triangular* ($U_{jk} = 0$ for $k > j$).

In turns, (14.3) is equivalent to the two triangular systems

$$L\underline{y} = \underline{b} \qquad U\underline{x} = \underline{y}.$$

The former can be solved by *forward substitution*

$$y_1 = \frac{b_1}{L_{11}} \qquad y_j = \frac{1}{L_{jj}}\left(b_j - \sum_{k=1}^{j-1} L_{jk}y_k\right),$$
$$j = 2,\dots,d-1,$$

while the latter can be solved by *backward substitution* (with a similar recursive formula).

The factorization of matrices, such as the LU-factorization, are important in the efficient solution of linear systems, as there can be a significant reduction in the number of flops required to get the solution compared with Gaussian elimination. This is particularly relevant, when we are required to solve many problems for the same matrix A, but different vectors $\underline{b}$. Gauss elimination roughly requires $2d^3/3$ operations, whereas forward and backward substitutions only require d^2 flops each. See [2, Chapter 4] and references therein.

The pivot elements of a given system are all different from zero if, and only if, each one of the $r \times r$ upper-left minors of A, $[A_{jk}]^r_{jk=1}$, are nonsingular. Otherwise, methods called *pivoting*, which involve reordering either the rows or the columns (*partial pivoting*) – or both (*full*), can be employed to produce an LU-factorization of the matrix. The Gauss elimination with partial pivoting (either by row or by column) always leads to nonzero pivot elements that produce a nonsingular LU-factorization of A if and only if $\det A \neq 0$. In fact, the pivoting technique can also be implemented to reduce errors caused by numerical rounding, which occur whenever the multipliers m_{jk} are small.

The LU-factorization of a matrix is not unique. When A has a given structure (e.g., symmetric), it is possible to find a factorization consistent with this structure, see Table 14.3. Banded matrices are of particular interest in applications. If the bandwidth is not very large, the factorization and backward/forward substitution steps for such matrices can be achieved in a comparatively small number of operations.

Table 14.3 Most common factorization methods for the solution of linear systems. [3, Chapter 3].

Name	Factorization	Description	flops	Comments / pitfalls
LU	$A = LU$	L is lower triangular and U is upper triangular	$2d^3/3$	With pivoting, it applies to any $A \in \mathbb{C}^{d\times d}$
Cholesky	$A = LL^\dagger$	L is lower triangular with positive diagonal entries	$d^3/3$	Only for $A \in \mathbb{R}^{d\times d}$ hermitian positive definite
LDL	$A = LDL^\dagger$	L is lower triangular and D is diagonal	$d^3/3$	For $A \in \mathbb{R}^{d\times d}$ general hermitian
Banded	As above	Factorization of A having bandwidth b: $A_{jk} = 0$ $\lvert j-k\rvert > b$	$2b^2d$	Gain only for matrices of small bandwidth

Example 14.4 [Default Matlab and Octave linear solvers] See [1, Section 5.8]. The default command for solving linear systems in the computer languages Matlab and Octave is the *backslash* "\" command. Each time it is invoked, specific algorithms are used, depending on the structure of the matrix A. These algorithms are all based on direct methods.

14.3.2 Iteration Methods for Linear Problems

Iterative methods for linear system are closely related with their nonlinear counterpart. An iteration method for the solution of a linear system is called *stationary*, if for given initial $\underline{x}_0$,

$$\underline{x}_{n+1} = B\underline{x}_n + \underline{c}, \qquad n = 1, \ldots$$

The *iteration matrix* B depends on A and determines the method, and

$$\underline{c} = (I - B)A^{-1}\underline{b}.$$

Splitting the matrix $A = P - R$, where P is usually called a *preconditioner* and $R = P - A$, gives $B = I - P^{-1}A$ and $\underline{c} = P^{-1}\underline{b}$. As (14.3) is equivalent to the system

$$P\underline{x} = R\underline{x} + \underline{b},$$

the static method can be written as

$$P\underline{\delta}_n = \underline{b} - A\underline{x}_n \qquad \underline{x}_{n+1} = \underline{x}_n + \alpha\underline{\delta}_n,$$

where $\alpha \neq 0$ is a parameter chosen so that convergence might be improved.

The iteration matrix B should be constructed so that each iteration can be computed efficiently. In particular, it is necessary that the system associated to the preconditioner is solved quickly. Writing $A = D + E + F$ where D is the diagonal part of A, E is lower triangular with zeros on the diagonal and F is upper triangular with zeros on the diagonal, give rise to the standard methods as described in Table 14.4.

The residual vector satisfies

$$\|\underline{e}_{n+1}\| \leq \rho(B)\|\underline{e}_n\|, \qquad n = 1, \ldots,$$

where $\rho(B)$ is the *spectral radius*: the largest of the moduli of the eigenvalues of B. Hence the iteration method will be convergent for any choice of $\underline{x}_0$, only if $\rho(B) < 1$. When $0 < w < 1$, the SOR method can converge when Gauss–Seidel does not. When $w = 1$, the two methods are the same. When

Table 14.4 The most common iteration stationary methods for the solution of linear systems [3, Chapter 4].

Name	P, α	Comments / pitfalls
Jacobi	D, 1	See below.
Gauss–Seidel	$D - E$, 1	Usually faster than Jacobi.
Successive over relaxation (SOR)	$\frac{1}{w}D - E$, 1	Only converges in general for $1 < w < 2$.
Static Richardson	any, any	Choice of α can improve convergence.

$1 < w < 2$, SOR can converge faster than Gauss–Seidel. If A is a positive definite hermitian matrix, then SOR converges for all $0 < w < 2$. If, in addition, A is tridiagonal, (bandwidth $b = 2$), then the optimal choice of w in the SOR method is

$$w = \frac{2}{1 - \sqrt{1 - \tilde{\rho}}},$$

where $\tilde{\rho}$ is the spectral radius of the matrix B of the corresponding Gauss–Seidel method. See [6, 7] for details.

Example 14.5 [The PageRank algorithm] See [8]. The notion of ranking web pages arises in commercial and noncommercial online search tools. The computer algorithm known as *PageRank* developed by Larry Page from the software company Google was fairly popular among website administrators until 2009. A simplified model of how this algorithm works is illustrated as follows. Consider as a small example the set of eight web pages with the hyperlinks as shown in Figure 14.1.

Suppose that a random walk moves through the web network by two rules: 85% of the time follow one of the hyperlinks from the page it is on by assigning equal weight to each and picking one at random with probability $1/H_{jj}$, then 15% of the time skip randomly to another website anywhere else on the network. Denoting by x_j the probability that the page j is visited, we have

$$\underline{x} = \underbrace{\delta C^{t} H^{-1} \underline{x}}_{\text{first rule}} + \underbrace{\frac{1-\delta}{d}\underline{e}}_{\text{second rule}}$$

$$\iff \quad \left(I - \delta C^{t} H^{-1}\right) \underline{x} = \frac{1-\delta}{d}\underline{e}$$

here $\delta = 0.85$, C is the *connection matrix* associated to the network, H is the corresponding *links-out* diagonal matrix and $\underline{e}$ is the vector with all entries equal to 1.

The Jacobi method is convenient for the solution of this linear system when d is very large for two main reasons. On the one hand, it can be proved to converge: $\rho(B) = \delta$ in this case. On the other hand, any of the iterates $\underline{x}_n$ preserves the property of being a probability vector (nonnegative entries adding exactly up to 1), if $\underline{x}_0$ is a probability vector.

The iteration formula for the stationary methods can be generalized to

$$P\underline{\delta}_n = \underline{b} - A\underline{x}_n \qquad \underline{x}_{n+1} = \underline{x}_n + \alpha_n \underline{\delta}_n,$$

where $\alpha_n \neq 0$ is a parameter chosen so that convergence might be improved. In this case, the technique is some times called *nonstationary* or *dynamic Richardson method*. The *residual vector*

$$\underline{r}_n = \underline{b} - A\underline{x}_n$$

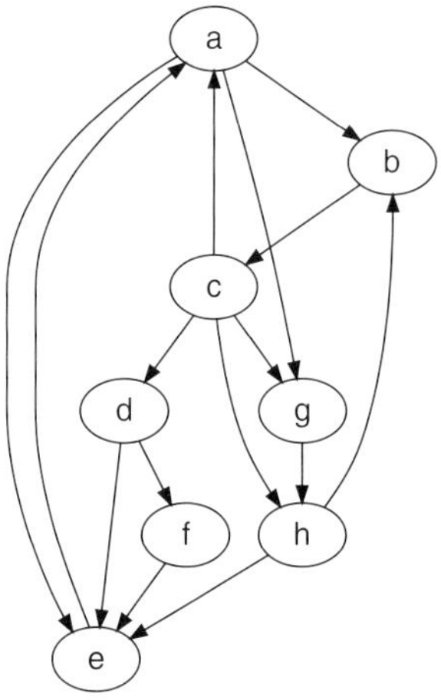

Order the pages as (a,b,c,d,e,f,g,h)

- Page 1 links to pages 2, 5, and 7
- Page 2 links to page 3
- Page 3 links to page 1, 4, 7, and 8
- Page 4 links to page 5 and 6
- Page 5 links to page 1
- Page 6 links to page 5
- Page 7 links to page 8
- Page 8 links to pages 2 and 5

$$\underbrace{C = \begin{bmatrix} 0 & 1 & 0 & 0 & 1 & 0 & 1 & 0 \\ 0 & 0 & 1 & 0 & 0 & 0 & 0 & 0 \\ 1 & 0 & 0 & 1 & 0 & 0 & 1 & 1 \\ 0 & 0 & 0 & 0 & 1 & 1 & 0 & 0 \\ 1 & 0 & 0 & 0 & 0 & 0 & 0 & 0 \\ 0 & 0 & 0 & 0 & 1 & 0 & 0 & 0 \\ 0 & 0 & 0 & 0 & 0 & 0 & 0 & 1 \\ 0 & 1 & 0 & 0 & 1 & 0 & 0 & 0 \end{bmatrix}}_{\text{Connection matrix}} \qquad \underbrace{H = \begin{bmatrix} 3 &&&&&&& \\ & 1 &&&&&& \\ && 4 &&&&& \\ &&& 2 &&&& \\ &&&& 1 &&& \\ &&&&& 1 && \\ &&&&&& 1 & \\ &&&&&&& 2 \end{bmatrix}}_{\text{Links-out matrix}}$$

Figure 14.1 Small example ($d = 8$) of a web network.

clearly measures how close $\underline{x}_n$ is to the actual solution of the system. As $P\underline{\delta}_n = \underline{r}_n$, $\underline{\delta}_n$ is called in this context the *preconditioned residual*. Recalling Example 14.3, if we pick $P = I$ and

$$\alpha_n = \frac{\underline{r}_n^{\mathrm{t}} \cdot \underline{r}_n}{\underline{r}_n^{\mathrm{t}} \cdot A\underline{r}_n},$$

we recover the *gradient method*. Thus the stationary and nonstationary methods can be viewed as generalizations of the descent method for the minimization of quadratic forms, where the descent direction is taken to be the preconditioned residual. See [1, §4.3] and references therein.

As mentioned in Example 14.3, the *conjugate gradient (CG) method* picks $\underline{\delta}_n$ in a slightly more sophisticated fashion. Beginning with initial $\underline{x}_0$ and $\underline{r}_0 = \underline{\delta}_0 = \underline{b} - A\underline{x}_0$, we set

$$\begin{aligned} \alpha_n &= \frac{\underline{r}_n^{\mathrm{t}} \cdot \underline{\delta}_n}{\underline{\delta}_n^{\mathrm{t}} \cdot A\underline{\delta}_n} & \underline{x}_{n+1} &= \underline{x}_n + \alpha_n \underline{\delta}_n, \\ && && n = 0, \ldots \\ \beta_{n+1} &= \frac{\underline{r}_{n+1}^{\mathrm{t}} \cdot \underline{\delta}_n}{\underline{\delta}_n^{\mathrm{t}} \cdot A\underline{\delta}_n} & \underline{\delta}_{n+1} &= \underline{r}_{n+1} + \beta_{n+1}\underline{\delta}_n, \end{aligned}$$

If A is hermitian and positive definite, both the gradient and the CG method converge for any initial choice of $\underline{x}_0$. Replace the second row of this iteration formula by

$$P\underline{y}_{n+1} = \underline{r}_{n+1} \qquad \beta_{n+1} = \frac{\underline{y}_{n+1}^{\mathrm{t}} \cdot \underline{\delta}_n}{\underline{\delta}_n^{\mathrm{t}} \cdot A\underline{\delta}_n}$$
$$\underline{\delta}_{n+1} = \underline{y}_{n+1} + \beta_{n+1}\underline{\delta}_n,$$

for P hermitian and positive definite, and we obtain the *preconditioned conjugate gradient (PCG) method.*

When all the iterations of the CG or the PCG method are computed exactly, they will converge to the true solution of the system in exactly d steps. Unfortunately, in practice, the numerical rounding will prevent this from occurring.

14.3.3 Computing Eigenvalues of Finite Matrices

As for the case of the finite system of linear equations, the numerical estimation of the *eigenvalues* λ and *eigenvectors* $\underline{x}$, associated to the eigenvalue problem (14.4) can also be achieved by means of minimization methods. The role of the *Rayleigh quotient,*

$$R(\underline{x}) = \frac{\underline{x}^{\mathrm{t}} \cdot A\underline{x}}{\underline{x}^{\mathrm{t}} \cdot \underline{x}}$$

in these methods may be illustrated on a simple case, as follows.

Suppose that A is a 3×3 hermitian matrix with eigenvalues $\lambda_1 \leq \lambda_2 \leq \lambda_3$ and eigenvectors $\underline{u}_1, \underline{u}_2, \underline{u}_3 \in \mathbb{C}^3$ with unit Euclidean norm. These eigenvalues may be written as extremal problems involving $R(\underline{x})$. Indeed, any vector in $\mathbb{C}^3$ expands as $\underline{x} = \sum x_j \underline{u}_j$, so

$$R(\underline{x}) = \frac{\sum \lambda_j |x_j|^2}{\sum |x_j|^2}.$$

Then

$$\lambda_1 = \min\{R(\underline{u})\}, \ \lambda_2 = \min\{R(\underline{u}) \,:\, \underline{u} \perp \underline{u}_1\},$$
$$\lambda_3 = \min\{R(\underline{u}) \,:\, u \perp \underline{u}_1, \underline{u}_2\}.$$

Moreover, λ_j are stationary points of the map $R : \mathbb{R}^3 \longrightarrow \mathbb{R}$.

Note that, if we employ a steepest descent method, essentially no prior knowledge or calculation of the eigenvectors is required to get λ_1. We can, in fact, characterize λ_2 also, without explicit information about the eigenvectors. If $S \subset \mathbb{C}^3$ is an arbitrary two-dimensional space, there always exists a nonzero vector $\underline{x} \in S$ such that $\underline{x} \perp \underline{u}_1$. As $R(\underline{x}) \geq \lambda_2$, we gather that $\max_{\underline{u} \in S} R(\underline{u}) \geq \lambda_2$ and

$$\lambda_2 = \min_{\dim S=2} \max_{\underline{u} \in S} R(\underline{u}).$$

A similar argument shows that

$$\lambda_3 = \min_{\dim S=3} \max_{\underline{u} \in S} R(\underline{u}).$$

In turns, bounds for the eigenvalues can be found from $R(u)$ without much prior information about the eigenvectors.

The above procedure can be extended to matrices of any size, and in fact to infinite-dimensional linear *operators*. The following result is of fundamental importance and it is known as the min–max principle. It was first investigated over one hundred years ago by Lord Rayleigh and Walter Ritz in connection with acoustics, but in its current form is due to Courant and Fischer. See [9, Section 4.5] or [10, Section XIII.2].

Theorem 14.2 (Min–max principle) *Let $A \in \mathbb{C}^{d \times d}$ be a hermitian matrix and let its eigenvalues be ordered nondecreasingly as $\lambda_1 \leq \cdots \leq \lambda_d$. Then*

$$\lambda_j = \min_{\dim S=j} \max_{\underline{u} \in S} R(\underline{u}).$$

In particular, $\lambda_1 = \min R(\underline{u})$ and $\lambda_d = \max R(\underline{u})$.

For certain practical problems, knowledge of all the eigenvalues might not be required and quite often only the largest or smallest eigenvalue is needed. This is highly relevant, for example, in the context of quantum mechanics.

Example 14.6 [Validation of the Lamb shift] See [10, Section XIII.2]. The ground state energy of hydrogen can be found analytically. The same cannot be said for helium, even though the ground state energy of a helium ion can be found exactly from the hydrogen model. In the early days of quantum mechanics, Hylleraas computed an upper approximation of the helium ground state energy by hand. The calculation involved finding the smallest eigenvalue of a 6×6 matrix and it was regarded as important support for the confirmation of the correctness of quantum mechanics. With the advent of computers, finding eigenvalues of larger matrices was made possible. Early approximations involving linear systems of sizes 39 by Kinoshita and later 1078 by Perkeris, were crucial in order to test agreement of the Lamb shift with experimental data. The min–max principle played a crucial role in this process.

Assume that $A \in \mathbb{R}^{d \times d}$ has eigenvalues such that

$$|\lambda_d| > |\lambda_{d-1}| \geq \cdots \geq |\lambda_1|.$$

Note that the largest eigenvalue in modulus is forced to be simple. The *power method* is an iterative procedure to approximate λ_d and $\underline{u}_d$. For initial nonzero vectors $\underline{x}_0$ and $\underline{y}_0 = \underline{x}_0 / \|\underline{x}_0\|$, define

$$\begin{cases} \underline{x}_{n+1} = A\underline{y}_n \\ \underline{y}_{n+1} = \dfrac{\underline{x}_{n+1}}{\|\underline{x}_{n+1}\|} \\ \lambda_{d,n+1} = R\left(\underline{y}_{n+1}\right) \end{cases}, \qquad n = 0, \ldots$$

Then the nth iterate vector is such that $\underline{y}_n = \beta_n A^n \underline{y}_0$ for suitable scalar β_n. As it turns, this recursive formula ensures that $\underline{y}_n$ aligns with the eigenvector direction $\underline{u}_d$ and

$$|\lambda_{d,n} - \lambda_d| \sim \left|\frac{\lambda_{d-1}}{\lambda_d}\right|^n \quad \text{(generic matrices)},$$

$$|\lambda_{d,n} - \lambda_d| \sim \left|\frac{\lambda_{d-1}}{\lambda_d}\right|^{2n} \quad \text{(hermitian matrices)}$$

as $n \to \infty$. Therefore $\lambda_{d,n} \to \lambda_d$, whenever the above conditions on the largest modulus eigenvalue hold true. See [3, Section 5.3] and [11, pp. 406–407].

A slight modification of the power method allows calculation of the eigenvalue of A that is smallest in modulus, under the condition $0 < |\lambda_1| < |\lambda_2|$ and no constraints on $|\lambda_d|$. For the *inverse power method*, we begin with initial vectors as before, but then iterate

$$\begin{cases} A\underline{x}_{n+1} = \underline{y}_n \\ \underline{y}_{n+1} = \dfrac{\underline{x}_{n+1}}{\|\underline{x}_{n+1}\|} \\ \mu_{d,n} = R(\underline{y}_{n+1}) \end{cases}, \qquad n = 0, \ldots$$

This is nothing more than the old power method applied to the matrix A^{-1}, whose eigenvalues are $\mu_j = 1/\lambda_{d-j+1}$, and so swapped in modulus order with respect to the eigenvalues of A. From the convergence of the power method and the above condition, $\mu_{d,n} \to \mu_d = \lambda_1^{-1}$, so we have a way of computing λ_1.

For any $\mu \in \mathbb{C}$ not an eigenvalue of A, the matrix $B = (A - \mu I)$ has its eigenvalues equal to $(\mu - \lambda_j)^{-1}$. Then, replacing the matrix A with the matrix $(A - \mu I)$ in the *inverse shifted power method* will allow computing the eigenvalue of A that is closest to μ.

This idea can also be applied to refine the computation of λ_d, by changing μ at each step of the iteration by the better guess $\mu_{d,n-1}$. It is remarkable that, despite the fact that $(A - \mu_{d,n})$ is near singular, this procedure might allow convergence under certain circumstances. One instance of this, with a different normalization, was already encountered in Example 14.2.

14.4 Approximation of Continuous Data

We now show how to represent continuous mathematical quantities by means of a finite number of degrees of freedom. In turns this is of natural importance for numerical approximation. A fairly complete account on the ideas discussed in this section can be found in [3, Chapter 8] and references therein.

14.4.1 Lagrange Interpolation

Let $f : [\alpha, \beta] \longrightarrow \mathbb{R}$ be a given continuous function and let

$$\alpha \leq x_0 < \cdots < x_N \leq \beta$$

be $N+1$ points called *interpolation nodes* of the interval $[\alpha, \beta]$. There is a unique *Lagrange interpolating polynomial* of degree N, $L_N f \in \mathbb{P}_N$, satisfying

$$L_N f(x_k) = f(x_k), \qquad k = 0, \dots, N.$$

Indeed, the family of N *characteristic polynomials*

$$\ell_k(x) = \prod_{\substack{j=0\\ j\neq k}}^{N} \frac{x - x_j}{x_k - x_j}, \qquad k = 0, \dots, N,$$

forms a basis of $\mathbb{P}_N$, which ensures that we can write

$$L_N f(x) = \sum_{k=0}^{N} f(x_k)\ell_k(x)$$

in a unique manner. The latter formula is often referred to as the *Lagrange form* of the corresponding interpolating polynomial.

The most economical way of computing the interpolating polynomial is not the one given by the Lagrange form. A more neat procedure is set by the *Newton form* of the interpolating polynomial. Set $L_0 f(x) = f(x_0)$ and

$$L_k f(x) = L_{k-1} f(x) + p_k(x), \qquad k = 1, \dots, N,$$

where $L_k f \in \mathbb{P}_k$ is the Lagrange interpolating polynomial for f at the points $\{x_j\}_{j=0}^k$ and $p_k \in \mathbb{P}_k$ depends on $\{x_j\}_{k=0}^k$ and only one unknown coefficient. Then

$$p_k(x) = a_k w_k(x) \quad \text{for} \quad w_k(x) = \prod_{j=0}^{k-1}(x - x_j).$$

Thus

$$a_0 = f(x_0), \qquad a_k = \frac{f(x_k) - L_{k-1} f(x_k)}{w_k(x_k)} = \sum_{j=0}^{k} \frac{f(x_j)}{w'_{k+1}(x_k)}.$$

In turns, we see that we can construct recursively a_k in such a way that

$$L_N f(x) = \sum_{k=0}^{N} a_k w_k(x).$$

This formula is known as the *Newton divided difference formula* and the coefficients a_k are the *kth divided differences.*

14.4.2 The Interpolation Error

The *interpolation error function*

$$\varepsilon_N(x) = f(x) - L_N f(x)$$

naturally measures how well the Lagrange interpolating polynomial approximates the function f. If $f \in C^r(\alpha, \beta)$, then it can be shown that [12, 13]

$$\varepsilon_N(x) = \frac{f^{(N+1)}(\xi)}{(N+1)!} w_{N+1}(x) \qquad (14.5)$$

for some $\xi \in (\alpha, \beta)$. Note that from the definition, it follows that

$$\varepsilon_N(x) = a_{N+1} w_{N+1}(x),$$

where the $(N+1)$th divided difference is obtained by considering as interpolation nodes the set $\{x_j\}_{j=0}^N \cup \{x\}$.

We now highlight one remarkable drawback of general polynomial interpolation. Consider a hierarchy of interpolation nodes of the segment $[\alpha, \beta]$,

$$\mathcal{X} = \{\Xi_N\}_{N=1}^{\infty}, \qquad \Xi_N = \{x_{N,k}\}_{k=0}^{N}.$$

There always exists a continuous function $f : [\alpha, \beta] \longrightarrow \mathbb{R}$, such that the Lagrange interpolating polynomial on Ξ_N does not capture f uniformly in the full segment $[\alpha, \beta]$ as $N \to \infty$. To be precise,

$$\lim_{N\to\infty} \max_{x\in[\alpha,\beta]} |\varepsilon_N(x)| \neq 0.$$

In fact, such a function can some times be constructed explicitly. See [14].

Example 14.7 [Runge's phenomenon] Let

$$f(x) = \frac{1}{1+x^2}, \qquad x \in [-5, 5].$$

Suppose that the partition Ξ_N of the segment $[-5, 5]$ is made out of equally spaced nodes. Then

$$\lim_{N\to\infty} \max_{x\in[-5,5]} |\varepsilon_N(x)| = \infty.$$

By contrast, if the partition Ξ_N of the segment $[-5, 5]$ is made out of the *Chebyshev—Gauss–Lobatto nodes,*

$$x_{N,k} = \frac{\alpha+\beta}{2} + \frac{\beta-\alpha}{2}\tilde{x}_k$$
$$\tilde{x}_k = -\cos\left(\frac{\pi k}{N}\right), \qquad k = 0, \ldots, N,$$

then

$$\lim_{N\to\infty} \max_{x\in[-5,5]} |\varepsilon_N(x)| = 0.$$

14.4.3 Hermite Interpolation

In the case of *Hermite interpolation,* rather than values of the function at the nodes, higher-order derivatives are matched by the given polynomial. Suppose that we wish to match

$$f^{(r)}(x_k) \quad \text{for} \quad r = 0, \ldots, m_k \quad \text{and} \quad k = 0, \ldots, N.$$

Let $d = \sum_{k=0}^{N}(m_k + 1)$. There exists a unique polynomial $H_{d-1} \in \mathbb{P}_{d-1}$ of order $d-1$, the *Hermite interpolating polynomial,* such that

$$H_{d-1}^{(r)}(x_k) = f^{(r)}(x_k) \quad \text{for} \quad r = 0, \ldots, m_k \quad \text{and} \quad k = 0, \ldots, N.$$

A concrete formula for the Hermite interpolating polynomial is given by

$$H_{d-1}(x) = \sum_{k=0}^{N} \sum_{r=0}^{m_k} f_k^{(r)}(x_k) K_{kr}(x),$$

where

$$K_{kr}^{(p)}(x_j) = \begin{cases} 1 & k = j \text{ and } r = p \\ 0 & \text{otherwise.} \end{cases}$$

Moreover, the error formula for Hermite interpolation reads [3, Section 8.4]

$$\varepsilon_N(x) = f(x) - H_{d-1}(x) = \frac{f^{(d)}(\xi)}{d!}\mathfrak{h}_d(x)$$
$$\mathfrak{h}_d(x) = \prod_{k=0}^{N}(x - x_k)^{m_k+1} \tag{14.6}$$

for a given $\xi \in [\alpha, \beta]$.

Example 14.8 [Hermite quadrature formulae] Polynomial interpolation can be used in order to derive numerical methods for the approximation of definite integrals of the form

$$I(f) = \int_a^b f(x)\,\mathrm{d}x.$$

Hermite polynomial interpolation gives rise to "corrected" versions of classical formulae for integration, such as the trapezoidal rule. Assuming that the $2N + 2$ values of $f(x_k)$ and $f'(x_k)$ are given at nodes $\{x_k\}_{k=0}^N$ of the segment $[a, b]$. The corresponding Hermite interpolant polynomial is

$$H_{2N+1}f(x) = \sum_{j=0}^{N} \left(f(x_j)\mathfrak{l}_j(x) + f'(x_j)\mathfrak{m}_j(x)\right)$$

for

$$\mathfrak{l}_j(x)\left(1 - \frac{w''_{N+1}(x_j)}{w'_{N+1}(x_j)}(x - x_j)\right)\ell_j(x)^2$$
$$\text{and}\quad \mathfrak{m}_j(x) = (x - x_j)\ell_j(x)^2$$

(the definition of w_j and ℓ_j is given in Section 14.4.1). If we integrate on $[a, b]$, we derive the quadrature formula

$$I_N(f) = \sum_{j=0}^{N}\left(\alpha_j f(x_j) + \beta_j f'(x_j)\right),$$
$$\alpha_j = I(\mathfrak{l}_j) \quad \text{and} \quad \beta_j = I(\mathfrak{m}_j),$$

which is an approximation of $I(f)$. For $N = 1$, we obtain the *corrected trapezoidal formula*

$$I_1(f) = \frac{b-a}{2}\left(f(a) + f(b)\right) + \frac{(b-a)^2}{12}\left(f'(a) + f'(b)\right),$$

which has higher accuracy than the approximation of $I(f)$ by the classical trapezium rule.

14.4.4 Piecewise Polynomial Interpolation

As described in Example 14.7, a possible way to avoid the Runge phenomenon is by considering a nonuniform distribution of interpolation nodes. One other possibility is to use to *piecewise polynomial interpolation*. The latter forms the background of the *finite element method* [15, 16].

Let $f : [a, b] \longrightarrow \mathbb{R}$ be a continuous function. The idea behind piecewise polynomial interpolation is to fix a partition of the segment $[a, b]$ into subsegments whose interiors are disjoint from one another, and conduct polynomial interpolation of a given fixed order in each one of these subsegments. Let $\Xi_h = \{\mathcal{J}_j\}_{j=1}^N$ be a partition of $[a, b]$ into N subintervals,

$$[a, b] = \bigcup_{j=1}^{N} \mathcal{J}_j, \qquad \mathcal{J}_j = [x_{j-1}, x_j],$$

whose maximum length is

$$h = \max_{1\le k\le N}(x_j - x_{j-1}).$$

For $q \in \mathbb{N}$, let

$$\mathcal{V}(q, \Xi_h) = \{w \in C^0(a, b) : w|_{\mathcal{J}_j} \in \mathbb{P}_q(\mathcal{J}_j),\ j = 1, \ldots, N\}$$

be the space of continuous functions on $[a, b]$ which are polynomials of degree no larger than q in each $\mathcal{J}_j$. The map

$$\Pi_{h,q} : C^0(a, b) \longrightarrow \mathcal{V}(q, \Xi_h)$$

such that $\Pi_{h,q} f|_{\mathcal{J}_j}$ is the interpolating polynomial (e.g., Lagrange or Hermite) on suitable nodes of the subsegments $\mathcal{J}_j$, is usually called the *piecewise polynomial interpolant* associated to Ξ_h.

By applying the error formulas (14.5) or (14.6) in each one of the subsegments of Ξ_h, it can be shown that there exists a constant $c(q) > 0$ independent of f such that

$$\max_{x\in[a,b]} |f(x) - \Pi_{h,q}(x)| \leq c(r)h^{q+1} \max_{x\in[a,b]} |f^{(q+1)}(x)|$$

for any $f \in C^{q+1}(a, b)$. Moreover, $\Pi_{h,q} f$ turns out to approach f also in the mean square sense. Indeed, denote by $L^2(a, b)$ the complex Hilbert space of all Lebesgue square integrable functions with the standard inner product $\langle f, g\rangle = \int_a^b f(x)\overline{g(x)}\, dx$ and norm $\|f\| = \langle f, f\rangle^{1/2}$. For a proof of the following result in a more general context, see, for example, [15, Theorem 3.1.6].

Theorem 14.3 *Let $q \geq 1$ and assume that $r = 0, \ldots, q+1$. There exists a constant $c(q) > 0$ independent of h ensuring that, if*

$$\sum_{r=0}^{q+1} \|f^{(r)}\| < \infty,$$

then

$$\|f^{(r)} - (\Pi_{h,q} f)^{(r)}\| \leq c(q)h^{q+1-r}\|f^{(q+1)}\|.$$

Example 14.9 [The Hermite elements of order three] The *Hermite element of order three* are piecewise polynomials of order $q = 3$ with prescribed values of f and f' at the interpolation nodes. Let the functions

$$\psi(x) = (|x| - 1)^2(2|x| + 1) \quad \text{and} \quad \omega(x) = x(|x| - 1)^2$$

be defined on $-1 \leq x \leq 1$, see Figure 14.2. Then

$$\psi, \omega \in \mathcal{V}(3, \Xi_1), \qquad \Xi_1 = \{[-1, 0], [0, 1]\}.$$

Both these functions have continuous derivatives, if extended by 0 to $\mathbb{R}$, and

$$\psi(0) = 1,\ \psi(\pm 1) = \psi'(0) = \psi'(\pm 1) = 0$$
$$\omega'(0) = 1,\ \omega(\pm 1) = \omega(0) = \omega'(\pm 1) = 0.$$

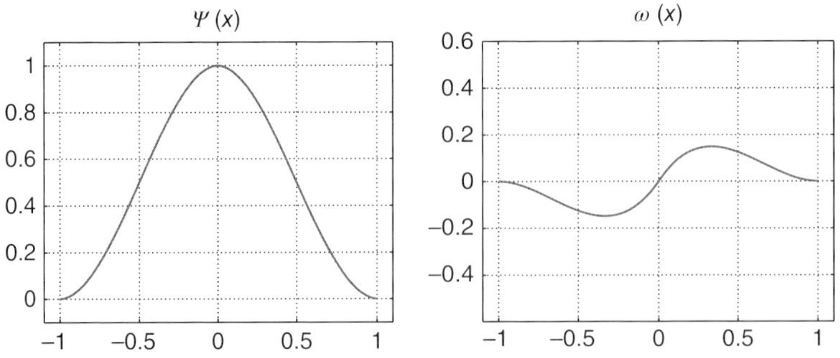

Figure 14.2 Profile of the functions for the Hermite elements of order three in one dimension.

A basis of the linear space $\mathcal{V}(3, \Xi_h)$ where Ξ_h a uniform partition of a generic interval $[a, b]$ is obtained by translation and dilation. That is

$$\Pi_{h,3} f(x) = \sum_{j=0}^{N} \left(f(x_j)\phi_j(x) + f'(x_j)\omega_j(x) \right),$$

for

$$\phi_j(x) = \phi\left(\frac{x - x_j}{h}\right) \qquad \omega_j(x) = \omega\left(\frac{x - x_j}{h}\right), \qquad j = 0, \dots, N.$$

14.5 Initial Value Problems

We now examine in some detail numerical solutions of the general *Cauchy problem*

$$\begin{aligned} &\frac{\mathrm{d}}{\mathrm{d}t}\underline{y}(t) = g(t, \underline{y}(t)), \qquad t_0 < t < T, \\ &\underline{y}(t_0) = \underline{y}_0, \end{aligned} \tag{14.7}$$

where $g : [t_0, \infty) \times \mathbb{R}^d \longrightarrow \mathbb{R}^d$ is a continuous function and $T > t_0$.

Using arguments involving fixed-point theorems, we know quite general conditions ensuring the existence and uniqueness of solutions of the above evolution problem [17]. Indeed, (14.7) will admit a unique *local solution* in a suitably small neighborhood $\mathcal{N} \subset \mathcal{I} \times \mathcal{J}$ of $(t_0, \underline{y}_0)$, if the function g is *Lipschitz continuous in the space variable,*

$$\|g(t, \underline{y}_2) - g(t, \underline{y}_1)\| \le L\|\underline{y}_2 - \underline{y}_1\|, \quad \text{for all } t \in \mathcal{I} \quad \text{and} \quad \underline{y}_1, \underline{y}_2 \in \mathcal{J}. \tag{14.8}$$

Moreover, it will admit a unique *global solution,* if the constant L can be chosen uniformly for $\mathcal{J} = \mathbb{R}$.

We will be mostly concerned with the case $d = 1$ which in itself possesses all the features of the higher-dimensional case. See [3, Chapter 11] and references therein.

14.5.1 One-Step Methods

In order to establish numerical procedures for the solution of the Cauchy problem, we begin by partitioning the segment $[t_0, t_0 + T]$ into a number N of subsegments of *step size* $h > 0$. That is

$$[t_0, t_0 + T] = \bigcup_{n=0}^{N-1} [t_n, t_{n+1}],$$

where the *nodes* t_n are given by $t_0 + nh$. Let

$$y_n = y(t_n).$$

We seek for approximations u_n of the true solution value y_n at the nodes. We set

$$g_n = g(t_n, u_n).$$

For the *one-step methods,* the approximation u_{n+1} at the $(n + 1)$th node depends solely on u_n. In their general formulation, these methods can be written as

$$u_{n+1} = u_n + h\Phi(t_n, u_n, g_n; h) \qquad u_0 = y_0, \qquad n = 0, \dots, N - 1,$$

where Φ is often referred to as the *incremental function.*

By writing

$$y_{n+1} = y_n + h\Phi_h(t_n, y_n, g(t_n, y_n)) + h\tau_{n+1}(h), \qquad n = 0, \dots, N - 1,$$

where $\tau_n(h)$ is a correction term, we define the *local truncation error,*

$$\tilde{\varepsilon}_n(h) = |\tau_n(h)|,$$

at the node t_n. If the incremental function satisfies the condition

$$\lim_{h\to 0} \Phi_h(t_n, y_n, g(t_n, y_n)) = g(t_n, y_n),$$

by expanding y in Taylor series at that node, we can show that [3, Section 11.3]

$$\lim_{h\to 0} \tilde{\varepsilon}_n(h) = 0, \qquad n = 0, \dots, N-1.$$

In fact, it can also be shown that the *global truncation error*

$$\tilde{\varepsilon}(h) = \max_{n=0,\dots,N-1} |\tilde{\varepsilon}_n(h)|$$

does satisfy

$$\lim_{h\to 0} \tilde{\varepsilon}(h) = 0$$

under the above condition, ensuring the *consistency* of the one-step method. The *order* of a consistent method is the largest power $q \in \mathbb{N}$, such that

$$\lim_{h\to 0} \frac{\tilde{\varepsilon}(h)}{h^q} < \infty.$$

In Table 14.5, we show common choices of incremental functions alongside the corresponding order of approximation. The method is *explicit*, if u_{n+1} can be computed directly from u_j for $j \leq n$, otherwise it is called *implicit*. An iteration formula in the case of the implicit methods should be derived for each individual Cauchy problem.

14.5.2 Multistep Methods

In the case of the *multistep methods* the approximated solution at the $(n+1)$th node is computed by means of a formula involving $u_n, \dots, u_{n+1-q}$. The smallest value of q for which this is possible is the *number of steps* of the method.

Example 14.10 [the midpoint method] The midpoint rule for differentiation gives rise to the *midpoint method*, in which

$$u_{n+1} = u_{n-1} + 2hg_n, \qquad n = 1, \dots, N-1.$$

Here $u_0 = y_0$ and u_1 is to be determined. This is an example of an explicit multistep method comprising two steps.

An *implicit* two-step method is the following.

Example 14.11 [the Simpson's method] The Simpson's rule for integration gives rise to the *Simpson's method*. The iteration formula reads

$$u_{n+1} = u_{n-1} + \frac{h}{3}\left(g_{n-1} + 4g_n + g_{n+1}\right), \quad n = 1, \dots, N-1.$$

The *linear multistep methods* represent an important class of procedures for the efficient solution of the Cauchy problem.

Table 14.5 Simple one-step methods for the numerical solution of the Cauchy problem.

Name	$\Phi_h(t_n, u_n, g_n)$	Order	Type
Forward Euler	g_n	1	Explicit
Backward Euler	g_{n+1}	1	Implicit
Crank–Nicholson	$\frac{g_n+g_{n+1}}{2}$	2	Implicit
Heun	$\frac{g_n+g(t_{n+1},u_n+hg_n)}{2}$	2	Explicit

See [18]. Set real coefficients $a_0, \dots, a_{q-1}$ and $b_{-1}, \dots, b_{q-1}$, such that $a_{q-1}b_{q-1} \neq 0$. Set the iteration formula

$$u_{n+1} = \sum_{k=0}^{q-1} a_k u_{n-k} + h \sum_{k=-1}^{q-1} b_k g_{n-k}, \quad n = 1, \dots, N-1. \tag{14.9}$$

The parameter q is the numbers of steps of the method and the coefficients (a_k, b_k) fully characterize it. If $b_{-1} = 0$, the method is explicit, otherwise it is implicit.

The local truncation error for multistep methods can be defined from the recursion

$$h\tilde{\varepsilon}_{n+1}(h) = y_{n+1} - \sum_{k=0}^{q-1} a_k y_{n-k} - h \sum_{k=-1}^{q-1} b_k y'_{n-j},$$

where $y'_j = \frac{\mathrm{d}}{\mathrm{d}t} y(t_j)$; that is, $h\tilde{\varepsilon}_{n+1}(h)$ is the error generated at the node t_{n+1}, if we substitute the analytical solution of the initial-value problem into the scheme (14.9). The definition of consistency and order of the method mentioned above are also available for multistep methods.

By choosing $a_0 = 1$ and all other $a_j = 0$ we obtain the *Adams methods*. Integrating (14.7), yields

$$y(t) - y_0 = \int_{t_0}^{t} g(s, y(s))\,\mathrm{d}s.$$

If we approximate this integral on the segment $[t_n, t_{n+1}]$ by the integral of the interpolating polynomial of g on q distinct nodes, particular Adams methods can be derived which, by construction, turn out to be consistent. Fixing as interpolation nodes $t_n, \dots, t_{n-q+1}$, gives the *Adams–Bashforth* methods, which are explicit, as they always lead to $b_{-1} = 0$. Fixing as interpolation nodes $t_{n+1}, \dots, t_{n-q+2}$, gives the *Adams–Moulton* methods, which are implicit.

For q steps, the Adams–Bashforth method can be shown to be of order q.

Example 14.12 [Adams–Bashforth methods] Examples of Adams–Bashforth methods include the following.

1. For $q = 1$, we recover the forward Euler method.
2. For $q = 2$, the interpolating polynomial is

   $$P(t) = g_n + (t - t_n)\frac{g_{n-1} - g_n}{t_{n-1} - t_n}.$$

   Thus

   $$\int_{t_n}^{t_{n+1}} P(s)\,\mathrm{d}s = \frac{h}{2}(3g_n - g_{n-1}).$$

   Hence the iteration formula of the Adams–Bashforth method with two steps requires

   $$(b_0, b_1) = \left(\frac{3}{2}, -\frac{1}{2}\right).$$

3. For $q = 3$, we get in a similar fashion an iteration formula for the Adams–Bashforth method with three steps, requiring

   $$(b_0, b_1, b_2) = \left(\frac{23}{12}, -\frac{4}{3}, \frac{5}{12}\right).$$

4. For $q = 4$, the iteration formula requires

   $$(b_0, b_1, b_2, b_3) = \left(\frac{55}{24}, -\frac{59}{24}, \frac{37}{24}, -\frac{3}{8}\right).$$

The Adams–Moulton methods can be derived similarly. Table 14.6 shows the corresponding coefficients for small q.

For $q + 1$ steps, the Adams–Moulton methods can be proved to be of order $q + 1$. This statement is a consequence

Table 14.6 Adams–Moulton methods.

q	$(b_{-1},\dots,b_{q-1})$
0	1
1	$\left(\frac{1}{2},\frac{1}{2}\right)$
2	$\left(\frac{5}{12},\frac{2}{3},-\frac{1}{12}\right)$
3	$\left(\frac{3}{8},\frac{19}{24},-\frac{5}{24},\frac{1}{24}\right)$
4	$\left(\frac{251}{720},\frac{323}{360},-\frac{11}{30},\frac{53}{360},-\frac{19}{720}\right)$

of the following remarkable result [3, Theorem 11.3].

Theorem 14.4 *The multistep method with iteration formula* (14.9) *is consistent if, and only if,*

$$\sum_{k=0}^{q-1} a_k = 1 \qquad \sum_{k=0}^{q-1} b_k - \sum_{k=-1}^{q-1} k a_k = 1.$$

Moreover, if the solution $y \in C^{r+1}(t_0, T)$, *then the method is of order r if, and only if, additionally*

$$\sum_{k=0}^{q-1}(-k)^j a_k + j\sum_{k=-1}^{q-1}(-k)^{j-1} b_k = 1,\quad j = 2,\dots,r.$$

14.5.3 Runge–Kutta Methods

An alternative approach to multistep methods, is the one adopted by the *Runge–Kutta methods*. These are one-step methods that involve several evaluations of $g(t,y)$ on the subsegment $[t_n, t_{n+1}]$. Their iteration formula reads

$$u_{n+1} = u_n + h\sum_{k=1}^{s} b_k R_k, \quad n = 1,\dots,N-1,$$

where

$$R_k = g\left(t_n + c_k h, u_n + h\sum_{k=1}^{s} a_{jk} R_k\right),\quad j = 1,\dots,s.$$

Here s is called the number of *stages* of the method. See [19] and the references therein. The parameters completely determine the method with the condition $c_j = \sum_{k=1}^{s} a_{jk}$ and are represented neatly by the *Butcher array*

$$\begin{array}{c|c} \underline{c} & A \\ \hline & \underline{b}^{t} \end{array} \quad \text{for} \quad A = [a_{jk}]_{jk=1}^{s} \quad \underline{b} = (b_k)_{k=1}^{s} \quad \text{and} \quad \underline{c} = (c_k)_{k=1}^{s}.$$

A Runge–Kutta method is explicit if, and only if, the matrix A is lower triangular. Implicit methods are quite difficult to derive in general, because the computation of R_k involves solving a nonlinear system of size s.

Example 14.13 [Some Runge–Kutta methods with a small number of stages] The Butcher arrays

$$\begin{array}{c|c} 1 & 1 \\ \hline & 1 \end{array} \qquad \begin{array}{c|c} \frac{1}{2} & \frac{1}{2} \\ \hline & 1 \end{array} \qquad \begin{array}{c|cc} 0 & 0 & 0 \\ 1 & \frac{1}{2} & \frac{1}{2} \\ \hline & \frac{1}{2} & \frac{1}{2} \end{array}$$

correspond, respectively, to the backward Euler, the implicit midpoint, and the Crank–Nicolson methods.

It is possible to construct implicit or explicit Runge–Kutta methods of any large order. One of the most widely used Runge–Kutta method is the one with four stages determined by the Butcher array

$$
\begin{array}{c|cccc}
0 & & & & \\
\frac{1}{2} & \frac{1}{2} & & & \\
\frac{1}{2} & 0 & \frac{1}{2} & & \\
1 & 0 & 0 & 1 & \\
\hline
 & \frac{1}{6} & \frac{1}{3} & \frac{1}{3} & \frac{1}{6}
\end{array}
$$

It can be derived in a similar fashion as for the Adams methods. It is explicit and convergent to the fourth order with respect to h. An implicit method with two stages that is also convergent to fourth order is given by the array

$$
\begin{array}{c|cc}
\frac{3-\sqrt{3}}{6} & \frac{1}{4} & \frac{3-2\sqrt{3}}{12} \\
\frac{3+\sqrt{3}}{6} & \frac{3+2\sqrt{3}}{12} & \frac{1}{4} \\
\hline
 & \frac{1}{2} & \frac{1}{2}
\end{array}
$$

We have exclusively dealt above with the case of the scalar Cauchy problem. When confronted with the case $d > 1$, two possible approaches are available. We either apply to each individual equation one of the scalar methods, or we write the iteration formula for the method in vector form.

Example 14.14 [Airy's equation] The Airy functions have a wide range of applications in mathematical physics. The solution of the Airy equation that generates the Airy functions,

$$
\frac{\mathrm{d}^2}{\mathrm{d}t^2}u(t) = tu(t), \qquad t > 0,
$$

$$
u(0) = 1, \qquad \frac{\mathrm{d}}{\mathrm{d}t}u(0) = 0,
$$

can be achieved by reducing this equation to the system

$$
\frac{\mathrm{d}}{\mathrm{d}t}\underline{y}(t) = g(t, \underline{y}(t)), \quad t > 0 \quad \Bigg| \quad \underline{y}(t) = \begin{bmatrix} u(t) \\ \frac{\mathrm{d}}{\mathrm{d}t}u(t) \end{bmatrix}
$$

$$
\underline{y}(0) = \begin{bmatrix} 1 \\ 0 \end{bmatrix} \qquad \Bigg| \quad g(t, \underline{y}) = \begin{bmatrix} y_2 \\ ty_1 \end{bmatrix}.
$$

Even though analytical methods can be applied to the study of Airy functions, numerical methods as described above are better suited for the estimation of numerical values of the Airy functions.

14.5.4 Stability and Global Stability

A scheme on a given fixed bounded segment will certainly be regarded as *stable*, if small perturbations of the data render bounded perturbations of the numerical solution in the regime $h \to 0$. This idea can be made more precise as follows. Assume that v_n is an approximated solution produced by the numerical method in question, applied to a *perturbation* of the original problem, and that σ_n is the size of this perturbation at the step n. The numerical method will be called *zero-stable*, if for fixed $T > t_0$, there exists $h_0, C_T, \varepsilon_0 > 0$ such that for all $0 < h \le h_0$ and $0 < \varepsilon \le \varepsilon_0$,

$$
|\sigma_n| \le \varepsilon \quad \text{implies} \quad |u_n - v_n| \le C_T\varepsilon
$$

$$
\text{for all} \quad n = 0, \dots, \max\{n : t_n \le t_0 + T\}.
$$

Example 14.15 [Consistency implies zero-stability for one-step methods] If g satisfies (14.8), a consistent one-step method for the solution of the corresponding Cauchy problem will be zero-stable. Moreover, in this case, the constant C_T is proportional to $\mathrm{e}^{(T-t_0)L}$.

Let

$$
\begin{aligned}
\mathfrak{p}(r) &= r^q - \sum_{k=0}^{q-1} a_k r^{q-1-k} \\
&= (r - r_0) \cdots (r - r_{q-1})
\end{aligned}
$$

be the *characteristic polynomial* associated to the coefficients of the multistep method

(14.9). The latter will be zero-stable, if and only if all $|r_k| \leq 1$ and those with $|r_k| = 1$ are all simple roots. The latter is often called the *root condition* of the method. See Table 14.7.

Theorem 14.5 (Lax–Ritchmyer) *A numerical method for the solution of the Cauchy problem that is consistent, will also be convergent if and only if it is zero-stable.*

A complete proof of this result can be found in [3, Section 11.6.3].

Another notion of stability is concerned with the behavior of the scheme, when the corresponding Cauchy problem is posed on an infinite segment. For *global stability*, the methods are tested against each other by means of the *reference equation*

$$\frac{\mathrm{d}y}{\mathrm{d}t} = \lambda y, \qquad t > 0,$$
$$y(0) = 1$$

for all given $\lambda \in \mathbb{C}$. This reference equation has as exact solution $y(t) = \mathrm{e}^{\lambda t}$. The *region of absolute stability* of an iterative scheme is the subset of the complex plane defined as

$$\mathcal{R} = \{h\lambda \in \mathbb{C} : |u_n| \to 0 \quad t_n \to \infty\}.$$

That is, the region of values of the product $h\lambda$, such that the numerical scheme leads to a solution decaying at infinity.

Example 14.16 [Absolute stability of the forward Euler method] The iteration formula of the forward Euler method for the solution of the reference equation is

$$u_{n+1} = h\lambda u_n, \qquad u_0 = 1, \qquad n = 0, \ldots$$

Induction allows finding the exact solution to this recursion,

$$u_n = (1 + h\lambda)^n, \qquad n = 0, \ldots$$

For inclusion in $\mathcal{R}$, we therefore require in this case that $|1 + h\lambda| < 1$. Hence,

$$\mathcal{R} = \left\{ \mathrm{Re}(h\lambda) < 0 \quad \text{and} \quad 0 < h < -\frac{2\,\mathrm{Re}(\lambda)}{|\lambda|^2} \right\},$$

which is a circle of radius 1 centered at $(-1, 0)$.

For a general multistep method, let

$$\mathfrak{q}(r) = \sum_{k=-1}^{q-1} b_k r^{q-1-k}.$$

The boundary of the region of absolute stability is characterized by the identity [3, Section 11.6.4]

$$\partial\mathcal{R} = \{h\lambda \in \mathbb{C} : \mathfrak{q}(r)h\lambda = \mathfrak{p}(r) \quad \text{for some} \quad |r| = 1\}.$$

14.6 Spectral Problems

In order to illustrate the numerical solution to infinite-dimensional spectral problems, we consider the reference eigenvalue equation corresponding to the one-dimensional Schrödinger Hamiltonian. Given a real-valued continuous potential $V : (a, b) \longrightarrow \mathbb{R}$, the aim is to determine

Table 14.7 Regions of absolute convergence for the most common one-step methods.

Method	$\mathcal{R}$
Forward Euler	$\{\lvert 1 + z\rvert < 1\}$
Backward Euler	$\mathbb{C} \setminus \{\lvert 1 - z\rvert \leq 1\}$
Crank–Nicolson	$\{\mathrm{Re}\, z < 0\}$
Heun	$\{\lvert 1 + z + \frac{z^2}{2}\rvert < 1\}$

approximately the solution of the equation

$$-\frac{d^2}{dx^2}y(x) + V(x)y(x) = \lambda y(x), \quad a < x < b,$$
$$y(a) = y(b) = 0. \tag{14.10}$$

Here $a < b$ and either one of them or both can be infinite. By a solution of (14.10), we mean an eigenvalue $\lambda \in \mathbb{R}$ and a corresponding nonzero eigenfunction $y \in L^2(a, b)$.

This simple equation possesses all the features of the more complicated multiparticle settings.

14.6.1 The Infinite-Dimensional min–max Principle

The left hand side of (14.10) gives rise to a *Hamiltonian*,

$$H : D(H) \longrightarrow L^2(a, b),$$

which is a self-adjoint operator defined on the dense *domain* $D(H) \subset L^2(a, b)$, dependent on the potential. Formally,

$$H = -\frac{d^2}{dx^2} + V.$$

The eigenvalue equation is equivalent to the weak equation

$$\mathfrak{h}(y, w) = \lambda\langle y, w\rangle, \qquad \text{for all} \quad w \in D(\mathfrak{h}),$$

where the energy functional $\mathfrak{h} : D(\mathfrak{h}) \times D(\mathfrak{h}) \longrightarrow \mathbb{C}$ is given by

$$\mathfrak{h}(y, w) = \langle Hy, w\rangle = \int_a^b \left(y'(x)\overline{w'(x)} + V(x)y(x)\overline{w(x)}\right) dx.$$

Here $D(\mathfrak{h}) \subset L^2(a, b)$ is another suitable dense domain containing $D(H)$ that renders the form *closed*.

By mimicking the finite-dimensional setting, let

$$\lambda_j(H) = \min_{\substack{S \subset D(\mathfrak{h}) \\ \dim S = j}} \max_{w \in S} R(w), \qquad j = 1, \ldots,$$

where the Rayleigh quotient

$$R(w) = \frac{\mathfrak{h}(w, w)}{\langle w, w\rangle}.$$

Then we have an analogous version of Theorem 14.2, [9, Section 4.5] or [10, Section XIII.2].

Theorem 14.6 (Min–max principle) for (14.10) *Assume that the potential V is such that $R(w) \geq c$ for all $w \in D(\mathfrak{h})$. Then $b \leq \lambda_j(H) \leq e(H)$ are the eigenvalues of* (14.10) *lying below*

$$e(H) = \lim_{j\to\infty} \lambda_j(H).$$

Moreover, the corresponding nonzero eigenfunctions $y_j \in D(\mathfrak{h})$ satisfy

$$\lambda_j(H) = \min_{w \perp \{y_k\}_{k=1}^{j-1}} R(w) = R(y_j).$$

By construction, $\lambda_j(H)$ are nondecreasing in j. The limit $e(H) > b$ might or might not be infinity and it is the bottom of the *essential spectrum* of H.

Numerical methods for computing upper bounds for the $\lambda_j(H)$ can then be derived by means of the *Galerkin method*. Pick a linear subspace $\mathcal{L} \subset D(\mathfrak{h})$ of dimension d with a linearly independent basis $\{b_j\}_{j=1}^d$. Assemble the matrices

$$K = [\mathfrak{h}(b_j, b_k)]_{jk=1}^d \quad \text{and} \quad M = [\langle b_j, b_k\rangle]_{jk=1}^d.$$

Then the eigenvalues $\mu_j(H, \mathcal{L})$ of the finite-dimensional linear eigenvalue problem

$$K\underline{u} = \lambda M\underline{u}$$

are such that $\mu_j(H, \mathcal{L}) \geq \lambda_j(H)$.

14.6.2 Systems Confined to a Box

Different methods arise by constructing families of subspaces $\mathcal{L}$ in different ways. We illustrate the standard way of implementing the Galerkin method on an artificial example, where all quantities involved can be found analytically.

Example 14.17 [The free particle in a box] Let $V(x) = 0$, $a = 0$ and $b > 0$. The solution of (14.10) is given explicitly by

$$\lambda_j(H) = \frac{j^2\pi^2}{b^2}, \qquad y_j(x) = \sin\left(\frac{j\pi}{b}x\right).$$

In this case, $\mathrm{D}(\mathfrak{h}) = H_0^1(0, b)$, the Sobolev space of all absolutely continuous functions vanishing at 0 and b.

Assume that $\mathcal{L}$ is the space of Lagrange finite elements of order 1. A basis for $\mathcal{L}$ is given by the piecewise linear continuous functions

$$b_j(x) = \begin{cases} \dfrac{x - x_{j-1}}{x_j - x_{j-1}} & x_{j-1} \leq x \leq x_j \\ \dfrac{x_{j+1} - x}{x_{j+1} - x_j} & x_j \leq x \leq x_{j+1} \\ 0 & \text{otherwise} \end{cases}$$
$$j = 1, \ldots, N-1,$$

where $h = b/N$ and $x_j = jh$. Note that b_0 and b_N are not considered here, as the trial functions in $\mathcal{L}$ should satisfy the Dirichlet boundary conditions.

We can find the eigenvalues $\mu_j(H, \mathcal{L})$ explicitly. Let

$$\tilde{M} = \begin{bmatrix} 4 & 1 & & & \\ 1 & 4 & 1 & & \\ & 1 & \ddots & \ddots & \\ & & \ddots & 4 & 1 \\ & & & 1 & 4 \end{bmatrix}$$

and

$$\tilde{K} = \begin{bmatrix} 2 & -1 & & & \\ -1 & 2 & -1 & & \\ & -1 & \ddots & \ddots & \\ & & \ddots & 2 & -1 \\ & & & -1 & 2 \end{bmatrix}.$$

Then $K = (1/h)\tilde{K}$ and $M = (h/6)\tilde{M}$. The eigenvalues of the Toeplitz matrix

$$T = \begin{bmatrix} 0 & 1 & & & \\ 1 & 0 & 1 & & \\ & 1 & \ddots & \ddots & \\ & & \ddots & 0 & 1 \\ & & & 1 & 0 \end{bmatrix}$$

are $\tau_j = 2\cos(j\pi/N)$, $j = 1, \ldots, N-1$. As the commutation relation $\tilde{M}\tilde{K} = \tilde{K}\tilde{M}$ holds true,

$$\mu_j(H, \mathcal{L}) = \frac{6N^2(2 - \tau_j)}{b^2(4 + \tau_j)}.$$

In the explicit calculation made in the previous example, $\mu_j(H, \mathcal{L}) \downarrow \lambda_j(H)$ and

$$\lim_{N\to\infty} \frac{\mu_j(H, \mathcal{L}) - \lambda_j(H)}{N^2} = \frac{j^4\pi^4}{12b^2}, \quad j = 1, \ldots, N-1.$$

In other words, the approximated eigenvalues miss the exact eigenvalues by a factor proportional to h^2.

Let Ξ_h be an equidistant partition of the segment $[a, b]$ into N subintervals

$$\mathcal{J}_j = [x_{j-1}, x_j]$$

of length $h = b - a/N$. Let

$$\mathcal{V}(q, \Xi_h) = \{w \in C^1(a,b) \,:\, w|_{\mathcal{J}_j} \in \mathbb{P}_q(\mathcal{J}_j),\ 1 \le j \le N,\ w(a) = w(b) = 0\}$$

be the space of piecewise differentiable functions on Ξ that are polynomials of degree q in each $\mathcal{J}_j$ and satisfy the boundary conditions of (14.10). See [16, Theorem 6.1].

Theorem 14.7 *Let a and b be fixed and both finite. Assume that the potential V is continuous. There exists $c(q) > 0$, such that*

$$\lambda_j(H) \le \mu_j(H, \mathcal{V}_h(q, \Xi)) \le \lambda_j(H) + c(q)h^{2q}\lambda_j(H)^{q+1}$$

for all h sufficiently small.

14.6.3 The Case of Unconfined Systems

When the segment (a, b) is of infinite length, Theorem 14.6 is still holds as long as $\mathcal{L}$ is contained in $\mathrm{D}(\mathfrak{h})$. One possible approach is to assume that $\mathcal{L}$ is made out of Gaussian-type functions.

Another possible approach is to observe that, generally, the eigenfunctions of (14.10) are exponentially decaying at $\pm\infty$ whenever the potential is sufficiently regular. Let $L > 0$ and denote by $\lambda_j(H, L)$ the eigenvalues of (14.10) for $a = -L$ and $b = L$.

Theorem 14.8 *Let $j \in \mathbb{N}$ be fixed and assume that the potential V is continuous. There exist constants $c_j > 0$ and $a_j > 0$ both independent of L, such that*

$$\lambda_j(H) < \lambda_j(H, L) < \lambda_j(H) + c_j \mathrm{e}^{-a_j L}.$$

A proof of this result follows from the known exponential decay of the eigenfunctions of one-dimensional Schrödinger operators, see, for example, [20, Chapter 3.3] and references therein.

Further Reading

For a more comprehensive survey on the subjects of numerical solutions of nonlinear systems and numerical optimization, see [3, Chapter 7]. In particular, we have ignored completely the subject of constrained optimization and Lagrange multiplier techniques [21–23].

A substantial body of research has been devoted to the sensitivity analysis of the solution of linear systems by direct methods and the effect of rounding errors. For details on this, see [11, 24, 25]. The PCG method is a particular realization of the more general Krylov methods that can be applied to nonhermitian matrices [6, 26, 27]. Of particular importance are the full orthogonalization method (FOM) and the generalized minimum residual (GMRES) method, which converge extremely fast for the right class of matrices.

Computation of all the eigenvalues of a matrix via QR factorization or singular-value decomposition is a well-studied subject [11, 28]. Much research has also been conducted on the more general matrix polynomials, in particular, their factorization properties and the corresponding generalized eigenvalue problem [29]. The remarkable theory of computation with matrix functions has also been recently examined [30].

Reference on topics such as approximation theory by orthogonal polynomials can be found in [3, Chapter 10] and references therein. Classical references on the finite element method are [15, 16, 31]. A survey on the theory of curves and surface fitting can be found in [32].

An important class of methods for the solution of Cauchy problems that we have not mentioned here is the class of predictor–corrector methods [3, Section 11.7]. For further reading on the numerical solution of PDEs both evolutionary and stationary, see [33, 34] and references therein.

A fairly complete account on the theory of eigenvalue computation for differential equations including estimation of *a priori* and *a posteriori* complementary bounds for eigenvalues can be found in [35]. Computation of eigenvalues in gaps of the essential spectrum is a much harder problem; see [36] and references therein.

References

1. Quarteroni, A., Saleri, F., and Gervasio, P. (2010) *Texts in Computational Science and Engineering, Scientific Computing with MATLAB and Octave*, vol. 2, Springer-Verlag, Berlin.
2. Ridgway Scott, L. (2011) *Numerical Analysis*, Princeton University Press, Princeton, NJ.
3. Quarteroni, A., Sacco, R., and Saleri, F. (2007) *Numerical Mathematics, Texts in Applied Mathematics*, vol. 37, 2nd edn, Springer-Verlag, Berlin.
4. Ortega, J.M. and Rheinboldt, W.C. (2000) *Iterative Solution of Nonlinear Equations in Several Variables, Classics in Applied Mathematics*, vol. 30, Society for Industrial and Applied Mathematics (SIAM), Philadelphia, PA.
5. Dennis, J.E. Jr., and Schnabel, R.B. (1996) *Numerical Methods for Unconstrained Optimization and Nonlinear Equations, Classics in Applied Mathematics*, vol. 16, Society for Industrial and Applied Mathematics (SIAM), Philadelphia, PA.
6. Hackbusch, W. (1994) *Iterative Solution of Large Sparse Systems of Equations, Applied Mathematical Sciences*, vol. 95, Springer-Verlag, New York.
7. Young, D.M. (2003) *Iterative Solution of Large Linear Systems*, Dover Publications, Inc., Mineola, NY.
8. Langville, A.N. and Meyer, C.D. (2012) *Google's PageRank and Beyond: The Science of Search Engine Rankings*, Princeton University Press, Princeton, NJ.
9. Davies, E.B. (1995) *Spectral Theory and Differential Operators, Cambridge Studies in Advanced Mathematics*, vol. 42, Cambridge University Press, Cambridge.
10. Reed, M. and Simon, B. (1978) *Methods of Modern Mathematical Physics. IV. Analysis of Operators*, Academic Press, New York.
11. Golub, G.H. and Van Loan, C.F. (2013) *Matrix Computations. Johns Hopkins Studies in the Mathematical Sciences. Johns Hopkins*, 4th edn, University Press, Baltimore, MD.
12. Wendroff, B. (1966) *Theoretical Numerical Analysis*, Academic Press, New York.
13. Davis, P.J. (1975) *Interpolation and Approximation*, Dover Publications, Inc., New York.
14. Erdős, P. (1961) Problems and results on the theory of interpolation. II. *Acta Math. Acad. Sci. Hungar.*, **12**, 235–244.
15. Ciarlet, P.G. (2002) *The Finite Element Method for Elliptic Problems, Classics in Applied Mathematics*, vol. 40, Society for Industrial and Applied Mathematics (SIAM), Philadelphia, PA.
16. Strang, G. and Fix, G. (2008) *An Analysis of the Finite Element Method*, 2nd edn, Wellesley-Cambridge Press, Wellesley, MA.
17. Coddington, E.A. and Levinson, N. (1955) *Theory of Ordinary Differential Equations*, McGraw-Hill Book Company, London.
18. Lambert, J.D. (1991) *Numerical Methods for Ordinary Differential Systems*, John Wiley & Sons, Ltd., Chichester, The initial value problem.
19. Butcher, J.C. and Wanner, G. (1996) Runge-Kutta methods: some historical notes. *Appl. Numer. Math.*, **22**(1-3), 113–151.
20. Simon, B. (1982) Schrödinger semigroups. *Bull. Amer. Math. Soc. (N.S.)*, 7(3), 447–526.
21. Avriel, M. (2003) *Nonlinear Programming*, Dover Publications, Inc., Mineola, NY.
22. Bertsekas, D.P. (1982) Constrained optimization and Lagrange multiplier methods, in *Computer Science and Applied Mathematics*, Academic Press, Inc., New York.
23. Canon, M.D., Cullum, C.D. Jr., and Polak, E. (1970) *Theory of Optimal Control and Mathematical Programming*, McGraw-Hill Book Co., New York.

24. Datta, B.N. (2010) *Numerical Linear Algebra and Applications*, 2nd edn, Society for Industrial and Applied Mathematics (SIAM), Philadelphia, PA.
25. Stewart, G.W. (1973) *Introduction to Matrix Computations*, Academic Press, New York, London.
26. Axelsson, O. (1994) *Iterative Solution Methods*, Cambridge University Press, Cambridge.
27. Saad, Y. (2003) *Iterative Methods for Sparse Linear Systems*, 2nd edn, Society for Industrial and Applied Mathematics, Philadelphia, PA.
28. Wilkinson, J.H. (1988) *The Algebraic Eigenvalue Problem, Monographs on Numerical Analysis*, The Clarendon Press - Oxford University Press, New York.
29. Gohberg, I., Lancaster, P., and Rodman, L. (1982) *Matrix Polynomials*, Academic Press, Inc., New York.
30. Higham, N.J. (2008) *Functions of Matrices*, Society for Industrial and Applied Mathematics (SIAM), Philadelphia, PA.
31. Brenner, S.C. and Ridgway Scott, L. (2008) *The Mathematical Theory of Finite Element Methods, Texts in Applied Mathematics*, vol. 15, 3rd edn, Springer, New York.
32. Lancaster, P. and Šalkauskas, K. (1986) *Curve and Surface Fitting*, Academic Press, Inc., London.
33. Quarteroni, A. and Valli, A. (1994) *Numerical Approximation of Partial Differential Equations, Springer Series in Computational Mathematics*, vol. 23, Springer-Verlag, Berlin.
34. Ern, A. and Guermond, J.-L. (2004) *Theory and Practice of Finite Elements, Applied Mathematical Sciences*, vol. 159, Springer-Verlag, New York.
35. Weinberger, H.F. (1974) *Variational Methods for Eigenvalue Approximation*, Society for Industrial and Applied Mathematics, Philadelphia, PA.
36. Boulton, L., Boussaïd, N., and Lewin, M. (2012) Generalised Weyl theorems and spectral pollution in the Galerkin method. *J. Spectr. Theory*, **2**(4), 329–354.

15 Mathematical Transformations

Des McGhee, Rainer Picard, Sascha Trostorff, and Marcus Waurick

15.1 What are Transformations and Why are They Useful?

A transformation[1] is a process that turns objects in one area into objects of a different area in such a way that no information is lost and the original can be reconstructed without loss. In mathematical terms, this could be described as an invertible mapping from one topological space onto another such that both the mapping and its inverse are continuous. Transformations are advantageous if terms, concepts, and ideas are considered to be more elementary or intuitive in one area and less elementary or intuitive in the other. A simple example is the use of different maps (i.e., different projections) of the Earth's surface. An angle-preserving map is useful for ocean navigation, whereas an area-preserving map is useful for comparing the size of different islands.

Another more mathematical example is the principal axis transformation, which brings the equation of an ellipse into a standard (canonical) form where, for example, the axes sizes, eccentricity, and foci are easily read off or deduced from formula. This amounts to the diagonalization of a real (2×2)−matrix A

$$A = U^* D U, \tag{15.1}$$

where U is an orthogonal (real) (2×2)−matrix, i.e.

$$U^{-1} = U^*.$$

Here U^* is simply the transpose matrix to U.

1) There is a linguistic subtlety regarding the standard mathematical terminology. Transformation is a process and the result of the transformation applied to a particular object is its transform. In the literature, these terms are almost always used as synonyms and we shall not attempt to rectify this situation. The device performing the transformation is a transformer or transformator. We will not use the term "transformer" as it might lead to confusion with other areas of life, see, for example, [1]. The term "transformator" is rarely used anyway, so we shall follow the standard practice and use "transformation" to describe both process and device (compare map/mapping, operator/operation or projector/projection for similar terminological difficulties).

Mathematical Tools for Physicists, Second Edition. Edited by Michael Grinfeld.

Physicists should be familiar with the ubiquitous Laplace and Fourier transformations (including the sine and cosine transformations). Probably the most familiar application of these transformations is to convert "hard" differential equations into "easier" algebraic equations. Once the latter are solved, the solutions of the original differential equations are obtained by applying the inverse transformation.

The choice of transformation is determined by the differential operator involved including consideration of its domain determined by initial or boundary conditions. The Laplace transformation L defined by

$$(Lf)(s) = \int_{\mathbb{R}_+} \mathrm{e}^{-st} f(t)\,\mathrm{d}t$$

is appropriate for initial-value problems on the half-line $[0, \infty[$. Thus, for some suitable range of values of s, the function $t \mapsto \mathrm{e}^{-st} f(t)$ must be integrable over $[0, \infty[$, a property that will contribute to the determination of an appropriate space of functions in which the problem can be considered. The salient property is that the Laplace transformation converts differentiation into multiplication by the variable subject to the function value at $t = 0$ being known, the easily derived formula being

$$(Lf')(s) = s(Lf)(s) - f(0)$$

(derived using integration by parts and assuming that $\mathrm{e}^{-st} f(t) \to 0$ as $t \to \infty$). This generalizes to higher derivatives as

$$(Lf^{(n)})(s) = s^n (Lf)(s) - \sum_{r=0}^{n-1} s^{n-1-r} f^{(r)}(0).$$

The Fourier transformation F defined by[2)]

$$(Ff)(x) = \frac{1}{\sqrt{2\pi}} \int_{\mathbb{R}} \exp\left(-\mathrm{i}xy\right) f\left(y\right)\,\mathrm{d}y \tag{15.2}$$

is often appropriate for boundary-value problems on $\mathbb{R}$ where solutions are required to decay to zero at $\pm\infty$. Once again, the requirement for the above integral to exist contributes to the definition of appropriate spaces in which to consider problems. Clearly, the integral exists for any function that is integrable on $\mathbb{R}$, that is, for all $L^1(\mathbb{R})$-functions, because for all $x, y \in \mathbb{R}$ we have $|\exp\left(-\mathrm{i}xy\right)| = 1$. The Fourier transformation can also be defined on all of $L^2(\mathbb{R})$ where it is traditionally referred to as the *Fourier–Plancherel transformation* (see [2] and Section 15.4). For the Fourier transformation, the result for differentiation of an n–times continuously differentiable function is

$$(Ff^{(n)})(x) = (\mathrm{i}x)^n (Ff)(x),$$

that is, differentiation is transformed into multiplication by i times the variable.

The Fourier sine and cosine transformations are defined respectively by[3)]

$$(\tilde{F}_{\sin} f)(x) = \sqrt{\frac{2}{\pi}} \int_{\mathbb{R}_+} \sin(xy) f(y)\,\mathrm{d}y,$$

$$(\tilde{F}_{\cos} f)(x) = \sqrt{\frac{2}{\pi}} \int_{\mathbb{R}_+} \cos(xy) f(y)\,\mathrm{d}y$$

for suitable functions f on $[0, \infty[$ such that the integrals exist. The differentiation formulae for twice continuously differentiable function f on $[0, \infty[$ with $f(x) \to 0$ as x tends to infinity are

$$(\tilde{F}_{\sin} f'')(x) = -x^2 (\tilde{F}_{\sin} f)(x) + \sqrt{\frac{2}{\pi}} x f(0),$$

$$(\tilde{F}_{\cos} f'')(x) = -x^2 (\tilde{F}_{\cos} f)(x) - \sqrt{\frac{2}{\pi}} f'(0),$$

2) There are various, essentially equivalent, definitions of the Fourier Transformation. We shall discuss some issues later.

3) The rationale for the ~ notation here is explained in Section 15.6.

making these transformations useful typically for the canonical second-order partial differential equations (Laplace, wave, and diffusion equations) on a space domain $[0, \infty[$ where, respectively, the function or its derivative is given as boundary data at zero (along with decay to zero at ∞).

Such applications make clear the need for a transformation to be a continuous one-to-one mapping with a continuous inverse. If not one-to-one, then the inverse transformation would not exist and uniqueness of the inverse transform would be lost: more than one function purporting to solve the differential equation would correspond to any solution of the algebraic equation. If the mapping or its inverse were not continuous, then any small error (rounding error in calculation or measurement error in experiment) would result in large, uncontrollable errors in transforms or inverse transforms, respectively.

We shall consider these transformations and others in some mathematical generality in the following sections. We shall not however rehearse standard applications of mathematical transformations to typical problems of mathematical physics because there is no shortage of appropriate sources of such material. Instead, our aim is threefold:

1. to shed light on why mathematical transformations work by clarifying what governs the choice of transformation that may be appropriate to a particular problem;
2. to show that various transformations widely used in physics and engineering are closely related; and
3. to introduce sufficient mathematical rigor so that the application of mathematical transformations to entities such as the Dirac delta and its derivatives that are so useful in modeling point sources and impulses, are fully understood.

We set the scene by considering an appropriate setting in which to develop a deeper understanding of the role and usefulness of mathematical transformations. We focus attention on linear mappings between Hilbert spaces H_0 and H_1. A transformation is a linear mapping

$$T : H_0 \to H_1$$

such that

$$T^{-1} : H_1 \to H_0$$

exists (i.e., T is a bijection, a one-to-one and onto map) and both T and T^{-1} are continuous, that is, $T \in \mathcal{B}\left(H_0, H_1\right)$ and $T^{-1} \in \mathcal{B}\left(H_1, H_0\right)$ (see [2]). In other words, T is a linear homeomorphism. Of particular interest are transformations T that preserve the Hilbert space structure, that is, which are unitary. Unitary transformations are isometric, that is, norm preserving, which means that for all $x \in H_0$

$$|Tx|_{H_1} = |x|_{H_0}.$$

A useful first example is given by coordinate transformations. Let $\phi : S_0 \to S_1$ be a smooth bijection between open sets S_0, $S_1 = \operatorname{Ran}(\phi)$ in $\mathbb{R}^n$, $n \in \{1, 2, 3, \dots\}$ such that the determinant $\det(\partial\phi)$ of the Jacobian $\partial\phi$ is bounded and has a bounded inverse. Such a coordinate transformation induces a unitary transformation between the spaces $L^2\left(S_k\right) := L^2\left(S_k, \mathrm{d}\lambda, \mathbb{C}\right)$ of (equivalence classes of) complex-valued, square-integrable functions on $\mathbb{R}^n$ vanishing outside S_k, $k \in \{0, 1\}$. (Here λ denotes the standard Lebesgue measure.) Indeed, for all $f, g \in L^2\left(S_1\right)$, the substitution formula for integrals yields

$$\begin{aligned}
&\langle f|g\rangle_{L^2(S_1)}\\
&= \int_{S_1} \bar{f}\, g \, d\lambda\\
&= \int_{S_1} \overline{f(y)}\, g(y)\; dy\\
&= \int_{S_0} \overline{f(\phi(x))}\, g(\phi(x))\; |\det((\partial\phi)(x))|\; dx.
\end{aligned}$$

The latter can be re-written as

$$\int_{S_0} \overline{\sqrt{|\det((\partial\phi)(x))|} f(\phi(x))}\, \sqrt{|\det((\partial\phi)(x))|} g(\phi(x))\; dx$$

or

$$\left\langle \sqrt{|\det((\partial\phi)(\mathbf{m}))|} f\circ\phi \middle| \sqrt{|\det((\partial\phi)(\mathbf{m}))|} g\circ\phi \right\rangle_{L^2(S_0)},$$

where $\sqrt{|\det((\partial\phi)(\mathbf{m}))|}$ is a suggestive notation for the multiplication operator mapping f to $x \mapsto \sqrt{|\det((\partial\phi)(x))|} f(x)$, that is, $T_{\sqrt{|\det((\partial\phi))|}}$ in the notation of [2]. This shows that

$$\begin{array}{ccc}
L^2(S_1) & \to & L^2(S_0)\\
f & \mapsto & \sqrt{|\det((\partial\phi)(\mathbf{m}))|}\, f\circ\phi
\end{array} \tag{15.3}$$

is an isometry and, in fact, it is also onto and therefore is a unitary transformation.

We shall focus mainly on some particular unitary transformations obtained by continuous extension from integral expressions, because they are particularly relevant in applications to physics [3–7].

15.2 The Fourier Series Transformations

15.2.1 The Abstract Fourier Series

The first example is the so-called abstract Fourier expansion giving rise to the abstract Fourier series transformation. It may come as a surprise to some that Fourier series expansions are essentially transformations as defined above.

Let o be an orthonormal set in a Hilbert space H. Then

$$x = \sum_{e\in o} \langle e|x\rangle_H \; e \tag{15.4}$$

for all x in the linear envelope (linear span, linear hull) of o, which is the smallest linear subspace of H containing o. The orthonormal set o is called *total* or *complete* (or an orthonormal basis) if the linear envelope of o is dense in H, [2]. In case of an orthonormal basis, we have a series expansion of the form (15.4) for every $x \in H$. The series expansion holds independently of any ordering of o, which is why we chose the summation notation in (15.4). For a separable Hilbert space, an orthonormal basis is finite or countably infinite and so there is a parametric "enumeration" $(e_\alpha)_{\alpha\in S}$ of $o = \{e_\alpha | \alpha \in S\}$ and we may write $\sum_{\alpha\in S} \langle e_\alpha|x\rangle_H \; e_\alpha$. For $S = \{0,\ldots,N\}$, $S = \mathbb{N}$ or $S = \mathbb{Z}$, the respective notations $\sum_{\alpha=0}^{N} \langle e_\alpha|x\rangle_H \; e_\alpha$, $\sum_{\alpha=0}^{\infty} \langle e_\alpha|x\rangle_H \; e_\alpha$ and $\sum_{\alpha=-\infty}^{\infty} \langle e_\alpha|x\rangle_H \; e_\alpha$ are commonly used. The associated unitary transformation (as stated by Parseval's equality, [2]) is the abstract Fourier series transformation

$$\begin{aligned}
&H \to \ell^2(o) := \left\{ f : o \to \mathbb{C} \,\middle|\, \sum_{e\in o} |f(e)|^2 < \infty \right\},\\
&x \mapsto \left(\langle e|x\rangle_H\right)_{e\in o},
\end{aligned}$$

which associates with each element $x \in H$ its sequence of so-called Fourier coefficients. The inverse of this transformation is the reconstruction of the original $x \in H$ from its (abstract) Fourier coefficients

$$\begin{aligned}
&\ell^2(o) \to H,\\
&(\phi_e)_{e\in o} \mapsto \sum_{e\in o} \phi_e \, e.
\end{aligned}$$

15.2.2 The Classical Fourier Series

Of course, the French baron Jean-Baptiste-Joseph Fourier (March 21, 1768 – May 16, 1830) never envisioned such a general concept of orthogonal expansion. He had a very specific orthonormal basis in mind. In today's language, he was concerned with the orthonormal basis $\left\{1/\sqrt{2\pi}\exp(\mathrm{i}k\,\cdot\,)\,|k \in \mathbb{Z}\right\}$ in the Hilbert space $L^2(]-\pi,\pi]) = L^2(]-\pi,\pi],\mathrm{d}\lambda,\mathbb{C})$. Equivalently, we could have no scalar factor in front of the exponential and have $\{\exp(\mathrm{i}k\,\cdot\,)\,|k \in \mathbb{Z}\}$ as an orthonormal basis in $L^2(]-\pi,\pi],1/2\pi\mathrm{d}\lambda,\mathbb{C})$. We shall, however, stay with the first option.

The classical Fourier expansion gives rise to the transformation, that is, a unitary mapping,

$$F_\# : L^2(]-\pi,\pi]) \to \ell^2(\mathbb{Z}),$$
$$f \mapsto \left(\hat{f}(k)\right)_{k\in\mathbb{Z}},$$

where the Fourier coefficients $\hat{f}(k)$ are given by

$$\hat{f}(k) = \frac{1}{\sqrt{2\pi}}\int_{]-\pi,\pi]} \exp(-\mathrm{i}ky)\, f(y)\; \mathrm{d}y,$$

for each $k \in \mathbb{Z}$ and the inverse transformation, $F_\#^{-1} = F_\#^*$ is given by

$$F_\#^{-1} : \ell^2(\mathbb{Z}) \to L^2(]-\pi,\pi]),$$
$$\left(\hat{f}(k)\right)_{k\in\mathbb{Z}} \mapsto f,$$

where

$$f(x) = \frac{1}{\sqrt{2\pi}}\sum_{k=-\infty}^{\infty} \hat{f}(k)\;\exp(\mathrm{i}kx)$$

for almost every $x \in]-\pi,\pi]$. Thus, for all $f \in L^2(]-\pi,\pi])$,

$$f(x) = \frac{1}{2\pi}\sum_{k=-\infty}^{\infty}\int_{]-\pi,\pi]} \exp(-\mathrm{i}ky)\, f(y)\;\mathrm{d}y\;\exp(\mathrm{i}kx)$$

for almost all $x \in]-\pi,\pi]$.

A unitary equivalence of an operator A to a multiplication-by-argument operator in some suitable function space is called a *spectral representation* of A.[4] The main point of interest is that $F_\#$ is a spectral representation associated with differentiation.

Indeed, if $\partial_\#$ denotes the differentiation operator in $L^2(]-\pi,\pi])$ with domain defined by periodic boundary conditions then

$$\partial_\# = F_\#^*\,\mathbf{im}\,F_\#$$

showing that applying $\partial_\#$ in $L^2(]-\pi,\pi])$ is unitarily equivalent to i times the multiplication by the argument $\mathbf{m} : \left(a_k\right)_{k\in\mathbb{Z}} \mapsto \left(k\,a_k\right)_{k\in\mathbb{Z}}$ in $\ell^2(\mathbb{Z})$. In other words, $F_\#$ is a spectral representation for $\frac{1}{\mathrm{i}}\partial_\#$. This property is used to turn the problem of solving a linear differential equation with constant coefficients $a_k \in \mathbb{C}$, $k \in \{0,\ldots,N\}$, $N \in \mathbb{N}$,

$$p\left(\partial_\#\right)u := \sum_{k=0}^{N} a_k\partial_\#^k u = f$$

under a periodic boundary condition on $[-\pi,\pi]$ into the question of calculating $F_\#^*\left((1/p(\mathrm{i}k))\hat{f}(k)\right)_{k\in\mathbb{Z}}$, which is always possible if the polynomial function $x \mapsto p(\mathrm{i}x)$ has no integer zeros.

The rescaling $(h > 0)$ or rescaling with reflection $(h < 0)$ of $]-\pi,\pi]$ to $]-|h|\,\pi,|h|\,\pi]$, $h \in \mathbb{R}\setminus\{0\}$, is the coordinate transformation

4) In as much as multiplication by a diagonal matrix can be interpreted as multiplication-by-argument operator if the argument ranges over the eigenvalues of the matrix A, (15.1) states that the unitary matrix U is a spectral representation of A.

$$\mathbb{R} \to \mathbb{R}$$
$$t \mapsto ht$$

inducing by (15.3) a unitary rescaling transformation σ_h given by

$$\sigma_h : L^2(\mathbb{R}) \to L^2(\mathbb{R}), \qquad (15.5)$$
$$f \mapsto \sqrt{|h|} f(h\cdot)$$

with $\sigma_{1/h} = \sigma_h^*$ as inverse. This provides a neat way of obtaining Fourier series expansion results for functions on different length intervals. For $L^2(]-|h|\pi, |h|\pi])$, the unitary Fourier series transformation is then[5]

$$F_\# \sigma_h : L^2(]-|h|\pi, |h|\pi]) \to \ell^2(\mathbb{Z}),$$
$$f \mapsto \left(\widehat{\sigma_h f}(k)\right)_{k\in\mathbb{Z}}$$

with

$$\begin{aligned}\widehat{\sigma_h f}(k) &= \left\langle \frac{1}{\sqrt{2\pi}} \exp(\mathrm{i}k\cdot) | \sigma_h f \right\rangle_{L^2(]-\pi,\pi])}, \\ &= \left\langle \sigma_h^* \frac{1}{\sqrt{2\pi}} \exp(\mathrm{i}k\cdot) | f \right\rangle_{L^2(]-|h|\pi,|h|\pi])}, \\ &= \left\langle \frac{1}{\sqrt{2\pi|h|}} \exp(\mathrm{i}k\cdot/h) | f \right\rangle_{L^2(]-|h|\pi,|h|\pi])}.\end{aligned}$$

The set $\{1/\sqrt{2\pi|h|}\exp(\mathrm{i}k\cdot/h) | k \in \mathbb{Z}\}$ is an orthonormal basis in $L^2(]-|h|\pi, |h|\pi])$, because unitary transformations map orthonormal bases to orthonormal bases. Conversely,

$$f(x) = \frac{1}{2\pi|h|} \sum_{k\in\mathbb{Z}} \alpha_k \exp\left(\frac{\mathrm{i}kx}{h}\right)$$

for almost all, where

$$\alpha_k = \left\langle \exp\left(\frac{\mathrm{i}k}{h}\cdot\right) | f \right\rangle_{L^2(]-|h|\pi,|h|\pi])}$$

for all $k \in \mathbb{Z}$" at the end of the sentence.

5) Compositions of unitary mappings are again unitary.

It may be interesting to note that by choosing $h = \frac{N+1}{\pi}$, $N \in \mathbb{N}$, and evaluating pointwise at integer values from $-N$ to $N+1$, we have that

$$\left\{ \left(\frac{1}{\sqrt{2(N+1)}} \exp\left(\frac{\pi \mathrm{i} sk}{N+1}\right) \right)_{s\in\{-N,\dots,N+1\}} \middle| k \in \{-N,\dots,N+1\} \right\}$$

is a complete orthonormal set in $\mathbb{C}^{2(N+1)}$. This follows from the fact that $\{\exp(\pi \mathrm{i}s/(N+1)) | s \in \{-N,\dots,N+1\}\}$ is the set of roots of unity, that is, the solution set of the equation

$$z^{2(N+1)} = 1. \qquad (15.6)$$

For more details we refer to [8].

Thus we obtain the so-called discrete Fourier transformation

$$\left(\frac{1}{\sqrt{2(N+1)}} \exp\left(\frac{\pi \mathrm{i}(jk)}{(N+1)}\right) \right)_{k,j\in\{-N,\dots,N+1\}}$$

as a matrix defining a unitary map in $\mathbb{C}^{2(N+1)}$.

Noting that

$$\frac{1}{\sqrt{2(N+1)}} \sum_{s=-N}^{N+1} \exp\left(\frac{-\pi \mathrm{i}ks}{(N+1)}\right) f_s \qquad (15.7)$$

may serve as a crude approximation of the kth Fourier coefficient of $\sigma_{(N+1)/\pi} f$ associated with the step function

$$f = \sum_{s=-N}^{N+1} f_s \chi_{](k-1),k]} \in L^2(]-(N+1),(N+1)]),$$

where $f_s \in \mathbb{C}$ for $s \in \{-N,\dots,N+1\}$, we see that the discrete Fourier transformation can be utilized for numerical purposes. Noting that (15.7) is actually a polynomial evaluated at $\exp(-\pi \mathrm{i}k/(N+1)) \in \mathbb{C}$, for numerical purposes one may apply efficient polynomial evaluation strategies, such as Horner's scheme (or rule), to obtain what is known as the *fast Fourier transformation* (FFT) [9].

15.2.3 The Fourier Series Transformation in $L^2(S_{\mathbb{C}}(0,1))$

By the complex coordinate transformation

$$\begin{aligned}]-\pi,\pi] &\to S_{\mathbb{C}}(0,1) \\ t &\mapsto \exp(\mathrm{i}t) \end{aligned}$$

of the interval $]-\pi,\pi]$ onto the unit circle $S_{\mathbb{C}}(0,1)$ in $\mathbb{C}$ centered at the origin, the Fourier series transformation may also be considered as acting on $L^2(S_{\mathbb{C}}(0,1))$. This induces the unitary transformation

$$\begin{aligned} \Pi_1 : L^2(S_{\mathbb{C}}(0,1)) &\to L^2(]-\pi,\pi]) \\ \psi &\mapsto \Pi_1\psi, \end{aligned}$$

where

$$(\Pi_1\psi)(t) := \psi(\exp(\mathrm{i}t)).$$

This is easily seen to be an isometry. Indeed, noting that for the line element $\mathrm{d}s(z)$ on $S_{\mathbb{C}}(0,1) = \{\exp(\mathrm{i}t) \mid t \in]-\pi,\pi]\}$, we have

$$\mathrm{d}s(\exp(\mathrm{i}t)) = \mathrm{d}t,$$

and it follows that

$$\begin{aligned} \int_{S_{\mathbb{C}}(0,1)} |\psi(z)|^2 \,\mathrm{d}s(z) &= \int_{-\pi}^{\pi} |\psi(\exp(\mathrm{i}t))|^2 \,\mathrm{d}t \\ &= \int_{-\pi}^{\pi} \left|(\Pi_1\psi)(t)\right|^2 \mathrm{d}t. \end{aligned}$$

By this transformation, it follows that $(z \mapsto 1/\sqrt{2\pi z^k})_{k\in\mathbb{Z}}$ is an orthonormal basis for $L^2(S_{\mathbb{C}}(0,1))$ and we have the unitary mapping

$$\begin{aligned} F_{\#}\Pi_1 : L^2(S_{\mathbb{C}}(0,1)) &\to \ell^2(\mathbb{Z}), \\ \psi &\mapsto \left(F_{\#}(\psi(\exp(\mathrm{i}\,\cdot\,)))(k)\right)_{k\in\mathbb{Z}}, \end{aligned}$$

as the corresponding Fourier series transformation. Thus for all $\psi \in L^2(S_{\mathbb{C}}(0,1))$,

$$\psi(z) = \frac{1}{\sqrt{2\pi}} \sum_{k=-\infty}^{\infty} \hat{\psi}(k)\, z^k$$

for almost all $z \in S_{\mathbb{C}}(0,1)$ where

$$\hat{\psi}(k) = \frac{1}{\sqrt{2\pi}} \int_{S_{\mathbb{C}}(0,1)} z^{-k}\psi(z)\,\mathrm{d}s(z).$$

This result is simply a restatement of the classical Fourier series transformation for periodic functions because such functions may be understood as functions defined on the unit circle.

15.3 The z-Transformation

Rather than taking the unit circle, we may repeat the discussion of the previous section on a circle $S_{\mathbb{C}}(0,r) \subseteq \mathbb{C}$ of radius $r \in \mathbb{R}_{>0}$ centered at the origin. Thus, we consider the transformation

$$\begin{aligned}]-\pi,\pi] &\to S_{\mathbb{C}}(0,r) \\ t &\mapsto r\exp(\mathrm{i}t) \end{aligned}$$

of the interval $]-\pi,\pi]$ onto the circle $S_{\mathbb{C}}(0,r)$. The induced unitary transformation is now

$$\begin{aligned} \Pi_r : L^2(S_{\mathbb{C}}(0,r)) &\to L^2(]-\pi,\pi]) \\ \psi &\mapsto \Pi_r\psi, \end{aligned}$$

where

$$(\Pi_r\psi)(t) := \sqrt{r}\psi(r\exp(\mathrm{i}t)).$$

This time, the line element $\mathrm{d}s(z)$ on the circle $S_{\mathbb{C}}(0,r) = \{r\exp(\mathrm{i}t) \mid t \in]-\pi,\pi]\}$ is

$$\mathrm{d}s\left(r\exp\left(\mathrm{i}t\right)\right) = r\,\mathrm{d}t,$$

and therefore

$$\int_{S_{\mathbb{C}}(0,r)} |\psi(z)|^2\,\mathrm{d}s(z) = \int_{-\pi}^{\pi} |\psi(r\exp(\mathrm{i}t))|^2\, r\,\mathrm{d}t = \int_{-\pi}^{\pi} \left|\Pi_r\psi(t)\right|^2\,\mathrm{d}t$$

showing that $\Pi_r : L^2(S_{\mathbb{C}}(0,r)) \to L^2(]-\pi,\pi])$ is indeed an isometry. It follows that $\left(z \mapsto 1/\sqrt{2\pi r}(z/r)^k\right)_{k\in\mathbb{Z}}$ forms an orthonormal basis of $L^2(S_{\mathbb{C}}(0,r))$ and we have the unitary mapping $F_\#\,\Pi_r$, that is,

$$F_\#\,\Pi_r : L^2\left(S_{\mathbb{C}}(0,r)\right) \to \ell^2(\mathbb{Z}),\quad \psi \mapsto \left(\widetilde{\psi}(k)\right)_{k\in\mathbb{Z}},$$

with $\widetilde{\psi}(k) := (F_\#(\sqrt{r}\psi(r\exp(\mathrm{i}\,\cdot\,))))(k)$ for $k\in\mathbb{Z}$ as the corresponding Fourier series transformation. The inverse transformation, that is, the reconstruction of ψ, is given by

$$\psi(z) = \frac{1}{\sqrt{2\pi r}}\sum_{k\in\mathbb{Z}} r^{-k}\widetilde{\psi}(k)\,z^k$$

for almost all $z\in S_{\mathbb{C}}(0,r)$. This latter formula gives rise to the so-called z-transformation. Consider an exponentially weighted ℓ^2-type space: $r^{-\mathbf{m}}[\ell^2(\mathbb{Z})] := \{(a_k)_{k\in\mathbb{Z}} \mid \sum_{k\in\mathbb{Z}} |a_k|^2 r^{2k} < \infty\}$. For $r\in\mathbb{R}_{>0}$, we define

$$Z_r : r^{-\mathbf{m}}\left[\ell^2(\mathbb{Z})\right] \to L^2\left(S_{\mathbb{C}}(0,r)\right),\quad \left(a_k\right)_{k\in\mathbb{Z}} \mapsto \frac{1}{\sqrt{2\pi r}}\sum_{k\in\mathbb{Z}} a_k\,z^k.$$

Thus we have a family of z-transformations[6] parameterized by $r\in\mathbb{R}_{>0}$. The image of $\left(a_k\right)_{k\in\mathbb{Z}}$ under such a z–transformation is frequently referred to as its generating function. In applications, it is often the case that we have $a_k = 0$ for $k<0$.

Depending on the sequence $\left(a_k\right)_{k\in\mathbb{Z}}$ it may happen that its generating function $\Phi := Z_r\left(\left(a_k\right)_{k\in\mathbb{Z}}\right)$ has an analytic extension to an annulus $R_{\mathbb{C}}\left(0,r_1,r_2\right) := B_{\mathbb{C}}\left(0,r_2\right)\setminus\overline{B_{\mathbb{C}}\left(0,r_1\right)}$, $r_2>r_1>0$. Then methods from the theory of analytic functions are available for further consideration of the properties of the sequence $\left(a_k\right)_{k\in\mathbb{Z}}$ and for comparison with other such sequences in overlapping annuli. Typical applications are to difference equations, which recursively define such sequences $\left(a_k\right)_{k\in\mathbb{Z}}$ (usually with $a_k=0$ for $k<0$). The sequence can then be identified as the coefficients in the Laurent expansion of an analytic function Φ (its generating function).

15.4 The Fourier–Laplace Transformation

The Fourier–Laplace transformation is a generalization of the Fourier–Plancherel transformation $F : L^2(\mathbb{R}) \to L^2(\mathbb{R})$ defined by (15.2) where the integral over $\mathbb{R}$ is understood as the L^2-limit of the integral over $[-n,n]$, the so-called "limit in the mean", see [2]. Given $\varrho\in\mathbb{R}$, define

$$H_{\varrho,0}(\mathbb{R}) := L^2\left(\mathbb{R},\ \exp(-2\varrho\mathbf{m})\,\mathrm{d}\lambda,\mathbb{C}\right)$$

where the inner product of $H_{\varrho,0}(\mathbb{R})$ is

$$\left(f,g\right) \mapsto \int_{\mathbb{R}} \overline{f(t)}\,g(t)\,\exp(-2\varrho t)\,\mathrm{d}t.$$

The Fourier–Laplace transformation is a family of unitary maps

$$L_\varrho := F\exp(-\varrho\mathbf{m}) : H_{\varrho,0}(\mathbb{R}) \to L^2(\mathbb{R}),$$

6) The z-transformation is usually considered without the normalizing factor $1/\sqrt{2\pi r}$. This factor is needed to produce a unitary z-transformation.

for $\rho \in \mathbb{R}$ that is,

$$\begin{aligned}
&\left(L_\varrho f\right)(x)\\
&= \frac{1}{\sqrt{2\pi}} \int_{\mathbb{R}} \exp\left(-\mathrm{i}xy\right) \exp\left(-\varrho y\right) f\left(y\right)\, \mathrm{d}y\\
&= \frac{1}{\sqrt{2\pi}} \int_{\mathbb{R}} \exp\left(-\left(\mathrm{i}x+\varrho\right) y\right) f\left(y\right)\, \mathrm{d}y\\
&= \frac{1}{\sqrt{2\pi}} \int_{\mathbb{R}} \exp\left(-\mathrm{i}\left(x-\mathrm{i}\varrho\right) y\right) f\left(y\right)\, \mathrm{d}y\\
&=: \hat{f}\left(x-\mathrm{i}\varrho\right),
\end{aligned}$$

where $\hat{f} = Ff$, the Fourier–Plancherel transform of f.

It is clear that $\exp(-\varrho\mathbf{m})$ defined by $(\exp(-\varrho\mathbf{m})\,\psi)(x) = \exp(-\varrho x)\,\psi(x)$ for almost all $x \in \mathbb{R}$, is a unitary mapping of $H_{\varrho,0}(\mathbb{R})$ onto $L^2(\mathbb{R})$ and so the Fourier–Laplace transformation is unitary because it is the composition of two unitary maps. A Fourier–Laplace transform is frequently interpreted as an element of the form $\hat{f}(-\mathrm{i}\,\cdot\,)$ in $L^2(\mathrm{i}[\mathbb{R}]+\varrho)$ given by

$$\begin{aligned}
&\mathrm{i}[\mathbb{R}]+\varrho \to \mathbb{C}\\
&\quad p \mapsto \frac{1}{\sqrt{2\pi}} \int_{\mathbb{R}} \exp\left(-py\right) f\left(y\right)\, \mathrm{d}y\\
&\qquad = \hat{f}\left(-\mathrm{i}p\right).
\end{aligned}$$

Here $L^2(\mathrm{i}[\mathbb{R}]+\varrho)$ is a Hilbert space with inner product

$$(f,g) \mapsto \int_{\mathrm{i}[\mathbb{R}]+\varrho} \overline{f(z)}\, g(z)\, \mathrm{d}s(z)$$

induced by the complex transformation

$$x \mapsto \mathrm{i}x + \varrho$$

yielding

$$\mathrm{d}s\left(\mathrm{i}x+\varrho\right) = \mathrm{d}x.$$

The importance of the Fourier–Laplace transformation lies in the property that it is a spectral representation associated with the operator

$$\partial_\varrho := \exp\left(\varrho\mathbf{m}\right)\left(\partial+\varrho\right)\exp\left(-\varrho\mathbf{m}\right)$$

representing differentiation in $H_{\varrho,0}(\mathbb{R})$, where ∂ represents ordinary differentiation in $L^2(\mathbb{R})$. Indeed, we have the unitary equivalence[7)]

$$\partial_\varrho = L_\varrho^*\, \left(\mathrm{i}\mathbf{m}+\varrho\right)\, L_\varrho. \tag{15.8}$$

Note that $H_{0,0}(\mathbb{R}) = L^2(\mathbb{R})$ and $L_0 = F$, the Fourier–Plancherel transformation. Observing that $L^2(]-\pi,\pi])$ may be considered as a subspace of $L^2(\mathbb{R})$ of (equivalence classes of) functions vanishing (almost everywhere) outside of $]-\pi,\pi]$, we have for the Fourier series transformation $F_\#$ discussed above that

$$\begin{aligned}
F_\# f &= \left(\left(Ff\right)(k)\right)_{k\in\mathbb{Z}}\\
&= \left(\left(L_0 f\right)(k)\right)_{k\in\mathbb{Z}}
\end{aligned}$$

for all $f \in L^2(]-\pi,\pi]) \subseteq L^2(\mathbb{R})$. As already noted for the Fourier transformation and in the context of the Fourier series transformation, there are also various, essentially equivalent, versions of the Fourier–Laplace transformation in use and in applying "standard results" from tables, care needs to be taken to ensure that formulae are correct for the version of the transformation in use. The Fourier–Laplace transformation variant[8)]

$$\tilde{L}_\varrho := \sigma_{2\pi} L_\varrho \tag{15.9}$$

7) This could be rephrased as saying that L_ϱ is a spectral representation for $(1/\mathrm{i})\left(\partial_\varrho - \varrho\right) = (1/\mathrm{i})\partial_\varrho + \mathrm{i}\varrho$.

8) In general, the Fourier–Laplace transformation interacts with rescaling, (15.5), in the following way

$$\sigma_L L_\varrho = L_{\varrho/L}\sigma_{1/L},\ L \in \mathbb{R}\setminus\{0\};$$

that is, smaller scale is turned into larger scale and vice versa.

has the advantage that neither $\tilde{L}_\varrho$ nor its inverse $\tilde{L}_\varrho^{-1} = \tilde{L}_\varrho^*$ have a numerical factor in front of the integral expression defining it. This results in simplifications of various formulae. In contrast, $\tilde{L}_\varrho$ yields a more complicated unitary equivalence to differentiation of the form

$$\partial_\varrho = \tilde{L}_\varrho^*(2\pi \mathrm{i}\mathbf{m} + \rho)\tilde{L}_\varrho,$$

which may not be desirable. The case $\varrho = 0$ leads to a variant of the Fourier transformation, $\mathcal{F} := \tilde{L}_0$ given by

$$(\mathcal{F}f)(x) = \int_{\mathbb{R}} \exp\left(-\mathrm{i}2\pi xy\right) f\left(y\right)\, \mathrm{d}y,$$

which is popular in particular areas of physics and electrical engineering. We shall, however, maintain the use of the Fourier–Laplace transformation version L_ϱ defined above and the special case of the Fourier–Plancherel transformation $F = L_0$.

15.4.1 Convolutions as Functions of ∂_ϱ

15.4.1.1 Functions of ∂_ϱ

A spectral representation allows us to straightforwardly define functions of an operator. Generalizing (15.8), we have for any polynomial P

$$P\left(\partial_\varrho\right) = L_\varrho^* P\left(\mathrm{i}\mathbf{m} + \varrho\right) L_\varrho.$$

This suggests the definition of more general functions of ∂_ϱ as

$$S(\partial_\varrho) := L_\varrho^* S(\mathrm{i}\,\mathbf{m} + \varrho) L_\varrho, \qquad (15.10)$$

where S can be fairly general (and even matrix-valued, in which case the transformation is carried out entry by entry). In the case $\varrho = 0$, functions of ∂ are known as filters in the context of signal processing. Owing to the unitary equivalence between ∂_ϱ and $\partial + \varrho$ via the unitary transformation $\exp\left(-\varrho\mathbf{m}\right) : H_{\varrho,0}\left(\mathbb{R}\right) \to L^2\left(\mathbb{R}\right)$, we obtain the unitary equivalence

$$S\left(\partial_\varrho\right) = \exp\left(\varrho\mathbf{m}\right)\; S\left(\partial + \varrho\right)\; \exp\left(-\varrho\mathbf{m}\right),$$

where

$$S\left(\partial + \varrho\right) = F^*\, S(\mathrm{i}\,\mathbf{m} + \varrho)\, F. \qquad (15.11)$$

This indicates that it would suffice to consider only the Fourier–Plancherel transformation, that is, the case $\varrho = 0$, taking the exponential weight $\exp\left(-\varrho\,\cdot\,\right)$ into account separately.

The concept of a function of ∂_ϱ is extremely powerful because many models discussed in so-called linear systems theory; see [10, 11], can be understood as such a function $S(\partial_\varrho)$ acting on an input u resulting in an output $S(\partial_\varrho)u$. Characteristic properties of such systems are that their input–output relations are linear and translation invariant, that is, sums, multiples, and shifts of input result in the same sums, multiples, and (up to scaling) shifts of output. Indeed, translation τ_h defined by

$$\tau_h f := f(\,\cdot\, + h)$$

is itself a function of ∂_ϱ in the above sense. Indeed, for suitable f we may calculate

$$\begin{aligned}
&\exp(h\,(\mathrm{i}x + \varrho))\,(L_\rho f)(x)\\
&= \frac{1}{\sqrt{2\pi}} \int_{\mathbb{R}} \exp\left(-\left(\mathrm{i}x + \varrho\right)\left(y - h\right)\right) f\left(y\right) \mathrm{d}y\\
&= \frac{1}{\sqrt{2\pi}} \int_{\mathbb{R}} \exp\left(-\left(\mathrm{i}x + \varrho\right) s\right) f\left(s + h\right)\, \mathrm{d}s\\
&= \left(L_\varrho \tau_h f\right)(x)
\end{aligned}$$

showing that

$$\tau_h = L_\varrho^{-1} \exp(h\,(\mathrm{i}\mathbf{m} + \varrho))\, L_\varrho = \exp(h\partial_\varrho).$$

It is interesting to note that by substituting the series expansion for the exponential function, this formula (if applied to a suitable function f) is easily recognized as nothing but the Taylor expansion of f about a point x.

As another example of functions of ∂_ϱ, we consider fractional calculus [12]. For $\varrho \in \mathbb{R}_{>0}$ we obtain ∂_ϱ^{-1} as forward causal integration. Indeed, for integrable functions f in $H_{\varrho,0}(\mathbb{R})$, we have

$$\left(\partial_\varrho^{-1} f\right)(t) = \int_{-\infty}^{t} f(s)\,\mathrm{d}s,\ t \in \mathbb{R}. \qquad (15.12)$$

Fractional powers of ∂_ϱ^{-1} can now be defined by

$$\partial_\varrho^{-\alpha} := L_\varrho^* (\mathrm{i}\mathbf{m} + \varrho)^{-\alpha} L_\varrho,\ \alpha \in [0, 1[\,,$$

and[9]

$$\partial_\varrho^{-\beta} := \partial_\varrho^{-\lfloor\beta\rfloor} \partial_\varrho^{\lfloor\beta\rfloor-\beta},\ \beta \in \mathbb{R}. \qquad (15.13)$$

Note that $\beta - \lfloor\beta\rfloor \in [0, 1[$ and $z \mapsto z^\alpha$ for $\alpha \in]-1, 0[$ is the principle root function, that is, with $z = |z| \exp\left(\mathrm{i}\arg(z)\right), \arg(z) \in]-\pi, \pi]$, we have

$$z^\alpha := |z|^\alpha \exp\left(\mathrm{i}\,\alpha \arg(z)\right).$$

For positive β, (15.13) is the fractional integral and, for negative β, equation (15.13) defines fractional differentiation. Here, $\varrho \in \mathbb{R}_{>0}$ is important because for $\varrho \in \mathbb{R}_{<0}$, we obtain the backward causal integral, and for $\varrho = 0$, the inverse of differentiation does not exist; there are the candidates corresponding to the forward or the backward causal situation both of which are unbounded linear operators.

In other cases, or following the observation (15.11), the choice $\varrho = 0$ is of particular interest. An example of a function of ∂ (i.e., in the case $\varrho = 0$) is the so-called Hilbert transformation

$$\begin{aligned}\mathrm{H} &:= F^* \frac{1}{\mathrm{i}} \operatorname{sgn}(\mathbf{m})\, F,\\ &= \frac{1}{\mathrm{i}} \operatorname{sgn}\left(\frac{1}{\mathrm{i}}\partial\right),\\ &= \mathrm{i} \operatorname{sgn}(\mathrm{i}\partial).\end{aligned}$$

We see that H is a unitary mapping in $L^2(\mathbb{R})$ because $(1/\mathrm{i})\operatorname{sgn}(\mathbf{m})$ is a multiplication operator on $L^2(\mathbb{R})$ with $\left|(1/\mathrm{i})\operatorname{sgn}(\mathbf{m})\right| = 1$ and

$$\mathrm{H}^2 = -1.$$

Consequently,

$$\mathrm{H}^* = \mathrm{H}^{-1} = -\mathrm{H}$$

and so H is skew-selfadjoint and unitary in $L^2(\mathbb{R})$.

15.4.1.2 Convolutions

Assuming that S allows for a Fourier–Laplace transformation by L_ϱ and $\left(\widehat{S}\right)\left((1/\mathrm{i})\partial_\varrho\right)$ is a function of ∂_ϱ in the above sense, we define

$$S * g := \sqrt{2\pi}\left(\widehat{S}\right)\left(\frac{1}{\mathrm{i}}\partial_\varrho\right) g, \qquad (15.14)$$

which is well defined for any g in the domain of $\sqrt{2\pi}\left(\widehat{S}\right)\left((1/\mathrm{i})\partial_\varrho\right)$. In many cases, $S * g$ is – at least for "well-behaved" functions $g \in H_{\varrho,0}(\mathbb{R})$ – actually given as an integral expression, which is then a so-called convolution integral:

9) Here $\lfloor\beta\rfloor$ denotes the floor function evaluated at $\beta \in \mathbb{R}$, which is the largest integer less than or equal to β, that is, $\lfloor\beta\rfloor := \sup\{k \in \mathbb{Z} \mid k \le \beta\}$. Note that $-\lfloor\beta\rfloor = \lceil-\beta\rceil$ for $\beta \in \mathbb{R}$, $\lceil\gamma\rceil := \inf\{k \in \mathbb{Z} \mid \gamma \le k\}$ for $\gamma \in \mathbb{R}$.

$$(S * g)(x) = \int_{-\infty}^{\infty} S(x-y)\,g(y)\,\mathrm{d}y \quad (15.15)$$

(hence the name convolution operator for $S * := \sqrt{2\pi}\left(\hat{S}\right)\left((1/\mathrm{i})\partial_\varrho\right)$). In accordance with (15.14), for such convolution integrals, the so-called convolution theorem holds

$$L_\varrho(S * g) = \sqrt{2\pi}\left(L_\varrho S\right)(\mathbf{m})\left(L_\varrho g\right) = \sqrt{2\pi} L_\varrho S \cdot L_\varrho g. \quad (15.16)$$

By the above definition, this always holds, even if there is no actual integral expression involved.

To illustrate this, let us first consider the fractional integral. For $\varrho \in \mathbb{R}_{>0}$, we obtain

$$\partial_\varrho^{-\alpha} f = \Phi * f,$$

where

$$\Phi(t) := \frac{1}{\Gamma(\alpha)} \chi_{\mathbb{R}_{>0}}(t)\ t^{\alpha-1}.$$

For sufficiently well-behaved functions $f \in H_{\varrho,0}(\mathbb{R})$ (so that the integral exists), this convolution (in the general sense) can actually be written as a convolution integral[10)]

$$\left(\partial_\varrho^{-\alpha} f\right)(t) = \frac{1}{\Gamma(\alpha)} \int_{\mathbb{R}} \chi_{\mathbb{R}_{>0}}(t-s)\ (t-s)^{\alpha-1} f(s)\ \mathrm{d}s = \frac{1}{\Gamma(\alpha)} \int_{-\infty}^{t} (t-s)^{\alpha-1} f(s)\ \mathrm{d}s,\ t \in \mathbb{R}.$$

In particular, the case $\alpha = 1/2$ is referred to as the *Abel transformation* which generates a unitary transformation. Indeed, the mapping

$$\partial_\varrho^{-1/2} : H_{\varrho,0}(\mathbb{R}) \to H_{\varrho,0}(\mathbb{R})$$

is clearly an isometry if the first space $H_{\varrho,0}(\mathbb{R})$ is equipped with the inner product

$$(f, g) \mapsto \left\langle \partial_\varrho^{-1/2} f \middle| \partial_\varrho^{-1/2} g \right\rangle_{\varrho,0}.$$

Denoting the smallest Hilbert space with this inner product containing $H_{\varrho,0}(\mathbb{R})$ by $H_{\varrho,-1/2}(\mathbb{R})$, the Abel transformation $A_\varrho : H_{\varrho,-1/2}(\mathbb{R}) \to H_{\varrho,0}(\mathbb{R})$, defined as the continuous extension of $\partial_\varrho^{-1/2}$, is unitary. As A_ϱ is essentially a fractional integral, it follows that $A_\varrho^* = A_\varrho^{-1}$ is related to the fractional derivative $\partial_\varrho^{1/2} = \partial_\varrho \partial_\varrho^{-1/2}$. The continuous extension of $\partial_\varrho^{1/2}$ to a mapping from $H_{\varrho,0}(\mathbb{R})$ to $H_{\varrho,-1/2}(\mathbb{R})$ corresponds to A_ϱ^*.

The integral representation of $A_\varrho f$, when $f = 0$ on $\mathbb{R}_{<0}$ and is suitably well behaved, is

$$\left(A_\varrho f\right)(t) = \sqrt{\frac{1}{\pi}} \chi_{\mathbb{R}_{>0}}(t) \int_0^t (t-s)^{-1/2} f(s)\ \mathrm{d}s,\ t \in \mathbb{R}.$$

Letting $r^{-2} = t$ and substituting $s = w^{-2}$ yields

$$r^{-1}\left(A_\varrho f\right)\left(r^{-2}\right) = 2\sqrt{\frac{1}{\pi}} r^{-1} \int_r^{\infty} \left(r^{-2} - w^{-2}\right)^{-1/2} f\left(w^{-2}\right)\ w^{-3}\,\mathrm{d}w = 2\sqrt{\frac{1}{\pi}} \int_r^{\infty} \left(w^2 - r^2\right)^{-1/2} \left(f\left(w^{-2}\right) w^{-3}\right)\ w\,\mathrm{d}w.$$

The function

$$r \mapsto 2\sqrt{\frac{1}{\pi}} \int_r^{\infty} \left(w^2 - r^2\right)^{-1/2} g(w)\ \ w\,\mathrm{d}w$$

is frequently referred to as the Abel transform of g. Introducing suitably weighted spaces this can (by the unitarity of A_ϱ) also be realized as a unitary transformation.

10) Frequently, there is an implicit assumption that f vanishes for negative arguments in which case we get the convolution integral expression

$$\frac{1}{\Gamma(\alpha)} \chi_{\mathbb{R}_{>0}}(t) \int_0^t (t-s)^{\alpha-1} f(s)\ \mathrm{d}s,\ t \in \mathbb{R}.$$

15.4.2 The Fourier–Plancherel Transformation

Focusing on the case $\varrho = 0$, we may consider the Fourier–Plancherel transformation as a unitary mapping in $L^2(\mathbb{R})$ in its own right. Every unitary operator has its spectrum on the unit circle $S_{\mathbb{C}}(0, 1)$. The spectrum of the Fourier–Plancherel transformation $L_0 = F$ is particularly simple. It is pure point spectrum and consists of the four points in $\{1, \mathrm{i}, -1, -\mathrm{i}\}$. The corresponding orthonormal basis of eigensolutions is given by[11)]

$$\left(\frac{2^{-k/2}}{\sqrt{k!}} (\mathbf{m} - \partial)^k \gamma\right)_{k \in \mathbb{N}},$$

where γ is the Gaussian distribution function defined by $\gamma(x) = \pi^{-1/4} \exp\left(-x^2/2\right)$, normalized such that $|\gamma|_{L^2(\mathbb{R})} = 1$. These have the form

$$\frac{2^{-k/2}}{\sqrt{k!}} (\mathbf{m} - \partial)^k \gamma = P_k(\mathbf{m}) \gamma,$$

where P_k is, up to a renormalization constant, the Hermite polynomial of degree k, $k \in \mathbb{N}$. Thus, the spectral representation takes on the simple form discussed in general in Section 15.2.1 and the Fourier–Plancherel transform $\hat{f} = Ff$ of $f \in L^2(\mathbb{R})$ takes the form

$$\sum_{k=0}^{\infty} \mathrm{e}^{\mathrm{i}k\pi/2} \left\langle \frac{(\mathbf{m} - \partial)^k}{2^k k!} \gamma \middle| f \right\rangle_{L^2(\mathbb{R})} (\mathbf{m} - \partial)^k \gamma.$$

11) The adjoint of the operator $1/\sqrt{2}(\mathbf{m} - \partial)$ is $1/\sqrt{2}(\mathbf{m} + \partial)$ and their product

$$\frac{1}{2}(\mathbf{m} + \partial)(\mathbf{m} - \partial) = \frac{1}{2}\left(\left(\mathbf{m}^2 - \partial^2\right) + 1\right),$$

which is the (quantum-mechanical) harmonic oscillator, has the same eigensolutions as the Fourier–Plancherel transformation with associated point spectrum $\mathbb{N}$.

Recalling that a spectral representation of an operator allows the definition of functions of the operator, we may define functions of the Fourier–Plancherel transformation F. A function S defined on the spectrum $\{1, \mathrm{i}, -1, -\mathrm{i}\}$ then gives rise to an operator $S(F)$ given by

$$S(F)f = \sum_{k=0}^{\infty} S\left(\mathrm{e}^{\mathrm{i}k\pi/2}\right) \left\langle \frac{(\mathbf{m} - \partial)^k}{2^k k!} \gamma \middle| f \right\rangle_{L^2(\mathbb{R})} (\mathbf{m} - \partial)^k \gamma.$$

For example, we may define fractional Fourier–Plancherel transformations F^α, $\alpha \in \mathbb{R}$, by

$$F^\alpha f := \sum_{k=0}^{\infty} \mathrm{e}^{\mathrm{i}k\alpha\pi/2} \left\langle \frac{(\mathbf{m} - \partial)^k}{2^k k!} \gamma \middle| f \right\rangle_{L^2(\mathbb{R})} (\mathbf{m} - \partial)^k \gamma,$$

which have in recent years found many applications in physics and engineering (see [13]).

15.5 The Fourier–Laplace Transformation and Distributions

We can extend the meaning of the $L^2(\mathbb{R})$-inner-product, which we now denote simply by $\langle \cdot | \cdot \rangle$, by introducing generalized functions f (also called *distributions*, see [14]) as linear functionals. Initially, for simplicity, consider the so-called space of "test functions," $C_0^\infty(\mathbb{R})$, of infinitely differentiable functions on $\mathbb{R}$ that have compact support, that is, they vanish outside a bounded interval, and let $f : C_0^\infty(\mathbb{R}) \to \mathbb{C}$, $\phi \mapsto f(\phi)$ be a linear functional. We introduce the "inner product" notation

$$\langle f \mid \phi \rangle := f(\phi).$$

Using this notation, the linearity of f is expressed by

$$\langle f \mid \phi + \alpha\,\psi \rangle = \langle f \mid \phi \rangle + \alpha\,\langle f \mid \psi \rangle$$

for all complex numbers α and all $\phi, \psi \in C_0^\infty \mathbb{R}$. Clearly, for any $f \in L^2\mathbb{R}$ the inner product $\langle f|\phi\rangle$ defines a generalized function (i.e., a complex-valued linear functional defined for $\phi \in C_0^\infty \mathbb{R}$), but for example $\exp(\mathrm{i}p\,\cdot\,)$, $p \in \mathbb{R}$, although not an element of $L^2\mathbb{R}$, also defines a generalized function via

$$\begin{aligned}\langle \exp(\mathrm{i}p\,\cdot\,)|\phi\rangle &:= \int_{\mathbb{R}} \overline{\exp(\mathrm{i}px)}\,\phi(x)\,\mathrm{d}x \\ &= \sqrt{2\pi}\hat{\phi}(p)\end{aligned}$$

for all $\phi \in C_0^\infty \mathbb{R}$. Further, a generalized function, which is *not* given by any classical function, is $\delta_{\{\omega\}}$ defined by

$$\delta_{\{\omega\}}(\phi) = \left\langle \delta_{\{\omega\}} \mid \phi \right\rangle := \phi(\omega) \text{ for all } \phi \in C_0^\infty \mathbb{R}.$$

This distribution samples, that is, evaluates, the test function ϕ at the point $\omega \in \mathbb{R}$. The special choice $\delta := \delta_{\{0\}}$ is the so-called Dirac-δ-distribution. As it can only be distinguished from 0 by testing with $\phi \in C_0^\infty \mathbb{R}$ satisfying $\phi(0) \neq 0$, we often say $\delta = 0$ on $\mathbb{R} \setminus \{0\}$. However, δ is (by definition) not identically zero as a functional. It is particularly useful in many contexts to have $\delta_{\{\omega\}}\ (= \tau_{-\omega}\delta)$ as a mathematical model for physical phenomena that are vanishingly small in space or time such as a mass point, an elementary charge, a light point, a short impulse, or a particle at location ω (see Section 15.9 for the particularly interesting higher-dimensional case).

An important class of distributions is the space of so-called tempered distributions S' (see [14]). With sufficient care, another space of distributions $\exp(\varrho\mathbf{m})\left[S'\right]$ (distributions which after multiplication by $\exp(-\varrho\mathbf{m})$, are in S') can be established for which the concept of derivative and Fourier–Laplace transformation can be generalized by defining, for $f \in \exp(\varrho\mathbf{m})\left[S'\right]$,

$$\begin{aligned}\langle \partial f\,|\phi\rangle &:= -\langle f\,|\partial\phi\rangle\,, \\ \left\langle L_\varrho f\,|\phi\right\rangle &:= \left\langle f \mid L_{-\varrho}^{*}\phi\right\rangle, \\ \left\langle L_\varrho^{*} f\,|\phi\right\rangle &:= \left\langle f \mid L_{-\varrho}\phi\right\rangle\end{aligned} \tag{15.17}$$

for all $\phi \in C_0^\infty \mathbb{R}$.

As an example of generalized differentiation, for the characteristic function $\chi_{]0,\infty[}$ of the interval $]0,\infty[$, we get

$$\partial\chi_{]0,\infty[} = \delta;$$

that is, the generalized derivative of the characteristic function of $]0,\infty[$ is the Dirac-δ-distribution.

The Fourier–Laplace transform of the distribution $\delta_{\{\omega\}}$ is – according to (15.17) –

$$L_\varrho\delta_{\{\omega\}} = \frac{1}{\sqrt{2\pi}}\exp\left(-\mathrm{i}\left(\,\cdot\, - \mathrm{i}\varrho\right)\omega\right). \tag{15.18}$$

Conversely, we have

$$L_\varrho^{*}\ \exp(-\mathrm{i}\omega\,\cdot\,) = \sqrt{2\pi}\exp(\varrho\omega)\,\delta_{\{\omega\}}. \tag{15.19}$$

The latter fact can be used to detect oscillatory behavior of frequency ω that may be hiding in seemingly random data. The oscillation would show up in the approximately Fourier–Laplace transformed data as a “peak” or “spike” at the point ω.

The concept of functions of ∂_ϱ and, in particular, the concept of convolution can also be carried over to exponentially weighted tempered distributions in $\exp(\varrho\mathbf{m})\left[S'\right]$. This will be illustrated by a few applications.

15.5.1 Impulse Response

The importance of the Dirac-δ-distribution is that a translation-invariant, linear system $S(\partial_\varrho)$ is completely described by its impulse response $S(\partial_\varrho)\delta$. The general response $S(\partial_\varrho)f$ is given by convolution with the impulse response

$$S(\partial_\varrho)f = (S(\partial_\varrho)\delta) * f.$$

In the case that $S(\partial_\varrho)$ is the solution operator $S(\partial_\varrho) = P(\partial_\varrho)^{-1}$ of the differential equation $P(\partial_\varrho)u = f$, where P is a polynomial, the impulse response is also known as the *fundamental solution* (or *Green's function*), cf. [14].

15.5.2 Shannon's Sampling Theorem

Another distribution of particular interest is the (sampling or) comb distribution $\text{III} := \sum_{k\in\mathbb{Z}} \delta_{\{\sqrt{2\pi}k\}}$, which takes samples at equidistant points (distance $\sqrt{2\pi}$). This distribution is reproduced by the Fourier–Plancherel transformation

$$F\text{III} = \text{III}, \tag{15.20}$$

which is a way of stating the Poisson summation formula:

$$F\sum_{k\in\mathbb{Z}} \delta_{\{\sqrt{2\pi}k\}} = \frac{1}{\sqrt{2\pi}}\sum_{k=-\infty}^{+\infty} \exp\left(-\mathrm{i}\sqrt{2\pi}k\,\cdot\right) = \sum_{k=-\infty}^{+\infty} \delta_{\{\sqrt{2\pi}k\}}.$$

Note that for the first equality we have used (15.18).

If $f \in L^2(\mathbb{R})$ and $f = 0$ outside of $[-\sqrt{\pi/2}, \sqrt{\pi/2}]$, then $\text{III} * f$ is simply the periodic extension of f from the interval $[-\sqrt{\pi/2}, \sqrt{\pi/2}]$, to all of $\mathbb{R}$. Therefore, by "cutting-off" with the characteristic function $\Pi := \chi_{[-\sqrt{\pi/2},\sqrt{\pi/2}]}$, we recover

$$f = \Pi(\text{III} * f). \tag{15.21}$$

As we shall see, the so-called Shannon sampling theorem follows from (15.20). In fact, applying (15.21) to $\hat{f} \in L^2(\mathbb{R})$ with $\hat{f} = Ff = 0$ outside of $[-\sqrt{\pi/2}, \sqrt{\pi/2}]$ and applying the inverse Fourier transformation we get

$$\begin{aligned} f &= F^*(\Pi(\text{III} * \hat{f})),\\ &= \text{sinc}\left(\sqrt{\pi/2}\,\cdot\right) * \left(\sum_{n=-\infty}^{+\infty} f(\sqrt{2\pi}n)\,\delta_{\{\sqrt{2\pi}n\}}\right),\\ &= \sum_{n=-\infty}^{+\infty} f(\sqrt{2\pi}n)\,\text{sinc}\left(\sqrt{\pi/2}\left(\,\cdot\, - \sqrt{2\pi}n\right)\right), \end{aligned} \tag{15.22}$$

which is Shannon's theorem. Here we have used the fact that $(F^*\Pi)(t) = \text{sinc}(\sqrt{\pi/2}t)$ and $\text{sinc}(t) := \sin(t)/t$.

Any function f with $\hat{f}$ vanishing outside a bounded interval is called *band limited.* As any such function can be easily rescaled to have $\hat{f} = 0$ outside of $[-\sqrt{\pi/2}, \sqrt{\pi/2}]$, (15.22) shows that any band-limited function f can be completely recovered from equidistant sampling.

By interchanging the role of f and $\hat{f}$ in the Shannon sampling theorem, we have

$$\hat{f} = \text{sinc}\left(\sqrt{\frac{\pi}{2}}\,\cdot\right) * \left(\sum_{n=-\infty}^{+\infty} \hat{f}(\sqrt{2\pi}n)\,\delta_{\{\sqrt{2\pi}n\}}\right) \tag{15.23}$$

and after applying the inverse Fourier transformation we obtain a scaled variant of the Fourier series expansion

$$f = \sum_{n=-\infty}^{+\infty} \hat{f}(\sqrt{2\pi}n)\,\Pi\,\exp\left(\mathrm{i}\sqrt{2\pi}n\,\cdot\right) \tag{15.24}$$

of a function $f \in L^2(\mathbb{R})$ with $f = 0$ outside of $[-\sqrt{\pi/2}, \sqrt{\pi/2}]$. Note that

$\{(2\pi)^{-1/4}\Pi \exp(\mathrm{i}\sqrt{2\pi}n\cdot) \mid n \in \mathbb{Z}\}$ is an orthonormal set in $L^2(\mathbb{R})$, that is, all elements are normalized and pairwise orthogonal, which allows a Fourier series expansion in the sense of (15.24) for every $f \in L^2(\mathbb{R})$ with $f = 0$ outside of $[-\sqrt{\pi/2}, \sqrt{\pi/2}]$.

As the Fourier–Plancherel transformation is unitary, it follows that the Fourier–Plancherel transform of the orthonormal basis

$$\left\{(2\pi)^{-1/4}\,\Pi \exp(\mathrm{i}\sqrt{2\pi}n\cdot) \mid n \in \mathbb{Z}\right\}$$

of $L^2\left(]-\sqrt{\pi/2}, \sqrt{\pi/2}]\right)$ is also an orthonormal basis. Thus,

$$\left\{(2\pi)^{-1/4}\,\mathrm{sinc}\left(\sqrt{\pi/2}\left(\cdot - \sqrt{2\pi}n\right)\right)\middle| n \in \mathbb{Z}\right\}$$

is also an orthonormal set and indeed an orthonormal basis for all band-limited functions in $L^2(\mathbb{R})$ with band in $[-\sqrt{\pi/2}, \sqrt{\pi/2}]$.

15.6 The Fourier-Sine and Fourier-Cosine Transformations

As a unitary operator, the Fourier–Plancherel transformation F is also normal with real part $F_{\cos} := \mathfrak{Re}\, F = (1/2)(F + F^*)$, the Fourier cosine transformation and negative imaginary part $F_{\sin} := -\mathfrak{Im}\, F = -(1/2\mathrm{i})(F - F^*)$, the Fourier sine transformation. Being real and imaginary parts of a normal operator, the Fourier cosine and the Fourier sine transformations are selfadjoint and therefore their spectra $\sigma(F_{\cos})$ and $\sigma(F_{\sin})$ are real. We already know that

$$\mathfrak{Re}\,\sigma(F) = \{+1, -1\} \subseteq \sigma(F_{\cos}),$$
$$\mathfrak{Im}\,\sigma(F) = \{+1, -1\} \subseteq \sigma(F_{\sin}).$$

The relation between $F_{\cos}$, $F_{\sin}$, and F yields further insight. We have (with σ_{-1} as the reflection at the origin)

$$\begin{aligned} F_{\cos} &= \frac{1}{2}(F + F^*),\\ &= \frac{1}{2}(1 + \sigma_{-1})\,F,\\ &= F\,\frac{1}{2}(1 + \sigma_{-1}),\\ &= \frac{1}{2}(1 + \sigma_{-1})\,F^*,\\ &= F^*\,\frac{1}{2}(1 + \sigma_{-1});\\ F_{\sin} &= -\frac{1}{2\mathrm{i}}(F - F^*),\\ &= \frac{1}{2}(1 - \sigma_{-1})\,\mathrm{i}F,\\ &= \mathrm{i}F\,\frac{1}{2}(1 - \sigma_{-1}). \end{aligned} \tag{15.25}$$

It is not hard to see that $1/2(1 + \sigma_{-1})$ and $(1/2)(1 - \sigma_{-1}) = 1 - \frac{1}{2}(1 + \sigma_{-1})$ are the orthogonal projections onto the (almost everywhere) even and odd functions in $L^2(\mathbb{R})$, respectively. Therefore, we also have $0 \in \sigma(F_{\cos})$ and $0 \in \sigma(F_{\sin})$. Moreover, we calculate with (15.25)

$$F_{\sin}F_{\cos} = F_{\cos}F_{\sin} = 0, \tag{15.26}$$
$$F_{\sin}F_{\sin} = \frac{1}{2}(1 - \sigma_{-1}), \tag{15.27}$$
$$F_{\cos}F_{\cos} = \frac{1}{2}(1 + \sigma_{-1}). \tag{15.28}$$

Thus, we see that $F_{\cos}$, $F_{\sin}$ are unitary on the subspaces $\frac{1}{2}(1 \pm \sigma_{-1})\left[L^2(\mathbb{R})\right]$, respectively. This, along with the selfadjointness of $F_{\cos}$ and $F_{\sin}$, allows us to deduce that

$$\begin{aligned} \sigma(F_{\cos}) = \sigma(F_{\sin}) &= P\sigma(F_{\cos})\\ &= P\sigma(F_{\sin}) = \{0, +1, -1\}. \end{aligned}$$

We can identify $\frac{1}{2}(1 \pm \sigma_{-1})\left[L^2(\mathbb{R})\right]$ with $L^2(\mathbb{R}_{>0})$ via the unitary transformations $E_\pm : L^2(\mathbb{R}_{>0}) \to \frac{1}{2}(1 \pm \sigma_{-1})\left[L^2(\mathbb{R})\right]$

defined by

$$(E_\pm f)(x) := \frac{1}{\sqrt{2}} \begin{cases} f(x) & \text{for } x \in \mathbb{R}_{>0} \\ \pm f(-x) & \text{for } x \in \mathbb{R}_{\leq 0}, \end{cases}$$

so that, for $f \in L^2(\mathbb{R}_{>0})$, $\sqrt{2}E_+ f$ is the even extension of f and $\sqrt{2}E_- f$ is the odd extension of f. The calculation

$$\begin{aligned} |E_\pm \varphi|^2_{L^2(\mathbb{R})} &= \int_{\mathbb{R}} |(E_\pm \varphi)(x)|^2 \,\mathrm{d}x, \\ &= \frac{1}{2} \int_{\mathbb{R}_{>0}} |\varphi(x)|^2 \,\mathrm{d}x \\ &\quad + \frac{1}{2} \int_{\mathbb{R}_{<0}} |\varphi(-x)|^2 \,\mathrm{d}x, \\ &= \int_{\mathbb{R}_{>0}} |\varphi(x)|^2 \,\mathrm{d}x = |\varphi|^2_{L^2(\mathbb{R}_{>0})}, \end{aligned}$$

for all $\varphi \in L^2(\mathbb{R}_{>0})$ shows that $E_\pm$ are unitary and so the mappings

$$\begin{aligned} E_+^* F_{\cos} E_+ &: L^2(\mathbb{R}_{>0}) \to L^2(\mathbb{R}_{>0}), \\ E_-^* F_{\sin} E_- &: L^2(\mathbb{R}_{>0}) \to L^2(\mathbb{R}_{>0}) \end{aligned}$$

are also unitary. Here, $E_\pm^* = E_\pm^{-1}$ are defined by

$$\begin{aligned} E_\pm^* : \frac{1}{2}(1 \pm \sigma_{-1}) \left[L^2(\mathbb{R})\right] &\to L^2(\mathbb{R}_{>0}), \\ \phi &\mapsto \sqrt{2}\,\phi|_{\mathbb{R}_{>0}}. \end{aligned}$$

For "nice" functions, for example, for $\phi \in C_0^\infty(\mathbb{R}_{>0})$, the space of smooth functions vanishing outside of a compact set in $\mathbb{R}_{>0}$, we get the following well-known integral representations discussed in Section 15.1:

$$\begin{aligned} (\tilde{F}_{\cos}\phi)(x) &:= (E_+^* F_{\cos} E_+ \phi)(x) \\ &= \sqrt{\frac{2}{\pi}} \int_{\mathbb{R}_{>0}} \cos(xy)\,\phi(y)\,\mathrm{d}y, \\ (\tilde{F}_{\sin}\phi)(x) &:= (E_-^* F_{\sin} E_- \phi)(x) \\ &= \sqrt{\frac{2}{\pi}} \int_{\mathbb{R}_{>0}} \sin(xy)\,\phi(y)\,\mathrm{d}y. \end{aligned}$$

The unitary transformations $\tilde{F}_{\cos}$ and $\tilde{F}_{\sin}$ are spectral representations for $|\partial|$ with no or Dirichlet type boundary condition, respectively, at the origin. As noted in Section 15.1, they can be used to discuss problems involving the second-order differential operator $-\partial^2 = |\partial|^2$ with Neumann or Dirichlet boundary conditions at the origin, respectively.

15.7 The Hartley Transformations $H_\pm$

As the Fourier transformation F is unitary, it follows immediately that $\exp(\mathrm{i}\,\pi/4)\,F$ is also unitary. Considering the real and imaginary part of $\exp(\mathrm{i}\,\pi/4)\,F$ leads to an interesting situation:

$$\begin{aligned} \mathfrak{Re}\,(\exp(\mathrm{i}\,\pi/4)\,F) &= \frac{1}{\sqrt{2}}(F_{\cos} + F_{\sin}), \\ \mathfrak{Im}\,(\exp(\mathrm{i}\,\pi/4)\,F) &= \frac{1}{\sqrt{2}}(F_{\cos} - F_{\sin}), \end{aligned}$$

and therefore, for $H_\pm := F_{\cos} \pm F_{\sin}$, we have

$$\begin{aligned} H_+ &= \mathfrak{Re}\left(\sqrt{2}\,\exp(\mathrm{i}\,\pi/4)\,F\right), \\ H_- &= \mathfrak{Im}\left(\sqrt{2}\,\exp(\mathrm{i}\,\pi/4)\,F\right) \end{aligned}$$

and so for all $\varphi \in L^2(\mathbb{R})$,

$$\begin{aligned} |H_\pm \varphi|^2_{L^2(\mathbb{R})} &= |(F_{\cos} \pm F_{\sin})\varphi|^2_{L^2(\mathbb{R})} \\ &= |F_{\cos}\varphi|^2_{L^2(\mathbb{R})} \\ &\quad \pm 2\mathfrak{Re}\,\langle F_{\cos}\varphi | F_{\sin}\varphi \rangle_{L^2(\mathbb{R})} \\ &\quad + |F_{\sin}\varphi|^2_{L^2(\mathbb{R})} \\ &= |F_{\cos}\varphi|^2_{L^2(\mathbb{R})} + |F_{\sin}\varphi|^2_{L^2(\mathbb{R})} \\ &= |F\varphi| = |\varphi|^2_{L^2(\mathbb{R})}. \end{aligned}$$

The transformations $H_\pm$, which are unitary and selfadjoint, are known as *Hartley transformations*. In particular,

$$\sigma(H_{\pm}) = P\sigma(H_{\pm}) = \{+1, -1\}$$

and

$$\begin{aligned}
H_+H_- &= H_-H_+\\
&= F_{\cos}F_{\cos} - F_{\sin}F_{\sin}\\
&= \frac{1}{2}(1+\sigma_{-1}) - \frac{1}{2}(1-\sigma_{-1})\\
&= \sigma_{-1},\\
H_+H_+ &= H_-H_-\\
&= F_{\cos}F_{\cos} + F_{\sin}F_{\sin}\\
&= \frac{1}{2}(1+\sigma_{-1}) + \frac{1}{2}(1-\sigma_{-1})\\
&= 1.
\end{aligned}$$

On $C_0^\infty(\mathbb{R})$, we have

$$\left(H_{\pm}\varphi\right)(\omega) = \frac{1}{\sqrt{2\pi}}\int_{\mathbb{R}}(\cos(\omega t) \pm \sin(\omega t))\,\varphi(t)\,\mathrm{d}t$$

and

$$\left(H_-\varphi\right)(\omega) = \left(H_+\varphi\right)(-\omega)$$

for $\omega \in \mathbb{R}$ or

$$H_- = \sigma_{-1}H_+.$$

As the Fourier–Plancherel transformation has an extension to tempered distributions, so have the Hartley transformations.

The Hartley transformations are real, that is, they commute with complex conjugation, which may be an advantage in applications; see [15] for more details. On the downside, the Hartley transformations $H_{\pm}$ are spectral representations for the rarely used operators $\pm\sigma_{-1}\mathrm{i}\partial$. Hidden periodic behavior is revealed in a more complicated fashion:

$$\begin{aligned}
H_{\pm}\cos(\omega\,\cdot\,) &= \sqrt{\frac{\pi}{2}}\left(\delta_{\{\omega\}} + \delta_{\{-\omega\}}\right),\\
H_{\pm}\sin(\omega\,\cdot\,) &= \pm\sqrt{\frac{\pi}{2}}\left(\delta_{\{\omega\}} - \delta_{\{-\omega\}}\right),
\end{aligned}$$

indicating that period $2\pi/\omega$ behavior in a function will show up as spikes at $\pm\omega$ in the Hartley transforms of the function.

15.8 The Mellin Transformation

Let us consider the transformation

$$\begin{aligned}
\ln : \mathbb{R}_{>0} &\to \mathbb{R}\\
x &\mapsto \ln(x).
\end{aligned}$$

Analogous to the reasoning leading up to (15.3), this transformation induces a unitary mapping

$$\begin{aligned}
\Lambda : L^2(\mathbb{R}) &\to \sqrt{\mathbf{m}}\left[L^2\left(\mathbb{R}_{>0}\right)\right]\\
&:= L^2\left(\mathbb{R}_{>0}, \mathbf{m}^{-1}\mathrm{d}\lambda, \mathbb{C}\right)\\
f &\mapsto f\circ\ln
\end{aligned}$$

with

$$\begin{aligned}
\Lambda^* = \Lambda^{-1} : \sqrt{\mathbf{m}}\left[L^2\left(\mathbb{R}_{>0}\right)\right] &\to L^2(\mathbb{R})\\
f &\mapsto f\circ\exp.
\end{aligned}$$

Unitarity is easily confirmed by the computation done in Section 15.1. The unitary mapping $M := F\Lambda^* : \sqrt{\mathbf{m}}\left[L^2\left(\mathbb{R}_{>0}\right)\right] \to L^2(\mathbb{R})$ is known as the *Mellin transformation*. We find the unitary equivalence

$$\begin{aligned}
\mathbf{m}\partial &= \Lambda\partial\Lambda^*\\
&= \Lambda F^*\mathrm{i}\mathbf{m}\,F\Lambda^*
\end{aligned}$$

and so that M is a spectral representation for $(1/\mathrm{i})\mathbf{m}\partial$. For well-behaved (to ensure that the integrals exist) $f \in \sqrt{\mathbf{m}}\left[L^2\left(\mathbb{R}_{>0}\right)\right]$ and $g \in L^2(\mathbb{R})$, we get

$$\begin{aligned}
(Mf)(x) &= \frac{1}{\sqrt{2\pi}}\int_{\mathbb{R}}\exp(-\mathrm{i}xy)f(\exp(y))\,\mathrm{d}y\\
&= \frac{1}{\sqrt{2\pi}}\int_{\mathbb{R}_{>0}} s^{-\mathrm{i}x-1}f(s)\,\mathrm{d}s,\ x\in\mathbb{R},
\end{aligned}$$

and

$$\left(M^{*}g\right)(t)=\frac{1}{\sqrt{2\pi}}\int_{\mathbb{R}}\exp\left(\mathrm{i}\,\ln(t)\,y\right)g\left(y\right)\mathrm{d}y$$
$$=\frac{1}{\sqrt{2\pi}}\int_{\mathbb{R}}t^{\mathrm{i}y}g\left(y\right)\,\mathrm{d}y,\ t\in\mathbb{R}_{>0}.$$

15.9 Higher-Dimensional Transformations

Following the rationale that most of the classical integral transformations are spectral representations of differential operators or functions of such (see [16]), to identify applicable higher-dimensional transformations, we should be looking for spectral representations associated with partial differential operators and investigating the corresponding operator function calculus associated with such a spectral representation. For many applications, however, there is a somewhat easier approach to higher-dimensional transformations, which is based on tensor product structures (see [2]), implied by separation-of-variables arguments.

The idea is roughly as follows. A tensor product $X_1\otimes\cdots\otimes X_n$, $n\in\mathbb{N}$, of complex function spaces X_k, $k\in\{1,\ldots,n\}$, can be considered to be the completion with respect to an appropriate metric of the space of all linear combinations of functions in product (separable) form $u_1\otimes\cdots\otimes u_n$ given by

$$\left(x_1,\ldots,x_n\right)\mapsto\prod_{k=0}^{n}u_k\left(x_k\right)$$

with $u_k\in X_k$, $k\in\{1,\ldots,n\}$. If the spaces X_k, $k\in\{1,\ldots,n\}$, are Hilbert spaces then so is $X_1\otimes\cdots\otimes X_n$ with the inner product $\langle u_1\otimes\cdots\otimes u_n|v_1\otimes\cdots\otimes v_n\rangle_{X_1\otimes\cdots\otimes X_n}$ of two such functions $u_1\otimes\cdots\otimes u_n, v_1\otimes\cdots\otimes v_n\in X_1\otimes\cdots\otimes X_n$ of product form given by

$$\prod_{k=1}^{n}\left\langle u_k|v_k\right\rangle_{X_k},$$

that is, the product of the inner products. Densely defined closed linear operators $A_k:D\left(A_k\right)\subseteq X_k\to Y_k$, where Y_k, $k\in\{1,\ldots,n\}$, are also Hilbert spaces can now be combined to give an operator denoted by $A_1\otimes\cdots\otimes A_n$ between tensor product spaces $X_1\otimes\cdots\otimes X_n$ and $Y_1\otimes\cdots\otimes Y_n$ defined (see [2]) as the closed linear extension of the mapping defined on separable functions by

$$u_1\otimes\cdots\otimes u_n\mapsto A_1u_1\otimes\cdots\otimes A_nu_n.$$

Applying this for example to n copies of the Fourier–Laplace transformation yields the n-dimensional Fourier–Laplace transformation. Indeed, because

$$\exp\left(\left(z_1,\ldots,z_n\right)\left(w_1,\ldots,w_n\right)\right)$$
$$:=\exp\left(\sum_{k=1}^{n}z_kw_k\right)$$
$$=\exp\left(z_1w_1\right)\cdots\exp\left(z_nw_n\right),$$

for $\nu=\left(\nu_1,\ldots,\nu_n\right)\in\mathbb{R}^n$ and $H_{\nu,0}\left(\mathbb{R}^n\right):=\left\{f\,\middle|\exp\left(-\nu\mathbf{m}\right)f\in L^2\left(\mathbb{R}^n\right)\right\}=\bigotimes_{k=1}^{n}H_{\nu_k,0}\left(\mathbb{R}\right)$, we have that the n-dimensional Fourier–Laplace transformation, $n\in\mathbb{N}_{\geq 2}$, can be understood as a repeated one-dimensional Fourier transformation, that is, with $\mathbf{m}=\left(\mathbf{m}_1,\ldots,\mathbf{m}_n\right)$ as the n-tuple of multiplication by the respective argument

$$(L_\nu f)(p)$$
$$=\frac{1}{(2\pi)^{n/2}}\int_{-\infty}^{\infty}\exp(-\mathrm{i}\,x_1p_1)\cdots$$
$$\cdots\int_{-\infty}^{\infty}\exp(-\mathrm{i}\,x_np_n)\exp\left(-\nu\mathbf{m}\right)f(x)\,\mathrm{d}x_n$$
$$\cdots\mathrm{d}x_2\,\mathrm{d}x_1,$$

for well-behaved $f\in H_{\nu,0}\left(\mathbb{R}^n\right)$; in other terms

$$L_\nu=L_{\nu_1}\otimes\cdots\otimes L_{\nu_n}.$$

As tensor products of unitary operators are unitary, it follows that $L_\nu : H_{\nu,0}(\mathbb{R}^n) \to L^2(\mathbb{R}^n)$ is a unitary transformation. As a consequence, many of the above considerations, such as the extension to distribution spaces, can be carried over immediately to the higher-dimensional case.

For example, the n-dimensional Dirac-δ-distribution $\delta = \delta_{\{0\}}$ simply takes a single sample at the origin in $\mathbb{R}^n$ and similarly the shifted version $\delta_{\{x\}}$ samples at $x \in \mathbb{R}^n$. The latter has now $1/(2\pi)^{n/2}\exp(-\mathrm{i}x\cdot)$ as its Fourier–Plancherel transform, that is, the Fourier–Laplace transform for $\nu = (0,\ldots,0) \in \mathbb{R}^n$ (compare [2]). Also, $\text{III} := \sum_{x\in\mathbb{Z}^n} \delta_{\{\sqrt{2\pi}x\}}$, which takes samples at every point of $\mathbb{R}^n$ with coordinates in $\sqrt{2\pi}[\mathbb{Z}]$, still satisfies

$$F\text{III} = \text{III}. \tag{15.29}$$

Thus, in particular, corresponding variants of the Shannon sampling theorem and of the Fourier series expansion hold. The higher-dimensional Fourier–Plancherel transformation finds its applications in higher-dimensional, translation-invariant, linear systems, [11]. A particularly prominent application is in optics (see, e.g., [10, 17, 18].

In fact, the solution theory of general linear partial differential equations and systems with constant coefficients can be based on higher-dimensional Fourier–Laplace transformation strategies, see [14].

The scaling behavior of the Fourier–Plancherel transformation finds its generalization in the following. Let $(\sigma_A f)(x) := \sqrt{|\det(A)|} f(Ax)$, $x \in \mathbb{R}^n$, for a non-singular real $(n\times n)$-matrix A. Then the generalization of the one-dimensional rescaling property, see footnote 8, is

$$\sigma_{(A^{-1})^*}F = F\sigma_A.$$

If A is orthogonal, that is, $A^* = A^{-1}$, then σ_A and F commute. This implies that rotational symmetries are preserved by the Fourier–Plancherel transformation, a fact that is frequently used in applications.

15.10 Some Other Important Transformations

In this final section, we consider a few other mathematical transformations that have found important application in physics or engineering.

15.10.1 The Hadamard Transformation

The Hadamard transformation, also called the Hadamard–Rademacher–Walsh, the Hadamard–Walsh, or the Walsh transformation, is an example of a higher-dimensional discrete transformation defined for functions in $\mathbb{C}^{\{0,1\}^n} := \{f | f : \{0,1\}^n \to \mathbb{C}\}$ as a mapping $W : \mathbb{C}^{\{0,1\}^n} \to \mathbb{C}^{\{0,1\}^n}$ given by[12)]

$$W\varphi(x) := \sum_{k\in\{0,1\}^n} \varphi(k)(-1)^{\langle k|x\rangle_{\{0,1\}^n}}.$$

The Hadamard transformation is used in the analysis of digital devices.

If we choose a particular enumeration $e : \{0,\ldots 2^n - 1\} \to \{0,1\}^n$ of $\{0,1\}^n$, that is, e is bijective (one-to-one and onto), then, using the standard inner products for complex-valued mappings defined on finite sets, we get that

$$E : \mathbb{C}^{\{0,1\}^n} \to \mathbb{C}^{2^n}$$
$$\varphi \mapsto \varphi\circ e$$

12) Compare higher-dimensional Fourier series transformation (and the above discussion of the discrete Fourier transformation at the end of Section 15.2.2 for the one-dimensional case).

is a unitary map. Thus, EWE^* yields a matrix representation in terms of so-called Hadamard matrices[13] $\left((-1)^{\langle e(s)|e(t)\rangle_{\{0,1\}^n}}\right)_{s,t\in\{0,\ldots,2^n-1\}}$.

15.10.2 The Hankel Transformation

The (one-dimensional) Hankel transformation H_0 is defined, for suitable functions f on $]0,\infty[$, by

$$(\mathrm{H}_0 f)(s) = \int_0^\infty r f(r) J_0(sr)\,\mathrm{d}r,$$

where J_0 denotes the Bessel function of the first kind $x \mapsto \sum_{r=0}^{\infty}(-1)^r \left(\frac{x}{2}\right)^{2r}/(r!)^2$.

Let $\phi\colon \mathbb{R}^2 \to \mathbb{C}$ be circularly symmetric, which means that the function $(r,\theta) \mapsto \phi(r\cos(\theta), r\sin(\theta))$, obtained by a polar coordinate transformation of ϕ, is independent of θ, that is,

$$\phi(r\cos(\theta), r\sin(\theta)) = f(r) := \phi(r,0)$$

for every $\theta \in \mathbb{R}$. The Hankel transform of f can be found by using polar coordinates to evaluate $\hat{\phi}$, the two-dimensional Fourier–Plancherel transform of ϕ:

$$\hat{\phi}(s,0) = \int_0^\infty r f(r) J_0(sr)\,\mathrm{d}r.$$

Now, due to the circular-symmetry-preserving property of the Fourier–Plancherel transformation, we have $\hat{\phi}(x) = \hat{\phi}(s,0)$ for all $x \in \mathbb{R}^2$ with $|x| = |s|$. Thus, the Hankel transformation H_0 is essentially the two-dimensional Fourier–Plancherel transformation restricted to the circularly symmetric function.

To find the precise spaces between which H_0 is unitary, we need to be more detailed. As circularly symmetric functions are uniquely determined by their values on $\mathbb{R}_{>0} \times \{0\}$, we find, using polar coordinates,

$$\begin{aligned}
&\langle \phi|\psi\rangle_{L^2(\mathbb{R}^2)} \\
&= \int_{\mathbb{R}^2} \overline{\phi(x)}\,\psi(x)\,\mathrm{d}x \\
&= 2\pi \int_{\mathbb{R}_{>0}} \overline{\phi(r,0)}\,\psi(r,0)\,r\,\mathrm{d}r \\
&= \left\langle \sqrt{2\pi\mathbf{m}}\phi(\cdot,0)\,\middle|\,\sqrt{2\pi\mathbf{m}}\psi(\cdot,0)\right\rangle_{L^2(\mathbb{R}_{>0})}.
\end{aligned}$$

Thus, $\phi \mapsto \phi(\,\cdot\,,0)$ is a unitary mapping s_0 from

$$\left\{\phi \in L^2\left(\mathbb{R}^2\right) \mid \sigma_U\phi = \phi \text{ for all rotations } U\right\}$$

to the weighted L^2-type space

$$\frac{1}{\sqrt{2\pi\mathbf{m}}}\left[L^2\left(\mathbb{R}_{>0}\right)\right] := \left\{f \mid \sqrt{2\pi\mathbf{m}}f \in L^2\left(\mathbb{R}_{>0}\right)\right\}.$$

Consequently, the Hankel transformation $\mathrm{H}_0 := s_0 F s_0^*$ is unitary in the Hilbert space $1/\sqrt{2\pi\mathbf{m}}\left[L^2\left(\mathbb{R}_{>0}\right)\right]$.

The significance of the Hankel transformation lies in particular in the fact that it is a spectral representation for $\sqrt{-\mathbf{m}^{-1}\partial\mathbf{m}\partial}$, where $-\mathbf{m}^{-1}\partial\mathbf{m}\partial$ is a self-adjoint realization of the Bessel operator in $1/\sqrt{2\pi\mathbf{m}}\left[L^2\left(\mathbb{R}_{>0}\right)\right]$ with Neumann boundary condition at 0.

15.10.3 The Radon Transformation

Finally, we consider the Radon transformation, which is used, for instance, in the field of tomography, cf. [19]. This transformation also yields a spectral representation of a differential operator. The Radon transform of $f : \mathbb{R}^2 \to \mathbb{C}$ is formally given by

$$Rf(\alpha,\sigma) = \frac{1}{\sqrt{2\pi}} \int_{\{x\in\mathbb{R}^2 | x\cdot y(\alpha)=\sigma\}} f\,\mathrm{d}s, \quad (15.30)$$

13) Hadamard matrices are orthogonal matrices, that is, real unitary matrices, with entries ± 1.

where $\alpha \in]-\pi,\pi], \sigma \in \mathbb{R}$ and $y(\alpha) \in S^1$, the unit sphere in $\mathbb{R}^2$, given by $y(\alpha) := \begin{pmatrix} \cos\alpha \\ \sin\alpha \end{pmatrix}$. To analyze this, we employ a parameterization of $\{x \in \mathbb{R}^2 | x \cdot y(\alpha) = \sigma\}$ as the line given by

$$\begin{pmatrix} x_1 \\ x_2 \end{pmatrix} = \begin{pmatrix} \cos\alpha & -\sin\alpha \\ \sin\alpha & \cos\alpha \end{pmatrix} \begin{pmatrix} \sigma \\ t \end{pmatrix}$$

for $t \in \mathbb{R}$. Substituting this into the integral expression of R yields

$$Rf(\alpha,\sigma) = \int_{\mathbb{R}} f\left(\begin{pmatrix} \cos\alpha & -\sin\alpha \\ \sin\alpha & \cos\alpha \end{pmatrix} \begin{pmatrix} \sigma \\ t \end{pmatrix}\right) \mathrm{d}t.$$

Using the equality

$$\left((1 \otimes F) Rf\right)(\alpha, r) = (Ff)\left(r \begin{pmatrix} \cos\alpha \\ \sin\alpha \end{pmatrix}\right)$$

for each $\alpha \in]-\pi,\pi], r \in \mathbb{R}$, we compute

$$\begin{aligned}
&|f|^2_{L^2(\mathbb{R}^2)} \\
&= |Ff|^2_{L^2(\mathbb{R}^2)} \\
&= \int_{\mathbb{R}^2} |(Ff)(x)|^2 \,\mathrm{d}x \\
&= \int_{\mathbb{R}_{>0}} \int_{]-\pi,\pi]} \left|(Ff)\left(r \begin{pmatrix} \cos\alpha \\ \sin\alpha \end{pmatrix}\right)\right|^2 \mathrm{d}\alpha \; r \,\mathrm{d}r \\
&= \frac{1}{2} \int_{\mathbb{R}} \int_{]-\pi,\pi]} \left|(Ff)\left(s \begin{pmatrix} \cos\alpha \\ \sin\alpha \end{pmatrix}\right)\right|^2 \mathrm{d}\alpha \; |s| \,\mathrm{d}s \\
&= \frac{1}{2} \int_{\mathbb{R}} \int_{]-\pi,\pi]} \left|\left((1 \otimes F) Rf\right)(\alpha, s)\right|^2 \mathrm{d}\alpha \; |s| \,\mathrm{d}s \\
&= \frac{1}{2} \int_{\mathbb{R}} \int_{]-\pi,\pi]} \left|\left((1 \otimes F) \sqrt{|\partial_2|} Rf\right)(\alpha, s)\right|^2 \mathrm{d}\alpha \,\mathrm{d}s \\
&= \frac{1}{2} \int_{\mathbb{R}} \int_{]-\pi,\pi]} \left|\left(\sqrt{|\partial_2|} Rf\right)(\alpha, s)\right|^2 \mathrm{d}\alpha \,\mathrm{d}s \\
&= \frac{1}{2} \left|\sqrt{|\partial_2|} Rf\right|^2_{L^2(]-\pi,\pi]\times\mathbb{R})}.
\end{aligned}$$

Thus, R defined by the integral expression (15.30) can be extended by continuity to all of $L^2(\mathbb{R}^2)$ and then

$$\begin{aligned}
\frac{1}{\sqrt{2}} R : L^2(\mathbb{R}^2) &\to H_{1/2}\left(|\partial_2|,]-\pi,\pi] \times \mathbb{R}\right) \\
f &\mapsto \frac{1}{\sqrt{2}} Rf
\end{aligned}$$

is unitary. Here $H_{1/2}\left(|\partial_2|,]-\pi,\pi] \times \mathbb{R}\right)$ is the completion[14] of $D(\partial_2)$ with respect to the norm induced by the inner product

$$(f,g) \mapsto \left\langle |\partial_2|^{1/2} f \,\middle|\, |\partial_2|^{1/2} g \right\rangle_{L^2(]-\pi,\pi]\times\mathbb{R})}.$$

We now define a mapping $R^\diamond : H_{-1/2}(|\partial_2|,]-\pi,\pi] \times \mathbb{R}) \to L^2(\mathbb{R}^2)$ by the duality relation

$$\langle R^\diamond g | f \rangle_{L^2(\mathbb{R}^2)} = \langle g | Rf \rangle_{L^2(]-\pi,\pi]\times\mathbb{R})},$$

where the inner product on the right-hand side is utilized as a so-called duality pairing, which is the continuous extension of $\langle \cdot | \cdot \rangle_{L^2(]-\pi,\pi]\times\mathbb{R})}$ to $H_{-1/2}(|\partial_2|,]-\pi,\pi] \times \mathbb{R}) \times H_{1/2}(|\partial_2|,]-\pi,\pi] \times \mathbb{R})$.

Inverting our initial parameterization, we see that for well-behaved g, we have

$$\begin{aligned}
&\int_{\mathbb{R}} \int_{\mathbb{R}} \overline{g(\alpha,\sigma)} f\left(\begin{pmatrix} \cos\alpha & -\sin\alpha \\ \sin\alpha & \cos\alpha \end{pmatrix} \begin{pmatrix} \sigma \\ t \end{pmatrix}\right) \mathrm{d}\sigma \,\mathrm{d}t \\
&= \int_{\mathbb{R}} \int_{\mathbb{R}} \overline{g\left(\alpha, x_1\cos\alpha + x_2\sin\alpha\right)} \\
&\quad f(x_1,x_2) \,\mathrm{d}x_1 \,\mathrm{d}x_2
\end{aligned}$$

and so, at least formally, we see that

$$\begin{aligned}
&\left(R^\diamond g\right)(x_1,x_2) \\
&\quad = \int_{]-\pi,\pi]} g\left(\alpha, x_1\cos\alpha + x_2\sin\alpha\right) \mathrm{d}\alpha.
\end{aligned}$$

The dual operator $R^\diamond$ is again the continuous extension of the mapping defined for

14) That is the construction – already used several times – to obtain the smallest Hilbert space with the prescribed inner product containing $D(\partial_2)$.

well-behaved functions by the latter integral expression. Then

$$\frac{1}{2}R^{\diamond}\,|\partial_2|\,Rf = f$$

and so $R^{-1} = 1/\sqrt{2}\left(\frac{1}{\sqrt{2}}R\right)^{-1} = \frac{1}{\sqrt{2}}(\frac{1}{\sqrt{2}}R)^{*} = \frac{1}{2}R^{*}$ is given by

$$\frac{1}{2}R^{\diamond}|\partial_2| \;:\; H_{1/2}(|\partial_2|,]-\pi,\pi] \times \mathbb{R}) \to L^2(\mathbb{R}^2). \tag{15.31}$$

Using the Hilbert transformation H, we find

$$\begin{aligned}|\partial_2| &= |\mathrm{i}\partial_2| \\ &= \mathrm{i}\partial_2 \,\mathrm{sgn}\,(\mathrm{i}\partial_2) \\ &= \partial_2\, \mathcal{H}_2\,,\end{aligned}$$

where $\mathcal{H}_2$ denotes the continuous extension of $1 \otimes \mathrm{H}$ to $H_{1/2}\left(|\partial_2|,]-\pi,\pi] \times \mathbb{R}\right)$, the inverse Radon transform may be rewritten as

$$R^{-1} = \frac{1}{2}R^{\diamond}\partial_2\,\mathcal{H}_2.$$

Note here that $\mathcal{H}_2$ commutes with ∂_2 because it is a function of ∂_2 defined via the operator function calculus.

References

1. Bay, M. and Hasbro (2007) *Transformers.* Science fiction action film, Paramount Pictures, DreamWorks SKG.
2. Functional Analysis. This handbook, Chapter 13.
3. Ronald, N.B. (1983) *The Fourier Transform and Its Applications,* 2nd edn, 3rd printing, International Student Edition. Auckland etc.: McGraw-Hill International Book Company.
4. Weisstein, N. (1980) The joy of Fourier analysis, in *Visual Coding and Adaptibility* (ed. C.S. Harris), Erlbaum, Hillsdale, NJ.
5. Akhiezer, N.I. and Glazman, I.M. (1981) in *Theory of Linear Operators in Hilbert Space, Monographs and Studies in Mathematics,* 9, 10, Vols I, II (eds Transl. from the 3rd Russian ed. by E.R. Dawson and W.N. Everitt), Publishers in association with Scottish Academic Press, Edinburgh, Boston, MA, Pitman Advanced Publishing Program, London, Melbourne.
6. Zemanian, A.H. (1987) *Generalized Integral Transformations (Unabridged republ. of the 1968 orig. publ. by Interscience, New York),* vol. XVI, Dover Publications, Inc., New York, 300 p.
7. Beckenbach, E.F. and Hestenes, M.R. (eds) (1962) *Modern Mathematics for the Engineer, Second Series* (University of California, Engineering Extension Series), McGraw-Hill Book Company Inc., New York, Toronto, London.
8. Akansu, A.N. and Agirman-Tosun, H. (2010) Generalized Discrete Fourier Transform With Nonlinear Phase. *IEEE Trans. Signal Process.*, **58**(9), 4547–4556.
9. Brigham, E.O. (1988) *The Fast Fourier Transform and Applications,* Prentice Hall, Englewood Cliffs, NJ.
10. Gaskill, J.D. (1978) *Linear Systems, Fourier Transformations and Optics,* John Wiley & Sons, Inc., New York, Chichester, Brisbane, Toronto.
11. Norman, F.M. (1981) Lectures on linear systems theory. *J. Math. Psychol.*, **23**, 1–89.
12. McBride, A.C. (1979) *Fractional Calculus and Integral Transforms of Generalized Functions, Pitman Research Notes in Mathematics,* vol. 31, Pitman Advanced Publishing Program, San Francisco, CA, London, Melbourne, ISBN: 0-273-08415-1.
13. Ozaktas, H.M., Zalevsky, Z., and Alper, M. (2001) *The Fractional Fourier Transform with Applications in Optics and Signal Processing, Series in Pure and Applied Optics,* John Wiley & Sons, Inc, ISBN: 0-471-96346-1.
14. Partial Differential Equations. This handbook, Chapter 16.
15. Bracewell, R.N. (1986) *The Hartley Transform, Oxford Engineering Science Series,* vol. 19, Oxford University Press, New York, Clarendon Press, Oxford.
16. Picard, R. (1989) *Hilbert Space Approach to Some Classical Transforms,* John Wiley & Sons, Inc., New York.

17. Goodman, J.W. (1968) *Introduction to Fourier Optics*, McGraw-Hill, New York.
18. Taylor, C.A. (1978) *Images: A Unified View of Diffraction and Image Formation with All Kinds of Radiation*, Wykeham Publ., London, Basingstoke.
19. Ramm, A.G. and Katsevich, A.I. (1996) *The Radon Transform and Local Tomography*, CRC Press Inc.

16
Partial Differential Equations

Des McGhee, Rainer Picard, Sascha Trostorff, and Marcus Waurick

16.1
What are Partial Differential Equations?

Partial differential equations (PDEs) provide the essential mathematical models of physical phenomena that vary in time and space. They are equations involving one or more unknown function(s) of (more than one) independent variables and their partial derivatives. In standard applications, we would often expect one variable, t, for time and three independent space variables, Cartesian coordinates x, y, z or polar coordinates r, θ, ϕ or some other standard three-dimensional coordinate system. Of course, sometimes steady-state models are considered so there is no dependence on t and/or symmetries allow lower-dimensional models to be studied so that only one or two space variables are required. On the other hand, physics often involves mathematical models that introduce higher-dimensional "space" (inverted commas because this may not be physical space) requiring independent variables $x_1, x_2, x_3, \ldots, x_n$, $n > 3$, so that, along with time, t, there are $n+1$ independent variables.

A core objective is to determine the unknown functions that satisfy the PDE in order to predict the behavior of the physical system being modeled. The physical phenomenon being modeled will normally determine an appropriate domain in which the PDE is to be solved, that is, a "space"-domain $D \subseteq \mathbb{R}^n$ and, if there is time dependence, a time interval $I \subseteq \mathbb{R}$ in which the PDE is to be satisfied. In addition to the PDE itself, there will often be some boundary conditions that the unknown functions have to satisfy on the boundary ∂D of D and, if time is involved, some initial conditions to be satisfied by the unknown functions at some initial time $t = t_0$. Often, an analytic solution of the PDE may not be available but qualitative information derived from a careful analysis of the system, for example, existence and uniqueness of the solution, may allow numerical methods to be developed and applied to approximate the solution to an acceptable degree of accuracy.

As elementary notions of differentiation require functions to be continuous, one might initially think that the importance of PDEs is restricted to phenomena that can

Mathematical Tools for Physicists, Second Edition. Edited by Michael Grinfeld.

be modeled as continuous (indeed smooth) processes. However, PDEs also play a fundamental role in modeling discrete physical phenomena, most notably in quantum mechanics via the Schrödinger equation. In addition, mathematical generalization allows the precise definition of derivatives of discontinuous functions, and indeed of generalized functions such as the Dirac-delta, so that PDEs are often appropriate mathematical models even when the unknown functions to be found are not expected to be smooth or even continuous.

In full generality, that is, where the PDE involves an arbitrary function of the unknowns and their derivatives, there is little that can be said about properties of solutions or methods of solution. However special cases, particularly those that are of central interest in applications, have been dealt with extensively. All the core physical phenomena, such as sound, heat, electromagnetism, elasticity, fluid flow, and so on are modeled by PDEs and these models have been well studied. In this chapter, therefore, we shall not dwell on the standard theory and methods of solution of the classical models of mathematical physics referring the reader to standard texts at elementary level (e.g., [1–4]) or advanced level (e.g., [5–8]). Rather, we shall concentrate on structural matters concerning PDEs that are necessary for reasonable models of physical phenomena and precise notions of solvability. We shall however illustrate the theory using well-known (systems of) equations from mathematical physics.

Historically, linear models of physical phenomena have been developed and extensively studied. Progressively, more and more nonlinear effects are being taken into account but, because the analysis of nonlinear PDEs is based on a deep understanding of the linear case, the latter will be the focus of our discussion.

A linear PDE is of the form

$$\begin{aligned}&\textstyle\sum_{\alpha_0=0}^{m_0}\cdots\sum_{\alpha_n=0}^{m_n} a_{(\alpha_0,\dots,\alpha_n)}(t,x)\\ &\times\left(\partial_0^{\alpha_0}\cdots\partial_n^{\alpha_n}u\right)(t,x)=f(t,x),\end{aligned}$$

where (t,x) ranges in a subset[1] $I\times D\subseteq\mathbb{R}^{1+n}$, that is, $t\in I\subseteq\mathbb{R}$, $x=(x_1,\dots,x_n)\in D\subseteq\mathbb{R}^n$. In many physical applications we have, as discussed above, $n=3$ and t plays the role of time. The symbol ∂_k denotes the partial derivative with respect to the variable at position $k\in\{0,1,\dots,n\}$ in the list $(t,x_1,\dots,x_n)$. The coefficient functions $a_{(\alpha_0,\dots,\alpha_n)}$ and the "source term" f are commonly assumed to be given and the main task is to find solutions u that are usually real- or complex-valued functions defined on $I\times D$. To emphasize that we are looking for functions as solutions and not really for values at a specific point, we usually drop the reference to the variables $(t,x)=(t,x_1,\dots,x_n)$. Also, introducing the standard multi-index notation $(\partial_0,\dots,\partial_n)^{(\alpha_0,\dots,\alpha_n)}$ for $\partial_0^{\alpha_0}\cdots\partial_n^{\alpha_n}$, abbreviating $\partial:=(\partial_0,\dots,\partial_n)$, $\alpha:=(\alpha_0,\dots,\alpha_n)$, $m:=(m_0,\dots,m_n)$, and writing 0 for $(0,\dots,0)$, we get the more compact form

$$P(\partial)u:=\sum_{\alpha=0}^{m}a_\alpha\partial^\alpha u=f. \qquad (16.1)$$

We could also drop the summation range altogether, because we could implicitly think of the coefficients a_α as being zero if they are not in the prescribed summation range.

1) Here, we use the superscript $1+n$ to draw attention to the distinction between the time variable and the space variable(s). We revert later to the superscript $n+1$ when we have variables $(x_0,x_1,\dots,x_n)$, that is, there is no time dependence or a suitable variable to regard as "time-like" has not yet been identified (see Section 16.2.1).

We may think of the coefficients a_α not just as scalar valued but possibly matrix valued, all of the same matrix size $s \times r$, with $u = (u_1, u_2, \dots, u_r)$ and $f = (f_1, f_2, \dots, f_s)$. In this way, (16.1) describes a system of PDEs as well as a single PDE in one unified form. Whichever the case, we shall simply refer to (16.1) as a PDE. A single PDE is a special matrix-valued case with 1×1-matrices as coefficients and generally linear algebra teaches us that we should not expect a unique solution to a linear system of equation if the coefficients are not square matrices and so we shall assume this throughout.

When can such a problem be considered as a reasonable model of a physical phenomenon? In general, the answer to this question is provided by the three Hadamard conditions:

1. Existence. There should be at least one solution for every right-hand side.
2. Uniqueness. There should be at most one solution.
3. Robustness. The solution should depend continuously on the given data.

The first condition may imply a constraint on the admissible data, that is, the prescription of a suitable function space that f must belong to. Satisfying the second condition may require constraints to the admissible set of solutions, that is, the definition of a suitable function space in which to seek solutions u. The third condition guarantees that small errors in data, which will always be present in any physical measurement, result in only small errors in solutions; without this requirement, the solution of the PDE may provide no meaningful insight to reality.

Satisfying all three conditions commonly results in generalizations of solution concepts that takes us beyond the classical understanding of differentiation because the function spaces required to ensure the Hadamard requirements are spaces of generalized functions (distributions). Problems satisfying Hadamard's requirements are called *well-posed*, those which do not are called *ill-posed*. However, for mathematical models of physical phenomena that are in the first instance ill-posed problems, there are often physically justifiable modifications that can be made to obtain a problem that is well-posed.

In general, establishing well-posedness for a PDE of the form (16.1) is not an easy task. We therefore start with the restrictive assumption that the coefficients a_α are constant.

16.2 Partial Differential Equations in $\mathbb{R}^{n+1}$, $n \in \mathbb{N}$, with Constant Coefficients

Given a PDE with constant coefficients, the issue of well-posedness can be reduced to algebraic characterizations via the application of an integral transformation. For $\nu \in \mathbb{R}^{n+1}$, the unitary mapping [2)]

$$L_\nu : H_{\nu,0}\left(\mathbb{R}^{n+1}\right) \to L^2\left(\mathbb{R}^{n+1}\right)$$

generated by the integral expression

$$\begin{aligned}(L_\nu \varphi)(x) &= \hat{\varphi}(x - \mathrm{i}\nu) \\ &= \frac{1}{(2\pi)^{(n+1)/2}} \int_{\mathbb{R}^{n+1}} \exp\left(-(\mathrm{i}x + \nu)\cdot y\right) \varphi(y)\, \mathrm{d}y\end{aligned}$$

is known as the Fourier–Laplace transformation, see [9].

2) Recall from [9] that $H_{\nu,0}\left(\mathbb{R}^{n+1}\right)$ is the space of (equivalence classes of) functions f such that $x \mapsto \exp(-\nu \cdot x)\, f(x)$ is square integrable. The inner product is

$$(f, g) \mapsto \int_{\mathbb{R}^{n+1}} \overline{f(x)}\, g(x)\, \exp(-2\nu \cdot x)\, \mathrm{d}x.$$

A Cartesian product $X_0 \times \cdots \times X_N$ of Hilbert spaces, equipped with the inner product X_k, $k \in \{0, \ldots, N\}$, $\langle v|w\rangle_{\bigoplus_{k\in\{0,\ldots,N\}} X_k} := \sum_{k\in\{0,\ldots,N\}} \langle v_k \mid w_k \rangle_{X_k}$ for all $v = (v_0, \ldots, v_N)$, $w = (w_0, \ldots, v_N)$ $\in X_0 \times \cdots \times X_N$ is a Hilbert space, called the **(finite) direct sum** of $(X_k)_{k\in\{0,\ldots,N\}}$ and is denoted by $\bigoplus_{k\in\{0,\ldots,N\}} X_k$ or $X_0 \oplus \cdots \oplus X_N$, see [10]. The elements $(x_0, \ldots, x_N)$ of such a direct sum will also be denoted by $x_0 \oplus \cdots \oplus x_N$ or – more suggestively – by a column matrix

$$\begin{pmatrix} x_0 \\ \vdots \\ x_N \end{pmatrix}.$$

Similarly, a $(K+1) \times (L+1)$-matrix of elements in Hilbert spaces can be regarded as a $(K+1) \cdot (L+1)$-dimensional vector and hence an element in the direct sum of the appropriate $(K+1) \cdot (L+1)$ Hilbert spaces.

Applying this construction, we see that we can easily extend the Fourier–Laplace transformation to vectors or indeed matrices with components in $H_{\nu,0}\left(\mathbb{R}^{n+1}\right)$ by componentwise application. Moreover, this matrix convention makes the action of a differential operator $P(\partial)$, where P is a polynomial with matrix coefficients, rather intuitive as a form of matrix multiplication. Indeed, for a polynomial matrix P, that is, a matrix with scalar polynomials as entries,

$$L_\nu P(\partial) = P(\mathbf{im} + \nu) L_\nu. \tag{16.2}$$

Here $\mathbf{m}$ symbolizes "multiplication-by-the-argument" and $P(\mathbf{im} + \nu)$ denotes a multiplication operator defined by $P(\mathbf{im} + \nu)(g)(x) = P(ix + \nu) g(x)$, [10]. The multiplication on the right-hand side is, for g in an appropriate direct sum space, to be understood as an ordinary matrix product. Indeed, if $P(\partial)$ has $(J+1) \times (K+1)$ – matrix coefficients, then it is clear that g must be either a $(K+1)$ – column vector with entries in $H_{\nu,0}\left(\mathbb{R}^{n+1}\right)$, that is, $g \in \bigoplus_{k\in\{0,\ldots,K\}} H_{\nu,0}\left(\mathbb{R}^{n+1}\right)$, or possibly a $(K+1) \times (L+1)$ – matrix with entries in $H_{\nu,0}\left(\mathbb{R}^{n+1}\right)$, that is, $g \in \bigoplus_{l\in\{0,\ldots,L\}} \bigoplus_{k\in\{0,\ldots,K\}} H_{\nu,0}\left(\mathbb{R}^{n+1}\right)$. We shall, however, rarely note matrix sizes and simply write $g \in H_{\nu,0}\left(\mathbb{R}^{n+1}\right)$, because the matrix sizes should be clear from the context.

Then the application of the Fourier–Laplace transformation to the PDE (16.1) and using (16.2) results in the algebraic problem

$$P(\mathbf{im} + \nu) L_\nu u = L_\nu f$$

for the unknown $L_\nu u$. If this can be solved, then the solution of the PDE can be obtained by applying the inverse transformation.

16.2.1 Evolutionarity

Before focusing on the issue of constructing solutions, we briefly discuss how to detect a direction of a possible time variable in a differential expression $P(\partial)$. This can be a problem because changes of variables made, for example, to simplify the appearance of a PDE may result in the direction of evolution being somewhat hidden.

A fundamental (and from our point of view characterizing) property of a mathematical model

$$P(\partial)u = f$$

of a physical phenomenon is that it should be causal, that is the effect of data f on the solution u should not precede (in time) the application of the data. Therefore, if the polynomial $\det(P(z))$ has no zeros in a half-space of $\mathbb{C}^{n+1}$ of the form

$$\mathrm{i}\left[\mathbb{R}^{n+1}\right] + \left[\mathbb{R}_{\geq \varrho_0}\right] \nu_0$$

for some $\varrho_0 \in \mathbb{R}_{>0}$, $\nu_0 \in \mathbb{R}^{n+1}$, $\left|\nu_0\right| = 1$, then we call the direction ν_0 evolutionary or time-like because in the direction of ν_0 we have causality in the sense that for all $a \in \mathbb{R}$, if the data f vanishes on the half-space $\left[\mathbb{R}_{<a}\right] \nu_0 + \left\{\nu_0\right\}^{\perp}$ with interior unit normal ν_0 –that is, the set $\left\{x \in \mathbb{R}^{n+1} | x = s\nu_0 + y,\ s < a,\ \nu_0 \cdot y = 0\right\}$ – then the solution $P(\partial)^{-1} f$ also vanishes on the same half-space. The partial differential operator $P(\partial)$ is then called evolutionary in direction ν_0. If $P(\partial)$ has a direction in which it is evolutionary then we briefly say $P(\partial)$ is evolutionary; otherwise we say $P(\partial)$ is nonevolutionary. If $P(\partial)$ is evolutionary in both direction ν_0 and $-\nu_0$ we say that $P(\partial)$ is reversibly evolutionary (in direction ν_0). If $P(\partial)$ is only evolutionary in direction ν_0 but not in direction $-\nu_0$, then we say $P(\partial)$ is irreversibly evolutionary.

In $\mathbb{R}^2$, for a second-order scalar differential operator $p(\partial) = p(\partial_1, \partial_2)$ with real coefficients, a more common name for irreversibly evolutionary is parabolic, reversibly evolutionary coincides with hyperbolic, and nonevolutionary with elliptic. This classical terminology is based on the fact that the three situations correspond to $\left\{x \in \mathbb{R}^2 | p(x) = 0\right\}$ being respectively a parabola, a hyperbola, or an ellipse. These classifications have various generalizations to arbitrary dimensions (leaving, however, many partial differential operators, such as the Schrödinger operator, unclassified).

Note that $\left(1 - \partial_1^2 - \partial_2^2\right)$ is nonevolutionary if considered in $\mathbb{R}^2$ (static), but it is reversibly evolutionary in $\mathbb{R}^{1+2}$ in direction $e_0 = (1, 0, 0)$ (quasi-static).

The direction of evolutionarity ν_0 will be used as the weight occurring in the exponential weighted space $H_{\nu,0}(\mathbb{R}^{n+1})$, the space where we will seek for solutions, via $\nu = \varrho\nu_0$ for $\varrho > 0$ sufficiently large and the Fourier–Laplace transformation L_ν will be used to transform the PDE problem into an algebraic one using (16.2) as outlined above. In a nonevolutionary case, we may choose $\nu = 0$ and apply the more familiar Fourier transformation $F = L_0$.

16.2.2
An Outline of Distribution Theory

As indicated above, we need to consider a concept of generalized functions, so-called distributions. A classical development of a theory of distributions relies on the introduction of the Schwartz space $\mathcal{S}(\mathbb{R}^{n+1})$ of rapidly decreasing smooth functions $f : \mathbb{R}^{n+1} \to \mathbb{C}$ defined by

$$\mathcal{S}(\mathbb{R}^{n+1}) := \left\{ f \in C_\infty(\mathbb{R}^{n+1}) \mid \bigwedge_{\alpha,\beta \in \mathbb{N}^{n+1}} \sup_{x \in \mathbb{R}^{n+1}} |x^\alpha \partial^\beta f(x)| < \infty \right\}$$

endowed with the topology defined by

$$f_k \to 0 \text{ in } \mathcal{S}(\mathbb{R}^{n+1}) \text{ as } k \to \infty \iff \bigwedge_{\alpha,\beta \in \mathbb{N}^{n+1}} \sup_{x \in \mathbb{R}^{n+1}} |x^\alpha \partial^\beta f_k(x)| \to 0 \text{ as } k \to \infty.$$

A *(tempered) distribution* is then an element of $\mathcal{S}'(\mathbb{R}^{n+1})$, the space of continuous linear functionals on $\mathcal{S}(\mathbb{R}^{n+1})$. Examples of tempered distributions are $\delta_{\{\omega\}}$, the Dirac-δ-functionals of evaluation of a function $f \in \mathcal{S}(\mathbb{R}^{n+1})$ at a point $\omega \in \mathbb{R}^{n+1}$, that is, $\delta_{\{\omega\}} f := f(\omega)$. Other examples are functionals $\delta_{\mathcal{M}}$ of "evaluation" on a manifold $\mathcal{M} \subseteq \mathbb{R}^{n+1}$. These are defined by $\delta_{\mathcal{M}} f := \int_{\mathcal{M}} f$. Frequently, it is convenient to apply functionals $\mathcal{S}'(\mathbb{R}^{n+1})$ to elements in a dense subset of $\mathcal{S}(\mathbb{R}^{n+1})$ such as $C_0^\infty(\mathbb{R}^{n+1})$, the space of infinitely differentiable functions on $\mathbb{R}^{n+1}$ that have compact support, that is, vanish outside of a bounded set.

It is remarkable to note that there is a more intrinsic way of constructing the space of tempered distributions than the ad hoc definition from above, see, for example, [Example 2.2] [11]. For simplicity, we only treat the one-dimensional case. For this, consider the position operator $\mathbf{m}$ of multiplication-by-argument, that is, $(\mathbf{m}f)(x) := xf(x)$, and the momentum operator ∂, that is, $\partial f(x) = f'(x)$ for suitable f and $x \in \mathbb{R}$, realized as operators in $L^2(\mathbb{R})$. Then, define the annihilation operator $\mathcal{D} := (1/\sqrt{2})(\mathbf{m} + \partial)$ and the creation operator $\mathcal{D}^* = \frac{1}{\sqrt{2}}(\mathbf{m} - \partial)$. Realizing that $\gamma : x \mapsto \exp\left(-x^2/2\right)$ is an eigenfunction of the harmonic oscillator $\mathcal{D}^*\mathcal{D}$ with eigenvalue 1 and using the density of the linear hull of the Hermite functions $\{\Gamma_k \mid k \in \mathbb{N}\}$, where $\Gamma_k := (1/|\mathcal{D}^k\gamma|_{L^2(\mathbb{R})})\mathcal{D}^k\gamma$ for all $k \in \mathbb{N}$, we can determine the action of any power of $\mathcal{D}^*\mathcal{D}$ by knowing only how $\mathcal{D}^*\mathcal{D}$ acts on $\{\Gamma_k \mid k \in \mathbb{N}\}$. In this way, $\mathcal{S}(\mathbb{R}) = \bigcap_{k\in\mathbb{N}} D\left((\mathcal{D}^*\mathcal{D})^k\right)$. A sequence $(\phi_n)_{n\in\mathbb{N}}$ in $\mathcal{S}(\mathbb{R})$ then, by definition, converges to some $\phi \in \mathcal{S}(\mathbb{R})$ if for all $k \in \mathbb{N}$ we have $(\mathcal{D}^*\mathcal{D})^k \phi_n \to (\mathcal{D}^*\mathcal{D})^k \phi$ in $L^2(\mathbb{R})$. Endow $D\left((\mathcal{D}^*\mathcal{D})^k\right)$ with the scalar product $(\phi, \psi) \mapsto \langle (\mathcal{D}^*\mathcal{D})^k \phi | (\mathcal{D}^*\mathcal{D})^k \psi\rangle$, where $\langle\cdot|\cdot\rangle$ denotes the usual L^2-scalar product. Then $D\left((\mathcal{D}^*\mathcal{D})^k\right)$ becomes a Hilbert space. The space of tempered distributions on $\mathbb{R}$ can then be written as

$$\mathcal{S}'(\mathbb{R}) = \bigcup_{k\in\mathbb{N}} D\left((\mathcal{D}^*\mathcal{D})^k\right)'.$$

Thus, for a tempered distribution $f \in \mathcal{S}'(\mathbb{R})$, the following continuity estimate holds:

$$\bigvee_{k\in\mathbb{N}, C>0} \bigwedge_{\phi\in\mathcal{S}(\mathbb{R})} |\langle f|\phi\rangle| = |f(\phi)| \le C|(\mathcal{D}^*\mathcal{D})^k \phi|_{L^2(\mathbb{R})}.$$

Indeed, if we choose $f = \delta_{\{0\}}$ given by $\langle\delta_{\{0\}}|\phi\rangle := \phi(0)$ for $\phi \in \mathcal{S}(\mathbb{R})$ we estimate

$$\begin{aligned} |\phi(0)| &= \left| \int_{-\infty}^{0} (1+t)\phi'(t)\frac{1}{1+t}\,\mathrm{d}t \right| \\ &\le |(1+\mathbf{m})\partial\phi|_{L^2(\mathbb{R})} \\ &\le C|\mathcal{D}^*\mathcal{D}\phi|_{L^2(\mathbb{R})} \end{aligned}$$

for some constant $C > 0$.

Observing that the Schwartz space $\mathcal{S}(\mathbb{R}^{n+1})$ is also an algebra, that is, for any two functions $\phi, \psi \in \mathcal{S}(\mathbb{R}^{n+1})$, the product $\phi \cdot \psi$ also lies in $\mathcal{S}(\mathbb{R}^{n+1})$, we can define the product of a tempered distribution $f \in \mathcal{S}'(\mathbb{R}^{n+1})$ and a function $\phi \in \mathcal{S}(\mathbb{R}^{n+1})$ by setting

$$\langle\phi \cdot f|\psi\rangle := \langle f|\overline{\phi} \cdot \psi\rangle$$

for every $\psi \in \mathcal{S}(\mathbb{R}^{n+1})$. The resulting functional $\phi \cdot f$ is also a tempered distribution. Moreover, because the convolution $\phi * \psi$ of two functions $\phi, \psi \in \mathcal{S}(\mathbb{R}^{n+1})$, given by

$$(\phi * \psi)(x) := \int_{\mathbb{R}^{n+1}} \phi(x-y)\psi(y)\,\mathrm{d}y$$

is again an element of $\mathcal{S}(\mathbb{R}^{n+1})$, we define the convolution of a tempered distribution f with a function $\phi \in \mathcal{S}(\mathbb{R}^{n+1})$ by

$$\langle f * \phi|\psi\rangle := \langle f|\overline{\sigma_{-1}\phi} * \psi\rangle \qquad (16.3)$$

for all $\psi \in \mathcal{S}(\mathbb{R}^{n+1})$. Here, $\sigma_{-1}\phi$ is given by $(\sigma_{-1}\phi)(x) = \phi(-x)$ for $x \in \mathbb{R}^{n+1}$. The resulting functional $f * \phi$ is again a tempered distribution.

One of the most important properties of the Schwartz space $\mathcal{S}(\mathbb{R}^{n+1})$ is that the Fourier transformation F defines a bijection on it. This fact allows to extend the Fourier transformation to tempered distributions by defining

$$\langle Ff \,|\, \psi\rangle := \langle f \,|\, F^{*}\psi\rangle$$

for all $f \in \mathcal{S}'(\mathbb{R}^{n+1})$, $\psi \in \mathcal{S}(\mathbb{R}^{n+1})$.

16.2.3
Integral Transformation Methods as a Solution Tool

As illustrated above, the Fourier–Laplace transformation can be utilized to reformulate a linear PDE with constant coefficients as an equation involving (matrix-valued) polynomials only, which is easier to handle. Moreover, it is sometimes possible to write a solution for $P(\partial)u = f$ as a certain superposition of so-called fundamental solutions. In order to compute such fundamental solutions, one has to extend the Fourier–Laplace transformation to distributions.

For $\nu \in \mathbb{R}^{n+1}$ the Fourier–Laplace transformation has a continuous extension to distributions in the space

$$\exp(\nu\mathbf{m})\left[\mathcal{S}'\left(\mathbb{R}^{n+1}\right)\right] := \left\{\exp(\nu\mathbf{m})f \,|\, f \in \mathcal{S}'\left(\mathbb{R}^{n+1}\right)\right\},$$

that is the space of distributions $\exp(\nu\mathbf{m})f$ given by

$$\langle \exp(\nu\mathbf{m})f \,|\, \psi\rangle := \langle f \,|\, \exp(-\nu\mathbf{m})\psi\rangle,$$

for all functions ψ such that $\exp(-\nu\mathbf{m})\psi \in \mathcal{S}(\mathbb{R}^{n+1})$. For such a distribution, the Fourier–Laplace transformation is given by

$$\begin{aligned}\langle L_\nu \exp(\nu\mathbf{m})f \,|\, \phi\rangle &:= \langle \exp(\nu\mathbf{m})f \,|\, L_\nu^{*}\phi\rangle \\ &= \langle \exp(\nu\mathbf{m})f \,|\, \exp(\nu\mathbf{m})F^{*}\phi\rangle \\ &= \langle f \,|\, F^{*}\psi\rangle \\ &= \langle Ff \,|\, \phi\rangle\end{aligned}$$

for $\phi \in \mathcal{S}(\mathbb{R}^{n+1})$, that is,

$$\begin{aligned}L_\nu : \exp(\nu\mathbf{m})\left[\mathcal{S}'(\mathbb{R}^{n+1})\right] &\to \mathcal{S}'(\mathbb{R}^{n+1}) \\ \exp(\nu\mathbf{m})f &\mapsto Ff.\end{aligned}$$

Let us consider an example. Obviously, $\delta_{\{0\}} \in \exp(\nu\mathbf{m})\left[\mathcal{S}'\left(\mathbb{R}^{n+1}\right)\right]$ for every $\nu \in \mathbb{R}^{n+1}$ with $\delta_{\{0\}} = \exp(\nu\mathbf{m})\delta_{\{0\}}$, the right-hand side $\delta_{\{0\}}$ being interpreted in $\mathcal{S}'(\mathbb{R}^{n+1})$. Thus, we obtain $L_\nu\delta_{\{0\}} = F\delta_{\{0\}}$ and

$$\begin{aligned}\langle F\delta_{\{0\}} \,|\, \phi\rangle &= \langle \delta_{\{0\}} \,|\, F^{*}\phi\rangle \\ &= \frac{1}{(2\pi)^{(n+1)/2}} \int_{\mathbb{R}^{n+1}} \phi(x)\,\mathrm{d}x\end{aligned}$$

for each $\phi \in \mathcal{S}(\mathbb{R}^{n+1})$, showing

$$L_\nu\delta_{\{0\}} = F\delta_{\{0\}} = \frac{1}{(2\pi)^{(n+1)/2}}. \tag{16.4}$$

With this we can now address the possibility of representing the solution u of the problem

$$P(\partial)u = f,$$

for given f and a suitable (matrix-valued) polynomial P as a convolution.

A Green's tensor G is a solution of the equation

$$P(\partial)\,G = \delta_{\{0\}}, \tag{16.5}$$

where the right-hand side is the diagonal matrix having $\delta_{\{0\}}$ as diagonal entries. It is a rather remarkable fact that if $\det(P)$ is not the zero polynomial, then such a Green's tensor always exists (even if

$x \mapsto \det(P(ix+\nu))$ has zeros and so $x \mapsto P(ix+\nu)^{-1}$ is not everywhere defined) but it may not be uniquely determined (division problem).

Applying the Fourier–Laplace transformation L_ν to (16.5) and using (16.4), we arrive at

$$P(\mathbf{im}+\nu)\,L_\nu G = \frac{1}{(2\pi)^{(n+1)/2}}.$$

Note that because $\delta_{\{0\}}$ and $\frac{1}{(2\pi)^{(n+1)/2}} := (\phi \mapsto \int_{\mathbb{R}^{n+1}} \frac{1}{(2\pi)^{(n+1)/2}}\phi)$ is a tempered distribution, so is – entry by entry – G.

As the transformed equation involves only polynomial multiplication, solving a (system of) PDE(s) reduces to a linear algebra issue of inverting matrices.

Now, the Green's tensor can be utilized to compute solutions u for the (inhomogeneous) equation

$$P(\partial)u = f.$$

Indeed, if $L_\nu G$ can be used to define a multiplication operator $\left(L_\nu G\right)(\mathbf{m})$ on f, which may require severe constraints on f, then

$$\begin{aligned} f &= L_\nu^* \frac{1}{(2\pi)^{(n+1)/2}}\,(2\pi)^{(n+1)/2}\,L_\nu f \\ &= (2\pi)^{(n+1)/2}\,L_\nu^* P(im+\nu)\left(L_\nu G\right)(\mathbf{m})L_\nu f \\ &= P(\partial)\;(2\pi)^{(n+1)/2}\,L_\nu^*\left(L_\nu G\right)(\mathbf{m})L_\nu f. \end{aligned}$$

One usually writes

$$G * f := (2\pi)^{(n+1)/2}\,L_\nu^*\left(L_\nu G\right)(\mathbf{m})L_\nu f$$

and speaks of a convolution because, for certain "good" right-hand sides f, $G*f$ can be written as a convolution-type integral. This definition of convolution coincides with the definition given in (16.3). Thus we get the existence of a solution in the form

$$u = G * f.$$

Typical examples of such Green's tensors are the fundamental solutions of the Poisson equation $\Delta u = f$ in $\mathbb{R}^3$ and the wave equation $(\partial_0^2 - \Delta)u = f$ in $\mathbb{R}^{1+3}$. The Green's tensor for the former is

$$\mathbb{R}^3 \setminus \{0\} \ni x \mapsto -\frac{1}{4\pi|x|},$$

and for the latter

$$\mathbb{R} \setminus \{0\} \ni t \mapsto \frac{1}{4\pi\sqrt{2}}\frac{1}{t}\delta_C(t,\cdot),$$

where δ_C is the distribution of integration over the "forward light cone" $C = \{(t,x) \in \mathbb{R}^{1+3} | t > 0, |x| = t\}$. Thus, a solution for $\Delta u = f$ is

$$u(x) = -\int_{\mathbb{R}^3} \frac{1}{4\pi|x-y|} f(y)\mathrm{d}y \quad (x \in \mathbb{R}^3)$$

and for $(\partial_0^2 - \Delta)u = f$ in $\mathbb{R}^{1+3}$ is

$$u(t,x) = \int_{\mathbb{R}^3} \frac{1}{4\pi|y|} f(t-|y|, x-y)\mathrm{d}y$$

assuming well-behaved data f.

There is a convenient mechanism for obtaining a Green's tensor for systems if the Green's tensor for an associated scalar problem is known. Consider an $(N+1)\times(N+1)$ matrix operator $P(\partial)$. By the Caley–Hamilton theorem, we know that $P(\partial)$ satisfies its minimal polynomial. In other words, (treating ∂ as if it were an array of numbers) there is a (scalar) polynomial $q_P(\lambda) := \sum_{k=0}^{d} c_k(\partial)\lambda^k$ (of smallest [3] degree) with scalar polynomials c_k such that

$$q_P(P(\partial)) = \sum_{k=0}^{d} c_k(\partial)\,P(\partial)^k = 0.$$

3) If minimality of degree is not wanted, $q_P(\lambda)$ can be replaced by the characteristic polynomial $q_P(\lambda) := \det(P(\partial) - \lambda)$.

If we have a Green's tensor (fundamental solution) g for $c_0(\partial)$ then, identifying g with the $(N+1)\times(N+1)$ diagonal matrix with g as diagonal entries, we obtain

$$\delta_{\{0\}} = c_0(\partial)g = -\sum_{k=1}^{d} c_k(\partial)P(\partial)^k g$$
$$= P(\partial)\left(-\sum_{s=0}^{d-1} c_{s+1}(\partial)P(\partial)^s g\right).$$

This calculation shows that $G = -\sum_{s=0}^{d-1} c_{s+1}(\partial)P(\partial)^s g$ is a Green's tensor for $P(\partial)$.

As an example, we consider an extended[4] system based on the equations of magnetostatics:

$$P(\partial) := \begin{pmatrix} 0 & \nabla^\top \\ -\nabla & \nabla\times \end{pmatrix}$$
$$:= \begin{pmatrix} 0 & \partial_1 & \partial_2 & \partial_3 \\ -\partial_1 & 0 & -\partial_3 & \partial_2 \\ -\partial_2 & \partial_3 & 0 & -\partial_1 \\ -\partial_3 & -\partial_2 & \partial_1 & 0 \end{pmatrix}.$$

We find

$$P(\partial)^2 = -\left(\partial_1^2 + \partial_2^2 + \partial_3^2\right)\begin{pmatrix} 1 & 0 & 0 & 0 \\ 0 & 1 & 0 & 0 \\ 0 & 0 & 1 & 0 \\ 0 & 0 & 0 & 1 \end{pmatrix}$$

showing that $q_P(\lambda) = \lambda^2 + \Delta$ so that $c_0(\partial) = \Delta$. Thus, with $g = -1/4\pi|\cdot|$ as the associated fundamental solution, we obtain

$$G = -P(\partial)g$$

as a Green's tensor for $P(\partial) = \begin{pmatrix} 0 & \nabla^\top \\ -\nabla & \nabla\times \end{pmatrix}$. Thus, we have for $y \in \mathbb{R}^3 \setminus \{0\}$

$$G(y) = -\frac{1}{4\pi|y|^3}\begin{pmatrix} 0 & y^\top \\ -y & y\times \end{pmatrix}.$$

Hence, a solution of

$$\nabla^\top H = 0$$
$$\nabla \times H = J$$

can, for suitable divergence-free J, be given in an obvious block matrix notation as $\begin{pmatrix} 0 \\ H(x) \end{pmatrix}$ in terms of a componentwise convolution integral as

$$\int_{\mathbb{R}^3} \frac{1}{4\pi|x-y|^3}\begin{pmatrix} 0 & -(x-y)^\top \\ (x-y) & -(x-y)\times \end{pmatrix}\begin{pmatrix} 0 \\ J(y) \end{pmatrix}\mathrm{d}y,$$

that is,

$$H(x) = -\int_{\mathbb{R}^3} \frac{1}{4\pi|x-y|^3}(x-y)\times J(y)\,\mathrm{d}y,$$

which is the well-known Biot–Savart formula.

The generality of the above solution concepts (more details can be found in [11]) is mathematically pleasing. It turns out, however, that the particular PDEs and systems of interest in applications are from a rather small subclass, see [12]. As a first simplifying observation, we note that we may assume that we have a specific evolutionary direction, which, by a simple rotation, we always may assume to be $v_0 = (1, 0, \ldots, 0)$, that is, the direction of the "time" variable t. By introducing higher time derivatives as new unknowns, we may also assume that the differential operators of interest are first

4) This extension is a technicality to obtain a formally invertible differential operator matrix.

order in time, that is, they are of the form

$$P(\partial_0, \dots, \partial_n) = \partial_0 M_0 + M_1 + P_0(\partial_1, \dots, \partial_n) \tag{16.6}$$

with $P_0(0) = 0$, i.e. $P(0) = M_1$ and $P(1, 0, \dots, 0) = M_0$. Indeed, typical problems of mathematical physics are in this form.

As an example, let us consider the equations of acoustics

$$\partial_0 p + \sigma p + \nabla^\top v = f \qquad (\mu, \sigma, \kappa \in \mathbb{R}_{\geq 0})$$

combined with a Newton law of the form

$$\partial_0(\mu v) + \kappa v + \nabla p = 0,$$

where $\nabla = \begin{pmatrix} \partial_1 \\ \partial_2 \\ \partial_3 \end{pmatrix}$, and $\nabla^\top = \begin{pmatrix} \partial_1 & \partial_2 & \partial_3 \end{pmatrix}$.

These can be combined in a block matrix notation to give an equation of the form

$$P(\partial) \begin{pmatrix} p \\ v \end{pmatrix} = \begin{pmatrix} f \\ 0 \end{pmatrix}$$

involving the partial differential operator

$$\begin{aligned} &P(\partial_0, \partial_1, \partial_2, \partial_3) \\ &= \partial_0 \begin{pmatrix} 1 & (0\ 0\ 0) \\ \begin{pmatrix} 0 \\ 0 \\ 0 \end{pmatrix} & \begin{pmatrix} \mu & 0 & 0 \\ 0 & \mu & 0 \\ 0 & 0 & \mu \end{pmatrix} \end{pmatrix} \\ &+ \begin{pmatrix} \sigma & (0\ 0\ 0) \\ \begin{pmatrix} 0 \\ 0 \\ 0 \end{pmatrix} & \begin{pmatrix} \kappa & 0 & 0 \\ 0 & \kappa & 0 \\ 0 & 0 & \kappa \end{pmatrix} \end{pmatrix} \\ &+ P_0(\partial_1, \partial_2, \partial_3), \end{aligned} \tag{16.7}$$

where

$$P_0(\partial_1, \partial_2, \partial_3) := \begin{pmatrix} 0 & \nabla^\top \\ \nabla & \begin{pmatrix} 0 & 0 & 0 \\ 0 & 0 & 0 \\ 0 & 0 & 0 \end{pmatrix} \end{pmatrix}. \tag{16.8}$$

It is only after elimination of the velocity field v (assuming $\partial_0 \mu + \kappa \neq 0$) that we obtain a familiar wave equation in pressure p alone:

$$\begin{aligned} &\mu \partial_0^2 p + (\kappa + \mu\sigma)\, \partial_0 p + \kappa\sigma p - \Delta p \\ &\quad = \partial_0 \mu f + \kappa f =: g. \end{aligned} \tag{16.9}$$

Note that if $\mu = 0$, this turns into a diffusion equation

$$\kappa \partial_0 p + \eta p - \Delta p = g \qquad (\eta := \kappa\sigma) \tag{16.10}$$

(also known in different contexts as a Fokker–Planck equation or a heat equation). This is a typical situation in applications: what are essentially the same mathematical (systems of) equations may have different names (and units) and interpretations in different contexts.

The limit case $\kappa = 0$, $\eta = 1$, now leads to an invertible elliptic problem. The case $\kappa = 0$, $\eta = 0$, gives Poisson's equation, imposing further constraints on g and leading to nonuniqueness. Indeed, there are infinitely many polynomial solutions in $\operatorname{Ker}(\Delta)$, the so-called harmonic polynomials. Modifying this degenerate case to achieve a well-posed problem is the subject of potential theory and involves the introduction of growth constraints on the distributional solution p and decay conditions for the right-hand side g.

It turns out that, in this first-order-in-time form, problems of interest are characterized by a single property of fundamental importance: strict positive definiteness in $H_{v,0}(\mathbb{R}^{n+1})$, in the nonevolutionary case for a particular v (say $v = 0$) or, for the evolutionary case, for all $v = \varrho v_0$ in the time direction $v_0 = (1, 0, \dots, 0)$ and all sufficiently large $\varrho \in \mathbb{R}_{>0}$. Indeed, partial differential operators $P(\partial)$ of the form (16.6), predominant in applications, when realized in $H_{\nu,0}\left(\mathbb{R}^{n+1}\right)$, all satisfy

$$\mathfrak{Re}\,\langle u|P(\partial)\,u\rangle_{\nu,0} \geq c_0\,\langle u|u\rangle_{\nu,0} = c_0\,|u|^2_{\nu,0} \tag{16.11}$$

for some $c_0 \in \mathbb{R}_{>0}$ and all $u \in H_{\nu,0}\left(\mathbb{R}^{n+1}\right)$ such that $P(\partial)\,u \in H_{\nu,0}\left(\mathbb{R}^{n+1}\right)$. By the Cauchy–Schwarz inequality, we have

$$\mathfrak{Re}\,\langle u|P(\partial)\,u\rangle_{\nu,0} \leq |u|_{\nu,0}\,|P(\partial)\,u|_{\nu,0}$$

and so we read off

$$|P(\partial)\,u|_{\nu,0} \geq c_0\,|u|_{\nu,0}\,,$$

which implies invertibility of $P(\partial)$ and continuity of the inverse. The same result follows for the adjoint differential operator $P(\partial)^*$. Since by the projection theorem (see [10]),

$$H_{\nu,0}\left(\mathbb{R}^{n+1}\right) = \overline{\mathrm{Ran}\,(P(\partial))}^{\perp} \oplus \overline{\mathrm{Ran}\,(P(\partial))}$$

with $\overline{\mathrm{Ran}\,(P(\partial))}^{\perp} = \mathrm{Ker}\left(P(\partial)^*\right) = \{0\}$ (because $P(\partial)^*$ is invertible), we get

$$\left|P(\partial)^{-1}f\right|_{\nu,0} \leq c_0^{-1}\,|f|_{\nu,0}$$

for all $f \in \mathrm{Ran}\,(P(\partial))$ and, by continuous extension, for all $f \in \overline{\mathrm{Ran}\,(P(\partial))} = H_{\nu,0}\left(\mathbb{R}^{n+1}\right)$. Thus, the Hadamard requirements are satisfied: we have existence and uniqueness of solution for every right-hand side in $H_{\nu,0}\left(\mathbb{R}^{n+1}\right)$ and continuous dependence on the data. As $\nu = (\varrho, 0, \ldots, 0)$, we have

$$H_{\nu,0}\left(\mathbb{R}^{n+1}\right) = H_{\varrho,0}\,(\mathbb{R}) \otimes L^2\,(\mathbb{R}^n), \tag{16.12}$$

where the tensor product (see [10]) on the right-hand side may more intuitively be understood as $H_{\varrho,0}$–functions on $\mathbb{R}$ with values in $L^2\,(\mathbb{R}^n)$, that is, we may consider $H_{\nu,0}\left(\mathbb{R}^{n+1}\right)$ as $H_{\varrho,0}\left(\mathbb{R}, L^2\,(\mathbb{R}^n)\right)$ noting that

$$\langle u|v\rangle_{\nu,0} = \int_{\mathbb{R}} \langle u\,(t)\,|v\,(t)\rangle_{L^2(\mathbb{R}^n)}\,\exp\,(-2\varrho t)\,\mathrm{d}t =: \langle u|v\rangle_{\varrho,0,0}\,. \tag{16.13}$$

In the above example of the equations of acoustics, (16.11) does not impose any additional constraints on P_0 because, in this case, $\mathfrak{Re}\left\langle u|P_0\left(\partial_1, \partial_2, \partial_3\right)u\right\rangle_{\nu,0}$ happens to vanish as an integration by parts calculation confirms. In other words, $A := P_0\left(\partial_1, \partial_2, \partial_3\right)$ is skew-self-adjoint, that is, $A = -A^*$ (A is skew-self-adjoint if and only if iA is self-adjoint, compare [10]). That the spatial operator A is skew-self-adjoint is not exceptional but rather the general situation in mathematical physics due to the typical Hamiltonian structure of A as a block operator matrix of the form $\begin{pmatrix} 0 & -C^* \\ C & 0 \end{pmatrix}$ where C is a closed, densely defined Hilbert space operator. We shall explore this more deeply in the next section.

16.3 Partial Differential Equations of Mathematical Physics

As discussed in the previous section, strict positive definiteness is at the heart of the solution theory for PDEs of mathematical physics. Generalization to the case of phenomena confined to an open region Ω in $\mathbb{R}^n$ (boundary value problems) and to media with properties varying with their location in space (variable coefficients) is straightforward. The typical Hamiltonian structure is usually preserved by a suitable choice of boundary conditions. Limit cases such as static or stationary solutions may need specific strategies (e.g., decay constraints or radiation conditions). Generic cases, however, enjoy the same simple structure making them accessible to a unified solution theory.

The abstract solution theory for a general strictly positive definite, closed and densely defined linear operator Q with a strictly positive definite adjoint Q^* follows as above by simply replacing $P(\partial)$ by Q. Somewhat miraculously, this simple idea suffices to understand the main PDEs of mathematical physics. We assume that the time direction is fixed and ∂_0 is the associated derivative (i.e., $t = x_0$ and ∂_0 is the time derivative). To leave the spatial part as general as possible, we merely assume that we are dealing with suitable functions on $\mathbb{R}$ (the time domain) with values in a Hilbert space X. In comparison with (16.12), (16.13) above, this amounts to replacing $L^2(\mathbb{R}^n)$ by an arbitrary Hilbert space so that we are concerned with a solution theory in a Hilbert space $H_{\varrho,0}(\mathbb{R}, X)$ with inner product

$$\langle u|v\rangle_{\varrho,0,0} := \int_{\mathbb{R}} \langle u(t)\,|\,v(t)\rangle_X \,\exp(-2\varrho t)\,\mathrm{d}t. \tag{16.14}$$

Thus we have an "ordinary" differential equation on $\mathbb{R}$ in a Hilbert space X.

Following the structure introduced in (16.6), we focus on the particular class of abstract differential operators

$$Q = \partial_0 M_0 + M_1 + A, \tag{16.15}$$

where A is the canonical extension, to the time-dependent case, of a closed and densely defined linear differential operator $A_0 : D(A_0) \subseteq X \to X$, that is, $(Au)(t) = A_0 u(t)$ for (almost every) $t \in \mathbb{R}$, and similarly M_0, M_1 are the canonical extensions to the time-dependent case of a bounded linear operator in X. From the observation at the end of the last section, we may focus our interest on the case where A_0 and consequently A is skew-self-adjoint. Indeed, the typical shape of A_0 is in a block matrix operator form $\begin{pmatrix} 0 & -C^* \\ C & 0 \end{pmatrix}$. Frequently, this is not obvious in historical formulations of the equations of mathematical physics. For example, the Schrödinger equation, involving an operator of the form

$$\partial_0 + \mathrm{i}L,$$

where L is nonnegative and self-adjoint, appears not to fit in the above setting. However, separating real and imaginary part of the equation

$$(\partial_0 + \mathrm{i}L)\,u = f$$

yields the system

$$\left(\partial_0 + \begin{pmatrix} 0 & -L \\ L & 0 \end{pmatrix}\right)\begin{pmatrix} \mathfrak{Re}\,u \\ \mathfrak{Im}\,u \end{pmatrix} = \begin{pmatrix} \mathfrak{Re}\,f \\ \mathfrak{Im}\,f \end{pmatrix},$$

where

$$A = \begin{pmatrix} 0 & -L \\ L & 0 \end{pmatrix}$$

has the required block matrix form because L is self-adjoint.

Given that A is skew-self-adjoint, the strict positive definiteness of Q and Q^* (see (16.15)) reduces to that of $\partial_0 M_0 + M_1$, which is equivalent to requiring

$$\langle u|\varrho M_0 + \mathfrak{Re}\,M_1 u\rangle_{\varrho,0,0} \geq c_0 \langle u|u\rangle_{\varrho,0,0} \tag{16.16}$$

for all $u \in H_{\varrho,0}(\mathbb{R}, X)$ and all sufficiently large $\varrho \in \mathbb{R}_{>0}$ as a constraint on the self-adjoint, bounded linear operators M_0, $\mathfrak{Re}\,M_1 := \frac{1}{2}(M_1 + M_1^*)$ in X. Note that considering $\mathfrak{Re}\,M_1$ has little to do with the entries of M_1 being real or complex. For example, take M_1 to be some simple 2×2 matrices:

$$\mathfrak{Re}\begin{pmatrix} 1 & -1 \\ 0 & 1 \end{pmatrix} = \frac{1}{2}\begin{pmatrix} 2 & -1 \\ -1 & 2 \end{pmatrix},$$

$$\mathfrak{Re}\begin{pmatrix} \mathrm{i} & -\mathrm{i} \\ 0 & 1 \end{pmatrix} = \frac{1}{2}\begin{pmatrix} 0 & -\mathrm{i} \\ \mathrm{i} & 2 \end{pmatrix}.$$

If there is a need for more general material laws, we may, by a simple perturbation argument, include arbitrary additional bounded linear operators on $H_{\varrho,0}(\mathbb{R}, X)$. Such an operator may result from a linear operator $\mathbb{M}$ mapping X-valued step functions (i.e., linear combinations of functions of the form $\chi_{_I} \otimes h := t \mapsto \chi_{_I}(t)\, h$, where χ_I denotes the characteristic function of a bounded interval $I \subseteq \mathbb{R}$ and $h \in X$) into $\bigcap_{\varrho \geq \varrho_0} H_{\varrho,0}(\mathbb{R}, X)$. Assuming the estimate

$$|\mathbb{M}\phi|_{\varrho,0,0} \leq c(\varrho)\, |\phi|_{\varrho,0,0}$$

for all X-valued step functions ϕ for suitable constants $c(\varrho)$ satisfying

$$\limsup_{\varrho \to \infty} c(\varrho) < c_0, \tag{16.17}$$

we can continuously extend $\mathbb{M}$ to an operator $\widetilde{M} : H_{\varrho,0}(\mathbb{R}, X) \to H_{\varrho,0}(\mathbb{R}, X)$. The operator $\widetilde{M}$ can model time-dependent material laws or memory effects. Let us treat an example. For $h < 0$, we consider the operator $\mathbb{M} := \tau_h$ of time translation, mapping an X-valued step function u to the step function $t \mapsto u(t + h)$. Then for $\varrho > 0$, we compute

$$|\tau_h u|^2_{\varrho,0,0} = \int_{\mathbb{R}} |u(t+h)|^2 \mathrm{e}^{-2\varrho t}\, \mathrm{d}t = \mathrm{e}^{2\varrho h} |u|^2_{\varrho,0,0}$$

showing that

$$\limsup_{\varrho \to \infty} c(\varrho) = \lim_{\varrho \to \infty} \mathrm{e}^{2\varrho h} = 0 < c_0,$$

because $h < 0$.

A modified Q would then be of the form $\partial_0 M_0 + M_1 + \widetilde{M} + A$. Abbreviating $M_0 + \partial_0^{-1}\left(M_1 + \widetilde{M}\right)$ as $\mathcal{M}\left(\partial_0^{-1}\right)$, we can rewrite this generalized version of Q simply as $\partial_0 \mathcal{M}\left(\partial_0^{-1}\right) + A$ and so we may reformulate the problem to be solved as finding $U, V \in H_{\varrho,0}(\mathbb{R}, X)$ such that

$$\partial_0 V + AU = F, \tag{16.18}$$

where the "material law"

$$V = \mathcal{M}\left(\partial_0^{-1}\right) U \tag{16.19}$$

is satisfied.

This perturbation argument can actually be made more useful by refining the strict positive definiteness condition. An integration by parts calculation yields

$$\begin{aligned} &\mathfrak{Re}\left\langle U \middle| \left(\partial_0 M_0 + M_1\right) U \right\rangle_{\varrho,0,0} \\ &\quad = \varrho \langle U | M_0 U \rangle_{\varrho,0,0} + \left\langle U \middle| \left(\mathfrak{Re}\, M_1\right) U \right\rangle_{\varrho,0,0} \\ &\quad \geq d_1 \varrho \,\langle PU | PU \rangle_{\varrho,0,0} \\ &\qquad + d_2 \langle (1-P)\, U | (1-P)\, U \rangle_{\varrho,0,0}, \end{aligned}$$

where P is the orthogonal projector onto $\operatorname{Ran}(M_0)$, $d_1 > 0$ depends on the positive definiteness constant for M_0 restricted to the subspace $\operatorname{Ran}(M_0)$, and $d_2 > 0$ depends on the positive definiteness constant for $\mathfrak{Re}\, M_1$ restricted to the subspace $\operatorname{Ker}\left(M_0\right) = \left(\operatorname{Ran}(M_0)\right)^{\perp}$. Thus, more general perturbations $\widetilde{M}$ can be considered if $\widetilde{M}P$ is a bounded linear operator and $\widetilde{M}(1-P)$ is a bounded linear operator with a bound $\left\|\widetilde{M}(1-P)\right\|$ less than d_2 uniformly for all sufficiently large ϱ.

16.4
Initial-Boundary Value Problems of Mathematical Physics

As we have indicated, the equations of mathematical physics have not in general been formulated as the above type of first-order systems and it can be a formidable task to rewrite the system to display the generic Hamiltonian structure. As an example, we discussed the system of acoustics. What is the impact of our abstract considerations on this particular example?

The system already has the required form with very simple operators M_0, M_1, which are just multiplication by diagonal matrices. For the spatial operator

$$A := \begin{pmatrix} 0 & \nabla^\top \\ \nabla & \begin{pmatrix} 0 & 0 & 0 \\ 0 & 0 & 0 \\ 0 & 0 & 0 \end{pmatrix} \end{pmatrix} \tag{16.20}$$

to be skew-self-adjoint a suitable boundary condition on the underlying domain $\Omega \subseteq \mathbb{R}^3$ must be chosen. Vanishing of the pressure distribution p on the boundary (Dirichlet boundary condition) or the normal component of the velocity field (Neumann boundary condition for p) are typical choices resulting in skew-self-adjointness of A.

From a mathematical perspective, we can have general operators M_0, M_1, which may be such that there is little chance of recovering anything close to a well-known wave equation. A simple special case would be that all entries in the (not necessarily block diagonal) operator matrices M_0, M_1 are actually bounded multiplication operators in space, that is, we consider coefficients in the usual sense, which are varying in space. Non-block-diagonal situations have been investigated in recent times in studies of metamaterials, see for example, [13, 14].

To illustrate some issues for the problem class under consideration, we look at a very specific "wave equation" more closely. Of course, for computational purposes, it may be interesting to have more or less explicit representation formulae for the solutions. Such formulae, however, cannot be expected in the general case. Even in the block diagonal case, an example such as

$$\partial_0 \begin{pmatrix} \mathbb{P}_{\Omega_1} & 0 \\ 0 & \mathbb{P}_{\Omega_2} \end{pmatrix} + \begin{pmatrix} \mathbb{P}_{\Omega\setminus\Omega_1} & 0 \\ 0 & \mathbb{P}_{\Omega\setminus\Omega_2} \end{pmatrix} + A$$

for measurable sets $\Omega_1, \Omega_2 \subseteq \Omega$, where $\mathbb{P}_S$ denotes the orthogonal projector generated by multiplication with the characteristic function χ_S of a subset $S \subseteq \mathbb{R}^n$, illustrates the problems involved. In $(\Omega \setminus \Omega_1) \cap (\Omega \setminus \Omega_2)$ the time derivative vanishes (elliptic case), in $\Omega_1 \cap \Omega_2$ we have reversible wave propagation (hyperbolic case) and everywhere else, that is, in $(\Omega_1 \setminus \Omega_2) \cup (\Omega_2 \setminus \Omega_1)$, we have a system associated with a diffusion (parabolic case). However, the complete system is covered by our general framework.

Returning to our general form $\partial_0 M_0 + M_1 + \widetilde{M} + A$ with A from (16.20), we note that the crucial concept of causality characterizing evolutionary processes is maintained for this abstract class in the sense that, for all $a \in \mathbb{R}$, if the right-hand side (as an X-valued function) vanishes on the open interval $]-\infty, a[$ then so does the solution. Owing to time translation invariance (which is satisfied for M_0, M_1, A and which we assume to hold for $\widetilde{M}$), it suffices to consider $a = 0$. The above yields that, in a suitable sense, the term $M_0 U$ associated with the solution U for a given right-hand side F vanishing on $\mathbb{R}_{\leq 0}$ must – as a result its continuity – vanish on $\mathbb{R}_{\leq 0}$ and so, in particular, at time 0. Note that, if M_0 has a nontrivial null space, U itself may be discontinuous and so, although $U(0-) = 0$,

$U(0)$ and consequently $U(0+)$ may not be defined. From the perspective of classical initial value problems, U satisfies the initial condition

$$\left(M_0 U\right)(0+) = 0. \tag{16.21}$$

Thus, the solution U of $\left(\partial_0 M_0 + M_1 + \widetilde{M} + A\right) U = F$ with F vanishing on $(-\infty, 0]$ satisfies homogeneous initial conditions. How can we implement nonzero initial data? We think of nonvanishing initial data as a jump in the solution occurring at time 0. Noting that the derivative of a constant is 0, we have

$$\partial_0 M_0 \left(U - \chi_{\mathbb{R}_{>0}} \otimes U_0\right) = \partial_0 M_0 U \text{ on } \mathbb{R}_{>0}.$$

Thus, we may consider

$$\begin{aligned}&\partial_0 M_0 \left(U - \chi_{\mathbb{R}_{>0}} \otimes U_0\right) \\ &\quad + M_1 U + \widetilde{M} U + A U = F\end{aligned} \tag{16.22}$$

as the proper formulation for the initial value problem

$$\begin{aligned}\partial_0 M_0 U + M_1 U + \widetilde{M} U + A U = F \text{ on } \mathbb{R}_{>0}, \\ M_0 U(0+) = M_0 U_0.\end{aligned}$$

Recall that $\widetilde{M}$ is a sufficiently small perturbation in the sense of property (16.17). As $\partial_0 \chi_{\mathbb{R}_{>0}}$ is the Dirac-δ-distribution $\delta_{\{0\}}$, problem (16.22) is formally equivalent to

$$\begin{aligned}&\left(\partial_0 M_0 + M_1 + \widetilde{M} + A\right) U \\ &\quad = F + \delta_{\{0\}} \otimes M_0 U_0.\end{aligned} \tag{16.23}$$

This yields the interpretation that initial data correspond to Dirac $\delta_{\{0\}}$ distribution-type sources. However, rather than discussing possible generalizations of our above solution theory to distributions, we prefer to reformulate (16.22) as a problem for finding $V := U - \chi_{\mathbb{R}_{>0}} \otimes U_0$. Assuming $U_0 \in D\left(A_0\right)$, we obtain

$$\begin{aligned}&\left(\partial_0 M_0 + M_1 + \widetilde{M} + A\right) V \\ &\quad = F - \chi_{\mathbb{R}_{>0}} \otimes \left(M_1 + A\right) U_0 \\ &\qquad - \widetilde{M}\left(\chi_{\mathbb{R}_{>0}} \otimes U_0\right) =: G,\end{aligned}$$

which is covered by the above solution theory.

Alternatively, introducing the perturbation argument in a different way, we have

$$\begin{aligned}V &= \left(U - \chi_{\mathbb{R}_{>0}} \otimes U_0\right) \\ &= \left(\partial_0 M_0 + M_1 + A\right)^{-1} G \\ &\quad - \left(\partial_0 M_0 + M_1 + A\right)^{-1} \widetilde{M} V \\ &=: T(V),\end{aligned}$$

where T, by our assumptions, is a contraction, that is,

$$|T(V)|_{\varrho,0,0} \leq q\,|V|_{\varrho,0,0}$$

for some $q < 1$ independent of $V \in H_{\varrho,0}(\mathbb{R}, X)$ for all sufficiently large ϱ. Hence, by the contraction mapping theorem, the solution V can be found iteratively:

$$V = \lim_{k\to\infty} T^k\left(V_0\right)$$

for arbitrary choice of $V_0 \in H_{\varrho,0}(\mathbb{R}, X)$. The solution U of the initial value problem (16.22) is then

$$U = V + \chi_{\mathbb{R}_{>0}} \otimes U_0.$$

Let us return to the simpler problem with $\widetilde{M} = 0$. In this case, the underlying media, assumed to satisfy (16.16), can be roughly categorized by properties of M_0, M_1 as follows.

$$\text{Lossless media: } \mathfrak{Re}\ M_1 = 0;$$

then, for media that are not lossless, we have,

$$\text{Lossy media: } \mathfrak{Re}\ M_1 \geq 0;$$

$$\text{Gainy media: } \mathfrak{Re}\ M_1 \leq 0;$$

$$\text{Chiral media: } \mathfrak{Im}\ M_1 \neq 0.$$

The case $\widetilde{M} = 0$ also exhibits "energy conservation." Indeed, if F vanishes above a time threshold t_0, then we have the following pointwise relation for the solution U:

$$\frac{1}{2}\langle U(b)\,|M_0 U(b)\rangle_X + \int_a^b \langle U(s)\,|\mathfrak{Re}\,M_1 U(s)\rangle_X \,\mathrm{d}s = \frac{1}{2}\langle U(a)\,|M_0 U(a)\rangle_X$$

for $b \geq a \geq t_0$.

While we have concentrated so far on the acoustic case (16.20), the underlying considerations are completely general. Clearly ∇ can be considered in higher (or lower) dimensions. Further, we could for example also replace ∇ by some other differential operator C (and $\nabla^\top$ by $-C^*$) to obtain other well-known systems of equations from mathematical physics. To illustrate, we shall consider two other cases more closely.

16.4.1
Maxwell's Equations

First, we consider Maxwell's equations in the isotropic, homogeneous case. These can be written in 2×2–block matrix operator form

$$\left(\partial_0 M_0 + M_1 + A\right)\begin{pmatrix} E \\ H \end{pmatrix} = \begin{pmatrix} -J_0 \\ 0 \end{pmatrix}, \tag{16.24}$$

where

$$M_0 := \begin{pmatrix} \begin{pmatrix} \varepsilon & 0 & 0 \\ 0 & \varepsilon & 0 \\ 0 & 0 & \varepsilon \end{pmatrix} & \begin{pmatrix} 0 & 0 & 0 \\ 0 & 0 & 0 \\ 0 & 0 & 0 \end{pmatrix} \\ \begin{pmatrix} 0 & 0 & 0 \\ 0 & 0 & 0 \\ 0 & 0 & 0 \end{pmatrix} & \begin{pmatrix} \mu & 0 & 0 \\ 0 & \mu & 0 \\ 0 & 0 & \mu \end{pmatrix} \end{pmatrix},$$

$$M_1 := \begin{pmatrix} \begin{pmatrix} \sigma & 0 & 0 \\ 0 & \sigma & 0 \\ 0 & 0 & \sigma \end{pmatrix} & \begin{pmatrix} 0 & 0 & 0 \\ 0 & 0 & 0 \\ 0 & 0 & 0 \end{pmatrix} \\ \begin{pmatrix} 0 & 0 & 0 \\ 0 & 0 & 0 \\ 0 & 0 & 0 \end{pmatrix} & \begin{pmatrix} 0 & 0 & 0 \\ 0 & 0 & 0 \\ 0 & 0 & 0 \end{pmatrix} \end{pmatrix},$$

($\varepsilon, \mu \in \mathbb{R}_{>0}$, $\sigma \in \mathbb{R}_{\geq 0}$) and

$$A := \begin{pmatrix} \begin{pmatrix} 0 & 0 & 0 \\ 0 & 0 & 0 \\ 0 & 0 & 0 \end{pmatrix} & \nabla\times \\ -\nabla\times & \begin{pmatrix} 0 & 0 & 0 \\ 0 & 0 & 0 \\ 0 & 0 & 0 \end{pmatrix} \end{pmatrix}, \tag{16.25}$$

where $\nabla\times := \begin{pmatrix} 0 & -\partial_3 & \partial_2 \\ \partial_3 & 0 & -\partial_1 \\ -\partial_2 & \partial_1 & 0 \end{pmatrix}$. By a suitable choice of boundary condition, which commonly is the vanishing of the tangential component of the electric field E on the boundary of the underlying open set $\Omega \subseteq \mathbb{R}^3$ (electric boundary condition, boundary condition of total reflexion), the operator A can be established as skew-self-adjoint in a six-component $L^2(\Omega)$-space. The material laws $\mathcal{M}\left(\partial_0^{-1}\right)$ can be arbitrary as long as (16.16) is satisfied. Thus, in this general formulation, the most complex materials (chiral media, metamaterials) become accessible.

16.4.2 Viscoelastic Solids

The system of linearized viscoelasticity is commonly presented as

$$\operatorname{Div} T + f = \varrho_0\, \partial_0^2 u,$$

where u denotes the displacement field, $T = \left(T_{jk}\right)_{j,k\in\{1,2,3\}}$ is the stress tensor, ϱ_0 is the mass density and $\operatorname{Div} T := \left(\sum_{k=1}^{3} \partial_k T_{jk}\right)_{j\in\{1,2,3\}}$ denotes the tensorial divergence (in Cartesian coordinates).

With $v := \partial_0 u$, we first derive from the definition

$$\mathcal{E} := \operatorname{Grad} u,$$

where $\operatorname{Grad} u := \frac{1}{2}\left(\partial \otimes u + (\partial \otimes u)^{\top}\right)$ denotes the symmetric part of the Jacobi matrix $\partial \otimes u$, another first-order dynamic equation

$$\partial_0 \mathcal{E} = \operatorname{Grad} v.$$

We can formally summarize the resulting system in the form

$$\partial_0 \begin{pmatrix} w \\ \mathcal{E} \end{pmatrix} + \begin{pmatrix} 0 & -\operatorname{Div} \\ -\operatorname{Grad} & 0 \end{pmatrix} \begin{pmatrix} v \\ T \end{pmatrix} = \begin{pmatrix} f \\ 0 \end{pmatrix}$$

with $w = \varrho_0\, v$. The system is completed by linear material relations of various forms linking $\mathcal{E}$ and T. We follow the presentation in [15].

16.4.2.1 The Kelvin–Voigt Model

This class of materials is characterized by a material relation of the form

$$T = C\mathcal{E} + D\partial_0 \mathcal{E}, \tag{16.26}$$

where the elasticity tensor C and the viscosity tensor D are assumed to be modeled as bounded, self-adjoint, strictly positive definite mappings in a Hilbert space X_{sym} of $L^2(\Omega)$-valued, symmetric 3×3-matrices, with the inner product induced by the Frobenius norm

$$(\Phi, \Psi) \mapsto \int_{\Omega} \operatorname{trace}\left(\Phi(x)^{*}\, \Psi(x)\right)\, \mathrm{d}x .$$

By a suitable choice of boundary condition, that is, domain for $\begin{pmatrix} 0 & -\operatorname{Div} \\ -\operatorname{Grad} & 0 \end{pmatrix}$, we can, as in previous cases, achieve skew-self-adjointness in the Hilbert space

$$X := L^2(\Omega) \oplus L^2(\Omega) \oplus L^2(\Omega) \oplus X_{\mathrm{sym}}.$$

For sake of definiteness, let us consider the vanishing of the displacement velocity at the boundary of the domain Ω containing the medium. With this choice of operator domain $D(A)$,

$$A := \begin{pmatrix} 0 & -\operatorname{Div} \\ -\operatorname{Grad} & 0 \end{pmatrix}$$

is skew-self-adjoint in $H_{\varrho,0}(\mathbb{R}, X)$.

For $D \geq c_0 > 0$, we obtain from (16.26) that

$$\begin{aligned} \mathcal{E} &= \left(C + D\partial_0\right)^{-1} T, \\ &= \partial_0^{-1}\left(\partial_0^{-1} C + D\right)^{-1} T, \end{aligned}$$

which amounts to a "material law" of the form

$$\begin{pmatrix} w \\ \mathcal{E} \end{pmatrix} = \begin{pmatrix} \varrho_0 & 0 \\ 0 & 0 \end{pmatrix} \begin{pmatrix} v \\ T \end{pmatrix} + \partial_0^{-1}\left(\begin{pmatrix} 0 & 0 \\ 0 & D^{-1} \end{pmatrix} + \widetilde{M}\right) \begin{pmatrix} v \\ T \end{pmatrix}$$

with

$$\widetilde{M} = \partial_0^{-1} \begin{pmatrix} 0 & 0 \\ 0 & -\left(\partial_0^{-1} C + D\right)^{-1} C D^{-1} \end{pmatrix}.$$

This is the so-called Kelvin–Voigt model of viscoelasticity. The case $C = 0$ leads to a system for a purely viscous behavior (Newton model). On the other hand, if C is strictly positive definite, then the limit case $D = 0$ leads to the standard system for elastic solids.

16.4.2.2 The Poynting–Thomson Model (The Linear Standard Model)

The linear standard model or Poynting–Thomson model is based on a generalization of the Maxwell model involving another coefficient operator R and has the form

$$\partial_0 \mathcal{E} + R\,\mathcal{E} = C^{-1} \partial_0 T + D^{-1} T.$$

Solving for $\mathcal{E}$, this yields

$$\begin{aligned} \mathcal{E} &= \left(\partial_0 + R\right)^{-1} \left(C^{-1}\partial_0 + D^{-1}\right) T = C^{-1} T \\ &\quad + \partial_0^{-1} \left(1 + R\,\partial_0^{-1}\right)^{-1} \left(D^{-1} - RC^{-1}\right) T \end{aligned}$$

leading to a slightly more complex material law

$$\begin{aligned} \begin{pmatrix} w \\ \mathcal{E} \end{pmatrix} &= \begin{pmatrix} \varrho_0 & 0 \\ 0 & C^{-1} \end{pmatrix} \begin{pmatrix} v \\ T \end{pmatrix} \\ &\quad + \partial_0^{-1} \left(\begin{pmatrix} 0 & 0 \\ 0 & D^{-1} - RC^{-1} \end{pmatrix} + \widetilde{M} \right) \begin{pmatrix} v \\ T \end{pmatrix} \end{aligned}$$

with

$$\widetilde{M} = \partial_0^{-1} \begin{pmatrix} 0 & 0 \\ 0 & -R\left(1 + R\,\partial_0^{-1}\right)^{-1} \left(D^{-1} - RC^{-1}\right) \end{pmatrix}.$$

16.5 Coupled Systems

An understanding of the specific mathematical form that material laws must take allows for a transparent discussion on how to couple different physical phenomena to obtain a suitable evolutionary problem. Here we shall focus on the abstract structure of coupled systems (compare [16] for a discussion of coupled systems of mathematical physics).

Without coupling, systems of interest can be combined simply by writing them together in diagonal block operator matrix form:

$$\partial_0 \begin{pmatrix} V_0 \\ \vdots \\ \vdots \\ V_n \end{pmatrix} + A \begin{pmatrix} U_0 \\ \vdots \\ \vdots \\ U_n \end{pmatrix} = \begin{pmatrix} f_0 \\ \vdots \\ \vdots \\ f_n \end{pmatrix},$$

where

$$A = \begin{pmatrix} A_0 & 0 & \cdots & 0 \\ 0 & \ddots & & \vdots \\ \vdots & & \ddots & 0 \\ 0 & \cdots & 0 & A_n \end{pmatrix}$$

inherits the skew-self-adjointness in $X = \bigoplus_{k=0}^{n} X_k$ from its skew-self-adjoint diagonal entries $A_k : D\left(A_k\right) \subseteq X_k \to X_k$, $k \in \{0, \ldots, n\}$. The combined material law takes the simple Block-diagonal form

$$\begin{pmatrix} V_0 \\ \vdots \\ \vdots \\ V_n \end{pmatrix} = \begin{pmatrix} \mathcal{M}_{00}\left(\partial_0^{-1}\right) & 0 & \cdots & 0 \\ 0 & \ddots & & \vdots \\ \vdots & & \ddots & 0 \\ 0 & \cdots & 0 & \mathcal{M}_{nn}\left(\partial_0^{-1}\right) \end{pmatrix} \begin{pmatrix} U_0 \\ \vdots \\ \vdots \\ U_n \end{pmatrix}.$$

Coupling between the phenomena now can be modeled by introducing suitable off-diagonal entries. The full material law now is of the familiar form

$$\begin{pmatrix} V_0 \\ \vdots \\ \vdots \\ V_n \end{pmatrix} = \mathcal{M}\left(\partial_0^{-1}\right) \begin{pmatrix} U_0 \\ \vdots \\ \vdots \\ U_n \end{pmatrix}$$

with

$$\mathcal{M}\left(\partial_0^{-1}\right) := \begin{pmatrix} \mathcal{M}_{00}\left(\partial_0^{-1}\right) & \cdots & \cdots & \mathcal{M}_{0n}\left(\partial_0^{-1}\right) \\ \vdots & \ddots & & \vdots \\ \vdots & & \ddots & \vdots \\ \mathcal{M}_{n0}\left(\partial_0^{-1}\right) & \cdots & \cdots & \mathcal{M}_{nn}\left(\partial_0^{-1}\right) \end{pmatrix}.$$

We consider some applications.

16.5.1 Thermoelasticity

As a first illustration, we consider the general thermoelastic system

$$\partial_0 V + \begin{pmatrix} 0 & \mathrm{Div} & 0 & 0 \\ \mathrm{Grad} & 0 & 0 & 0 \\ 0 & 0 & 0 & \nabla^\top \\ 0 & 0 & \nabla & 0 \end{pmatrix} \begin{pmatrix} v \\ T \\ \theta \\ Q \end{pmatrix} = \begin{pmatrix} f \\ 0 \\ g \\ 0 \end{pmatrix},$$

where we assume that the domains of operators containing the spatial derivatives are such that we model, for example, Dirichlet boundary conditions for the displacement velocity and the temperature in order to maintain skew-self-adjointness of

$$\begin{pmatrix} 0 & \mathrm{Div} & 0 & 0 \\ \mathrm{Grad} & 0 & 0 & 0 \\ 0 & 0 & 0 & \nabla^\top \\ 0 & 0 & \nabla & 0 \end{pmatrix}.$$

The material law is of the form

$$V = \mathcal{M}\left(\partial_0^{-1}\right) \begin{pmatrix} v \\ T \\ \theta \\ Q \end{pmatrix},$$

where $\mathcal{M}\left(\partial_0^{-1}\right)$ is given by

$$\begin{pmatrix} \varrho_0 & 0 & 0 & 0 \\ 0 & C^{-1} & C^{-1}\Gamma & 0 \\ 0 & \Gamma^* C^{-1} & w + \Gamma^* C^{-1}\Gamma & 0 \\ 0 & 0 & 0 & q_0 + q_2\left(\alpha + \beta\partial_0\right)^{-1} \end{pmatrix}.$$

Via symmetric row and column operations, this can be reduced to the Block-diagonal form

$$\begin{pmatrix} \varrho_0 & 0 & 0 & 0 \\ 0 & C^{-1} & 0 & 0 \\ 0 & 0 & w & 0 \\ 0 & 0 & 0 & q_0 + q_2\left(\alpha + \beta\partial_0\right)^{-1} \end{pmatrix}$$

and so the issue of classifying the material is simplified. For example, the issue of strict positive definiteness of

$$M_0 = \begin{pmatrix} \varrho_0 & 0 & 0 & 0 \\ 0 & C^{-1} & C^{-1}\Gamma & 0 \\ 0 & \Gamma^* C^{-1} & w + \Gamma^* C^{-1}\Gamma & 0 \\ 0 & 0 & 0 & q_0 \end{pmatrix}$$

hinges on the strict positive definiteness of ϱ_0, C, w, q_0.

For $q_0 = 0$, the above system is known as a type-3 thermoelastic system. With $\alpha = 0$, we obtain the special case of thermoelasticity with second sound, that is, with the Cataneo modification of heat transport. The so-called type-2 thermoelastic system results by letting $q_2 = 0$.

We point out that the well-known Biot system, see, for example, [17–19], which describes consolidation of a linearly elastic porous medium, can be reformulated so that, up to physical interpretations, it has the same form as the thermoelastic system. The coupling operator Γ of thermoelasticity is, in the poroelastic case, given as $\Gamma = \begin{pmatrix} \alpha & 0 & 0 \\ 0 & \alpha & 0 \\ 0 & 0 & \alpha \end{pmatrix}$, where α is a coupling parameter.

16.5.2 Piezoelectromagnetism

As a second class of examples we consider the coupling of elastic and electromagnetic wave propagation. Here, we have a system of the form

$$\partial_0 V + A \begin{pmatrix} v \\ T \\ E \\ H \end{pmatrix} = \begin{pmatrix} f \\ 0 \\ -J \\ 0 \end{pmatrix},$$

where, by a suitable choice of boundary conditions, A will be a skew-self-adjoint block operator matrix of the form

$$A = \begin{pmatrix} 0 & \mathrm{Div} & 0 & 0 \\ \mathrm{Grad} & 0 & 0 & 0 \\ 0 & 0 & 0 & \nabla\times \\ 0 & 0 & -\nabla\times & 0 \end{pmatrix}.$$

This system needs to be completed by suitable material relations. A simple piezoelectromagnetic model is described by

$$V = \begin{pmatrix} \varrho_0 & 0 & 0 & 0 \\ 0 & C^{-1} & C^{-1}d & 0 \\ 0 & d^*C^{-1} & \varepsilon + d^*C^{-1}d + \partial_0^{-1}\sigma & 0 \\ 0 & 0 & 0 & \mu \end{pmatrix} \begin{pmatrix} v \\ T \\ E \\ H \end{pmatrix}$$

with the coupling given by a bounded linear mapping d from $L^2(\Omega) \oplus L^2(\Omega) \oplus L^2(\Omega)$ to X_{sym}.

Following [20], we obtain a more complicated coupling mechanism. Adding a conductivity term, the coupling is initially described in the form

$$\begin{aligned} T &= C\,\mathcal{E} - dE - qH, \\ D &= d^*\mathcal{E} + \varepsilon E + eH + \partial_0^{-1}\sigma E, \\ B &= q^*\,\mathcal{E} + e^*E + \mu H. \end{aligned}$$

Domain and range spaces for the additional bounded, linear coefficient operators q and e are clear from these equations and, for sake of brevity, we shall not elaborate on this. As has been already noted, for a proper formulation we need to solve for $\mathcal{E}$ to obtain suitable material relations. We find

$$\begin{aligned} \mathcal{E} &= C^{-1}T + C^{-1}dE + C^{-1}qH, \\ D &= d^*C^{-1}T + \left(\varepsilon + d^*C^{-1}d\right)E \\ &\qquad + d^*C^{-1}qH + eH + \partial_0^{-1}\sigma E, \\ B &= q^*\,C^{-1}T + q^*\,C^{-1}dE \\ &\qquad + q^*C^{-1}qH + e^*\,E + \mu H. \end{aligned}$$

Thus, we obtain the material law

$$V = \mathcal{M}\left(\partial_0^{-1}\right) \begin{pmatrix} v \\ T \\ E \\ H \end{pmatrix}$$

with

$$\mathcal{M}\left(\partial_0^{-1}\right) = \begin{pmatrix} \varrho_0 & 0 & 0 & 0 \\ 0 & C^{-1} & C^{-1}d & C^{-1}q \\ 0 & d^*C^{-1} & \left(\varepsilon + d^*C^{-1}d\right) + \partial_0^{-1}\sigma & d^*C^{-1}q + e \\ 0 & q^*C^{-1} & q^*C^{-1}d + e^* & \mu + q^*C^{-1}q \end{pmatrix}.$$

Via symmetric row and column operations, we obtain the block-diagonal operator matrix

$$\begin{pmatrix} \varrho_0 & 0 & 0 & 0 \\ 0 & C^{-1} & 0 & 0 \\ 0 & 0 & \varepsilon + \partial_0^{-1}\sigma & 0 \\ 0 & 0 & 0 & \mu - e^*\varepsilon^{-1}e \end{pmatrix}.$$

Thus, the given form of material relations leads to a material law in the above sense if in addition to the strict positive definiteness of the self-adjoint bounded operators ϱ_0, C, ε, and μ we require

$$\mu \geq \mu_0 + e^*\varepsilon^{-1}e$$

for some constant $\mu_0 \in \mathbb{R}_{>0}$. Again, we emphasize that, without additional effort, more complex material relations in the sense of the above theory could be considered.

16.5.3 The Extended Maxwell System and its Uses

It has been shown [21] that Maxwell's equations in an open domain $\Omega \subseteq \mathbb{R}^3$ may be formulated as a formally coupled system of PDEs. For this, we introduce the formal operator matrices

$$A_{\mathrm{D}} := \begin{pmatrix} 0 & \nabla^\top & 0 & 0 \\ \nabla & 0 & 0 & 0 \\ 0 & 0 & 0 & 0 \\ 0 & 0 & 0 & 0 \end{pmatrix},$$

$$A_{\mathrm{N}} := \begin{pmatrix} 0 & 0 & 0 & 0 \\ 0 & 0 & 0 & 0 \\ 0 & 0 & 0 & \nabla \\ 0 & 0 & \nabla^\top & 0 \end{pmatrix},$$

$$A_{\mathrm{E}} := \begin{pmatrix} 0 & 0 & 0 & 0 \\ 0 & 0 & -\nabla\times & 0 \\ 0 & \nabla\times & 0 & 0 \\ 0 & 0 & 0 & 0 \end{pmatrix},$$

subject to boundary conditions: A_{D} inherits homogeneous Dirichlet boundary conditions for ∇, A_{N} is endowed with homogeneous Neumann boundary conditions for $\nabla^\top$, and the lower left operator in the operator matrix A_{E} carries the electric boundary condition. Now, assuming for simplicity that all material parameters are set to one, it turns out that Maxwell's equation $\left(\partial_0 + A_{\mathrm{E}}\right) u = f$ can be written as

$$\left(\partial_0 + A_{\mathrm{E}} + A_{\mathrm{D}} + A_{\mathrm{N}}\right) U = F,$$

where $U = (0, u, 0)$ and $F = \widetilde{f} + \partial_0^{-1} \left(A_{\mathrm{N}} + A_{\mathrm{D}}\right)\widetilde{f}$ with $\widetilde{f} = (0, f, 0)$.

This so-called extended Maxwell's equation gives a link to the Dirac equation. Indeed, we compute

$$\begin{aligned} &\left(\partial_0 + A_{\mathrm{E}} + A_{\mathrm{D}} + A_{\mathrm{N}}\right)\left(\partial_0 - \left(A_{\mathrm{E}} + A_{\mathrm{D}} + A_{\mathrm{N}}\right)\right) \\ &= \partial_0^2 - \left(A_{\mathrm{E}} + A_{\mathrm{D}} + A_{\mathrm{N}}\right)^2 \\ &= \partial_0^2 - \Delta. \end{aligned}$$

Thus, the extended Maxwell system $\left(\partial_0 + A_{\mathrm{E}} + A_{\mathrm{D}} + A_{\mathrm{N}}\right)$ corresponds to a 0-mass Dirac equation.

By appropriate permutation of rows and columns the more general mass 1 Dirac equation admits the form

$$\left(\partial_0 + \begin{pmatrix} 0 & -S^* \\ S & 0 \end{pmatrix} + A_{\mathrm{E}} + A_{\mathrm{D}} + A_{\mathrm{N}}\right) V = G,$$

where

$$S = \begin{pmatrix} 0 & \begin{pmatrix} 0 & 0 & 1 \end{pmatrix} \\ \begin{pmatrix} 0 \\ 0 \\ 1 \end{pmatrix} & \begin{pmatrix} 0 & 1 & 0 \\ -1 & 0 & 0 \\ 0 & 0 & 0 \end{pmatrix} \end{pmatrix}.$$

Finally, the Maxwell–Dirac system also shares a similar form. Consider the system

$$\left(\partial_0 + A\right)\begin{pmatrix} 0 \\ E \\ H \\ 0 \end{pmatrix} = f,$$

$$\left(\partial_0 + M_1 + A\right)\Psi = g, \qquad (16.27)$$

$$\left(\partial_0 - A\right)\begin{pmatrix} \varphi \\ \alpha \\ 0 \\ 0 \end{pmatrix} = \begin{pmatrix} 0 \\ E \\ H \\ 0 \end{pmatrix},$$

where $A = \left(A_{\mathrm{E}} + A_{\mathrm{D}} + A_{\mathrm{N}}\right)$ and f and g are suitable source terms. As ∂_0 and A commute, the first and third equations can be combined to give

$$\left(\partial_0^2 - A^2\right)\begin{pmatrix} \varphi \\ \alpha \\ 0 \\ 0 \end{pmatrix} = f,$$

and trivially we have

$$\left(A\partial_0 - \partial_0 A\right)\begin{pmatrix} \varphi \\ \alpha \\ 0 \\ 0 \end{pmatrix} = 0.$$

These last two equations can be written as

$$\left(\partial_0 + \begin{pmatrix} 0 & A \\ A & 0 \end{pmatrix}\right)\begin{pmatrix} \partial_0\begin{pmatrix} \varphi \\ \alpha \\ 0 \\ 0 \end{pmatrix} \\ -A\begin{pmatrix} \varphi \\ \alpha \\ 0 \\ 0 \end{pmatrix} \end{pmatrix} = \begin{pmatrix} f \\ 0 \end{pmatrix}.$$

Inserting the second equation from (16.27) gives

$$\left(\partial_0 + \widetilde{M}_1 + \begin{pmatrix} 0 & 0 & A \\ 0 & A & 0 \\ A & 0 & 0 \end{pmatrix}\right)U = \begin{pmatrix} f \\ g \\ 0 \end{pmatrix},$$

where

$$\widetilde{M}_1 = \begin{pmatrix} 0 & 0 & 0 \\ 0 & M_1 & 0 \\ 0 & 0 & 0 \end{pmatrix} \quad \text{and}$$

$$U = \begin{pmatrix} U_0 \\ U_1 \\ U_2 \end{pmatrix} = \begin{pmatrix} \partial_0\begin{pmatrix} \varphi \\ \alpha \\ 0 \\ 0 \end{pmatrix} \\ \Psi \\ -A\begin{pmatrix} \varphi \\ \alpha \\ 0 \\ 0 \end{pmatrix} \end{pmatrix}.$$

Coupling occurs here in an essential way via a suitable quadratic nonlinear dependence of f and g on the solution U. The middle component $U_1 = \Psi$ is the Dirac field, the electromagnetic field can be recovered as $U_0 + U_2 = \begin{pmatrix} 0 \\ E \\ H \\ 0 \end{pmatrix}$, and the so-called potential can be obtained by integrating $U_0 = \partial_0\begin{pmatrix} \varphi \\ A \\ 0 \\ 0 \end{pmatrix}$.

References

1. Constanda, C. (2010) *Solution Techniques for Elementary Partial Differential Equations*, Taylor and Francis.
2. Haberman, R. (2004) *Applied Partial Differential Equations*, Prentice Hall.

3. Weinberger, H.F. (1995) *A First Course in Partial Differential Equations*, Dover, New York.
4. Zauderer, E. (1998) *Partial Differential Equations of Applied Mathematics*, Wiley-Interscience, New York.
5. Garabedian, P.R. (1964) *Partial Differential Equations*, John Wiley & Sons , Inc., New York.
6. Hormander, L. (1990) *The Analysis of Linear Partial Differential Operators I*, Springer-Verlag.
7. Leis, R. (1986) *Initial Boundary Value Problems in Mathematical Physics*, John Wiley and Sons Ltd and B. G. Teubner, Stuttgart.
8. Showalter, R.E. (1977) *Hilbert Space Methods for Partial Differential Equations*, Pitman.
9. Mathematical Transformations and Their Uses. This handbook, Chapter 15.
10. Exner, P. Functional Analysis. This handbook, Chapter 13.
11. Picard, R. and McGhee, D. (2011) *Partial Differential Equations: A Unified Hilbert Space Approach, de Gruyter Expositions in Mathematics 55*. de Gruyter, Berlin, xviii.
12. Picard, R. (2009) A structural observation for linear material laws in classical mathematical physics. *Math. Methods Appl. Sci.*, **32** (14), 1768–1803.
13. Lindell, I.V., Sihvola, A.H., Tretyakov, S.A., and Viitanen, A.J. (1994) *Electromagnetic waves in chiral and bi-isotropic media*, Artech House, Boston, MA and London.
14. Lakhtakia, A. (1994) *Beltrami Fields in Chiral Media*, World Scientific, Singapore.
15. Bertram, A. (2005) *Elasticity and Plasticity of Large Deformations: An Introduction*, Springer, Berlin.
16. Bednarcyk, B.A. (2002) A Fully Coupled Micro/Macro Theory for Thermo-Electro-Magneto-Elasto-Plastic Composite Laminates. Technical Report 211468, NASA.
17. Biot, M.A. (1941) General theory of three-dimensional consolidation. *J. Appl. Phys., Lancaster, PA*, **12**, 155–164.
18. Showalter, R.E. (2000) Diffusion in poro-elastic media. *J. Math. Anal. Appl.*, **251** (1), 310–340.
19. Bear, J. and Bachmat, Y. (1990) *Introduction to Modelling of Transport Phenomena in Porous Media, Theory and Applications of Transport in Porous Media 4*, Kluwer Academic Publishers, Dordrecht etc.
20. Pan, E. and Heyliger, P.R. (2003) Exact solutions for magneto-electro-elastic laminates in cylindrical bending. *Int. J. Solids Struct.*, **40** (24), 6859–6876.
21. Picard, R. (1984) On the low frequency asymptotics in electromagnetic theory. *J. Reine Angew. Math.*, **354**, 50–73.

17
Calculus of Variations

Tomáš Roubíček

17.1
Introduction

The history of the calculus of variations dates back several thousand years, fulfilling the ambition of mankind to seek lucid principles that govern the Universe. Typically, one tries to identify scalar-valued functionals having a clear physical interpretation, for example, time, length, area, energy, and entropy, whose extremal (critical) points (sometimes under some constraints) represent solutions of the problem in question. Rapid development was initiated between the sixteenth and nineteenth centuries when practically every leading scholar, for example, J. Bernoulli, B. Bolzano, L. Euler, P. Fermat, J.L. Lagrange, A.-M. Legendre, G.W. Leibniz, I. Newton, K. Weierstrass and many others, contributed to variational calculus; at that time, the focus was rather on one-dimensional problems cf. also [1–3]. There has been progress through the twentieth century, which is still continuing, informed by the historically important project of Hilbert [4], Problems 19, 20, and 23] and accelerated by the development of functional analysis, theory of partial differential equations, and efficient computational algorithms supported by rigorous numerical analysis and computers of ever-increasing power. Modern methods allow simple formulations in abstract spaces where technicalities are suppressed, cf. Section 17.2, although concrete problems ultimately require additional tools, cf. Section 17.3. An important "side effect" has been the development of a sound theory of optimization and optimal control and of its foundations, convex and nonsmooth analysis.

17.2
Abstract Variational Problems

Variational problems typically deal with a real-valued functional $\Phi : V \to \mathbb{R}$ on an abstract space V that is equipped with a linear structure to handle variations and a topological structure to handle various continuity/stability/localization concepts. In the simplest and usually sufficiently general scenario, V is a *Banach space*[1] [5] or,

1) A linear space equipped with a norm $\|\cdot\|$, that is, $0 \le \|u+v\| \le \|u\|+\|v\|$, $\|u\|=0 \Rightarrow u=0$, $\|\lambda u\| = \lambda\|u\|$ for any $\lambda\geq 0$ and $u, v \in V$, is

Mathematical Tools for Physicists, Second Edition. Edited by Michael Grinfeld.

in physics, often even a *Hilbert space*.[2] The Banach space structure allows us to define basic notions, such as linearity, continuity, and convexity: Φ is called *continuous* if $\Phi(u_k) \to \Phi(u)$ for any $u_k \to u$, *convex* if $\Phi(\lambda u + (1-\lambda)v) \leq \lambda\Phi(u) + (1-\lambda)\Phi(v)$ for any $u, v \in V$ and $0 \leq \lambda \leq 1$, *concave* if $-\Phi$ is convex, or *linear* if it is convex, concave, and $\Phi(0) = 0$.

Yet it should be pointed out that the linear structure imposed on a problem is the result of our choice; it serves rather as a mathematical tool used to define variations or laws of evolution, or to devise numerical algorithms, and so on. Often, this choice is rather artificial, especially if it leads to nonquadratic or even nonconvex functionals possibly with nonlinear constraints.

17.2.1 Smooth (Differentiable) Case

The Banach space structure allows further to say that Φ is *directionally differentiable* if the directional derivative at u in the direction of (variation) v, defined as

$$\mathrm{D}\Phi(u, v) = \lim_{\varepsilon \searrow 0} \frac{\Phi(u + \varepsilon v) - \Phi(u)}{\varepsilon}, \quad (17.1)$$

exists for any $u, v \in V$, and is *smooth* if it is directionally differentiable and $\mathrm{D}\Phi(u, \cdot) : V \to \mathbb{R}$ is a linear continuous functional; then the *Gâteaux differential* $\Phi'(u) \in V^*$, with V^* being the *dual space*,[3] is defined by

$$\left\langle \Phi'(u), v \right\rangle = \mathrm{D}\Phi(u, v). \quad (17.2)$$

If $\Phi' : V \to V^*$ is continuous, then Φ is called *continuously differentiable*. Furthermore, $u \in V$ is called a *critical point* if

$$\Phi'(u) = 0, \quad (17.3)$$

which is an abstract version of the *Euler–Lagrange equation*. In fact, (17.3) is a special case of the abstract operator equation

$$A(u) = f \quad \text{with } A : V \to V^*, f \in V^*, \quad (17.4)$$

provided $A = \Phi' + f$ for some *potential* Φ whose existence requires some symmetry of A: if A itself is Gâteaux differentiable and *hemicontinuous*,[4] it has a potential if, and only if, it is *symmetric*, that is,

$$\left\langle [A'(u)](v), w \right\rangle = \left\langle [A'(u)](w), v \right\rangle \quad (17.5)$$

for any $u, v, w \in V$; up to a constant; this potential is given by the formula

$$\Phi(u) = \int_0^1 \left\langle A(\lambda u), u \right\rangle \mathrm{d}\lambda\,. \quad (17.6)$$

Equation (17.3) is satisfied, for example, if Φ attains its minimum[5] or maximum at u. The former case is often connected with a *minimum-energy principle* that is assumed

called a *Banach space* if it is complete, that is, any Cauchy sequence $\{u_k\}_{k\in\mathbb{N}}$ converges: $\lim_{\max(k,l)\to\infty} \|u_k - u_l\| = 0$ implies that there is $u \in V$ such that $\lim_{k\to\infty} \|u_k - u\| = 0$; then we write $u_k \to u$.

2) This is a Banach space V whose norm makes the functional $V \to \mathbb{R} : u \mapsto \|u+v\|^2 - \|u-v\|^2$ linear for any $v \in V$; in this case, we define the scalar product by $(u, v) = \frac{1}{4}\|u+v\|^2 - \frac{1}{4}\|u-v\|^2$.

3) The dual space V^* is the Banach space of all linear continuous functionals f on V with the norm $\|f\|_* = \sup_{\|u\|\leq 1} \langle f, u\rangle$, with the duality pairing $\langle \cdot, \cdot \rangle : V \times V^* \to \mathbb{R}$ being the bilinear form defined by $\langle f, u\rangle = f(u)$.

4) This is a very weak mode of continuity, requiring that $t \mapsto \langle A(u+tv), w\rangle$ is continuous.

5) The proof is simple: suppose $\Phi(u) = \min \Phi(\cdot)$ but $\Phi'(u) \neq 0$, then for some $v \in V$ we would have $\langle \Phi'(u), v\rangle = \mathrm{D}\Phi(u, v) < 0$ so that, for a sufficiently small $\varepsilon > 0$, $\Phi(u+\varepsilon v) = \Phi(u) + \varepsilon\left\langle \Phi'(u), v\right\rangle + o(\varepsilon) < \Phi(u)$, a contradiction.

to govern many steady-state physical problems. The existence of solutions to (17.3) can thus often be based on the existence of a minimizer of Φ, which can rely on the *Bolzano – Weierstrass theorem*, which states that a *lower* (resp. upper) *semicontinuous* functional[6] on a *compact*[7] set attains its minimum (resp. maximum).

In infinite-dimensional Banach spaces, it is convenient to use this theorem with respect to *weak* convergence*: assuming $V = (V')^*$ for some Banach space V' (called the *pre-dual*), we say that a sequence $\{u_k\}_{k\in\mathbb{N}}$ converges weakly* to u if $\lim_{k\to\infty}\langle u_k, z\rangle = \langle u, z\rangle$ for any $z \in V'$. If V^* is taken instead of V', this mode of convergence is called *weak convergence*. Often $V' = V^*$ (such spaces are called *reflexive*), and then the weak* and the weak convergences coincide. The Bolzano – Weierstrass theorem underlies the *direct method*,[8] invented essentially in [6], for proving existence of a solution to (17.3). We say that Φ is *coercive* if $\lim_{\|u\|\to\infty} \Phi(u)/\|u\| = +\infty$.

Theorem 17.1 (Direct method) [9] *Let V have a pre-dual and $\Phi : V \to \mathbb{R}$ be weakly* lower semicontinuous, smooth, and coercive. Then* (17.3) *has a solution.*

6) Lower semicontinuity of Φ means that $\liminf_{k\to\infty} \Phi(u_k) \geq \Phi(u)$ for any sequence $\{u_k\}_{k\in\mathbb{N}}$ converging (in a sense to be specified) to u; more precisely, this is sequential lower semicontinuity, but we will confine ourselves to the sequential concept throughout the chapter, which is sufficiently general provided the related topologies are metrizable.

7) A set is compact if any sequence has a converging (in the same sense as used for the semicontinuity of the functional) subsequence.

8) This means that no approximation and subsequent convergence is needed.

9) The proof relies on coercivity of Φ, which allows for a localization on bounded sets and then, due to weak* compactness of convex closed bounded sets in V, on the Bolzano – Weierstrass theorem.

AS continuous convex functionals are also weakly* lower semicontinuous, one gets a useful modification:

Theorem 17.2 (Direct method II) *Let V have a pre-dual and let $\Phi : V \to \mathbb{R}$ be continuous, smooth, coercive, and convex. Then* (17.3) *has a solution.*

If Φ is furthermore *strictly convex* in the sense that $\Phi(\lambda u+(1-\lambda)v) < \lambda\Phi(u) + (1-\lambda)\Phi(v)$ for any $u \neq v$ and $0 < \lambda < 1$, then (17.3) has at most one solution.

We say that a nonlinear operator $A : V \to V^*$ is *monotone* if $\langle A(u)-A(v), u-v\rangle \geq 0$ for any $u, v \in V$. Monotonicity of a potential nonlinear operator implies convexity of its potential, and then Theorem 17.2 implies the following.

Theorem 17.3 *Let V be reflexive and $A : V \to V^*$ be monotone, hemicontinuous, coercive in the sense that $\lim_{\|u\|\to\infty}\langle A(u), u\rangle = \infty$, and possess a potential. Then, for any $f \in V^*$,* (17.4) *has a solution.*

In fact, Theorem 17.3 holds even for mappings not having a potential but its proof, due to Brézis [7], then relies on an approximation and on implicit, nonconstructive fixed-point arguments.

The solutions to (17.3) do not need to represent the global minimizers that we have considered so far. Local minimizers, being consistent with physical principles of minimization of energy, would also serve well. The same holds for maximizers. Critical points may, however, have a more complicated saddle-like character. One intuitive example is the following: let the origin, being at the level 0, be surrounded by a range of mountains all of height $h > 0$ at distance ρ from the origin, but assume that there is at least one location v beyond that circle, which has lower altitude. Going from the origin to v, one is tempted to minimize climbing and takes a mountain pass. The Ambrosetti–Rabinowitz *mountain pass theorem* [8] says that there is such a mountain pass and Φ' vanishes there. More rigorously, we have Theorem 17.4.

Theorem 17.4 (Mountain pass) *Let Φ be continuously differentiable, satisfy the Palais–Smale property*[10] *and satisfy the following three conditions:*

$$\Phi(0) = 0, \tag{17.7a}$$

$$\exists \rho, h > 0\colon \ \|u\| = \rho \ \Rightarrow \ \Phi(u) \geq h, \tag{17.7b}$$

$$\exists v \in V\colon \quad \|v\| > \rho, \qquad \Phi(v) < h. \tag{17.7c}$$

Then Φ has a critical point $u \neq 0$.

A similar assertion relies on a Cartesian structure, leading to a *von Neumann's saddle-point theorem.*

Theorem 17.5 (Saddle point)[11] *Let $V = Y \times Z$ be reflexive, $\Phi(y,\cdot) : Z \to \mathbb{R}$ be concave continuous and $\Phi(\cdot,z) : Y \to \mathbb{R}$ be convex continuous for any $(y,z) \in Y \times Z$, $\Phi(\cdot,\bar{z}) : Y \to \mathbb{R}$ and let $-\Phi(\bar{y},\cdot) : Z \to \mathbb{R}$ be coercive for some $(\bar{y},\bar{z}) \in Y \times Z$. Then there is $(y,z) \in Y \times Z$ so that*

$$\forall \tilde{y} \in Y \forall \tilde{z} \in Z : \ \Phi(\tilde{y},z) \geq \Phi(y,z) \geq \Phi(y,\tilde{z})$$

and, if Φ is smooth, then $\Phi'(y,z) = 0$.

10) More specifically, $\{\Phi(u_k)\}_{k\in\mathbb{N}}$ bounded and $\lim_{k\to\infty} \|\Phi'(u_k)\|_{V^*} = 0$ imply that $\{u_k\}_{k\in\mathbb{N}}$ has a convergent subsequence.

11) The proof is nonconstructive, based on a fixed-point argument, see, for example, [9, Theorems 9D and 49A with Prop. 9.9]. The original von Neumann's version [10] dealt with the finite-dimensional case only.

17.2.2
Nonsmooth Case

For $\Phi : V \mapsto \mathbb{R}\cup\{+\infty\}$ convex, we define the *subdifferential* of Φ at u as

$$\partial\Phi(u) = \Big\{ f \in V^*;\ \forall v \in V :\ \Phi(v) + \langle f, u-v\rangle \geq \Phi(u) \Big\}. \tag{17.8}$$

If Φ is Gâteaux differentiable, then $\partial\Phi(u) = \{\Phi'(u)\}$, hence this notion is indeed a generalization of the conventional differential. Instead of the abstract Euler–Lagrange equation (17.3), it is natural to consider the abstract inclusion $0 \in \partial\Phi(u)$. More generally, assuming $\Phi = \Phi_0 + \Phi_1$ with Φ_0 smooth and Φ_1 convex, instead of (17.3), we consider the *inclusion*

$$\partial\Phi_1(u) + \Phi_0'(u) \ni 0. \tag{17.9}$$

In view of (17.8), this inclusion can equally be written as a *variational inequality*

$$\forall v \in V :\ \ \Phi_1(v) + \langle \Phi_0'(u), v-u\rangle \geq \Phi_1(u). \tag{17.10}$$

Theorems 17.1 and 17.2 can be reformulated, for example, as follows.

Theorem 17.6 *Let V have a pre-dual and let $\Phi_0 : V \to \mathbb{R}$ be weakly* lower semicontinuous and smooth, $\Phi_1 : V \mapsto \mathbb{R}\cup\{+\infty\}$ convex and lower semicontinuous, and let $\Phi_0 + \Phi_1$ be coercive. Then* (17.9) *has a solution.*[12]

Introducing the *Fréchet subdifferential*

$$\partial_{\mathrm{F}}\Phi(u) = \Big\{ f \in V^*;\ \liminf_{\|v\|\to 0} \frac{\Phi(u+v) - \Phi(u) - \langle f, v\rangle}{\|v\|} \geq 0 \Big\}, \tag{17.11}$$

12) The proof relies on existence of a minimizer of $\Phi_0 + \Phi_1$ as in Theorem 17.1; then one shows that any such a minimizer satisfies (17.9).

the inclusion (17.9) can be written simply as $\partial_F \Phi(u) \ni 0$; in fact, a calculus for Fréchet subdifferentials can be developed for a wider class of (sometimes called *amenable*) functionals than that considered in (17.9), cf. [11, 12].

Example 17.1 Let us consider the *indicator function* δ_K of a set $K \subset V$ defined as

$$\delta_K(u) = \begin{cases} 0 & \text{if } u \in K, \\ +\infty & \text{if } u \notin K. \end{cases} \qquad (17.12)$$

Clearly, δ_K is convex or lower semicontinuous if (and only if) K is convex or closed, respectively. Assuming K convex closed, it is not difficult to check that $\partial\delta_K(u) = \{f \in V^*;\ \forall v \in K :\ \langle f, v-u\rangle \le 0\}$ if $u \in K$, otherwise $\partial\delta_K(u) = \emptyset$. The set $\partial\delta_K(u)$ is called the *normal cone* to K at u; denoted also by $N_K(u)$. For the very special case $\Phi_1 = \delta_K$, the variational inequality (17.10) (i.e. here also $\Phi_0'(u) \in -N_K(u)$) represents the problem of finding $u \in K$ satisfying

$$\forall v \in K :\ \left\langle \Phi_0'(u), v-u \right\rangle \ge 0. \qquad (17.13)$$

17.2.3 Constrained Problems

In fact, we saw in Example 17.1 a variational problem for Φ_0 with the constraint formed by a convex set K. Sometimes, there still is a need to involve constraints of the type $R(u) = 0$ (or possibly more general $R(u) \le 0$) for a nonlinear mapping $R : V \to \Lambda$ with Λ a Banach space that is possibly ordered; we say that Λ is *ordered* by "$\ge$" if $\{\lambda \ge 0\}$ forms a closed convex *cone*[13] in Λ. Then the constrained minimization problems reads as follows:

13) A cone C is a set such that $a\lambda \in C$ whenever $\lambda \in C$ and $a \ge 0$.

$$\text{Minimize } \Phi(u) \text{ subject to } R(u) \le 0,\ u \in K. \qquad (17.14)$$

Let us define the *tangent cone* $T_K(u)$ to K at u as the closure of $\cup_{a\ge 0} a(K-u)$. For $A : V \to \Lambda$ linear continuous, the *adjoint operator* $A^* : \Lambda^* \to V^*$ is defined by $\langle A^*\lambda^*, u\rangle = \langle \lambda^*, Au\rangle$ for all $\lambda^* \in \Lambda^*$ and $u \in V$. Assuming R to be smooth, the first-order necessary optimality *Karush—Kuhn–Tucker*[14] *condition* takes the following form:

Theorem 17.7 (First-order condition) *Let* $u \in V$ *solve* (17.14) *and let*[15]

$$\exists \tilde{u} \in T_K(u) :\quad [R'(u)](\tilde{u}) < 0 \qquad (17.15)$$

hold. Then there exists $\lambda^* \overset{*}{\ge} 0$[16] *such that*[17]

$$\left\langle \lambda^*, R(u) \right\rangle = 0 \qquad \textit{and} \qquad (17.16a)$$

$$\Phi'(u) + R'(u)^*\lambda^* + N_K(u) \ni 0. \qquad (17.16b)$$

The condition (17.15) is called the *Mangasarian–Fromovitz constraint qualification,* while (17.16a) is called the *complementarity* (or sometimes *orthogonality* or *transversality*) *condition* and the triple

$$R(u) \le 0, \quad \lambda^* \overset{*}{\ge} 0, \quad \langle \lambda^*, R(u)\rangle = 0 \quad (17.17)$$

is called a *complementarity problem*. Defining the *Lagrangean* by

$$\mathscr{L}(u, \lambda^*) = \Phi(u) + \lambda^* \circ R(u), \qquad (17.18)$$

14) Conditions of this kind were first formulated in Karush's thesis [13] and later independently in [14].

15) The inequality "<" in (17.15) means that a neighborhood of $[R'(u)](\tilde{u})$ still lies in the cone $\{\lambda \le 0\}$.

16) The so-called dual ordering $\overset{*}{\ge}$ on Λ^* means that $\lambda^* \overset{*}{\ge} 0$ if, and only if, $\langle \lambda^*, v\rangle \ge 0$ for all $v \ge 0$.

17) The linear operator $R'(u)^* : \Lambda^* \to V^*$ is adjoint to $R'(u) : V \to \Lambda$ and (17.16b) is meant in V^*.

we can write the inclusion (17.16b) simply as $\mathscr{L}'_u(u,\lambda^*) + N_K(u) \ni 0$. The optimality condition *à la* Example 17.1 for maximization of $\mathscr{L}(u,\cdot) : \Lambda^* \to \mathbb{R}$ over the cone $\{\lambda^* \overset{*}{\geq} 0\}$ is simply $R(u) \leq 0$.

If R is a *convex mapping*[18] and K is a convex set, then (17.15) is equivalent to the simpler *Slater constraint qualification*: $\exists u_0 \in K : R(u_0) < 0$. If Φ is also convex, then (17.16) represents the first-order sufficient optimality condition in the sense that if (17.16) is satisfied, u solves (17.14). Moreover, the couple (u,λ^*) represents a saddle point for $\mathscr{L}$ on the set $K \times \{\lambda^* \overset{*}{\geq} 0\}$, and its existence can be proved by using Theorem 17.5.

Minimization problems without the constraint $R(u) \leq 0$ may be much easier to solve in specific cases. In particular, one can explicitly calculate the value $D(\lambda^*) = \min_{u\in K} \mathscr{L}(u,\lambda^*)$. The functional $D : \Lambda^* \to \mathbb{R} \cup \{-\infty\}$ is concave and

$$\text{maximize } D(\lambda^*) \text{ subject to } \lambda^* \overset{*}{\geq} 0 \quad (17.19)$$

is called the *dual problem*. The supremum of (17.19) is always below the infimum of (17.14). Under additional conditions, they can be equal to each other, and (17.19) has a solution λ^* that can serve as the multiplier for (17.16). For duality theory, see, for example, [12, Chapter 12].

In the general nonconvex case, (17.16) is no longer a sufficient condition and construction of such conditions is more involved. A prototype is a *sufficient second-order condition* that uses the *approximate critical cone* C_ε:

$$C_\varepsilon(u) = \{h \in T_K(u);\ \Phi'(u)h \leq \varepsilon\|h\|,\ \operatorname{dist}\big(R'(u)h, T_{-D}(R(u))\big) \leq \varepsilon\|h\|\}$$

for some $\varepsilon > 0$:

18) In this Banach-valued case, convexity means $R(\lambda u+(1-\lambda)v) \leq \lambda R(u) + (1-\lambda)R(v)$ for any $u, v \in V$ and $0 \leq \lambda \leq 1$ with $\leq$ referring to the ordering in Λ.

Theorem 17.8 (Second-order condition) *Let Φ and R be twice differentiable and let the first-order necessary condition* (17.16) *with a multiplier $\lambda^* \overset{*}{\geq} 0$ hold at some u and let*

$$\exists \varepsilon, \delta > 0\ \forall h \in C_\varepsilon(u) : \quad \mathscr{L}''_u(u,\lambda^*)(h,h) \geq \delta\|h\|^2. \quad (17.20)$$

Then u is a local minimizer for (17.14).

A very special case is when $R \equiv 0$ and $K = V$: in this unconstrained case, $N_K = \{0\}$, $C_\varepsilon = V$, and (17.16) and (17.20) become, respectively, the well-known classical condition $\Phi'(u) = 0$ and $\Phi''(u)$ is positive definite.

17.2.4 Evolutionary Problems

Imposing a linear structure allows us not only to define differentials by using (17.1) and (17.2) but also to defining the derivatives $\mathrm{d}u/\mathrm{d}t$ of trajectories $t \mapsto u(t) : \mathbb{R} \to V$.

17.2.4.1 Variational Principles

Minimization of the energy Φ is related to a gradient flow, that is, a process u evolving in time, governed by the gradient Φ' in the sense that the velocity $\mathrm{d}u/\mathrm{d}t$ is always in the direction of steepest descent $-\Phi'$ of Φ. Starting from a given initial condition u_0 and generalizing it for a time-dependent potential $\Phi - f(t)$ with $f(t) \in V^*$, one considers the initial-value problem (a *Cauchy problem*) for the *abstract parabolic equation*:

$$\frac{\mathrm{d}u}{\mathrm{d}t} + \Phi'(u) = f(t), \qquad u(0) = u_0. \quad (17.21)$$

It is standard to assume $V \subset H$, with H a Hilbert space, this embedding being *dense*[19] and continuous. Identifying H with its own dual, we obtain a *Gelfand-triple* $V \subset H \subset V^*$. Then, with the coercivity/growth assumption

$$\exists \epsilon > 0: \quad \epsilon \|u\|_V^p \le \Phi(t,u) \le \frac{1+\|u\|_V^p}{\epsilon} \tag{17.22}$$

for some $1 < p < +\infty$ and fixing a time horizon $T > 0$, the solution to (17.21) is sought in the affine manifold

$$\left\{ v \in L^p(I;V);\ v(0) = u_0,\ \frac{\mathrm{d}v}{\mathrm{d}t} \in L^{p'}(I;V^*) \right\} \tag{17.23}$$

with $I = [0, T]$, where $L^p(I; V)$ stands for a Lebesgue space of abstract functions with values in a Banach space (here V), which is called a *Bochner space*.

By continuation, we obtain a solution u to (17.21) on $[0, +\infty)$. If Φ is convex and f is constant in time, there is a relation to the variational principle for $\Phi - f$ in Section 17.2.1: the function $t \mapsto [\Phi - f](u(t))$ is nonincreasing and convex, and $u(t)$ converges weakly as $t \to \infty$ to a minimizer of $\Phi - f$ on V.

The variational principle for (17.21) on the bounded time interval I uses the functional $\mathfrak{F}$ defined by

$$\mathfrak{F}(u) = \int_0^T \Phi(t, u(t)) + \Phi^*\left(t, f(t) - \frac{\mathrm{d}u}{\mathrm{d}t}\right) - \langle f(t), u(t)\rangle\,\mathrm{d}t + \frac{1}{2}\|u(T)\|_H^2, \tag{17.24}$$

where $\Phi^*(t,\cdot): V^* \to \mathbb{R} \cup \{+\infty\}$ is the *Legendre conjugate* to $\Phi(t,\cdot): V \to \mathbb{R} \cup \{+\infty\}$ defined by

$$\Phi^*(t,f) = \sup_{v \in V} \langle f, v\rangle - \Phi(v); \tag{17.25}$$

19) A subset is dense if its closure is the whole space, here H.

the construction $\Phi(t,\cdot) \mapsto \Phi^*(t,\cdot)$ is called the *Legendre transformation*. Omitting t for the moment, Φ^* is convex and

$$\Phi^*(f) + \Phi(v) \ge \langle f, v\rangle, \tag{17.26}$$

which is *Fenchel's inequality*. If Φ, resp. Φ^*, is smooth, the equality in (17.26) holds if, and only if, $f \in \Phi'(v)$, resp. $v \in [\Phi^*]'(f)$. Moreover, if $\Phi(\cdot)$ is lower semicontinuous, it holds $\Phi^{**} = \Phi$.

The infimum of $\mathfrak{F}$ on (17.24) is equal to $\frac{1}{2}\|u_0\|_H^2$. If u from (17.23) minimizes $\mathfrak{F}$ from (17.24), that is, $\mathfrak{F}(u) = \frac{1}{2}\|u_0\|_H^2$, then u solves the Cauchy problem (17.21); this is the *Brezis–Ekeland–Nayroles principle* [15, 16]. It can also be used in the direct method, see [17] or [18, Section 8.10]:

Theorem 17.9 (Direct method for (17.21)**)** *Let $\Phi: [0,T] \times V \to \mathbb{R}$ be a Carathéodory function such that $\Phi(t,\cdot)$ is convex, both $\Phi(t,\cdot)$ and $\Phi^*(t,\cdot)$ are smooth,* (17.22) *holds, $u_0 \in H$, and $f \in L^{p'}(I;V^*)$. Then $\mathfrak{F}$ from* (17.24) *attains a minimum on* (17.23) *and the (unique) minimizer solves the Cauchy problem* (17.21).

One can consider another side-condition instead of the initial condition, for example, the periodic condition $u(0) = u(T)$, having the meaning that we are seeking periodic solutions with an a priori prescribed period T assuming f is periodic with the period T. Instead of (17.21), one thus considers

$$\frac{\mathrm{d}u}{\mathrm{d}t} + \Phi'(u) = f(t), \quad u(0) = u(T). \tag{17.27}$$

Then, instead of (17.23), solutions are sought in the linear (in fact, Banach) space

$$\left\{ v \in L^p(I;V);\ v(0) = v(T),\ \frac{\mathrm{d}v}{\mathrm{d}t} \in L^{p'}(I;V^*) \right\}. \tag{17.28}$$

The direct method now uses, instead of (17.24), the functional

$$\mathfrak{F}(u) = \int_0^T \Phi(t, u(t)) - \langle f(t), u(t)\rangle + \Phi^*\Big(t, f(t) - \frac{\mathrm{d}u}{\mathrm{d}t}\Big)\,\mathrm{d}t, \quad (17.29)$$

and an analog of Theorem 17.9 but using (17.28) and (17.29); the minimum is 0 and the minimizer need not be unique, in general.

Often, physical and mechanical applications use a convex (in general, nonquadratic) *potential of dissipative forces* $\Psi : H \to \mathbb{R} \cup \{+\infty\}$ leading to a *doubly nonlinear Cauchy problem*:

$$\Psi'\Big(\frac{\mathrm{d}u}{\mathrm{d}t}\Big) + \Phi'(u) = f(t), \quad u(0) = u_0. \quad (17.30)$$

In fact, the hypothesis that (here abstract) dissipative forces, say $A(\mathrm{d}u/\mathrm{d}t)$, have a potential needs a symmetry of A, cf. (17.5), which has been under certain conditions justified in continuum-mechanical (even anisothermal) linearly responding systems (so that the resulting Ψ is quadratic); this is *Onsager's* (or *reciprocal*) *symmetry* condition [19],[20] cf. [20, Section 12.3]. Sometimes, (17.30) is also equivalently written as

$$\frac{\mathrm{d}u}{\mathrm{d}t} = [\Psi^*]'\big(f(t) - \Phi'(u)\big), \quad u(0) = u_0, \quad (17.31)$$

where Ψ^* again denotes the conjugate functional, that is, here $\Psi^*(v^*) = \sup_{v\in H}\langle v^*, v\rangle - \Psi(v)$. If Ψ is also proper in the sense that $\Psi > -\infty$ and $\Psi \not\equiv +\infty$, then $[\Psi^*]' = [\Psi']^{-1}$, which was used in (17.31). For $\Psi = \frac{1}{2}\|\cdot\|_H^2$, we get $\mathrm{d}u/\mathrm{d}t = f - \Phi'(u)$, cf. (17.21). Thus, for $f = 0$, (17.31) represents a *generalized gradient flow*. For a general f, a *Stefanelli's variational principle* [21] for (17.30) employs the functional

$$\mathfrak{F}(u, w) = \bigg(\int_0^T \Psi\Big(\frac{\mathrm{d}u}{\mathrm{d}t}\Big) - \Big\langle f, \frac{\mathrm{d}u}{\mathrm{d}t}\Big\rangle + \Psi^*\big(w\big)\,\mathrm{d}t + \Phi(u(T)) - \Phi(u_0)\bigg)^+ + \int_0^T \Phi(u) - \langle f - w, u\rangle + \Phi^*(f - w)\,\mathrm{d}t \quad (17.32)$$

to be minimized on the affine manifold

$$\Big\{(u, w) \in L^\infty(I; V);\quad u(0) = u_0,\quad \frac{\mathrm{d}u}{\mathrm{d}t} \in L^q(I; H),\quad w \in L^{q'}(I; H)\Big\}, \quad (17.33)$$

where $1 < q < +\infty$ refers to a coercivity/growth condition for Ψ. On the set (17.33), $\mathfrak{F}(u, w) \geq 0$ always holds, and $\mathfrak{F}(u, w) = 0$ means that $w = \Psi'(\mathrm{d}u/\mathrm{d}t)$ and $f - w = \Phi'(u)$ a.e. (almost everywhere) on I, that is, u solves (17.30).

Another option is to use the conjugation and Fenchel inequality only for Ψ, which leads to[21]

$$\mathfrak{G}(u) = \int_0^T \Psi\Big(\frac{\mathrm{d}u}{\mathrm{d}t}\Big) + \Psi^*\big(f - \Phi'(u)\big) + \Big\langle \frac{\mathrm{d}f}{\mathrm{d}t}, u\Big\rangle\,\mathrm{d}t + \Phi(u(T)) \quad (17.34)$$

to be minimized on a submanifold $\{u = w\}$ of (17.33). The infimum is $\Phi(u_0) - f(0) + f(T)$ and any minimizer u is a solution to (17.30). Sometimes, this is known under the name *principle of least dissipation*, cf. [22] for Ψ quadratic. The relation

$$\mathfrak{G}(u) = \Phi(u_0) - f(0) + f(T) \quad (17.35)$$

20) A Nobel prize was awarded to Lars Onsager in 1968 "for the discovery of the reciprocal relations bearing his name, which are fundamental for the thermodynamics of irreversible processes."

21) Here (17.26) reads as $\Psi(\mathrm{d}u/\mathrm{d}t) + \Psi^*(f - \Phi'(u)) \geq \langle f - \Phi'(u), \mathrm{d}u/\mathrm{d}t\rangle = \langle f, \mathrm{d}u/\mathrm{d}t\rangle - [\mathrm{d}\Phi/\mathrm{d}t](u)$, from which (17.34) results by integration over $[0, T]$.

is sometimes called *De Giorgi's formulation* of (17.30); rather than for existence proofs by the direct method, this formulation is used for various passages to a limit. Note that for f constant, the only time derivative involved in (17.34) is $\Psi(\mathrm{d}u/\mathrm{d}t)$, which allows for an interpretation even if V is only a metric space and thus $\mathrm{d}u/\mathrm{d}t$ itself is not defined, which leads to a theory of *gradient flows in metric spaces*, cf. [23, Theorem 2.3.3].

A combination of (17.27) and (17.30) leads to

$$\Psi'\left(\frac{\mathrm{d}u}{\mathrm{d}t}\right) + \Phi'(u) = f(t), \qquad u(T) = u(0),$$

and the related variational principle uses $\mathfrak{F}$ from (17.32) but with $\Phi(u(T))-\Phi(u_0)$ omitted, to be minimized on the linear manifold (17.33) with u_0 replaced by $u(T)$.

Many physical systems exhibit oscillatory behavior combined possibly with attenuation by nonconservative forces having a (pseudo)potential Ψ, which can be covered by the (Cauchy problem for the) abstract second-order evolution equation

$$\mathcal{T}'\frac{\mathrm{d}^2u}{\mathrm{d}t^2} + \Psi'\left(\frac{\mathrm{d}u}{\mathrm{d}t}\right) + \Phi'(u) = f(t),$$
$$u(0) = u_0, \quad \frac{\mathrm{d}u}{\mathrm{d}t}(0) = v_0, \qquad (17.36)$$

where $\mathcal{T} : H \to \mathbb{R}$ is the positive (semi)definite quadratic form representing the kinetic energy. The *Hamilton variational principle* extended to dissipative systems says that the solution u to (17.36) is a critical point of the integral functional

$$\int_0^T \mathcal{T}\left(\frac{\mathrm{d}u}{\mathrm{d}t}\right) - \Phi(u) + \langle f-\mathfrak{f}, u\rangle \,\mathrm{d}t \qquad (17.37)$$

with a nonconservative force $\mathfrak{f} = \Psi'(\mathrm{d}u/\mathrm{d}t)$ considered fixed on the affine manifold $\{u\in L^\infty(I;V);\ \mathrm{d}u/\mathrm{d}t\in L^\infty(I;H),\ \mathrm{d}^2u/\mathrm{d}t^2\in L^2(I;V^*),\ u(0)=u_0,\ \mathrm{d}u/\mathrm{d}t = v_0\}$, cf. [24].

17.2.4.2 Evolution Variational Inequalities

For nonsmooth potentials, the above evolution equations turn into inclusions. Instead of the Legendre transformation, we speak about the *Legendre–Fenchel transformation*. For example, instead of $[\Psi^*]' = [\Psi']^{-1}$, we have $\partial\Psi^* = [\partial\Psi]^{-1}$. Note that variational principles based on $\mathfrak{F}$ from (17.24), (17.29), or (17.32) do not involve any derivatives of Φ and Ψ and are especially designed for nonsmooth problems, and also $\mathfrak{G}$ from (17.34) allows for Ψ to be nonsmooth. For example, in the case of (17.30), with convex but nonsmooth Ψ and Φ, we have the *doubly nonlinear inclusion*

$$\partial\Psi\left(\frac{\mathrm{d}u}{\mathrm{d}t}\right)+\partial\Phi(u)\ni f(t),\ u(0) = u_0, \quad (17.38)$$

and $\mathfrak{F}(u,w) = 0$ in (17.32) and (17.33) means exactly that $w\in\partial\Psi(\mathrm{d}u/\mathrm{d}t)$ and $f-w\in\partial\Phi(u)$ hold a.e. on I,[22)] which (in the spirit of Section 17.2.2) can be written as a system of two variational inequalities for u and w:

$$\forall v:\quad \Psi(v) + \left\langle w, v-\frac{\mathrm{d}u}{\mathrm{d}t}\right\rangle \geq \Psi\left(\frac{\mathrm{d}u}{\mathrm{d}t}\right), \qquad (17.39a)$$

$$\forall v:\quad \Phi(v) + \langle f-w, v-u\rangle \geq \Phi(u). \qquad (17.39b)$$

For a systematic treatment of such multiply nonlinear inequalities, see [25].

In applications, the nonsmoothness of Ψ occurs typically at 0 describing activation phenomena: the abstract driving force $f-\partial\Phi(u)$ must pass a threshold, that is, the boundary of the convex set $\partial\Psi(0)$, in order to trigger the evolution of u. Often, any

22) The idea behind the principle in (17.32) and (17.33) is to apply two Fenchel inequalities to (17.38) written as $w\in\partial\Psi(\mathrm{d}u/\mathrm{d}t)$ and $f-w\in\partial\Phi(u)$.

rate dependence is neglected, and then Ψ is degree-1 positively homogeneous.[23] In this kind of *rate-independent* case, $\Psi^* = \delta_{\partial\Psi(0)}$, while $\Psi = \delta^*_{\partial\Psi(0)}$, and the De Giorgi formulation (17.35) leads to the *energy equality*

$$E(T, u(T)) + \int_0^T \Psi\left(\frac{du}{dt}\right) dt = E(0, u_0) - \int_0^T \left\langle \frac{df}{dt}, u\right\rangle dt$$
$$\text{for } E(t,u) = \Phi(u) - \langle f(t), u\rangle \quad (17.40a)$$

together with $f(t)-\Phi'(u(t))\in\partial\Psi(0)$ for a.a. (almost all) $t\in[0,T]$; here, in accordance with (17.35), we assume Φ to be smooth for the moment. This inclusion means $\Psi(v) - \langle f - \Phi'(u), v\rangle \geq \Psi(0) = 0$ and, as Φ is convex, we obtain the *stability condition*,[24]

$$\forall t\in[0,T]\ \forall v\in V: \quad E(t, u(t)) \leq E(t, v) + \Psi(v-u(t)). \quad (17.40b)$$

Moreover, in this rate-independent case, $\partial\Psi^* = N_{\partial\Psi(0)}$ and (17.31) reads $du/dt \in N_{\partial\Psi(0)}(f - \Phi'(u))$. By (17.13), it means that $\langle du/dt, v - f + \Phi'(u)\rangle \leq 0$ for any $v\in \partial\Psi(0)$, that is,

$$\max_{v\in\partial\Psi(0)} \left\langle \frac{du}{dt}, v\right\rangle = \left\langle \frac{du}{dt}, f - \Phi'(u)\right\rangle, \quad (17.41)$$

which says that the dissipation due to the driving force $f - \Phi'(u)$ is maximal compared to all admissible driving forces provided the rate du/dt is kept fixed; this is the *maximum dissipation principle.*

In fact, (17.40) does not contain Φ' and thus works for Φ convex nonsmooth, too. Actually, (17.40) was invented in [26], where it is called the *energetic formulation* of (17.38), cf. also [27].

23) This means $\Psi(\lambda v) = \lambda\Psi(v)$ for any $\lambda \geq 0$.

24) By convexity of Φ, we have $\Phi(v) \geq \Phi(u) + \langle \Phi'(u), v-u\rangle$, and adding it with $\Psi(v-u) - \langle f-\Phi'(u), v-u\rangle \geq 0$, we get (17.40b).

17.2.4.3 Recursive Variational Problems Arising by Discretization in Time

The variational structure related to the potentials of Section 17.2.4.1 can be exploited not only for formulation of "global" in time-variational principles, but, perhaps even more efficiently, to obtain recursive (incremental) variational problems when discretizing the abstract evolution problems in time by using some (semi) implicit formulae. This can serve as an efficient theoretical method for analyzing evolution problems (the *Rothe method,* [28]) and for designing efficient conceptual algorithms for numerical solution of such problems.

Considering a uniform partition of the time interval with the time step $\tau > 0$ with T/τ integer, we discretize (17.21) as

$$\frac{u_\tau^k - u_\tau^{k-1}}{\tau} + \Phi'(u_\tau^k) = f(k\tau), \qquad k = 1, \dots, \frac{T}{\tau}, \quad u_\tau^0 = u_0. \quad (17.42)$$

This is also known as the *implicit Euler formula* and u_τ^k for $k = 1, \dots, T/\tau$ approximate respectively the values $u(k\tau)$. One can apply the direct method by employing the recursive variational problem for the functional

$$\Phi(u) + \frac{1}{2\tau}\|u-u_\tau^{k-1}\|_H^2 - \langle f(k\tau), u\rangle \quad (17.43)$$

to be minimized for u. Obviously, any critical point u (and, in particular, a minimizer) of this functional solves (17.42) and we put $u = u_\tau^k$. Typically, after ensuring existence of the approximate solutions $\{u_\tau^k\}_{k=1}^{T/\tau}$, a priori estimates have to be derived[25] and then convergence as $\tau \to 0$ is to be proved by

25) For this, typically, testing (17.42) (or its difference from $k-1$ level) by u_τ^k or by $u_\tau^k - u_\tau^{k-1}$ (or $u_\tau^k - 2u_\tau^{k-1} + u_\tau^{k-2}$) is used with Young's and (discrete) Gronwall's inequalities, and so on.

various methods.[26] Actually, Φ does not need to be smooth and, referring to (17.11), we can investigate the set-valued variational inclusion $\mathrm{d}u/\mathrm{d}t+\partial_{\mathrm{F}}\Phi(u) \ni f$.

In specific situations, the fully implicit scheme (17.42) can be advantageously modified in various ways. For example, in case $\Phi = \Phi_1 + \Phi_2$ and $f = f_1 + f_2$, one can apply the *fractional-step method*, alternatively to be understood as a Lie–Trotter (or *sequential*) *splitting* combined *with the implicit Euler formula*:

$$\frac{u_\tau^{k-1/2} - u_\tau^{k-1}}{\tau} + \Phi_1'(u_\tau^{k-1/2}) = f_1(k\tau), \tag{17.44a}$$

$$\frac{u_\tau^{k} - u_\tau^{k-1/2}}{\tau} + \Phi_2'(u_\tau^{k}) \ni f_2(k\tau), \tag{17.44b}$$

with $k = 1, \dots, T/\tau$. Clearly, (17.44) leads to two variational problems that are to be solved in alternation.

Actually, we have needed rather the splitting of the underlying operator $A = \Phi_1' + \Phi_2' : V \to V^*$ and not of its potential $\Phi = \Phi_1 + \Phi_2 : V \to \mathbb{R}$. In case $\Phi : V = Y \times Z \to \mathbb{R}$, $u = (y, z)$ and $f = (g, h)$ where (17.21) represents a system of two equations

$$\frac{\mathrm{d}y}{\mathrm{d}t} + \Phi_y'(y, z) = g, \quad y(0) = y_0, \tag{17.45a}$$

$$\frac{\mathrm{d}z}{\mathrm{d}t} + \Phi_z'(y, z) = h, \quad z(0) = z_0, \tag{17.45b}$$

with Φ_y' and Φ_z' denoting partial differentials, one can thus think also about the splitting $\Phi'-f = (\Phi_y'-g, \Phi_z'-h) = (\Phi_y'-g, 0) + (0, \Phi_z'-h)$. Then the fractional method such as (17.44) yields a *semi-implicit scheme*[27]

$$\frac{y_\tau^k - y_\tau^{k-1}}{\tau} + \Phi_y'(y_\tau^k, z_\tau^{k-1}) = g(k\tau), \tag{17.46a}$$

$$\frac{z_\tau^k - z_\tau^{k-1}}{\tau} + \Phi_z'(y_\tau^k, z_\tau^k) = h(k\tau), \tag{17.46b}$$

again for $k = 1, \dots, T/\tau$. Note that the use of z_τ^{k-1} in (17.46a) decouples the system (17.46), in contrast to the fully implicit formula which would use z_τ^k in (17.46a) and would not decouple the original system (17.45). The underlying variational problems for the functionals $y \mapsto \Phi(y, z_\tau^{k-1}) + \frac{1}{2\tau}\|y-y_\tau^{k-1}\|^2 - \langle g(k\tau), y\rangle$ and $z \mapsto \Phi(y_\tau^k, z) + \frac{1}{2\tau}\|z-z_\tau^{k-1}\|^2 - \langle h(k\tau), z\rangle$ represent recursive alternating variational problems; these particular problems can be convex even if Φ itself is not; only separate convexity[28] of Φ suffices. Besides, under certain relatively weak conditions, this semi-implicit discretization is "numerically" stable; cf. [18, Remark 8.25]. For a convex/concave situation as in Theorem 17.5, (17.46) can be understood as an iterative *algorithm of Uzawa*'s type (with a damping by the implicit formula) for finding a saddle point.[29]

Of course, this decoupling method can be advantageously applied to nonsmooth situations and for u with more than two components, that is, for systems of more than two equations or inclusions. Even more, the splitting as in (17.45) may yield a variational structure of the decoupled incremental problems even if the original problem of the type $\mathrm{d}u/\mathrm{d}t + A(u) \ni 0$ itself does not have it. An obvious example for this is $A(y, z) = (\Phi_1'(\cdot, z)](y), \Phi_2'(y, \cdot)](z))$, which does not need to satisfy the symmetry (17.5) if $\Phi_1 \neq \Phi_2$ although the corresponding semi-implicit scheme

26) Typically, a combination of the arguments based on weak lower semicontinuity or compactness is used.

27) Indeed, in (17.44), one has $u_\tau^{k-1} = (y_\tau^{k-1}, z_\tau^{k-1})$, $u_\tau^{k-1/2} = (y_\tau^k, z_\tau^{k-1})$, and eventually $u_\tau^k = (y_\tau^k, z_\tau^k)$.

28) This means that only $\Phi(y, \cdot)$ and $\Phi(\cdot, z)$ are convex but not necessarily $\Phi(\cdot, \cdot)$.

29) This saddle point is then a steady state of the underlying evolution system (17.45).

(17.46) still possesses a "bi-variational" structure.

Similarly to (17.42), the doubly nonlinear problem (17.38) uses the formula

$$\partial\Psi\Big(\frac{u_\tau^k - u_\tau^{k-1}}{\tau}\Big) + \partial\Phi(u_\tau^k) \ni f(k\tau) \quad (17.47)$$

and, instead of (17.43), the functional

$$\Phi(u) + \tau\Psi\Big(\frac{u - u_\tau^{k-1}}{\tau}\Big) - \big\langle f(k\tau), u\big\rangle. \quad (17.48)$$

Analogously, for the second-order doubly nonlinear problem (17.36) in the nonsmooth case, that is, $\mathscr{T}\,'\mathrm{d}^2u/\mathrm{d}t^2 + \partial\Psi\big(\mathrm{d}u/\mathrm{d}t\big) + \partial\Phi(u) \ni f(t)$, we would use

$$\mathscr{T}\,'\frac{u_\tau^k - 2u_\tau^{k-1} + u_\tau^{k-2}}{\tau^2} + \partial\Psi\Big(\frac{u_\tau^k - u_\tau^{k-1}}{\tau}\Big) + \partial\Phi(u_\tau^k) \ni f(k\tau) \quad (17.49)$$

and the recursive variational problem for the functional

$$\Phi(u) + \tau\Psi\Big(\frac{u - u_\tau^{k-1}}{\tau}\Big) - \big\langle f(k\tau), u\big\rangle + \tau^2\mathscr{T}\Big(\frac{u - 2u_\tau^{k-1} + u_\tau^{k-2}}{\tau^2}\Big). \quad (17.50)$$

The fractional-step method and, in particular, various semi-implicit variants of (17.47) and (17.49) are widely applicable, too.

17.3 Variational Problems on Specific Function Spaces

We now use the abstract framework from Section 17.2 for concrete variational problems formulated on specific function spaces.

17.3.1 Sobolev Spaces

For this, we consider a bounded domain $\Omega \subset \mathbb{R}^d$ equipped with the Lebesgue measure, having a smooth boundary $\Gamma := \partial\Omega$. For $1 \le p < \infty$, we will use the standard notation

$$L^p(\Omega;\mathbb{R}^n) = \Big\{u : \Omega \to \mathbb{R}^n \text{ measurable}; \int_\Omega |u(x)|^p\,\mathrm{d}x < \infty\Big\}$$

for the *Lebesgue space*; the addition and the multiplication understood pointwise makes it a linear space, and introducing the norm

$$\|u\|_p = \Big(\int_\Omega |u(x)|^p\,\mathrm{d}x\Big)^{1/p}$$

makes it a Banach space. For $p = \infty$, we define $\|u\|_\infty = \operatorname{ess\,sup}_{x\in\Omega}|u(x)| = \inf_{N\subset\Omega,\,\mathrm{meas}_d(N)=0}\,\sup_{x\in\Omega\setminus N}|u(x)|$. For $1<p<\infty$, $L^p(\Omega;\mathbb{R}^n)$ is reflexive. For $1\le p<\infty$, $L^p(\Omega;\mathbb{R}^n)^* = L^{p'}(\Omega;\mathbb{R}^n)$ with $p' = p/(p-1)$ if the duality is defined naturally as $\langle f,u\rangle = \int_\Omega f(x)\cdot u(x)\,\mathrm{d}x$. For $p = 2$, $L^p(\Omega;\mathbb{R}^n)$ becomes a Hilbert space. For $n = 1$, we write for short $L^p(\Omega)$ instead of $L^p(\Omega;\mathbb{R})$.

Denoting the kth order gradient of u by $\nabla^k u = (\partial^k/\partial x_{i_1}\cdots\partial x_{i_k} u)_{1\le i_1,\dots,i_k\le d}$, we define the *Sobolev space* by

$$W^{k,p}(\Omega;\mathbb{R}^n) = \Big\{u\in L^p(\Omega;\mathbb{R}^n); \nabla^k u\in L^p(\Omega;\mathbb{R}^{n\times d^k})\Big\},$$

normed by $\|u\|_{k,p} = \sqrt[p]{\|u\|_p^p + \|\nabla^k u\|_p^p}$.

If $n = 1$, we will again use the shorthand notation $W^{k,p}(\Omega)$. If $p = 2$, $W^{k,p}(\Omega;\mathbb{R}^n)$ is a Hilbert space and we will write $H^k(\Omega;\mathbb{R}^n) = W^{k,2}(\Omega;\mathbb{R}^n)$. Moreover, we occasionally use a subspace of $W^{k,p}(\Omega;\mathbb{R}^n)$

with vanishing traces on the boundary Γ, denoted by

$$W_0^{k,p}(\Omega;\mathbb{R}^n) = \big\{u \in W^{k,p}(\Omega;\mathbb{R}^n);\ \nabla^l u = 0 \text{ on } \Gamma,\ l = 0,\dots,k-1\big\}. \tag{17.51}$$

To give a meaningful interpretation to traces $\nabla^l u$ on Γ, this boundary has to be sufficiently regular; roughly speaking, piecewise C^{l+1} is enough.

We denote by $C^k(\bar{\Omega})$ the space of *smooth functions* whose gradients up to the order k are continuous on the closure $\bar{\Omega}$ of Ω. For example, we have obviously embeddings $C^k(\bar{\Omega}) \subset W^{k,p}(\Omega) \subset L^p(\Omega)$; in fact, these embeddings are dense.

An important phenomenon is the compactifying effect of derivatives. A prototype for it is the Rellich – Kondrachov theorem, saying that $H^1(\Omega)$ is *compactly*[30] embedded into $L^2(\Omega)$. More generally, we have

Theorem 17.10 (Compact embedding)
For the Sobolev critical exponent

$$p^* \begin{cases} = dp/(d-p) & \textit{for } p < d, \\ \in [1,+\infty) \textit{ arbitrary} & \textit{for } p = d, \\ = +\infty & \textit{for } p > d, \end{cases}$$

the embedding $W^{1,p}(\Omega) \subset L^{p^*-\epsilon}(\Omega)$ *is compact for any* $0 < \epsilon \le p^*-1$.

Iterating this theorem, we can see, for example, that, for $p < d/2$, the embedding $W^{2,p}(\Omega) \subset L^{[p^*]^*-\epsilon}(\Omega)$ is compact; note that $[p^*]^* = dp/(d-2p)$.

Another important fact is the compactness of the trace operator $u \mapsto u|_\Gamma$:

Theorem 17.11 (Compact trace operator)
For the boundary critical exponent

$$p^\sharp \begin{cases} = (dp-p)/(d-p) & \textit{for } p < d, \\ \in [1,+\infty) \textit{ arbitrary} & \textit{for } p = d, \\ = +\infty & \textit{for } p > d, \end{cases}$$

the trace operator $u \mapsto u|_\Gamma : W^{1,p}(\Omega) \subset L^{p^\sharp-\epsilon}(\Gamma)$ *is compact for any* $0 < \epsilon \le p^\sharp-1$.

For example, the trace operator from $W^{2,p}(\Omega)$ is compact into $L^{[p^*]^\sharp-\epsilon}(\Gamma)$.[31]

17.3.2 Steady-State Problems

The above abstract functional-analysis scenario gives a lucid insight into concrete variational problems leading to boundary-value problems for quasilinear equations in divergence form which is what we will now focus on. We consider a bounded domain $\Omega \subset \mathbb{R}^d$ with a sufficiently regular boundary Γ divided into two disjoint relatively open parts Γ_D and Γ_N whose union is dense in Γ. An important tool is a generalization of the superposition operator, the *Nemytskiĭ mapping* $\mathcal{N}_a$, induced by a *Carathéodory*[32] *mapping* $a : \Omega \times \mathbb{R}^n \to \mathbb{R}^m$ by prescribing $[\mathcal{N}_a(u)](x) = a(x,u(x))$.

Theorem 17.12 (Nemytskiĭ mapping)
Let $a : \Omega \times \mathbb{R}^n \to \mathbb{R}^m$ *be a Carathéodory mapping and* $p, q \in [1,\infty)$. *Then* $\mathcal{N}_a$ *maps* $L^p(\Omega;\mathbb{R}^n)$ *into* $L^q(\Omega;\mathbb{R}^m)$ *and is continuous if, and only if, for some* $\gamma \in L^q(\Omega)$ *and* $C < \infty$, *we have that*

$$|a(x,u)| \le \gamma(x) + C|u|^{p/q}.$$

30) This means that the embedding is a compact mapping in the sense that weakly converging sequences in $H^1(\Omega)$ converge strongly in $L^2(\Omega)$.

31) To see this, we use Theorem 17.10 to obtain $W^{2,p}(\Omega) \subset W^{1,p^*-\epsilon}(\Omega)$, and then Theorem 17.11 with $p^* - \epsilon$ in place of p.

32) The Carathéodory property means measurability in the x-variable and continuity in all other variables.

17.3.2.1 Second Order Systems of Equations

First, we consider the integral functional

$$\Phi(u) = \int_{\Omega} \varphi(x, u, \nabla u)\,\mathrm{d}x + \int_{\Gamma_\mathrm{N}} \phi(x, u)\,\mathrm{d}S \tag{17.52a}$$

involving Carathéodory integrands $\varphi : \Omega\times\mathbb{R}^n\times\mathbb{R}^{n\times d} \to \mathbb{R}$ and $\phi : \Gamma_\mathrm{N}\times\mathbb{R}^n \to \mathbb{R}$. The functional Φ is considered on an affine closed manifold

$$\{u \in W^{1,p}(\Omega;\mathbb{R}^n);\ u|_{\Gamma_\mathrm{D}} = u_\mathrm{D}\} \tag{17.52b}$$

for a suitable given u_D; in fact, existence of $\bar{u}_\mathrm{D} \in W^{1,p}(\Omega;\mathbb{R}^n)$ such $u_\mathrm{D} = \bar{u}_\mathrm{D}|_{\Gamma_\mathrm{D}}$ is to be required. Equipped with the theory of $W^{1,p}$-Sobolev spaces,[33] one considers a p-polynomial-type coercivity of the highest-order term and the corresponding growth restrictions on the partial derivatives φ'_F, φ'_u, and ϕ'_u with some $1 < p < \infty$, that is,

$$\varphi(x, u, F){:}F \geq \epsilon|F|^p + |u|^\epsilon - \frac{1}{\epsilon}, \tag{17.53a}$$

$$\exists\gamma \in L^{p'}(\Omega):\ |\varphi'_F(x, u, F)| \leq \gamma(x) + C|u|^{(p^*-\epsilon)/p'} + C|F|^{p-1}, \tag{17.53b}$$

$$\exists\gamma \in L^{p^{*\prime}}(\Omega):\ |\varphi'_u(x, u, F)| \leq \gamma(x) + C|u|^{p^*-1-\epsilon} + C|F|^{p/p^{*\prime}}, \tag{17.53c}$$

$$\exists\gamma \in L^{p^{\sharp\prime}}(\Gamma):\ |\phi'_u(x, u)| \leq \gamma(x) + C|u|^{p^\sharp-1-\epsilon} \tag{17.53d}$$

for some $\epsilon > 0$ and $C < \infty$; we used F as a placeholder for ∇u. A generalization of Theorem 17.12 for Nemytskiǐ mappings of several arguments says that (17.53b) ensures just continuity of $\mathscr{N}_{\varphi'_F} : L^{p^*-\epsilon}(\Omega;\mathbb{R}^n) \times L^p(\Omega;\mathbb{R}^{n\times d}) \to L^{p'}(\Omega;\mathbb{R}^{n\times d})$, and analogously also (17.53c) works for $\mathscr{N}_{\varphi'_u}$, while (17.53d) gives continuity of $\mathscr{N}_{\phi'_u} : L^{p^\sharp-\epsilon}(\Gamma;\mathbb{R}^n) \to L^{p^{\sharp\prime}}(\Gamma;\mathbb{R}^n)$. This, together with Theorems 17.10 and 17.11, reveals the motivation for the growth conditions (17.53b–d).

For $\epsilon \geq 0$, (17.53b–d) ensures that the functional Φ from (17.52a) is *Gâteaux differentiable* on $W^{1,p}(\Omega;\mathbb{R}^n)$. The abstract *Euler–Lagrange equation* (17.3) then leads to the integral identity

$$\int_{\Omega} \varphi'_{\nabla u}(x, u, \nabla u){:}\nabla v + \varphi'_u(x, u, \nabla u){\cdot}v\,\mathrm{d}x + \int_{\Gamma_\mathrm{N}} \phi'_u(x, u){\cdot}v\,\mathrm{d}S = 0 \tag{17.54}$$

for any $v \in W^{1,p}(\Omega;\mathbb{R}^n)$ such that $v|_{\Gamma_\mathrm{D}} = 0$; the notation " : " or " · " means summation over two indices or one index, respectively. Completed by the Dirichlet condition on Γ_D, this represents a *weak formulation* of the *boundary-value problem* for a system of second-order elliptic quasilinear equations:[34]

$$\operatorname{div}\varphi'_{\nabla u}(u, \nabla u) = \varphi'_u(u, \nabla u) \quad \text{in } \Omega, \tag{17.55a}$$

$$\varphi'_{\nabla u}(u, \nabla u){\cdot}\vec{n} + \phi'_u(u) = 0 \quad \text{on } \Gamma_\mathrm{N}, \tag{17.55b}$$

$$u|_\Gamma = u_\mathrm{D} \quad \text{on } \Gamma_\mathrm{D}, \tag{17.55c}$$

where x-dependence has been omitted for notational simplicity. The conditions (17.55b) and (17.55c) are called the *Robin* and the *Dirichlet boundary conditions*, respectively, and (17.55) is called the *classical formulation* of this boundary-value problem. Any $u \in C^2(\bar{\Omega};\mathbb{R}^n)$ satisfying (17.55) is called a *classical solution*, while $u \in W^{1,p}(\Omega;\mathbb{R}^n)$ satisfying (17.54) for any $v \in W^{1,p}(\Omega;\mathbb{R}^n)$ such that $v|_{\Gamma_\mathrm{D}} = 0$ is

33) More general nonpolynomial growth and coercivity conditions would require the theory of Orlicz spaces instead of the Lebesgue ones, cf. [9, Chapter 53].

34) Assuming sufficiently smooth data as well as u, this can be seen by multiplying (17.55a) by v, using the Green formula $\int_\Omega(\operatorname{div} a)v + a{\cdot}v\,\mathrm{d}x = \int_\Gamma(a\cdot\vec{n})v\,\mathrm{d}S$, and using $v = 0$ on Γ_D and the boundary conditions (17.55b) on Γ_N.

called a *weak solution*; note that much less smoothness is required for weak solutions.

Conversely, taking general Carathéodory integrands $a : \Omega\times\mathbb{R}^n\times\mathbb{R}^{n\times d} \to \mathbb{R}^{n\times d}$, $b : \Gamma_{\rm N}\times\mathbb{R}^n \to \mathbb{R}^n$, and $c : \Omega\times\mathbb{R}^n\times\mathbb{R}^{n\times d} \to \mathbb{R}^n$, one can consider a boundary-value problem for a system of second-order elliptic quasilinear equations

$$\operatorname{div} a(u,\nabla u) = c(u,\nabla u) \quad \text{in } \Omega, \tag{17.56a}$$

$$a(u,\nabla u)\cdot\vec{n} + b(u) = 0 \quad \text{on } \Gamma_{\rm N}, \tag{17.56b}$$

$$u|_{\Gamma_{\rm D}} = u_{\rm D} \quad \text{on } \Gamma_{\rm D}. \tag{17.56c}$$

Such a problem does not need to be induced by any potential Φ; nevertheless, it possesses a weak formulation as in (17.54), namely, $\int_\Omega a(u,\nabla u){:}\nabla v + c(u,\nabla u)\cdot v\,\mathrm{d}x + \int_{\Gamma_{\rm N}} b(u)\cdot v\,\mathrm{d}S = 0$ for any "variation" v as in (17.54), and related methods are sometimes called variational in spite of absence of a potential Φ. The existence of such a potential requires a certain *symmetry* corresponding to that in (17.5) for the underlying nonlinear operator $A : W^{1,p}(\Omega;\mathbb{R}^n) \to W^{1,p}(\Omega;\mathbb{R}^n)^*$ given by $\langle A(u),v\rangle = \int_\Omega a(u,\nabla u){:}\nabla v + c(u,\nabla u)\cdot v\,\mathrm{d}x + \int_{\Gamma_{\rm N}} b(u)\cdot v\,\mathrm{d}S$, namely,

$$\frac{\partial a_{il}(x,u,F)}{\partial F_{jk}} = \frac{\partial a_{kj}(x,u,F)}{\partial F_{li}}, \tag{17.57a}$$

$$\frac{\partial a_{il}(x,u,F)}{\partial u_j} = \frac{\partial c_j(x,u,F)}{\partial F_{li}}, \tag{17.57b}$$

$$\frac{\partial c_j(x,u,F)}{\partial u_l} = \frac{\partial c_l(x,u,F)}{\partial u_j} \tag{17.57c}$$

for all $i,k = 1,\ldots,d$ and $j,l = 1,\ldots,n$ and for a.a. $(x,u,F)\in\Omega\times\mathbb{R}^n\times\mathbb{R}^{n\times d}$, and also

$$\frac{\partial b_j(x,u)}{\partial u_l} = \frac{\partial b_l(x,u)}{\partial u_j}. \tag{17.57d}$$

for all $j,l = 1,\ldots,n$ and for a.a. $(x,u) \in \Gamma\times\mathbb{R}^n$. Note that (17.57a–c) just means a symmetry for the Jacobian of the mapping $(F,u) \mapsto (a(x,u,F), c(x,u,F)) : \mathbb{R}^{n\times d}\times\mathbb{R}^d \to \mathbb{R}^{n\times d}\times\mathbb{R}^d$ while (17.57d) is the symmetry for the Jacobian of $b(x,\cdot) : \mathbb{R}^n \to \mathbb{R}^n$.

Then (17.6) leads to the formula (17.52a) with

$$\varphi(x,u,F) = \int_0^1 a(x,\lambda u,\lambda F){:}F + c(x,\lambda u,\lambda F)\cdot u\,\mathrm{d}\lambda, \tag{17.58a}$$

$$\phi(x,u) = \int_0^1 b(x,\lambda u)\cdot u\,\mathrm{d}\lambda. \tag{17.58b}$$

Relying on the minimization-of-energy principle described above, which is often a governing principle in steady-state mechanical and physical problems, and on Theorem 17.1 or 17.2, one can prove existence of weak solutions to the boundary-value problem by the *direct method*; cf. e.g. [29–32]. Theorem 17.2 imposes a strong (although often applicable) structural restriction that $\varphi(x,\cdot,\cdot) : \mathbb{R}^n\times\mathbb{R}^{n\times d} \to \mathbb{R}$ and $\phi(x,\cdot) : \mathbb{R}^n \to \mathbb{R}$ are convex for a.a. x.

Yet, in general, Theorem 17.1 places fewer restrictions on φ and ϕ by requiring only weak lower semicontinuity of Φ. The precise condition (i.e., sufficient and necessary) that guarantees such semicontinuity of $u \mapsto \int_\Omega \varphi(x,u,\nabla u)\,\mathrm{d}x$ on $W^{1,p}(\Omega;\mathbb{R}^n)$ is called *$W^{1,p}$-quasiconvexity*, defined in a rather nonexplicit way by requiring

$$\forall x\in\Omega\ \forall u\in\mathbb{R}^n\ \forall F\in\mathbb{R}^{n\times d} : \ \varphi(x,u,F) = = \inf_{v\in W_0^{1,p}(O;\mathbb{R}^d)} \int_O \frac{\varphi(x,u,F+\nabla v(\xi))}{\operatorname{meas}_d(O)}\,\mathrm{d}\xi,$$

where $O\subset\mathbb{R}^d$ is an arbitrary smooth domain. This condition cannot be verified efficiently except for very special cases, unlike, for example, polyconvexity which is a (strictly) stronger condition. Subsequently, another type of convexity, called *rank-one convexity*, was introduced by Morrey [33] by requiring

$\lambda \mapsto \varphi(x, u, F+\lambda a\otimes b) : \mathbb{R} \to \mathbb{R}$ to be convex for any $a\in\mathbb{R}^d$, $b\in\mathbb{R}^n$, $[a\otimes b]_{ij} = a_i b_j$. For smooth $\varphi(x, u, \cdot)$, rank-one convexity is equivalent to the *Legendre–Hadamard condition* $\varphi''_{FF}(x, u, F)(a\otimes b, a\otimes b) \geq 0$ for all $a\in\mathbb{R}^n$ and $b\in\mathbb{R}^d$. Since Morrey's [33] introduction of quasiconvexity, the question of its coincidence with rank-one convexity had been open for many decades and eventually answered negatively by Šverák [34] at least if $n \geq 3$ and $d \geq 2$. Weak lower semicontinuity of the boundary integral $u \mapsto \int_\Omega \phi(x, u)\,\mathrm{d}S$ in (17.52a) does not entail any special structural condition because one can use compactness of the trace operator, cf. Theorem 17.11. Here, Theorem 17.1 leads to the following theorem:

Theorem 17.13 (Direct method) *Let* (17.53) *hold with $\epsilon > 0$, let $\varphi(x, u, \cdot)$ be quasiconvex, and let $u_{\mathrm{D}} \in W^{1-1/p,p}(\Gamma_{\mathrm{D}}; \mathbb{R}^n)$.*[35] *Then* (17.54) *has a solution, that is, the boundary-value problem* (17.55) *has a weak solution.*

For $n = d$, an example for a quasiconvex function is $\varphi(x, u, F) = \mathfrak{f}(x, u, F, \det F)$ with a convex function $\mathfrak{f}(x, u, \cdot, \cdot) : \mathbb{R}^{d\times d} \times \mathbb{R} \to \mathbb{R}$. The weak lower semicontinuity of Φ from (17.52a) is then based on the weak continuity of the nonlinear mapping induced by $\det : \mathbb{R}^{d\times d} \times \mathbb{R} \to \mathbb{R}$ if restricted to gradients, that is,

$$u_k \to u \text{ weakly in } W^{1,p}(\Omega; \mathbb{R}^d) \quad \Rightarrow \quad \det\nabla u_k \to \det\nabla u \text{ weakly in } L^{p/d}(\Omega), \tag{17.59}$$

which holds for $p > d$; note that nonaffine mappings on Lebesgue spaces such as $G \mapsto \det G$ with $G \in L^p(\Omega; \mathbb{R}^{d\times d}) \to L^{p/d}(\Omega)$ can be continuous[36] but not weakly continuous, so (17.59) is not entirely trivial. Even less trivial, it holds for $p = d$ locally (i.e., in $L^1(K)$ for any compact $K \subset \Omega$) if $\det\nabla u_k \geq 0$.[37] Invented by Ball [36], such functions $\varphi(x, u, \cdot)$ are called *polyconvex*, and in general this property requires

$$\varphi(x, u, F) = \mathfrak{f}\Big(x, u, (\mathrm{adj}_i F)_{i=1}^{\min(n,d)}\Big) \tag{17.60}$$

for some $\mathfrak{f} : \Omega \times \mathbb{R}^n \times \prod_{i=1}^{\min(n,d)} \mathbb{R}^{\kappa(i,n,d)} \to \mathbb{R} \cup \{\infty\}$ such that $\mathfrak{f}(x, u, \cdot)$ is convex, where $\kappa(i, n, d)$ is the number of all minors of the ith order and where $\mathrm{adj}_i F$ denotes the determinants of all $(i\times i)$-submatrices. Similarly, as in (17.59), we have that $\mathrm{adj}_i\nabla u_k \to \mathrm{adj}_i\nabla u$ weakly in $L^{p/i}(\Omega; \mathbb{R}^{\kappa(i,n,d)})$ provided $p > i \leq \min(n, d)$, and Theorem 17.13 directly applies if $\mathfrak{f}$ from (17.60) gives φ satisfying (17.53a–c).

Yet, this special structure allows for much weaker restrictions on φ if one is concerned with the minimization of Φ itself rather than the satisfaction of the Euler–Lagrange equation (17.54):

Theorem 17.14 (Direct method, polyconvex) *Let φ be a normal integrand*[38] *satisfying* (17.53a) *with $\varphi(x, u, \cdot) : \mathbb{R}^{n\times d} \to \mathbb{R} \cup \{\infty\}$ polyconvex, and let $u_{\mathrm{D}} \in W^{1-1/p,p}(\Gamma_{\mathrm{D}}; \mathbb{R}^n)$. Then the minimization problem* (17.52) *has a solution.*

Obviously, polyconvexity (and thus also quasi- and rank-one convexity) is weaker than usual convexity. Only for $\min(n, d) = 1$, all mentioned modes coincide with usual convexity of $\varphi(x, u, \cdot)$.

35) Without going into detail concerning the so-called Sobolev–Slobodetskiĭ spaces with fractional derivatives, this condition means exactly that u_{D} allows an extension onto Ω belonging to $W^{1,p}(\Omega; \mathbb{R}^n)$.

36) For $p \geq d$, Theorem 17.12 gives this continuity.

37) Surprisingly, not only $\{\det\nabla u_k\}_{k\in\mathbb{N}}$ but even $\{\det\nabla u_k \ln(2+\det\nabla u_k)\}_{k\in\mathbb{N}}$ stays bounded in $L^1(K)$, as proved by S. Müller in [35].

38) This means φ is measurable but $\varphi(x, \cdot, \cdot)$ is only lower semicontinuous.

Example 17.2 [Oscillation effects.] A simple one-dimensional counterexample for *nonexistence* of a solution due to oscillation effects is based on

$$\Phi(u) = \int_0^L \Big(\big(\frac{du}{dx}\big)^2 - 1\Big)^2 + u^2 \, dx \qquad (17.61)$$

to be minimized for $u \in W^{1,4}([0,L])$. A minimizing sequence $\{u_k\}_{k\in\mathbb{N}}$ is, for example,[39]

$$u_k(0) = \frac{1}{k}, \quad \frac{du_k}{dx} = \begin{cases} 1 \text{ if } \sin(kx)>0, \\ -1 \text{ otherwise.} \end{cases} \qquad (17.62)$$

Then $\Phi(u_k) = \mathcal{O}(1/k^2) \to 0$ for $k\to\infty$, so that $\inf\Phi = 0$. Yet, there is no u such that $\Phi(u) = 0$.[40] We can observe that Theorem 17.1 (resp. Theorem 17.2) cannot be used due to lack of weak lower semicontinuity (resp. convexity) of Φ which is due to nonconvexity of the *double-well potential* density $F \mapsto \varphi(x,u,F) = (|F|^2-1)^2 + u^2$; cf. also (17.105) below for a "fine limit" of the fast oscillations from Figure 17.1.

Example 17.3 [Concentration effects.] The condition that V in Theorems 17.1 and 17.2 has a pre-dual, is essential. A simple one-dimensional counterexample for *nonexistence* of a solution in the situation where V is not reflexive and even does not have any pre-dual, is based on

$$\Phi(u) = \int_{-1}^{1} (1+x^2)\Big|\frac{du}{dx}\Big| \, dx + (u(-1)+1)^2 + (u(1)-1)^2 \qquad (17.63)$$

for $u \in W^{1,1}([-1,1])$. If u were a minimizer, then u must be nondecreasing (otherwise, it obviously would not be optimal), and we can always take some "part" of the nonnegative derivative of this function and add the corresponding area in a neighborhood of 0. This does not affect $u(\pm1)$ but makes $\int_{-1}^{1}(1+x^2)|du/dx|\,dx$ lower, contradicting the original assumption that u is a minimizer. In fact, as $1+x^2$ in (17.63) attains its minimum only at a single point $x = 0$, any minimizing sequence $\{u_k\}_{k\in\mathbb{N}}$ is forced to concentrate its derivative around $x = 0$. For example considering, for $k \in \mathbb{N}$ and $\ell \in \mathbb{R}$, the sequence given by

$$u_k(x) = \frac{\ell kx}{1 + k|x|} \qquad (17.64)$$

yields $\Phi(u_k) = 2\ell + 2(\ell-1)^2 + \mathcal{O}(1/k^2)$. Obviously, the sequence $\{u_k\}_{k\in\mathbb{N}}$ will minimize Φ provided $\ell = 1/2$; then $\lim_{k\to\infty}\Phi(u_k) = 3/2 = \inf\Phi$; see Figure 17.2. On the other hand, this value $\inf\Phi$ cannot be achieved, otherwise such u must have simultaneously $|du/dx| = 0$ a.e. and $u(\pm1) = \pm1/2$, which is not possible.[41] A similar effect occurs for $\varphi(F) = \sqrt{1+|F|^2}$ for which $\int_\Omega \varphi(\nabla u)\,dx$ is the area of the parameterized hypersurface $\{(x,u(x));\ x \in \Omega\}$ in $\mathbb{R}^{d+1}$. Minimization of such a functional is known as the *Plateau variational problem.* Hyper-*surfaces of minimal area* typically do not exists in $W^{1,1}(\Omega)$, especially if Ω is not convex and the concentration of the gradient typically occurs on Γ rather than inside Ω, cf. e.g. [37, Chapter V].

Example 17.4 [Lavrentiev phenomenon.] Coercivity in Theorems 17.1 and 17.2 is also essential even if Φ is bounded from below. An innocent-looking one-dimensional

39) Actually, $u_k(0) \neq 0$ was used in (17.62) only for a better visualization on Figure 17.1.

40) Indeed, then both $\int_0^L((du/dx)^2-1)^2dx$ and $\int_0^L u^2dx$ would have to be zero, so that $u = 0$, but then also $du/dx = 0$, which however contradicts $\int_0^L((du/dx)^2-1)^2dx = 0$.

41) This is because of the concentration effect. More precisely, the sequence $\{du_k/dx\}_{k\in\mathbb{N}} \subset L^1(-1,1)$ is not uniformly integrable.

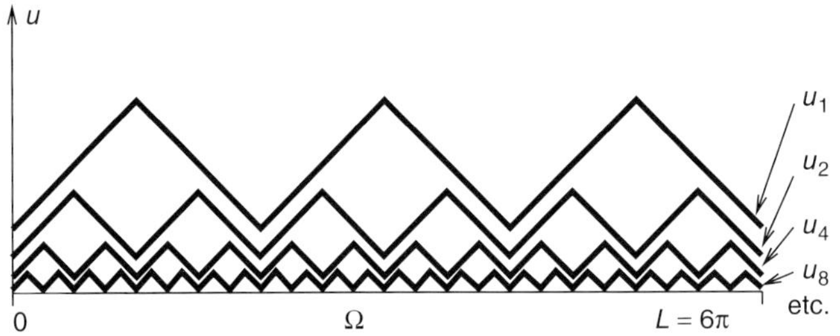

Figure 17.1 A minimizing sequence (17.62) for Φ from (17.61) whose gradient exhibits faster and faster spatial oscillations.

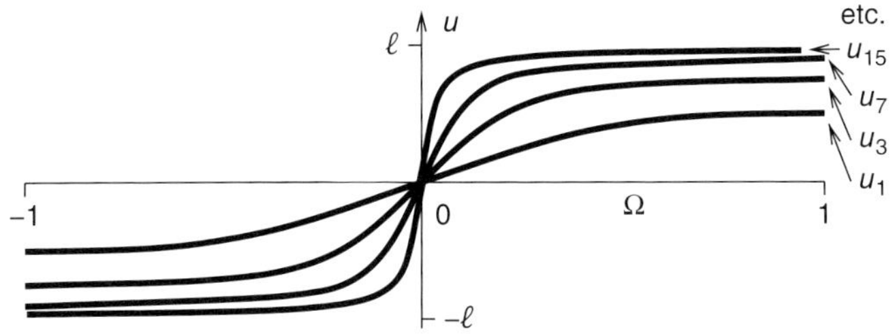

Figure 17.2 A minimizing sequence (17.64) for Φ from (17.63) whose gradient concentrates around the point $x = 0$ inside Ω.

counterexample for nonexistence of a solution in the situation where V is reflexive and $\Phi \geq 0$ is continuous and weakly lower semicontinuous is based on

$$\Phi(u) = \int_0^1 (u^3 - x)^2 \left(\frac{\mathrm{d}u}{\mathrm{d}x}\right)^6 \mathrm{d}x$$
$$\text{subject to} \quad u(0) = 0, \ u(1) = 1, \qquad (17.65)$$

for $u \in W^{1,6}([0,1]) = V$. The minimum of (17.65) is obviously 0, being realized on $u(x) = x^{1/3}$. Such $u \in W^{1,1}([0,1])$, however, does not belong to $W^{1,6}([0,1])$ because $|\mathrm{d}u/\mathrm{d}x|^6 = 3^{-6}x^{-4}$ is not integrable owing to its singularity at $x = 0$. Thus (17.65) attains the minimum on $W^{1,p}([0,1])$ with $1 \leq p < 3/2$ although Φ is not (weakly lower semi-) continuous and even not finite on this space, and thus abstract Theorem 17.1 cannot be used. A surprising and not entirely obvious phenomenon is that the infimum (17.65) on $W^{1,6}([0,1])$ is positive, that is, greater than the infimum on $W^{1,p}([0,1])$ with $p < 3/2$; this effect was first observed in [38], cf. also, e.g. [1, Section 4.3.]. Note that $W^{1,6}([0,1])$ is dense in $W^{1,p}([0,1])$ but one cannot rely on $\Phi(u_k) \to \Phi(u)$ if $u_k \to u$ in $W^{1,p}([0,1])$ for $p < 6$; it can even happen that $\Phi(u) = 0$ while $\Phi(u_k) \to \infty$ for $u_k \to u$, a *repulsive effect*, cf. [39, Section 7.3]. Here $\varphi(x, u, \cdot)$ is not uniformly convex, yet the Lavrentiev phenomenon can occur even for uniformly convex φ's, cf. [40].

17.3.2.2 Fourth Order Systems

Higher-order problems can be considered analogously but the complexity of the problem grows with the order. Let us therefore use for illustration fourth-order problems only, governed by an integral functional

$$\Phi(u) = \int_\Omega \varphi(x, u, \nabla u, \nabla^2 u)\,dx + \int_{\Gamma_N} \phi(x, u, \nabla u)\,dS \qquad (17.66a)$$

involving Carathéodory integrands $\varphi : \Omega\times\mathbb{R}^n\times\mathbb{R}^{n\times d}\times\mathbb{R}^{n\times d\times d} \to \mathbb{R}$ and $\phi : \Gamma_N\times\mathbb{R}^n\times\mathbb{R}^{n\times d} \to \mathbb{R}$. The functional Φ is considered on an affine closed manifold

$$\left\{u\in W^{2,p}(\Omega;\mathbb{R}^n);\ u|_{\Gamma_D} = u_{D,1},\ \frac{\partial u}{\partial \vec{n}}|_{\Gamma_D} = u_{D,2}\right\} \qquad (17.66b)$$

for a given suitable $u_{D,1}$ and $u_{D,2}$. Instead of (17.54), the abstract Euler–Lagrange equation (17.3) now leads to the integral identity:

$$\int_\Omega \Big(\varphi'_{\nabla^2 u}(x, u, \nabla u, \nabla^2 u)\vdots\nabla^2 v + \varphi'_{\nabla u}(x, u, \nabla u, \nabla^2 u):\nabla v + \varphi'_u(x, u, \nabla u, \nabla^2 u)\cdot v\Big)dx + \int_{\Gamma_N}\Big(\phi'_{\nabla u}(x, u, \nabla u)\cdot\frac{\partial v}{\partial \vec{n}} + \phi'_u(x, u, \nabla u)\cdot v\Big)\,dS = 0 \qquad (17.67)$$

for any $v\in W^{2,p}(\Omega;\mathbb{R}^n)$ such that $v|_{\Gamma_D} = 0$ and $\partial u/\partial\vec{n}|_{\Gamma_D} = 0$; the notation " $\vdots$ " stands for summation over three indices. Completed by the Dirichlet conditions on Γ_D, this represents a *weak formulation* of the *boundary-value problem* for a system of fourth-order elliptic quasilinear equations

$$\operatorname{div}^2 \varphi'_{\nabla^2 u}(u, \nabla u, \nabla^2 u) - \operatorname{div}\varphi'_{\nabla u}(u, \nabla u, \nabla^2 u) + \varphi'_u(u, \nabla u, \nabla^2 u) = 0 \text{ in } \Omega, \qquad (17.68a)$$

with two natural (although quite complicated) boundary conditions prescribed on each part of the boundary, namely,

$$\big(\operatorname{div}\varphi'_{\nabla^2 u}(u,\nabla u,\nabla^2 u) - \varphi'_{\nabla u}(u,\nabla u,\nabla^2 u)\big)\cdot\vec{n} + \operatorname{div}_S\big(\varphi'_{\nabla^2 u}(u, \nabla u, \nabla^2 u)\vec{n}\big) - \big(\operatorname{div}_S\vec{n}\big)\big(\vec{n}^\top\varphi'_{\nabla^2 u}(u, \nabla u, \nabla^2 u)\vec{n}\big) + \phi'_{\nabla u}(u, \nabla u) = 0 \quad \text{on } \Gamma_N, \qquad (17.68b)$$

$$\varphi'_{\nabla^2 u}(u, \nabla u, \nabla^2 u):(\vec{n}\otimes\vec{n}) + \phi'_u(u, \nabla u) = 0 \quad \text{on } \Gamma_N, \qquad (17.68c)$$

$$u|_\Gamma = u_{D,1}, \quad \frac{\partial u}{\partial\vec{n}}\Big|_\Gamma = u_{D,2} \quad \text{on } \Gamma_D. \qquad (17.68d)$$

Again, (17.68) is called the *classical formulation* of the boundary-value problem in question, and its derivation from (17.67) is more involved than in Section 17.3.2.1. One must use a general decomposition $\nabla v = \partial v/\partial\vec{n} + \nabla_S v$ on Γ with $\nabla_S v = \nabla v - \partial v/\partial\vec{n}$ being the *tangential gradient* of v. On a smooth boundary Γ, one can use another (now $(d-1)$-dimensional) Green-type formula on tangent spaces:[42)]

$$\int_\Gamma a:(\vec{n}\otimes\nabla v)\,dS = \int_\Gamma \big(\vec{n}^\top a\vec{n}\big)\frac{\partial v}{\partial\vec{n}} + a:(\vec{n}\otimes\nabla_S v)\,dS = \int_\Gamma \big(\vec{n}^\top a\vec{n}\big)\frac{\partial v}{\partial\vec{n}} - \operatorname{div}_S\big(a\vec{n}\big)v + \big(\operatorname{div}_S\vec{n}\big)\big(\vec{n}^\top a\vec{n}\big)v\,dS, \qquad (17.69)$$

where $a = \varphi'_{\nabla^2 u}(x, u, \nabla u, \nabla^2 u)$ and $\operatorname{div}_S = \operatorname{tr}(\nabla_S)$ with $\operatorname{tr}(\cdot)$ being the trace of a $(d-1)\times(d-1)$-matrix, denotes the $(d-1)$-dimensional *surface divergence* so that $\operatorname{div}_S\vec{n}$ is (up to a factor $-1/2$) the mean curvature of the surface Γ. Comparing the variational formulation as critical points of (17.66) with the weak formulation (17.67) and with the classical formulation (17.68), one can see that although formally all formulations are equivalent to each other, the advantage of the variational formulations such as (17.66) in its simplicity is obvious.

42) This "surface" Green-type formula reads $\int_\Gamma w:((\nabla_S v)\otimes\vec{n})\,dS = \int_\Gamma(\operatorname{div}_S\vec{n})(w:(\vec{n}\otimes\vec{n}))v - \operatorname{div}_S(w\cdot\vec{n})v\,dS$.

As in (17.56), taking general Carathéodory integrands $a : \Omega\times\mathbb{R}^n\times\mathbb{R}^{n\times d}\times\mathbb{R}^{n\times d\times d} \to \mathbb{R}^{n\times d\times d}$, $b : \Omega\times\mathbb{R}^n\times\mathbb{R}^{n\times d}\times\mathbb{R}^{n\times d\times d} \to \mathbb{R}^{n\times d}$, $c : \Omega\times\mathbb{R}^n\times\mathbb{R}^{n\times d}\times\mathbb{R}^{n\times d\times d} \to \mathbb{R}^n$, $d : \Gamma_{\mathrm{N}}\times\mathbb{R}^n\times\mathbb{R}^{n\times d} \to \mathbb{R}^{n\times d}$, and finally $e : \Gamma_{\mathrm{N}}\times\mathbb{R}^n\times\mathbb{R}^{n\times d} \to \mathbb{R}^n$, one can consider a boundary-value problem for a system of fourth-order elliptic quasilinear equations:

$$\operatorname{div}^2 a(u,\nabla u,\nabla^2 u) - \operatorname{div} b(u,\nabla u,\nabla^2 u) + c(u,\nabla u,\nabla^2 u) = 0 \quad \text{in } \Omega, \tag{17.70a}$$

with the boundary conditions (17.68d) and

$$a(u,\nabla u,\nabla^2 u):(\vec{n}\otimes\vec{n}) + d(u,\nabla u) = 0 \quad \text{on } \Gamma_{\mathrm{N}}, \tag{17.70b}$$

$$\big(\operatorname{div} a(u,\nabla u,\nabla^2 u) - b(u,\nabla u,\nabla^2 u)\big)\cdot\vec{n} + \operatorname{div}_{\mathrm{S}}\big(a(u,\nabla u,\nabla^2 u)\vec{n}\big) - \big(\operatorname{div}_{\mathrm{S}}\vec{n}\big)\big(\vec{n}^\top a(u,\nabla u,\nabla^2 u)\vec{n}\big) + e(u,\nabla u) = 0 \quad \text{on } \Gamma_{\mathrm{N}}. \tag{17.70c}$$

Existence of a potential for which the boundary-value problem (17.70) would represent the Euler–Lagrange equation (17.3) requires the *symmetry* for the Jacobians of the mapping $(H,F,u) \mapsto (a(x,u,F,H), b(x,u,F,H), c(x,u,F,H)) : \mathbb{R}^{n\times d\times d}\times\mathbb{R}^{n\times d}\times\mathbb{R}^d \to \mathbb{R}^{n\times d\times d}\times\mathbb{R}^{n\times d}\times\mathbb{R}^d$ and of the mapping $(F,u) \mapsto (d(x,u,F), e(x,u,F)) : \mathbb{R}^{n\times d}\times\mathbb{R}^d \to \mathbb{R}^{n\times d}\times\mathbb{R}^d$; we used H as a placeholder for $\nabla^2 u$. The formula (17.58) then takes the form:

$$\varphi(x,u,F,H) = \int_0^1 a(x,\lambda u,\lambda F,\lambda H):H + b(x,\lambda u,\lambda F,\lambda H):F + c(x,\lambda u,\lambda F,\lambda H)\cdot u \,\mathrm{d}\lambda, \tag{17.71a}$$

$$\phi(x,u,F) = \int_0^1 d(x,\lambda u,\lambda F):F + e(x,\lambda u,\lambda F)\cdot u \,\mathrm{d}\lambda. \tag{17.71b}$$

Analogously to Theorem 17.13, one can obtain existence of weak solutions by the direct method under a suitable coercivity/growth conditions on φ and an analogue of quasiconvexity of $\varphi(x,u,F,\cdot) : \mathbb{R}^{n\times d\times d} \to \mathbb{R}$, and $u_{\mathrm{D},1} \in W^{2-1/p,p}(\Gamma_{\mathrm{D}};\mathbb{R}^n)$ and $u_{\mathrm{D},2} \in W^{1-1/p,p}(\Gamma_{\mathrm{D}};\mathbb{R}^n)$.

So far, we considered two Dirichlet-type conditions (17.68d) on Γ_{D} (dealing with zeroth- and first-order derivatives) and two Robin-type conditions (17.68b,c) on Γ_{N} (dealing with second- and third-order derivatives). These arise either by fixing both $u|_\Gamma$ or $\partial u/\partial\vec{n}|_\Gamma$ or neither of them, cf. (17.66b). One can, however, think about fixing only $u|_\Gamma$ or only $\partial u/\partial\vec{n}|_\Gamma$, which gives other two options of natural boundary conditions, dealing with zeroth- and second-order derivatives or first- and third-order derivatives, respectively.

The other two combinations, namely the zeroth- and the third-order derivatives or the first- and the second-order derivatives, are not natural from the variational viewpoint because they overdetermine some of the two boundary terms arising in the weak formulation (17.67).

17.3.2.3 Variational Inequalities

Merging the previous Sections 17.3.2.1 – 17.3.2.2 with the abstract scheme from Sections 17.2.2 – 17.2.3, has important applications. Let us use as illustration $\Phi_0 = \Phi$ from (17.52) and $\Phi_1(v) = \int_\Omega \zeta(v)\,\mathrm{d}x + \int_{\Gamma_{\mathrm{N}}} \xi(v)\,\mathrm{d}S$ as in Remark 17.1, now with some convex $\zeta,\xi : \mathbb{R}^n \to \mathbb{R}\cup\{+\infty\}$. In view of the abstract inequality (17.10), the weak formulation (17.54) gives the variational inequality

$$\int_\Omega \zeta(v) + \varphi'_{\nabla u}(x,u,\nabla u):\nabla(v-u) + \varphi'_u(x,u,\nabla u)\cdot(v-u)\,\mathrm{d}x + \int_\Gamma \xi(v) + \phi'_u(x,u)\cdot(v-u)\,\mathrm{d}S$$

$$\geq \int_\Omega \zeta(u)\,\mathrm{d}x + \int_\Gamma \xi(u)\,\mathrm{d}S \quad (17.72)$$

for any $v \in W^{1,p}(\Omega;\mathbb{R}^n)$ such that $v|_{\Gamma_D} = 0$. The passage from the weak formulation to the classical boundary-value problem analogous to (17.54) ⟶ (17.55) leads to the differential inclusion on Ω:

$$\operatorname{div}\varphi'_{\nabla u}(u,\nabla u) - \varphi'_u(u,\nabla u) - \partial\zeta(u) \ni 0 \quad \text{in } \Omega, \quad (17.73a)$$

with another differential inclusion in the boundary conditions

$$\varphi'_{\nabla u}(u,\nabla u)\cdot\vec{n} + \phi'_u(u) + \partial\xi(u) \ni 0 \quad \text{on } \Gamma_N, \quad (17.73b)$$

$$u|_\Gamma = u_D \quad \text{on } \Gamma_D. \quad (17.73c)$$

There is an extensive literature on mathematical methods in variational inequalities, cf. e.g. [18, 41–44].

17.3.3 Some Examples

Applications of the previous general boundary-value problems to more specific situations in continuum physics are illustrated in the following examples.

17.3.3.1 Nonlinear Heat-Transfer Problem

The steady-state temperature distribution θ in an anisotropic nonlinear heat-conductive body $\Omega \subset \mathbb{R}^3$ is described by the balance law

$$\operatorname{div} j = f \quad \text{with}$$

$$j = -\kappa(\theta)\mathbb{K}\nabla\theta \quad \text{on } \Omega, \quad (17.74a)$$

$$\vec{n}\cdot j + b(\theta) = g \quad \text{on } \Gamma, \quad (17.74b)$$

where $b(\cdot) > 0$ is a boundary heat outflow, g the external heat flux, f the bulk heat source, and with the heat flux j governed by the Fourier law involving a symmetric positive definite matrix $\mathbb{K} \in \mathbb{R}^{d\times d}$ and a nonlinearity $\kappa : \mathbb{R} \to \mathbb{R}^+$. In terms of a and c in (17.56a), we have $n = 1$, $a(x,u,F) = \kappa(u)\mathbb{K}F$ and $c(x,u,F) = f(x)$ and the symmetry (17.57b) fails, so that (17.74) does not have the variational structure unless κ is constant. Yet, a simple rescaling of θ, called the *Kirchhoff transformation*, can help: introducing the substitution $u = \widehat{\kappa}(\theta) = \int_0^\theta \kappa(\vartheta)\,\mathrm{d}\vartheta$, we have $j = -\mathbb{K}\nabla u$ and (17.74) transforms to

$$\operatorname{div}(\mathbb{K}\nabla u) + f = 0 \quad \text{on } \Omega, \quad (17.75a)$$

$$\vec{n}^\top\mathbb{K}\nabla u + b(\widehat{\kappa}^{-1}(u)) = g \quad \text{on } \Gamma, \quad (17.75b)$$

which already fits in the framework of (17.52) with $\varphi(x,u,F) = \frac{1}{2}F^\top\mathbb{K}F - f(x)u$ and $\phi(x,u) = \widetilde{b}(u) - g(x)u$ where $\widetilde{b}$ is a primitive of $b \circ \widehat{\kappa}^{-1}$. Eliminating the nonlinearity from the bulk to the boundary, we thus gain a variational structure at least if $f \in L^{6/5}(\Omega)$ and $g \in L^{4/3}(\Gamma)$ [43] and thus, by the direct method, we obtain existence of a solution $u \in H^1(\Omega)$ to (17.75) as well as a possibility of its efficient numerical approximation, cf. Section 17.4.1; then $\theta = \widehat{\kappa}^{-1}(u)$ yields a solution to (17.74). Our optimism should however be limited because, in the heat-transfer context, the natural integrability of the heat sources is only $f \in L^1(\Omega)$ and $g \in L^1(\Gamma)$, but this is not consistent with the variational structure if $d > 1$:

Example 17.5 [Nonexistence of minimizers] Consider the heat equation $-\operatorname{div}\nabla u = 0$ for $d = 3$ and with, for simplicity, zero Dirichlet boundary conditions, so that the underlying variational problem is to minimize $\int_\Omega \frac{1}{2}|\nabla u|^2 - fu\,\mathrm{d}x$ on $H^1_0(\Omega)$. Yet,

43) If $d = 3$, this integrability of the heat sources is necessary and sufficient to ensure the functional $(u \mapsto \int_\Omega fu\,\mathrm{d}x + \int_\Gamma gu\,\mathrm{d}S)$ to belong to $H^1(\Omega)^*$.

$$\inf_{u\in H_0^1(\Omega)} \int_\Omega \frac{1}{2}|\nabla u|^2 - fu\,\mathrm{d}x = -\infty, \qquad (17.76)$$

whenever $f\in L^1(\Omega)\backslash L^{6/5}(\Omega)$.[44)]

17.3.3.2 Elasticity at Large Strains

A prominent application of multidimensional ($d > 1$) vectorial ($n > 1$) variational calculus is to elasticity under the hypothesis that the stress response on the deformation gradient is a gradient of some potential; such *materials* are called *hyperelastic*. The problem is very difficult especially at large strains where the geometry of the standard Cartesian coordinates may be totally incompatible with the largely deformed geometry of the specimen Ω.

Here, u will stand for the deformation (although the usual notation is rather y) and we will consider $n = d$ and $\varphi = \varphi(F)$ taking possibly also the value $+\infty$. The ultimate requirement is frame indifference, that is, $\varphi(RF) = \varphi(F)$ for all $R \in \mathbb{R}^{d\times d}$ in the special orthogonal group SO(d). One could try to rely on Theorem 17.13 or 17.14. In the former case, one can consider non-polyconvex materials and one also has the Euler–Lagrange equation (17.54) at one's disposal, but the growth condition (17.53b) does not allow an infinite increase of energy when the volume of the material is locally shrunk to 0, that is, we cannot satisfy the condition

$$\varphi(F) \to +\infty \qquad \text{if } \det F \to 0+ \qquad (17.77a)$$

$$\varphi(F) = +\infty \qquad \text{if } \det F \le 0. \qquad (17.77b)$$

An example for a polyconvex frame-indifferent φ satisfying (17.77) is the *Ogden material*

$$\varphi(F) = a_1 \mathrm{tr}(F^\top F)^{b_1} + a_2|\mathrm{tr}(\mathrm{cof}(F^\top F))|^{b_2} + \gamma(\det F) \qquad (17.78)$$

with $a_1, a_2, b_1, b_2 > 0$ and $\gamma : \mathbb{R}_+ \to \mathbb{R}$ convex such that $\gamma(\delta) = +\infty$ for $\delta \le 0$ and $\lim_{\delta\to 0_+} \gamma(\delta) = +\infty$, and where $\mathrm{cof}(A) = (\det A)A^\top$, considering $d = 3$, and where $\mathrm{tr}\,A = \sum_{i=1}^d A_{ii}$. Particular Ogden materials are Mooney–Rivlin materials with $\varphi(F) = |F|^2 + |\det F|^2 - \ln(\det F)$ or compressible neo-Hookean materials with $\varphi(F) = a|F|^2 + \gamma(\det F)$. The importance of polyconvexity is that the existence of energy-minimizing deformation can be based on Theorem 17.14, which allows us to handle local nonpenetration

$$\det(\nabla u) > 0 \qquad \text{a.e. on } \Omega \qquad (17.79a)$$

involved in φ via (17.77). Handling of non-penetration needs also the global condition

$$\int_\Omega \det(\nabla u)\,\mathrm{d}x \le \mathrm{meas}_d\big(u(\Omega)\big); \qquad (17.79b)$$

(17.79) is called the *Ciarlet–Nečas nonpenetration condition* [45]. Whether it can deal with quasiconvex but not polyconvex materials remains an open question, however cf. [46].

An interesting observation is that polyconvex materials allow for an energy-controlled stress $|\varphi'(F)F^\top| \le C(1 + \varphi(F))$ even though the so-called Kirchhoff stress $\varphi'(F)F^\top$ itself does not need to be bounded. This can be used, for example, in sensitivity analysis and to obtain modified Euler–Lagrange equations to overcome the possible failure of (17.54) for such materials, cf. [46]. It is worth noting that even such spatially homogeneous, frame-indifferent, and polyconvex materials can exhibit the *Lavrentiev phenomenon*, cf. [47].

Another frequently used ansatz is just a quadratic form in terms of the

44) The proof is very simple: $f \notin H_0^1(\Omega)^*$ means $\|f\|_{H_0^1(\Omega)^*} = \sup_{u\in W_0^{1,\infty}(\Omega),\ \|u\|_{1,2}\le 1} \int_\Omega fu\,\mathrm{d}x = +\infty$, which further means $\int_\Omega fu_k\,\mathrm{d}x \to +\infty$ for some $u_k \in W_0^{1,\infty}(\Omega)$ such that $\int_\Omega \frac{1}{2}|\nabla u_k|^2\,\mathrm{d}x \le \frac{1}{2}$, so that $\int_\Omega \frac{1}{2}|\nabla u_k|^2 - fu_k\,\mathrm{d}x \to -\infty$.

Green–Lagrange strain tensor

$$E = \tfrac{1}{2}F^\top F - \tfrac{1}{2}\mathbb{I}. \qquad (17.80)$$

An example is an isotropic material described by only two elastic constants; in terms of the *Lamé constants* μ and λ, it takes the form

$$\varphi(F) = \tfrac{1}{2}\lambda|\mathrm{tr}E|^2 + \mu|E|^2, \qquad (17.81)$$

and is called a *St.Venant–Kirchhoff's material*, cf. [48, Volume I, Section 4.4]. If $\mu > 0$ and $\lambda > -\frac{2}{d}\mu$, φ from (17.81) is coercive in the sense of $\varphi(F) \geq \epsilon_0|F|^4 - 1/\epsilon_0$ for some $\epsilon_0 > 0$ but not quasiconvex (and even not rank-one convex), however. Therefore, existence of an energy-minimizing deformation in $W^{1,4}(\Omega;\mathbb{R}^d)$ is not guaranteed.

A way to improve solvability for non-quasiconvex materials imitating additional interfatial-like energy is to augment φ with some small convex higher-order term, for example, $\varphi_1(F,H) = \varphi(F) + \sum_{i,k,l,m,n=1}^d \mathbb{H}_{klmn}(x)H_{ikl}H_{imn}$ with $\mathbb{H}$ a (usually only small) fourth-order positive definite tensor, and to employ the fourth-order framework of Section 17.3.2.2. This is the idea behind the mechanics of *complex* (also called *nonsimple* or a special *micropolar*) *continua*, cf. e.g., [49].

17.3.3.3 Small-Strain Elasticity, Lamé System, Signorini Contact

Considering the deformation y and displacement $u(x) = y(x) - x$, we write $F = \nabla y = \nabla u + \mathbb{I}$ and, for $|\nabla u|$ small, the tensor E from (17.80) is

$$E = \frac{F^\top F - \mathbb{I}}{2} = \frac{(\nabla u)^\top + \nabla u + (\nabla u)^\top \nabla u}{2} = \tfrac{1}{2}(\nabla u)^\top + \tfrac{1}{2}\nabla u + o\big(|\nabla u|\big), \qquad (17.82)$$

which leads to the definition of the linearized-strain tensor, also called *small-strain tensor*, as $e(u) = \frac{1}{2}(\nabla u)^\top + \frac{1}{2}\nabla u$. In fact, a vast amount of engineering or also, for example, geophysical models and calculations are based on the small-strain concept. The specific energy in homogeneous materials is then $\varphi : \mathbb{R}^{d\times d}_{\mathrm{sym}} \to \mathbb{R}$ where $\mathbb{R}^{d\times d}_{\mathrm{sym}} = \{A \in \mathbb{R}^{d\times d};\ A^\top = A\}$. In linearly responding materials, φ is quadratic. An example is an isotropic material; in terms of Lamé's constants as in (17.81), it takes the form

$$\varphi_{\mathrm{Lame}}(e) = \tfrac{1}{2}\lambda|\mathrm{tr}\, e|^2 + \mu|e|^2. \qquad (17.83)$$

Such φ_{Lame} is positive definite on $\mathbb{R}^{d\times d}_{\mathrm{sym}}$ if $\mu > 0$ and $\lambda > -(2d/\mu)$. The positive definiteness of the functional $\int_\Omega \varphi_{\mathrm{Lame}}(e(u))\,\mathrm{d}x$ is a bit delicate as the rigid-body motions (translations and rotations) are not taken into account by it. Yet, fixing positive definiteness by suitable boundary conditions, coercivity can then be based on the *Korn inequality*

$$\forall v \in W^{1,p}(\Omega;\mathbb{R}^d),\ v|_{\Gamma_{\mathrm{D}}} = 0 : \quad \|v\|_{W^{1,p}(\Omega;\mathbb{R}^d)} \leq C_p\|e(v)\|_{L^p(\Omega;\mathbb{R}^{d\times d}_{\mathrm{sym}})} \qquad (17.84)$$

to be used for $p = 2$; actually, (17.84) holds for $p > 1$ on connected smooth domains with Γ_{D} of nonzero measure, but notably counterexamples exist for $p = 1$. Then, by the direct method based on Theorem 17.2, one proves existence and uniqueness of the solution to the *Lamé system* arising from (17.55) by considering $\varphi(x,u,F) = \varphi_{\mathrm{Lame}}(\frac{1}{2}F^\top + \frac{1}{2}F)$ and $\phi(x,u) = g(x)u$:

$$\mathrm{div}\,\sigma + f = 0 \quad \text{in } \Omega \quad \text{with } \sigma = 2\mu e(u) + \lambda(\mathrm{div}\,u)\mathbb{I}, \qquad (17.85a)$$

$$\sigma\vec{n} = g \quad \text{on } \Gamma_{\mathrm{N}}, \qquad (17.85b)$$

$$u|_\Gamma = u_{\mathrm{D}} \quad \text{on } \Gamma_{\mathrm{D}}. \qquad (17.85c)$$

The Lamé potential (17.83) can be obtained by an asymptotic expansion of an

Ogden material (17.78), see [48, Volume I, Theorem 4.10-2].

An illustration of a very specific variational inequality is a *Signorini (frictionless) contact* on a third part Γ_C of the boundary Γ; so now we consider Γ divided into three disjoint relatively open parts Γ_D, Γ_N, and Γ_C whose union is dense in Γ. This is a special case of the general problem (17.72) with $\zeta \equiv 0$ and $\xi(x,u) = 0$ if $u{\cdot}\vec{n}(x) \le 0$ on Γ_C, otherwise $\xi(x,u) = +\infty$. In the classical formulation, the boundary condition on Γ_C can be identified as

$$\left.\begin{aligned} &(\sigma\vec{n}){\cdot}\vec{n} \le 0,\\ &u{\cdot}\vec{n} \le 0,\\ &\big((\sigma\vec{n}){\cdot}\vec{n}\big)\big(u{\cdot}\vec{n}\big) = 0 \end{aligned}\right\} \quad \text{on } \Gamma_C, \qquad (17.85d)$$

$$(\sigma\vec{n}) - ((\sigma\vec{n}){\cdot}\vec{n})\vec{n} = 0 \quad \text{on } \Gamma_C; \qquad (17.85e)$$

note that (17.85d) has a structure of a complementarity problem (17.17) for the normal stress $(\sigma\vec{n}){\cdot}\vec{n}$ and the normal displacement $u{\cdot}\vec{n}$, while (17.85e) is the equilibrium condition for the tangential stress.

This very short excursion into a wide area of contact mechanics shows that it may have a simple variational structure: In terms of Example 17.1, the convex set K is a cone (with the vertex not at origin if $u_D \ne 0$):

$$K = \big\{u{\in}H^1(\Omega;\mathbb{R}^n);\ u|_\Gamma = u_D \text{ on } \Gamma_D,\ u{\cdot}\vec{n} \le 0 \text{ on } \Gamma_C\big\} \qquad (17.86)$$

and then (17.85) is just the classical formulation of the first-order condition (17.13) for the simple problem

$$\text{minimize } \Phi(u) = \int_\Omega \frac{\lambda}{2}|\mathrm{div}\, u|^2 + \mu|e(u)|^2 \mathrm{d}x$$
$$\text{subject to } u{\in}K \text{ from (17.86).}$$

As in Section 17.3.2.2, it again demonstrates one of the advantages of the variational formulation as having a much simpler form in comparison with the classical formulation.

17.3.3.4 Sphere-Valued Harmonic Maps

Another example is a minimization problem with $\varphi(x,u,F) = \omega(u) + h(x){\cdot}u + \frac{\epsilon}{2}|F|^2$ and the nonconvex constraint $|u| = 1$, that is,

$$\begin{aligned} &\text{minimize } \Phi = \int_\Omega \omega(u) + \frac{\epsilon}{2}|\nabla u|^2 - h{\cdot}u \,\mathrm{d}x\\ &\text{subject to } R(u) = |u|^2 - 1 = 0 \text{ on } \Omega, \end{aligned} \qquad (17.87)$$

which has again the structure (17.14) now with $V = K = H^1(\Omega;\mathbb{R}^3)$ and $\Lambda = L^2(\Omega)$ with the ordering by the trivial cone $\{0\}$ of "nonnegative" vectors. This may serve as a simplified model of static micromagnetism[45] in ferromagnetic materials at low temperatures so that the *Heisenberg constraint* $|u| = 1$ is well satisfied pointwise by the magnetization vector u. Alternatively, it may also serve as a simplified model of liquid crystals. The weak formulation of this minimization problem is

$$\int_\Omega \epsilon\nabla u{:}\nabla(v - (u{\cdot}v)u) + \big(\omega'(u) - h\big){\cdot}\big(v - (u{\cdot}v)u\big)\,\mathrm{d}x = 0 \qquad (17.88)$$

for any $v \in H^1(\Omega;\mathbb{R}^3)$ with $|v|^2 = 1$ a.e. on Ω; for $\omega \equiv 0$ cf. [50, Section 8.4.3]. The corresponding classical formulation then has the form

$$-\,\mathrm{div}(\epsilon\nabla u) + \omega'(u) - h = \big(|\nabla u|^2 + (\omega'(u) - h){\cdot}u\big)u \quad \text{in } \Omega, \qquad (17.89a)$$

$$\epsilon\nabla u{\cdot}\vec{n}\big|_\Gamma = 0 \quad \text{on } \Gamma. \qquad (17.89b)$$

45) In this case, $\omega : \mathbb{R}^3 \to \mathbb{R}$ is an anisotropy energy with minima at easy-axis magnetization and $\epsilon > 0$ is an exchange-energy constant, and h is an outer magnetic field. The demagnetizing field is neglected.

Comparing it with (17.16) with $N_K \equiv \{0\}$ and $R' = 2\times$identity on $L^2(\Omega;\mathbb{R}^3)$, we can see that $\lambda^* = \frac{1}{2}|\nabla u|^2 + \frac{1}{2}(\omega'(u)-h)\cdot u$ plays the role of the Lagrange multiplier for the constraint $|u|^2 = 1$ a.e. on Ω.

Let us remark that, in the above-mentioned micromagnetic model, R is more complicated than (17.87) and involves also the differential constraints

$$\operatorname{div}(h-u) = 0 \quad \text{and} \quad \operatorname{rot} h = j \qquad (17.90)$$

with j assumed fixed, which is the (steady state) Maxwell system where j is the electric current, and the minimization in (17.87) is to be done over the couples (u,h); in fact, (17.90) is considered on the whole $\mathbb{R}^3$ with j fixed and u vanishing outside Ω.

17.3.3.5 Saddle-Point-Type Problems

In addition to minimization principles, other principles also have applications. For example, for the usage of the mountain pass Theorem 17.4 for potentials such as $\varphi(u,F) = \frac{1}{2}|F|^2 + c(u)$ see [50, Section 8.5] or [51, Section II.6].

Seeking saddle points of Lagrangeans such as (17.18) leads to *mixed formulations* of various constrained problems. For example, the Signorini problem (17.85) uses $\Phi(u) = \int_\Omega \varphi_{\text{Lame}}(e(u))\,\mathrm{d}x + \int_{\Gamma_N} g\cdot u\,\mathrm{d}S$, $R : u \mapsto u\cdot\vec{n} : H^1(\Omega;\mathbb{R}^d) \to H^{1/2}(\Gamma_C)$, and $D = \{v \in H^{1/2}(\Gamma_C);\ \ v \le 0\}$. The saddle point $(u,\lambda^*) \in H^1(\Omega;\mathbb{R}^d) \times H^{-1/2}(\Gamma_C)$ with $u|_{\Gamma_D} = g$ and $\lambda^* \le 0$ on Γ_C exists and represents the mixed formulation of (17.85); then $\lambda^* = (\sigma\vec{n})\cdot\vec{n}$ and cf. also the Karush–Kuhn–Tucker conditions (17.16) with $N_K \equiv \{0\}$, cf. e.g. [37, 43].

Another example is a saddle point on $H_0^1(\Omega;\mathbb{R}^d) \times L^2(\Omega)$ for $\mathscr{L}(u,\lambda^*) = \int_\Omega \frac{1}{2}|\nabla u|^2 + \lambda^* \operatorname{div} u\,\mathrm{d}x$ that leads to the system

$$-\Delta u + \nabla\lambda^* = 0 \quad \text{and} \quad \operatorname{div} u = 0, \qquad (17.91)$$

which is the *Stokes system* for a steady flow of viscous incompressible fluid. The primal formulation minimizes $\int_\Omega \frac{1}{2}|\nabla u|^2\mathrm{d}x$ on $H_0^1(\Omega;\mathbb{R}^d)$ subject to $\operatorname{div} u = 0$. The Karush—Kuhn–Tucker conditions (17.16) with $R : u \mapsto \operatorname{div} u : H_0^1(\Omega;\mathbb{R}^d) \to L^2(\Omega)$ and the ordering of $D = \{0\}$ as the cone of nonnegative vectors gives (17.91).

17.3.4 Evolutionary Problems

Let us illustrate the *Brezis–Ekeland–Nayroles principle* on the initial-boundary-value problem for a quasilinear parabolic equation

$$\frac{\partial u}{\partial t} - \operatorname{div}\big(|\nabla u|^{p-2}\nabla u\big) = g \quad \text{in } Q, \qquad (17.92a)$$

$$u = 0 \quad \text{on } \Sigma, \qquad (17.92b)$$

$$u(0,\cdot) = u_0 \quad \text{in } \Omega \qquad (17.92c)$$

with $Q = [0,T]\times\Omega$ and $\Sigma = [0,T]\times\Gamma$. We consider $V = W_0^{1,p}(\Omega)$ equipped with the norm $\|u\|_{1,p} = \|\nabla u\|_p$, $\Phi(u) = (1/p)\|\nabla u\|_p^p$ and $\langle f(t), u\rangle = \int_\Omega g(t,x)u(x)\,\mathrm{d}x$. Let us use the notation $\Delta_p : W_0^{1,p}(\Omega) \to W^{-1,p'}(\Omega) = W_0^{-1,p}(\Omega)^*$ for the p-Laplacian; this means $\Delta_p u = \operatorname{div}(|\nabla u|^{p-2}\nabla u)$. We have that

$$\Phi^*(\xi) = \frac{1}{p'}\|\xi\|_{W^{-1,p'}(\Omega)}^{p'} = \frac{1}{p'}\|\Delta_p^{-1}\xi\|_{1,p}^{p} = \frac{1}{p'}\|\nabla\,\Delta_p^{-1}\xi\|_p^p\,.$$

Thus we obtain the following explicit form of the functional $\mathfrak{F}$ from (17.24):

$$\mathfrak{F}(u) = \int_Q \frac{1}{p}|\nabla u|^p + \Big(\frac{\partial u}{\partial t} - g\Big)u + \frac{1}{p'}\Big|\nabla\,\Delta_p^{-1}\Big(\frac{\partial u}{\partial t} - g\Big)\Big|^p \mathrm{d}x\mathrm{d}t + \int_\Omega \frac{1}{2}\big|u(T)\big|^2\,\mathrm{d}x. \qquad (17.93)$$

We can observe that the integrand in (17.93) is nonlocal in space, which is, in fact, an inevitable feature for parabolic problems, cf. [52].

Wide generalization to *self-dual problems* of the type $(Lx, Ax) \in \partial\mathscr{L}(Ax, Lx)$ are in [53], covering also nonpotential situations such as the Navier–Stokes equations for incompressible fluids and many others.

Efficient usage of the "global" variational principles such as (17.24), (17.32), or (17.34) for parabolic equations or inequalities is, however, limited to theoretical investigations. Of much wider applicability are recursive variational problems arising by implicit or various semi-implicit time discretization as in Section 17.2.4.3, possibly combined also with spatial discretization and numerical algorithms leading to computer implementations, mentioned in Section 17.4.1 below.

Other, nonminimization principles have applications in hyperbolic problems of the type $\rho\partial^2 u/\partial t^2 - \mathrm{div}\big(|\nabla u|^{p-2}\nabla u\big) = g$ where the *Hamilton principle* (17.37) leads to seeking a critical point of the functional $\int_0^T \int_\Omega \rho|\partial u/\partial t|^2 - (1/p)|\nabla u|^p + g{\cdot}u\,\mathrm{d}x\mathrm{d}t$.

17.4 Miscellaneous

The area of the calculus of variations is extremely wide and the short excursion above presented a rather narrow selection. Let us at least briefly touch a few more aspects.

17.4.1 Numerical Approximation

Assuming $\{V_k\}_{k\in\mathbb{N}}$ is a nondecreasing sequence of finite-dimensional linear subspaces of V whose union is dense, that is,

$$\forall v \in V\ \exists v_k \in V_k :\quad v_k \to v, \qquad (17.94)$$

we can restrict the original variational problem of V to V_k. Being finite dimensional, V_k possesses a basis and, in terms of coefficients in this basis, the restricted problem then becomes implementable on computers. This is the simplest idea behind numerical approximation, called the *Ritz method* [54] or, rather in a more general nonvariational context, also the *Galerkin method* [55]. This idea can be applied on an abstract level to problems in Section 17.2. In the simplest (conformal) version instead of (17.3), one is to seek $u_k \in V_k$ such that

$$\forall v \in V_k :\qquad \langle \Phi'(u_k), v\rangle = 0; \qquad (17.95)$$

such u_k is a critical point of the restriction of Φ on V_k, that is, of $\Phi : V_k \to \mathbb{R}$.

One option is to solve numerically the system of nonlinear equations (17.95) iteratively, for example, by the (quasi) Newton method. Yet, the variational structure can advantageously be exploited in a number of other options such as the conjugate-gradient or the variable-metric methods, cf. e.g. [56, Section 7]. Then approximate satisfaction of optimality conditions typically serves as a stopping criterion for an iterative strategy; in this unconstrained case, it is the residual in (17.95) that is to be small. For constrained problems, the methods of Section 17.2.3 can be adapted.

Application to concrete problems in Section 17.3.2 on function spaces opens further interesting possibilities. Typically, V_k are chosen as linear hulls of piecewise polynomial functions $\{v_{kl}\}_{l=1,\ldots,L_k}$ whose supports in Ω are only very small sets so that, for most pairs (l_1, l_2), we have that $\int_\Omega \nabla^i v_{kl_1} \vdots \nabla^j v_{kl_2}\,\mathrm{d}x = 0$; more precisely, this holds for each pair (l_1, l_2) for which $\mathrm{supp}(v_{kl_1}) \cap \mathrm{supp}(v_{kl_2}) = \emptyset$. For integral functionals Φ from (17.52) or (17.66a), the system of algebraic equations resulting from (17.95) is sparse, which facilitates its implementation and numerical solution on computers; this is the essence of the *finite-element method.*

Phenomena discussed in Examples 17.2–17.5 can make the approximation issue quite nontrivial. For Lavrentiev-type phenomena, see, for example, [39]. An example sure to lead to this phenomenon is the constraint $|u|^2 = 1$ in Section 17.3.3.4, which is not compatible with any polynomial approximation so that plain usage of standard finite-element approximation cannot converge.

All this can be applied to time-discretized evolution problems from Section 17.2.4.3, leading to implementable numerical strategies for evolution problems from Sections 17.2.4.1 and 17.2.4.2.

17.4.2
Extension of Variational Problems

Historically, variational problems have been considered together with the Euler–Lagrange equations in their classical formulations, that is, in particular, the solutions are assumed to be continuously differentiable. Here (17.55) or (17.70) are to be considered holding pointwise for $u \in C^2(\overline{\Omega};\mathbb{R}^d)$ and $C^4(\overline{\Omega};\mathbb{R}^d)$, respectively. Yet, such classical solutions do not need to exist[46] and thus the weak formulations (17.54) and (17.67) represent a natural extension (using the density $C^{2k}(\overline{\Omega}) \subset W^{k,p}(\Omega)$) of the original problems defined for smooth functions. The adjective "natural" here means the extension by continuity, referring to the continuity of $\Phi : C^{2k}(\overline{\Omega};\mathbb{R}^d) \to \mathbb{R}$ with respect to the norm of $W^{k,p}(\Omega;\mathbb{R}^d)$ provided natural growth conditions on φ and ϕ are imposed; for $k = 1$; see (17.53). Weak solutions thus represent a natural generalization of the concept of classical solutions.

In general, the method of extension by (lower semi)continuity is called *relaxation.* It may provide a natural concept of generalized solutions with some good physical meaning. One scheme, related to the minimization principle, deals with situations when Theorem 17.1 cannot be applied owing to the lack of weak* lower semicontinuity. The relaxation then replaces Φ by its *lower semicontinuous envelope* $\bar{\Phi}$ defined by

$$\bar{\Phi}(u) = \liminf_{v \to u \text{ weakly*}} \Phi(v). \tag{17.96}$$

Theorem 17.1 then applies to $\bar{\Phi}$ instead of the original Φ, yielding thus a generalized solution to the original variational problem. The definition (17.96) is only conceptual and more explicit expressions are desirable and sometimes actually available. In particular, if $n=1$ or $d=1$, the integral functional (17.52a) admits the formula

$$\bar{\Phi}(u) = \int_\Omega \varphi^{**}(x,u,\nabla u)\,\mathrm{d}x + \int_{\Gamma_N} \phi(x,u)\,\mathrm{d}S, \tag{17.97}$$

where $\varphi^{**}(x,u,\cdot) : \mathbb{R}^{n\times d} \to \mathbb{R}$ denotes the convex envelope of $\varphi(x,u,\cdot)$, that is, the maximal convex minorant of $\varphi(x,u,\cdot)$. In Example 17.2, $\bar{\Phi}$ is given by (17.97) with

$$\varphi^{**}(u,F) = \begin{cases} u^2 + \left(|F|^2 - 1\right)^2 & \text{if } |F| \geq 1, \\ u^2 & \text{if } |F| < 1, \end{cases}$$

cf. Figure 17.3, and with $\phi = 0$, and the only minimizer of $\bar{\Phi}$ on $W^{1,4}(\Omega)$ is $u = 0$, which is also a natural $W^{1,4}$-weak limit of all minimizing sequences for Φ, cf. Figure 17.1.

46) Historically the first surprising example for a minimizer of a (17.52) with $\varphi = \varphi(F)$ smooth uniformly convex, which was only in $W^{1,\infty}(\Omega;\mathbb{R}^m)$ but not smooth is due to Nečas [57], solving negatively the 19th Hilbert's problem [4] if $d > 1$ and $m > 1$ are sufficiently high. An example for even $u \notin W^{1,\infty}(\Omega;\mathbb{R}^m)$ for $d = 3$ and $m = 5$ is in [58].

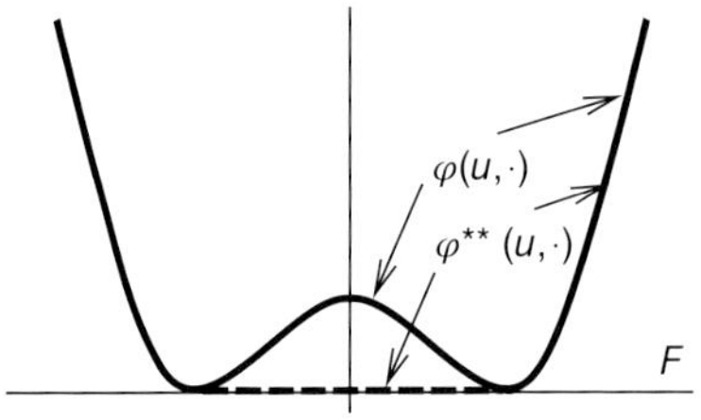

Figure 17.3 A convex envelope $\varphi^{**}(u,\cdot)$ of the double-well potential $\varphi(u,\cdot)$.

Fast oscillations of gradients of these minimizing sequences can be interpreted as *microstructure*, while the minimizers of $\bar{\Phi}$ bear only "macroscopical" information. This reflects the multiscale character of such variational problems. In general, if both $n > 1$ and $d > 1$, (17.97) involves the *quasiconvex envelope* $\varphi^\sharp(x,u,\cdot) : \mathbb{R}^{n\times d} \to \mathbb{R}$ rather than φ^{**}; this is defined by

$$\forall x \in \Omega\ \forall u \in \mathbb{R}^n\ \forall F \in \mathbb{R}^{n\times d} : \quad \varphi^\sharp(x,u,F) = \inf_{v\in W_0^{1,p}(O;\mathbb{R}^n)} \int_O \frac{\varphi(x,u,F+\nabla v(\tilde{x}))}{\text{meas}_d(O)}\,\mathrm{d}\tilde{x}\,;$$

this definition is independent of O but is only implicit and usually only some upper and lower estimates (namely, rank-one convex and polyconvex envelopes) are known explicitly or can numerically be evaluated.

To cope with both nonconvexity and with the unwanted phenomenon of nonexistence as in Example 17.2, one can consider *singular perturbations*, such as

$$\Phi_\varepsilon(u) = \int_\Omega \varphi(x,u,\nabla u) + \varepsilon \mathbb{H}\nabla^2 u \vdots \nabla^2 u\,\mathrm{d}x + \int_{\Gamma_\mathrm{N}} \phi(x,u)\,\mathrm{d}S \qquad (17.98)$$

with a positive definite fourth-order tensor $\mathbb{H}$ and small $\varepsilon > 0$; cf. also $\mathbb{H}$ in Section 17.3.3.2. Under the growth/coercivity conditions on φ and ϕ induced by (17.53) with $1 < p < 2^*$, (17.98) possesses a (possibly nonunique) minimizer $u_\varepsilon \in W^{2,2}(\Omega;\mathbb{R}^d)$. The parameter ε determines an internal length scale of possible oscillations of ∇u_ε occurring if $\varphi(x,u,\cdot)$ is not convex, cf. also Figure 17.4. As ε is usually very small, it makes sense to investigate the asymptotics when it approaches 0. For $\varepsilon \to 0$, the sequence $\{u_\varepsilon\}_{\varepsilon>0}$ possesses a subsequence converging weakly in $W^{1,p}(\Omega;\mathbb{R}^d)$ to some u and every such a limit u minimizes the relaxed functional

$$\bar{\Phi}(u) = \int_\Omega \varphi^\sharp(x,u,\nabla u)\,\mathrm{d}x + \int_{\Gamma_\mathrm{N}} \phi(x,u)\,\mathrm{d}S. \qquad (17.99)$$

The possible fast spatial oscillations of the gradient are smeared out in the limit.

To record some information about such oscillations in the limit, one should make a relaxation by continuous extension rather than only by weak lower semicontinuity. To ensure the existence of solutions, the extended space should support a compact topology which makes the extended functional continuous; such a relaxation is called a *compactification*. If the extended space also supports a convex structure (not necessarily coinciding with the linear structure of the original space), one can define variations, differentials, and the abstract Euler–Lagrange equality (17.13); then we speak about the *convex compactification* method, cf. [59].

A relatively simple example can be the relaxation of the micromagnetic problem (17.87)–(17.90) that, in general, does not have any solution if $\varepsilon = 0$ due to nonconvexity of the Heisenberg constraint $|u| = 1$. One can embed the set of admissible u's, namely $\{u \in L^\infty(\Omega;\mathbb{R}^3);\, u(x) \in S \text{ for a.a. } x\}$ with the sphere $S = \{|s| = 1\} \subset \mathbb{R}^3$ into a larger set $\mathscr{Y}(\Omega; S) = \{\nu = (\nu_x)_{x\in\Omega};\ \nu_x$ a probability[47] measure on S and $x \mapsto \nu_x$ weakly* measurable$\}$; the embedding is realized by the mapping $u \mapsto \nu = (\delta_{u(x)})_{x\in\Omega}$ where δ_s denotes here the Dirac measure supported at $s \in S$. The elements of $\mathscr{Y}(\Omega; S)$ are called *Young measures* [60][48] and this set is (considered as) a convex weakly* compact subset of $L^\infty_{w*}(\Omega; \mathrm{M}(S))$ where $\mathrm{M}(S) \cong C(S)^*$ denotes the set of Borel measures on S.[49] The problem (17.87)–(17.90) with $\varepsilon = 0$ then allows a continuous extension

$$\begin{aligned}
&\text{minimize } \bar{\Phi}(\nu, h) \\
&= \int_\Omega\int_S \omega(s) - h\cdot s\,\nu_x(\mathrm{d}s)\mathrm{d}x \\
&\text{subject to } \operatorname{div}(h-u) = 0, \quad \operatorname{rot} h = j, \\
&\nu \in \mathscr{Y}(\Omega; S), \\
&\text{with } u(x) = \int_S s\,\nu_x(\mathrm{d}s) \text{ for } x \in \Omega. \qquad (17.100)
\end{aligned}$$

The functional $\bar{\Phi}$ is a continuous extension of $(u, h) \mapsto \Phi(u, h)$ from (17.87), which is even convex and smooth with respect to the geometry of $L^\infty_{w*}(\Omega; \mathrm{M}(S)) \times L^2(\Omega; \mathbb{R}^3)$. Existence of solutions to the relaxed problem (17.100) is then obtained by Theorem 17.1 modified for the constrained case. Taking into account the convexity of $\mathscr{Y}(\Omega; S)$, the necessary and sufficient optimality conditions of the type (17.13) for (17.100) lead, after a disintegration, to a pointwise condition

$$\int_S \mathfrak{H}_h(x, s)\,\nu_x(\mathrm{d}s) = \max_{s\in S} \mathfrak{H}_h(x, s),$$
$$\text{with } \mathfrak{H}_h(x, s) = h(x)\cdot s - \omega(s) \qquad (17.101)$$

to hold for a.a. $x \in \Omega$ with h satisfying the linear constraints in (17.100), that is, $\operatorname{div}(h-u) = 0$, $\operatorname{rot} h = j$, and $u = \int_S s\,\nu(\mathrm{d}s)$. The integrand of the type $\mathfrak{H}_h$ is sometimes called a *Hamiltonian* and conditions like (17.101), the *Weierstrass maximum principle*, formulated here for the relaxed problem and revealed as being a standard condition of the type (17.13) but with respect to a nonstandard geometry imposed by the space $L^\infty_{w*}(\Omega; \mathrm{M}(S))$. The solutions to (17.100) are typically nontrivial Young measures in the sense that ν_x is not a Dirac measure. From the maximum principle (17.101), one can often see that they are composed from a weighted sum of a finite number of Dirac measures supported only at such $s \in S$ that maximizes $\mathfrak{H}_h(x, \cdot)$. This implies that minimizing sequences for the original problem (17.87)–(17.90) with $\varepsilon = 0$ ultimately must exhibit finer and finer spatial oscillations of u's; this effect is experimentally observed in ferromagnetic materials, see Figure 17.4.[50] In fact, a small parameter $\varepsilon > 0$ in the original problem (17.87)–(17.90) determines the lengthscale of magnetic domains and also the typical width of the walls between the domains. For the Young measure relaxation in micromagnetism see, for example, [61, 63].

47) The adjective "probability" means here a positive measure with a unit mass but does not refer to any probabilistic concept.

48) In fact, L.C. Young had already introduced such measures in 1936 in a slightly different language even before the theory of measure had been invented. For modern mathematical theory see, for example, [30, Chapter 8], [61, Chapter 6–8], [62, Chapter 2], or [59, Chapter 3].

49) "L^∞_{w*}" denotes "weakly* measurable" essentially bounded" mappings, and $L^\infty_{w*}(\Omega; \mathrm{M}(S))$ is a dual space to $L^1(\Omega; C(S))$, which allows to introduce the weak* convergence that makes this set compact.

50) Actually, the minimization-energy principle governs magnetically soft materials where

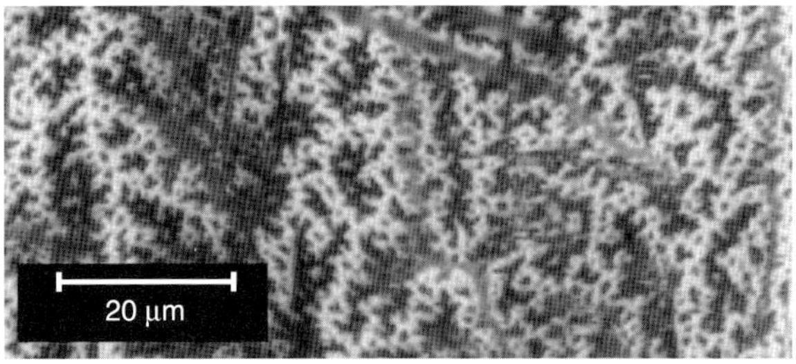

Figure 17.4 Fast spatial oscillations of the magnetization vector minimizing Φ from (17.87)–(17.90) with a double-well potential ω forming a fine microstructure in a ferromagnetic tetragonal single-crystal of NiMnGa with only one axis of easy magnetization normal to the observed surface. (Courtesy O. Heczko, Institute of Physics, ASCR.)

A relaxation by continuous extension of the originally discussed problem (17.52) is much more complicated because the variable exhibiting fast-oscillation tendencies (i.e., ∇u) is in fact subjected to some differential constraint (namely, $\mathrm{rot}\,(\nabla u) = 0$) and because, in contrast to the previous example, is valued on the whole $\mathbb{R}^{n\times d}$, which is not compact. We thus use only a subset of Young measures, namely,

$$\begin{aligned}\mathscr{G}^p(\Omega;\mathbb{R}^{n\times d}) = \Big\{ & \nu \in \mathscr{Y}(\Omega;\mathbb{R}^{n\times d});\\ & \exists u \in W^{1,p}(\Omega;\mathbb{R}^n):\\ & \nabla u(x) = \int_{\mathbb{R}^{n\times d}} F\,\nu_x(\mathrm{d}F)\ \forall_{\mathrm{a.a.}} x \in \Omega,\\ & \int_\Omega \int_{\mathbb{R}^{n\times d}} |F|^p\,\nu_x(\mathrm{d}F)\mathrm{d}x < \infty \Big\}.\end{aligned}$$

The relaxed problem to (17.52) obtained by continuous extension then has the form

$$\begin{aligned}&\text{minimize } \bar{\Phi}(\nu,h) = \int_\Omega \int_{\mathbb{R}^{n\times d}} \varphi(u,F)\,\nu_x(\mathrm{d}F)\mathrm{d}x + \int_\Gamma \phi(u)\,\mathrm{d}S\\ &\text{subject to } \nabla u(x) = \int_{\mathbb{R}^{n\times d}} F\,\nu_x(\mathrm{d}F)\ \ \forall_{\mathrm{a.a.}} x,\\ &(u,\nu) \in W^{1,p}(\Omega;\mathbb{R}^n) \times \mathscr{G}^p(\Omega;\mathbb{R}^{n\times d}).\end{aligned} \tag{17.102}$$

Proving existence of solutions to (17.102) is possible although technically complicated[51] and, moreover, $\mathscr{G}^p(\Omega;\mathbb{R}^{n\times d})$ is unfortunately not a convex subset of $L^\infty_{\mathrm{w}*}(\Omega;\mathrm{M}(\mathbb{R}^{n\times d}))$ if $\min(n,d) > 1$. Only if $n = 1$ or $d = 1$, we can rely on its convexity and derive Karush—Kuhn–Tucker-type necessary optimality conditions of the type (17.13) with λ^* being the multiplier to the constraint $\nabla u(x) = \int_{\mathbb{R}^{n\times d}} F\,\nu_x(\mathrm{d}F)$; the adjoint operator $[R']^*$ in (17.16) turns "∇" to "div." The resulted system takes the form

$$\begin{aligned}&\int_{\mathbb{R}^{n\times d}} \mathfrak{H}_{u,\lambda^*}(x,F)\,\nu_x(\mathrm{d}F) = \max_{F\in\mathbb{R}^{n\times d}} \mathfrak{H}_{u,\lambda^*}(x,F),\\ &\quad\text{with } \mathfrak{H}_{u,\lambda^*}(x,F) = \lambda^*(x)\cdot F - \varphi(x,u(x),F)\\ &\quad\mathrm{div}\,\lambda^* = \int_{\mathbb{R}^{n\times d}} \varphi'_u(u,F)\,\nu_x(\mathrm{d}F) \text{ on } \Omega,\\ &\quad\lambda^*\cdot\vec{n} + \phi'_u(u) = 0 \qquad \text{on } \Gamma,\end{aligned} \tag{17.103}$$

cf. [59, Chapter 5]. If $\varphi(x,u,\cdot)$ is convex, then there exists a standard weak solution u, that is, $\nu_x = \delta_{\nabla u(x)}$, and (17.103) simplifies to

the hysteresis caused by pinning effects is not dominant.

51) Actually, using compactness and the direct method must be combined with proving and exploiting that minimizing sequences $\{u_k\}_{k\in\mathbb{N}}$ for (17.52) have $\{|\nabla u_k|^p;\ k\in\mathbb{N}\}$ uniformly integrable if the coercivity (17.53a) with $p > 1$ holds.

$$\mathfrak{H}_{u,\lambda^*}(x,\nabla u(x)) = \max_{F\in\mathbb{R}^{n\times d}} \mathfrak{H}_{u,\lambda^*}(x,F),$$
$$\operatorname{div}\lambda^* = \varphi'_u(u,\nabla u) \qquad \text{on } \Omega,$$
$$\lambda^*\cdot\vec{n} + \phi'_u(u) = 0 \qquad \text{on } \Gamma. \quad (17.104)$$

One can see that (17.104) combines the *Weierstrass maximum principle* with a half of the Euler–Lagrange equation (17.55). If $\varphi(x,u,\cdot)$ is not convex, the oscillatory character of ∇u for minimizing sequences can be seen from (17.103) similarly as in the previous example, leading to nontrivial Young measures.

We can illustrate it on Example 17.2, where (17.103) leads to the system $\mathrm{d}\lambda^*/\mathrm{d}x = 2u$ and $\mathrm{d}u/\mathrm{d}x = \int_{\mathbb{R}} F\,\nu_x(\mathrm{d}F)$ on $(0,6\pi)$ with the boundary conditions $\lambda^*(0) = 0 = \lambda^*(6\pi)$ and with the Young measure $\{\nu_x\}_{0\le x\le 6\pi} \in \mathscr{Y}^4((0,6\pi);\mathbb{R})$ such that ν_x is supported on the finite set $\{F\in\mathbb{R};\ \lambda^*(x)F - (|F|^2-1)^2 = \max_{\widetilde{F}\in\mathbb{R}} \lambda^*(x)\widetilde{F} - (|\widetilde{F}|^2-1)^2\}$. The (even unique) solution of this set of conditions is

$$u(x) = 0, \quad \lambda^*(x) = 0, \quad \nu_x = \frac{1}{2}\delta_1 + \frac{1}{2}\delta_{-1} \qquad (17.105)$$

for $x \in (0,6\pi)$. This (spatially constant) Young measure indicates the character of the oscillations of the gradient in (17.62).

Having in mind the elasticity interpretation from Section 17.3.3.2, this effect is experimentally observed in some special materials, see Figure 17.5;[52] for modeling of such microstructure by nonconvex problems see, for example, [20, 62, 64–66].

Models based on relaxation of continuous extensions such as (17.100) or (17.102), are sometimes called *mesoscopical*, in contrast to the original problems such as (17.87)–(17.90) or (17.98) with small $\varepsilon > 0$, which are called *microscopical*, while the models using original spaces but lower semicontinuous extensions such as (17.97), which forget any information about fine microstructures, are called *macroscopical*.

52) Actually, Figure 17.5 refers to a multidimensional vectorial case (i.e., $d > 1$ and $n > 1$) where (17.103) is not available.

17.4.3 Γ-Convergence

We saw above various situations where the functional itself depends on a parameter. It is then worth studying convergence of such functionals. In the context of minimization, a prominent role is played by *Γ-convergence* introduced by De Giorgi [67], sometimes also called *variational convergence* or *epigraph convergence*, cf. also [68–70]. We say that the functional Φ is the Γ-limit of a sequence $\{\Phi_k\}_{k\in\mathbb{N}}$ if

$$\forall u_k \to u: \ \liminf_{k\to\infty} \Phi_k(u_k) \ge \Phi(u), \qquad (17.106a)$$

$$\forall u\in V\ \exists\{u_k\}_{k\in\mathbb{N}} \text{ with } u_k \to u: \ \limsup_{k\to\infty} \Phi_k(u_k) \le \Phi(u). \qquad (17.106b)$$

One interesting property justifying this mode of convergence is the following:

Theorem 17.15 (Γ-convergence.) *If $\Phi_k \to \Phi$ in the sense* (17.106) *and if u_k minimizes Φ_k, then any converging subsequence of $\{u_k\}_{k\in\mathbb{N}}$ yields, as its limit, a minimizer of Φ.*[53]

Identifying the Γ-limit (if it exists) in concrete cases can be very difficult. Few relatively simple examples were, in fact, already stated above.

53) The proof is simply by a contradiction, assuming that $\Phi(u) > \Phi(v)$ for $u = \lim_{l\to\infty} u_{k_l}$ and some $v \in V$ and using (17.106) to have a recovery sequence $v_k \to v$ so that $\Phi(v) = \liminf_{k\to\infty} \Phi_k(v_k) \ge \liminf_{k\to\infty} \Phi_k(u_k) \ge \Phi(u)$.

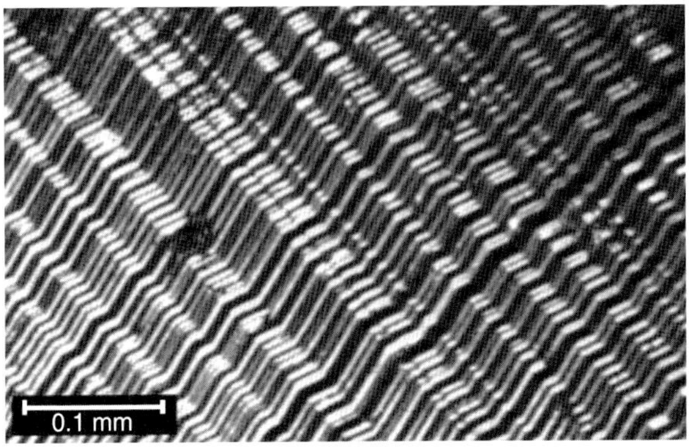

Figure 17.5 Oscillations of the deformation gradient minimizing Φ with a 6-well potential $\varphi(F)$: an orthorhombic martensite microstructure in a single-crystal of CuAlNi. (Courtesy H. Seiner, Inst. of Thermomechanics, ASCR.)

A simple example is the numerical approximation in Section 17.4.1 where we had the situation

$$V_k \subset V_{k+1} \subset V \quad \text{for } k \in \mathbb{N}, \text{ and} \tag{17.107a}$$

$$\Phi_k(v) = \begin{cases} \Phi(v) & \text{if } v \in V_k, \\ +\infty & \text{otherwise.} \end{cases} \tag{17.107b}$$

Let us further suppose that Φ is continuous with respect to the convergence used in (17.94). Then Φ_k Γ-converges to Φ.[54] Note that lower semicontinuity of Φ would not be sufficient for it, however.[55]

Another example of (17.106) with the weak topology we already saw is given by *singular perturbations*: the functionals $\Phi_\varepsilon : W^{1,p}(\Omega;\mathbb{R}^d) \to \mathbb{R} \cup \{+\infty\}$ defined by (17.98) for $u \in W^{1,p}(\Omega;\mathbb{R}^d) \cap W^{2,2}(\Omega;\mathbb{R}^d)$ and by $+\infty$ for $u \in W^{1,p}(\Omega;\mathbb{R}^d) \setminus W^{2,2}(\Omega;\mathbb{R}^d)$ Γ-converge, for $\varepsilon \to 0$, to Φ from (17.52) if $\varphi(x,u,\cdot)$ is quasiconvex, otherwise one should use $\bar{\Phi}$ from (17.99). Alternatively, if Φ_ε is extended to the Young measures by $+\infty$ if the Young measure is not of the form $\{\delta_{\nabla u(x)}\}_{x\in\Omega}$ for some $u \in W^{1,p}(\Omega;\mathbb{R}^d) \cap W^{2,2}(\Omega;\mathbb{R}^d)$, one Γ-converges as $\varepsilon \to 0$ to the functional from (17.102).[56] Similarly, (17.87)–(17.90) Γ-converges to (17.100) if $\epsilon \to 0$.

Other prominent applications of Γ-convergence are dimensional reduction from three-dimensional problems to one-dimensional (springs, rods, beams) or two-dimensional (membranes, thin films, shells, plates), or homogenization of composite materials with periodic structure, cf. e.g. [48, 71].

54) Indeed, (17.106a) holds because $\Phi_k \geq \Phi_{k+1} \geq \Phi$ due to (17.107a). For any $\hat{v} \in V$, there is $\hat{v}_k \in V_k$ such that $\hat{v}_k \to \hat{v}$. Then $\lim_{k\to\infty} \Phi_k(\hat{v}_k) = \lim_{k\to\infty} \Phi(\hat{v}_k) = \Phi(\hat{v})$ and also $\lim_{k\to\infty} \hat{v}_k = v$ in V so that $\{\hat{v}_k\}_{k\in\mathbb{N}}$ is a recovery sequence for (17.106b).

55) A simple counterexample is $\Phi = +\infty$ everywhere except some $v \in V \setminus \cup_{k\in\mathbb{N}} V_k$; then $\Phi_k \equiv \infty$ obviously does not Γ-converge to Φ.

56) Again, (17.106a) is simply due to $\Phi_\varepsilon \geq \bar{\Phi}$ and $\bar{\Phi}$ is lower semicontinuous. The construction of particular recovery sequences for (17.106b) is more involved, smoothing the construction of a recovery sequence for $\varphi^\sharp$ or for the minimizing gradient Young measure as in [61, 2].

Glossary

A lot of notions, definitions, and assertions are presented above. The following list tries to sort them according subjects or disciplines, giving the link to particular pages where the particular item is highlighted.

Topological notions:

Γ-convergence, p.581
continuous (weakly), p.552 (p.553)

compact mapping, p.563
compact set, p.553
dense, p.557
hemicontinuous mapping, p.552
lower semicontinuous, p.553
  envelope, p.577
variational convergence, p.581

Linear spaces, spaces of functions:
adjoint operator, p.555
Banach space, p.551
  ordered, p.555
  reflexive, p.553
Bochner space $L^p(I;V)$, p.557
boundary critical exponent $p^\sharp$, p.563
dual space, p.552
Gelfand triple $V \subset H \subset V^*$, p.557
Hilbert space, p.552
Lebesgue space L^p, p.562
pre-dual, p.553
smooth functions $C^k(\bar{\Omega})$, p.563
Sobolev critical exponent p^*, p.563
Sobolev space $W^{k,p}$, p.562
Young measures, p.579

Convex analysis:
cone, p.555
convex/concave, p.552
convex mapping, p.556
Fenchel inequality, p.557
indicator function δ_K, p.555
Legendre conjugate, p.557
Legendre transformation, p.557
Legendre – Fenchel transformation, p.559
linear, p.552
monotone, p.553
normal cone $N_K(u)$, p.555
polyconvexity, p.566
rank-one convexity, p.565
strictly convex, p.553
subdifferential ∂, p.554
tangent cone $T_K(u)$, p.555

Smooth analysis:
continuously differentiable, p.552
directionally differentiable, p.552
Fréchet subdifferential ∂_F, p.554
Gâteaux differential, pp.552, 564
smooth, p.552

Optimization theory:
adjoint system, p.585
constraint qualification
  Mangasarian-Fromovitz, p.555
  Slater, p.556
complementarity condition, p.555
critical point, p.552
dual problem, p.556
Euler – Lagrange equation, pp.552, 564
Karush – Kuhn – Tucker condition, p.555
Lagrangean $\mathscr{L}(u,\lambda^*)$, p.555
optimal control, p.585
  relaxed, p.585
sufficient 2nd-order condition, p.556
transversality condition, p.555

Variational principles and problems:
Brezis – Ekeland – Nayroles, pp.557,575
complementarity problem, p.555
Ekeland principle, p.584
Hamilton principle, pp.559,576
Lavrentiev phenomenon, pp.567,572
least dissipation principle, p.558
minimum-energy principle, p.552
maximum dissipation, p.560
nonexistence, pp.567,571
potential, p.552
  coercive, p.553
  double-well, p.567
  of dissipative forces, p.558
Palais – Smale property, p.554
Plateau minimal-surface problem, p.567
Pontryagin maximum principle, p.585
relaxation, p.577
  by convex compactification, p.578
singular perturbations, pp.578, 582
Stefanelli principle, p.558
symmetry condition, pp.552, 565, 570
  Onsager, p.558
variational inequality, p.554
Weierstrass maximum principle, p.579

Differential equations and inequalities:
abstract parabolic equation, p.556
boundary conditions, p.564
boundary-value problem, pp.564, 569
Carathéodory mapping, p.563
Cauchy problem, p.556
  doubly nonlinear, p.558
classical solution, p.564
doubly nonlinear inclusion, p.559
formulation
  classical, pp.564,569
  De Giorgi, p.559
  energetic, p.560
  mixed, p.575
  weak, p.564,569
generalized gradient flow, p.558
  in metric spaces, p.559
Legendre – Hadamard condition, p.566
Nemytskiĭ mapping, p.563

Further Reading

The convex/smooth setting with one objective functional on which we primarily focused in Section 17.2 can be extensively generalized to nonconvex and nondifferentiable cases and/or to multi-objective situations, including dualization schemes, optimality conditions, sensitivity analysis, generalized equilibria, and many others, cf. e.g. [9, 12, 37, 72–74]. Many proof techniques are based on the remarkable *Ekeland variational principle* saying that, for a Gâteaux differentiable functional Φ bounded from below on a Banach space V, holds that

$$\begin{aligned} &\forall u \in V,\ \varepsilon > 0: \quad \Phi(u) \le \inf \Phi + \varepsilon \\ &\Rightarrow \quad \exists v \in V: \quad \Phi(v) \le \Phi(u), \\ &\qquad \|v-u\| \le \sqrt{\varepsilon}, \quad \|\Phi'(v)\| \le \sqrt{\varepsilon}, \end{aligned}$$

See, for example, [11, 31, 37], in particular, also for a general formulation in metric spaces.

In concrete situations, solutions of variational problems often enjoy additional properties (typically, despite the counterexamples as [57, 58], some smoothness); there is an extensive literature in this direction of *regularity* of solutions, for example, [32, 50, 75].

There has been intensive effort leading to efficient and widely applicable methods to avoid the symmetry conditions (17.5), cf. also (17.57), based on the concept of monotonicity. Nonsymmetric nonlinear monotone-type operators (possibly generalized, for example, to pseudo-monotone operators or of the types (M) or (S), etc.) have been introduced on an abstract level in the work of Brézis [7], Minty [76], and others. Many monographs are available on this topic, also applied to concrete *nonsymmetric* quasilinear *equations* or inequalities, cf. e.g. [18, 50, 77, 78].

Even for situations conforming with the symmetry conditions of the type (17.57), Example 17.5 showed that sometimes variational methods even for linear boundary-value problem such as

$$-\mathrm{div}\nabla u = f \text{ on } \Omega, \quad \nabla u \cdot \vec{n} + u = g \text{ on } \Gamma,$$

are not compatible with natural physical demands that the right-hand sides f and g have an L^1-structure. This is why also *nonvariational methods* have been extensively developed. One method to handle general right-hand sides is Stampacchia's [79] transposition method, which has been analyzed for linear problems by Lions and Magenes [80]. Another general method is based on metric properties and contraction based on *accretivity* (instead of compactness and monotonicity) and, when applied to evolution problems, is connected with the theory of nonexpansive semigroups; from a very wide literature cf. e.g. the monographs by Showalter [78, Chapter 4], Vainberg [81, Chapter VII], or Zeidler [9, Chapter 57], or also [18, Chapter 3 and 9]. An estimation technique fitted with L^1-structure and applicable to thermal problems possibly coupled with mechanical or other physical systems, has been developed in [82], cf. also e.g. [18].

In fact, for $d = 1$ and $\Omega = [0, T]$, Section 17.3.2.1 dealt in particular with a very special *optimal control* problem of the *Bolza type*: minimize the objective $\int_0^T \varphi(t, u(t), v(t))\,\mathrm{d}t + \phi(T, u(T))$ for the initial-value problem for a simple (system of) ordinary differential equation(s) $\mathrm{d}u/\mathrm{d}t = v$, $u(0) = u_0$, with the *control* v being possibly subjected to a constraint $v(t) \in S$, with $t \in [0, T]$ playing the role of time. One can also think about generalization to (systems of) nonlinear ordinary differential equations of the type $\mathrm{d}u/\mathrm{d}t = f(t, u, v)$. If $\varphi(t, u, \cdot)$ is convex and $f(t, u, \cdot)$ is affine, one obtains existence of optimal control v and the corresponding response by the direct method as we did in Section 17.3.2.1. If fact, convexity of the so-called orientor field $Q(t, u) := \{(q_0, q_1);\ \exists s \in S(t) : q_0 \geq \varphi(t, u, s),\ q_1 = f(t, u, s)\}$ is decisive for existence of optimal control. In the general case, the existence is not guaranteed and one can make a relaxation as we did in (17.100) obtaining the *relaxed optimal control* problem

$$\begin{aligned} &\text{minimize } = \int_0^T \int_S \varphi(t, u(t), s)\, v_t(\mathrm{d}s)\mathrm{d}t \\ &\text{subject to } \frac{\mathrm{d}u}{\mathrm{d}t} = \int_S f(t, u(t), s)\, v_t(\mathrm{d}s) \\ &\qquad v \in \mathscr{Y}([0, T]; S). \end{aligned} \qquad (17.108)$$

The optimality conditions of the type (17.16) results in a modification of the Weierstrass maximum principle (17.103), namely,

$$\begin{aligned} &\int_S \mathfrak{H}_{u,\lambda^*}(t, s)\, v_t(\mathrm{d}s) = \max_{\tilde{s} \in S} \mathfrak{H}_{u,\lambda^*}(t, \tilde{s}) \quad \text{with} \\ &\mathfrak{H}_{u,\lambda^*}(t, s) = \lambda^*(t) \cdot f(t, u(t), s) - \varphi(t, u(t), s), \\ &\frac{\mathrm{d}\lambda^*}{\mathrm{d}x} + \int_S f_u'(t, u(t), s)^\top \lambda^*(t)\, v_t(\mathrm{d}s) \\ &\qquad = \int_S \varphi_u'(t, u(t), s)^\top\, v_t(\mathrm{d}s) \text{ on } [0, T], \\ &\lambda^*(T) = \phi_u'(T, u(T)). \end{aligned} \qquad (17.109)$$

The linear terminal-value problem in (17.109) for λ^* is called the *adjoint system*, arising from the adjoint operator in (17.16). Of course, if (by chance) the optimal control v of the original problem exists, then the first condition in (17.109) reads as $\mathfrak{H}_{u,\lambda^*}(t, v(t)) = \max_{\tilde{v} \in S} \mathfrak{H}_{u,\lambda^*}(t, \tilde{v})$. Essentially, this has been formulated in [83, 84] and later become known as the *Pontryagin maximum principle*, here in terms of the so-called relaxed controls. We can see that it is a generalization of the Weierstrass principle and can be derived as a standard Karush–Kuhn–Tucker condition but with respect to the convex geometry induced from the space of relaxed controls.[57] One can also consider optimal control of partial differential equations instead of the ordinary ones, cf. also [59]. There is a huge literature about optimal control theory in all usual aspects of the calculus of variations as briefly presented above, cf. e.g. [73, 85, 86].

Acknowledgments. The author is very thankful to Jan Malý for fruitful discussions (particularly with regard to Example 17.5), Alexander Mielke, and Jiří V. Outrata, and to Oleg Heczko and Hanuš Seiner for providing the experimental figures 17.4 and 17.5 . The institutional support RVO:61388998 (ČR) and the support from the grant 201/10/0357 (GA ČR) are also warmly acknowledged.

References

1. Buttazzo, G., Giaquinta, M., and Hildebrandt, S. (eds) (1998) *One-Dimensional Variational Problems*, Clarendon, Oxford.
2. Drake, G.W.F. (2005) Variational methods, in *Mathematical Tools for Physicists* (ed. G.L. Trigg), Wiley-VCH Verlag GmbH, Weinheim, pp. 619–656.
3. Jost, J. and Li-Jost, X. (1998) *Calculus of Variations*, Cambridge University Press, Cambridge.
4. Hilbert, D. (1901) Mathematische probleme. *Archiv d. Math. u. Physik*, **1**, 44–63, 213–237; (English Translation: (1902) *Bull. Am. Math. Soc.*, **8**, 437–479).
5. Banach, S. (1932) *Théorie des Opérations Linéaires*, M.Garasiński, Warszawa.
6. Tonelli, L. (1915) Sur un méthode directe du calcul des variations. *Rend. Circ. Mat. Palermo*, **39**, 233–264.
7. Brezis, H. (1968) Équations et inéquations non-linéaires dans les espaces vectoriel en dualité. *Ann. Inst. Fourier*, **18**, 115–176.

57) The original and rather technical method was based on the so-called needle variations, however.

8. Ambrosetti, A. and Rabinowitz, P.H. (1973) Dual variational methods in critical point theory and applications. *J. Funct. Anal.*, **14**, 349–380.
9. Zeidler, E. (1985) *Nonlinear Functional Analysis and Its Applications III: Variational Methods and Optimization*, Springer, New York.
10. von Neumann, J. (1928) Zur Theorie der Gesellschaftsspiele. *Math. Ann.*, **100**, 295–320.
11. Borwein, J.M. and Zhu, Q.J. (eds) (2005) *Techniques of Variational Analysis*, Springer, Berlin.
12. Rockafellar, R.T. and Wets, R.J.-B. (1998) *Variational Analysis*, Springer, Berlin.
13. Karush, W. (1939) *Minima of functions of several variables with inequalities as side conditions*. PhD thesis, Department of Mathematics - University of Chicago, Chicago, IL.
14. Kuhn, H. and Tucker, A. (1951) Nonlinear programming, *2nd Berkeley Symposium on Mathematical Statistics and Probability*, University of California Press, Berkeley, pp. 481–492.
15. Brezis, H. and Ekeland, I. (1976) Un prinicpe varationnel associeé à certaines équations paraboliques. *C. R. Acad. Sci. Paris*, **282**, 971–974 and 1197–1198.
16. Nayroles, B. (1976) Deux théorèmes de minimum pour certains systèmes dissipatifs. *C. R. Acad. Sci. Paris Sér. A-B*, **282**, A1035–A1038.
17. Roubíček, T. (2000) Direct method for parabolic problems. *Adv. Math. Sci. Appl.*, **10**, 57–65.
18. Roubíček, T. (2013)*Nonlinear Partial Differential Equations with Applications*, 2nd edn, Birkhäuser, Basel.
19. Onsager, L. (1931) Reciprocal relations in irreversible processes I. *Phys. Rev.*, **37**, 405–426; Part II, **38**, 2265–2279.
20. Šilhavý, M. (1997) *The Mechanics and Thermodynamics of Continuous Media*, Springer, Berlin.
21. Stefanelli, U. (2008) The Brezis-Ekeland principle for doubly nonlinear equations. *SIAM J. Control Optim.*, **47**, 1615–1642.
22. Onsager, L. and Machlup, S. (1953) Fluctuations and irreversible processes. *Phys. Rev.*, **91**, 1505–1512.
23. Ambrosio, L., Gigli, N., and Savaré, G. (2008) *Gradient Flows*, 2nd edn, Birkhäuser, Basel.

24. Bedford, A. (ed.) (1985) *Hamilton's Principle in Continuum Mechanics*, Pitman, Boston, MA.
25. Visintin, A. (1996) *Models of Phase Transitions*, Birkhäuser, Boston, MA.
26. Mielke, A. and Theil, F. (2004) On rate-independent hysteresis models. *Nonlin. Diff. Equ. Appl.*, **11**, 151–189.
27. Mielke, A. and Roubíček, T. (2014) *Rate-Independent Systems - Theory and Application*, Springer, New York, to appear.
28. Rothe, E. (1930) Zweidimensionale parabolische Randwertaufgaben als Grenzfall eindimensionaler Randwertaufgaben. *Math. Ann.*, **102**, 650–670.
29. Dacorogna, B. (1989) *Direct Methods in the Calculus of Variations*, Springer, Berlin.
30. Fonseca, I. and Leoni, G. (2007) *Modern Methods in the Calculus of Variations: L^p Spaces*, Springer, New York.
31. Giusti, E. (2003) *Direct Methods in Calculus of Variations*, World Scientific, Singapore.
32. Nečas, J. (1967) *Les Méthodes Directes en la Théorie des Equations Elliptiques*, Academia & Masson, Praha & Paris; (English Translation: Springer, Berlin, 2012).
33. Morrey, C.B. Jr. (1966) *Multiple Integrals in the Calculus of Variations*, Springer, Berlin.
34. Šverák, V. (1992) Rank-one convexity does not imply quasiconvexity. *Proc. R. Soc. Edinburgh Sect. A*, **120**, 185–189.
35. Müller, S. (1989) A surprising higher integrability property of mappings with positive determinants. *Bull. Am. Math. Soc.*, **21**, 245–248.
36. Ball, J.M. (1977) Convexity conditions and existence theorems in nonlinear elasticity. *Arch. Ration. Mech. Anal.*, **63** (4), 337–403.
37. Ekeland, I. and Temam, R. (1976) *Convex Analysis and Variational Problems*, North-Holland.
38. Lavrentiev, A. (1926) Sur quelques problémes du calcul des variations. *Ann. Mat. Pura Appl.*, **41**, 107–124.
39. Carstensen, C. and Ortner, C. (2010) Analysis of a class of penalty methods for computing singular minimizers. *Comput. Meth. Appl. Math.*, **10**, 137–163.
40. Ball, J.M. and Mizel, V.J. (1985) One-dimensional variational problems whose minimizers do not satisfy the Euler-Lagrange equation. *Arch. Ration. Mech. Anal.*, **90**, 325–388.
41. Baiocchi, C. and Capelo, A. (1984) *Variational and Quasivariational Inequalities*, John Wiley & Sons, Ltd, Chichester.
42. Duvaut, G. and Lions, J.-L. (1976) *Inequalities in Mechanics and Physics*, Springer, Berlin.
43. Hlaváček, I., Haslinger, J., Nečas, J., and Lovíšek, J. (1988) *Solution of Variational Inequalities in Mechanics*, Springer, New York.
44. Kinderlehrer, D. and Stampacchia, G. (1980) *An Introduction to Variational Inequalities and their Applications*, Academic Press, New York.
45. Ciarlet, P.G. and Nečas, J. (1987) Injectivity and self-contact in nonlinear elasticity. *Arch. Ration. Mech. Anal.*, **97**, 171–188.
46. Ball, J.M. (2002) Some open problems in elasticity, in *Geometry, Mechanics, and Dynamics* (ed. P. Newton, P. Holmes, A. Weinstein), Springer, New York, pp. 3–59.
47. Foss, M., Hrusa, W.J., and Mizel, V.J. (2003) The Lavrentiev gap phenomenon in nonlinear elasticity. *Arch. Ration. Mech. Anal.*, **167**, 337–365.
48. Ciarlet, P.G. (1988, 1997, 2000) *Mathematical Elasticity, Vol.I: Three-Dimensional Elasticity, Vol.II: Theory of Plates, Vol.III: Theory of Shells*, North-Holland, Amsterdam.
49. Eringen, A.C. (2002) *Nonlocal Continuum Field Theories*, Springer, New York.
50. Evans, L.C. (1998) *Partial Differential Equations*, AMS, Providence, RI.
51. Struwe, M. (1990) *Variational Methods: Applications to Nonlinear Partial Differential Equations and Hamiltonian Systems*, Springer, Berlin.
52. Hlaváček, I. (1969) Variational principle for parabolic equations. *Apl. Mat.*, **14**, 278–297.
53. Ghoussoub, N. (2009) *Self-Dual Partial Differential Systems and Their Variational Principles*, Springer, New York.
54. Ritz, W. (1908) Über eine neue Methode zur Lösung gewisser Variationsprobleme der mathematischen Physik. *J. Reine u. Angew. Math.*, **135**, 1–61.
55. Galerkin, B.G. (1915) Series development for some cases of equilibrium of plates and beams (In Russian). *Vestnik Inzhinierov Teknik*, **19**, 897–908.
56. Christara, C.C. and Jackson, K.R. (2005) Numerical methods, in *Mathematical Tools*

for Physicists (ed. G.L. Trigg), John Wiley & Sons, Inc., Weinheim, pp. 281–383.

57. Nečas, J. (1977) Example of an irregular solution to a nonlinear elliptic system with analytic coefficients and conditions of regularity, in *Theory of Nonlinear Operators Proceedings of Summer School (Berlin 1975)* (eds W. Muller, Berlin, Akademie-Verlag, pp. 197–206.
58. Šverák, V. and Yan, X. (2000) A singular minimizer of a smooth strongly convex functional in three dimensions. *Calc. Var.*, **10**, 213–221.
59. Roubíček, T. (1997) *Relaxation in Optimization Theory and Variational Calculus*, W. de Gruyter, Berlin.
60. Young, L.C. (1969) *Lectures on the Calculus of Variations and Optimal Control Theory*, W.B. Saunders, Philadelphia, PA.
61. Pedregal, P. (1997) *Parametrized Measures and Variational Principles*, Birkhäuser, Basel.
62. Pedregal, P. (2000) *Variational Methods in Nonlinear Elasticity*, SIAM, Philadelphia, PA.
63. Kružík, M. and Prohl, A. (2006) Recent developments in the modeling, analysis, and numerics of ferromagnetism. *SIAM Rev.*, **48**, 439–483.
64. Ball, J.M. and James, R.D. (1987) Fine phase mixtures as minimizers of energy. *Arch. Ration. Mech. Anal.*, **100**, 13–52.
65. Bhattacharya, K. (ed.) (2003) *Microstructure of MartenSite. Why it Forms and How it Gives Rise to the Shape-Memory Effect*, Oxford University Press, New York.
66. Müller, S. (1999) Variational models for microstructure and phase transitions, *Calculus of Variations and Geometric Evolution Problems*, Springer, Berlin, pp. 85–210.
67. De Giorgi, E. (1977.) Γ-convergenza e *G*-convergenza. *Boll. Unione Mat. Ital., V. Ser.*, **A 14**, 213–220.
68. Attouch, H. (1984) *Variational Convergence of Functions and Operators*, Pitman.
69. Braides, A. (ed.) (2002) *Γ-Convergence for Beginners*, Oxford University Press, Oxford.
70. Dal Maso, G. (1993) *An Introduction to Γ-Convergence*, Birkhäuser, Bostonm, MA.
71. Friesecke, G., James, R.D., and Müller, S. (2006) A hierarchy of plate models derived from nonlinear elasticity by Gamma-convergence. *Arch. Ration. Mech. Anal.*, **180**, 183–236.
72. Bonnans, J.F. and Shapiro, A. (eds) (2000) *Perturbation Analysis of Optimization Problems*, Springer-Verlag, New York.
73. Clarke, F.H. (1983) *Optimization and Nonsmooth Analysis*, Wiley.
74. Mordukhovich, B.S. (2006) *Variational Analysis and Generalized Differentiation I, II*, Springer, Berlin.
75. Malý, J. and Ziemer, W.P. (1997) *Fine Regularity of Solutions of Elliptic Partial Differential Equations*, American Mathematical Society, Providence, RI.
76. Minty, G. (1963) On a monotonicity method for the solution of non-linear equations in Banach spaces. *Proc. Natl. Acad. Sci. U.S.A.*, **50**, 1038–1041.
77. Lions, J.L. (1969) *Quelques Méthodes de Résolution des Problémes aux Limites non linéaires*, Dunod, Paris.
78. Showalter, R.E. (1997) *Monotone Operators in Banach Space and Nonlinear Partial Differential Equations*, AMS.
79. Stampacchia, G. (1965) Le problème de Dirichlet pour les équations elliptiques du second ordre à coefficients discontinus. *Ann. Inst. Fourier*, **15**, 189–258.
80. Lions, J.L. and Magenes, E. (1968) *Problèmes aux Limites non homoegènes et Applications*, Dunod, Paris.
81. Vainberg, M.M. (1973) *Variational Methods and Method of Monotone Operators in the Theory of Nonlinear Equations*, John Wiley & Sons, Inc., New York.
82. Boccardo, L. and Gallouet, T. (1989) Non-linear elliptic and parabolic equations involving measure data. *J. Funct. Anal.*, **87**, 149–169.
83. Boltyanskiĭ, V.G., Gamkrelidze, R.V., and Pontryagin, L.S. (1956) On the theory of optimal processes (In Russian). *Dokl. Akad. Nauk USSR*, **110**, 7–10.
84. Hestenes, M.R. (1950) *A general problem in the calculus of variations with applications to the paths of least time.* Technical Report 100, RAND Corp., Santa Monica, CA.
85. Ioffe, A.D. and Tikhomirov, V.M. (1979) *Theory of Extremal Problems*, North-Holland, Amsterdam.
86. Tröltzsch, F. (2010) *Optimal Control of Partial Differential Equations*, American Mathematical Society, Providence, RI.

Index

Mathematical Tools for Physicists, Second Edition. Edited by Michael Grinfeld.

g

x

y